The MOCVD *Challenge*

Second Edition

Electronic Materials and Devices Series

Series Editors:
Yongbing Xu, *University of York, UK*
Jean-Pierre Leburton, *University of Illinois at Chicago, USA*

This series seeks to publish books covering materials, properties and fabrication processes for electronic materials, nanostructures and devices for applications in solid state electronics, optoelectronics, photovoltaics, sensors and integrated systems. Including both theoretical and application oriented texts spanning physics, engineering and material science the series provides high quality texts for advanced students and researchers at the interface of these disciplines.

Published titles:

The MOCVD Challenge: A survey of GaInAsP-InP and GaInAsP-GaAs for photonic and electronic device applications, Second Edition
Manijeh Razeghi

Forthcoming titles:

Handbook of Modern Photovoltaics: Fundamentals, Engineering, and Applications
Siva Sivoththaman (Ed)

Handbook of Zinc Oxide and Related Materials: Vol 1 - Materials
Zhe Chuan Feng (Ed)

Handbook of Zinc Oxide and Related Materials: Vol 2 - Devices and Nano-Engineering
Zhe Chuan Feng (Ed)

The MOCVD *Challenge*

A survey of GaInAsP-InP and GaInAsP-GaAs for photonic and electronic device applications

Second Edition

Manijeh Razeghi
Northwestern University
Illinois, USA

CRC Press
Taylor & Francis Group
Boca Raton   London   New York

CRC Press is an imprint of the
Taylor & Francis Group, an **informa** business

A TAYLOR & FRANCIS BOOK

CRC Press
Taylor & Francis Group
6000 Broken Sound Parkway NW, Suite 300
Boca Raton, FL 33487-2742

First issued in paperback 2017

ISBN 13: 978-1-138-11493-7 (pbk)
ISBN 13: 978-1-4398-0698-2 (hbk)

Library of Congress Cataloging-in-Publication Data

Razeghi, M.
    The MOCVD challenge : a survey of GaInAsP-InP and GaInAsP-GaAs for photonic and electronic device applications / author, Manijeh Razeghi. -- 2nd ed.
        p. cm. -- (Electronic materials and devices series)
    "A CRC title."
    Includes bibliographical references and index.
    ISBN 978-1-4398-0698-2 (hardcover : alk. paper)
    1. Metal organic chemical vapor deposition. 2. Electronic apparatus and appliances--Materials. I. Title. II. Series.

TS695.R39 2010
621.381--dc22

2010026136

Visit the Taylor & Francis Web site at
http://www.taylorandfrancis.com

and the CRC Press Web site at
http://www.crcpress.com

To my parents, my educators, my family, and my past, present, and future students.

Contents

Foreword xvii

Preface xix

1. **Introduction to Semiconductor Compounds** 1
   1.1. Introduction 1
   1.2. III–V semiconductor alloys 1
       1.2.1. III–V binary compounds 1
       1.2.2. III–V ternary alloys 3
       1.2.3. III–V quaternary compounds 3
   1.3. III–V semiconductor devices 5
   1.4. Technology of multilayer growth 10
       1.4.1. Epitaxial technology 10
       1.4.2. Review of III–V heterostructures grown by LPE, MBE, and
               MOCVD 14
   References 19

2. **Growth Technology** 23
   2.1. Introduction 23
       2.1.1. Liquid-phase epitaxy 25
       2.1.2. Vapor-phase epitaxy 26
       2.1.3. Molecular beam epitaxy 27
   2.2. Metalorganic chemical vapor deposition 28
       2.2.1. Experimental details 29
       2.2.2. Reactor design 30
       2.2.3. Growth procedure 31
       2.2.4. Flow patterns 32
       2.2.5. Starting materials 34
   2.3. New non-equilibrium growth techniques 40
       2.3.1. Chemical beam epitaxy 40
       2.3.2. Atomic layer epitaxy 42
       2.3.3. Metalorganic atomic layer epitaxy 42
       2.3.4. Chloride atomic layer epitaxy 43
       2.3.5. Migration-enhanced epitaxy 43
   References 43

3. *In situ* **Characterization during MOCVD** 45
   3.1. Introduction 45
   3.2. Reflectance anisotropy and ellipsometry................................ 48
      3.2.1. Principles of RDS..49
      3.2.2. The optical setup and data acquisition system...............50
      3.2.3. Experimental procedure 52
      3.2.4. Adaptation to a MOCVD reactor 53
      3.2.5. Comparison with other RDS techniques 54
   3.3. Optimization of the growth of III–V binaries by RDS 55
      3.3.1. Effects of the MOCVD reactor design on the growth 55
      3.3.2. The III/V flux dependence of RDS signals 57
      3.3.3. The temperature dependence of RDS signals................58
      3.3.4. Discussion 59
   3.4. RDS investigation of III–V lattice-matched heterojunctions 62
      3.4.1. RDS observations of heterojunction growth 62
      3.4.2. Interpretations of the experimental observations................65
      3.4.3. Applications of RDS for growth monitoring................76
   3.5. RDS investigation of III–V lattice-mismatched structures........ 79
      3.5.1. Growth procedure of InAs/InP and InP/GaAs/Si 79
      3.5.2. RDS records during real-time growth 81
      3.5.3. Optical models 85
      3.5.4. The first stage of 3D growth..............................90
      3.5.5. Subsequent growth 93
      3.5.6. Discussion of optical models..............................95
      3.5.7. RDS monitoring of growth of GaInP/GaAs and GaInP/InP.......97
   3.6. Insights on the growth process 99
   References 101

4. *Ex situ* **Characterization Techniques**...................................... 105
   4.1. Introduction 106
   4.2. Chemical bevel revelation 106
      4.2.1. Principle of magnification 106
      4.2.2. Description 107
      4.2.3. Bath revelation 108
   4.3. Deep-level transient spectroscopy 111
      4.3.1. Deep levels..111
      4.3.2. Generation of capacitance transients.....................112
      4.3.3. Determination of trap parameters.........................116
   4.4. X-ray diffraction 120
      4.4.1. Introduction...120
      4.4.2. Bragg's law 120
      4.4.3. Bond's method 122
      4.4.4. Simple analysis of x-ray diffraction 123
      4.4.5. Dynamical x-ray diffraction theory simulation 126
      4.4.6. Structural characterization of GaAs–GaInP superlattices 129
   4.5. Photoluminescence 133

4.5.1. Fourier transform spectroscopy 134
4.5.2. Recombination processes 136
4.6. Electromechanical capacitance-voltage and photovoltage
spectroscopy 139
4.6.1. Electrochemical capacitance–voltage measurement 139
4.6.2. Photovoltage spectroscopy 144
4.7. Resistivity and Hall measurement 146
4.7.1. Resistivity measurement 146
4.7.2. Hall measurement 149
4.8. Thickness measurement 157
4.8.1. Ball polishing 157
4.8.2. Revelation with RCA solution 158
4.8.3. Rinsing procedure 158
4.8.4. Theory behind the thickness measurement 158
4.8.5. Other thickness measurement techniques 159
References 160

5. MOCVD Growth of GaAs Layers 163
5.1. Introduction 163
5.2. GaAs and related compounds band structure 163
5.3. MOCVD growth mechanism of GaAs and related
compounds 173
5.4. Experimental details 176
5.5. Incorporation of impurities in GaAs grown by MOCVD 182
5.5.1. Residual impurities 183
5.5.2. Carbon incorporation in GaAs grown by MOCVD 183
5.5.3. n-type GaAs 184
5.5.4. p-type GaAs 188
5.5.5. Erbium-doped GaAs 190
References 191

6. Growth and Characterization of the GaInP–GaAs System 193
6.1. Introduction 194
6.2. Growth details 195
6.3. Structural order in $Ga_xIn_{1-x}P$ alloys grown by MOCVD 196
6.4. Defects in GaInP layers grown by MOCVD 199
6.5. Doping behavior of GaInP 201
6.5.1. n-type doping 201
6.5.2. p-type doping 207
6.5.3. Conclusions 209
6.6. GaAs–GaInP heterostructures 210
6.6.1. Band offset measurements in the $GaAs/Ga_{0.51}In_{0.49}P$ system ... 210
6.6.2. Observation of the two-dimensional properties of the
electron gas in $Ga_{0.51}In_{0.49}P$/GaAs heterojunctions grown by
MOCVD 216
6.6.3. High-mobility GaInP–GaAs heterostructures 219

6.6.4. Electron spin resonance in the two-dimensional electron
gas of a GaAs–Ga$_{0.51}$In$_{0.49}$P heterostructure 223
6.6.5. Magnetotransport measurements in GaInP/GaAs
heterostructures using δ-doping 228
6.6.6. Persistent photoconductivity in Ga$_{0.51}$In$_{0.49}$P/GaAs
heterojunctions 231
6.6.7. Quantum and classical lifetimes in a Ga$_{0.51}$In$_{0.49}$P/GaAs
heterojunction...238
6.6.8. FIR magnetoemission study of the quantum Hall state and
the breakdown of the quantum Hall effect in GaAs–GaInP 242
6.7. Growth and characterization of GaInP–GaAs multilayers by
MOCVD.. 243
6.8. Optical and structural investigations of GaAs–GaInP
quantum wells and superlattices grown by MOCVD 248
6.9. Characterization of GaAs–GaInP quantum wells by auger
analysis of chemical bevels... 254
6.10. Evaluation of the band offsets of GaAs–GaInP multilayers
by electroreflectance 259
6.10.1. Study of the GaAs–GaInP superlattice...................................260
6.10.2. The study of the GaAs/GaInP three-quantum-well
structure...265
6.10.3. III–V intersubband infrared detectors 267
6.11. Intersubband hole absorption in GaAs–GaInP quantum
wells 270
References 273

7.   Optical Devices... 279
7.1. Electro-optical modulators 279
7.1.1. Introduction..279
7.1.2. Multiple-quantum-well modulators based on the quantum-
confined Stark effect 280
7.1.3. Superlattice modulators based on Wannier–Stark
localization 285
7.1.4. Perspectives..288
7.2. GaAs-based infrared photodetectors grown by MOCVD 289
7.2.1. Basic physics of photodetectors 289
7.2.2. II–VI material-based photodetectors 292
7.2.3. InAs$_{1-x}$Sb$_x$ materials...294
7.2.4. Intersubband GaAs quantum well photodetectors...................296
7.3. Solar cells and GaAs solar cells 304
References 309

8.   GaAs-Based Lasers... 313
8.1. Introduction 313
8.2. Basic physical concepts.. 314
8.2.1. Optical gain and feedback 315

8.2.2. Threshold current 316
8.2.3. Spectrum of longitudinal modes.............................318
8.2.4. Carrier and light confinement................................318
8.2.5. High-speed modulation 320
8.2.6. Quantum size effects 321
8.3. Laser structures.. 322
8.3.1. Threshold current density and transverse waveguiding...........323
8.3.2. Threshold current and lateral mode control...................329
8.4. New GaAs-based materials for lasers............................. 332
8.4.1. Strained InGaAs layer laser structures 332
8.4.2. AlGaInP visible lasers..339
8.4.3. GaInAsP/GaAs lasers...341
References 361

9. **GaAs-Based Heterojunction Electron Devices Grown by
MOCVD**.. **365**
9.1. Introduction 365
9.1.1. Heterojunctions 366
9.2. Heterostructure field-effect transistors (HFETs) 367
9.2.1. AlGaAs/GaAs MODFETs.....................................368
9.2.2. AlGaAs/InGaAs MODFETs 373
9.2.3. GaInP/GaAs MODFETs 374
9.2.4. MODFETs on silicon substrate 382
9.2.5. GaInP/GaAs heterostructure insulated gate field effect
transistors 383
9.3. Heterojunction bipolar transistors (HBTs) 386
9.3.1. Introduction..386
9.3.2. Frequency response and design considerations.......................389
9.3.3. AlGaAs/GaAs HBTs...391
9.3.4. GaInP/GaAs HBTs...392
References 395

10. **Optoelectronic Integrated Circuits (OEICs)**............................. **399**
10.1. Introduction 399
10.2. Material considerations 400
10.3. OEICs on silicon substrates... 401
10.4. The role of optoelectronic integration in computing 405
10.5. Examples of optoelectronic integration by MOCVD 410
10.5.1. Monolithic integration of photodetectors with transistors......411
10.5.2. Monolithic integration of photodetectors with optical
waveguides..411
10.5.3. Monolithic integration of optical modulator with lasers 412
10.5.4. Surface emitting laser diodes 412
References 412

11. InP–InP System: MOCVD Growth, Characterization, and Applications.. 415
11.1. Introduction 415
11.2. Energy band structure of InP 416
11.3. Growth and characterization of InP using TEIn 419
    11.3.1. Preparation of substrates 420
    11.3.2. Orientation effects...421
    11.3.3. Source-purity effects 422
    11.3.4. Material characterization....................................423
    11.3.5. Interfaces..436
11.4. Growth and characterization of InP using TMIn 437
11.5. Incorporation of dopants...................................... 442
    11.5.1. *p*-type..442
    11.5.2. *n*-type..444
11.6. Applications of InP epitaxial layers....................... 448
    11.6.1. Gunn diodes 448
References 451

12. GaInAs–InP System: MOCVD Growth, Characterization, and Applications 453
12.1. Introduction 454
12.2. Growth conditions 454
12.3. Optical and crystallographic properties, and impurity incorporation in GaInAs grown by MOCVD 461
    12.3.1. Sample preparation..461
    12.3.2. Electrical, crystallographic, and optical experiments 462
    12.3.3. Results and discussion..464
    12.3.4. Exciton line 466
    12.3.5. Determination of E_g (x_{Ga}) near $x_{Ga} = 47\%$.........470
    12.3.6. Donor–acceptor pair recombination 472
    12.3.7. Zn-doped samples 474
    12.3.8. Deep Fe and intrinsic defect levels in $Ga_{0.47}In_{0.53}As/InP$........475
    12.3.9. Comparison of transport, optical, and crystallographic properties........478
12.4. Shallow p^+ layers in GaInAs grown by MOCVD by mercury implantation 479
12.5. GaInAs–InP heterojunctions: Multiquantum wells and superlattices grown by MOCVD 482
    12.5.1. Growth technique...483
    12.5.2. Structural characterization of GaInAs–InP quantum wells grown by MOCVD........484
    12.5.3. Optical properties of GaInAs–InP quantum wells................489
    12.5.4. Room-temperature excitons in GaInAs–InP superlattices grown by MOCVD................493

12.5.5. Negative differential resistance at room temperature from
resonant tunneling in GaInAs/InP double-barrier
heterostructures 495
12.6. Magnetotransport in GaInAs–InP heterojunctions grown
by MOCVD.. 498
12.6.1. Shubnikov–de Haas and quantum Hall effects.................498
12.6.2. Observation of a two-dimensional hole gas in a
$Ga_{0.47}In_{0.53}As/InP$ heterojunction grown by MOCVD.............505
12.6.3. Precise quantized Hall resistance measurements in
$In_xGa_{1-x}As/InP$ heterostructures...511
12.6.4. Persistent photoconductivity and the quantized Hall effect
in $In_{0.53}Ga_{0.47}As/InP$ heterostructures grown by MOCVD 513
12.6.5. The effect of hydrostatic pressure on a $Ga_{0.47}In_{0.53}As/InP$
heterojunction with three electric subbands 516
12.6.6. Cyclotron resonance..522
12.6.7. Shallow-donor spectroscopy and polaron coupling in
$Ga_{0.47}In_{0.53}As$–InP grown by MOCVD....................................530
12.7. Applications of GaInAs–InP system grown by MOCVD 533
12.7.1. PIN photodetector 533
12.7.2. Field-effect transistor 535
12.7.3. GaInAs–InP optical waveguides 537
References 539

**13. GaInAsP–InP System: MOCVD Growth, Characterization,
and Applications 545**
13.1. Introduction 545
13.2. Growth conditions 546
13.2.1. Orientation effects...549
13.2.2. Carrier-gas effects...550
13.2.3. Pyrolysis-oven effects 550
13.2.4. Photoluminescence spectra...551
13.3. Characterization... 553
13.3.1. Calorimetric absorption and photoluminescence studies of
interface disorder in InGaAsP–InP quantum wells 553
13.3.2. Magnetotransport 557
13.3.3. Observation of quantum Hall effect in a GaInAsP–InP
heterostructure grown by MOCVD..564
13.3.4. Disorder of a $Ga_xIn_{1-x}As_yP_{1-y}$–InP quantum well by Zn
diffusion 568
13.3.5. Interface study of $Ga_xIn_{1-x}As_yP_{1-y}$–InP heterojunctions
grown by MOCVD by spectroscopic ellipsometry 573
13.4. Applications of GaInAsP–InP systems grown by MOCVD.. 583
13.4.1. Broad-area 1.2–1.6 μm $Ga_xIn_{1-x}As_yP_{1-y}$–InP DH lasers
grown by MOCVD...583
13.4.2. Buried-ridge-structure lasers grown by MOCVD 594
13.4.3. Distributed feedback lasers fabricated on material grown
completely by MOCVD 612

13.4.4. CW phase-locked array $Ga_{0.25}In_{0.75}As_{0.5}P_{0.5}$–InP high-
          power semiconductor laser grown by MOCVD 625
13.4.5. GaInAsP–InP quantum-well lasers..629
13.4.6. Buried waveguides in InGaAsP–InP material grown by
          MOCVD..634
References 636

**14. Strained Heterostructures: MOCVD Growth,
     Characterization, and Applications 641**
14.1. Introduction 641
14.2. Growth procedure and characterization................................ 642
     14.2.1. Growth procedure...642
     14.2.2. Secondary-ion mass spectrometry 644
     14.2.3. Auger analysis...646
     14.2.4. Etch-pit density..647
14.3. Growth of GaInAs–InP multiquantum wells on $Gd_3Ga_5O_{12}$
     garnet (GGG) substrates 649
14.4. Applications... 653
     14.4.1. Photonic devices based on strained layers.............................653
     14.4.2. Electronic devices—GaInP/GaInAs/InP MESFET 654
     14.4.3. Optoelectronic integrated circuits (OEIC)............................655
14.5. Monolayer epitaxy of $(GaAs)_n(InAs)_n$–InP by MOCVD 662
     14.5.1. Multiquantum wells of $(GaAs)_2(InAs)_2$/InP 664
References 666

**15. MOCVD Growth of III–V Heterojunctions and
     Superlattices on Silicon Substrates 669**
15.1. Introduction 669
     15.1.1. The difference in lattice parameter.......................................670
     15.1.2. The difference in lattice symmetry.......................................671
     15.1.3. The difference in thermal expansion coefficient 672
     15.1.4. Preparation and chemical etching of Si substrates 672
     15.1.5. The cross doping of the III–V epilayer with Si 673
15.2. MOCVD growth of GaAs on silicon.................................... 673
15.3. InP grown on silicon... 675
15.4. GaInAsP–InP grown on silicon 679
15.5. Applications... 683
     15.5.1. Room-temperature CW operation of a GaInAsP/InP
              ($\lambda = 1.15$ μm) light-emitting diode on silicon substrate 683
     15.5.2. GaInAsP/InP double heterostructure laser emitting at
              1.3 μm on silicon substrate..686
     15.5.3. CW operation of a $Ga_{0.25}In_{0.75}As_{0.5}P_{0.5}$/InP BRS laser on
              silicon substrate..690
     15.5.4. InGaAs–InP MQW on Silicon substrate 693
     15.5.5. GaInAs–InP PIN photodetector...694
References 698

16. Optoelectronic Devices Based on Quantum Structures 701
16.1. Introduction 701
16.2. GaAs- and InP-based quantum well infrared
    photodetectors (QWIP) 702
    16.2.1. GaAs-based QWIPs 704
    16.2.2. InP-based QWIPs 704
    16.2.3. InP-based nanopillar QWIPs 705
16.3. Self-assembled quantum dots, and quantum dot–based
    photodetectors 706
    16.3.1. Growth of InAs QDs on InP 707
    16.3.2. The quantum dot-in-well infrared photodetector (QDWIP)...710
    16.3.3. QDWIP device performance 711
    16.3.4. The QDIP structure 713
    16.3.5. Gain in QDIP and QDWIP 715
    16.3.6. QDIP performance 717
16.4. Quantum dot lasers 718
16.5. InP based quantum cascade lasers (QCLs) 720
    16.5.1. Quantum cascade laser growth 721
    16.5.2. Short wavelength QCLs 725
    16.5.3. Middle wavelength QCLs 726
    16.5.4. Long wavelength QCLs 726
    16.5.5. Photonic crystal distributed feedback QCLs 727
References 728

Appendices 731
A.1. Effect of substrate miscut on the measured superlattice
    period 733
A.2. Optimization of thickness and indium composition of
    InGaAs wells for 980 nm lasers 737
A.3. Energy levels and laser gains in a quantum well
    (GaInAsP): The "effective mass approximation" 743
A.4. Luttinger–Kohn Hamiltonian 747
A.5. Infrared detectors 751
A.6. Physical properties and safety information of
    metalorganics 757
MOCVD Challenge Volume 1: Original foreword 769
MOCVD Challenge Volume 1: Original introduction 771
MOCVD Challenge Volume 2: Original foreword 772
MOCVD Challenge Volume 2: Original introduction 773

Index 775

Foreword

Epitaxial technologies have come a long way, after taking years of considerable effort to make them work. We now see the results of this gigantic achievement in almost every electronic and optoelectronic device that we encounter — whether it be in the electronics that flood the consumer market, the communications infrastructure that is working to shrink the world, or in the specialized components used for defense and security. Industry is producing tens of millions of sophisticated components every year. Metal Organic Chemical Vapor Deposition (MOCVD) technology is an integral part of making it possible to realize such a diverse array of devices. With it large-scalability and fast turn-around, MOCVD has proven well adapted to industrialization and thus has driven the development of low-cost optoelectronics. This field grew immensely in the last several years, in particular due to the rapid development of diode lasers and LEDs, ranging from the ultraviolet to the infrared.

The author of this book, Manijeh Razeghi, is herself, one of the preeminent founders of semiconductor epitaxy. She was the driving force behind the development of MOCVD technology. Manijeh has been recognized all over the world for her ability to achieve astonishingly brilliant results with MOCVD technologies. I remember in 1981 when Manijeh first arrived at the Thomson-CSF laboratory (which I directed at that time). This was soon after completing her thesis, and she was already full of enthusiasm and energy, an attitude which she has retained up to now. She has always demonstrated a rare premonition of what is going to be important — an impressive number of her papers have led to opening new avenues for optoelectronics research. In her laboratory at Thomson-CSF, she created an extremely stimulating and passionate atmosphere, transferring her own passion for the science and technology to her collaborators and to her students. In 1991 she left Thomson-CSF to become a professor at Northwestern University; there she founded the Center for Quantum Devices and has built a new world-class laboratory from the ground up where she is still continuing to foster the same sort of passionate research. Today, she has former students all over the world, who have gained a strong technical background from working with Manijeh and can be seen continuing to disseminate that same passion for research.

Manijeh Razeghi's early work on MOCVD immediately attracted the attention of famous scientists such as Cyril Hilsum (who was, at the time, the director of research at RSRE Malvern), and the soon to be Nobel laureates, Claus von Klitzing and Daniel Tsui. They saw, in her work on InP, the opportunity to develop the brilliant new properties of the III–V materials, such as the direct band gap, the valley structures, and the low effective masses, which led to the Gunn–Hilsum effect, quantum Hall effect, the family of high mobility transistors, and the

xvii

development of world-record lasers. It was Cyril Hilsum who first urged Manijeh Razeghi to write the original edition of The MOCVD Challenge and discuss the MOCVD growth of InP and related compounds. Despite initial hesitation, she finally gave in, producing what is still one of the most comprehensive and informative books on the growth of InP-based heterostructures and devices that currently exist.

Since the first edition was published, Manijeh Razeghi has made a number of additional world class contributions, including the development of new material systems and the demonstration of optoelectronic devices covering the infrared, visible, and ultraviolet spectra. Of course, all of these accomplishments have the same fundamental basis, a deep understanding of semiconductor epitaxy. This new edition combines and updates the work of the first two volumes of *The MOCVD Challenge* and gives the reader a comprehensive, stimulating, and updated view into the development of this exciting and important field.

Erich Spitz

French Academy of Technology
French Academy of Science

Preface

This book represents the combined updated version of Volumes 1 and 2 of *The MOCVD Challenge*. The first volume started with an in-depth overview of the growth, characterization and applications of InP and its related compounds. At the time, many publishers had asked me to write a book about MOCVD; however, it was Dr. Cyril Hilsum, then a consultant editor of the Institute of Physics Publishing (IOP), who was most persistent. In short, I relented and wrote *The MOCVD Challenge Volume 1: A Survey of GaInAsP–InP for Photonic and Electronic Applications*. Toward the end of this work Dr. Hilsum asked about GaAs and its related compounds, saying that it was impossible to write a book on MOCVD without including GaAs. I told him that as I had only just started working with this material system, and that I was leaving this for Volume 2 because I wanted to gain further insight and experience before writing about it.

Five years later I had developed a deep understanding of GaAs-related materials and began work on the second volume; this new book was titled *The MOCVD Challenge Volume 2: A survey of GaInAsP–GaAs for photonic and electronic device applications*. The research charted much of the pioneering work that I completed in the growth of GaInAsP–GaAs in collaboration with and with the encouragement of many excellent scientists from around the world, especially Klaus von Klitzing's group. The aim of Volume 2 was to provide insight, detail, and an overview of the MOCVD growth process for GaAs and its related compounds. The GaAs–GaInP heterojunction has been of particular strategic interest as a robust and powerful alternative to GaAs–AlGaAs for devices with improved performance and reliability. Much of Volume 2 also focused on the growth of a sophisticated GaInAsP quaternary material, addressing issues such as crystal quality, optical properties, and composition control. The driving force behind my work was the reliable growth of high-quality phosphorus-based materials for low-dimensional electronic, photonic and optoelectronic integrated circuit (OEIC) devices. Naturally, a key element of this work is the application of GaAs and its related compounds. It is through the development of these important material systems that MOCVD has been demonstrated as the technique of choice for industrial mass production as well as for cutting-edge research.

In this new combined book, Chapter 1 is devoted to an introduction to semiconductor compounds and Chapter 2 describes the MOCVD growth process. Chapter 3 discusses *in situ* characterization for MOCVD growth — in particular, reflection difference spectroscopy (RDS). This technique is important because it is a direct analogue of the RHEED technique for MBE growth. The RDS technique has proved critical in the optimization of reactor design and improving the quality of the material and understanding the properties of heterojunction interfaces. Chapter 4 discusses *ex situ* characterization techniques. Chapter 5 covers the specifics of the growth of GaAs as the basis for Chapter 6 which covers the growth and characterization of the GaAs–GaInP system. Chapter 7 describes optical devices based on GaAs and related compounds whereas Chapter 8 details the specifics of GaAs-based laser diode structures. Chapter 9 discusses electronic devices, and finally Chapter 10 provides an overview of optoelectronic integrated circuits (OEICs). Most of this material is based on the original Volume 2. However, this new book combines and integrates material from both volumes where applicable and focuses on logical presentation of the material.

The second half of the book covers InP and related materials. Chapter 11 covers InP-InP systems, Chapters 12 and 13 cover GaInAs(P)-InP MOCVD growth characterization and application, Chapter 14 covers strained heterostructures, Chapter 15 covers the MOCVD growth of III–V heterojunctions and superlattices on silicon substrates and their applications. Most of this material is based on the original Volume 1. Chapter 16 is wholly new and contains an update on some of the most exciting recent achievements in this field.

The results reported in this combined book span more than 20 years of research. It would have been impossible without my excellent groups at Thomson-CSF in the Exploratory Material Laboratory and, more recently, at the Center for Quantum Devices at Northwestern University. I would like to thank all of my students and colleagues; this book is for them, and all of my future students and colleagues.

I am particularly indebted to Dr. Hilsum, who reviewed and edited Volume 1 of *The MOCVD Challenge* and large parts of Volume 2. He was the driving force behind this series. In addition, I am also grateful to Dr. Clivia Sotomayor Torres and Professor Greg Stillman who also reviewed and edited the original manuscripts. Their suggestions made an invaluable contribution to this work.

Since I joined Northwestern University in 1991 I have had the support and encouragement of many scientists, in particular Drs. Leo Esaki, Lester Eastman, George Wright, Yoon-Soo Park, Max Yoder, L.N. Durvasula, Ray Balcerak, Henry Everitt, Gail Brown, G. Witt, W. Mitchel, R. Burnham, John Fan, Mark Spitzer, R. Bredhauer, E. Spitz, G. Nuzillat, D. Kaplan and J.P. Duchemin. Moreover, the Center for Quantum Devices would not have come into existence except for the original vision of Northwestern University. I am grateful to Drs. A. Weber, H. Bienen, D. Cohen, W. Kern, W. Fischer, and A. Haddad for their unequivocal support since I joined Northwestern University. I am particularly indebted to Dean Jerome Cohen for convincing me that Northwestern University is the place to do research, for facilitating my move from Thomson-CSF, and for his continuing support and encouragement.

Manijeh Razeghi
Walter P. Murphy Professor of Electrical Engineering and Computer Science
Director Center for Quantum Devices
McCormick School of Engineering and Applied Science
Northwestern University, Evanston, Illinois

1. Introduction to Semiconductor Compounds

1.1.   Introduction
1.2.   III–V semiconductor alloys
    1.2.1.   III–V binary compounds
    1.2.2.   III–V ternary alloys
    1.2.3.   III–V quaternary compounds
1.3.   III–V semiconductor devices
1.4.   Technology of multilayer growth
    1.4.1.   Epitaxial technology
    1.4.2.   Review of III–V heterostructures grown by LPE, MBE and MOCVD

1.1. Introduction

The *MOCVD Challenge Volume I* concentrated on MOCVD growth of InP–GaInAsP multilayers for photonic and electronic devices. The second Volume concentrated on MOCVD growth of GaAs and related alloys for photonic and electronic devices. Both of these volumes have now been combined together and updated to form this new edition. This new book incorporates material from Volume 1 and Volume 2 and provides a state of the art review of methods of producing ultra thin accurately controlled epitaxial layers of $Ga_xIn_{1-x}As_yP_{1-y}$ binary ternary and quaternary semiconductor multilayers and microstructure on a large variety of substrates.

1.2. III–V semiconductor alloys

1.2.1. III–V binary compounds

Most of the III–V semiconductor compounds have the zinc blende structure. Table 1.1 shows the bandgap energy, lattice parameter, refractive index, dielectric constant, conduction band effective mass and valence band effective masses of some of the binary III–V compounds. These binary semiconductors are generally useful as substrates and some binary, ternary and quaternary alloys of these compounds have certain useful characteristics. Some of the ternary and quaternary alloys are lattice matched to certain binary compounds with different bandgap energies providing interesting physical properties.

III–V binary compound	Atomic number $\bar{Z}$	Lattice parameter (Å)	Bandgap energy (eV)	Refractive index ($\bar{n}$)	Effective mass			Dielectric constant ($\varepsilon/\varepsilon_0$)	Electron affinity χ (eV)
					(m_c/m_0)	(m_{hh}/m_0)	(m_{lh}/m_0)		
InSb	50	6.479 37	0.17	4.0	0.0145	0.44	0.016	17.7	4.69
InAs	41	6.0584	0.36	3.520	0.022	0.41	0.025	14.6	4.45
GaSb	41	6.09593	0.73	3.820	0.044	0.33	0.056	15.7	4.03
InP	32	5.868 75	1.35	3.450	0.078	0.8	0.012	12.4	4.4
GaAs	32	5.653 21	1.424	3.655	0.065	0.45	0.082	13.1	4.5
AlSb	32	6.133 5	1.58	3.400	0.39	0.5	0.11	14.4	3.64
AlAs	23	5.662 2	2.16	3.178	0.11	—	0.22	10.1	—
GaP	23	5.45117	2.26	3.452	0.35	0.86	0.14	11.1	4.0
AlP	14	5.451	2.45	3.027	—	0.63	0.20	—	—

Table 1.1. Physical constants of some III–V binary compounds at 300 K.

1.2.2. III–V ternary alloys

When more than one element from group III or group V is distributed randomly on group III or group V lattice sites, III–III–V or III–V–V ternary alloys can be achieved. The notation most frequently used is $III_xIII_{1-x}V$ or IIIV_yV_{1-y}. There are 18 possible ternary systems among the group III and group V elements of interest.

The bandgap energy $E_g(x)$ of a ternary compound varies with the composition x as follows:

Eq. (1.1) $\quad E_g(x) = E_g(0) + bx + cx^2$

where $E_g(0)$ is the bandgap energy of the lower-bandgap binary compound and c is the bowing parameter. The bowing parameter c can be theoretically determined [Van Vechten and Bergstresser 1970]. It is especially helpful to estimate c when experimental data are unavailable. The lattice constant a of ternary compounds can be calculated using Vegard's law. According to Vegard's law the lattice constant of the ternary alloys can be expressed as follows:

Eq. (1.2) $\quad a_{alloy} = xa_A + (1-x)a_B$

where a_A and a_B are the lattice constants of the binary alloys A and B. Vegard's law is obeyed quite well in most of the III–V ternary alloys. The compositional dependence of the energy gaps of various III–V ternary alloys at 300 K is given in Table 1.2 [Casey and Panish 1978].

Ternary	Direct energy gap E_g (eV)
$Al_xGa_{1-x}As$	$E_g(x) = 1.424 + 1.247x$
$Al_xIn_{1-x}As$	$E_g(x) = 0.360 + 2.012x + 0.698x^2$
$Al_xGa_{1-x}Sb$	$E_g(x) = 0.726 + 1.139x + 0.368x^2$
$Al_xGa_{1-x}Sb$	$E_g(x) = 0.172 + 1.621x + 0.43x^2$
$Ga_xIn_{1-x}P$	$E_g(x) = 1.351 + 0.643x + 0.786x^2$
$Ga_xIn_{1-x}As$	$E_g(x) = 0.360 + 1.064x$
$Ga_xIn_{1-x}Sb$	$E_g(x) = 0.172 + 0.139x + 0.415x^2$
GaP_xAs_{1-x}	$E_g(x) = 1.424 + 1.15x + 0.176x^2$
$GaAs_xSb_{1-x}$	$E_g(x) = 0.726 - 0.502x + 1.2x^2$
InP_xAs_{1-x}	$E_g(x) = 0.36 + 0.891x + 0.101x^2$
$InAs_xSb_{1-x}$	$E_g(x) = 0.18 - 0.41x + 0.58x^2$

Table 1.2. Compositional dependence of the energy gap in the III–V ternary solid solution at 300 K (after Casey and Panish [1978]).

1.2.3. III–V quaternary compounds

The interest in quaternary alloys has centered on their use in conjunction with binary and ternary alloys to form lattice-matched heterojunction structures with different bandgaps. The reduction of stress in $Al_xGa_{1-x}As$ As layers grown on GaAs substrates is illustrated by the introduction of small amounts of P to give the quaternary $Al_xGa_{1-x}P_yAs_{1-y}$. The $InP/Al_xGa_{1-x}P_yAs_{1-y}$ heterojunction serves as a successful example of a binary–quaternary lattice-matched system.

Ilegems and Panish [1974] calculated quaternary phase diagrams with the solid decomposed into ternary alloys: ABC, ACD, ABD and BCD (where A and B are group III elements, and C and D are group V elements). Jordan and Ilegems [Jordan et al. 1974] obtained equivalent formulations considering the solid as a mixture of binary alloys: AC, AD, BC and BD. Assuming a linear dependence on composition of lattice parameter a_{AC} for the binary AC, and similarly for the other lattice parameters, the lattice parameter of the alloy $A_xB_{1-x}C_yD_{1-y}$ is

Quaternary	Lattice-matched binary	Wavelength, λ (μm)
$Al_xGa_{1-x}P_yAs_{1-y}$	GaAs	0.8–0.9
$Al_xGa_{1-x}As_ySb_{1-y}$	InP	1
$Al_xGa_{1-x}As_ySb_{1-y}$	InAs	3
$Al_xGa_{1-x}As_ySb_{1-y}$	GaSb	1.7
$Ga_xIn_{1-x}P_yAs_{1-y}$	GaAs, InP	1–1.7
$Ga_xIn_{1-x}P_ySb_{1-y}$	InP, GaSb, AlSb	2
$In(P_xAs_{1-x})_y Sb_{1-y}$	AlSb, GaSb, InAs	2–4
$(Al_xGa_{1-x})_y In_{1-y}P$	GaAs, $Al_x Ga_{1-x}$ As	0.57
$(Al_xGa_{1-x})_y In_{1-y}As$	InP	0.8–1.5
$(Al_xGa_{1-x})_y In_{1-y}Sb$	AlSb	1.1–2.1

Table 1.3. Binary to quaternary III–V lattice-matched systems of multilayer heterostructures [Casey and Panish 1978].

Eq. (1.3) $a_0 = xya_{AC} + x(1-y)a_{AD} + (1-x)ya_{BC} + (1-x)(1-y)a_{BD}$

The bandgap energy determination is more complicated. However, if the bowing parameter (c) is neglected, the bandgap energy may be approximated from the bandgap of the binaries, assuming linear variation:

Eq. (1.4) $E_g = xyE_{AC} + x(1-y)E_{AD} + (1-x)yE_{BC} + (1-x)(1-y)E_{BD}$

The binary to quaternary III–V lattice-matched systems for multilayer heterostructures are listed in Table 1.3.

1.3. III–V semiconductor devices

The most important semiconductor multilayer devices fall into two groups:

(i)   optoelectronic devices (Table 1.4), and
(ii)   electronic devices (Table 1.5).

Device	Operational principle
Light-emitting diodes (LEDs) and laser diodes (LDs)	Convert electrical energy into optical radiation
Solar cells	Convert optical radiation into electrical energy
Modulators	Convert electrical signals into optical signals
Photodetectors (PDs)	Convert optical signals into electric current or voltage

Table 1.4. Main optoelectronic devices.

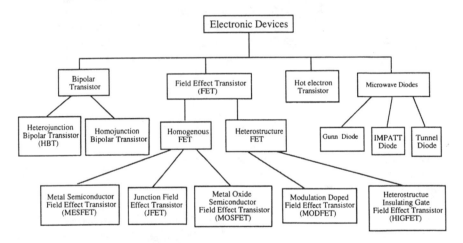

Table 1.5. Most commonly used electronic devices.

Electronic devices

In bipolar devices both electrons and holes are involved in the transport processes. The bipolar transistor in which there is interaction between two closely coupled *p–n* junctions is one of the most important semiconductor devices. The thyristor, which basically comprises three closely coupled *p–n* junctions in the form of a multilayered *p–n–p–n* structure, exhibits bistable characteristics and can be switched between a high-impedance, low-current OFF state, and a low-impedance, high-current ON state.

In unipolar devices only one type of carrier participates predominantly in the conduction mechanism. The junction field effect transistors (JFETs), and metal oxide semiconductor devices (MOSFETs) are the most important devices for very large-scale integrated (VLSI) circuits.

(1) Microwave diodes can be made with operating frequencies covering the range from about 0.1 GHz to 1000 GHz with corresponding wavelengths from 300 cm to 0.3 mm.

Tunnel diode devices are associated with quantum tunneling phenomena (a majority carrier effect). The tunneling time of carriers through the potential energy barrier is very short, permitting the use of tunnel devices well into the millimeter-wave region [Esaki 1958].

The IMPATT (impact ionization avalanche transit time) diode is one of the most powerful solid state sources of microwave power [Sze and Ryder 1971]. IMPATT diodes employ impact ionization and transit time properties of semiconductor structures to produce negative resistance at microwave frequencies. The Gunn diode discovered by Gunn [1963, 1964], is a transferred electron device. He found that a coherent microwave output was generated when a DC electric field in excess of a critical threshold value of several thousand volts per centimeter was applied across an *n*-type sample of GaAs or InP. Earlier, Hilsum [1962] and Ridley [Ridley and Watkins 1961] had proposed a theory of negative differential resistance, which is due to a field-induced transfer of conduction band electrons from a low-energy, high-mobility valley to higher-energy, low-mobility satellite valleys that could result in microwave oscillation. The transferred electron effect has been referred to as the Gunn effect or as the Hilsum–Ridley effect.

Optoelectronic devices

Optoelectronic and photonic devices are those in which photons play a major role. These devices can detect, generate and convert optical energy to electrical energy (photodetectors, PD) or vice versa (light-emitting diodes (LEDs) and laser diodes (LDs)). Optoelectronic and photonic devices can be divided into several groups: solar cells, photodetectors, semiconductor light-emitting diodes, laser diodes and optical modulators.

Solar cells convert optical radiation into electrical energy. As worldwide energy demand increases, natural resources of fossil fuels will become exhausted. The solar cell can be considered as a major candidate for obtaining energy from the sun. The solar cell was first developed by Chapin et al. [1954] using a diffused silicon *p–n* junction. To date, solar cells have been made from many III–V semiconductors.

(a) Photodetector devices convert the received optical power into an electrical current or voltage which is amplified and processed to deliver information in a useful form. In general, the performance of a photodetector depends on two parameters: the quantum efficiency (the number of electron–hole pairs collected per incident photon), and the energy gap of the material with a high absorption coefficient ($\simeq 10^3$ cm^{-1}). Above the band gap, III–V multilayer photodiodes can absorb efficiently in a layer only a few micrometers thick, leading to a small device structure and a short carrier transit time.

(b) The semiconductor light-emitting diode and laser diode devices convert electrical energy into optical radiation. The demonstration of the injection laser and continuous wave (CW) operation at room temperature had to await not only the basic invention of the laser and the advances in semiconductor physics of the 1950s, but also the development of suitable materials technology for the growth of several new semiconductors. Semiconductor lasers are considered to be the most important light sources for optical fiber communication systems.

The principal materials for semiconductor lasers were the direct energy gap compounds and alloys between elements of group III and group V of the periodic table. In a series of binary III–V compounds like InP, GaAs and GaP, increasing the atomic weight of the group III or group V element in general decreases the energy gap (E_g) and increases the refractive index $\bar{n}$ with exception of GaAs–AlAs and GaP–AlP, where the lattice parameter a increases. The crystalline solid solutions between these binary compounds, i.e. alloys, usually have properties intermediate between the end components as described in Eq. (1.3) and Eq. (1.4).

Heterostructure lasers are layered semiconductor structures in which the lattice parameter is usually held constant from layer to layer while E_g and $\bar{n}$ are varied. The quaternary solid solution systems provide continuously variable E_g and $\bar{n}$ for various compositions at constant a_0 and permit additional possible heterostructure laser semiconductors.

The active layer in a double heterostructure (DH) laser is the region where the light is generated (Fig. 1.1). The ratio of the optical intensity within the active layer to the total optical intensity is the confinement factor, and is used to represent the effect of the waveguide parameters on the threshold current density (Fig. 1.2). The DH laser provides a well defined dielectric waveguide and was the first injection laser structure to permit CW operation at room temperature.

Virtually all lasing semiconductors have direct bandgaps. This is to be expected since the radiative transition in a direct-bandgap semiconductor is a first-order process, and the transition probability is high. Some of the binary–ternary and binary–quaternary III–V lattice-matched systems for DH laser applications are listed in Table 1.6 and Table 1.7. The bandgap energy of the recombination region (active layer) controls the emission wavelength

Eq. (1.5) $$\lambda(\mu\mathrm{m}) = \frac{hc}{E_g} = \frac{1.24}{E_g\,(\mathrm{eV})}$$

where E_g is the energy gap, c is the velocity of light in vacuum and h is Planck's constant.

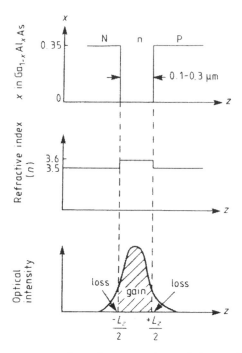

Fig. 1.1. Alloy composition, the refractive index and the confinement of light in the GaAs–GaAlAs double-heterostructure lasers.

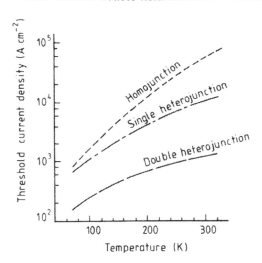

Fig. 1.2. Variation of threshold current density (Jth) versus temperature for a homojunction, single heterojunction and double heterojunction of GaAs–GaAlAs lasers.

Ternary	Lattice-matching binary substrate	Bandgap E_g (eV, 300 K)	Wavelength λ (μm, 300 K)	DH for laser
$Ga_{0.9}In_{0.1}Sb$	AlSb (indirect)	0.633	2.0	$AlSb/Ga_{0.9}In_{0.1}Sb/AlSb$
$InAs_{0.82}Sb_{0.18}$	AlSb	0.23	0.59	$AlSb/InAs_{0.82}Sb_{0.18}/AlSb$
$Ga_{1-x}Al_xAs$	GaAs	1.49–1.9	0.7–0.9	$Ga_{1-x}Al_xAs/GaAs/Ga_{1-x}Al_xAs$
$Ga_{0.49}In_{0.51}P$	GaAs	1.883	0.65	$Ga_{0.49}In_{0.51}P/GaAs/Ga_{0.49}In_{0.51}P$
$Al_{0.49}In_{0.51}P$	GaAs	2.3		$Al_{0.49}In_{0.51}P/Ga_{0.49}In_{0.51}P/Al_{0.49}In_{0.51}P$
$AlAs_{0.08}Sb_{0.92}$	GaSb	0.72	1.72	$AlAs_{0.08}Sb_{0.92}/GaSb/AlAs_{0.08}Sb_{0.92}$
$InAs_{0.91}Sb_{0.09}$	GaSb	0.287	4.3	$GaSb/InAs_{0.91}Sb_{0.09}/GaSb$
$Ga_{0.47}In_{0.53}As$	InP	0.75	1.7	$InP/Ga_{0.47}In_{0.53}As/InP$
$Al_{0.47}In_{0.53}As$	InP	1.46	0.85	$Al_{0.47}In_{0.53}As/InP/Al_{0.47}In_{0.53}As$
$AlAs_{0.16}Sb_{0.84}$	InAs	0.36	3.4	$AlAs_{0.16}Sb_{0.84}/InAs/AlAs_{0.16}Sb_{0.84}$
$GaAs_{0.08}Sb_{0.92}$	InAs	0.36		$GaAs_{0.08}Sb_{0.92}/InAs/GaAs_{0.08}Sb_{0.92}$

Table 1.6. Ternary III–V lattice matched to binary III–V systems for laser applications using multilayer heterostructures.

Substrate	Quaternary	Bandgap energy E_g (eV)	Wavelength λ (μm)
InAs	$In_xGa_{1-x}As_ySb_{1-y}$ $(1-y)=0.916(1-x)$	0.35–0.7	3.5–1.77
InP	$In_xGa_{1-x}As_yP_{1-y}$ $Y=2.16(1-x)$	0.75–1.35	1.7–0.9
GaAs	$In_xGa_{1-x}As_yP_{1-y}$ $(1-y)=2.04x$	1.42–1.9	0.87–0.65

Table 1.7. Quaternary III–V lattice matched to binary III–V systems for laser applications using multilayer heterostructures.

1.4. Technology of multilayer growth

Here we consider some of the important materials for microwave and optoelectronic devices, i.e. GaAs, $Ga_xAl_{1-x}As$ and $Ga_xIn_{1-x}P$. In the GaAs–GaAlAs system, the lattice constant of GaAlAs is almost identical to that of the GaAs substrate so there is no lattice mismatch problem. In the $Ga_xIn_{1-x}P$ system, the lattice constants of GaInP can vary widely and the quality of the epilayer depends on how closely the lattice constant of the GaInP is matched to that of the GaAs substrate. However, the technology behind liquid phase epitaxy (LPE), vapor phase epitaxy (VPE), molecular beam epitaxy (MBE) and metalorganic chemical vapor deposition (MOCVD) provides the means to achieve multilayer structures in this III–V system. Devices such as diode lasers, LEDs, detectors, FETs, phototransistors and waveguides fabricated with these two material systems have been prepared by these four growth processes.

1.4.1. Epitaxial technology

A wide range of semiconducting materials, particularly materials based on InP and GaAs, need to be grown as thin single-crystal films for a variety of applications. The discovery of quantum wells and superlattices has revolutionized the area of semiconductor technology in terms of new devices. These new devices require precise control and uniformity of thickness, excellent homogeneity, high purity, very sharp interfaces between the substrate and epitaxial layers, and low misfit dislocations in the epilayers. In the last few decades, epitaxial techniques have advanced to a level where such requirements can be easily met by a variety of growth techniques. Each technique has its own strengths and weaknesses.

(a) Liquid phase epitaxy

LPE involves precipitation of material from a supercooled solution onto an underlying substrate. The composition of the layers formed on the substrate depends mainly on the equilibrium phase diagram and, to a lesser extent, on the orientation of the substrate. There are basically three parameters that affect the growth in LPE. These are growth temperature, growth time and melt composition. The first LPE growth apparatus was developed by Nelson [1963]. The LPE reactor consists of a horizontal furnace system and a sliding graphite boat. The LPE apparatus is quite simple and excellent quality layers and high purity levels can be achieved.

The advantages of LPE are the simplicity of equipment, higher deposition rates and the high-purity layers that can be obtained. Background elemental impurities are eliminated due to the availability of high-purity metals and the inherent purification process that occurs during the liquid to solid phase transition. For example, during the purification stage of Al-containing material, oxygen in the system forms highly stable Al_2O_3 on the surface of the liquid which prevents oxygen incorporation into the epitaxial layer.

The limitations of LPE technology are the poor thickness uniformity of the epitaxial layers and difficulty in growing multilayer structures with extremely abrupt interfaces due to the high growth rate and meltback effect. Another limitation is the difficulty in growing certain materials. For example, in alloys containing both Al and In, the high Al distribution coefficient leads to much difficulty in growing alloys with a high composition of In.

(b) Vapor phase epitaxy

VPE, like LPE, is a thermodynamic equilibrium growth technique. The growth of InP-based materials can be obtained in a fused-silica reactor composed of two zones set at different temperatures by using a multielement furnace that surrounds the reactor. Hydrogen or hydrogen chloride (HCl) are often used as a carrier gas, and arsine (AsH_3) and phosphine (PH_3) as arsenic (As) and phosphorus (P) sources, respectively. Pure indium and gallium metal can be used as group III elemental sources. In the first zone of the reactor (source zone), which is held at 750°C to 800°C, the gaseous species to be transported are synthesized following the reaction

Eq. (1.6) $In + HCl \rightarrow InCl + \frac{1}{2} H_2$

In the second zone (deposition zone), which has a temperature range 650–750 °C, growth occurs via the reaction

Eq. (1.7) $\frac{1}{4} P_2 + InCl + \frac{1}{2} H_2 \rightarrow InP + HCl$

One of the major advantages of VPE over LPE is the possibility of localized epitaxy. This means that if one grows InP-based material on an InP substrate consisting of a stripe of SiO_2 or Si_3N_4, there is monocrystal growth on the substrate, but no growth on the dielectric (SiO_2 or Si_3N_4) surface, which could be important for monolithic integrated circuits on InP substrates.

The disadvantages of VPE include the potential for hillock and haze formation, and interfacial decomposition during the preheat stage. It is very difficult, if not impossible, to grow superlattice structures. Alternating layers are normally obtained by physically moving the substrate back and forth between reactor tubes. This approach is unattractive when compared with techniques such as MBE or MOCVD where the transport of source materials rather than the substrates is controlled.

(c) Molecular beam epitaxy

The MBE process involves the evaporation of elemental sources at a controlled rate onto a crystalline substrate surface held at a suitable temperature under ultrahigh-vacuum (UHV) conditions. Therefore, it is a high-vacuum technique where beams of evaporated molecules or atoms are focused onto the substrate. The use of UHV introduces two advantages. Firstly, atoms and molecules reach the growth surface in a very clean condition. Secondly, the growth process can be monitored *in situ* by diagnostic techniques as the crystal grows one atomic layer at a time [Parker 1985]. *In situ* diagnostic techniques such as reflection high-energy electron diffraction (RHEED), Auger electron spectroscopy (AES), x-ray photoelectron spectroscopy (XPS), low-energy electron diffraction (LEED), secondary-ion mass spectroscopy (SIMS) and ellipsometry are used.

MBE has attracted much interest as an excellent crystal growth technology for the production of complex and varied structures, especially for GaAs-based multilayer structures. This is due to its capability for extremely precise control over layer thickness and doping profile. However, the disadvantage of MBE is that it is expensive because of the need for UHV apparatus. Another major problem is the difficulty in growing phosphorus-containing materials such as InP and GaInAsP. Phosphorus is found to bounce around in the system, eventually collecting in the vacuum pumps. In addition, the growth of alloys containing both As and P is particularly difficult.

(d) Metalorganic chemical vapor deposition

Growth of III–V compounds from organometallic and hydride sources was first reported in 1960 by Didchenko et al. [1960]. In their experiment, deposition of InP from trimethylindium (TMIn) and phosphine (PH_3) was obtained in a closed-tube system. Later, Manasevit [1968, 1971, Manasevit et al. 1971] established the possibility of depositing many common compound semiconductors from organometallic materials and he coined the term metalorganic chemical vapor deposition (MOCVD). It is also called metalorganic vapor phase epitaxy (MOVPE). This technique for epitaxial growth of compound semiconductors using organometallic precursors has advanced rapidly during the past decades. In recent years, MOCVD has established itself as an important epitaxial crystal growth technique, yielding high-quality low-dimensional structures for fundamental semiconductor physics research and useful electronic and photonic semiconductor devices.

In MOCVD, chemically active species interact to produce a corresponding epitaxial layer either in the vapor phase or on a solid surface of the substrate. In particular, growth of semiconductor III–V compounds results from introducing metered amounts of group III alky Is and group V hydrides into a quartz tube which contains a substrate placed on an heated susceptor. The hot susceptor has a catalytic effect on the decomposition of the gaseous products and the growth therefore primarily takes place on this hot surface. MOCVD is attractive because of its relative simplicity compared to other growth methods. It can produce heterostructures, multiquantum wells and superlattices with very abrupt transitions in composition as well as in doping profiles in continuous growth by rapid changes of the gas composition in the reaction chamber.

The technique is also attractive because of its ability to grow uniform layers, low-background doping density, sharp interfaces and the potential for large-scale commercial applications. MOCVD can be used to prepare multilayer structures with thicknesses as thin as a few atomic layers. This allows the study and incorporation into device applications of the two-dimensional electron gas (2DEG) [Delahaye et al. 1986, Guldner et al. 1986, Razeghi et al. 1986], two-dimensional hole gas (2DHG) [Razeghi et al. 1986, Rogers et al. 1986] and transport and quantum size effects (QSE) [Razeghi 1983] in a variety of III–V compound semiconductors, heterojunctions and multilayers [Razeghi et al. 1986, Nicholas et al. 1985]. It also makes it possible to "engineer the bandgap" by growing a predetermined alloy composition and doping profile [Razeghi et al. 1986]. As a result, an entirely new class of electronic and photonic devices is realized [Razeghi et al. 1987]. Another recent advance is the ability to grow strained-layer superlattices, in which the crystal lattices of the two materials are not very closely matched and there is a built-in strain in each layer [Osbourn 1982].

The main disadvantage of the MOCVD growth technique is the use of large quantities of poisonous gases such as AsH_3 and PH_3. However, the recent development of less hazardous precursors for MOCVD has made it a very powerful technique for epitaxial growth.

(e) Metalorganic molecular beam epitaxy

Low-pressure MOCVD techniques have been shown to provide high-quality films by eliminating the parasitic reactions normally observed in atmospheric reactors. This idea of low pressure was further extended to high vacuum to grow high-quality GaAs, GaAsP and GaInAs films [Fraas 1981].

Since conventional MBE has the disadvantage of difficulty in the growth of phosphide compounds, the solid sources for the group V element, usually As and P, are replaced with the gaseous sources AsH_3 and PH_3, giving rise to gas source molecular beam epitaxy (GSMBE). In addition, even group III solid elements in MBE have been replaced by simple metalorganic compounds—metalorganic molecular beam epitaxy (MOMBE) [Vodjani 1982]. It is also called chemical beam epitaxy (CBE) and metalorganic chemical beam deposition (MOCBD).

Compared to MOCVD growth, the MOCBD technique has the advantages of MBE, such as *in situ* surface diagnostic techniques which provide monitoring of *in situ* etching and the removal of oxides before growth, and is compatible with other high-vacuum thin-film processing.

GSMBE looks more promising, since it uses lower quantities of phosphine and also it reduces the problem of using chemical waste disposal systems such as scrubbers normally used in MOCVD.

(f) Atomic layer epitaxy

Atomic layer epitaxy (ALE) is a new growth technique with control at the monolayer level. ALE was originally proposed by Suntola [Suntola et al. 1980] as a novel mode of preparing thin films of ZnS by evaporative deposition. ALE operates by growing complete layers on top of each other, similarly to evaporative deposition. Thus there are two basic variants: heated elemental source materials and sequential surface exchange reactions between compound reactants. By definition, ALE is based on chemical reactions at the solid surface of a substrate, to which the reactants are transported alternately as pulses of neutral molecules or atoms, either as chopped beams in high vacuum or as switched streams of vapor possibly with an inert carrier gas. The incident pulse reacts directly and chemically only with the outermost atomic layer of the substrate. The film therefore grows stepwise, a single monolayer per pulse, provided that at least one complete monolayer coverage of a constituent element, or of a chemical compound containing it, is formed before the next pulse is allowed to react with the surface.

In metalorganic atomic layer epitaxy (MOALE) [Nishizawa and Kurabayashi 1986, Doi et al. 1986] hydrides are used as the starting materials and instead of alkyls, chloride is used along with hydrides. The technique is called chloride atomic layer epitaxy (chloride-ALE) [Matsumoto and Usui 1986]. Kobayashi et al. [1985] introduced a small amount of AsH₃ during the TEGa exposure step in order to grow η-type GaAs films. They called this technique flow-rate modulation epitaxy (FME). The major advantages of ALE, especially digital ALE, where the thickness grown is insensitive to any analog quantities such as source gas pressure, growth temperature and growth time, is the ability to obtain large-area growth with monolayer control of thickness and composition.

(g) Migration-enhanced epitaxy

Rapid migration of the evaporated materials on the growing surface is essential to the growth of high-quality epitaxial layers. In conventional MBE growth of GaAs and AlAs layers, migrating materials on the growing surface are Ga–As and Al–As molecules rather than Ga and Al atoms, respectively, because these layers are grown under arsenic-stabilized conditions. The migration of these molecules on the surface is very slow, especially at low temperatures. Therefore, lowering the substrate temperature considerably deteriorates the crystal quality of grown layers.

Horikoshi proposed a new mode of MBE growth which makes it possible to grow high-quality GaAs and AlAs layers at very low substrate temperatures (~ 200 °C) by enhancing the migration of the materials evaporated on the growing surface [Horikoshi 1986]. This method has been termed migration-enhanced epitaxy (MEE). It is based on the very rapid migration of Ga and Al atoms on the growing surface, and on the GaAs surface, much more rapidly than Ga–As and Al–As molecules, and the fact that they migrate very actively even at temperatures as low as 200 °C. They showed that high-quality GaAs and AlAs could be grown at very low temperatures by alternately supplying Ga (Al) atoms and arsenic molecules to the substrate surface. Applying the MEE method, 1–2 μm thick GaAs layers were grown at a substrate temperature of 200 °C. These layers showed efficient photoluminescence from the band-edge excitons at 4.2 K. AlAs–GaAs single-quantum-well structures with 3–6 nm widths were also grown at 200 °C by MEE. These structures showed photoluminescence due to electronic transitions between quantized levels in the wells, indicating a reasonable quality of AlAs [Horikoshi 1986].

1.4.2. Review of III–V heterostructures grown by LPE, MBE, and MOCVD

Table 1.8 shows recent achievements in the growth of modulation-doped structures and superlattices of GaAs–Ga$_x$Al$_{1-x}$As using different growth techniques. Similar information for the Ga$_x$In$_{1-x}$As$_y$P$_{1-y}$–InP lattice-matched system is presented in Table 1.9.

The improvement of the threshold current density of injection in GaAs–Ga$_x$Al$_{1-x}$As lasers, using double-heterostructure and quantum well active layers is illustrated in Table 1.10, and Table 1.11 indicates the best threshold current density (J_{th}) of GaInAsP–InP double-heterostructure lasers emitting at 1.3 μm produced by different growth techniques (MBE, LPE, VPE and LP-MOCVD). From the results illustrated in Table 1.8 – Table 1.11, one can draw the following conclusions.

Description	Year/Method	Reference
Observation of quantum size effect in GaAs–GaAlAs multiquantum wells	1974 MBE	Dingle et al. [1974]
Experimental observation of 2DEG detected by Shubnikov–de Haas magnetoresistance oscillations of the conduction electrons in uniformly Sn-doped MBE-grown GaAs–GaAlAs superlattice	1977 MBE	Chang et al. [1977]
GaAs–GaAlAs heterostructures and superlattices with modulated silicon doping	1978 MBE	Dingle et al. [1978]
Observation of 2DEG in LPE-grown GaAs–GaAlAs heterojunctions	1979 LPE	Tsui and Logan [1979]
Modulation-doped GaAs–GaAlAs heterojunctions by MBE with highest mobilities μ (2 K) = 11.7×10^6 cm$^2\cdot$V$^{-1}\cdot$s^{-1}	1988 MBE	Pfeiffer et al. [1989] Foxon et al. [1989]
Modulation-doped GaAs–GaAlAs heterojunctions by MOCVD with highest mobilities μ (2 K) = 7×10^5 cm$^2\cdot$V$^{-1}\cdot$s^{-1}	1989 MOCVD	Frijkink et al. [1991]
GaAs with highest mobilities μ (40 K) = 3.35×10^5 cm$^2\cdot$V$^{-1}\cdot$s^{-1}	1989 LP-MOCVD	Razeghi et al. [1989]

Table 1.8. *Development of the growth of modulation-doped structures and superlattices in the GaAs–Ga$_x$Al$_{1-x}$As lattice-matched system.*

Description	Year/Method	Reference
Observation of quantum size effect (QSE) in the multiple thin layers of Ga$_x$In$_{1-x}$As$_y$P$_{1-y}$ (λ = 1.1 μm)	1977 LPE	Rezek et al. [1977]
Observation of 2DEG in modulation-doped Ga$_{0.47}$In$_{0.53}$As–InP	1982 LP-MOCVD	Razeghi et al. [1982]
Modulation-doped heterostructures of Ga$_{0.47}$In$_{0.53}$As–InP superlattices	1983 LP-MOCVD	Razeghi et al. [1983b]
Observation of QSE in multi-ultrathin QW structures of Ga$_{0.47}$In$_{0.53}$As–InP (with well width of GaInAs of 200, 100, 50 and 25 Å)	1983 LP-MOCVD	Razeghi et al. [1983c]
Observation of QSE and the 2DEG in heterojunctions of GaInAsP–InP λ = 1.3 μm	1984 LP-MOCVD	Razeghi and Duchemin [1984]
Growth of Ga$_{0.25}$In$_{0.75}$As$_{0.5}$P$_{0.5}$–InP superlattices (10 wells of GaInAsP of 75 Å with InP barrier of 75 Å)	1984 LP-MOCVD	Duchemin and Razeghi [1984]
Growth of superlattices and modulation-doped GaInAs/InP	1983 VPE	Komeno et al. [1983]
GaInAs–InP heterostructure with highest mobilities (700,000 cm$^2\cdot$V$^{-1}\cdot$s^{-1})	1987 LP-MOCVD	Razeghi et al. [1987]

Table 1.9. *Development of the growth of modulation-doped structures and superlattices in the Ga$_x$In$_{1-x}$As$_y$P$_{1-y}$ lattice-matched system.*

(a) GaAs–GaAlAs systems

The smoothest and sharpest layers have been demonstrated by MBE [Gossard 1979] and MOCVD [Frijkink 1991]. The highest mobilities have been achieved with the MBE technique, but the mobility enhancement observed with LPE [Tsui and Logan 1979] and MOCVD [Hersee et al. 1982] is also very appreciable.

For optoelectronic devices based on the GaAs–GaAlAs system, high-quality quantum well (QW) lasers, MQW lasers, graded index waveguides and separate confinement heterostructure (GRIN-SCH) lasers have been achieved by MBE and by MOCVD [Hersee et al. 1982, Burnham et al. 1983, Kasemset et al. 1982]. Both techniques showed extremely low threshold current densities, J_{th}, of 100 to 300 A·cm^{-2}. Consequently, MOCVD and MBE are potentially useful for the growth of high-quality multilayers (MQW and SL) of GaAs–GaAlAs for optoelectronic and microwave devices. Considerations of economics or convenience may determine the choice of crystal growth technique.

(b) GaAs–GaInP systems

The interest in $Ga_{0.51}In_{0.49}P$ lattice matched to GaAs substrate is due to its potential for optoelectronic and microwave devices. Because of its wide direct bandgap ($E_g = 1.9$ eV), this material provides a visible wavelength operation ($\lambda = 0.65$ μm) for light-emitting and laser diodes at room temperature. The GaInP–GaAs system is very promising for high-speed circuit applications. The large valence band discontinuity ($\Delta E_v = 0.24$ eV) offers good device performance for n–p–n heterojunction bipolar transistors (HBTs), and p-channel FETs. The most important feature of the $Ga_xIn_{1-x}P$–GaAs heterostructure is that the crossover of its direct (Γ) and indirect (X) conduction bands lies at $x = 0.74$, which is far from the lattice-matched composition ($x = 0.51$). In the case of $Al_xGa_{1-x}As$, the crossover of direct and indirect conduction bands is around $x = 0.45$. Donor-related deep traps, the so-called DX centers, were observed in $Al_xGa_{1-x}As$ and their presence was particularly important for $x > 0.3$, i.e. for a composition close to the $x = 0.45$ crossover point. Trap activation energies were found to follow the minimum of the indirect L band. Fig. 1.3(a) and (b) compare the composition dependence of direct and indirect gaps of GaInP and GaAlAs.

The presence of DX centers in highly n-doped GaInP–GaAs heterostructures will be smaller than that in GaAlAs–GaAs layers. The DX center is a deep donor level related to the crystal lattice rather than a defect and its activation energies follow one of the indirect conduction bands in III–V compounds. Once the composition of a ternary compound is near or over the crossover point of the bandgap transition, DX centers will start to affect the device electrical properties. For those compositions below the crossover point, DX centers lie above the minimum of the conduction band and no deep trap effect will be observed. Since this type of deep trap is quite common in all III–V semiconductors, a large separation between the lattice-matched composition and that of the crossover point helps to eliminate the DX center problems. Therefore, $Ga_{0.51}In_{0.49}P$ allows operation without significant donor-related deep traps.

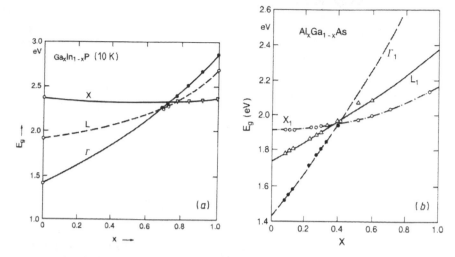

Fig. 1.3. (a) $Ga_xIn_{1-x}P$: composition dependence of direct and indirect gaps of $Ga_xIn_{1-x}P$ at 10 K (full symbols: piezoresistance measurements for compositions near the band crossover; open circles: experimental values for InP and GaP). For the indirect edges the energy values correspond to $E_{g-ind} - Eb + \hbar\omega_{phonon}$, where Eb is the exciton binding energy, (b) $Al_xGa_{1-x}As$: composition dependence of direct and indirect gaps of AlGaAs and of a deep donor level obtained by conductivity and Hall effect measurements.

Studies by deep-level transient spectroscopy (DLTS) [Watanabe and Ohba 1987] have shown that there are no detectable deep levels in *n*-doped GaInP layers, and that the concentration of electron traps is lower than 10^{13} cm^{-3}. In the absence of deep traps in doped GaInP layers, this material provides an option to replace the AlGaAs layer of HEMTs, HIGFETs and HBT in high-speed circuit applications [Chan et al. 1989].

(c) GaInAsP–InP systems

The first volume of this series concentrated on the survey of GaInAsP–InP for photonic and electronic applications. The smoothest and sharpest layers of GaInAsP–InP have been demonstrated by LP-MOCVD [Razeghi and Duchemin 1984]. The highest mobilities for modulation-doped GaInAs–InP have been obtained from LP-MOCVD- and VPE-grown materials [Razeghi et al. 1983a, Komeno et al. 1983]. The lowest threshold current density of a GaInAsP–InP double heterostructure emitting at 1.3 µm has been achieved with LP-MOCVD [Razeghi et al. 1983c, Razeghi 1985].

For optoelectronic GaInAsP–InP DH lasers, LPE, VPE and LP-MOCVD are potentially useful for the growth of high-quality DH lasers emitting between 1.2 and 1.6 µm.

(d) GaInAlP–GaAs systems

A useful material for short-wavelength optical devices is the quaternary alloy $Ga_{1-x-y}In_yAl_xP$ lattice matched to GaAs. This alloy lattice matched to GaAs has a direct bandgap between 1.9 eV ($x = 0$) and 2.3 eV ($x = 0.7$). This system is important for the fabrication of visible-, red-light-emitting lasers (λ = 630–680 nm), and for heterojunction bipolar transistor applications.

Suzuki et al. [1988a, b, c] have reported that the bandgap energy of $(Al_xGa_{1-x})_{0.5}In_{0.5}P$, at a fixed value of x grown by MOCVD, exhibits a variation in energy gap of more than 100 meV, which can be related to the (1 1 1) ordering on the group III sublattice. They showed that the degree of ordering in this system is related to the MOCVD growth conditions such as growth temperature, growth rate, V/III ratio and GaAs substrate orientation. High-quality $Ga_xIn_yAl_{1-x-y}P$–GaAs heterojunctions, quantum wells and superlattices have been grown by Valster et al. [1991], Mpaskoutas et al. [1991] and Watanabe and Ohba [1987].

Watanabe and Ohba [1987] studied the conduction-band discontinuity ΔE_c and interface charge density σ for GaAs–$In_{0.5}(Ga_{1-x}Al_x)_{0.5}P$ heterojunctions, grown by MOCVD. They investigated the dependence of ΔE_c and σ on composition x for $0 < x < 1$. They also studied $Ga_{0.5}In_{0.5}P/In_{0.5}Al_{0.5}P$ and found that the valence-band discontinuity ΔE_v for GaAs/$In_{0.5}(Ga_{1-x}Al_x)_{0.5}P$ is a linear function of x and is larger than ΔE_c. They also found that σ for GaAs/$In_{0.5}(Ga_{1-x}Al_x)_{0.5}P$ is one order of magnitude larger than that for $In_{0.5}Al_{0.5}P/In_{0.5}Ga_{0.5}P$ and GaAs/GaAlAs. Kondo et al. [1991] reported the MOCVD growth of $(Al_{0.7}Ga_{0.3})_{0.5}In_{0.5}P$–$Ga_xIn_{1-x}P$ strained single-quantum-well structures on GaAs substrate. The growth temperature was 710 °C and the growth pressure was 76 Torr. The V/III ratio was 200 and the growth rate was 1.3 $\mu m \cdot h^{-1}$. By changing x between 0.38 and 0.68 the biaxial misfit strain (ε) was varied between +1.0% and −1.24%. The photoluminescence (PL) microscopy showed that 100 Å thick GaInP layers are free of misfit dislocations at $|\varepsilon| < 1.5\%$. The PL linewidth of AlGaInP–GaInP strained single quantum wells (SSQWs) was as narrow as 9–12 meV at 4 K, independent of misfit strain, which confirms the structural perfection of SSQWs.

Growth technique	J_{th} (A cm^{-2})	Cavity length (μm)	Reference
LPE	770	1500	Itaya et al. [1979]
VPE	980	400	Mizutani et al. [1980]
MBE	1000	400	Tsang et al. [1982]
LP-MOCVD	430	400	Razeghi et al. [1983a]

Table 1.10. The best J_{th} of GaInAsP–InP DH lasers emitting at 1.3 μm.

Homostructure GaAs p–n junction laser $J_{th} \approx 10^5 A\ cm^{-2}$	1968 LPE	Hayashi et al. [1969]
Single heterostructure of GaAs–GaAlAs p–n heterojunction lasers $J_{th} \approx 4 \times 10^3\ A\ cm^{-2}$	1969 LPE	Alferov et al. [1969]
Double heterostructure of $Ga_xAl_{1-x}As/GaAs/Ga_xAl_{1-x}As$ lasers $J_{th} \approx 2 \times 10^3\ A\ cm^{-2}$	1970 LPE	Hayashi and Panish [1970]
The use of a graded aluminum fraction (x) in the $Ga_xAl_{1-x}As$ optical cavity surrounding a thin active region (GRIN-SCH) lasers $J_{th} \approx 250\ A\ cm^{-2}$	1982 MOCVD	Tsang et al. [1982]
The lowest threshold current density of GRIN-SCH lasers $J_{th} \approx 120\ A\ cm^{-2}$	1982 MOCVD	Hersee et al. [1982]
Multiquantum well (MQW) of $GaAs/Ga_xAl_{1-x}As$ lasers $J_{th} \approx 170\ A\ cm^{-2}$	1982 MOCVD MBE	Dupuis and Dapkus [1978] Burnham et al. [1983] Hiyamizu [1982]

Table 1.11. The improvement of J_{th} of injection in GaAs–Ga_xAl_{1-x}As lasers using DH and QW active layers.

References

Alferov, Z.I., Andreev, V.M., Korol'kov, V.I., and Portnoi, E.L. *Sov. Phys.-Semicond.* **2** 843, 1969.

Burnham, R.D., Streifer, W., Paoli, T.L., and Holonyak, N.J. *IOOC-83 (Tokyo)*, 1983.

Casey, H.C. and Panish, M.B. *Heterostructure Laser* (New York: Academic), 1978.

Chan, Y.J., Pavlidis, D., Razeghi, M., and Omnes, F. 1989 *Gallium Arsenide and Related Compounds (Inst. Phys. Conf. Ser. 106)* ed Ikoma, T. and Watanabe, H. (Bristol: Institute of Physics) p 891, 1989.

Chang, L.L., Sakaki, M., Chang, C.A., and Esaki, L. *Phys. Rev. Lett.* **38** 1489, 1977.

Chapin, D.M., Fuller, C.S., and Pearson, G.L. *J. Appl. Phys.* **25** 676, 1954.

Delahaye, F., Dominguez, D., Alexandre, F., Andre, J.P., Hirtz, J.P., and Razeghi, M. *Metrologia* **22** 103, 1986.

Didchenko, R., Alix, J.E., and Toeniskoetter, R.H. *J. Inorg. Nucl. Chem.* **14** 35, 1960.

Dingle, R., Stormer, H., Gossard, A.C., and Weigmann, W. *Appl. Phys. Lett.* **33** 665, 1978.

Dingle, R., Weigmenn, W., and Henry, C.H. *Phys. Rev. Lett.* **38** 1489, 1974.

Doi, A., Aoyagi, Y., and Namba, S. *Appl. Phys. Lett.* **48** 1787, 1986.

Dupuis, R.D. and Dapkus, P.D. *Gallium Arsenide and Related Compounds (Inst. Phys. Conf. Ser. 45)* ed Wolfe, C.M. (Bristol: Institute of Physics), 1978.

Esaki, L. *Phys. Rev.* **109** 603, 1958.

Foxon, C.T., Harris, J.J., Hilton, D., Hewett, J., and Roberts, C. *Semicond. ScL Technol.* **4** 582, 1989.

Fraas, L.M. *J. Appl. Phys.* **52** 6939, 1981.

Frijkink, P.M., Nicolas, J.L., and Suchet, P. *J. Crystal Growth* **107** 166

Gossard, A.C. 1979 *Thin Solid Films* **57** 3, 1991.

Guldner, Y., Vieren, J.P., Voos, M., Delahaye, F., Dominguez, D., Hirtz, J.P., and Razeghi, M. *Phys. Rev.* B **23** 3990, 1986.

Gunn, J.B. *Solid State Commun.* **1** 88, 1963.

Gunn, J.B. *IBM J. Res. Dev.* **8** 141, 1964.

Hayashi, I. and Panish, M.B. Device Research Conf. (Seattle, WA), 1970.

Hayashi, I., Panish, M.B. and Foy, P.W. *IEEE J. Quantum Electron.* **QE-5** 211, 1969.

Hersee, S.D., Baldy, M., Assenat, P., de Cremoux, B., and Duchemin, J.P. *Electron. Lett.* **18** 870, 1982.

Hilsum, C. *Proc. IRE* **5** 185, 1962.

Hiyamizu, S. *Oyo Buturi* **51** 942 (in Japanese), 1982.

Horikoshi, L., *The Physics and Fabrication of Microstructures*, eds. Kelly, M.J. and Weisbuch, C., Springer-Verlag, Berlin, 1986.

Ilegems, M. and Panish, M.B. *J. Phys. Chem. Solids* **35** 409, 1974.

Itaya, Y., Katoyama, S., and Sumatsu, Y. *Electron. Lett.* **15** 123, 1979.

Jordan, A.S. and Ilegems, M. *J. Phys. Chem. Solids* **36** 329, 1974.

Kasemset, D., Mong, C.S., Patel, N.B., and Dapkus, P.D. *J. Appl. Phys. Lett.* **41** 912, 1982.

Kobayashi, N., Makimoto, T., and Horikoshi, Y. *Japan. J. Appl. Phys.* **24** L962, 1985.

Komeno, J., Takikawa, M., and Ozeki, M. *Electron. Lett.* **19** 473, 1983.

Kondo, M., Domen, K., Anayama, C., Tanahashi, T., and Nakajima, K. *J. Crystal Growth* **107** 578, 1991.

Manasevit, H.M. *Appl. Phys. Lett.* **12** 156, 1968.

Manasevit, H.M. *J. Electrochem. Soc.* **118** 647, 1971.

Manasevit, H.M., Erdmann, F.M., and Simpson, W. I *J. Electrochem. Soc.* **118** 1864, 1971.

Matsumoto, T. and Usui, A. *Abstracts of 1986 Fall Meeting of Japan Soc. Appl. Phys.*, 1986.

Mizutani, T., Yoshida, M., Usai, A., Watanabe, H., Yuasa, T., and Hayashi, I. *Japan. J. Appl. Phys.* **19** L113, 1980.

Mpaskoutas, M., Morhaim, C., Patillon, J.N., Andre, J.P., Valster, A., and Rusch, J.J. *J. Crystal Growth* **107** 192, 1991.

Nelson, H. *RCA Rev.* **19** 603, 1963.

Nicholas, R.J., Brunel, L.C., Huant, S., Karrai, K., Portal, J.C., Brummell, M.A. Razeghi, M., Chang, K.Y., and Cho, A.Y. *Phys. Rev. Lett.* **55** 8883, 1985.

Nishizawa, J. and Kurabayashi, T. *J. Crystallogr. Soc. Japan* **28** 133, 1986.

Osbourn, G.C. *J. Vac. ScL Technol.* **21** 469, 1982.

Parker, E.H.C. *The Technology and Physics of Molecular Beam Epitaxy* (New York: Plenum), 1985.

Pfeiffer, L., West, K.W., Stormer, H.L., and Baldwin, K.W. *Appl. Phys. Lett.* **55** 1888, 1989

Razeghi, M. *Rev. Tech.* Thomson-CSF **15** 1, 1983.

Razeghi, M. *Lightwave Technology for Communication* ed. W.T. Tsang (New York: Academic), 1985

Razeghi, M. and Duchemin J P *J. Crystal Growth* **70** 145, 1984.

Razeghi, M., Duchemin, J.P., Portal, J.C., Dmowski, L., Remeni, G., Nicholas, R.J. and Briggs, A. *Appl. Phys. Lett.* **48** 712, 1986.

Razeghi, M., Hersee, S., Blondeau, R., Hirtz, P., and Duchemin, J.P. *Electron. Lett.* **19** 336, 1983a

Razeghi, M., Hirtz, P., Blondeau, R., and Duchemin, J.P. *Electron. Lett.* **19** 481, 1983b

Razeghi, M., Hirtz, J.P., Ziemelis, U.D., Delagande, C., Etienne, B., and Voos, M. *Appl. Phys. Lett.* **43** 586, 1983c

Razeghi, M., Hosseini, A., Vilcot, J.P., and Decoster, D. *Gallium Arsenide and Related Compounds* (Bristol: Institute of Physics) ρ 625, 1987.

Razeghi, M., Omnes, F., Nagle, J., Defour, M., Acher, D., and Bove, P. *Appl. Phys. Lett.* **55** 1677, 1989.

Razeghi, M., Poisson, M.A., Larivain, J.P., de Cremoux, B., and Duchemin, J.P. *Electron. Lett.* **18** 339, 1982.

Rezek, E.A., Holonyak, N.J., Vojak, B.A., Stillman, G.E., Rossi, J.A., and Keune, D. L *Appl Phys. Lett.* **31** 288, 1977.

Ridley, B.K. and Watkins, T.B. *Proc. Phys. Soc. London* **78** 293, 1961.

Rogers, D.C., Nicholas, R.J., Portal, J.C., and Razeghi, M. *Semicond. Sci. Technol.* **1** 350, 1986.

Suntola, T., Atron, J., Pakkala, A., and Linfors, S. *SID 80 Digest* 108, 1980.

Suzuki, T., Gomyo, A., Hino, I., Kobayashi, K., Kawata, S., and Lijina, S. *Japan. J. Appl. Phys.* **27** L1549, 1988a.

Suzuki, T., Gomyo, A., and Iijima, S. *J. Crystal Growth* **93** 389, 1988b.

Suzuki, T., Gomyo, A., Ijima, S., Kobayashi, K., Kawata, S., Hino, I., and Yuasa, T. *Japan. J. Appl. Phys.* **27** 2098, 1988c.

Sze, S.M. and Ryder, R.M. *Proc. IEEE* **59** 1140, 1971.

Tsang, W.T., Remhart, F.K., and Ditzenberg, J.A. *Electron. Lett.* **18** 75, 1982.

Tsui, D.C. and Logan, R.A. *Appl. Phys. Lett.* **15** 35, 1979.

Valster, A., Liedenbaum, C.T., Finke, M.N., Severens, A.L., Boermans, M.B, Vandenhoudt, D.W., and Bulle-Lieuwma, C.T. *J. Crystal Growth* **107** 403, 1991.

Van Vechten, J.A. and Bergstresser, T.K. *Phys. Rev.* B **1** 3361, 1970.

Vodjani, N. *These de Doctorat 3e Cycle* University Paris XI, 1982

Watanabe, M.O., and Ohba, Y., *Appl. Phys. Lett.* **50** 906, 1987.

2. Growth Technology

2.1.   Introduction
    2.1.1.   Liquid-phase epitaxy
    2.1.2.   Vapor-phase epitaxy
    2.1.3.   Molecular beam epitaxy
2.2.   Metalorganic chemical vapor deposition
    2.2.1.   Experimental details
    2.2.2.   Reactor design
    2.2.3.   Growth procedure
    2.2.4.   Flow patterns
    2.2.5.   Starting materials
2.3.   New non-equilibrium growth techniques
    2.3.1.   Chemical beam epitaxy
    2.3.2.   Atomic layer epitaxy
    2.3.3.   Metalorganic atomic layer epitaxy
    2.3.4.   Chloride atomic layer epitaxy
    2.3.5.   Migration-enhanced epitaxy

2.1. Introduction

During the past 10 years extensive theoretical and experimental studies on low-dimensional structures (LDS) of InP, GaAs and related compounds have been performed in two distinct contexts.

(1)   Low-dimensional, meaning, less than three dimensions: the properties of the two-dimensional electron or hole gas in different III–V semiconductor heterojunctions is now a mature, even aging subject.
(2)   Low in the sense of small: as in the physics of semiconductors, where the feature size of the sample is small compared with the intrinsic quantum length scale associated with carriers in semiconductors. This is given by the de Broglie wavelength, denoted as $\lambda = h/p$ where h is Planck's constant and p the carrier momentum, typically given by $p^2/2m^* = kT$, where m* is the effective mass of the carrier, k is Boltzmann's constant and T is the temperature.

Low-dimensional structures have also become essential elements in modern device technology.

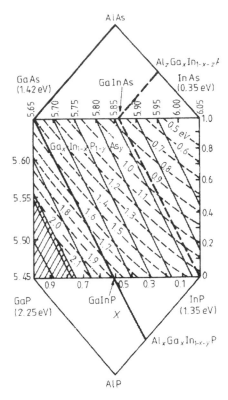

Fig. 2.1. The x–y compositional plane for quaternary III–V alloys at 300 K. The x–y coordinate of any point in the plane gives the compositions. The full lines are lattice parameters. The broken curves are direct energy gap values.

Fig. 2.1 shows the x–y compositional plane for ternary and quaternary alloys lattice-matched to an InP substrate. Each corner has a binary III–V material. The mixture of binaries at each side gives the ternary III–III–V or III–V–V alloys, and the combination of binaries or ternaries inside gives the quaternary alloys. The broken curves represent equal energy gaps, and the full lines represent equal lattice parameters. Two systems are very interesting for optoelectronic and microwave devices.

One of them is ternaries and quaternaries lattice-matched to a GaAs substrate, for which the energy gap varies between 1.43 and 2.4 eV. In this system, all of the ternaries and quaternaries have an energy gap higher than GaAs binary crystal, so in heterostructures the electrons and holes are usually confined in the binary material.

The other system is composed of ternaries and quaternaries lattice-matched to an InP substrate, for which the energy gap varies between 0.75 and 1.35 eV. In this system all of the ternaries and quaternaries have an energy gap lower than InP, so electrons and holes are confined in the ternary or quaternary alloys.

There is now a need for a wide range of semiconducting materials, especially InP, GaAs and related compounds, to be grown as thin single-crystal films. Thickness has to be controlled, increasingly more accurately, and uniformity of thickness can be vital. Epitaxial layers as thin as 10 Å or less are now needed. Excellent homogeneity and purity are required. It is desirable to have no misfit dislocations present, and to have a very sharp boundary between the substrate and epitaxial layers. A major goal for solid state physics and solid state technology is the perfection of materials and substrate in which charge carriers have long lifetime, low scattering, high mobilities and controlled densities: such materials have allowed the elucidation of the electronic structure of solids and the development of semiconductor electronic and photonic devices.

To grow heterojunctions, quantum wells (QW) and superlattices of InP-and GaAs-based materials, different growth techniques are available. In this chapter, we first describe briefly other methods currently used, then describe the metalorganic chemical vapor deposition method and finally discuss recent developments of new methods.

2.1.1. Liquid-phase epitaxy

Liquid-phase epitaxy (LPE) involves the precipitation of material from a supercooled solution onto an underlying substrate. The composition of the layer formed on the substrate depends mainly on the equilibrium phase diagram and to a lesser extent on the orientation of the substrate.

Growth of InP–GaInAsP has been started mainly by LPE, the LPE reactor consisting of a horizontal furnace system and a sliding graphite boat. There are basically three parameters that influence growth in LPE: the melt composition, the growth temperature and the growth time. The first reported growth apparatus was developed by Nelson [1963].

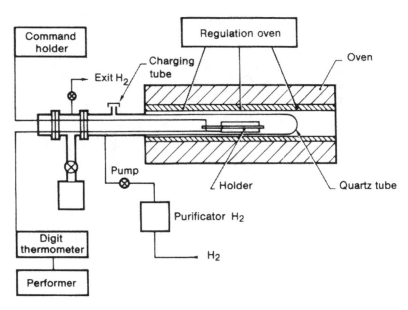

Fig. 2.2. Principles of a liquid phase epitaxy system.

The advantages of LPE for the growth of InP-based materials are simplicity of equipment, higher deposition rates and elimination of parasitic reactions due to use of reactive gases and their reactive products, which are often toxic and explosive.

The limitations of LPE technology are non-uniformity of the layer thickness, roughness of surface morphology and smaller wafer sizes. Owing to the high growth rate and melt-back effect, it is very difficult to grow multiquantum-well (MQW) or superlattice (SL) structures with $L_z \leq 100$ Å using LPE technology. Fig. 2.2 shows a schematic diagram of an LPE reactor.

2.1.2. Vapor-phase epitaxy

The vapor-phase epitaxy (VPE) method was first demonstrated by Tietjen and Amick, [Tietjen et al. 1966] for the growth of GaAsP alloys. The growth of InP-based materials can be obtained in a fused-silica reactor composed of two zones set at different temperatures by using a multi-element furnace that surrounds the reactor. Hydrogen is often used as a carrier gas, and arsine (AsH_3) and phosphine (PH_3) as arsenic (As) and phosphorus (P) sources. Pure indium and gallium metal can be used as group III element sources. In the first zone of the reactor, called the source zone, held at a temperature T_s lying in the range 750–800 °C, the gaseous species to be transported are synthesized, following the reaction

Eq. (2.1)
$$In + HCl \rightarrow InCl + \tfrac{1}{2} H_2$$
$$PH_3 \rightarrow \tfrac{1}{4} P_4 + \tfrac{3}{2} H_2$$

In the second zone, called the deposition zone, with a temperature T_D in the range 650–750 °C, growth occurs via the reaction

Eq. (2.2) $\tfrac{1}{4} P_4 + InCl + H_2 \rightarrow InP + HCl$

VPE is a thermodynamic equilibrium growth technique (like LPE).

The advantages of VPE over LPE are a high degree of flexibility in introducing dopant into the material as well as control of composition gradients by accurate flow metering.

Disadvantages include potential for hillock and haze formation, and interfacial decomposition during the preheat stage. It is very difficult to grow MQW or SL structures with $L_z \leq 50$ Å using VPE technology. One of the major advantages of VPE over LPE is the possibility of localized epitaxy. This means that if one grows InP-based material on an InP substrate consisting of a stripe of SiO_2 or Si_3N_4, there is monocrystal growth on the substrate, but no growth on the oxide (SiO_2 or Si_3N_4) surface, which could be important for monolithic integrated circuits on InP substrates. Fig. 2.3 shows a schematic illustration of a VPE reactor.

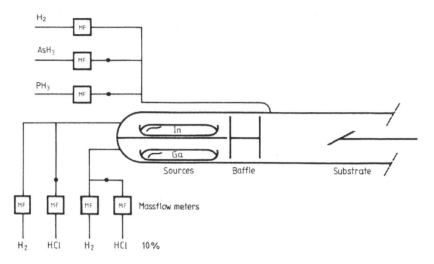

H₂

AsH₃

PH₃

Fig. 2.3. Schematic illustration of a vapor phase epitaxy reactor.

2.1.3. Molecular beam epitaxy

The molecular beam epitaxy (MBE) process involves the reaction of one or more thermal beams of atoms and molecules of the III–V elements with a crystalline substrate surface held at a suitable temperature under ultra-high-vacuum (UHV) conditions. In 1958, using multiple beams, Gunther [1958], described the growth of III–V materials; his films were grown on glass substrates and hence were polycrystalline. In 1968, Davey and Pankey [Davey et al. 1968] and Arthur [1968] grew monocrystalline GaAs films by MBE.

Since MBE is essentially a UHV evaporation technique, the growth process can be controlled *in situ* by the use of equipment such as pressure gauge, mass spectrometer and electron diffraction facility located inside the MBE reactor. The MBE growth chamber can contain other components for surface analytical techniques, including reflection high-energy electron diffraction (RHEED) (the first information about the dynamics of film growth by MBE was obtained by Neave and Joyce [Neave et al. 1983] by studying RHEED oscillations), Auger electron spectroscopy AES (another non-destructive analytical method for characterization of the initial substrate surface, verification of surface accumulation of dopant elements during epitaxial growth, and determination of the change in the relative ratio of the constituent elements of reconstructed surface structures), x-ray photoelectron spectroscopy (XPS)), low-energy electron diffraction (LEED), electron spectroscopy for chemical analysis (ESCA), secondary-ion mass spectroscopy (SIMS) (a powerful tool for surface and bulk material-composition analysis for pre- and post-deposition analysis during MBE growth) and ellipsometry, which can all be used as *in situ* surface diagnostic techniques during MBE growth, due to the UHV growth conditions.

MBE has attracted much interest as an excellent crystal growth technology, especially for GaAs-based multilayer structures, because of its extremely precise control over layer thickness and doping profile, and the high uniformity of the epitaxial layer over a large area of a substrate (>3 inch diameter).

The disadvantages of MBE are that it is expensive, and that difficulties have been reported with *p*-type doping by Zn [Noyanuma et al. 1975] and with the growth of phosphorus-bearing alloys such as InP and GaInAsP. Fig. 2.4 shows an MBE reactor.

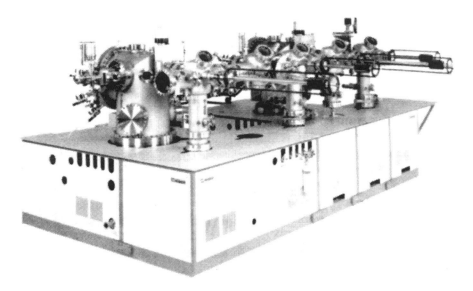

Fig. 2.4. Photograph of an MBE reactor.

2.2. Metalorganic chemical vapor deposition

Metalorganic chemical vapor deposition (MOCVD) has established itself as a unique and important epitaxial crystal growth technique yielding high-quality low-dimensional structures (LDS) for fundamental semiconductor physics research and useful semiconductor devices, both electronic and photonic. The growth of semiconductor III–V compounds results from introducing metered amounts of the group III alkyls and the group V hydrides into a quartz tube that contains a substrate placed on radiofrequency (RF) heated carbon susceptor. The hot susceptor has a catalytic effect on the decomposition of the gaseous products and the growth therefore takes place primarily at this hot surface. MOCVD is attractive because of its relative simplicity compared to other growth methods. It can produce heterostructures, multiquantum wells (MQW) and superlattices (SL) with very abrupt switch-on and switch-off transitions in composition as well as in doping profiles in continuous growth by rapid changes of the gas composition in the reaction chamber.

The technique is attractive in its ability to grow uniform layers, its low background doping density and sharp interfaces, and in the potential for commercial applications. MOCVD can prepare multilayer structures with thicknesses as thin as a few atomic layers.

The main disadvantage of the MOCVD growth technique is the use of a high quantity of poisonous gases such as ASH_3 and PH_3. By comparison to MBE, it is very difficult to do *in situ* characterization during MOCVD growth (see Fig. 2.5).

Fig. 2.5. Photograph of an MOCVD reactor.

2.2.1. Experimental details

The InP-based semiconductor alloys defined to have an energy gap $E_g \leq 2.0$ eV discussed in this book are listed in Table 2.1, together with properties relevant to their crystal growth and some of their physical parameters. They all melt at high temperatures and have a strong tendency to decompose well below their melting points. As a consequence, controlled melt growth of these compounds is difficult, and most efforts have been directed towards vapor growth as a means for producing high-quality heterojunctions, quantum wells and superlattices.

Compound	Energy gap (eV)	Refractive index n	Electron effective mass, m_e^*	Heavy-hole effective mass, m_h^*
InP	1.35	3.45	0.08	0.56
$Ga_{0.25}In_{0.75}As_{0.5}P_{0.5}$	0.95	3.52	0.053	0.5
$Ga_{0.40}In_{0.60}AS_{0.85}P_{0.15}$	0.80	3.55	0.045	0.5
$Ga_{0.47}In_{0.53}As$	0.75	3.56	0.041	0.5

Table 2.1. Some of the physical parameters of InP-based materials lattice-matched to InP (lattice parameter 5.869 Å).

Elemental vapor-phase epitaxy usually requires growth temperatures within the range 700–900 °C. It has been recognized increasingly that high-temperature growth thermodynamically favors the formation of a multitude of impurity–defect complexes which act as electron traps and pin the Fermi level far from both the conduction and valence band edges of these compounds, and hence the material exhibits compensating properties. The lower growth temperature (500–630 °C) for LP-MOCVD of InP-based materials, a feature shared with MBE, has stimulated substantial interest in this preparative route as a potential way of either eliminating or controlling detrimental impurities and defects. This control is particularly important where the material is the active medium in optoelectronic devices.

A review by Ludowise [1985] presented practical information about the apparatus and technique as well as more fundamental details concerning the interrelation of the several parameters that control deposition and material quality.

LP-MOCVD was first applied to the growth of InP using the metal alkyl triethylindium ((C_2H_5)$_3$In, TEIn) and phospine (PH_3) as the source of phosphorus by Duchemin et al. [1979]. A review by Razeghi [1985] described advances in the field prior to 1985. The developments since then for the growth of InP-based materials for optoelectronic and microwave devices form the basis of the present chapter.

2.2.2. Reactor design

The most important part of the growth apparatus is the deposition chamber. There are two basic reactor types, a vertical design originated by Manasevit and Simpson [Manasevit et al. 1971], in which the gas flow is perpendicular to the substrate surface, and the horizontal version developed by Bass [1975], in which the gas flow is parallel to the substrate surface. Both types have been used successfully at low and atmospheric pressures using several different substrate heating methods involving radiofrequency induction, resistance, radiant and laser heating. Fig. 2.6 shows schematic drawings of vertical and horizontal reaction chambers.

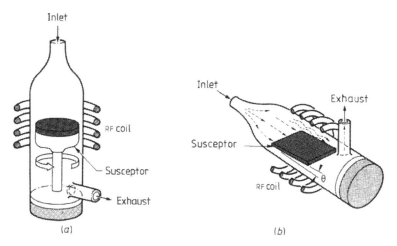

Fig. 2.6. Schematic drawings of (a) vertical and (b) horizontal reaction chambers (after Ludowise [1985]).

The advantages of a low-pressure MOCVD process are:

(1)   elimination of parasitic nucleations in the gas phase;
(2)   reduction of out-diffusion (i.e. the solid-state diffusion of impurities from the substrate through active layers or from one active layer to another);
(3)   reduction of autodoping (i.e. the doping of an epitaxial layer by volatile impurities that originate from the substrate);
(4)   improvement of the interface sharpness and impurity profiles;
(5)   lower growth temperature;
(6)   thickness uniformity and compositional homogeneity; and
(7)   elimination of memory effect.

The presence of vortices (behind the susceptor) and dead volumes (sharp corners in reactor inlet) will act as sources of unwanted materials which cannot be removed easily. So during the growth of a sharp heterojunction or a steep doping profile, by switching a flow with another chemical composition, the original composition is still present in the vortices and trapped gases. The slow out-diffusion from these parts will smear out the doping or heterojunction profiles. This is called the memory effect. We can eliminate such trapped gases by rapid evacuation, i.e. working at low pressure.

In relation to reactor design, the main advantage of the horizontal reactor is considered to be the uniform gas flow achieved over a slightly angled susceptor (7–15°), which leads to uniform deposition as the reactants are depleted from the gas flow.

The most important feature in the growth of III–V semiconductor materials is the arrangement for mixing the gases in order to inhibit the pre-reaction between the constituent gas flow. The reactor design has concentrated on introducing the gases to the reactor separately. Good mixing between the metal alkyls and hydrides through a delivery tube near to the heated substrate is vital. So the use of a carrier gas (H_2 or N_2) which eliminates the pre-reaction enables good mixing to take place and allows the efficient gas mixer designed for III–V semiconductor growth to be utilized.

2.2.3. Growth procedure

The exact chemical decomposition pathways in MOCVD are not yet clearly understood. The nature of the reactions is in part determined by the dynamics of the gas, i.e. the velocity and temperature profiles in the vicinity of the susceptor and the subsequent concentration and thermal gradients that are established. The reaction pathways, of course, are also strongly influenced by the choice of precursor chemicals.

The MOCDV growth of InP, GaAs and related compounds can be explained by the rapid transport of reactive species from the pipeline to the deposition zone by forced flow (in this case, the control of flow dynamics, flow mixing and the adjustment of the flow to geometrical effects and temperature changes are essential). Then the transport of reactive species takes place in the deposition zone to the hot substrates by diffusion (knowledge about development of concentration profiles, the effect of thermal gradients on diffusion and the effect of annihilation or creation of molecular species on diffusion is crucial).

Since the growth rate in MOCVD is limited by mass transport of the group III growth component, flow dynamics coupled with diffusion govern the deposition rates and thereby the gas-phase depletion. A profound knowledge of flow patterns and concentration profiles in the reactor is essential for optimization of reactor construction. In this respect two important factors are homogeneous growth on large surface areas and minimization of gas memory effects, which are essential in the growth of InP-based hetero-junctions and superlattices with sharp interfaces. To achieve these, the flow regime must be laminar with no turbulence in order to achieve control over the growth process, and it must develop its pattern in a controlled way when it enters the reactor. Also, when the gas is heated to the process temperature, no instabilities due to natural convection (buoyancy) in the boundary layer must occur. The term "boundary layer" is used here to define the regions of rapidly increasing compositional, thermal or momentum gradients perpendicular to the substrate.

2.2.4. Flow patterns

One of the important factors in the MOCVD reactor is the flow pattern. We can specify the macroscopic gas movements in the MOCVD reactor (Fig. 2.7) as follows:

(1)   laminar flow versus turbulence;
(2)   diffusion versus free convection (buoyancy).

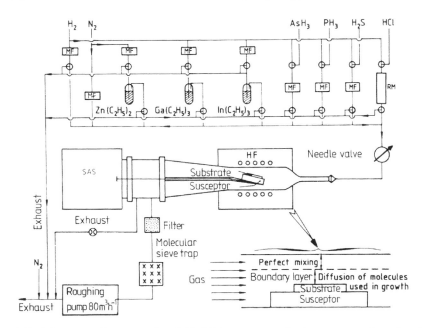

Fig. 2.7. Schematic diagram of an MOCVD reactor and tube cross section showing laminar and mixed flow zones.

In order to specify the circumstances in which the different types of flow occur, we need to introduce the concept of the Reynolds number for the first case and the Rayleigh number for the second case.

The Reynolds number R_e (dimensionless) of reactor flow is

Eq. (2.3) $R_e \simeq \rho \bar{V} d / \eta$

where d is the diameter of the tube (m), $\bar{V}$ the average flow-rate (m·s^{-1}), ρ the density (kg·m^{-3}) and η the dynamic viscosity (kg·m^{-1}·s^{-1}) of the gas.

When R_e is small (less than 100) the flow regime is laminar. In laminar flow, the velocity at a fixed position is always the same. Each element of the reactive species travels smoothly along a simple well denned path [Prandt et al. 1934]. Each element starting at the same place follows the same path.

When R_e is high, the flow becomes turbulent, and none of these features is retained. The flow develops a highly random character with rapid irregular fluctuations of velocity in both space and time. In this case, an element of gas flow follows a highly irregular distorted path. Different elements starting at the same place follow different paths, since the pattern of irregularities is changing all the time [Shapiro 1961].

The Rayleigh number R_a (dimensionless) of reactor flow is

Eq. (2.4) $R_a \simeq g \alpha C_p \rho^2 h^3 \Delta T / \eta K$

where α is the coefficient of thermal expansion (K^{-1}), g the gravitational constant (9.81 m·s^{-2}), C_p the specific heat (J·kg^{-1}·K^{-1}), ρ the density (kg·m^{-3}), h the free height above the susceptor (m), $\Delta T = T$(susceptor) $- T$(reactor wall), η the dynamic viscosity (kg·m^{-1}·s^{-1}) and K the thermal conductivity (J m^{-1} s^{-1} K^{-1}). When $R_a \leq 1700$, the gas is stable; for $R_a > 1700$, free convection occurs.

At a high Rayleigh number, free convection occurs, which affects the mass transfer, growth rate and homogeneity. Usually convection occurs between hot substrate and cold reactor wall. The cause of convection is the action of the gravitational field on the density variations associated with temperature variations. The heavy cold gas is situated above the light hot gas. If the former moves downwards and the latter upwards, there is a release of potential energy which can provide kinetic energy for motion. There is thus the possibility that the equilibrium will be unstable.

From the kinetic gas theory, it follows that both the dynamic viscosity η and the thermal conductivity K are independent of reactor pressure. But the density ρ is pressure-dependent, which means that $R_e \propto \rho$ and $R_a \propto \rho^2$. So, the consequence is that lower pressures stabilize the convection behavior of the gas flows, and laminar flows are more easily obtained. For more details on models of flow dynamics, the reader is referred to the book by Tritton [1977].

In conclusion, for the growth of high-quality III–V semiconductor materials, with sharp interfaces, and to avoid parasitic reactions in the gas phase and condensation at the inlet of the deposition zone of the reactor, giving variations from day to day and from system to system, the following remarks on reactor design are crucial.

    (1) Laminar flows free of convection should exist by:
        (a)   using a horizontal reactor,
        (b)   working at low pressure and
        (c)   decreasing the reactor diameter.
    (2) No temperature gradient should be present across the susceptor.
    (3) Eliminate the memory effect:
        (a)   the geometry of the reactor is such that no vortices can develop;
        (b)   no dead volumes are present inside the reactor; and
        (c)   elimination of sharp corners in reactor inlet, where the laminar flow can go by without having a strong interaction, and also behind the susceptor.

2.2.5. Starting materials

Alkyls of the group II and III metals, and hydrides of the group V and VI elements, are usually used as precursor species in MOCVD. Dilute vapors of these chemicals are transported at or near room temperature to a hot zone where a pyrolysis reaction occurs. For example, gallium arsenide is formed by heating trimethylgallium ($(CH_3)_3Ga_2$) or triethylgallium ($(C_2H_5)_3Ga_2$) and arsine over a suitable substrate, usually a single crystal of GaAs or Si, at a temperature around 600 °C:

Eq. (2.5)
$$(C_2H_5)_3\,Ga_2 + AsH_3 \xrightarrow{\;H_2\;} GaAs + nC_2H_6$$
$$(CH_3)_3\,Ga_2 + AsH_3 \xrightarrow{\;H_2\;} GaAs + nCH_4$$

Problems may arise through the use of group III alkyls with group V hydrides because reactions may occur between the two species in the cold gas during transport to the hot growth zone.

    When the alkyl radical is big, such as triethyl ($(C_2H_5)_3$), the temperature of decomposition is low, and in this case the growth temperature will be low, but there is the possibility of parasitic reaction in the gas phase. When the alkyl radical is small, such as trimethyl ($(CH_3)_3$), it becomes more stable, and the temperature of decomposition is high. The reaction described by Eq. (2.5) can be generalized for the III–V compounds to

Eq. (2.6) $R_3M + EH_3 \rightarrow ME + 3RH$

where M is the group III element, such as Al, Ga or In, E is the group V element, such as Sb, P or As, and $R = CH_3$ or C_2H_5. This is a simplified form of the reaction and it ignores any intermediate steps that may occur. If more than one reaction occurs, such as

Eq. (2.7)
$$xR_3M + (1-x)R_3M' + EH_3 \rightarrow M_xM'_{1-x}E + 3RH$$
$$x(C_2H_5)_3\,In_2 + (1-x)(C_2H_5)_3\,Ga_2 + AsH_3 \xrightarrow{\;H_2\;} In_xGa_{1-x}As + nC_2H_6$$

or

Eq. (2.8)

$$xR_3M + (1-x)R_3M' + yEH_3 + (1-y)E'H_3 \xrightarrow{H_2} M_xM'_{1-x}E_yE'_{1-y} + nRH$$

$$x(C_2H_5)_3 In_2 + (1-x)(C_2H_5)_3 Ga_2 + yAsH_3 + (1-y)PH_3 \xrightarrow{H_2}$$

$$In_xGa_{1-x}As_yP_{1-y} + nC_2H_6$$

the alloy composition x and y is determined by the relative rates of the two reactions, which in turn depend upon a number of factors including gaseous diffusion, unknown intermediate steps, thermodynamics (sometimes) and reactor gas dynamics.

In general, the alkyl and hydride reactions inside the reactor can be described as

Eq. (2.9) $R_3M + EH_3 \rightarrow R_3M - EH_3$

The strength of the M–E bond is a measure of the stability of the complex R_3M–EH_3. Ludowise [Ludowise 1985] proposed four cases for the progression of this reaction: (a) no reaction occurs; (b) formation of a volatile compound; (c) formation of a non-volatile compound; and (d) progression to additional reactions which produce non-volatile compounds. Cases (a) and (b) are the trivial cases since vapor transport still proceeds as desired. When nonvolatile products are formed (cases (c) and (d)), the carrier gas can be depleted of one or more reactants, resulting in a lowered growth rate or shifted alloy composition. Two examples are the reaction of TEIn and AsH_3 or TEIn and PH_3. At the relatively low concentrations ($\leq 10^{-4}$ mole fraction) and high gas velocities used in an MOCVD reactor, the reactions may only progress a few percent towards completion before the reactants reach the pyrolysis zone. The observation of a particular reaction depends upon the specifics of the reactor geometry. This may account for the current disagreement over the severity or even the existence of a reaction between trimethylindium (TMIn) and PH_3 or AsH_3.

The alternative for MOCVD growth of III–V compounds involves using adducts. An adduct is a compound formed between a group III alkyl such as MR_3, which acts as an electron-acceptor molecule (Lewis acid), and a group V alkyl such as ER_3, which acts as an electron-donor molecule (Lewis base), according to

Eq. (2.10) $MR_3 + ER_3 \rightarrow R_3MER_3$

e.g.

Eq. (2.11) $(CH_3)_3 In + (CH_3)_3 P \rightarrow (CH_3)_3 - In - P - (CH_3)_3$

The idea of using adducts as precursors for both the group III and group V elements was proposed by Harrison and Tompkins [Harrison et al. 1962] for the growth of InSb and GaAs. Benz et al. [1981] used TMIn (($CH_3)_3In$) and TMP (($CH_3)_3P$) for the growth of InP material.

Later, Dietze et al. [1981] and Moss and Evans [Moss et al. 1981] developed methods for generating adducts within the reactor. Since the group V alkyls are, in general, stronger Lewis bases than the corresponding hydrides, the stable adducts can be used conveniently to inhibit case (d) reactions of the group III alkyl with group V hydrides. Adducts have a vapor pressure significantly lower than either of the individual components, however, and tend to be solids at room temperature. Adducts are less pyrophoric, are easier to handle than the alkyls and may be easier to purify. Because adducts are low-vapor-pressure solids, heated valves, lines and other equipment are often used to avoid condensation within the reactor. For more details on adducts, the reader is referred to the review by Moss and Evans [1981].

Table 2.3 lists the vapor pressures and some other physical properties of most of the commonly used alkyls. The chemical names are usually abbreviated to a four-letter acronym; for example, trimethylgallium becomes TMGa. All of the group III alkyls are pyrophoric and react vigorously on contact with air or water. The alkyls are colorless liquids at room temperature except for the white solids TMIn and Cp_2Mg. The alkyls are monomeric in the vapor except for TMAl, which is dimeric. Because of their reactivity with oxygen, the alkyls must be handled under an inert atmosphere or vacuum. A number of commercial vendors supply the alkyls in electronic-grade purity. Rigorous handling procedures and cleanliness measures are necessary to preserve the level of purity. With the exception of a few workers who have used fractional distillation to improve the purity, most laboratories use the as-received materials.

MOCVD Sources			
Group III	Group V	p-type dopant	n-type dopant
$Ga(C_2H_5)_3$	AsH_3	$(C_2H_5)_2Zn$	SH_2
$Ga(CH_3)_3$			SiH_4
$In(C_2H_5)_3$	PH_3		
$In(CH_3)_3$			

Table 2.2 Starting materials.

The compounds used as sources of the group III and group V elements in this book are listed in Table 2.2. Their vapor pressures as a function of temperature are given in Fig. 2.8. Their chemical properties are given in appendix A.6.

Triethylindium (TEIn), trimethylindium (TMIn) and triethylgallium (TEGa) have been used as group III sources. The hydrides, pure arsine (AsH_3) and pure phosphine (PH_3), have been used as group V sources. Diethylzinc (DEZn) is used for p-type doping and the sulfides, H_2S or silane (SiH_4), have been used for n-type doping. Pure hydrogen (H_2) and pure nitrogen (N_2) have been used as the carrier gas. The presence of N_2 is necessary to avoid the parasitic reaction between TEIn and AsH_3 or PH_3. The presence of H_2 is necessary to avoid the deposition of carbon. TEIn, TMIn and TEGa are contained in stainless-steel bubblers, which are held in controlled temperature baths at 31 and 0 °C, respectively. An accurately metered flow of nitrogen (N_2) for TEIn and purified H_2 for TEGa and TMIn is passed through the appropriate bubbler. To ensure that the source material remains in the vapor form, the saturated vapor that emerges from the bottle is immediately diluted by a flow of H_2. The mole fraction, and thus the partial pressure, of the source species is lower in the mixture and is prevented from condensing in the stainless-steel pipework. The gas piping is made primarily of $\frac{1}{4}$ or $\frac{1}{8}$ inch stainless steel, and Swagelok or VCR fittings can be used. Commercially available alkyls are delivered in stainless-steel containers provided with valves and an inlet diptube assembly (see appendix A.6) to allow the bubbling of carrier gas, which is high-purity palladium-diffused hydrogen. Flows are monitored by mass flow controllers. These containers are connected by stainless steel tubes and supported in thermostatically controlled baths in the range −40 to +150 °C. For safety information, see also appendix A.6. Fig. 2.9 shows different alkyl containers.

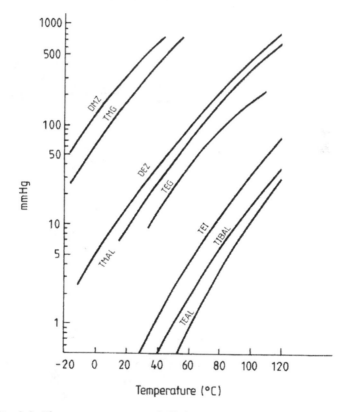

Fig. 2.8. The vapor pressure of alkyl sources of group III as a function of temperature.

The flow-rates of the hydrides, H_2 and N_2, are controlled by mass flow controllers within 0.2%. The metal alkyl or hydride flow can be injected either into the reactor or into the waste line by using the three-way valves (see the appendix). In each case, the source flow is first switched into the waste line to establish the flow-rate and then switched into the reactor.

Compound	Formula	Abbreviation	Melting point (°C)	Boiling point (°C)	Log10P (mmHg) (T in kelvin)	Range (°C)
Group II sources						
Diethylberyllium	$(C_2H_5)_2Be$	DEBe	12	194	$7.59 - 2200/T$	
Dimethylberyllium	$(CH_3)_2Be$	DMBe				
Bis(cyclopentadienyl) magnesium	$(C_5H_5)_2Mg$	Cp$_2$Mg	176		$25.14 - 2.18 \ln T - 4198/T$	
Group IIB sources[b]						
Dimethylzinc	$(CH_3)_2Zn$	DMZn	−42	46	$7.802 - 1560/T$	
Diethylzinc	$(C_2H_5)_2Zn$	DEZn	−28	118	$8.280 - 2190/T$	
Dimethylcadmium	$(CH_3)_2Cd$	DMCd	−4.5	105.5	$7.764 - 1850/T$	
Group III sources						
Trimethylaluminum	$(CH_3)_3Al$	TMAl	15.4	126	$7.3147 - 1534.1/(T - 53)$	17–100
Triethylaluminum	$(C_2H_5)_3Al$	TEAl	−58	194	$10.784 - 3625/7\cdot$	110–140
Trimethylgallium	$(CH_3)_3Ga$	TMGa	−15.8	55.7	$8.07 - 1703/T$	
Triethylgallium	$(C_2H_5)_3Ga$	TEGa	−82.3	143	$8.224 - 2222/T$	50–80
Trimethylindium	$(CH_3)_3In$	TMIn	88.4	133.8	$10.520 - 3014/T$	
Triethylindium	$(C_2H_5)_3In$	TEIn	−32	184	1.2	44
					3	53
					12	83

Compound	Formula	Abbreviation	Melting point (°C)	Boiling point (°C)	Log10P (mmHg) (T in kelvin)	Range (°C)
Group IV sources						
Tetramethylgermanium	$(CH_3)_4Ge$	TMGe	-88	43.6	139	0
Tetramethyltin	$(CH_3)_4Sn$	TMSn	-53	78	$7.495 - 1620/T$	
Tetraethyltin	$(C_2H_5)_4Sn$	TESn	-112	181		
Group V sources						
Trimethylphosphorus	$(CH_3)_3P$	TMP	-85	37.8	$7.7329 - 1512/T$	
Triethylphosphorus	$(C_2H_5)_3P$	TEP	-88	127	$7.86 - 2000/T$	18–78.2
Trimethylarsenic	$(CH_3)_3As$	TMAs	-87.3	50–52	$7.7119 - 1563/T$	
Triethylarsenic	$(C_2H_5)_3As$	TEAs	-91	140	15.5	37
Trimethylantimony	$(CH_3)_3Sb$	TMSb	-86.7	80.6	$7.7280 - 1709/T$	
Triethylantimony	$(C_2H_5)_3Sb$	TESb	-98	116	17	75
Group VI sources[c]						
Diethylselenium	$(C_2H_5)_2Se$	DESe	–	108.00		
Dimethyltellurium	$(CH_3)_2Te$	DMTe	-10	82.00	$7.97 - 1865/T$	
Diethyltellurium	$(C_2H_5)_2Te$	DETe	–	137–138	$7.99 - 2093/T$	

Table 2.3 Physical properties of some organometallics used in MOCVD[a].

[a] After Ludowise [1985].
[b] Used as p-type dopants.
[c] Used as n-type dopants.

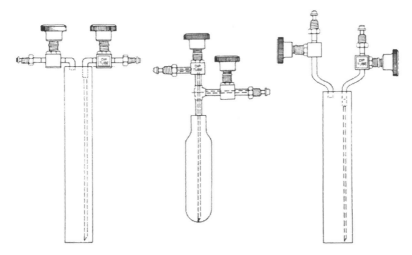

Fig. 2.9. The commercially available stainless steel alkyl containers with valves and an inlet dip-tube assembly.

2.3. New non-equilibrium growth techniques

2.3.1. Chemical beam epitaxy

For the growth of InP-based material with MBE, on the one hand, and for doing *in situ* characterization with MOCVD, on the other, there have been several modifications to the conventional non-equilibrium MBE and MOCVD growth techniques in the past few years. These have enhanced their versatility and increased their potential for the growth of InP and related compounds with extremely thin layers and abrupt interfaces.

In the case of MOCVD, the use of low-pressure MOCVD for the growth of InP was made by Duchemin et al. [1979]. It was shown that for silicon growth the use of low pressure reduced the absorption of hydrogen on the growing surface, thus permitting the growth of good-quality silicon at a lower temperature than is possible at atmospheric pressure. Fraas [1981] reported that epitaxial GaAs, GaAsP and GaInAs films with good surface morphologies and electrical properties could be grown by reacting combinations of triethylgallium (TEGa), triethylindium (TEIn), arsine and phosphine in a high-vacuum system.

In the case of MBE, Panish [1980] replaced the conventional condensed sources for the group V elements, usually As and P, with sources that decompose AsH_3 and PH_3, and called it gas-source molecular beam epitaxy (GSMBE).

Vodjdani [1982] and Tsang [1984] replaced the group III elements such as In and Ga used as the condensed source in conventional MBE by simple organometallic compounds. This extension of GSMBE was called metalorganic molecular beam epitaxy (MOMBE), chemical beam epitaxy (CBE) or metalorganic chemical beam deposition (MOCBD).

Fig. 2.10. Photograph of a CBE reactor.

The CBE reactor (Fig. 2.10) employed the gas-handling system of MOCVD and the growth chamber of MBE. The main advantages of this combination by comparison to the conventional MBE are:

(1)   The capability to grow high-quality InP-based material [Panish et al. 1986, Tsang 1984].

(2)   Elimination of oval defects even at high growth rates.

(3)   High-quality InP-based material at high growth rates.

(4)   The flexibility of source changes.

(5)   The versatility of using different alkyl and hydride sources.

(6)   Good homogeneity and composition uniformity over large areas of substrate, due to a single group III (and group V if desired) beam.

The main advantages of the above combination compared to MOCVD growth techniques are:

(1)   Use of at least 100 times less AsH_3 and PH_3 for the growth of GaAs- and InP-based materials.

(2)   Elimination of parasitic reaction in the gas phase, because of UHV conditions.

(3)   The possibility of using different *in situ* surface diagnostic techniques.

(4)   Improvement of homogeneity and composition uniformity and reproducibility of InP-based materials over large areas of substrate.

(5)   *In situ* etching and removal of oxide during the growth of InP on a silicon substrate.

(6)   Reduction of memory effect during *p*-type doping.

(7)   Reduction of boundary-layer thickness on the hot substrate surface.

(8)   Compatibility with other high-vacuum thin-film processing techniques such as plasma etching, metal evaporation, ion beam etching and ion implantation.

(9)   The possibility to do localized epitaxy.

No deposition was observed on a SiO_2 film [Cho and Ballamy 1975], [Takahashi 1984]. This is in contrast with conventional MBE or MOCVD where polycrystalline GaAs deposits on the SiO_2 masked region. It was found that monocrystalline GaAs was grown on the bare substrate with narrow-stripe patterns and that the growth did not take place on the large SiO_2 covered areas. This phenomenon had already been observed with VPE. The growth mechanism that causes this remarkable feature is not clear at present; this feature may be due to a surface-catalyzed growth process.

2.3.2. Atomic layer epitaxy

For future integrated circuits (IC) with monolayer controlled multilayer structure, one needs a more advanced epitaxial technique, which has atomic scale, high uniformity of thickness and composition, and perfect selective epitaxy, and is free from any defects such as spitting, dislocations, or hillocks. Atomic layer epitaxy (ALE) is a new growth technique with control at the monolayer level. ALE was originally proposed by Suntola et al. [1980], and Suntola et al. [1984], who described a novel mode of evaporative deposition for preparing thin films of zinc sulfide. ALE operates by growing complete layers on top of each other, there being two basic variants based on evaporative deposition, relying on heated elemental source materials, or chemical vapor deposition, relying on sequential surface exchange reactions between compound reactants. By definition, ALE is based on chemical reactions at the solid surface of a substrate, to which the reactants are transported alternately as pulses of neutral molecules or atoms, either as chopped beams in high vacuum, or as switched streams of vapor possibly with an inert carrier gas. The incident pulse reacts directly and chemically only with the outermost atomic layer of the substrate. The film therefore grows stepwise, a single monolayer per pulse, provided that at least one complete monolayer coverage of a constituent element, or of a chemical compound containing it, is formed before the next pulse is allowed to react with the surface.

2.3.3. Metalorganic atomic layer epitaxy

In this technique (MO-ALE) alkyls and hydrides are used for the starting materials. Nishizawa et al. [1986], and Doi et al. [1986] used MO_ALE for the growth of GaAs layers and they showed the GaAs to be *p*-type, with hole concentrations of 10^{19} cm^{-3} and mobilities of 60–200 C·m^2·V^{-1}·S^{-1}. Kobayashi et al. [1985] introduced a small amount of AsH_3 during the TEGa exposure step. They call this technique flow-rate modulation epitaxy (FME). They succeeded in the growth of *n*-type GaAs having carrier concentrations of 10^{14}–10^{16} cm^{-3} and mobilities as high as 42,000 C·m^2·V^{-1}·S^{-1} at 77 K. Additional AsH_3 introduction with TEGa suppresses carbon contamination and explains the gas properties [Makimoto et al. 1986].

2.3.4. Chloride atomic layer epitaxy

In this technique, chloride and hydride are used as starting materials. GaAs layers grown by GaCl-ALE are *n*-type with a carrier concentration of 6×10^{15} cm^{-3} and mobility of 16,000 C·m^2·V^{-1}·S^{-1} at 77 K. Matsumoto and Usui [1986] showed using photoluminescence at 5 K that the layers are compensated by a carbon acceptor. The electron trap EL$_2$ was not detected in the GaCl-ALE sample. This is in contrast with conventional VPE or MOCVD samples which exhibit EL$_2$ in the concentration range 10^{14}–10^{15} cm^{-3}.

The greatest advantage of ALE, especially digital ALE (where the thickness grown is insensitive to any analog quantities such as source gas pressure, growth temperature and growth time), is to obtain large-diameter integrated-circuit wafers with monolayer control of thickness and composition.

2.3.5. Migration-enhanced epitaxy

Rapid migration of the evaporated materials on the growing surface is essential to the growth of high-quality epitaxial layers. In conventional MBE growth of GaAs and AlAs layers, migrating materials on the growing surface are Ga–As and Al–As molecules rather than Ga and Al atoms, respectively, because these layers are grown under arsenic-stabilized conditions. The migration of these molecules on the surface is very slow, especially at low temperatures. Therefore, lowering the substrate temperature considerably deteriorates the crystal quality of grown layers.

Horikoshi [1986] proposed a new mode of MBE growth which makes it possible to grow high-quality GaAs and AlAs layers at very low substrate temperatures (~ 200 °C) by enhancing the migration of the materials evaporated on the growing surface. This method has been termed migration-enhanced epitaxy (MEE). It is based on the very rapid migration of Ga and Al atoms on the growing surface, and on the alternate supply of Ga(Al) atoms and arsenic molecules to the growing surface. RHEED intensity observations revealed that Ga and Al atoms migrate on the GaAs surface much more rapidly than Ga–As and Al–As molecules, and that they migrate very actively even at temperatures as low as 200 °C. They showed that high-quality GaAs and AlAs could be grown at very low temperatures by alternately supplying Ga(Al) atoms and arsenic molecules to the substrate surface. Applying the MEE method, 1–2 μm thick GaAs layers were grown at a substrate temperature of 200 °C. These layers showed efficient photoluminescence (PL) from the band-edge excitons at 4.2 K. AlAs–GaAs single-quantum-well structures with 3–6 nm widths were also grown at 200 °C by MEE. These structures showed PL due to electronic transitions between quantized levels in the wells, indicating a reasonable quality of AlAs [Horikoshi 1986].

References

Arthur, J.R., *J. Appl. Phys.* **39** 4032, 1968.

Bass, S.J., *J. Cryst. Growth* **31** 172, 1975.

Benz, K.W., Renz, H., Wiedlein, J., and Pilkuhn, N.H., *J. Electron. Mater.* **10** 185, 1981

Cho, A.Y. and Ballamy, W.C., *J. Appl. Phys.* **46** 783, 1975.

Davey, J.E. and Pankey, T.J. *J. Appl. Phys.* **39** 1941, 1968.

Dietze, W.T., Ludowise, M.J., and Copper, C.B., *Electron. Lett.* **17** 698, 1981.

Doi, A., Aoyagi, Y., and Namba, S., *Appl. Phys. Lett.* **48** 1787, 1986.

Duchemin, J.P., Bonnet, M., Koelsch, F., and Huighe, D., *J. Electrochem. Soc.* **126**, 1979.

Fraas, L.M., *J. Appl. Phys.* **52** 6939, 1981.

Gunther, K.G., *Patent* 2937816, 1958.

Harrison, B.C. and Tompkins, E.H., *Inorg. Chem.* **1** 951, 1962

Horikoshi, L., *The Physics and Fabrication of Microstructures* (Les Houches), 1986.

Kobayashi, N., Makimoto, T., and Horikoshi, Y., *Japan J. Appl. Phys.* **24** L962, 1985.

Ludowise, M.I., *J. Appl. Phys.* **58** R311134, 1985.

Makimoto, T., Kobayashi, N., and Horikoshi, Y., *Japan J. Appl. Phys.* **25** L513, 1986.

Manasevit, H.M., and Simpson, W.I., *J. Electrochem. Soc.* **118** C291, 1971.

Matsumoto, T. and Usui, A., Fall Meeting *Japan. Soc. Appl. Phys. Abstracts,* 1986.

Moss, R.H. and Evans, J.S., *J. Cryst. Growth* **55** 129, 1981.

Neave, J.H., Joyce, B.A., Dobson, P.J., and Norton, N., *Appl. Phys. A* **31** 1, 1983.

Nelson, H., *RCA Rev.* **19** (24) 603, 1963.

Nishizawa, J. and Kurabayashi, T., *J. Crystallogr. Soc. Japan* **28** 133, 1986.

Noyanuma, M. and Takahashi, K., *Appl. Phys. Lett.* **27** 342, 1975.

Panish, M.B., *J. Electrochem. Soc.* **127** 2729, 1980.

Panish, M.B. and Hamm, R.A., *J. Cryst. Growth* **78** 445, 1986.

Prandtl, L., and Tietgens, O.C., *Applied Hydro- and Aeromechanics* (New York: McGraw-Hill), 1934.

Razeghi, M., *Semicond. Semimet.* **22** 299, 1985.

Shapiro, A.H., *Shape and Flow* (London: Heinemann), 1961.

Suntola, T., Atron, J., Pakkala, A., and Linfors, S., *SID* **80**, Digest 108, 1980.

Suntola, T., 16th *Int. Conf. on Solid State Devices and Materials Tokyo*, Extended Abstract, p 647, 1984.

Takahashi, K., *Proc. Research and Development Association for Future Electron Devices Tokyo*, paper I-2, 1984.

Tietjen, J.J., and Amick, J.A., *J. Electrochem. Soc.* **113** 724, 1966.

Tritton, D.J., *Physical Fluid Dynamics* (Wokingham: Van Nostrand Reinhold), 1977.

Tsang, W.T., *Appl. Phys. Lett.* **45** 1234, 1984.

Tsang, W.T., *Appl. Phys. Lett.* **48** 511, 1986.

Vodjdani, N., *Thèse de Doctoral 3eme Cycle Université Paris XI*, 1982.

3. *In situ* Characterization during MOCVD

3.1.   **Introduction**
3.2.   **Reflectance anisotropy and ellipsometry**
    3.2.1.   Principles of RDS
    3.2.2.   The optical setup and data acquisition system
    3.2.3.   Experimental procedure
    3.2.4.   Adaptation to a MOCVD reactor
    3.2.5.   Comparison with other RDS techniques
3.3.   **Optimization of the growth of III–V binaries by RDS**
    3.3.1.   Effects of the MOCVD reactor design on the growth
    3.3.2.   The III/V flux dependence of RDS signals
    3.3.3.   The temperature dependence of RDS signals
    3.3.4.   Discussion
3.4.   **RDS investigation of III–V lattice-matched heterojunctions**
    3.4.1.   RDS observations of heterojunction growth
    3.4.2.   Interpretations of the experimental observations
    3.4.3.   Applications of RDS for growth monitoring
3.5.   **RDS investigation of III–V lattice-mismatched heterojunctions**
    3.5.1.   Growth procedure of InAs/InP and InP/GaAs/Si
    3.5.2.   RDS records during real-time growth
    3.5.3.   Optical models
    3.5.4.   The first stage of 3D growth
    3.5.5.   Subsequent growth
    3.5.6.   Discussion of optical models
    3.5.7.   RDS monitoring of growth of GaInP/GaAs and GaInP/InP
3.7.   **Insights on the growth process**

3.1. Introduction

Molecular beam epitaxy (MBE) benefits from many *in situ* characterization techniques because of its high-vacuum environment. For instance, reflection high-energy electron diffraction (RHEED) is used for optimization of MBE growth conditions. Very few techniques are available for *in situ* characterization of MOCVD because the development of such techniques is hampered by a number of difficulties. These include reactor geometry considerations, inability to use ultrahigh-vacuum techniques, and the deposition of reaction products on the reactor walls during growth. In comparison with MBE and metalorganic MBE (MOMBE), the *in situ* control of growth parameters is still in a premature state.

The kinetic data needed to understand the first step in growth include measurements of precursor decomposition, the search for and measurement of new products generated in the reactions, the influence of operation conditions on these data and monitoring the growing surfaces as RHEED does in MOMBE.

Mass spectroscopy [Lee et al. 1988, Mashita et al. 1986] has been widely used to sample species from the reaction zone. The mass spectrometer is usually at some distance from the sampling region. Therefore the sampled gas volume is studied under different conditions (pressures and temperatures) than those prevailing in the reactor. The sampled molecules undergo a very large number of collisions between the sampling point and the detection chamber. This implies that short-lived reactive species, e.g. free radicals, are likely to be removed by wall reactions and gas-phase reactions before the sample is analyzed. Thus mass spectrometry is not a true *in situ* characterization tool.

Since techniques with electron beams are not available in MOCVD, photons are used as probes. They can easily penetrate through the gas phase to the important regions near the substrate and can be chosen such that they do not modify the molecules studied.

For optical diagnostics, ultraviolet (UV), visible (VIS) [Haigh 1983] and near-infrared (NIR) probes are most often used to study the vibrational and electronic spectra of gas species respectively. IR laser absorption spectroscopy is applicable to a large number of molecular species, for example, to characterize gaseous chemical intermediates [Gaskill et al. 1988] and to monitor the vibrational frequencies of gaseous species [Butler et al. 1986], but the technique is hampered by averaging over the probe volume. This can make interpretation difficult for MOCVD systems where there are generally considerable spatial variations in concentrations and temperature.

Opto-galvanic detection and laser-induced fluorescence [Donnelly and Karlicek 1982] spectroscopy require a tunable laser to tune into either a fluorescent or an ionizing transition. They provide very sensitive and selective detection. However, opto-galvanic detection has not been applied to MOCVD and the potential of laser-induced fluorescence has been shown only in a few cases [Richter et al. 1991].

Raman scattering (RS), either linear (spontaneous) or non-linear, is experimentally elaborate. It always exhibits good temporal and spatial resolution. While spontaneous RS has a low sensitivity, the non-linear method coherent anti-Stokes Raman scattering (CARS) yields an adequate signal at typical MOCVD partial pressures.

All of the techniques mentioned above are applied in gas-phase diagnostics. They do not provide direct information about the growing surface. There is also a clear need for the development and use of diagnostic tools as monitoring devices in the routine growing equipment. In comparison with MBE and MOMBE, the on-line control and accurate presetting of growth parameters is still in a premature state. One of the techniques for monitoring the growth surface is quasielectric light scattering (QLS) which has been used to characterize surface defects and surface roughness [Olson and Kibbler 1986]. It is particularly sensitive to extended, correlated surface roughness of the kind typically found in morphologically unstable crystal growth systems.

Optical second-harmonic generation has been applied to monitor the dynamic surface chemical processes for low-pressure MOCVD. Preliminary results showed that the measurements depend exclusively upon surface rather than gas-phase processes. This work illustrated the potential for a second-harmonic generation technique as a monitoring tool for real-time growth of MOCVD. In principle, real-time ellipsometry is an apparently well adapted technique for the purpose of monitoring the growing surface of MOCVD because it is non-destructive, compatible with high pressure and various reactor geometries [Aspnes 1988b, Drevillon 1989, Collins 1989]. Ellipsometry analyses the polarization-dependent reflection properties of a light beam under oblique incidence and data are directly correlated to the dielectric function of the structure under investigation. Although it contains some information about the sample surface, it is very difficult to separate the surface contribution from the much larger bulk contribution. In order to yield information on the growing surface, UV light must be used to decrease the penetration depth to about 100 Å, thus the deposition on the reactor wall can prevent any prolonged real-time study.

The above limitations can be overcome using a technique called either reflectance anisotropy (RA) or reflectance difference spectroscopy (RDS) [Aspnes 1985, Aspnes and Studna 1985]. RDS measures the optical anisotropy of a sample under normal incidence. More precisely, the technique involves illuminating the sample with polarization-modulated light, and detecting the polarization of the reflected light. Thus, RDS is analogous to ellipsometry, in that oblique incidence is used. Because of the cubic symmetry of the III–V semiconductors, the bulk is nearly isotropic, while the regions of lower symmetry, like the surface and interfaces, can be anisotropic. In the case of the (1 0 0) surfaces of III–V semiconductors, the contribution from the bulk is expected to vanish [Agranovich and Ginzburg 1984]. Therefore, as compared to ellipsometry, RDS is expected to be directly sensitive to the surface (and interfaces) whatever the wavelength of the light. Also RDS is sensitive to both the chemical and structural state of the surface. Moreover the presence of a film deposited on the reactor walls does not affect RDS.

RDS has been adopted for real-time monitoring of MOCVD growth in several laboratories since 1985. A strong RDS "chemical" peak related to Ga near 2.5 eV was observed [Colas et al. 1989] in MOCVD growth of GaAs using arsine and trimethylgallium as the sources of As and Ga. It seems that the RDS can follow the relative surface coverages of Ga and As on (0 0 1) GaAs in MOCVD, and it was suggested that 2.5 eV (near the 2.41 eV, 514.5 nm Ar laser line) is a suitable energy to modify MOCVD growth processes with light [Aoyagi et al. 1987]. RDS has been applied to study, in real time, surface reconstruction changes, as well as catalytic effects of surfaces on the decomposition of adsorbed species from various reactants [Colas et al. 1988]. For example, Colas et al. found a difference in the response of TMGa and TMAl and hence in the decomposition mechanisms of these two molecules. A Ga response is immediately visible during the TMGa pulse and increases during the interruption pulse in hydrogen, whereas an Al response from TMAl is only visible when arsine is supplied [Colas et al. 1988]. MOCVD processes are less characterized than MBE and are potentially more complex. Colas et al. [1991] reported a quantitative study of the catalytic effect of the GaAs surface on the decomposition of TMGa and the growth process suggested that decomposition of an organic molecule occurs, e.g., TMGa in the case of GaAs. In addition, this decomposition may occur either on the surface or in the gas phase which contains both group III (TMGa) and group V (arsine) reactant molecules in conventional growth. Anisotropic changes at the growing surface can have a variety of origins including the surface roughness, which depends on sample history as well as growth conditions. It is possible to separate their individual surface-chemical effects and the difference pathways determining the kinetic limit of growth on (0 0 1) GaAs.

In 1990, Jonsson et al. [1990] reported the first measurements of growth oscillations in high-vacuum metalorganic vapor phase epitaxy. They found that the frequency of the optically detected growth oscillations is proportional to the flux of TEGa and to the growth rate. By comparing the oscillations obtained by RDS with those seen in RHEED, it was found that RHEED and RDS oscillations had the same period. Since it is well known that the RHEED oscillations correspond to the growth of exactly one monolayer, the RDS oscillations proved to be useful for the calibration of growth rates, for optimizing the buffer layer growth and for eliminating pressure transients which result in a varying growth rate at the initiation of growth. In a similar experiment, Samuelson et al. [1991] reported the monolayer saturation phenomenon, when too much TEGa was injected into the chamber, and by using the oscillation periodicity, they measured the kinetics of regeneration of As-stabilized surfaces.

The RDS technique appears particularly attractive for a variety of new applications including the *in situ* study of crystal growth by MOCVD [Colas et al. 1989, Acher et al. 1990b, c, Koch et al. 1990, 1991].

3.2. Reflectance anisotropy and ellipsometry

Despite its novelty, RDS belongs to the well known family of polarization-modulation techniques, like ellipsometry and dichroism. RDS is the determination of the difference between normal-incidence reflectance for light polarized along the two principal axes of the surface. Thus, an RDS instrument can be considered as a normal incidence ellipsometer. However, compared to the other polarization-modulation techniques, RDS measurements are only sensitive to the surface of the material, even though the light penetrates deep into the bulk. It was recently shown that, because of its lower symmetry, the surface of nominally isotropic crystals can display a measurable anisotropy [Aspnes 1985]. In particular, the bulk contribution is expected to vanish in the case of (1 0 0) surfaces of III–V semiconductors [Agranovich and Ginzburg 1984, Acher 1990], however the surface anisotropy generally induces small contributions to the RDS signal. Relative surface anisotropy ranging from 10^{-4} to 10^{-3} has typically been reported. This can be a strong limitation when dealing with real-time measurements because of the presence of noise induced by the experimental environment (rotating pumps, vibrations). As a consequence, the most successful *in situ* applications are performed using a fast photoelastic modulator (40–50 kHz) [Aspnes et al. 1988a, Paulsson et al. 1990, Colas et al. 1989, Acher et al. 1990b, c, Koch et al. 1990, 1991].

3.2.1. Principles of RDS

The numerous similarities between ellipsometry and RDS will be emphasized in this section for a better understanding of the RDS technique. Consider the two optical eigenaxes of a sample, x and y. In general, the ellipsometric technique measures the ratio between the two corresponding complex reflectances r_x and r_y [Azzam and Bashara 1977]

Eq. (3.1) $\qquad \rho = r_x / r_y = \tan \Psi \exp(i\Delta)$

where Δ is the optical path difference between x- and y-direction light. Considering oblique incidence, the optical eigenaxes are the directions parallel (p) and perpendicular (s) to the plane of incidence. Thus ellipsometry is the measurement of r_p/r_s.

In the case of a crystal surface, the eigenaxes can be deduced from symmetry considerations. For instance [0 1 1] and [0 1 $\bar{1}$] are the eigenaxes of the (1 0 0) surfaces of semiconductors and r_{011} and $r_{01\bar{1}}$ are the reflectances associated with these directions. RDS consists in the measurement of the relative difference:

Eq. (3.2) $\qquad r_a = \dfrac{r_{001} - r_{01\bar{1}}}{r}$

In the case of small anisotropies, Eq. (3.2) can be approximated to

Eq. (3.3) $\qquad r_a \approx 2\Psi' + i\Delta'$

where $\Psi = \Psi' - \pi/4$ and $\Delta' = \Delta - \pi$. When large anisotropies are considered [Acher et al. 1990b], r is generally replaced by $r_{01\bar{1}}$ in the denominator of Eq. (3.2) and then

Eq. (3.4) $\qquad r_a = \tan \Psi \exp(i\Delta') - 1$

At this point, let us reconsider the various ellipsometry techniques. Phase-modulated ellipsometry (PME) is based on the use of a photoelastic modulator generating a periodic phase shift $\delta(t) = A_m \cdot \sin(\omega t)$ ($\omega = 50$ Hz) between orthogonal amplitude components of the transmitted beam. Finally, the detected intensity takes the general form [Jasperson and Schnatterly 1969, Drevillon et al. 1982, Acher et al. 1989]:

Eq. (3.5) $\qquad I(t) = I[I_0 + I_s \sin \delta(t) + I_c \cos \delta(t)]$

where I_0, I_s and I_c are trigonometric functions of Ψ and Δ. Particular configurations of the optical elements allow simple determination of Ψ and Δ. In the case illustrated in Fig. 3.1, known in PME as configuration III, one obtains:

Eq. (3.6)
$$I_0 = 1$$
$$I_s = (\pm)_p (\pm)_A \sin 2\Psi \sin \Delta \approx -(\pm)_p (\pm)_A \Delta'$$
$$I_c = (\pm)_p (\pm)_M \cos 2\Psi \approx -(\pm)_p (\pm)_M \Psi'$$

which, in the small-anisotropy approximation, gives

Eq. (3.7)
$$I_s \approx -(\pm)_p (\pm)_A \Delta'$$
$$I_c \approx -(\pm)_p (\pm)_M 2\Psi'$$

where $(\pm)A = 1$ if $A = +45°$, and -1 if $A = -45°$, etc. Eq. (3.3) and Eq. (3.6) show that I_s and I_c are related to the imaginary and real part of the RDS signal, respectively. The optical configuration corresponding to Fig. 3.1 is generally used in RDS measurements.

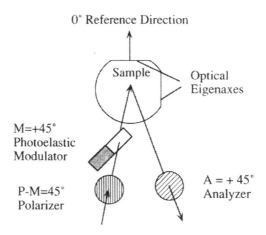

Fig. 3.1. Relative azimuths of the principal optical components of the RDS spectrometer (after Acher [1990]).

3.2.2. The optical setup and data acquisition system

The optical setup of the RDS spectrometer is presented in Fig. 3.2. The light source can be either a 75 W xenon lamp or a low-power laser. A mechanical shutter, included between the light source and the optical fiber, allows evaluation of the DC background. In order to increase the compactness of the spectrometer, optical fibers are used in both arms. Since the beam goes through a fixed polarizer before being modulated and through an analyzer before being detected, PME is insensitive to any polarization effect due to the optical fibers [Drevillon et al. 1989]. The polarizer and the analyzer are Glan-Taylor polarizers. The photoelastic modulator consists of a fused silica bar submitted to a periodical stress (50 kHz). The polarizer, the modulator and the analyzer are mounted on 1 min precision rotators. The angle between the modulator and the polarizer is fixed at 45°. The modulator and the analyzer can be automatically rotated by means of stepping motors. The spot size to the sample can be focused to less than $1mm^2$. The energy of the light is analyzed by a double-grating monochromator. In the simplest setup, a photomultiplier is used as detector. In this case, the available wavelength range is 230 to 830 nm. However, this range can easily be extended towards near IR by using narrow-gap semiconductor detectors. More generally, it can be pointed out that this setup is compatible with the simultaneous use of four detectors.

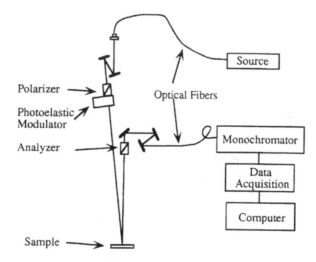

Fig. 3.2. Schematic diagram of the reflectance anisotropy spectrometer (after Acher [1990]).

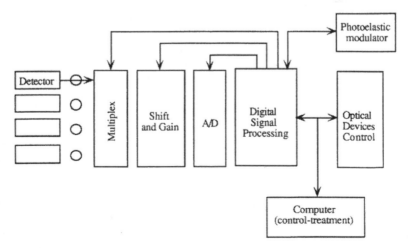

Fig. 3.3. Schematic diagram of the data acquisition system (after Acher [1990]).

A data acquisition system is presented in Fig. 3.3. It can be considered as a new version of the numerical Fourier transform processing system previously used [Drevillon and Karlicek 1982]. This new system is already used in phase-modulated ellipsometry [Drevillon et al. 1989]. The read-out system consists of a gain-controlled amplifier and a 14-bit digital–analog offset which are used to allow a correct matching between the detected signal and the analog–digital converter (ADC). A fast ADC (12-bit, 1 MHz) is then used to synchronously convert the signal at a frequency of 16ω. The Fourier analysis of the signal is then performed using a digital signal processor specially dedicated to the fast Fourier transform computation. Finally, the DC component together with the four first harmonics of the signal (S_ω, $S_{2\omega}$, $S_{3\omega}$ and $S_{4\omega}$) are continuously transmitted to a personal computer (PC) using a first-in–first-out register. The maximum data acquisition rate is fixed at 1 kHz. Moreover, the digital signal processor (DSP) allows on-line control of the modulation amplitude A_m, using the higher harmonics, $S_{3\omega}$ or $S_{4\omega}$, following a calibration procedure described in detail elsewhere [Acher et al. 1989, Drevillon et al. 1989]. Finally, the PC is devoted to the overall monitoring of the RDS spectrometer. In particular, it controls the optical devices (shutter, detector power supply, monochromator and stepping motors).

The data acquisition system allows further external connections, as previously described [Drevillon et al. 1989]. In particular, the data acquisition system can easily deal with a second low-frequency modulation. This can be very useful for extension to infrared because of the presence of a chopper. This facility can also be used to record RDS measurements performed with rotating samples, as has been suggested recently [Aspnes et al. 1990]. Furthermore, as mentioned above, four analog channels can be recorded simultaneously. These measurements can correspond to different light detectors or external parameters like control process information.

3.2.3. Experimental procedure

The analogy between RDS and PME allows a simple transposition of the calibration procedure used in ellipsometry. This procedure, based on the multiple-harmonic model, is described in detail by Acher et al. [1989]. In fact, the calibration procedure appears simpler in the case of RDS. In particular, the RDS optical setup acts in a transmission configuration when isotropic samples are considered such as amorphous materials or (1 1 1) surfaces of semiconductors. Moreover, as compared to ellipsometry, RDS allows the use of reference samples with well known reflectances like polarizers and birefringent plates.

The determination of the angles Ψ and Δ from the Fourier analysis is performed following the procedure used in PME. From Eq. (3.4), it can be easily shown that I_s and I_c are related to the ratios S_ω/S_0 and $S_{2\omega}/S_0$, respectively. Finally the useful quantities are deduced from Eq. (3.2) or Eq. (3.6).

In practical applications of RDS, the angle of incidence cannot strictly be zero (see Fig. 3.2). In the present case, the angle of incidence is found to be less than $3°$. Nevertheless, a parasitic ellipsometric contribution cannot be ignored. The orientation of the incidence plane as referred to the sample eigenaxis is defined by an angle Ω. The influence of the small angle of incidence can be evaluated from the general expressions for I_0, I_s and I_c in PME. It can be shown that the ellipsometric contribution vanishes if $\Omega = \pm 45°$.

Another imperfection of the RDS technique may be a possible misalignment of the sample eigenaxes, θ as referred to the reference position. This effect can be evaluated from the general expressions for I_0, I_s and I_c. In the case of small anisotropies, one obtains:

Eq. (3.8)
$$I_0 = 1 + (\pm)_A 2\Psi' \sin 2\theta$$
$$I_s = -(\pm)_p (\pm)_A \Delta' \cos 2\theta$$
$$I_c = -(\pm)_p (\pm)_M \cos 2\Psi' \cos 2\theta$$

Eq. (3.8) shows that a weak misalignment leads to second-order effects in S_ω/S_0 and $S_{2\omega}/S_0$. Moreover, Eq. (3.8) shows that the eigenaxes of any sample can be identified by the determination of the minima of S_ω/S_0 and $S_{2\omega}/S_0$ as functions of θ.

3.2.4. Adaptation to a MOCVD reactor

The RDS spectrometer can be directly adapted to various deposition chambers. Fig. 3.4 shows the RDS spectrometer coupled to a low-pressure MOCVD [Razeghi 1989] reactor chamber for real-time III–V crystal growth investigations. This diagram reveals that this adaptation can be realized without any modification of the MOCVD reactor. The light beam is not affected by the presence of the RF heating system. Nevertheless the MOCVD quartz tube does not allow UV light transmission below 3400 Å.

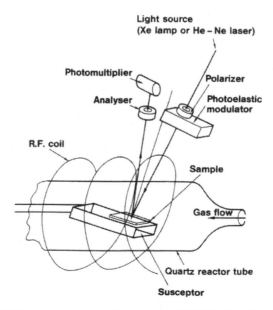

Fig. 3.4. The RDS spectrometer adapted to a MOCVD chamber (after Razeghi [1989]).

In any *in situ* application of RDS, the reactor windows can be submitted to inhomogeneous strains inducing birefringence effects that perturb the RDS measurements. In the case of ultrahigh-vacuum applications, low-strain windows can be used to overcome this experimental problem [Studna et al. 1989]. However, such a solution cannot be extended to MOCVD reactors based on the use of a cylindrical quartz tube. In order to evaluate this effect, the tube can be considered as two birefringence plates (input and output of the optical beam) with orientations θ_i (resp θ_o) and phase shifts δ_i (resp δ_o). Considering small anisotropies, it can be shown from Eq. (3.8) that to the first order in Δ, δ_i and δ_o

Eq. (3.9)
$$I_s = (\pm)_p (\pm)_A [\Delta' + \delta_i \cos 2\theta_i + \delta_o \cos 2\theta_o]$$
$$I_c = -(\pm)_p (\pm)_M 2\Psi'$$

Thus, the tube birefringence only affects the imaginary part of the RDS signal. This explains why during real-time investigations the noise level on Re r_a is generally lower than 10^{-4} while it is one order of magnitude higher on Im r_a. Therefore, both the real and the imaginary part of the RDS signal can only be measured *in situ* when large anisotropies are investigated [Acher et al. 1990b].

Finally, it has to be noted that in MOCVD reactors the substrate holder is slightly tilted (5–10°) as referred to the horizontal direction [Razeghi 1989]. Thus the light beam penetrates into the quartz tube at an oblique angle of incidence Φ. This induces a contribution to the real part of the RDS signal $2 \cdot \Phi^2[(n - l)/n]^2$ where Φ is measured with respect to the normal incidence and n is the refractive index of the tube. This constant contribution is found to be 4×10^{-3} for $\Phi = 8°$.

3.2.5. *Comparison with other RDS techniques*

The RDS system presented above can be compared to other RDS setups based upon modulation techniques. The first RDS measurements were performed using rotating samples [Aspnes 1985, Acosta-Ortiz and Lastras-Martinez 1987]. The detected intensity was measured using a lock-in amplifier. This technique is not sensitive to the phase shift at the reflection of the sample as defined by Eq. (3.2) and Eq. (3.3). Moreover, this technique is incompatible with growth techniques involving fixed samples.

Then, the adoption of a rotating analyzer ellipsometer to RDS was proposed for *in situ* applications with fixed samples [Aspnes et al. 1987]. It allows the determination of the real part of the RDS signal. In the case of real-time studies, the main limitation comes from the relatively low frequency of the mechanical rotation of the analyzer (50–100 Hz). Nevertheless *in situ* measurements of anisotropies in the 10^{-3} range were reported.

An RDS setup based on the use of two detectors has also been successfully used [Briones and Horikoshi 1990]. The sample is illuminated with polarized light tilted at 45° with respect to the optical eigenaxes. Then the two orthogonal polarizations are separated and detected. *In situ* measurements were performed. Nevertheless, the absence of modulation can be a limitation for more general real-time investigations.

Other normal-incidence RDS techniques have also been reported [Azzam 1977, 1981]. They consist of strictly normal-incidence techniques, using semi-transparent plates. Signal processing allows the determination of Re r_a and Im r_a^2. Thus, they do not appear to be well adapted to small anisotropy measurements (Im $r_a^2 \approx 10^{-7}, 10^{-8}$). Therefore these last techniques do not provide a crucial improvement as compared to the sample rotation setup.

3.3. Optimization of the growth of III–V binaries by RDS

RDS has been used in InAs growth by MOCVD and has been operated under different growth conditions to optimize the growth parameters. The effects of growth rate, III/V ratio and substrate temperature on the RDS signal were investigated. The difference in the optical anisotropy between the growing and non-growing surface, δ, was studied. This study also showed the possibility of monitoring transient flow perturbations on the growing surface due to the switching of gas flows.

The majority of the measurements were taken with light of energy 2.28 eV using a 0.5 mW He–Ne laser. A 75 W xenon lamp and a Jobin–Yvon H25 monochromator were used for the spectroscopic study. The fully digital acquisition system was set at an acquisition rate of 0.08 s per point, and the noise level of the measured signal using the laser was lower than 10^{-4}.

The measurements were taken on a 1 μm thick InAs buffer layer grown on either InAs or InP(IOO) substrates misoriented 2° toward $[01\bar{1}]$; the results were identical on the different substrates. In all cases the wafer surface was mirror-like after the growth. Unless otherwise stated, the following growth parameters were used: a growth rate of 250 Å·min^{-1}, a substrate temperature of 480 °C, a H_2 flow rate of 100 cm^3·min^{-1} in the trimethylindium (TMI) bubbler and an AsH$_3$ flow rate of 10 cm^3·min^{-1}.

3.3.1. Effects of the MOCVD reactor design on the growth

For InAs growth under optimal conditions, a significant change in the RDS signal was observed when the TMI flow was switched on and off, while maintaining a constant AsH$_3$ flow rate; this is shown in Fig. 3.5. It is clear from the figure that the steady-state optical anisotropy levels of the InAs surface under AsH$_3$ or during growth are stable and reproducible. The time constant associated with the switching off is very fast (<0.2 s), and then the signal stabilizes by approximately 2×10^{-3} with respect to the level of the growing surface. In contrast, when the TMI is switched on, the RDS signal exhibits a damped oscillation behavior before reaching its steady-state value.

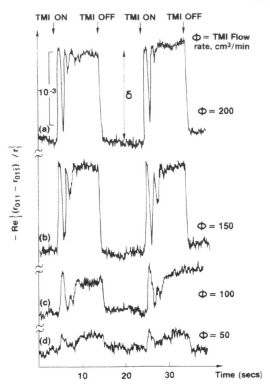

Fig. 3.5. RDS signals for InAs growth at 480 °C, where the TMI bubbler flow was switched on and off at the times indicated by the arrows. The different records were taken using the indicated TMI bubbler flows. The AsH_3 flow was maintained at $15\ cm^3 \cdot min^{-1}$ in each case (after Koch et al. [1990]).

It is interesting to note that the 1.5 s period of these signal oscillations is independent of growth rate, so that it does not correspond to the deposition of one monolayer; it is also independent of temperature. These observations suggest that the oscillations are due to perturbations in gas flows in the reactor, rather than a growth phenomenon. After intentionally changing gas flows, this was indeed found to be the case. The oscillations were observed whenever the gas flow in the organometallic line was increased, whether by switching on the TMI flow or simply by increasing the hydrogen flow while maintaining a constant TMI flow. Fig. 3.6 shows that oscillations are observed when the TMI flow is switched on, and also when an excess hydrogen flow in the organometallic line is switched on and off. This was found not to be due to the hydrogen itself, since there was no effect on the RDS signal when the extra hydrogen was sent through the line used for hydrides. There is a difference between the hydride line and the organometallic line; however, the hydride line is not regulated by a valve, whereas there is a needle valve between the organometallic line and the reactor tube, used to regulate the pressure in the organometallic line. Apparently this valve takes a few seconds to regulate a change in the pressure in the organometallic line; as the valve opens and closes during this transient period, more and less TMI is supplied to the reactor, even though the TMI flow set by the reactor control panel is kept fixed. Therefore the V/III ratio of the gas flows changes, and as described below, this gives rise to a measurable RDS signal change. Oscillations are therefore not seen when the TMI flow is stopped, consistent with this phenomenon. Razeghi and Acher [1989] showed that it was possible to cancel this effect by changing the flow rates of the various hydrogen flows used. These observations are important, because they show that RDS can be very useful in determining the effects of the MOCVD apparatus on growth, and therefore can aid in optimizing reactor design. In fact, RDS is the only method that can measure such reactor effects on the wafer surface. Particularly for the abrupt interfaces required in devices, especially quantum wells, this level of control during the first few seconds of growth is essential.

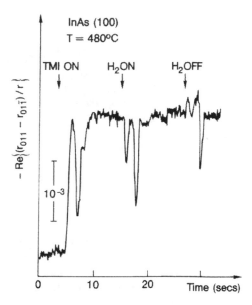

Fig. 3.6. RDS signal changes for InAs at 480°C upon introduction of TMI and H_2 into the reactor, at the times indicated by the arrows (after Koch et al. [1990]).

3.3.2. The III/V flux dependence of RDS signals

Fig. 3.5 shows that the steady-state optical anisotropy level of the InAs surface under AsH_3 is significantly different from that during growth. This difference, referred to as δ, was found to be very sensitive to the growth parameters, including the V/III ratio, growth rate and temperature. Fig. 3.5 shows a series of such measurements for a variety of different TMI flow rates. The value of δ clearly depends on the TMI flow rate.

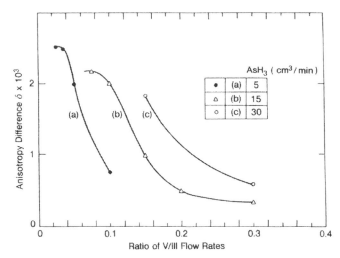

Fig. 3.7. A compilation of δ, the difference between the RDS signals for the growing and the static AsH₃-stabilized surface, as a function of V/III flow ratio. The different curves were measured using the indicated As H₃ flow rates (after Koch et al. [1990]).

Fig. 3.7 shows a compilation of δ values as a function of the V/III gas flow ratio; each curve shows the results for a given AsH_3 flow rate. It is clear that δ increases with decreasing V/III ratio until it reaches a saturation level. At very high V/III ratios, only small δ values can be measured. Other measurements show that δ changes in the same manner when the V/III ratio is changed by reducing the AsH_3 flow while keeping the TMI flow constant. Clearly this reflects the change in ratio of the element III and element V species on the surface when the V/III ratio of the gas flows is changed. Comparison of the different curves shows that δ depends not only on the V/III ratio, but also on the growth rate (for a fixed V/III ratio); it increases with TMI flow rate. It is interesting to note that, for a non-growing InAs surface at 480 °C under AsH_3, increasing the AsH_3 flow rate from 2 to 30 $cm^3 \cdot min^{-1}$ had no effect on the RDS signal. This indicates that the non-growing surface is AsH_3 stabilized.

3.3.3. The temperature dependence of RDS signals

The value of δ is also temperature dependent. Fig. 3.8 shows a compilation of δ as a function of temperature for several sets of gas flow rates. In each case δ is roughly constant at low temperatures (decreasing very slightly with decreasing temperature) and then drops to nearly zero by 520 °C. Comparison of the curves shows that the curves shift to higher temperatures for lower V/III flux ratios, and also for higher growth rate with the same AsH_3 flux. Also apparent from the figure is that the measured anisotropy difference in the lower temperature range is approximately 2.5×10^{-3}, regardless of the gas flow rates.

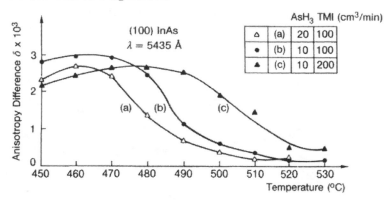

Fig. 3.8. The temperature dependence of δ, the difference between the RDS signals for the growing and the static AsH₃-stabilized surface. The different curves were measured using the indicated gas flow rates (after Koch et al. [1990]).

3.3.4. Discussion

These data show that the RDS signal is dependent on V/III ratio, TMI flow rate and temperature. The observed δ means that the growing surface has a different optical anisotropy than the AsH₃-stabilized, non-growing surface. These observations raise interesting questions about the nature of the growing InAs surface. It is not known whether the surface is terminated by As, hydrogen, organometallic groups or molecular species containing In and As. Unlike RHEED, which is sensitive only to the structure of the surface reciprocal lattice, RDS can also distinguish among various chemical structures. There are a number of different factors that influence the RDS signal, including surface roughness, surface reconstruction, surface stoichiometry and chemisorbed or physisorbed species on the surface.

"Islands" nucleating and coalescing for each monolayer formed have been used to explain RHEED oscillations during MBE [Harris et al. 1981] and metalorganic molecular-beam epitaxy (MOMBE) [Tsang et al. 1987] growth. However, they do not explain the RDS signals, as shown in Fig. 3.5 and Fig. 3.8. The surface may roughen in another manner during MOCVD growth. As Aspnes [1985] has shown, information about surface roughness can be obtained from the spectral dependence of the RDS signal. Fig. 3.9 shows the value of δ as a function of light energy, for InAs growth at 480 °C. There is a maximum in the range of 2.30–2.38 eV, and δ then falls to zero at approximately 3.26 eV. The contribution of surface roughness is expected to be larger with UV light than with infrared [Aspnes 1985]. Since this is not consistent with the results of Fig. 3.7, surface roughness as the dominant contribution to the RDS signal can be ruled out.

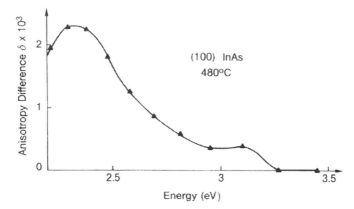

Fig. 3.9. The spectral dependence of the value of δ, the difference between the RDS signals for the growing and non-growing AsH₃-stabilized surface for InAs growth at 480 °C. The gas flow rates were 100 cm³·min⁻¹ for the TMI bubbler and 10 cm³·min⁻¹ for AsH₃ (after Koch et al. [1990]).

While surface roughness does not explain the measured δ, another possibility is that δ indicates a change in the surface reconstruction when the growth is stopped; since a change in surface reconstruction would change the surface symmetry, it could be detected by RDS. The optimal nucleation of InAs/GaAs by MBE occurs under In-rich conditions [Schaffer et al. 1983]. Although this cannot be verified in the case of MOCVD growth, the optimal growth conditions may also correspond to In-rich growth. In this case, δ would be due to a change from an In-stabilized to an As-stabilized surface. However, the In-stabilized (4 × 2) to As-stabilized (2 × 4) transition seen in MBE growth of InAs [Schaffer et al. 1983, Foxon and Joyce 1978, Sugiyama 1986] changes abruptly with temperature, all other factors being constant, whereas Fig. 3.7 shows a slower temperature dependence. Therefore surface reconstruction changes cannot be the only contribution to the RDS signal observed.

The results of Fig. 3.6, which indicate a decrease in δ with V/III ratio, suggest that In-related species or reaction products cause the change in optical anisotropy. The temperature dependence of δ, shown in Fig. 3.7, could be explained by the change in the V/III ratio of the species at the surface of the wafer due to temperature effects. Less efficient ASH₃ cracking in the lower-temperature range (450–470 °C) would lead to an increase in the density of In species on the surface and therefore a larger δ would be observed. The shift of the curves to higher temperature with lower V/III ratio is consistent with this explanation as well. For the temperature range studied excess evaporation of As species should not be a problem.

It is not evident, however, what chemical species are on the surface, and also what gives rise to the optical anisotropy. It is quite likely that organometallic groups are present on the surface when the film is growing [Stringfellow 1989] which can give rise to optical anisotropy. The decrease in δ at high temperature would then mean that the adsorbed species are no longer present on the surface, so that the surface optical anisotropy is the same during growth and without growth.

As Razeghi and Acher [1989] have reported, starting and stopping the flow of TMI changes the anisotropy of the surface under certain conditions. Starting and stopping the TMI and AsH_3 flows, individually or in combination, gives further insight into the processes occurring at the surface. Fig. 3.10 shows the result of an experiment of this type on an InAs surface at 480 °C with different types of gas flow. The measurement at time $t = 0$ is of the growing InAs surface with a TMI flow of 100 $cm^3 \cdot min^{-1}$ and an AsH_3 flow of 10 $cm^3 \cdot min^{-1}$.

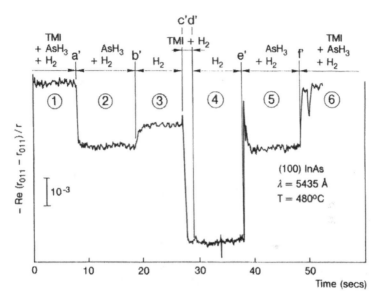

Fig. 3.10. r_a signal of InAs at 480 °C under a variety of different gas flows in the reactor, as indicated (after Koch et al. [1990]).

When the TMI flow is cut at time a', the signal changes abruptly by 2×10^{-3}. When the AsH_3 flow is stopped the signal again changes, this time toward the level of the growing InAs surface. Unlike the abrupt changes at the start and stop of growth, the time dependence of this change is more gradual, suggesting arsenic evaporation from the surface [Foxon and Joyce 1978]. As a result, the surface would be more In rich. When TMI alone is switched on for 2 s at time c' (corresponding to approximately two or three monolayers by extrapolation from the measured growth rate), the signal again changes abruptly, but surprisingly toward the direction of the AsH_3-stabilized surface, rather than the expected change toward the growing surface level. When the TMI is stopped and only hydrogen flows through the reactor, the signal level is unchanged, indicative of a passivated surface; when AsH_3 alone is switched on (time e'), there is an initial transient and then the signal stabilizes at the same level as it was in region 2 of the figure, after the TMI flow had been stopped for the first time. This shows that the surface had not undergone an irreversible change when the AsH_3 flow was cut, at least as far as can be determined by RDS. Finally, when the growth is resumed, the signal returns to its original level.

The RDS signal changes in Fig. 3.10 can be explained as follows. At time b', arsenic presumably desorbs from the surface. The resulting change in optical anisotropy may be due to a surface reconstruction change or simply a change in the surface stoichiometry without further structural changes. When the surface under hydrogen is exposed to TMI without AsH_3 at time c', the signal changes in the direction toward that of the AsH_3-stabilized surface rather than that of the surface with both AsH_3 and TMI. This means that the signal is due to other factors than simply the relative quantities of indium and arsenic on the surface; the most likely contributions are chemisorbed species and/or surface reconstruction changes. By using atomic layer epitaxy (ALE) for the MOMBE growth of GaAs, Chiu [1989] found that after the deposition of a Ga monolayer, the surface reconstructs in a manner different from the 4×2 Ga-stabilized surface of GaAs: 4×6 and 4×8 reconstructions can appear. Analogous changes may account for this observation. The signal "spikes" at time e' as shown in Fig. 3.10, when AsH_3 is again sent to the reactor. This is suggestive of AsH_3 reacting with the In species on the surface. The signal then stabilizes when the growth is complete.

3.4. RDS investigation of III–V lattice-matched heterojunctions

3.4.1. RDS observations of heterojunction growth

When heterojunctions are grown by MOCVD, the RDS signal changes abruptly at the onset of the growth of the new material as shown in Fig. 3.11 for the growth of two different $Ga_xIn_{1-x}As_yP_{1-y}$ materials on InP using light of wavelength 5435 Å. The signal recorded during the beginning of the measurement is that of the growing InP surface: at $t = 2.8$ s the InP growth was stopped and the quaternary growth was simultaneously started. The subsequent signal is that of the growing quaternary relative to the initial InP signal level. Each heterojunction is lattice matched, but the compositions and bandgaps are different. For $Ga_{0.25}In_{0.75}As_{0.5}P_{0.5}$, the bandgap is $\lambda = 1.3$ μm whereas for $Ga_{0.5}In_{0.5}As_{0.9}P_{0.1}$, $\lambda = 1.5$ μm. In each case, there is an abrupt signal change upon growth of the quaternary followed by a rise to a maximum. The signal then undergoes various changes before reaching a final level after about 10 s. Note that the signal changes do not correspond to the time to grow a monolayer, as in the case of RHEED oscillations.

There are a number of differences between the RDS signals of the two samples. The slope of the initial transient is the same in the two cases, but the time to reach the first maximum is 0.6 s for the 1.5 μm alloy and 1.3 s for the 1.3 μm alloy. The subsequent signal changes occur somewhat more rapidly for the 1.5 μm quaternary which also has a higher growth rate than that of the 1.3 μm quaternary. The signal for the 1.3 μm alloy oscillates around the value of approximately -6×10^{-4} with respect to the InP signal level. In contrast, the signal for the 1.5 μm alloy changes sign from negative to positive and then approaches a level near that of InP.

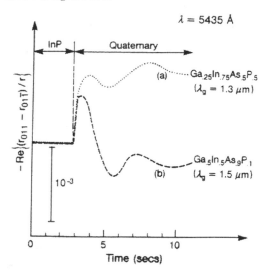

Fig. 3.11. RDS records of two different lattice-matched $Ga_xIn_{1-x}As_yP_{1-y}/InP$ heterojunctions using light of wavelength 5435 Å (after Koch et al. [1991]).

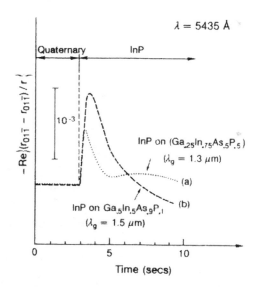

Fig. 3.12. RDS records of two different lattice-matched $InP/Ga_xIn_{1-x}As_yP_{1-y}$ heterojunctions using light of wavelength 5435 Å (after Koch et al. [1991]).

For the reverse sequence, the growth of InP on each of the two quaternaries, the signal evolution is quite different. This is shown in Fig. 3.12, again using 5435 Å light. In this case, each signal changes abruptly by a value of the order of -5×10^{-4} and then returns toward the $Ga_xIn_{1-x}As_yP_{1-y}$ level. Only one large peak is observed in each case. The peak height for the 1.5 μm alloy is consistently larger than that of the 1.3 μm alloy even though the same material, InP, is growing during the time measurement. This is due to the fact that the RDS signal is measuring the optical anisotropy of both the surface and hetero-interface, the latter being different in the two cases.

The form of each RDS record shown in Fig. 3.11 and Fig. 3.12 is reproducible: the time of the initial signal transient, the changes in the following few seconds and the final level reached after these changes were found to be characteristic of the wide variety of lattice-matched heterojunctions studied by Koch et al. [1991]. One can therefore refer to these characteristic signal changes as the RDS "signature" of each heterojunction.

The RDS signature of heterojunctions depends on the light wavelength used for the measurement. Fig. 3.13 shows a comparison between the RDS signatures taken with a green (5435 Å) and a red (6328 Å) laser for the $InP \rightarrow Ga_{0.25}In_{0.75}As_{0.5}P_{0.5}$ transition. While in both cases an abrupt change in the signal level is observed at the onset of the quaternary growth there are a number of differences between the two signatures. The most important one is that the signal changes resembling oscillations, observed using green light, are not seen with red light. In addition, the difference between the signal levels of InP and $Ga_{0.25}In_{0.75}As_{0.5}P_{0.5}$ after 10 s is roughly twice as large with the red laser as with the green laser.

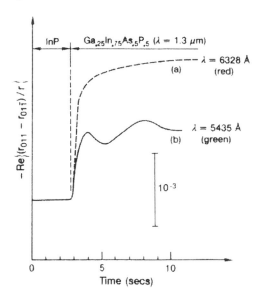

Fig. 3.13. A comparison of the RDS signatures for $Ga_{0.25}In_{0.75}As_{0.5}P_{0.5}/InP$ growth using light of wavelength (a) $\lambda = 6328$ Å and (b) $\lambda = 5435$ Å (after Koch et al. [1991]).

Although the RDS signals appear to stabilize after several seconds, in fact they slowly evolve over the course of several minutes. This is shown in Fig. 3.14 for GaInP growth on GaAs. The signal exhibits a damped oscillation behavior, with amplitude changes of the order of 10^{-3}. This is in contrast to lattice-mismatched growth [Acher et al. 1990b] where signal changes of the order of several tenths are observed during the first few minutes of growth. As will be discussed, the damped oscillation behavior seen in Fig. 3.14 is most likely due to interference effects; this continues until the epilayer is sufficiently thick that light reaching the interface is absorbed in the film. Therefore, the signal level reached after the buried interface oscillations have been damped out is the surface anisotropy of the second layer relative to that of the first layer; this level is different from the apparent steady-state level reached after the first few seconds of growth.

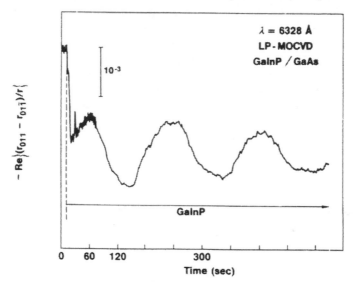

Fig. 3.14. The RDS record for GaInP/GaAs epitaxy during several minutes of growth using light of wavelength 6328 Å (after Koch et al. [1991]).

3.4.2. Interpretations of the experimental observations

There may be a variety of structural and chemical contributions to the RDS signals during heterojunction growth. The origins of optical anisotropy of materials have been investigated by Aspnes [Aspnes 1985, Aspnes and Studna 1985, Aspnes et al. 1988a, 1989] and others [Acosta-Ortiz and Lastras-Martinez 1987, Manghi et al. 1989, Berkovits et al. 1987]. Contributions to surface and interface anisotropy include chemical bonding at the surface and interface, anisotropic roughness and the presence of chemi- and physisorbed species on the surface.

The optical anisotropy of interfaces may depend on the bonding sequence that occurs. In the case of a perfectly abrupt InAlAs/InP interface, for example, the atomic sequence in the growth direction may be either ... P–In–P–/–(In, Al)–As–... or ...–P–In–A–As–(Al, In)–As; the associated RDS contribution is expected to be different for the two cases. In addition, the electric field at the interface, screening effects or interfacial roughness can contribute to the signal. Thus, RDS is expected to be sensitive to interface quality as will be discussed subsequently.

At the beginning of growth, both the surface and interfacial anisotropy are measured, and gradually the interfacial contribution becomes negligible for thick films. Surface processes at the very beginning of growth can be different from those after several seconds. Thus, the contributions to the RDS signal vary, depending on the growth time.

(a) Short-time-scale changes (less than two seconds)

The nature of the RDS signature can be analyzed by comparing the normal growth sequence with the effect of the group V gas change alone. This is shown in Fig. 3.15. Each part of the figure shows the standard RDS signature of a heterojunction (stopping the first-layer growth while simultaneously beginning the second-layer growth) compared with the signal change when the first-layer growth is stopped and only the group V gas of the second material is sent to the reactor. In each case the signal change is in the same sense (negative) for both the growth and for the group V change. The comparison between GaInAs grown on InP and AsH_3 alone over InP is shown in Fig. 3.15(a). The two signatures are clearly different, because there are signal changes resembling oscillations for the growing material, but not for the case of AsH_3 alone over InP. However, the initial transient time to reach the first feature in the GaInAs/InP signature, 0.1 s, is identical to the time to reach the steady-state value in the AsH_3-Only signature.

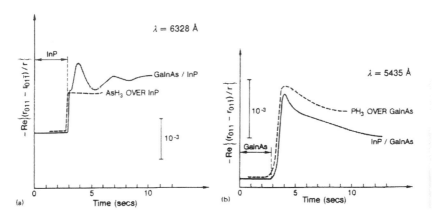

Fig. 3.15. Comparisons of the RDS signals between the standard heterojunction growth and the change in the group V species only, (a) GaInAs/InP as a function of AsH_3 over InP and (b) InP/GaInAs as a function of PH_3 over GaInAs (after Koch et al. [1991]).

In Fig. 3.15(*b*) for InP/GaInAs and PH_3/GaInAs, the time to reach the maximum is nearly the same in the two cases and the signal maxima roughly occur at the same amplitude. RDS signatures at $\lambda = 5435$ and 6328 Å for GaInP/GaAs as a function of PH_3/GaAs also show similar behavior: comparable initial transient times, but differences after this transient. These results suggest that during heterojunction growth the change in the group V species on the surface causes the initial RDS transient, and the subsequent signal oscillations are due to processes occurring during the growth.

Another rapid RDS signal change can occur when the group V gas flow is stopped over a non-growing surface. Fig. 3.16 shows the RDS signal change occurring when the PH_3 flow (500 cm^3·min^{-1}) over a non-growing InP surface is stopped; the substrate temperature is maintained at 540 °C. In 1.5 s, the signal amplitude changes by 10^{-3}, followed by a slower change and then the signal remains fairly constant. Upon restarting the PH_3 flow after 1 min there is an abrupt (0.3 s) return to the former signal level where the signal remains. The initial signal change is most likely phosphorus loss from the surface. The fact that the signal returns to its former level once the PH_3 flow is restarted indicates that the surface has not been damaged by the lack of PH_3, at least according to the detection sensibility of RDS. The origin of these signal changes can be both the stoichiometry change and the reconstruction change that is likely to occur, although this cannot be established with the present understanding of RDS without RHEED.

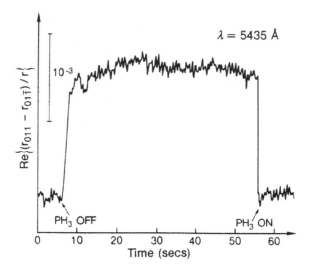

Fig. 3.16. RDS signal change occurring when the PH_3 flow over a non-growing InP surface is stopped and restarted. The substrate temperature is 540 °C (after Koch et al. [1991]).

The last measurement suggests that an important application of RDS can be determining whether the surface is stabilized by the element III or V species. This is done with RHEED for MBE and MOMBE, where the majority of III–V compounds are grown under group V–rich conditions. RHEED is used to determine what temperature and group V flux are needed to maintain a V-rich surface; this is judged by the reconstruction change from 4×2 (Ga rich) to 2×4 (As rich) in GaAs, for example. By repeating the process shown in Fig. 3.16 for a variety of different growth conditions, RDS may also be used in this manner.

Another factor which affects the first two seconds of a heterojunction signature is the manner in which the interface growth is done. Acher et al. [1991] found that slight delays in gas arrival to the substrate, for example, can affect the signature; this was done intentionally in order to study the different effects of the group III and group V species on the initial RDS signal change. This is shown in Fig. 3.17 for the InP→GaInAs transition. Fig. 3.17(a) shows the standard RDS signature for the normal growth procedure, whereas Fig. 3.17(b) shows the effect of stopping the PH_3

gas flow for 1 s before starting the GaInAs growth while maintaining a TMI flow. An extra peak appears in the RDS signal. This signal change is positive, in contrast to the negative signal change seen for the standard InP/GaInAs transition. When the PH_3 gas flow is stopped while introducing AsH_3 and TMI into the reactor for 1 s, Fig. 3.17(c), the extra dip in the signal is in the negative direction. Other similar experiments, such as briefly switching on the group III gases of the second layer before the group V gas, show the same effect: changing the group V species causes a signal change in the same sense as that observed for the normal growth procedure, while inducing a group III–rich surface by changing the group III gas causes the opposite sign change.

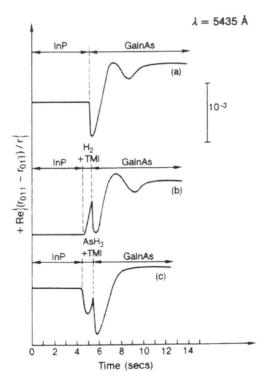

Fig. 3.17. Effects of slight delays in switching the gas during GaInAs/InP heterojunction growth, (a) Standard RDS signature, (b) The effect of stopping the PH_3 gas flow for 1 s before starting the GaInAs growth while maintaining a TMI flow, (c) The effect of stopping the PH_3 gas flow while introducing AsH_3 and TMI into the reactor for 1 s (after Koch et al. [1991]).

These observations again support the idea that the initial rapid signal change is due to changes associated with the group V species on the surface. The transient time is typically less than 1 s and shorter than the time for the growth of one monolayer. Therefore, the signal change is not simply due to the change in optical anisotropy when going from one group V–terminated material to another. Instead, a rapid process is occurring, such as the replacement of P atoms on the surface by As atoms for the case of AsH$_3$ flow over InP, as shown in Fig. 3.15(a). This possibility is supported by the results of an experiment similar to that shown in Fig. 3.16. In that case, Acher et al. [Acher 1990, Acher et al. 1990b] introduced an AsH$_3$ flow instead of only H$_2$ over the non-growing InP surface for several seconds. They observed RDS signal changes that were much larger than those in Fig. 3.16; in addition, the reflectivity changed significantly.

(b) RDS changes occurring from two to 10 seconds of growth

An interesting feature of RDS signatures are the signal changes that are typically observed for about 10 s before the signal reaches an apparently stabilized level. Acher et al. [1991] found that although these signal changes do not correspond to monolayer growth, the characteristic times of the signal changes depend on the growth rate. Fig. 3.18 shows a series of RDS signatures for Ga$_{0.47}$In$_{0.53}$As/InP lattice-matched growth for different growth rates and ASH$_3$ flow rates. The signatures obtained with normal gas flow rates, the growth rate divided by two and the growth rate divided by three are shown in Fig. 3.18(a), (b) and (c) respectively. In Fig. 3.18(c), both the growth rate and the ASH$_3$ flow rate are divided by two from the standard conditions. In all cases, the initial signal transient time is 0.4 s. The characteristic times for the subsequent signal changes, however, are longer when using lower gas flow rates. The time between the first minimum and the first maximum compared to that of the standard (curve (a)) is inversely proportional to the group III gas flow. The times between the other "oscillations" also increase with decreasing flow rate although not by exact factors of two or three. In contrast, there is only a slight difference between the RDS signals where group III as well as group V gas flows were reduced by half (Fig. 3.18(c)) compared to when only the group III gas flows were reduced by half (Fig. 3.18(b)). The growth of GaInP/GaAs was also investigated in the same manner, and again signal changes were found to be inversely proportional to the group III gas flow rate. These results, as well as those of Fig. 3.15, suggest that the signal changes are related to the growth process. However, since the RDS signal depends on the thickness of the material, the results in Fig. 3.18 may indicate that the optical anisotropy is due to volume effects, such as the electric field in the bulk originating from the interface.

The signal changes occurring during the first few seconds are not related to oscillations previously observed [Koch et al. 1990] when beginning the growth of a film on a non-growing substrate. As shown for the case of InAs growth [Koch et al. 1990], the valve regulating the pressure in the organometallic line fluctuates at the beginning of the growth, and this causes a fluctuation in the quantity of organometallic species reaching the substrate surface; in this case, damped sinusoidal signal changes are seen, with a period that is independent of the gas flow rates.

For the Ga$_{0.47}$In$_{0.53}$As→InP transition, Fig. 3.19, signal oscillations are not observed after the initial signal change and the signature shows essentially no effect arising from the change in gas flows. As for the InP→Ga$_{0.47}$In$_{0.53}$As transition, the initial transient time is the same regardless of growth rate.

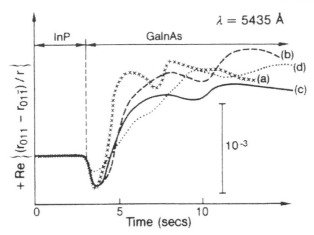

Fig. 3.18. A series of RDS signatures for $Ga_{0.47}In_{0.51}As/InP$ lattice-matched growth using different growth rates and group V flow rates. The growth runs were performed with (a) the standard flow rates, (b) group III flow rates divided by two, (c) both group III and V flow rates divided by two and (d) group III flow rates divided by three (after Koch et al. [1991]).

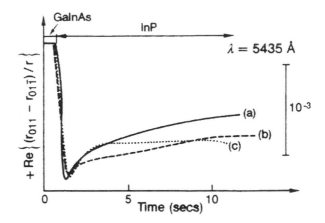

Fig. 3.19. A series of RDS signatures for $InP/Ga_{0.47}In_{0.53}As$ lattice-matched growth using different growth rates and group V flow rates. The growths were performed with (a) the standard flow rates, (b) the group III flow rates divided by two and (c) the group III flow rates divided by three (after Koch et al. [1991]).

Koch et al. [1991] also compared Al- and Ga-containing materials which have nearly identical lattice constants. Fig. 3.20 shows the signatures for GaInAs/InP and AlInAs/InP growth. As expected, the initial signal transient times are nearly identical for the two types of material both for red (Fig. 3.20(a)) and green (Fig. 3.20(b)) light as expected since the group V species used are the same in both cases. In addition, although GaInAs and AlInAs have quite different optical and electrical properties, the RDS signatures for the two are fairly similar, especially when using green light. This again suggests that the optical anisotropy is due to surface effects, such as reconstruction and chemisorbed species, rather than volume.

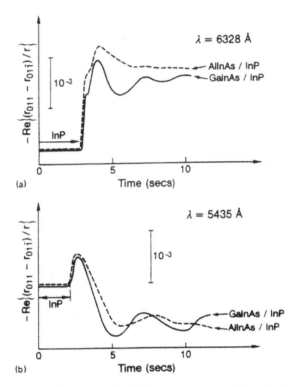

Fig. 3.20. A comparison between the RDS signatures of AlInAs/InP and GaInAs/InP using (a) $\lambda = 6328$ Å and (b) $\lambda = 5435$ Å (after Koch et al. [1991]).

The origins of these signal changes are not clear. The signal level reached after about 10 s is a measure of the surface anisotropy of the GaInP, combined with the effects of the GaInP/GaAs interface anisotropy relative to the value of the GaAs signal level at time $t = 0$. The fact that there is no observation of a change from the GaAs level to the GaInP level, but rather signal oscillations, suggests that some sort of transient processes occur at the beginning of the heterojunction growth. One possibility is that the sticking coefficients of the different group III species in a ternary or quaternary may be significantly different on the surface of the "substrate" than during the steady-state growth of the layer. For example, for GaInAs/InP growth, the flow rates of TEGa and TMI are adapted to give a lattice-matched GaInAs layer. However, the relative sticking coefficients of the two species may be different on a P-terminated surface compared to an As-terminated surface, so that the material grown for the first few monolayers may be rich in one of the group III species. Such a situation may be possible, because the use of organometallic sources leads to complicated reactions in the gas phase and at the surface: there may be a number of intermediate reaction species on the surface during the growth and desorption of group III species at each step of the reaction. RHEED oscillation and modulated-beam mass spectrometry measurements of MOMBE growth, for example, show that gallium incorporation in GaAs [Martines and Whitehouse 1989] is quite different from that in GaP [Maurel et al. 1990]. Clearly, further studies of surface kinetics and the interfacial structure are necessary.

Slight differences in the arrival rates between the two gases due to reactor geometry, for instance, may also affect the first few monolayers but not the thick-layer growth. Such explanations are consistent with the fact that Acher et al. [1991] did not observe signal oscillations during the growth of binary material on an alloy as shown in Fig. 3.12. However, it is possible that oscillations may be observed for binaries when using other light wavelengths, based on the data in Fig. 3.13 which show that the heterojunction signatures depend on the light wavelength. Another possibility, arising from photoluminescence data [Razeghi et al. 1990], is that the growth rate changes during the first few seconds of growth. Any of these explanations would mean that the signal oscillations are due to non-optimal growth. This would not be the case if the signal changes were due to volume effects but such contributions as will be shown later are negligible. Clearly, further understanding of the RDS signatures can yield important information about heterojunction optimization.

(c) Long-time-scale changes

As was shown in Fig. 3.14 the RDS signal shows a damped oscillation behavior over the course of several minutes. A model for this behavior has been developed ascribing the damped oscillations to light interference upon the growing layer, as well as the anisotropy of the buried interface. The model is based on the following facts. The growing layer acts as a Fabry–Pérot cavity. The reflection and transmission coefficients of the surface and interface are slightly anisotropic, and so the conditions of constructive and destructive interference depend slightly on the polarization. For Fabry–Pérot effects, the pseudoperiod τ can be related to the growth rate v by writing that the optical path length increases by λ during a pseudoperiod:

Eq. (3.10) $$\frac{v\tau}{2n} = \lambda$$

The material is modeled as a thin epilayer on a substrate as shown in Fig. 3.21, and 2×2 matrices are used to consider the reflection and transmission of the polarized light at the surface and interface using the method of Abeles [Azzam and Bashara 1977]. The surface and interface are treated as films with a very small thickness e compared to the light wavelength λ, and their optical properties are calculated to first-order development in e/λ as in McIntyre and Aspnes [1971]. The epilayer and substrate "bulk" materials are assumed to be isotropic. The surface and interface are assumed to have the same optical eigenaxes $[0\ 1\ 1]$ and $[0\ 1\ \bar{1}]$. The calculation leads to

Eq. (3.11)

$$\frac{r_{01\bar{1}} - r_{011}}{r} = \frac{4j\pi}{\lambda}\frac{N_0}{\varepsilon_1 - \varepsilon_0}e_s\delta\varepsilon_s + \frac{16j\pi}{\lambda}\frac{N_0\varepsilon_1}{(\varepsilon_1 - \varepsilon_0)(N_1 + N_2)^2}$$

$$\times\frac{X}{[1 + (r_{12}/r_{01})X](1 + r_{12}r_{01}X)}\left(e_i\delta\varepsilon_i\frac{\varepsilon_1 - \varepsilon_2}{\varepsilon_1 - \varepsilon_0}e_c\delta\varepsilon_c\right)$$

with

Eq. (3.12) $$X = \exp\left(-\frac{4j\pi N_1 d}{\lambda}\right)$$

where X is a damped oscillating function, $\delta\varepsilon_s$ is the difference between the surface dielectric constant associated with the direction of the two eigenaxes, the product $e_s\delta\varepsilon_s$ characterizes the surface anisotropy and $e_i\delta\varepsilon_i$ characterizes the interface anisotropy. The effective media model used for the interface and surface is somewhat simplistic since it neglects inhomogeneities. However, more detailed calculations on the optical properties of surfaces [Del Sole 1981] yield essentially the same results. In the present approach, N_0, N_1 and N_2 are the indices of refraction for the ambient, epilayer and substrate, respectively, ε_0, ε_1 and ε_2 are the dielectric constants of the ambient, epilayer and substrate, respectively, d is the thickness of the epilayer. r_{01} and r_{12} are the reflection coefficients of the surface and interface, respectively. t_{01} and t_{12} are the transmission coefficients of the surface and interface respectively. They can be expressed as a function of the index using the Fresnel relations [Azzam and Bashara 1977].

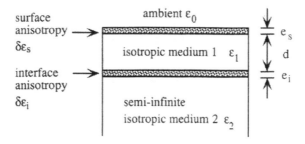

Fig. 3.21. Schematic diagram of the heterojunction model used in the calculations of the interference effects (after Koch et al. [1991]).

The real part of $\delta r/r$ gives the evolution of the RDS signal. The signal is the sum of three terms as shown in

Eq. (3.11). The first term is due to surface anisotropy without buried interface effects. The second and third terms indicate that the oscillations are due to the anisotropy of both the interface and surface with oscillations disappearing when layer thicknesses are large compared to the light penetration depth.

Experimentally, Acher et al. [1991] measured only variations and the real part of the anisotropy. Therefore one cannot determine $e_s \delta_s$ and $e_i \delta_i$ separately using

Eq. (3.11). However, for InP and AlInAs at the wavelength now used, $\varepsilon_1 \approx \varepsilon_2$, and so the prefactor $(\varepsilon_1 - \varepsilon_2)/ (\varepsilon_1 - \varepsilon_0)$ of

Eq. (3.11) is small. Therefore the oscillatory behavior is dominated by the interface anisotropy. This also means that $N_1 = N_2$ and $r_{12} \approx 0$.

Eq. (3.11) then simplifies to

Eq. (3.13)
$$\frac{r_{01\bar{1}} - r_{011}}{r} = \frac{4j\pi}{\lambda} \frac{N_0}{\varepsilon_1 - \varepsilon_0} \left(e_s \delta \varepsilon_s + X e_i \delta \varepsilon_i \right)$$

$e_i \delta \varepsilon_i$ therefore can be determined from the amplitude and phase of the observed oscillations.

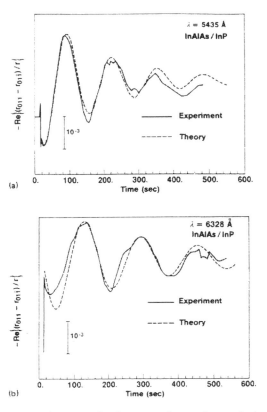

Fig. 3.22. A comparison between the theory and experimental observations for RDS records of InAlAs/InP growth over several minutes, (a) $\lambda = 5435$ Å and (b) $\lambda = 6328$ Å (after Koch et al. [1991]).

Using these results, the different features of the RDS signal of heterojunctions can be explained in the following manner. During the first few seconds of heterojunction growth ($X \approx 1$), the RDS signal is governed by changes in surface chemistry and hence by changes of the properties of the surface and interface. Then, the surface is expected to reach its steady-state level of anisotropy and the interface properties should stay the same. The RDS signal is then governed by the damped oscillation behavior described by Eq. (3.13). When the epilayer is sufficiently thick the light does not penetrate enough to be sensitive to interface effects, and the RDS signal is related simply to the surface anisotropy.

Calculations of this effect are in good agreement with experimental measurements using both green and red light as shown in Fig. 3.22(*a*) and (*b*) for the growth of InAlAs on InP. Using green light (Fig. 3.22(a)), the pseudoperiod of the damped oscillations is 131 s (± 6 s) [Acher et al. 1991]. Assuming the index of refraction of InAlAs is equal to that of InP, using Eq. (3.2) one finds that $v = 340$ Å·min^{-1} (± 16 Å·min^{-1}), which is in excellent agreement with *ex situ* thickness measurements.

For $\lambda = 5435$ Å, Fig. 3.22(*a*), the theoretical curve was obtained using an index of refraction of $N_1 = N_2 = 3.66 - 0.4$j [Palik 1985], and one finds that the value of the buried interface anisotropy is

Eq. (3.14) $e_i \delta_i = (-3 + 11j)$ Å

The difference between the surface anisotropy of InAlAs and InP is simply the real part of the difference between Eq. (3.13) for the two materials:

Eq. (3.15)   Re $\{\exp (1.8j)[(e_s \delta \varepsilon_s)_{\text{InAlAs}} - (e_s \delta \varepsilon_s)_{\text{InP}}]\} = 7$ Å

For $\lambda = 6328$ Å (Fig. 3.22(*b*)), the theoretical curve was obtained with an index of refraction of $N_1 = N_2 = 3.53 - 0.3$j, and one finds that the value of the buried interface anisotropy is

Eq. (3.16) $e_i \delta_i = (-9 + 4j)$ Å

The difference between the surface anisotropy of InAlAs and InP is

Eq. (3.17)   Re $\{\exp (1.75j)[(e_s \delta \varepsilon_s)_{\text{InAlAs}} - (e_s \delta \varepsilon_s)_{\text{InP}}]\} = 16$ Å

The results show that the theoretical calculation agrees well with the experimental measurement. However, the presence of bulk anisotropy will lead to similar signal oscillations. For the semiconductor (1 0 0) surface, the bulk is optically isotropic due to symmetry considerations. However, the electric field originating at the surface or interface may extend over quite long distances in the material relative to the light wavelength. This will give rise to bulk anisotropy [Acosta-Ortiz and Lastras-Martinez 1987]. If such an effect exists, the oscillation amplitude would first increase as the epilayer thickness increases (because there is more and more anisotropic material), and then decrease as less light penetrates to the interface. The bulk anisotropic effects will become more obvious when the difference between the optical indices of the epilayer and substrate is relatively large and when the absorption of the epilayer is not too strong. Calculations have shown that bulk effects are negligible.

3.4.3. Applications of RDS for growth monitoring

(a) Optimizing the MOCVD reactor geometry

One important goal of RDS is the assessment of interface quality during MOCVD growth. Fig. 3.23 shows the RDS signature of two GaAs–GaInP heterojunctions which differ significantly. The electron Hall mobilities at 77 K (30000 and 90,000 $cm^2 \cdot V^{-1} \cdot s^{-1}$) show that the two interfaces are quite different in quality [Razeghi et al. 1988b]. The growth conditions for these two GaInP/GaAs samples were the same except for the fact that the samples were in different reactor chambers. The higher-quality film was grown after modifying the reactor geometry. A major difference in their RDS signatures is the time constant of the initial variation, 2 s for sample (a) and 0.2 s for sample (b). According to the previous study of different heterojunctions, the initial variation of the RDS signal is mainly due to the change in element V source on the surface. Here, the relatively long time constant observed in Fig. 3.23(a) suggests that the switch from arsine to phosphine was not abrupt enough. These results therefore led to a modification of the geometry of the group V section in the gas panel to obtain fast initial transients (Fig. 3.23(b)). Conversely, the quality of the GaInP/GaAs films improved significantly. The evolution of the RDS signal during the period following the initial transient is not yet understood. However, it was found to be very sensitive to film composition and growth rate. Any delay in the switching of the different gases will change the behavior.

(b) Monitoring of superlattice growth

The sensitivity of RDS to small deviations from optimal conditions is extremely useful for monitoring complex structures. In the case of a quantum well or a superlattice, it is not possible to assess individually the quality of each interface using mobility measurements. However, observation of the RDS signature can be very useful. The RDS signature of the GaInP/GaAs interface was observed during the growth of several GaInP/GaAs/GaInP quantum wells (Fig. 3.24). For small GaAs thickness (10 s or less GaAs) the signature clearly departs from that of the high-quality interface of Fig. 3.23(b). This difference was greatly reduced by improving the switching sequence.

In Fig. 3.25, the RDS record of a GaInP–GaAs superlattice exhibits RDS features quite similar to Fig. 3.23(b). It also shows that the interfaces are reproducible. It is possible to compare the signatures of heterojunctions independently of the underlying structure only because RDS is not sensitive to bulk effects. This is an extremely useful feature of RDS when compared to other techniques like ellipsometry. Ellipsometric records of superlattices do not allow direct comparison of the different periods [Hottier et al. 1980].

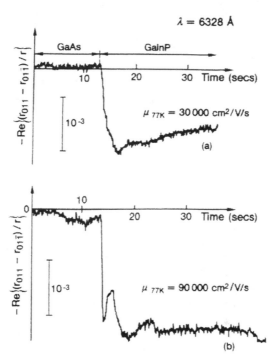

Fig. 3.23. RDS records during the MOCVD growth of GaInP/GaAs using two different reactor geometries. The Hall electron mobilities measured at 77 K are (a) 30,000 cm$^2 \cdot V^{-1} \cdot s^{-1}$, (b) 90,000 cm$^2 \cdot V^{-1} \cdot s^{-1}$ (after Koch et al. [1991]).

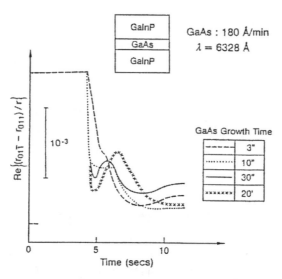

Fig. 3.24. RDS records of the GaInP/GaAs interface of quantum wells with different well thickness (after Acher et al. [1990a]).

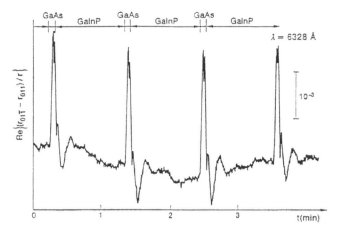

Fig. 3.25. RDS record of the growth of a GaInP–GaAs superlattice (after Acher et al. [1990a]).

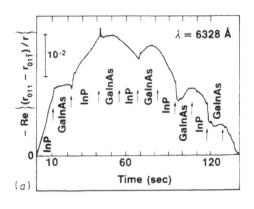

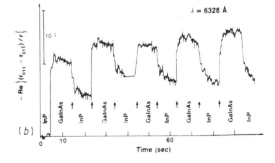

Fig. 3.26. RDS records of InP–GaInAs superlattices grown on InP/GaAs/Si. The superlattices were grown (a) immediately on GaAs/Si substrate, and (b) after growing 3600 Å of InP on GaAs/Si substrate (after Koch et al. [1991]).

Fig. 3.26(*a*) and (*b*), respectively, show the RDS record for InP–GaInAs superlattices grown on an InP/GaAs/Si structure at different stages of the InP growth. The superlattice shown in Fig. 3.26(*a*) was grown directly on the GaAs/Si substrate. The RDS signal varies irregularly with an amplitude as high as 10^{-2} indicating that the quality of this superlattice is quite poor due to the rough surface of lattice-mismatched growth. In contrast, the InP–GaInAs superlattice grown on InP/GaAs/Si substrate, with 3600 Å InP pregrown first, shows a good periodic structure with amplitude as low as 10^{-3}. As indicated in Fig. 3.26(*b*), the surface of InP has been smoothed. These figures show that, in contrast to most *ex situ* measurements, RDS is sensitive to each period in the superlattice. RDS can also be used to detect changes in superlattice growth, such as degradation over a period of time.

3.5. RDS investigation of III–V lattice-mismatched structures

GaAs on silicon and InP on GaAs on silicon structures are very attractive for device applications. A number of devices using this material system have been reported such as a GaInAsP–InP on silicon substrate laser emitting at 1.3 μm [Razeghi et al. 1988a] and a GaInAs–InP on GaAs/silicon photodiode [Razeghi et al. 1989]. The growth of lattice-mismatched III–V layers on other III–V substrates also has important applications. For example, since semi-insulating InAs substrates are not commercially available, GaAs and InP semi-insulating substrates can be used for InAs growth.

The use of RDS for *in situ* monitoring of the growth of InAs/InP, InP/GaAs/Si, GaInP/GaAs and GaInP/InP will be illustrated. The results will show that the anisotropic three-dimensional growth is responsible for the RDS signal changes. The detailed growth procedure of InAs/InP and InP/GaAs/Si will be discussed below. A model will be proposed and will be shown to agree well with experimental observations of the first stages of growth. As the roughness dimensions increase, precise modeling is not possible, but the broad features of RDS behavior are well understood.

3.5.1. Growth procedure of InAs/InP and InP/GaAs/Si

The lattice mismatch is 3.2% in the case of the InAs–InP system, and 3.8% for InP–GaAs. The growth of InAs was done directly on InP substrates. The substrate orientation was (1 0 0) with 2° off [0 1 1] unless otherwise stated. After introduction into the reactor, the substrates were heated under AsH_3 until they reached the growth temperature of 480 °C. The H_2 flow through the TMI bubbler was typically 100 $cm^3 \cdot min^{-1}$ and the AsH_3 flow was between 5 and 20 $cm^3 \cdot min^{-1}$. Under these conditions, the growth rate of InAs, deduced from thickness measurements performed on thick layers, was about 270 $Å \cdot min^{-1}$. Layer quality was assessed by x-ray diffraction. The x-ray diffraction pattern has well separated $K\alpha_1$ and $K\alpha_2$ diffraction peaks (Fig. 3.27) indicating a good structural quality. The typical Hall electron mobility at room temperature was about 11,500 $cm^2 \cdot V^{-1} \cdot s^{-1}$.

The MOCVD Challenge

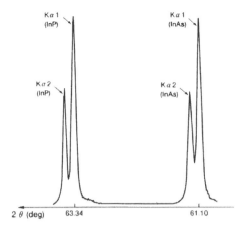

Fig. 3.27. *X-ray diffraction pattern of a 1 μm thick InAs layer grown on an InP
substrate (after Acher et al. [1990b]).*

The growth of InP/GaAs/Si structures has been described in detail elsewhere
[Razeghi et al. 1988c]. Some InP–GaInAs superlattices can be included in the InP to
reduce threading dislocation propagation. Lasers emitting at 1.3 μm in CW mode
were fabricated using these structures, indicative of excellent material quality
[Razeghi et al. 1988a].

RDS records of InP–GaInAs superlattices grown on GaAs/Si are shown in
Fig. 3.26. For both InAs/InP and InP/GaAs/Si, the sample becomes hazy during the
first minutes of growth, which indicates a three-dimensional (3D) growth. If the
growth is optimized, the haze disappears after 10 to 20 min. After 1 h of growth the
sample surface is mirror-like.

In contrast with the cases of lattice-matched structure, for lattice-mismatched
growth, the RDS signals are much larger. It is possible to record both the real and
imaginary part of the reflectance anisotropy with a small level of noise. Because the
perturbation due to the reactor tube adds an offset to the RDS signals, only the
variations of optical anisotropy are measured. The offset is of the order of 4×10^{-3}
which is large compared to surface chemistry changes, but negligible compared to
typical variations during the growth of lattice-mismatched materials. There is a
simple method to determine whether the measured RDS signals arise from parasitic
effects. Depending on the substrate orientation, the measured signal is either
$(r_{011} - r_{01\bar{1}})/r$ or its opposite $-(r_{011} - r_{01\bar{1}})/r$. Two identical experiments with the
substrate orientation differing by 90° should change the sign of the signal.
Conversely, when a material system is well known, the measured sign of the RDS
signal indicates the orientation [Acher et al. 1990c]. Fig. 3.28 illustrates the effects
of substrate orientation on RDS signal.

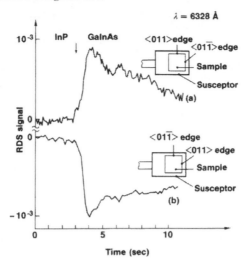

Fig. 3.28. RDS signal at λ = 6328 Å recorded during the growth of a GaInAs/InP heterojunction on two differently orientated substrates. The RDS signal is $\mathrm{Re}[(r_{011} - r_{01\bar{1}})/r]$ *for sample (a) and* $-\mathrm{Re}[(r_{011} - r_{01\bar{1}})/r]$ *sample (b) (after Razeghi and Acher [1989]).*

3.5.2. RDS records during real-time growth

Fig. 3.29 shows the RDS record of the growth of InAs on an InP substrate for 15 min using a 6328 Å light wavelength. The magnitude of the measured anisotropies is of the order of 5%. In contrast, the RDS signals measured for lattice-matched growth are generally about 10^{-3} [Aspnes 1988a, Acher et al. 1990c, Koch et al. 1990]. The occurrence of these large RDS signals is correlated with the occurrence of three dimensional (3D) growth, to a hazy surface appearance and a decrease in the measured reflectivity. The time scale of these anisotropy changes is also suggestive of structural evolution. By the end of 15 min growth, the RDS signal had returned to zero, the surface had smoothed and the reflectivity increased to a level comparable to the initial value. After 1 h, the growth was stopped, and the sample surface was mirror-like. Similar RDS records were obtained using different wavelengths, ranging from 3800 to 6328 Å.

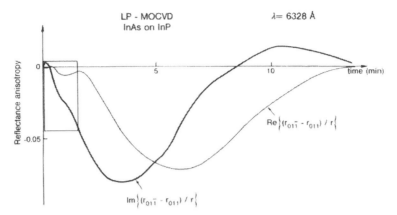

Fig. 3.29. RDS record of the growth of an InAs layer on an InP substrate using
$\lambda = 6328\ Å$ *light wavelength (Acher et al. [1990b]).*

The surface morphology of InAs on InP was observed using a scanning electron microscope (SEM). The growth was interrupted after 40 s (Fig. 3.30(*a*)), 2 min (Fig. 3.30(*b*)), 3 min 30 s (Fig. 3.30(*c*)) and 1 h (Fig. 3.30(*d*)). All the pictures were taken with the electron beam coming from the top of the pictures, and the samples were tilted by about 45°. The resolution in Fig. 3.30(*a*) is not sufficient to distinguish clearly the shape of the roughness. The lateral dimension of the roughness increases with deposition time and its geometry appears clearly on the other photos. It consists mainly of rectangular holes of different sizes but all with their edges parallel to [0 1 1] and [0 1 $\bar{1}$] . This observation is consistent with the prediction that they should be the preferred symmetry directions. The dispersion in size and shape is quite important in Fig. 3.30(*a*)–(*c*). These pictures do not reveal clear anisotropic patterns. The anisotropy may arise from the lateral dimensions of the holes, and also from their slope or their spacing. After 1 h growth, very few roughness patterns can be seen on the surface; one of these features is shown in Fig. 3.30(*d*). This is consistent with the fact that the sample is mirror-like after 1 h growth.

The dimensions of the roughness features increase with time (note the factor of three difference between the magnifications of Fig. 3.30(*c*) and (*d*)). One should point out that when the growth of the samples represented in Fig. 3.30(*a*)–(*c*) was stopped, the RDS signal continued to have an evolution as long as the sample was maintained at the growth temperature. This suggests that some mass transport occurs on the surface at that temperature. As a consequence, Fig. 3.30(*a*)–(*c*) may not show exactly the shape of the roughness under growth conditions. For the same reason, it is difficult to conduct spectroscopic studies on such samples. There is an evolution of properties determined through the RDS signal after stopping the growth which makes it difficult to compare kinetic and spectroscopic measurements. A fast cooling may be required to freeze the optical anisotropy.

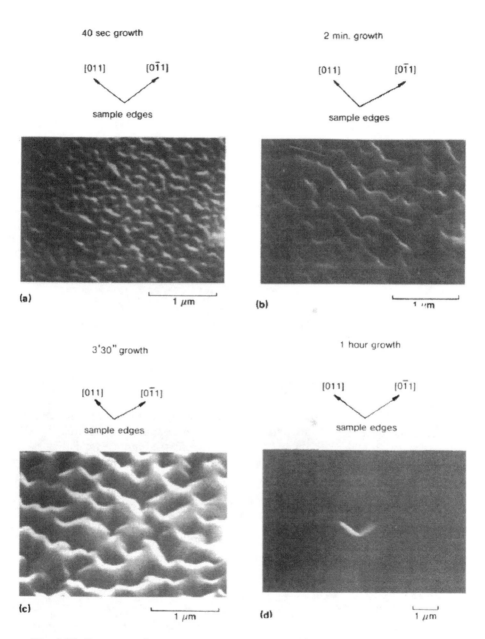

Fig. 3.30. Scanning electron microscope pictures of InAs on InP after: (a) 40 s
growth, (b) 2 min growth, (c) 3 min and 30 s growth and (d) 1 h growth (after Acher
et al. [1990b]).

Fig. 3.31 shows the RDS record using 5435 Å light of the growth of InP on a
GaAs on silicon substrate. In this case the optical anisotropy is extremely large,

sometimes with more than a factor of two difference between r_{011} and $r_{01\bar{1}}$. However, the broad features, the signs of the variations, are similar in Fig. 3.29 and Fig. 3.31. This suggests that the underlying mechanism responsible for the RDS signal evolution is the same, and is related to 3D growth, rather than effects associated with the specific material, such as chemical bonding.

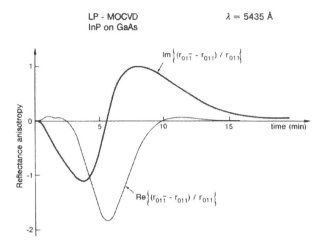

Fig. 3.31. RDS record of the growth of an InP layer on a GaAs on Si structure, using $\lambda = 5435$ Å light wavelength (after Acher et al. [1990b]).

For further discussion, it will be convenient to distinguish various stages in the RDS records. The initial transient corresponds to the first 5–10 s of growth. The experimental curve shown in Fig. 3.32 corresponds to the inset of Fig. 3.29, and it appears clearly that the initial behavior of the RDS signal differs from the behavior after 10 s growth. During this initial transient, the dominant contribution to RDS does not arise from surface roughness, as will be discussed further. In the first stage, corresponding to the first minute, the RDS record exhibits a rapid, nearly linear variation of $\mathrm{Im}[(r_{01\bar{1}} - r_{011})/r]$. In contrast, $\mathrm{Re}[(r_{01\bar{1}} - r_{011})/r]$ remains quite small. Fig. 3.29 shows a kind of oscillation with a time constant clearly different from the rest of the variations. The second stage corresponds to the rest of the growth, where the reflectance anisotropy shows a kind of damped oscillation behavior with large period and amplitude.

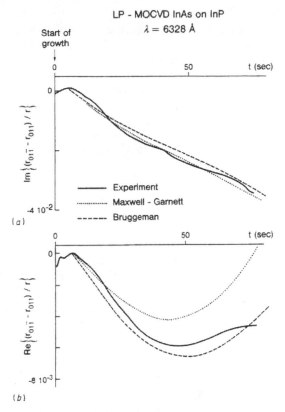

Fig. 3.32. Comparison between experiment (as per Fig. 3.29) and theory for the
RDS record using $\lambda = 6328$ Å of the first minute of growth of InAs on InP. (a)
$\mathrm{Im}[(r_{011} - r_{01\bar{1}})/r]$; (b) $\mathrm{Re}[(r_{011} - r_{01\bar{1}})/r]$. Note the change in scale between the two
figures. The Maxwell–Garnett fit was obtained with $F\,\delta q\,de/dt = 17\ \text{Å·min}^{-1}$. The
Bruggeman fit corresponds to $f = 0.7$, $q = 0.5$, $\delta q = 0.23$ and $de/dt = 100\ \text{Å·min}^{-1}$
(after Acher et al. [1990b]).

3.5.3. Optical models

It is clear from direct observation and SEM pictures (Fig. 3.30) that the surface of
the growing lattice-mismatched layer is rough during the first stages of growth. It is
suspected that this roughness accounts for the large reflectance anisotropies
observed.

Optical properties of rough surfaces are commonly treated using effective
medium theories (EMTs). Briefly, a rough surface can be approximated as a
homogeneous film, with a thickness corresponding to the height of the roughness,
and with a dielectric constant between that of the dense medium and that of the
ambient. Accordingly, a layer with a rough surface can be modeled as two layers: a
dense layer covered by a rough layer. A rough InAs/InP sample can be described as
a stack of three homogeneous layers: InP substrate, InAs dense layer and a rough
layer. The reflection coefficients of such a multilayer structure and its associated

optical anisotropy can easily be calculated using the Abeles formalism [Azzam and Bashara 1977], provided that the dielectric constant and thickness of all layers are known. The dielectric constants of GaAs, InP and InAs are tabulated [Palik 1985] and that of the rough layer evaluated using EMTs.

However, no reflectance anisotropy will arise from that model unless the shape of the roughness is anisotropic. It is well known that oblong particles may affect light polarization; this effect is known as form birefringence [Born and Wolf 1970]. It is possible to get from EMTs a quantitative evaluation of the form of birefringence associated with the anisotropic roughness, as shown below.

All EMTs yield the following form for the dielectric function of the effective medium [Aspnes 1983]

Eq. (3.18) $$\varepsilon_{\text{eff}} = \frac{q\varepsilon_0\varepsilon_1 + (1-q)\varepsilon_h\left[f\varepsilon_1 + (1-f)\varepsilon_0\right]}{(1-q)\varepsilon_h + q\left[f\varepsilon_1 + (1-f)\varepsilon_1\right]}$$

where ε_1 is the dielectric function of the dense medium, ε_0 is that of vacuum and f is the filling factor corresponding to the proportion of medium 1 making up this roughness. If $f < 0.5$, the roughness can be viewed as hills. If $f > 0.5$, the roughness can be viewed as a layer with depressions in it. The depolarization factor q is associated with the roughness shape, and accounts for the screening of the electric field within the effective medium. For anisotropic patterns, q depends on the orientation of the polarization, and one expects that $q_{011} \neq q_{01\bar{1}}$, leading to $\varepsilon_{011,\text{eff}} \neq \varepsilon_{01\bar{1},\text{eff}}$. If the roughness consists of long stripes, $q = 0$ for light polarized along the stripes and $q = 1$ for light polarization perpendicular to the stripes. The value of q associated with the main axes of ellipsoidal particles has been reported [Kittel 1976]. There is no straightforward derivation of q for patterns like that shown in Fig. 3.30. ε_h is the dielectric function of the background or "host" medium. In the Maxwell–Garnett theory (MG), $\varepsilon_h = \varepsilon_1$ or $\varepsilon_h = \varepsilon_0$, depending on the choice of prevalent medium. The Bruggeman theory takes $\varepsilon_h = \varepsilon_{\text{eff}}$. It is often preferred to MG, because it is self-consistent and does not attribute a particular role to either medium 0 or medium 1.

The present approach [Aspnes 1983] leads to a description of a sample using four media: the ambient, a rough anisotropic layer, a dense layer and the substrate. With the Abeles matrix formulation, it is very easy to calculate the optical properties of such multilayer systems numerically by computer. However, in order to get a better understanding of the observations, it is advantageous to use simpler models, if possible, with simple analytical solutions. The SEM observations and the thickness measurements suggest a growth process where both the dense layer and the top rough layer are growing.

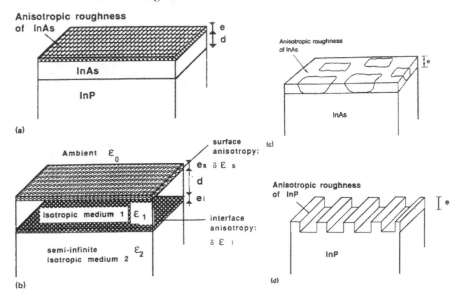

Fig. 3.33. Schematic view of the four models used to account for the observed RDS signal: (a) model with semi-infinite substrate, InAs dense layer and InAs rough anisotropic layer with thickness e « λ, (b) model with semi-infinite substrate, isotropic epilayer, anisotropic surface and interface, (c) model with semi-infinite InAs dense layer and InAs rough anisotropic layer and (d) model with semi-infinite layer and roughness consisting of infinite stripes. (d) corresponds to a particular case of model (c) (after Acher et al. [1990b]).

To simplify the calculation, one can consider different models for different growth stages of the RDS records. During the first stage (Fig. 3.33(*a*)), the roughness thickness *e* is very small compared to the light wavelength λ allowing a first-order development in *e*/λ. In the second stage of growth (Fig. 3.33(*c*)), the roughness should be treated without the small-thickness approximation. The effect of the substrate is ignored, because the dense layer is thick enough to absorb the light before it reaches the layer/substrate interface. This assumption is fully valid provided that the dense-layer thickness exceeds the penetration depth of light of a fixed wavelength in the material.

(a) Optical model for small roughness thickness

The model related to the first stage of growth is developed as follows. Fig. 3.33(*b*) shows schematically an isotropic layer growing on an isotropic substrate, with anisotropic surface and interface. The surface and interface are modeled as anisotropic layers with respective thicknesses e_s, e_i « λ. The Abeles matrix formulation and a first-order development in e/λ yield the reflectance anisotropy as shown in Eq. (3.8):

Eq. (3.19)

$$\frac{r_{01\bar{1}} - r_{011}}{r} = \frac{4j\pi}{\lambda} \frac{N_0}{\varepsilon_1 - \varepsilon_0} e_s \delta\varepsilon_s + \frac{16j\pi}{\lambda} \frac{N_0\varepsilon_1}{(\varepsilon_1 - \varepsilon_0)(N_1 + N_2)^2}$$

$$\times \frac{X}{[1 + (r_{12}/r_{01})X](1 + r_{12}r_{01}X)} \left(e_i \delta\varepsilon_i + \frac{\varepsilon_1 - \varepsilon_2}{\varepsilon_1 - \varepsilon_0} e_c \delta\varepsilon_c \right)$$

with

Eq. (3.20) $X = \exp\left(-\frac{4j\pi N_1 d}{\lambda}\right)$ $r_{12} = \frac{N_1 - N_2}{N_1 + N_2}$ $r_{01} = \frac{N_0 - N_1}{N_0 + N_2}$

and

Eq. (3.21) $\delta\varepsilon_s = \left(\varepsilon_{01\bar{1}} - \varepsilon_{011}\right)_{\text{surface}}$ $\delta\varepsilon_i = \left(\varepsilon_{01\bar{1}} - \varepsilon_{011}\right)_{\text{interface}}$

N_0, N_1 and N_2 are the optical indices of the ambient, epilayer and substrate, respectively, d is the thickness of the layer.

When the epilayer thickness is large enough, the influence of the substrate can be ignored, because all the light is absorbed before it reaches the substrate–epilayer interface. This corresponds to $X = 0$. In this case, Eq. (3.20) yields the known expression for the reflectance anisotropy of the surface of a bulk sample [Aspnes et al. 1989, McIntyre and Aspnes 1971, Del Sole and Selloni 1984]. It should be noticed that Eq. (3.20) was established using a time dependence of electromagnetic waves of exp(jωt) corresponding to a complex optical index given by $N = n - jk$ where n is the real index and k is the extinction coefficient. If the exp(jωt) convention is preferred, Eq. (3.20) should be replaced by its conjugate.

Fig. 3.33(a) corresponds to the case where the interface anisotropy $\delta\varepsilon_i$ is neglected and surface anisotropy is given by the EMTs. One can take surface roughness into account in the following way. For small anisotropies $\delta q = (q_{01\bar{1}} - q_{011}) \ll 1$ and for $\varepsilon_1 \gg \varepsilon_0 = 1$, a first-order development of MG with $\varepsilon_h = \varepsilon_1$ gives

Eq. (3.22) $\delta\varepsilon_s = -\delta q F \varepsilon_1$

where F is related to the roughness geometry by

Eq. (3.23) $F = f(1 - f)/(1 - qf)^2$

The RDS signal corresponding to the case in Fig. 3.33(a) with the Maxwell–Garnett description of roughness is obtained using Eq. (3.13) and Eq. (3.8), with $\varepsilon_1 \gg \varepsilon_0$.

Eq. (3.24)
$$\frac{r_{01\bar{1}}-r_{011}}{r}=\frac{4\mathrm{j}\pi}{\lambda}Fe\delta q\left(1+\frac{4X'}{\left(1-X'\right)^{2}}\right)$$

with

Eq. (3.25)
$$X'=r_{12}\exp\left(-\frac{4\mathrm{j}\pi N_{1}d}{\lambda}\right)\qquad r_{12}=\frac{N_{1}-N_{2}}{N_{1}+N_{2}}$$

where e is the roughness thickness, r_{12} is the reflection coefficient of the layer–substrate interface.

It is clear from Eq. (3.24) that the anisotropic roughness affects mainly the imaginary part of the reflectance anisotropy, as observed experimentally during the first stages of the growth shown in Fig. 3.29 and Fig. 3.32. The real part arises only from the term in X', which is a perturbation. It corresponds to light reflected at the substrate–dense-layer interface. As the dense-layer thickness increases, interference conditions between the light reflected at the surface and at the interface are modified. This leads to an oscillating behavior of $\mathrm{Re}[(r_{01\bar{1}}-r_{011})/r]$ as d increases.

The characteristics of roughness appear in the factor $Fe\,\delta q$. It is clearly not possible to obtain independently values for $f,\ q,\ e,\ \delta q$ from the experimental data. Indeed, the product $Fe\,\delta q$ is the only variable parameter that need be introduced in a fit of experimental measurement. It is important to emphasize that the relationship between the real and imaginary parts deduced from Eq. (3.24) is independent of roughness characteristics. For a given substrate and layer, if the growth rate of the dense layer is known, the model predicts the value of the real part of the reflectance anisotropy as a function of the imaginary part only. Therefore, the ability of the model to account for both components of the RDS signal is a good test of its validity.

If the Bruggeman theory is used instead of MG to calculate the value of $\delta\varepsilon_{s}$, one no longer obtains a simple analytic expression similar to Eq. (3.22). $\delta\varepsilon_{s}$ has to be evaluated numerically using Eq. (3.18). The RDS signal is computed using Eq. (3.20). Bruggeman is expected to be more accurate but both theories lead to similar results [Acher et al. 1991]. The discussion based on Eq. (3.24) is expected to remain essentially valid.

(b) Optical model for large roughness thickness

After both the rough and the dense layer have grown for some time, another model (Fig. 3.33(c)) should be used to describe the second stage. This is indeed a three-medium model, with an anisotropic intermediate medium treated using the MG or Bruggeman theory. The RDS signal associated with such a sample is easily computed as a function of roughness characteristics. The main features expected from such a model are the possibility to account for large anisotropies, and the interference-like features measured by RDS (Fig. 3.29 and Fig. 3.31). Clearly, the thicker the anisotropic medium, the greater the interaction with light, and therefore the larger the optical anisotropy. Constructive or destructive interference within this layer is possible only if the material is not too absorbing. Indeed, the effective medium described by the MG or Bruggeman theory is significantly less absorbing than the dense medium. The penetration depth of light in the rough layer exceeds the wavelength in this material. One can therefore expect to observe oscillating interference-like features on the RDS signal when the thickness of the roughness increases, even for light wavelengths strongly absorbed in the dense medium.

One should point out that the validity of EMTs depends on the dimensions of the rough features compared to the light wavelength. If the roughness consists of a series of parallel stripes of infinite length (Fig. 3.33(d)), the ratio between the lateral dimensions (stripe width and spacing) and the wavelength is the relevant criterion. This ratio should be small in order for the theories to be valid; discrepancies appear when this ratio exceeds 0.1. However, SEM images (Fig. 3.30) show that in fact roughness dimensions are comparable with light wavelength after 4 min growth of InAs on InP. Even if EMTs are used somewhat outside of their range of validity, they can still give a qualitative idea of the optical properties of a rough layer. One cannot expect to get quantitative information from RDS measurements after a few minutes of growth.

3.5.4. The first stage of 3D growth

Fig. 3.32, Fig. 3.34 and Fig. 3.35 present the RDS record of the first minute of InAs/InP growth and the corresponding fit using the model shown in Fig. 3.33(a). Fig. 3.32 is a detail of Fig. 3.29. After an initial transient of 5–10 s, $\mathrm{Im}[(r_{01\bar{1}} - r_{011})/r]$ shows a nearly linear variation. This suggests that the roughness thickness e increases linearly with time. Eq. (3.24) shows that the slope of the imaginary part of reflectance anisotropy is $F\,\delta q\,de/dt$. This is the only adjustable parameter used in the MG model to fit both the real and the imaginary part of $(r_{01\bar{1}} - r_{011})/r$. The evolution of the thickness d of the dense layer as a function of time has to be known in order to evaluate X' in Eq. (3.24). The growth rate of the dense InAs layer ($d(d)/dt$) was taken as 250 Å·min^{-1} for Fig. 3.32 and Fig. 3.34. This value was chosen to be slightly lower than the 270 Å·min^{-1} measured experimentally on thick samples. This is because only part of the growth contributes to the dense layer; another part contributes to increasing the roughness thickness. In the case of Fig. 3.35, where the growth rate was doubled, dd/dt was taken as 500 Å·min^{-1}. It can be noticed that a small error in the evaluation of $d(d)/dt$ would slightly change the period of the oscillation of $\mathrm{Re}\{(r_{01\bar{1}} - r_{011})/r\}$ but would not significantly affect the agreement observed between the experiment and the model.

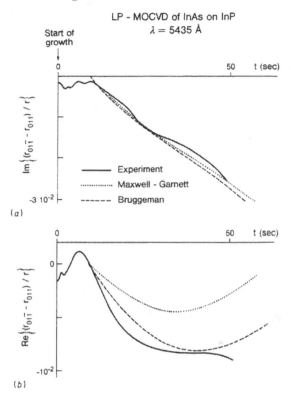

Fig. 3.34. *A comparison between experiment and theory for the RDS record using* $\lambda = 5435$ *Å of the first minute of growth of InAs on InP. The Maxwell–Garnett fit was obtained with* $F \, \delta q \, de/dt = 19.5$ *Å·min^{-1}. The Bruggeman fit corresponds to* $f = 0.7$, $q = 0.5$, $\delta q = 0.28$ *and* $de/dt = 100$ *Å·min^{-1} (after Acher et al. [1990b]).*

The simulations shown in Fig. 3.32, Fig. 3.34 and Fig. 3.35 using the Bruggeman theory were made using $f = 0.7$, $q = 0.5$, and by varying the shape anisotropy δq and the roughness growth rate de/dt. No attempt was made to find the best fit by allowing all parameters to vary. As discussed previously, it is not possible to obtain reliable evaluations for each parameter individually, but only for combinations of them. Using the MG theory, the model yields 17 Å·min^{-1}, 19.5 Å·min^{-1} and 35.5 Å·min^{-1} for the $F \, \delta q \, de/dt$ product for Fig. 3.32, Fig. 3.34 and Fig. 3.35, respectively. The corresponding values using Bruggeman theory are 11.5 Å·min^{-1}, 14 Å·min^{-1} and 24 Å·min^{-1}. This shows that the MG and Bruggeman theory yield comparable $F \, \delta q \, de/dt$ product for a given sample.

As seen from the figures, both models give comparable predictions. The model based on Bruggeman theory accounts very well for the behavior of the real part of the reflectance anisotropy, and is of the order of one magnitude difference between the real and the imaginary parts. The model agrees well with the measurements both at $\lambda = 5435$ Å (Fig. 3.34) and at $\lambda = 6328$ Å (Fig. 3.32). The $F \, \delta q \, de/dt$ products found in both cases are similar. Doubling the growth rate (Fig. 3.35) yields a nearly doubled $F \, \delta q \, de/dt$ product. The period of the oscillation of $\mathrm{Re}[(r_{01\bar{1}} - r_{011})/r]$ is shorter, in accordance with the higher growth rate of the dense layer. Fig. 3.32, Fig. 3.34 and Fig. 3.35 show that the model proposed gives a satisfactory description of experimental observations. The slope of $\mathrm{Im}[(r_{01\bar{1}} - r_{011})/r]$ gives a quantitative indication of the anisotropic thickness growth rate. The oscillatory behavior of $\mathrm{Re}[(r_{01\bar{1}} - r_{011})/r]$, as a function of dense-layer thickness predicted by Eq. (3.24), is actually observed. Other experiments have shown that this feature is observed also at the 5000 Å wavelength but is no longer seen with UV light. This is because the InAs epilayer is too absorbing in the UV and the quantity X' in Eq. (3.24) vanishes even for small epilayer thickness.

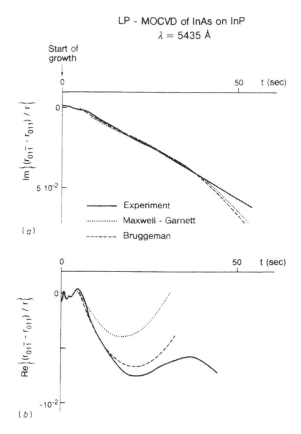

Fig. 3.35. Comparison between experiment and theory for the RDS record using $\lambda = 5435\ \mathring{A}$ of the first minute of growth of InAs on InP, for twice the usual growth rate. The Maxwell–Garnett fit was obtained with $F\ \delta q\ de/dt = 35.5\ \mathring{A}\cdot min^{-1}$. The Bruggeman fit corresponds to $f = 0.7$, $q = 0.5$, $\delta q = 0.8$ and $de/dt = 170\ \mathring{A}\cdot min^{-1}$ (after Acher et al. [1990b]).

The same model was also used for the growth of InP on GaAs. The sign of the reflection coefficient between InP and GaAs is the opposite of that of InAs on InP. Therefore, the sign of the initial variation of $\mathrm{Re}[(r_{01\bar{1}} - r_{011})/r]$ given by Eq. (3.24) should be opposite of that observed in the case of InAs on InP. This is indeed the case (Fig. 3.31). But the oscillatory behavior visible, especially in Fig. 3.33 and Fig. 3.35, is masked here by the variations of surface roughness. The assumption that $e \ll \lambda$ breaks down very quickly for the InP/GaAs samples.

The model does not account for the initial stage, corresponding to the first 5–10 s of the growth. It takes between 4 (Fig. 3.32, Fig. 3.35, and Fig. 3.34) for the imaginary part of the reflectance anisotropy to begin its linear variation. This shows that the variation of the RDS signal during this initial stage comes from a contribution other than anisotropic 3D growth.

3.5.5. Subsequent growth

After one minute of growth, the fit between the calculated and measured RDS signal evolution no longer holds for $\mathrm{Re}[(r_{01\bar{1}} - r_{011})/r]$. This is expected, because the model displayed in Fig. 3.33(a) is valid only for small roughness thickness, $e \ll \lambda$. One should use the model developed for the second stage (Fig. 3.33(c) and (d)), which works for finite roughness thickness.

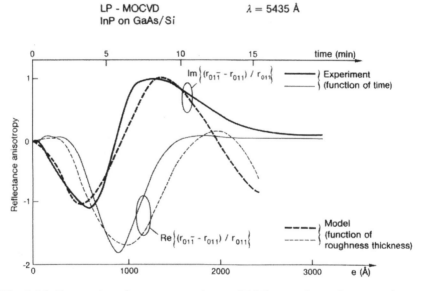

Fig. 3.36. Comparison between experimental RDS record as a function of time (same as Fig. 3.31), and theoretical prediction of model (Fig. 3.33(d)) as a function of roughness thickness. The filling factor was taken as $f = 0.5$ (after Acher et al. [1990b]).

This is shown in Fig. 3.36, for InP/GaAs/Si growth, where the calculated reflectance anisotropy is plotted as a function of roughness thickness, and compared with the RDS versus time record of Fig. 3.31. The calculated curve is obtained by approximating the roughness to be stripes (Fig. 3.33(d)), which is the most anisotropic shape ($\delta q = 1$). Both the MG and Bruggeman models are equivalent in that case. The filling factor was taken to $f = 0.5$. The theoretical curve as a function of roughness thickness e exhibits the same general features as the variations of measured signal as a function of time. There is agreement for the sign of the variation, the order of magnitude, the sign change of the imaginary part and the correspondence between the inflection points and the extrema of both components. This similarity between the theoretical curve as a function of e and the experimental record function of time suggests that roughness thickness e increases linearly with time. Differences exist between the observation and the model, but many reasons may account for this. The roughness shape, the filling factor and the growth rate of e may vary with time. Besides, as previously mentioned, the EMT is not used in its range of strict validity.

The growth of InAs on InP was investigated at different wavelengths, ranging from 3800 to 6328 Å. The broad features and the sign of the variations are the same at all wavelengths, except the initial oscillation of $\mathrm{Re}[(r_{01\bar{1}} - r_{011})/r]$, as mentioned before. This is consistent with a structural origin of the RDS behavior. The model of Fig. 3.33(c) was used to account for an RDS record of InAs/InP. Fig. 3.37 represents the calculated RDS signal for the growth of rough InAs/InP as a function of roughness thickness e. The roughness shape is kept constant ($f = 0.7$, $q = 0.5$, $\delta q = 0.16$). The model agrees quite well with the experimental record of RDS signal as a function of time (same record as the one detailed in Fig. 3.34). While it was not possible to determine separately the value of de/dt and the shape of the anisotropy δq during the first minute of growth, these two parameters have quite different effects on the RDS features for longer time scale. As a rough indication, the period of the oscillations of the RDS signal corresponds to constructive or destructive interference conditions, and is therefore related to the roughness thickness e; the magnitude of the signal at the extrema depends on the shape anisotropy δq.

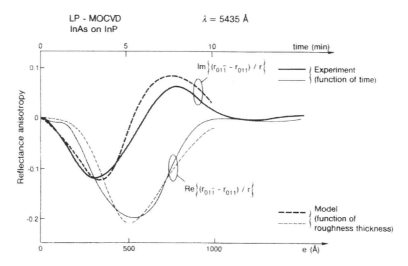

Fig. 3.37. Comparison between experimental RDS record as a function of time of InAs/InP, and theoretical prediction of the model (Fig. 3.33(c)) as a function of roughness thickness. The roughness was modeled using the Bruggeman theory, with $f = 0.7$, $\delta q = 0.16$ and $q = 0.5$ (after Acher et al. [1990b]).

The models used (Fig. 3.33(c) and (d)) assume that the influence of the substrate on the optical properties can be neglected. This assumption is valid provided that the penetration length of light in the material is smaller than the dense-layer thickness. This is true after several minutes of deposition, depending on the wavelength and the growth rate, but does not hold after only one minute of growth for the wavelengths used. However, as discussed above, the effects due to light reflected by the substrate–layer interface are small. Taking this effect into account during the second stage of the RDS signal would bring only small corrections.

Fig. 3.36 and Fig. 3.37 suggest that the roughness thickness keeps increasing with time. The corresponding growth rate of the roughness de/dt is 160 Å·min^{-1} in the case of InP/GaAs/Si (Fig. 3.36) and 100 Å·min^{-1} in the case of InAs/InP (Fig. 3.37). However, after 10 or 15 min growth, the surface improves, leading to a mirror-like surface after one hour of growth. This is consistent with the fact that the RDS signal returns to zero, instead of exhibiting more oscillations as the models would predict. The mechanisms that could lead to an increase both in apparent roughness thickness and in the inhibition of roughness have not been investigated. They are beyond the scope of the present modeling.

Various details of the models will now be considered in more depth. The anisotropy has been described only with the screening parameter anisotropy δq, and we shall try to relate this quantity to the shape of the roughness and to its anisotropy in the lateral dimension. The difficulty in finding a more accurate description for the optical properties of roughness with large dimensions will be emphasized. With our present understanding, RDS observations give some insights into the growth process. Information concerning the critical thickness of strained layers and the orientation of the anisotropic 3D patterns can be obtained. The influence of substrate misorientation as evidence of mass transport during annealing can be measured. With our present understanding, the information obtained using RDS can be of use for the optimization of the growth of lattice-mismatched semiconductors.

3.5.6. Discussion of optical models

The first stages of growth have been described successfully using the model of Fig. 3.33(a), and the agreement between experiment and theory is shown in Fig. 3.32, Fig. 3.34, and Fig. 3.35. The roughness features are small, and EMTs are clearly used within their range of validity. It is possible to give an idea of the correspondence between this parameter and the shape of the roughness in some cases. The sign of δq indicates that the screening is less effective along the [0 1 1] direction than along [0 1 $\bar{1}$]. This suggests that the roughness patterns are elongated in the [0 1 1] direction. For oblate spheroids, δq depends on the ratio between axial lengths c/a [Kittel 1976]; for example, $\delta q = 0.3$ corresponds to $c/a \approx 1.4$. This elongation is moderate but can account for the RDS behavior observed for InAs on InP (Fig. 3.37). The determination of the depolarization factor associated with a given particle shape is not straightforward, and no attempt was made to relate δq to SEM observations. The dispersion of roughness shape has not been taken into account in this calculation. It is possible to include it in a model [Borenszstein et al. 1988], but this would add more parameters.

In all the calculations, instead of high-temperature values of the optical index of InAs and InP, room-temperature values have been used. The main parameter affected by this simplification is the reflection coefficient at the substrate-layer interface r_{12}, which relates the variation of the real part of RDS signal to the variation of the imaginary part (Eq. (3.24)). However, r_{12} depends mainly on the difference between the indices, and should not be affected greatly by temperature, since the indices of both InAs and InP are expected to have similar variations with temperature.

No effects other than the structural contribution have been taken into account in this model. The contribution of chemisorbed species to the RDS signal is known to be in the 10^{-3} range, in particular for InAs. It may account for the variations of the RDS signal in the first seconds of growth, but clearly not for the larger variations observed afterwards. However, the effect of the strain should be considered. It is known that a lattice-mismatched layer is very strained at the beginning of growth, and gradually relaxes. If the deformation of the epilayer is purely tetragonal, no RDS contribution is expected to arise from the strain. If the relaxation process is different in the [0 1 1] and [0 1 $\bar{1}$] directions, as has been reported in the case of GaAs on silicon [Koch 1988], it is expected to have an influence on the RDS signal. The effect of strain on the optical properties of semiconductors has been studied using piezoreflectance measurements [Sell 1973, Camassel et al. 1975]; the effect is extremely wavelength dependent, increases by one order of magnitude, and has a sign change near the critical points E_0 and E_1. This probably rules out anisotropic strain as a major contribution to the RDS records of lattice-mismatched materials, because the observed features are essentially similar for two wavelengths. The influence of strain may be a key feature in understanding the evolution of the RDS signal in the first 5 or 10 s. Using available data on GaAs [Sell 1973] a strain anisotropy in the growth plane of 10^{-3} would lead to an RDS contribution less than 10^{-3} far below E_1. It would reach 5×10^{-3} in the E_1, $E_1 + \Delta_1$ region. Spectroscopic measurements would be necessary to assess this effect.

As previously mentioned, EMTs are based on the assumption that the dimensions of the roughness are very small compared to the light wavelength. This assumption clearly holds during the first minute, but is no longer true after 5 min. The use of EMTs to account for RDS behavior after 5 min of growth (Fig. 3.36 and Fig. 3.37) requires some further discussion. The reflectivity of the sample decreases by a factor of two to six after 10 min of 3D growth, before returning to a level comparable to its original value. EMTs fail to account for this evolution of the reflectivity. One reason is that they do not take into account light diffusion. When the typical size of roughness patterns approaches the light wavelength, part of the light is diffused, leading to a decrease in reflectivity and to a hazy appearance. But the reflectance anisotropy is not expected to be much affected by the diffused light since it contributes to decrease both r_{011} and $r_{01\bar{1}}$. Another important assumption which is no longer valid for thick roughness is the homogeneity along the growth direction. These factors may explain why EMTs account quite well for the behavior of the RDS signal $(r_{01\bar{1}} - r_{011})/r$, but completely fail to describe r_{011} and $r_{01\bar{1}}$ individually.

A more accurate model would be able to account for the optical properties of roughness with characteristic dimensions comparable to the wavelength. Much effort has been devoted to the study of scattering by anisotropic particles [Asano and Yamamoto 1975, Asano 1979, Barber and Wang 1978, Uzunoglu et al. 1976] leading very often to numerical solutions requiring long computations. The modeling of anisotropic patterns on a surface may be more difficult than that of isolated particles [Berreman 1970]. Besides, SEM pictures show that the roughness is not uniform, and a distribution of roughness characteristics should be used. This inhomogeneity may also induce partial depolarization of reflected light. A model based on the Fresnel–Kirchoff theory of diffraction accounts for the decrease in reflectivity of rough surfaces with lateral roughness dimensions that are large compared to the wavelength, and small roughness thickness [Ohlidal and Lucks 1972]. This model is only valid if the slope of the roughness features is very small, which does not seem to hold in the present case, and is unable to account for reflectance anisotropy under normal incidence. The difficulty in finding a model that would account both for reflectance measurements and characterization in polarized light has often been mentioned [Pickering et al. 1989].

3.5.7. RDS monitoring of growth of GaInP/GaAs and GaInP/InP

Fig. 3.38 shows the RDS records of growth of GaInP on two different substrates. In the case (a), the composition of GaInP is adjusted so that there is 0.6% compression stress in the epilayer. The variation of the imaginary part is in the 10^{-3} range, and is mainly due to noise. No anisotropic roughness is detected on the surface. In the case (b), the GaInP epilayer exhibits a 3.8% lattice mismatch with the InP substrate. The evolution of $\mathrm{Im}[(r_{01\bar{1}} - r_{011})/r]$ shows that the growth is three dimensional from the very first seconds of epilayer growth. The real part $\mathrm{Re}[(r_{01\bar{1}} - r_{011})/r]$ is not affected significantly by 3D growth, as expected from Eq. (3.15) because r_{12} is quite small. The thickness of the roughness is expected to increase at a rate de/dt with $F\ \delta q$ $de/dt \approx 24$ Å·min^{-1}.

On a larger time scale, the roughness thickness can increase so that the assumption $e \ll \lambda$ no longer holds. Then, both real and imaginary parts of $(r_{01\bar{1}} - r_{011})/r$ attain large values, sometimes larger than 0.1. However, if a proper growth procedure is followed, a smooth surface can be obtained and, as a consequence, the optical anisotropy returns to zero. RDS can give indications of whether the surface cures or not and at what rate. This information is particularly important since surface morphology is very sensitive to a slight perturbation of growth conditions. Fig. 3.39 illustrates this point. Two InAs layers were grown on InP substrates. Growth conditions were identical in both cases. The only difference consisted in the heating procedure, which was faster in (b) than in (a). It is clear from the RDS records that the roughness cures better in (a) than in (b). Therefore, the growth was stopped in the second case.

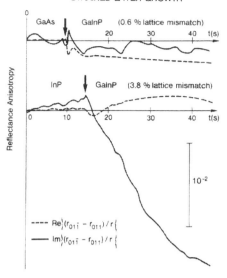

Fig. 3.38. RDS records of two strained layers, showing both the real and imaginary parts of the anisotropy: (a) GaInP on GaAs, with a lattice mismatch of 0.6% (compressive stress); (b) GaInP on InP, with a lattice mismatch of 3.8% (tensile stress) (after Acher et al. [1990a]).

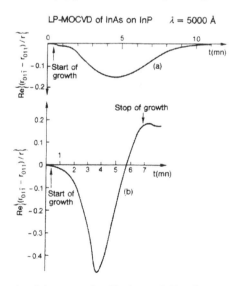

Fig. 3.39. RDS records of the growth of InAs on InP substrate (% lattice mismatch) Growth conditions are the same in both cases, but the heating procedure was faster in (b) than in (a) (after Acher et al. [1990a]).

3.6. Insights on the growth process

Using this understanding of the RDS technique, it is now possible to discuss the aspects of growth. Among the crystal-growth issues that can be dealt with is the assessment of critical thickness where 3D growth begins. During the initial stage of growth, RDS behavior does not correspond to anisotropic 3D growth. This could mean that the growth is two dimensional during the first 5–10 s. This corresponds to Stranski–Krastanov growth: growth of several layers of a complete InAs film, followed by island formation. After this initial stage, both the dense and the rough part of the film grow. The RDS signal evolution observed during the initial transient, corresponding to the first seconds of growth, may arise from a variety of contributions. The contribution of chemisorbed and physisorbed species are expected to be different in the case of InP or in the case of growing InAs. Atomic rearrangements at the surface or interface can affect the signal. Another contribution may arise from anisotropic strain in the layer, as discussed later.

The magnitude of the RDS signal for InAs/InP is very dependent on growth conditions, substrate cleaning and the heating procedure. It makes the quantitative RDS assessment of $F \, \delta q \, de/dt$ more suitable for *in situ* monitoring. The small difference between the values deduced from Fig. 3.32 and Fig. 3.34 corresponding to two different growths monitored using two different wavelengths may not arise from an unexpected spectroscopic dependence, but simply because the growth conditions were slightly different between the two experiments. The choice of wavelength for performing such *in situ* studies is not crucial. The same broad features are found at all wavelengths except the oscillation on $\mathrm{Re}[(r_{01\bar{1}} - r_{011})/r]$ during the first minute. It is not detectable for wavelengths too strongly absorbed in the dense medium.

The observations show that during the growth of both InAs/InP and InP/GaAs/Si, anisotropic 3D growth occurs. The magnitude of this effect is different for the two systems. Modeling suggests that roughness patterns are more anisotropic and grow faster in the case of InP/GaAs than in the case of InAs/InP.

In a preliminary study, the growth of InAs and InP was stopped while keeping the AsH$_3$ flux and temperature constant. Fig. 3.40 reports the RDS record of a 3 min 30 s growth of InAs on InP followed by about 1 min under AsH$_3$ without growth at the same temperature. This is the same sample as that used in Fig. 3.30(*c*). The RDS signal changes significantly during this annealing, suggesting that there was mass transport and a change in roughness characteristics. The change during the first 30 s is about 5% for $\mathrm{Re}[(r_{01\bar{1}} - r_{011})/r]$, and 20% for $d_{h_1 h_2 h_3}$, which corresponds to a variation rate larger than during growth. Then the optical anisotropy reaches a steady-state value. The change in RDS is expected to come from a change in roughness shape and characteristics since only structural contributions can account for this order of magnitude. The mechanism for such structural evolutions has not been investigated. It is important to note that stopping the growth and keeping the sample at the same temperature does not stop surface evolution. As previously mentioned, it makes a comparison between kinetic RDS data and spectroscopic RDS recorded after growth has been stopped more difficult. This is because the evolution of the optical properties of a sample during the process of stopping the growth and cooling down is not negligible.

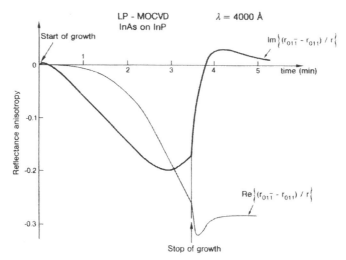

Fig. 3.40. *RDS record of the growth of an InAs layer on an InP substrate, using* $\lambda = 4000$ Å. *The growth is stopped after 3 min and 30 s, the* AsH_3 *flux and temperature being unchanged. Note the important change of RDS signal in the seconds following the growth interruption (after Acher et al. [1990b]).*

The models proposed for the first and second stages of the RDS records of roughness suggest that the roughness thickness continuously increases with time. It is also clear that after some time the roughness contribution disappears, indicating that the surface has smoothed. These two aspects are not contradictory if one supposes that the roughness thickness increases while its filling factor increases toward unity. Fig. 3.30(*a*)–(*d*) are consistent with roughness features increasing in size but becoming less and less numerous as a function of time. The mechanism needs further investigation.

Two factors contribute to the difference in optical properties of the [0 1 1] and [0 1 $\bar{1}$] directions. One is the chemical asymmetry between the two directions: the dangling bonds of the group V element are aligned along the [0 1 $\bar{1}$] direction. The other is the structural asymmetry due to the presence of steps on the surface. The steps are due to the slight tilt of the surface of the substrate relative to (1 0 0). For the samples misoriented toward [0 1 $\bar{1}$] that were used for this study, the steps are along the [0 1 1] direction. Steps are known to play an important role in the first stages of the growth of GaAs/Si [Koch 1988]: nucleation occurs at the step edges and GaAs islands are elongated along the step. In the present study, the sign of δq tells us that roughness is elongated along [0 1 1] which is also the direction of the steps. Preliminary experiments were conducted with (1 0 0) exact substrates, in order to assess the importance of the structural asymmetry of the substrate on the RDS behavior (Fig. 3.41). The RDS features related to 3D growth are quite different from those observed on substrates with 2° misorientation toward [0 1 $\bar{1}$]. This suggests that structural rather than chemical anisotropy of the surface of the substrate determines the privileged directions of anisotropic 3D growth. It shows the importance of substrate misorientation for the growth of lattice-mismatched materials. With our present understanding, the RDS record of Fig. 3.41 is difficult to comment on in more detail. It is not even clear that [0 1 1] and [0 1 $\bar{1}$] are the optical eigenaxes. The orientation of the commercially available substrates is not specified with a precision better than 0.5°, and optical eigenaxes may correspond to the orientation of the residual tilt. Further studies with different substrate misorientations are required to separate the influence of the chemical and structural asymmetry of the substrate on the roughness characteristics.

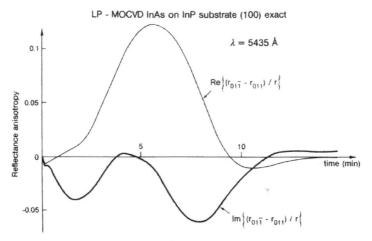

Fig. 3.41. RDS record of the growth of an InAs layer on an InP substrate with (1 0 0) exact surface orientation using $\lambda = 5435$ Å (after Acher et al. [1990b]).

References

Acher, O., *These de Doctoral* Universite Paris-Sud, 1990.

Acher, O., Benferhat, R., Drevillon, B., and Razeghi, M., *Proc. SPIE* **1361** 1156, 1990a.

Acher, O., Bigan, E., and Drevillon, B., *Rev. Sci. Instrum.* **60** 65, 1989.

Acher, O., Koch, S.M., Omnes, F., Defour, M., Razeghi, M., and Drevillon, B., *J. Appt. Phys.* **68** 3564, 1990b.

Acher, O., Omnes, F., Koch, S.M., Razeghi, M., Drevillon, B., and Bigan, E., *Revue Technique Thomson-CSF* (Paris: Gauthier-Villars), 1991.

Acher, O., Omnes, F., Razeghi, M., and Drevillon, B., *Mater Sci. Eng.* B **5** 223, 1990c.

Acosta-Ortiz, S.E. and Lastras-Martinez, A., *Solid State Commun.* **64** 809, 1987.

Agranovich, V.M., and Ginzburg, V.L., *Spatial Dispersion in Crystal Optics and Theory of Excitons* (Berlin: Springer), 1984.

Aoyagi, Y., Doi, A., Iwai, S., and Namba, S., *J. Vac. Sci. Technol.* B **5** 1460, 1987.

Asano, S., *Appl. Opt.* **18** 712, 1979.

Asano, S. and Yamamoto, G., *Appl. Opt.* **14** 29, 1975.

Aspnes, D.E., *J. Physique Coll.* C **10** 3, 1983.

Aspnes, D.E., *J. Vac. Sci. Technol.* B **3** 1498, 1985.

Aspnes, D.E., *Appl. Phys. Lett.* **52** 957, 1988a.

Aspnes, D.E., *Proc. SPIE* **946** 84, 1988b.

Aspnes, D.E., Chang, Y.C., Studna, A.A., Florez, L.T., Farrell, H.H., and Harbison, J.P., *Phys. Rev. Lett.* **64** 192, 1990.

Aspnes, D.E., Colas, E., Studna, A.A., Bhat, R., Koza, M.A., and Keramidas, V.G., *Phys. Rev. Lett.* **61** 2782, 1988a.

Aspnes, D.E., Harbison, J.P., Studna, A.A., and Florez, L.T., *Phys. Rev. Lett.* **59** 1687, 1987.

Aspnes, D.E., Harbison, J.P., Studna, A.A., Florez, L.T., and Kelly, M.K., *J. Vac. Sci. Technol.* B **6** 1127, 1988b.

Aspnes, D.E. and Studna, A.A., *Phys. Rev. Lett* **54** 1956, 1985.

Aspnes, D.E., Studna, A.A., Florez, L.T., Chang, Y.C., Harbison, J.P., Kelly, M.K., and Farrell, H.H., *J. Vac. Sci. Technol.* B **7** 901, 1989.

Azzam, R.A., *Opt. Commun.* **20** 405, 1977.

Azzam, R.A., *J. Opt. (Paris)* **12** 317, 1981.

Azzam, R.M. and Bashara, N.M., *Ellipsometry and Polarized Light* (Amsterdam: North-Holland), 1977.

Barber, P.W. and Wang, D.S., *Appl. Opt.* **17** 797, 1978.

Berkovits, V.L., Ivantsov, L.F., Makarenko, I.V., Minashvili, T.A., and Safarov, V.I., *Solid State Commun.* **64** 767, 1987.

Berreman, D.W., *J. Opt. Soc. Am.* **60** 499, 1970.

Borenszstein, Y., Jebari, M., Lopez-Rios, T., and Vuye, G., *Europhys. Lett.* **7** 617, 1988.

Born, M. and Wolf, E., *Principles of Optics* (New York: Pergamon), 1970.

Briones, F. and Horikoshi, Y., *J. Appl. Phys.* **29** 1014, 1990.

Butler, J.F., Bottka, N., Sillmon, R.S., and Gaskill, D.K., *J. Crystal Growth* **77** 163, 1986.

Camassel, J., Auvergne, D., and Mathieu, H., *J. Appl. Phys.* **46** 2683, 1975.

Chiu, T.H., *Appl. Phys. Lett.* **55** 1244, 1989.

Colas, E., Aspnes, D.E., Bhat, R., Studna, A.A., Harbison, J.P., Florez, L.T., Koza, M.A., and Keramidas, V.G., *J. Crystal Growth* **107** 47, 1991.

Colas, E., Aspnes, D.E., Bhat, R., Studna, A.A., Koza, M.A., and Keramidas, V.G., *J. Crystal Growth* **93** 931, 1988.

Colas, E., Aspnes, D.E., Bhat, R., Studna, A.A., Koza, M.A., and Keramidas, V.G., *J. Crystal Growth* **94** 613, 1989.

Collins, R.W., *Advances in Disordered Semiconductors* vol I (Singapore: World Scientific), 1989.

Del Sole, R., *Solid State Commun.* **37** 537, 1981.

Del Sole, R., and Selloni, A., *Solid State Commun.* **9** 825, 1984.

Donnelly, V.M. and Karlicek, R.F., *J. Appl. Phys.* **53** 6399, 1982.

Drevillon, B., *J. Non-Cryst. Solids* **114** 139, 1989.

Drevillon, B., Parey, J.Y., Stchakovsky, M., Benferhat, R., Josserand, T., and Schlayen, B., *Proc. SPIE* **1188** 174, 1989.

Drevillon, B., Perrin, J., Marbot, R., Violet, A., and Dalby, J.L., *Rev. Sci Instrum.* **53** 969, 1982.

Foxon, C.T. and Joyce, B.A., *J. Crystal Growth* **44** 75, 1978.

Gaskill, D.K., Kolubayev, V., Bottka, N., Sillmon, R.S., and Butler, J.E., *J. Crystal Growth* **93** 127, 1988.

Haigh, J., *J. Mater. Sci* **18** 1072, 1983.

Harris, J.J., Joyce, B.A., and Dobson, P.J., *Surf. Sci* **103** L90, 1981.

Hottier, F., Hallais, J., and Simondet, F., *J. Appl. Phys.* **51** 1599, 1980.

Jasperson, S.N. and Schnatterly, S.E., *Rev. Sci Instrum.* **40** 761, 1969.

Jonsson, J., Deppert, K., Jeppesen, S., Paulsson, G., Samuelson, L., and Schmit, P., *Appl. Phys. Lett.* **56** 2414, 1990.

Kittel, C., *Introduction to Solid State Physics* (New York: Wiley), 1976.

Koch, S.M., *PhD Thesis* Stanford University, 1988.

Koch, S.M., Acher, O., Omnes, F., Defour, M., Drevillon, B., and Razeghi, M., *J. Appl. Phys.* **68** 3364, 1990.

Koch, S.M., Acher, O., Omnes, F., Defour, M., Drevillon, B., and Razeghi, M., *J. Appl. Phys.* **69** 1389, 1991.

Lee, P.W., Omstead, T.R., McKenna, D.R., and Jensen, K.F. *J. Crystal Growth* **85** 165, 1987.

Lee, P.W., Omstead, T.R., McKenna, D.R., and Jensen, K.F., *J. Crystal Growth* **93** 134, 1988.

Manghi, F., Del Sole, R., Molinari, E., and Selloni, A., *Surf. Sci* **211/212** 518, 1989.

Martines, T. and Whitehouse, C.R., *Proc. 2nd Int. CBE Conf. (Houston),* 1989.

Mashita, M., Origuchi, S., Shimazu, M., Kamon, K., Mihara, M., and Ishii, M., *J. Crystal Growth* **77** 194, 1986.

Maurel, P., Bove, P., Garcia, J.C., and Razeghi, M., *Semicond. Sci Technol.* **5** 638, 1990.

McIntyre, J.E., and Aspnes, D.E., *Surf. Sci* **24** 417, 1971.

Ohlidal, I. and Lucks, F., *Opt. Acta* **19** 817, 1972.

Olson, J.M. and Kibbler, A., *J. Crystal Growth* **77** 182, 1986.

Palik, E.D., *Handbook of Optical Constants of Solids* (New York: Academic), 1985.

Paulsson, K., Deppert, S., Jeppeson, S., Jonsson, J., Samuelson, L., and Schmidt, P., *J. Crystal Growth* **105** 312, 1990.

Pickering, C., Greef, R., and Hodge, A.M., *Semicond. Sci Technol.* **4** 574, 1989.

Razeghi, M., *The MOCVD Challenge: Volume 1: A Survey of GaInAsP–InP for Photonic and Electronic Applications* (Bristol: Hilger), 1989.

Razeghi, M. and Acher, O., *Proc. NATO Studies 1989,* 1989.

Razeghi, M., Defour, M., Blondeau, R., Omnes, F., Maurel, P., Acher, O., Brillouet, F., Fan, J.C., and Salerno, J., *Appl. Phys. Lett.* **53** 2389, 1988a.

Razeghi, M., Defour, M., Omnes, F., Maurel, P., Bigan, E., Acher, O., Nagle, J., Brillouet, F., and Portal, J.C., *J. Crystal Growth* **93** 776, 1988b.

Razeghi, M., Defour, M., Omnes, F., Maurel, P., Chazelas, J., and Brillouet, F., *Appl. Phys. Lett.* **53** 725, 1988c.

Razeghi, M., Defour, M., Omnes, F., Nagle, J., Maurel, P., Acher, O., Huber, A., and Mijuin, D., *Mater. Res. Soc. Symp. Proc.* **126** 143, 1988d.

Razeghi, M., Machado, A., Koch, S.M., Acher, O., Omnes, F., and Defour, M., *SPIE Conf. (Aachen, 28 October–2 November 1990),* 1990.

Razeghi, M., Omnes, F., Blondeau, R., Maurel, P., Defour. M., Acher, O., Vassilakis, E., Mesquida, G., Fan, J.C., and Salerno, J., *Appl. Phys. Lett.* **65** 4066, 1989.

Richter, W., Kurpas, P., Lückerath, R., and Motzkus, M., *J. Crystal Growth* **107** 13, 1991.

Samuelson, L., Deppert, K., Jeppesen, S., Jonsson, J., Paulsson, G., and Schmidt, P., *J. Crystal Growth* **107** 68, 1991.

Schaffer, W.J., Lind, M.D., Kowalczyk, S.P., and Grant, R.W., *J. Vac Sci Technol* B **1** 688, 1983.

Sell, D.D., *Surf. Sci* **37** 876, 1973.

Studna, A.A., Aspnes, D.E., Florez, L.T., Wilkens, B.J., Harbison, J.P., and Ryan, R.E., *J Vac. Sci Technol.* **A 7** 3291, 1989.

Sugiyama, K., *J. Crystal Growth* **75** 435, 1986.

Tsang, W.T., Chiu, T.H., Cunningham, J.E., and Robertson, A., *Appl. Phys. Lett.* **50** 1376, 1987.

Uzunoglu, N., Evans, B.G., and Holt, A.R., *Electron. Lett.* **12** 312, 1976.

4. *Ex situ* Characterization Techniques

4.1.   **Introduction**
4.2.   **Chemical bevel revelation**
    4.2.1.   Principle of magnification
    4.2.2.   Description
    4.2.3.   Bath revelation
4.3.   **Deep-level transient spectroscopy**
    4.3.1.   Deep levels
    4.3.2.   Generation of capacitance transients
    4.3.3.   Determination of trap parameters
4.4.   **X-ray diffraction**
    4.4.1.   Introduction
    4.4.2.   Bragg's law
    4.4.3.   Bond's method
    4.4.4.   Simple analysis of x-ray diffraction
    4.4.5.   Dynamical x-ray diffraction theory simulation
    4.4.6.   Structural characterization of GaAs–GaInP superlattices
4.5.   **Photoluminescence**
    4.5.1.   Fourier transform spectroscopy
    4.5.2.   Recombination processes
4.6.   **Electromechanical capacitance-voltage and photovoltage spectroscopy**
    4.6.1.   Electrochemical capacitance–voltage measurement
    4.6.2.   Photovoltage spectroscopy
4.7.   **Resistivity and Hall measurement**
    4.7.1.   Resistivity measurement
    4.7.2.   Hall measurement
4.8.   **Thickness measurement**
    4.8.1.   Ball polishing
    4.8.2.   Revelation with RCA solution
    4.8.3.   Rinsing procedure
    4.8.4.   Theory behind the thickness measurement
    4.8.5.   Other thickness measurement techniques

4.1. Introduction

Ex situ measurements are very important to characterize material quality for predicting device performance and for optimizing growth conditions. This chapter introduces the most commonly used *ex situ* characterization techniques for routine measurement. Chemical bevel revelation and DLTS are used to obtain information about defects in the material. X-ray diffraction is a powerful tool for structural characterization. Photoluminescence is an optical characterization technique which provides information about optical properties such as bandgap energy, quantum-well energies, and deep-level energies. Electrochemical capacitance–voltage, photovoltage spectroscopy and Hall measurements can be used to obtain information about electrical properties, such as carrier concentration profiles, mobility and compensation rates. Various thickness-measurement techniques can be used to measure the thickness of epilayers or steps.

4.2. Chemical bevel revelation

The bevel revelation technique is a structural characterization tool developed for interface characterization of heterostructures and multilayers. It performs this task by magnifying and revealing the defect distribution of any grown layer using a combination of apparatus, that can create a bevel of a fraction of a degree, and chemical revelation baths. The system is simple to use and quick enough that it is viable as a standard characterization to complement Hall, photoluminescence and x-ray diffraction measurements.

The bevel revelation technique is capable of tracing any buildup of dislocations to its origins as well as detecting crystal breakdown through mismatch or other growth parameters. It is also used for determining substrate problems from polishing or cleaning deficiencies.

4.2.1. Principle of magnification

The idea behind magnification is that by creating a bevel, the effective distance between interfaces can be increased. As an example, if we take two layers of unit thickness, the distance between interfaces is increased by a factor of $\sqrt{2}$ and two by making a bevel of angle 45° and 30° respectively, as shown in Fig. 4.1.

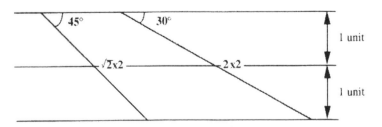

Fig 4 1. Principle of magnification in bevel revelation.

Taking this to the limit, angles of 0.005° or less, which equate to a magnification factor of 10000, can be achieved. This makes investigation of the fine layers much easier. To produce such slight gradients, a chemical etch is necessary since mechanical polishing will introduce extrinsic defects.

4.2.2. Description

These fine bevels are achieved using a special apparatus as shown in Fig. 4.2. The apparatus consists of a simple etch system with a sophisticated gas controlling system. Two vertically mounted, gas-tight Pyrex tubes are connected by a three-way tap capable of connecting either chamber with a drain or connecting the two chambers directly. A precisely regulated gas pumping system is used to pump the solutions from the right to the left chamber at a constant rate. The constant rate is assured by use of a mass flow controller. This is complemented by a switching manifold comprising pneumatic three-way valves, which are able to very rapidly reverse or stop the gas feed, minimizing the changeover time when the flow direction is reversed. The pumping gas is nitrogen which assures that sample oxidation is kept to a minimum.

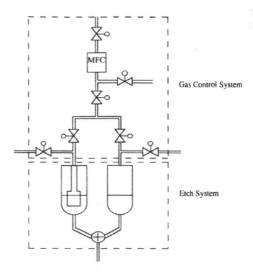

Fig. 4.2. Chemical bevel apparatus.

Prior to carrying out the experiment, it is sometimes necessary to put the samples through rigorous cleaning and deoxidation procedures for reproducible results. If a sample has an oxide layer on its surface the chemical attack will be uneven and unpredictable. As a result it gives a meaningless result.

Bromine/methanol solution is used as the etching solution for most III–V semiconductors and is placed in the chamber on the right. The sample is vertically attached on the holder of the chamber on the left above a protective layer of methanol. A protective resin is used to attach the sample and also cover half of its surface so as to leave a reference layer. The sample can be as small as 1 cm in length and 4 mm in width. The ability to use such small sized samples is an important factor in the basic research, where growth runs are made using substrate fragments of not greater than 10 cm^2 on which four or five destructive or non-destructive characterizations will be made and possibly devices processed.

By pumping the bromine solution in and out of the left-hand reservoir, a bevel is formed as a result of the bromine solution being in contact with the sample for longer at its base than at its head. A typical bevel of constant gradient with a parallel reference level is shown in Fig. 4.3. In this condition, the different refractive indices of the III–V compounds make it possible to see clearly the material interfaces. Changes in material appear as changes in brightness.

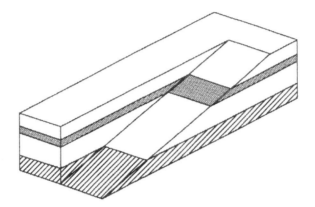

Fig. 4.3. A typical beveled sample with a reference layer.

The angle of the bevel can be controlled by varying the concentration of bromine/methanol solution and pumping speed. An increased concentration of bromine will produce a stronger attack and a greater bevel angle (smaller magnification). On the other hand, the pumping speed works in the opposite sense in that an increase in speed produces a shallower attack since the sample would remain in the dissolution mixture for a shorter time (larger magnification). Of the two, the pumping speed offers a more precise control of the bevel angle, mainly because the mass flow controller gives a very precise control of the pumping speed.

The etched samples are then studied under optical microscopes for more detail. Two microscopes are generally used to carry out the investigation. An ordinary contrast microscope is used when differentiating areas of different brightness and obvious features on the surface such as steps. A Nomarski-type microscope is used in situations where greater sensitivity to surface texture changes is required. A typical magnification is 250, which coupled with the bevel magnification will give an effective magnification factor in excess of a million. One advantage of this technique is that, by adjusting the magnification, an overall view as well as a small part of the material can be examined.

4.2.3. Bath revelation

Defects are revealed using a variety of baths and etching conditions. Revelation baths have a preferential rate of attack for certain crystal faces/planes. If a crystal structure is subjected to an environment which removes atoms from its surface, then the rate of removal around the defect may be different from that of the perfect surrounding matrix. The possible causes of this difference can be attributed to lattice distortion and strain field of the defect geometry of planes associated with the defect and the concentration of impurity atoms at the dislocations and defects, which change the chemical composition of the material in that region.

The type of etch feature formed depends on the attack of the bath. For those where material removal in the defects is more rapid than in the surroundings, etch pits result, while if etching is slower in the defects, hillocks will become apparent. Therefore, the defects are prominent and the level of dislocations due to the growth conditions or substrate deficiencies can be assessed. In addition, each bath requires that a certain thickness of the layer is etched away before these growth features are revealed. This thickness varies with the material to be characterized and the bath used. Through the use of a variety of baths and etching conditions, the weakness in the system can be reduced and even the finest layers can be characterized.

The baths currently being used most widely are AB, H, L and DSL baths. The properties and composition of these baths are as follows:

- AB bath [Abrahams and Buiocchi 1965] is a chemical mixture between $H_2O/AgNO_3/HF$ and H_2O/CrO_3. It preferentially attacks GaAs material but also etches related compounds at a slower rate. It is, however, not particularly suitable for the purpose of GaAs materials because of its high etch rate and the need for a strong light source for preferential etching.

- DSL bath [Weyher and Van de Ven 1983, 1986, 1988], a $HF/CrO_3/H_2O$ mixture, is being widely used as the replacement bath to AB.

- H bath [Huber et al. 1984], which consists of one volume of HBr and two volumes of H_3PO_4, attacks only InP and as it has an etch rate of about 1 μm min^{-1}, it can make successful revelations of very thin layers [Huber et al. 1983].

- L bath [Lourenco 1984], consisting of 200 ml H_2O, 16 g KOH, and 1 g $K_3Fe(CN)_6$, is a relatively recent bath which preferentially attacks (1 1 0) GaInAsP. It has also been found to be effective for InP, GaInAs and GaInP revelation. Due to its very low etch rate and strong preferentiality, when used under illumination, it is ideal for this characterization.

For all but H bath, a strong white light illumination is used to promote the preferential nature of the bath allowing the same degree of revelation with a much smaller attack.

Fig. 4.4 shows the final result of a GaAs/GaInP sample after using chemical bevel revelation. For this specific experiment, DSL bath was used for 4 min. A detailed defect distribution profile is obtained and interfaces are easily distinguishable. Fig. 4.4 also shows the overall view of the sample, but with higher optical magnification a small specific part of the sample can also be studied.

Such powerful, flexible and yet simple qualities of the bevel revelation technique have led to many research uses. Huber et al. [1987] used the technique to study the influence of As content on InP layer perfection. From the experiment, it was concluded that improvement of crystal quality seems to be critical in the case of LP-MOCVD if the As concentration is above 3×10^{19} cm^{-3}. Huber et al. [1984] also assessed the substrate quality by studying the depth of polishing damage as well as studying the quality of several quantum well (MQW) and superlattice (SL) structures such as GaInAs–InP MQW and Si δ-doped GaAs–AlGaAs SL. In the experiment with Si δ-doped GaAs–AlGaAs SL it was observed that the generation of some new defects occurs particularly in the δ-doped region of GaAs. This result was attributed to the formation of a small Si lattice when As vapor pressure is in excess, resulting in a locally disordered crystal.

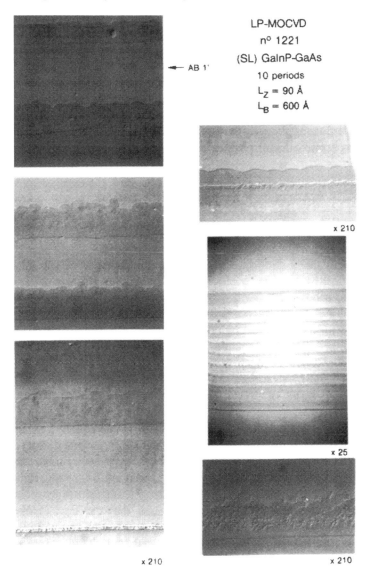

Fig. 4.4. View of GaAs–GaInP structure following chemical bevel revelation.

4.3. Deep-level transient spectroscopy

Deep levels play a vital role in semiconductor technology. They may be intentionally added to control material and device characteristics, but in many cases they are undesirable defects. Deep-level transient spectroscopy (DLTS), originally proposed by Lang [1974], is a capacitance transient thermal scanning characterization method used to identify such deep levels in semiconductors. The technique exploits the properties of a potential barrier such as a Schottky diode or *p–n* junction for the detection of these deep levels and it is on this fact that the high sensitivity and flexibility of DLTS is based.

DLTS displays the spectrum of traps as a function of temperature and this spectrum is used to extract information on energy, density and thermal emission properties of traps. The presence of each trap is indicated by a positive or negative peak as a function of temperature. The sign of each peak determines whether it is due to a majority- or minority-carrier trap and the height of the peaks indicates the respective trap concentration. The positions of the peaks are determined by the integrator gate settings and the thermal emission properties of the traps. With a suitable choice of experimental parameters, the thermal emission rate, activation energy, concentration profile, and capture rate of each trap can be measured. In addition, it is also capable of detecting both radiative and non-radiative recombination centers over a wide range of depths within the bandgap.

In the following section, the basic concepts of the DLTS technique and the underlying semiconductor physics are presented.

4.3.1. Deep levels

The electrical characteristics of semiconductors are controlled by foreign atoms or crystal defects sited within the regular semiconductor lattice. On a representative energy level diagram these atoms form discrete energy levels positioned in the forbidden bandgap. A certain amount of energy has to be gained or lost by a carrier positioned in a deep level in order to be excited to or from the bands.

Commonly, dopant atoms are intentionally added in order to make the semiconductor more conductive. These dopant atoms give rise to shallow levels positioned near to a band edge (within 0.1 eV) and are readily thermally ionized. The origin of deep levels is more complicated. They lie between 0.1 eV and mid-bandgap and are due to impurities and defects. Defect deep levels are the result of crystal imperfections, such as dislocations, stacking faults, precipitates, vacancies or interstitials. These imperfections may arise from: dopant atoms positioned substitutionally or interstitially, vacancies, host atoms on the wrong site in compound semiconductors (antisite defect) or damage induced by irradiation or ion-implantation. Impurity atoms may be contaminants from material growth or processing, oxygen or transition metals, for instance.

Deep levels may behave as carrier traps or as generation–recombination centers if they are near mid-bandgap. As traps, they can capture free carriers, thus compensating the shallow levels and reducing the effective doping density. This increases the resistivity of the material and could therefore have a detrimental effect. However, for very high levels of compensation the material acquires intrinsic-like properties. For example, III–V semiconductors may have resistivities greater than 10^7 $\Omega\cdot$cm and are said to be semi-insulating. Semi-insulating materials are essential for device substrates and therefore deep levels are deliberately introduced occasionally: chromium atoms are added to GaAs and Fe atoms are added to InP for this purpose.

Deep levels behaving as recombination centers may also be either beneficial or detrimental. They provide a path for the generation and recombination of electron–hole pairs across the entire bandgap. The performance of devices such as light emitting diodes (LEDs) and lasers requires radiative recombination; however, this process is degraded when deep levels form a parallel non-radiative recombination path. On the other hand, such paths also provide a means of controlling minority carrier lifetimes. Gold, for example, is commonly intentionally added to silicon for this purpose. In these cases, the deep levels are desirable and beneficial.

Deep levels may be characterized by three properties: the activation energy (E_T) which is related to the position of the level in the bandgap, its concentration (N_T) and its capture cross-section (σ) which provides a measure of the ability of the deep level to trap carriers. DLTS measures all of these properties directly.

4.3.2. Generation of capacitance transients

In order to obtain information about an impurity level in a depletion region, the capacitance transient, associated with the transition from thermal equilibrium of the trap level occupation following an initial non-equilibrium condition, is used. The change in capacitance is a result of applying a bias pulse to introduce carriers and therefore change the electron occupation of a trap from the steady-state condition. As the system returns to its equilibrium, the capacitance returns to its quiescent value.

The change in capacitance transient can be explained in more detail using an example involving a metal–semiconductor (Schottky) barrier. When a metal comes into contact with a semiconductor to form a Schottky barrier there is a transfer of free charge from the region of semiconductor next to the barrier to regions away from it. This results in a depletion (space charge) region, W, which is given by

Eq. (4.1) $$W = \sqrt{\frac{2\varepsilon(V_D - V)}{qN}}$$

where ε is the permittivity of the semiconductor, q the electronic charge, V_D the flat-band potential, V the applied potential across the barrier and N the effective doping density which should equal the shallow-level concentration compensated by any deep levels present.

The depletion region acts as a parallel plate capacitor of capacitance C given by

Eq. (4.2) $C = \varepsilon A / W$

where A is the area of the Schottky contact. The width of the depletion region and hence its capacitance as a function of the applied bias V is then given by

Eq. (4.3) $$C = \sqrt{\frac{q\varepsilon N A^2}{2(V_D - V)}}$$

Fig. 4.5(*a*) shows a Schottky barrier under reverse biased quiescent state. The Fermi level, the statistical energy level which governs the probability of any other level being empty or full, remains flat across the barrier. The semiconductor is assumed to be an *n*-type semiconductor with a single deep level behaving as a donor electron trap. The trap is empty and positively charged above the Fermi level and full and neutral below the Fermi level. The transient is produced in a controlled manner by applying a forward bias V_F to fill intentionally all the traps through the capture process shown in Fig. 4.5(*b*). The forward bias decreases the depletion width and so increases the capacitance as given by Eq. (4.2). When a reverse bias V_R is then applied, the depletion width instantaneously readjusts back to the quiescent width and as a result the capacitance decreases. In this new position some of the full deep levels find themselves above the Fermi level but the trapped electrons do not emit instantaneously because the response time of trapped electrons in the deep level is relatively slow and very temperature sensitive unlike free electrons. Therefore, the capacitance does not immediately return to its original value. The trapped electrons above the Fermi level are released by thermal emission at temperatures where sufficient excitation energy is present as shown in Fig. 4.5(*c*). The steady-state condition is reached when all deep levels above the Fermi level have lost their electrons. Fig. 4.5(*d*) shows the variation of the observed capacitance in Fig. 4.5(*a*), (*b*), and (*c*).

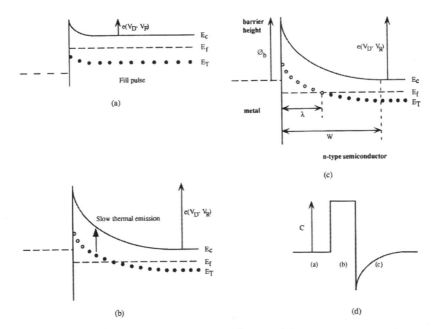

Fig. 4.5. Capacitance variation corresponding to: (a) quiescent reverse bias V_R applied across the Schottky barrier, (b) forward bias pulse V_F fills all the traps and (c) reverse bias pulse V_R slowly emits the trapped electrons in the deep levels.

The usual method for measuring the capacitance transient is the boxcar technique. The transient is measured conventionally between two sampling points at times t_1 and t_2. The time period between the sampling points is referred to as the rate window. Because deep-level emission is thermally activated, the time constant of the transient will change with temperature. As a result, as the temperature increases, the shape of the transient changes in a manner characteristic of a particular trap. ΔC will be small at high temperature since all the traps are readily ionized and small at low temperatures since few will be ionized. However, when the rate window matches the maximum thermal emission rate of the deep level, ΔC will be at a maximum. Hence an output of ΔC against temperature will show a DLTS peak (Fig. 4.6), which is characteristic of the trap being studied. The peak occurs at

Eq. (4.4) $$\frac{1}{e_n} = \frac{t_1 - t_2}{\ln(t_1 / t_2)}$$

Fig. 4.7 and Fig. 4.8 also illustrate, in four steps, how the capacitance transients vary between a majority- and minority-carrier trap. For simplicity of explanation, only those traps in the low-doped p side of an asymmetric n^+p diode will be considered and in both figures the band bending due to the junction electric field is omitted. Furthermore, only one emission rate will be considered for a given trap since for most centers one emission rate usually dominates. Generally, the electron emission rate for trap levels in the upper half of the bandgap is much higher than the hole emission, and similarly the hole emission rate is much higher than the electron emission rate for trap levels in the lower half of the bandgap.

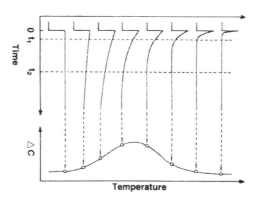

Fig. 4.6. The temperature development of a capacitance transient observed through a rate window produces a DLTS peak (after Lang [1974]).

Fig. 4.7 shows an electron (minority-carrier) trap, which is normally empty of electrons and thus capable of capturing them. Step I shows the reverse-biased quiescent state of the diode and in step II, an injection pulse is applied which momentarily injects electrons into the region of observation. At the same time, the injection pulse allows filling of the electron trap through the capture process.

After a reverse-biased pulse is applied, as in step III, the trapped electrons are released by thermal emission as shown in step IV. The steady-state condition is reached when all deep levels within the depletion region have lost their electrons. Fig. 4.7 also shows the typical time dependence involved in pulsed bias capacitance transients for minority-carrier traps.

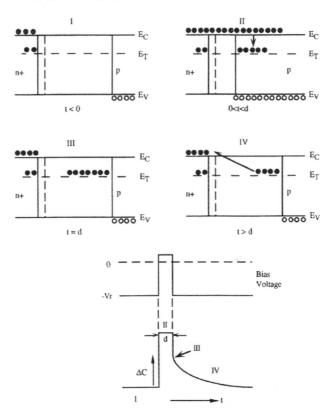

Fig. 4.7. Schematic summary of the emission and capture processes of a electron (minority-carrier) trap in p-type material and the resulting capacitance transient.

Fig. 4.8 shows the analogous majority-carrier pulse sequence for a hole (majority-carrier) trap. A hole trap is a trap which is normally full of electrons and is capable of capturing holes, i.e. recombination. The four processes previously discussed are shown. In this case, the pulse momentarily reduces the diode bias and introduces only holes (majority carriers) into the region of observation. The appropriate time dependence involved in pulse bias capacitance transients for majority-carrier traps is also shown in Fig. 4.8. It is therefore possible to distinguish from the capacitance transients whether the deep level is a minority- or a majority-carrier trap.

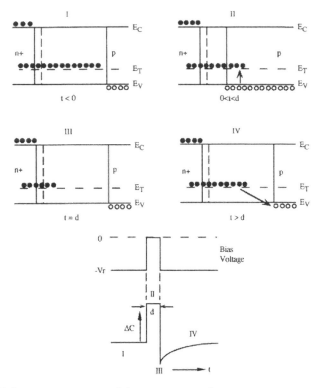

Fig. 4.8. Schematic summary of the emission and capture processes of a hole (majority-carrier) trap in p-type material and the resulting capacitance transient.

4.3.3. Determination of trap parameters

As mentioned before, the DLTS is capable of determining a number of trap parameters such as activation energy, concentration of traps, capture cross-section and capture rates. These are very important parameters in determining the quality of a material and performance of a device. Therefore, the principle behind the use of DLTS for determining these specific parameters will be covered in the following section.

(a) Trap activation energy

The most unique and essential quality of DLTS is the strong temperature dependence of the thermal emission rates. DLTS can observe a response peak of a transient only at the temperature where the trap emission rate is within an emission rate window. The transients are processed with the help of a signal average to provide an output signal, as a function of temperature.

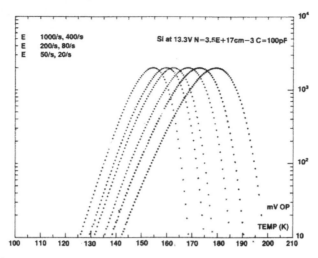

Fig. 4.9. An example of DLTS peaks acquired from a silicon sample.

To determine the energy level, different emission rate windows are selected for each thermal scan which moves the peak to different temperatures. Fig. 4.9 shows these multiple DLTS peaks of a silicon sample using a Polaron 4600 DLTS system. The emission rate is proportional to a Boltzmann factor k, and thus depends exponentially on the energy difference between the trap level and the conduction band, in the case of electron emission, and the trap level and the valence band, in the case of hole emission. The exact emission rate is given by

$$\text{Eq. (4.5)} \qquad e = \frac{N_0 \sigma V_{th}}{g} \exp\left(\frac{\Delta E}{kT}\right)$$

where N_0 is the effective density of states in the band associated with the trapped carriers, σ is the capture cross-section, V_{th} is the thermal velocity and g is the degeneracy of the level. If σ is independent of temperature, then the activation energy ΔE represents the energy of the trap from the band edge to which carriers are emitted.

In Eq. (4.5) the product $N_0 V_{th}$ is proportional to T^2, therefore the peak shift is used to construct an Arrhenius plot of $\ln(e/T^2)$ versus $1/T$ to calculate the activation energy from the slope of the plot. The activation energy ΔE is used to identify the trap present. ΔE can be viewed as the deep level's "label." However, there is no theoretical method of relating measured activation energies to any particular deep level. Therefore, in most cases, the deep level is identified by referring to published data. Fig. 4.10 shows an Arrhenius plot of the silicon sample of Fig. 4.9 and its comparison with the library of known deep levels.

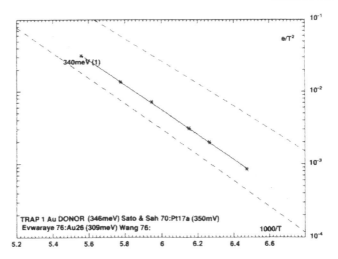

Fig. 4.10. The activation plot of the peaks in Fig. 4.9 and its comparison with the library of known Deep levels

(b) Trap concentration

The concentration of the trap (N_T) can be obtained directly from the capacitance change corresponding to completely filling the trap with a large enough pulse. The relationship for an electron trap in an n^+–p step junction is

Eq. (4.6) $$N_T = \frac{2\Delta C}{C}\left(N_A - N_D\right)$$

- where N_T is the trap concentration, ΔC is the capacitance change due to a saturating injection pulse, C is the capacitance of the diode under quiescent reverse-biased conditions and $N_A - N_D$ is the net acceptor concentration on the p side of the junction where the trap is observed. It should be stressed that these relationships apply accurately only under the following conditions:
- the trap and the doping density are uniformly distributed, with a field-independent emission rate
- only a single emission process occurs at the deep level
- $N_T \ll N_A - N_D$

(c) Capture cross-section

The third important parameter to be derived is the capture cross-section of the traps, σ_n or σ_p. The widespread practice is to extrapolate the Arrhenius plot to $T = \infty$ and calculate σ_n from the y-axis intercept. However, this method can give an apparent capture cross-section which is normally very different from the true cross-section.

Taking a thermodynamic approach, the effective capture cross-section $\sigma_{n,\text{eff}}$ can be given by [Schroder 1990]

Eq. (4.7) $\qquad \sigma_{n,\text{eff}} = \sigma_n X_n$

where σ_n is the true capture cross-section and X_n is defined as

Eq. (4.8) $\qquad X_n = \left(\dfrac{g_0}{g_1} \right) \exp\left(\dfrac{\Delta S_{na}}{k} \right)$

where g_0 is the degeneracy of the deep level unoccupied by an electron, g_1 is the degeneracy of the deep level occupied by one electron and ΔS_{na} is the entropy change due to atomic vibrational changes. As a result, effective cross-sections larger by factors of 50 or more than true cross-sections are not uncommon [Lang et al. 1980]. Further inaccuracies are introduced if σ is temperature or electric-field dependent. A good discussion of energy levels, enthalpies, entropies, capture cross-section, etc, can be found in the work of Lang et al. [1980] and further thermodynamic derivations can be found in the work of Thurmond [1975] and Van Vechten and Thurmond [1976].

A more accurate method of determining the capture cross-section is through the filling pulse method. The idea behind this method is that the capture rate varies depending on σ_n. If a Schottky barrier is under forward bias, all the majority-carrier traps are available for filling but for a certain time of the forward pulse only some of the traps will actually fill. Traps with a large capture cross-section are more likely to fill within a short pulse than those with a small capture cross-section.

To calculate σ_n the standard DLTS temperature scan is repeated at a fixed rate window, but with a varying fill pulse width. As the pulse width increases, the peak height also increases until it eventually reaches a maximum. At this point all the traps are able to fill within one pulse length (saturation). The final DLTS plot consists of peaks displaced vertically. In this manner the capture cross-section is determined from a capture, not an emission, process. The measurement is more difficult to implement because capture times are much shorter than emission times and the instrumentation is more demanding.

Standard DLTS processing is performed on the transient. The peak height is related to the fill pulse by

Eq. (4.9) $\qquad (1 - f) = \exp(-t / \tau)$

where f is the peak height for one pulse width/saturated peak height, τ is the characteristic filling time and t is the pulse width. The Capture cross-section is then calculated from the slope of $\ln(1 - f)$ plotted against t since the slope is $1/\tau$ and

Eq. (4.10) $\qquad \sigma_n = 1 / \tau V_{th} n$

where n is the effective doping density.

(d) Capture rate

Using the same filling pulse method as used for capture cross-section, the capture rate can be determined. The capture process into an initially empty trap is described by Miller et al. [1977]

Eq. (4.11) $N(t) = N_T \left[1 - \exp(-ct) \right]$

where $N(t)$ is the density of traps filled by a bias pulse of width t, N_T is the total trap density and c is the capture rate of the trap. $N(t)$ values are obtained from the peak heights at different widths of the filling pulse. N_T corresponds to the pulse width which saturates all the traps. The capture rate is then obtained from the plot of $\ln[(N_T - N(t))/N_T]$ against t.

4.4. X-ray diffraction

4.4.1. Introduction

X-ray diffraction is a well established and commonly used analytical tool for the non-destructive characterization of semiconductor structure. In principle, the interpretation of measured diffraction patterns is simple. Angular positions of the Bragg peaks of the epilayers and substrates are related to the lattice constant of the material. The alloy composition of the ternary alloy can be determined from the lattice constant by using Vegard's law.

An x-ray diffraction measurement, of course, can reveal a great deal of information about the structure of a compound semiconductor such as III–V or II–VI materials. In particular, the interference fringe structure accompanying the Bragg peak from a thin epilayer permits measurement of the layer thickness with an accuracy of one or two percent, depending on the quality of the material itself and the precision of the diffraction data. In this section we describe related theories behind x-ray diffraction and the simple analysis used for interpretation of the measured diffraction data; moreover, a dynamical x-ray diffraction simulation method based on the solution of Tagaki–Taupin equations is presented. An example of the characterization of superlattices is also included in the last part to describe the importance of dynamical simulation.

4.4.2. Bragg's law

Due to W. H. Bragg and W. L. Bragg [1913a, b] an improved experimental technique contributed greatly to the development of knowledge in the fields of the wave nature of x-rays and the periodic structure of crystals.

Crystal lattices can be modeled as three-dimensional gratings. A linear diffraction grating can be regarded as a special case of a one-dimensional grating. Based on this point of view a three-dimensional grating may be defined as a spatial distribution of matter for which the scattering power is a periodic function of three-dimensional space, i.e., $\Psi(r) = \Psi\ (r + L_1 a_1 + L_2 a_2 + L_3 a_3)$, where L_1, L_2 and L_3 are integers. With grating spaces a_1, a_2 and a_3, which are the periods of the different axes, the three-dimensional grating can thus be considered as consisting of three sets of one-dimensional linear gratings. To find the diffraction maximum according to the three-dimensional case, the Laue vector equations are employed as follows for each of the one-dimensional linear gratings:

$$a_1 \cdot \left(k_{H_1 H_2 H_3} - k_0 \right) = H_1$$

Eq. (4.12) $\qquad a_2 \cdot \left(k_{H_1 H_2 H_3} - k_0 \right) = H_2$

$$a_3 \cdot \left(k_{H_1 H_2 H_3} - k_0 \right) = H_3$$

where k_0 is the wavevector of the incident x-ray beam and $k_{H_1 H_2 H_3}$ is that of the diffracted beam. The quantities H_1, H_2 and H_3, being integers, are Miller indices of a family of lattice planes and are associated with each diffraction maximum. Eq. (4.12) can be further simplified as a vector equation (Laue vector equation):

Eq. (4.13) $\qquad k_H - k_0 = B_H$

In Eq. (4.13), the vector $B_H = B_{H_1 H_2 H_3} = H_1 b_1 + H_2 b_2 + H_3 b_3$ is normal to the $(H_1 H_2 H_3)$ planes and its length is equal to the reciprocal of the spacing between planes, where b_1, b_2, b_3 is the reciprocal unit vector set of a_1, a_2, a_3. Since $|k_H| = |k_0| = 1/\lambda$, the Laue vector equation expresses the fact that the only diffracted beam (k_H) formed is that shown in Fig. 4.11, namely one making a reflected angle θ equal to the incident angle θ. In other words, one may consider the diffracted beam to be produced by a reflection of the incident beam in the family of planes normal to B_H. Equating the magnitude of the two sides of Eq. (4.13), the Bragg equation can be derived

Eq. (4.14) $\qquad 2 \sin \theta_B / \lambda = 1 / d_H = n / d_h$

where θ_B is thus complementary to the angle of incidence or reflection and is called the Bragg glancing angle. The magnitude of the right-hand side of Eq. (4.13) is $1 / d_{H_1 H_2 H_3}$ or $n / d_{h_1 h_2 h_3}$ when H_1, H_2, H_3 have a common integral factor n, and h_1, h_2, h_3 are prime numbers and $d_{h_1 h_2 h_3}$, is the spacing between two consecutive planes between $(h_1 h_2 h_3)$ planes. The wavelength of the x-ray is of the order of 10^{-8} cm, whereas the smallest lattice period lies in the range $10^{-8} \sim 10^{-5}$ cm. Accordingly, it may be expected that the scattering angles for the diffraction maxima are large enough to be measured.

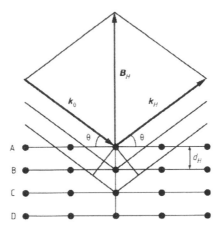

Fig. 4.11. Diffraction of x-rays by a section of a crystal with lattice planes spaced a distance d_H apart. The atoms are arranged on a set of parallel planes A, B, C,... normal to the plane of the drawing.

4.4.3. Bond's method

The Bond method [Bond 1960] presents a way to determine the lattice constant by measuring the Bragg angles at two complementary positions (180° azimuthal angle difference), which cancel the errors due to the misorientation of the sample and the x-ray systems. Lattice constants can be determined up to a few parts in a million for highly perfect crystals. This accuracy is comparable to the accuracy with which wavelengths of x-rays were determined. In the measurement, the axis of crystal rotation (θ) must be perpendicular to the plane of incidence within a few arc seconds. This alignment criterion may be discussed in terms of the non-orthogonality of the incident beam with the crystal rotation axis θ (or ω), called the beam tilt, and any lack of parallelism of the reflecting planes with the crystal rotation axis, called crystal tilt.

The main errors which may be caused by the system itself and non-optimized measurements include:

- angle reading error
- eccentricity error
- absorption error
- crystal tilt error
- axial divergence error (or beam tilt error).

Fig. 4.12 shows the spatial configuration of the diffractometer and various movements. Based on Bond's description, optimization is necessary to bring the scattering vector into the scattering plane. The scattering vector of a particular set of lattice planes is perpendicular to these planes. In the Bragg condition (Eq. (4.14)), the scattering vector bisects the incident and diffracted beams. The scattering plane is in a plane which is perpendicular to the goniometer rotation axis (see Fig. 4.13). Non-optimization results in a serious drop in intensity and resolution. Rotating the sample around Phi or alternatively around R can bring the scattering vector into the scattering plane. The better the alignment the more accurate the measurement.

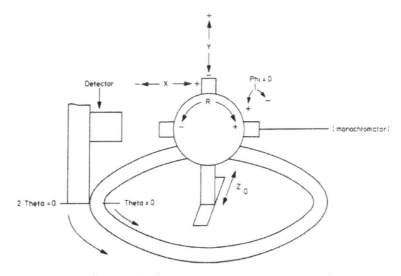

Fig. 4.12. Spatial configuration of the horizontal goniometer.

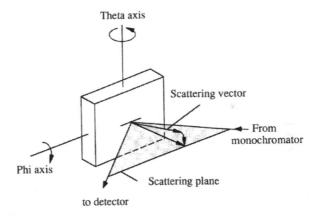

Fig. 4.13. Scattering geometry a crystal with diffraction planes roughly parallel to the surface.

4.4.4. Simple analysis of x-ray diffraction

(a) Interface quality determination

The diffraction patterns for the accurate determination of alloy composition or lattice mismatch are usually obtained on a x-ray diffractometer. The process of measuring a lattice parameter is a very indirect one, and is fortunately of such a nature that high precision is fairly easily obtainable by x-ray diffraction. The parameter a of a cubic structure substance is directly proportional to the spacing d of any particular set of lattice planes. If we measure the Bragg angle θ_B for this set of planes, we can use the Bragg law to determine d and, knowing d, we can calculate a.

The angular separation of the peaks can be related to the lattice mismatch through the differential form of the Bragg equation with respect to θ. For symmetric geometry, which is the case where the diffraction planes are parallel to the sample surface, we obtain

Eq. (4.15) $(\Delta a / a)_{\perp} = -\cot \theta_B \Delta \theta$

$(\Delta a/a)_{\perp}$ represents the lattice strain perpendicular to the crystal surface, which is related to the relaxed lattice mismatch $(\Delta a/a)_r$ by

Eq. (4.16) $(\Delta a / a)_r = (\Delta a / a)_{\perp} \left[(1-v)/(1+v) \right]$

provided the in-plane strain $(\Delta a/a)_{\parallel}$ parallel to the surface is assumed to be zero and v is Poisson's ratio for the layer material. This relaxed mismatch can then be related to the material composition, such as x in the ternary alloy $Ga_xIn_{1-x}P$, by assuming Vegard's law

Eq. (4.17) $a_{Ga_xIn_{1-x}P} = x a_{GaP} + (1-x) a_{InP}$

By measuring the lattice constant $a_{Ga_xIn_{1-x}P}$, the compositions x can be determined. Conversely, if we know the alloy compositions, the lattice constant of this ternary material can also be obtained precisely. Thus, once the angular separation between two peaks shown in the diffraction pattern is known, information on lattice mismatch, material quality and layer compositions can be acquired.

(b) Pendellosung fringes

The hyperfine structure observed in diffraction patterns near the substrate and/or layer peaks may be attributed to the interaction between incident and reflected waves through the whole epitaxial layer. In the reflection case (Bragg case) of the dynamical diffraction theory, the beating of two wave fields inside a crystal occurs between wave points on the same branch of the dispersion surface (Pendellosung). This type of wave interference is rarely observed and as such is more within the framework of a plane wave, rather than a spherical wave phenomenon. The known relation [Tapfer and Ploog 1986] between the angular spacing of the Pendellosung fringes $\Delta\theta$ and the thickness L of the whole epilayer is

Eq. (4.18) $L = \lambda |\gamma_h| / \Delta \theta \sin(2\theta_B)$

where $\gamma_h = \sin(\theta_B + \alpha)$ and α is the angle between the sample surface and diffraction plane. Eq. (4.18) can be simplified easily for symmetrical geometry

Eq. (4.19) $L = \lambda / (2\Delta\theta \cos\theta_B)$

Eq. (4.18) and Eq. (4.19) are also applicable to the period T of a superlattice, which is a periodic repeat of some layer structure with period T. The whole epilayer thickness is replaced by the superlattice period T, and the oscillation spacing is replaced by the angular distance between the satellite peaks which occur in x-ray diffraction curves from superlattices.

We conclude that, by using Eq. (4.18) or Eq. (4.19), the thickness of not only the superlattice periodicity but also some layer thickness of the multiquantum well can be determined; moreover, the thickness measurement can be carried out from measured symmetrical or asymmetrical diffraction patterns. These patterns can be simulated by using the dynamical x-ray diffraction theory described below.

(c) X-ray reflectivity

The specular x-ray reflectivity, or (0 0 0) Bragg reflection, can be interpreted simply within the framework which was developed for x-ray diffraction. For other Bragg reflection geometries like (0 0 4) and (0 0 2), the role played by the dimensions defining the layer within the layer plane has been ignored, so this implicitly assumes that the layer plane is infinitely extended in these directions. In the case of reflectivity, because of the low-angle geometry, information on the surface and interface roughness [Cowley and Ryan 1987] can be known, and, since this technique is independent of crystal periodicity and unit cells of materials, the dominant role is played by the shape of the area with different electron density as a function of depth. Therefore, x-ray reflectivity can be used to study liquids, amorphous or crystalline films. All that has to be considered is the electron density variation through layers. As mentioned before, although x-ray reflectivity can be explained within the framework developed there are some additional factors that must be considered, especially for superlattice structures, to understand the whole picture of this technique.

(i) X-ray refraction. Snell's law is the traditional approach used to relate the angle of incidence to the angle of refraction of electromagnetic waves travelling through media with different refractive indices n. For x-rays, the refractive index of most materials is slightly less than unity:

Eq. (4.20) $n_m = 1 - \delta - i\beta$

where i, which relates to the x-ray absorption, indicates the imaginary component of a complex number. From Snell's law, if the refractive index of the air/vacuum is taken to be unity, for total external reflection the refractive index n_m of the material is equal to cos θ_c, where θ_c, the incident angle between x-ray and interface, is the critical angle and is related to the density of the material and the wavelength of the x-rays. At θ_c, the penetration of the x-rays into the sample is very limited. Typically the penetration depth will be around 50 Å. As the incident angle is increased, the x-rays rapidly penetrate deeper into the material.

(ii) Effective layer thickness. The layer thickness measured directly from the raw data of a reflectivity curve appears to be thinner than its real thickness, since, due to the x-ray refraction, the effective path length of the x-rays in the material will be increased.

(iii) Scattered radiation effect. Strong interaction (between incident beam and scattered radiation) of the x-rays with the sample will happen within and close to the area where total external reflection occurs. Due to this interaction effect the complete optical theory of reflectivity using Maxwell's equations and the Fresnel formula must be introduced to describe the full behavior of x-ray reflectivity. Generally, within the range of twice the critical angle the strong interaction should be taken into account to avoid making serious measurement errors.

4.4.5. Dynamical x-ray diffraction theory simulation

The general basis and framework of the dynamical theory were developed soon after the discovery of x-ray diffraction in crystals. This theory, treated by Ewald [1916, 1917, 1920, 1921, 1924, 1925, 1958], took into account all wave interactions within the crystal. The measured diffraction pattern must be simulated by means of the dynamical diffraction theory in order to have more accurate and unambiguous information about the structure properties of the sample. As a consequence, the dynamical theory of x-ray diffraction was developed to take into account normal absorption as well as the interaction between incident and scattered radiation.

(a) The dynamical theory of x-ray diffraction

Normal absorption occurs for all directions of incidence while extinction is important only when the incident wavevector has such a value that the Laue vector equation (Eq. (4.13)) is exactly or very nearly fulfilled for one or more reciprocal lattice vectors $\boldsymbol{B}_H$. The incident x-ray wave also suffers changes when it enters the crystal; that is to say, diffracted waves depend upon the internal incident wave which in turn depends upon the diffracted waves and both waves form a coupled system. Hence, the incident wave inside and outside the crystal material can be represented by the expression:

Eq. (4.21) $\boldsymbol{D}_1(\boldsymbol{r}) = \boldsymbol{D}_0(\boldsymbol{r}) \exp[i(\omega_0 t - 2\pi \boldsymbol{\beta}_0 \cdot \boldsymbol{r})]$

where $\boldsymbol{r}$ is a point in the actual space, $\boldsymbol{D}(\boldsymbol{r})$ the displacement vector and $\boldsymbol{\beta}_0 \cdot \boldsymbol{r}$ a plane wave with $\boldsymbol{\beta}_0$ constant. Outside the crystal, i.e. in the vacuum, $\boldsymbol{\beta}_0 = \boldsymbol{k}_0$ and $\boldsymbol{\beta}_0 \cdot \boldsymbol{r}$ and $\boldsymbol{D}_0(\boldsymbol{r})$ are supposed to be real but space dependent, whereas the frequency ω_0 remains constant. On the contrary, inside the crystal the same function $\boldsymbol{\beta}_0 \cdot \boldsymbol{r}$ is maintained and $\boldsymbol{D}_0(\boldsymbol{r})$ is introduced in perturbations originated in the wave by the crystalline medium, i.e., $|\boldsymbol{\beta}_0|^2 = \varepsilon |\boldsymbol{k}_0|^2$, where the dielectric constant ε is a complex quantity and is equal to n_m^2, the square of the refractive index. Therefore, $D_0(r)$ becomes, in general, a complex quantity. On the other hand, the wave corresponding to the sum of incident and diffracted waves in the interior of the crystal may be written as:

Eq. (4.22) $D(r) = \sum_H D_H(r) \exp[i(\omega_0 t - 2\pi\beta_H \cdot r)]$

where $\beta_H = \beta_0 + B_H$ is the wavevector of a diffracted wave in the dynamical theory and the amplitude of this wave is expected to be negligibly small except when the Laue vector equation is satisfied. Obviously, the wave field represented in Eq. (4.22) is a set of coupled plane waves and H can take the value zero with $B_H = 0$. In the dynamical theory, the case in which the incident beam produces just one diffracted beam at a time is specifically treated. In other words, for a reciprocal lattice vector B_H, which lies relatively far away from where the Bragg geometry reflection should be, one may set the amplitude D_H to be zero.

(b) The fundamental equations for the general case

As mentioned earlier, the dielectric constant ε plays an important role in the dynamical theory of x-ray diffraction and may be expressed as follows:

Eq. (4.23) $\varepsilon = 1 + \Psi = 1 + \Psi' + i\Psi''$

The dielectric constant, however, is not a constant but a function of position having the periodicity of the crystal lattice. Since Ψ is periodic, this distribution function may be expressed as a Fourier series

Eq. (4.24) $\Psi(r) = \sum_H \Psi_H \exp(-i2\pi B_H \cdot r)$

where

Eq. (4.25) $\Psi_H = V^{-1} \int_V \Psi(r) \exp(i2\pi B_H \cdot r)\, dV$

and where H can take also the value zero with $B_H = 0$. $\Psi_{0,H} = -(e^2/mc^2)(\lambda^2/\pi V)F_{0,H}$ are the coefficients of Fourier expansion, where $F_{0,H}$ are the structure factors for the incident and diffracted beams and both include absorption, V is the unit cell volume and m the electron mass. Since the dielectric constant in the x-ray region is only slightly different from unity, the condition $\Psi \ll 1$ can be assumed and the refractive index of crystal medium n_m can be expressed as:

Eq. (4.26) $n_m = 1 + \Psi/2 = 1 + (\Psi' + i\Psi'')/2.$

This expression implies that the refractive index n_m becomes a complex and periodic function of position. Also, the imaginary part of Eq. (4.26) indicates that the scattering is accompanied by a true absorption corresponding to the emission of photoelectrons. In addition to the assumption $\Psi \ll 1$, the following conditions are assumed: (i) the magnetic permeability is unity and (ii) the current density is zero. According to the above assumptions, the Maxwell equation gives:

Eq. (4.27) $\nabla \times [\nabla \times (1 - \Psi)\boldsymbol{D}] = -\dfrac{1}{c^2}\dfrac{\partial^2 \boldsymbol{D}}{\partial t^2}$

where $1/\varepsilon$ is set to be $1 - \Psi$ and $\boldsymbol{D}$ represents the sum of incident and diffracted waves and is actually the one expressed in Eq. (4.22). By substituting Eq. (4.22) into Eq. (4.27), an equation with an infinite number of unknowns $\boldsymbol{D}_H$ is obtained. This equation can be further transformed into a system of an infinite number of equations with each one corresponding to the wave diffracted by the planes related to the reciprocal lattice vector $\boldsymbol{B}_H$. Therefore, when the Bragg condition is fulfilled for one reciprocal lattice vector $\boldsymbol{B}_H$ only, the incident wave and one diffracted wave of appreciable amplitude exist.

(c) Diffraction pattern simulation

Simulation of measured diffraction patterns can be accomplished through the use of a computer program developed by P. F. Fewster at Philips [Fewster and Curling 1987], which is based on the solution of the Takagi–Taupin equations [Takagi 1962, 1969, Taupin 1964] derived by the dynamical scattering approach described earlier with the introduction of a quantity $\alpha_H(\omega) = -2(\theta - \theta_B)\sin 2\theta_B$ measuring the deviation from the Bragg condition at each point of the crystal, where ω is the incident angle of the x-rays. The basic equations derived from Eq. (4.27) become

Eq. (4.28)
$$(i\lambda\gamma_H / \pi)(dD_H / dZ) = \Psi_0 D_H + C\Psi_H D_0 - \alpha_H(\omega) D_H$$
$$(i\lambda\gamma_0 / \pi)(dD_0 / dZ) = \Psi_0 D_0 + C\Psi_H D_H$$

where $D_{0,H}$ are complex amplitudes of the incident and diffracted beams, and Z is the depth into the crystal. $\gamma_{0,H} = \boldsymbol{n} \cdot \boldsymbol{k}_{0,H}$ are direction cosines of incident and diffracted beams, where $\boldsymbol{n}$ is the vector normal to the surface and $\boldsymbol{k}_{0,H}$ are wavevectors of the incident and diffracted beams, and λ is the wavelength of the x-ray source. The polarization factor C is unity for σ polarization and $\cos(2\theta_B)$ for π polarization. However, these two Laplacians are normally negligible as compared to the other terms of Eq. (4.28) once the curvature radius of the incident wave is larger than the absorption path through the crystal. Accordingly, if an amplitude ratio $X(z, \omega) = D_H/D_0$ is defined, a differential equation can be derived as follows:

Eq. (4.29) $\dfrac{\partial X}{\partial Z} = \dfrac{i\pi}{\lambda\gamma_0}\left\{\Psi_H X^2 + \left[\left(1 - \dfrac{\gamma_0}{|\gamma_H|}\right)\Psi_0 + \alpha_H(\omega)\dfrac{\gamma_0}{|\gamma_H|}\right]X - \dfrac{\gamma_0}{|\gamma_H|}\Psi_H\right\}$

and a simplified form may be expressed as:

Eq. (4.30) $\quad \partial X / \partial Z = iD\left(AX^2 + 2BX + C\right)$

where $A \simeq C$ for symmetrical reflections, B depends on $\theta - \theta_B$ and A, C, D depend on direction cosines and structure factors. For a layer of constant composition, the solution $X(z, \omega)$ can be obtained:

Eq. (4.31) $\quad X(z, \omega) = \dfrac{\left[Sx(Z,\omega) + i\left(Bx(Z,\omega) + C\right)\tan\left(DS(Z-z)\right)\right]}{\left(S - i\left(Ax(Z,\omega) + B\right)\tan\left(DS(Z-z)\right)\right)}$

with boundary condition $X = x$ at $z = Z$, by solving a standard integral which comes from the integration after separating Eq. (4.29) into partial fractions, where $S^2 = (B^2 - AC)$. Since the diffracted wave deep inside the crystal is almost zero, the solution at the surface of a thick substrate is derived by using the boundary condition $X = 0$ at $Z = \infty$. This amplitude ratio X can then be used as a boundary condition at the bottom of the first heteroepitaxial layer of depth Z and thickness t to calculate $X(Z - t, \omega)$ at the top of the first layer. The procedure is iterated from the first layer until the top of the surface layer is reached. The reflectivity is thus given by:

Eq. (4.32) $\quad R(\omega) = X(0, \omega) X^*(0, \omega)$

and the angular range (from ω_1 to ω_2) for simulation is based on the range used in the experiment. From Eq. (4.32) the calculated diffraction pattern (the fitted data) can be obtained and may be compared to the measured diffraction profile (the raw data). Without using a simulation program, especially for those harmonics extending to several degrees, it is difficult to extract accurate information on structures only from measured data.

4.4.6. Structural characterization of GaAs–GaInP superlattices

Many improvements have been made recently in understanding and controlling the purity, interface quality, optical and electronic properties of the GaAs–GaInP lattice-matched system grown by MOCVD. $Ga_{0.51}In_{0.49}P$ lattice matched to GaAs shows a number of unique and interesting features. Also, high-quality GaAs/$Ga_{0.51}In_{0.49}P$ heterostructures such as superlattices (SL) and multiquantum wells (MQW) have been grown by LP-MOCVD. Here, we describe the characterization of a superlattice structure by a high-resolution x-ray diffraction technique.

To attempt to fully characterize this GaAs/$Ga_{0.51}In_{0.49}P$ sample with high-resolution x-ray diffraction, a set of three Bragg geometries (0 0 4), (0 0 2) and (0 0 0) is selected for detailed study. The (0 0 4) reflection is the most commonly studied reflection for (0 0 1)-oriented III–V structures, being the point where all atomic planes within the sample scatter in phase. As such, the (0 0 4) reflection will be the most intense, and will exhibit features from the whole sample structure. For superlattice materials containing GaAs it is often very revealing to study the (0 0 2) reflection in conjunction with the traditional (0 0 4). At the (0 0 2) point, alternate atomic planes within the sample will scatter out of phase. So for materials like GaAs, where the average electron density of each atomic plane is very similar, the (0 0 2) reflection will be almost absent. In the case of GaAs–GaInP superlattices, this implies that the (0 0 2) reflection will exhibit features predominantly from the GaInP aspect of the superlattice. As an independent validation of any conclusions drawn from the (0 0 2) and (0 0 4) reflections, the x-ray reflectivity (0 0 0) was also measured. The comparison of x-ray reflectivity with x-ray diffraction can often provide vital clues as to the detailed layer structure as each is sensitive to different aspects of the sample. X-ray reflectivity is a probe of macroscopic electron density fluctuations, while x-ray diffraction is a probe of fluctuations in the crystal structure.

(a) Experimental results

The sample studied here is a 10-period GaAs–$Ga_xIn_{1-x}P$ superlattice grown on GaAs(1 0 0) substrate misoriented 2° off axis towards $\langle 0\ 1\ \bar{1} \rangle$ with an approximately 100 Å $Ga_xIn_{1-x}P$ ($x \sim 0.51$) barrier and 90 Å GaAs well [Razeghi and Omnes 1991]. A simple investigation of the raw data as measured for the (0 0 4), (0 0 2) and (0 0 0) reflections can provide lots of information for initial analysis. From the (0 0 4) data in Fig. 4.14(a) we can make a very good estimate of the superlattice period, 151 ± 9 Å, as evaluated from the measured separation of the superlattice satellites. The average superlattice mismatch of approximately -3116 ppm is simply derived from the separation between the substrate and the zero-order superlattice peaks. A closer inspection of the data reveals weak Pendellosung oscillations corresponding to a thickness of 1511 ± 90 Å. This thickness may be associated with the total superlattice thickness, which indicates that 10 periods of superlattice are involved in this structure. On the contrary, the (0 0 2) reflection in Fig. 4.14(b) exhibits a very simple structure, which enables us to extract the average superlattice mismatch, -3323 ppm, and superlattice period, 145 ± 9 Å, to a higher degree of confidence than from the (0 0 4). What is immediately obvious from the data is that the even-order satellites ±2 are missing, while the ±3 satellites are clearly visible. These features indicate that the GaAs and GaInP layers within the superlattice must be of a very similar size (approximately 75 Å). The asymmetry between the ±1 satellites further tells us that the superlattice unit cell is more complex than a simple GaAs/GaInP bilayer structure. According to the fine Pendellosung fringes shown in Fig. 4.14(c) a thickness of 1519 ± 90 Å also indicates that the number of periods is most likely to be 10. Furthermore, the reflectivity data in Fig. 4.14(d) exhibit features similar to the (0 0 4) and (0 0 2). What is most notable is the complete absence of any even-order satellites. This confirms the earlier observation that the GaAs and GaInP layers must be of a similar size, or to be more exact, that the superlattice unit cell is split into two areas of different electron density but very similar thickness of 75 Å. In addition, a strong broadening of the superlattice peaks, most likely resulting from thickness fluctuations, is in evidence as it moves away from the critical angle. The broadening is much more severe in this glancing angle measurement, and may indicate lateral inhomogeneity of the individual layer thicknesses averaged over the large illuminated surface area.

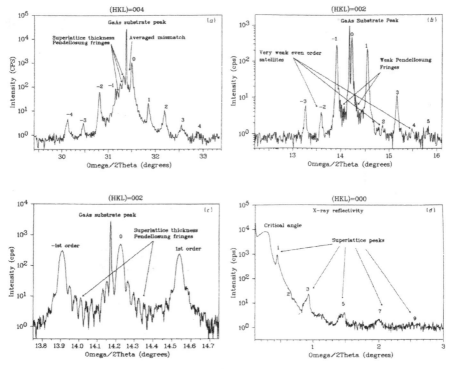

Fig. 4.14. *Measured x-ray diffraction patterns of a 10-period GaAs–Ga$_x$In$_{1-x}$P superlattice for various (h k l) reflections: (a) (0 0 4), (b) (0 0 2), (c) expanded view of (0 0 2), and (d) (0 0 0). The (0 0 0) reflection is measured as x-ray reflectivity.*

(b) Simulation results

Based upon the initial results and the information concerning the growth process, random thickness fluctuations are expected and additional layers may be present in the superlattice to generate the observed asymmetries. One monolayer, 10 Å, of Ga$_y$In$_{1-y}$P is expected to appear between Ga$_x$In$_{1-x}$P and GaAs. This additional layer allows a large range of density and mismatch to be covered with small variations in the y component. In Fig. 4.14(a)–(c), the simulated data, using the dynamical x-ray diffraction theory and the x-ray reflectivityas simulation programs, for the (0 0 4), (0 0 2) and (0 0 0) reflections are compared with the measured data. The agreement between the simulated and measured (0 0 2) reflection is remarkably consistent, including the matched FWHM of the GaInP-layer Bragg peak. All of the observed features are reproduced. In direct contrast, the simulated (0 0 4) data near the substrate peak are not well reproduced. The simulated reflectivity (Fig. 4.15(c)) is in quite close agreement with the measured curve. All of the observed features in the x-ray reflectivity are reproduced including the negative dip where the second-order peak should be. To reduce the satellite intensity in the simulation, the electron density difference between the GaAs and GaInP layer must be reduced. This is achieved by doping the GaAs layer with 15% of P. This does not result in a true representation of reality but in an illustration of the degree of density modification required to model accurately the data. The main difference between the two data sets is in the higher-order peak broadening of the superlattice peaks. This provides the information on random thickness fluctuations of this sample.

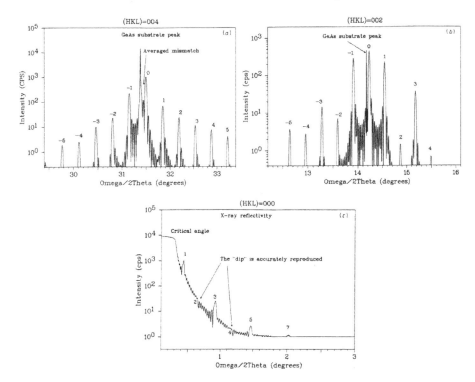

Fig. 4.15. Simulated satellite peak intensity profiles for various (h k l) reflections as a result of the parameter fitting, (a) (0 0 4), (b) (0 0 2), (c) (0 0 0). (a) and (b) are fitted by dynamical x-ray diffraction simulation and (c) is fitted by x-ray reflectivity simulation.

As a consequence, due to the above detailed analysis, the fitting of the simulated data to the measured data gives the closest match to the third-order satellite asymmetry and the GaAs/Ga$_{0.519}$In$_{0.481}$P matched system with a mismatch of $\Delta a/a_0 = -3.22 \times 10^{-4}$ is investigated. Without using the simulation programs, accurate structure parameters cannot be extracted. Information concerning the structure parameters of this sample is shown in Table 4.1.

hkl	Measurement			Simulation		
	004 (ppm)	002 (ppm)	000 (ppm)	004 (ppm)	002 (ppm)	000 (ppm)
$\Delta a/a_0$ ($Ga_xIn_{1-x}P$)	−6232	−6646	–	−322	−322	–
$\Delta a/a_0$ ($Ga_yIn_{1-y}P$)	–	–	–	−0.043	−0.043	–
Period thickness	151 ±9 Å	145 ±9 Å	150 Å	152 Å	144 Å	157 ±2 Å
GaAs thickness	–	–	–	76 Å	72 Å	78 ±2 Å
$Ga_xIn_{1-x}P$ thickness	–	–	–	66 Å	62 Å	73 ±2 Å
X	~0.56	~0.56	–	0.519	0.519	0.519
$Ga_yIn_{1-y}P$ thickness	–	–	–	10 Å	10 Å	5.1 ±0.1 Å
Y	–	–	–	0.82	0.82	0.82
Superlattice thickness	1511 ±90 Å	1519 ±90 Å	–	1520 Å	1440 Å	1570 ±20 Å
Periods	10	10	–	10	10	10
Superlattice FWHM	109 s	92.3 s	–	117 s	99.9 s	–
Surface roughness (GaAs)	–	–	–	–	–	6.7 ±0.1 Å
Surface roughness ($Ga_xIn_{1-x}P$)	–	–	–	–	–	5 Å (7.1 Å top layer)

Table 4.1. Information concerning the structure parameters of samples measured by x-ray diffraction and simulated by dynamical x-ray diffraction theory.

4.5. Photoluminescence

Luminescence is the term used to describe the non-equilibrium emission of radiation. Three processes are involved in the luminescence process: excitation, energy transfer and radiative transition of the electrons. When the light is absorbed as the excitation, the luminescence process is called photoluminescence (PL).

The study of photoluminescence spectroscopy, which is a technique to obtain the relation between the intensity and the wavelength of the emission of radiation, has become one of the most powerful techniques for the comprehensive and non-destructive assessment of materials since it yields a large amount of information about the electronic structure of the material [Bebb and Williams 1972].

Excess electron-hole pairs (EHPs) are generated when a semiconductor is illuminated by light with photon energy larger than the bandgap. The created EHPs will move or transfer their energy in the semiconductors, then recombine through various mechanisms, which can be radiative or non-radiative, such as the cascade phonon process and emission quench at defect centers etc.

The heart of the optical measurement is a spectrometer, which is used to discriminate between emission at different photon energies, resulting in a plot of light intensity as a function of energy. Multibeam interference is often the physical means used to achieve the spectrometer such as various grating monochromators. The dispersive prism is the oldest instrument to separate light into its component wavelengths. However, more and more, the Fourier transformation spectrometer is becoming the key spectrometer.

4.5.1. Fourier transform spectroscopy

The foundations of modern Fourier transform spectroscopy (FTS) were laid in the latter part of the nineteenth century by Michelson. Soon after that, Raleigh recognized the relationship between an interferogram and its spectrum by a Fourier transform. It was not until the advent of reasonably priced computers and the fast Fourier algorithm that interferometry began to be applied to spectroscopic measurements in the 1970s.

The basic optical component of Fourier transform spectrometers is the Michelson interferometer shown in simplified form in Fig. 4.16. Light from an infrared source is collimated and incident on a beam splitter. An ideal beam splitter creates two separate optical paths by reflecting 50% of the incident light and transmitting the remaining 50%. In one path the beam is reflected by a fixed-position mirror back to the beam splitter where it is partially reflected to the source and partially transmitted to the detector. In the other arm of the interferometer, the beam is reflected by the movable mirror that is translated back and forth and maintained parallel to itself. The beam from the movable mirror is also returned to the beam splitter where it too is partially transmitted back to the source and partially reflected to the detector. When the two beams are combined, the interference phenomenon is produced. Typically the movable mirror rides on an air bearing for good stability. Detectors can be a cooled HgCdTe or germanium photodiode, depending on the wavelength of interest.

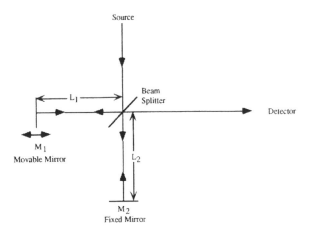

Fig. 4.16. Michelson interferometer.

The light intensity reaching the detector is the sum of the two beams. The two beams are in phase and reinforce each another when $L_1 = L_2$. When M_1 is moved, the optical path lengths are unequal, and an optical path difference d is introduced. If M_1 is moved a distance x, the retardation is $d = 2x$ since the light has to travel an additional distance x to reach the mirror and the same additional distance to reach the beam splitter.

Consider the output signal from the detector when the source emits a single frequency or wavelength. For $L_1 = L_2$ the two beams reinforce each other because they are in phase, $d = 0$, and the detector output is a maximum. If M_1 is moved by $x = \lambda/4$, the retardation becomes $d = 2x = \lambda/2$. The two wavefronts reach the detector 180° out of phase, resulting in destructive interference or zero output. For an additional $\lambda/4$ movement by M_1, $d = 1$ and constructive interference results again. The detector output — the interferogram — consists of a series of maxima and minima. The interferometer always retains its maximum at $x = 0$ where $L_1 = L_2$ because all wavelengths interfere constructively for that and only that mirror position. For $x \neq 0$, waves interfere destructively, and the interferogram amplitude decreases from its maximum as shown in the interferogram for a silicon wafer in Fig. 4.17. The strong maximum at $x = 0$ is known as the centerburst.

The measured quantity in Fourier transform spectroscopy is the interferogram. The interferogram, however, is of little direct interest. It is the spectral response that is of interest, which is calculated from the interferogram using the Fourier transformation

Eq. (4.33) $$B(f) = \int_{-\infty}^{\infty} I(x) \cos(2\pi x f)\, dx$$

where $B(f)$ is the source intensity and $I(x)$ is the intensity of the interferogram.

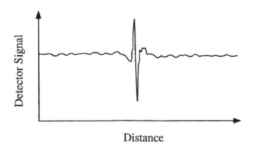

Fig. 4.17. A typical interferogram.

With the advent of computers of high calculation capacity and small size, all the calculations involved in a Fourier transform can be done in seconds. The data are stored in digital form and can be easily manipulated using different functions. In addition, many interferograms of a given sample can be collected and averaged to give a higher signal-to-noise ratio.

FTS has two major advantages over prism or grating monochromators. One is the multiplex gain advantage. The entire spectrum is observed during the measurement of FTS while only a small portion of the entire spectrum is measured at a time in the normal monochromator. So FTS has a signal-to-noise advantage of

$N^{1/2}$ for a spectra with N spectral elements if only noise other than the photon noise is considered in the detector. FTS can also acquire data in a far shorter time. The second advantage is the Jacquinot advantage or optical throughput gain advantage. Slits are used in a monochromator to collect a fraction of the spectra. So the optical throughput is only a portion of the incident light. There are no slits in FTS, so nearly 100% of the light is collected by the detector. This gives FTS nearly 100% throughput gain.

4.5.2. Recombination processes

When a sample is excited with an optical source with $hv > E_g$, EHPs are created and subsequently recombine by one of several mechanisms. At low temperature, recombination is dominantly through the annihilation of the exciton. An exciton is formed when Coulombic attraction between an electron–hole pair leads to an excited state in which an electron and a hole remain bound to each other in a hydrogen-like state [Wolfe and Mysyrowicz 1984]. When the exciton is formed by an electron in the conduction band and a hole in the valence band, the exciton is called a free exciton, since the exciton can move freely in the band with the electron and hole bound together. The energy of exciton recombination is slightly less than that of free EHP recombination due to the Coulombic attraction which results in a binding energy of the exciton E_x. With the exception of the free exciton, an exciton can be bound to some localized trap such as an ionized donor and acceptor and form a bound state at localized traps and impurity bands. This kind of exciton is called a bound exciton or Frank exciton. Typically, E_x is only about 30 meV which is comparable to the thermal energy at room temperature (26 meV), and in general no excitonic recombination is observable at room temperature. However, excitonic recombination dominates at low temperature. To obtain the maximum information, measurements at low temperature are required to decrease the thermal broadening of the emission peak and to eliminate the capture by thermally activated irradiation centers. For good PL output the majority of the recombination processes should be radiative. Fig. 4.18 shows five of the most commonly observed PL transitions [Smith 1981].

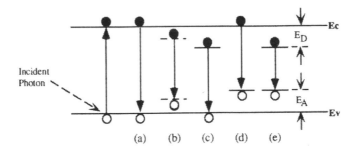

Fig. 4.18. Radiative transitions observed with photoluminescence [Smith 1981].

Band-to-band recombination (Fig. 4.18(*a*)) is the process whereby a free electron in the conduction band recombines with a free hole in the valence band and gives out radiation equivalent to the bandgap energy. It dominates at room temperature but it is rarely observed at low temperatures where excitonic recombination dominates. In materials that are sufficiently pure and at low temperature, *free excitons* form and recombine by emitting photons. The emitted photon energy in a direct bandgap semiconductor is given by [Pankove 1975]

Eq. (4.34) $hv = E_g - E_x$

where E_x is the free-exciton binding energy. In indirect bandgap semiconductors, momentum conservation requires the emission of a momentum conserving phonon, thus

Eq. (4.35) $hv = E_g - E_x - E_p$

where E_p is the phonon energy.

Bound-exciton recombination dominates over free-exciton recombination for less pure materials. The free exciton moves through the crystal and combines with a donor (Fig. 4.18(c)) to form an excitonic ion or bound exciton. If the donor is ionized, a donor ion and exciton complex (D^+, X) is formed. The electron bound to the donor travels in a wide orbit about the donor. Similarly, the neutral donor and exciton form a donor–exciton complex (D^0, X). Likewise, electrons combining with acceptors also form bound excitons (A^-, X), (A^0, X) (Fig. 4.18(d)).

Lastly, an electron on a neutral donor can recombine with a hole on a neutral acceptor, the well known donor–acceptor (D^0, A) recombination, illustrated in Fig. 4.18(e). The emission line has an energy modified by the Coulombic interaction between donors and acceptors given by

Eq. (4.36) $hv = E_g - (E_A + E_D) + \dfrac{q^2}{\varepsilon_r \varepsilon_0 r}$

where r is the distance between donor and acceptor.

Identification of the above recombinations depends on the careful consideration of the subtle competition between the above-mentioned mechanisms. For *p*-type semiconductors, (A^0, X) and (A^-, X) luminescence is stronger than the (D^0, X) and (D^+, X) luminescence. The opposite is true for *n*-type semiconductors. For the material of standard purity, the free-exciton luminescence is weaker than the bound-exciton luminescence. (D, A) recombination only exists at temperatures low enough so that dopants are not fully ionized. The energy position of the (D, A) peak is shifted to the lower-energy end when the semiconductor is heavily doped, due to the decrease of the average distance between the electrons in the donors and the holes in the acceptors. Another difference between (D, A) recombination and other recombination paths is that the full width at half maximum (FWHM) of the (D, A) recombination is typically $<3kT/2$ (k here is the Boltzman constant) and resemble slightly broadened delta functions. However, recombinations from other bound excitons are usually a few kT wide. The recombination between a band carrier and a deep center can also appear in the spectra which potentially gives us a method for identifying deep impurity species. Sometimes, phonon replicas of the main transitions also appear in the spectra.

Quantum-size effects become appreciable when the layer thickness becomes less than ~50 nm in III–V semiconductors, such as GaAs. This type of structure is called a quantum well, whereby quantum mechanics has to be employed to solve the equations yielding electronic states.

Since the pioneering work of Dingle [1975], a very large amount of literature has been published on the spectroscopic assessment of quantum-well structures. Photoluminescence spectroscopy is particularly important in the assessment of these structures.

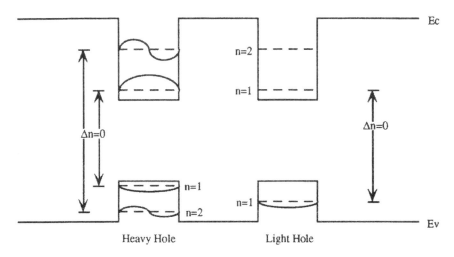

Fig. 4.19. Bound states and wavefunctions of quantized valence and conduction band states.

The bound energy and hole states in the quantum well are shown in Fig. 4.19 and absorption spectra for 400, 21, and 14 nm GaAs layers are shown in Fig. 4.20. Dingle [1975] proposed that the transitions were between the bound-hole and bound-electron states with equal quantum number n, following the $\Delta n = 0$ selection rule (see Fig. 4.19). It should be noted that this rule is generally valid but more recent work has shown that transitions can occur which violate the $\Delta n = 0$ selection rule.

Low-temperature PL measurements primarily detect $n = 1$ exciton recombination. PL provides a rapid assessment of well width, and the full width at half maximum (FWHM) of the luminescence peak is potentially a good measure of the quality of the interface since fluctuations in the well width produce considerable broadening, particularly in narrow wells. Bound exciton luminescence can also occur from donor and acceptors in the quantum wells. The binding energy depends both on the well width and the position of the impurity within the well.

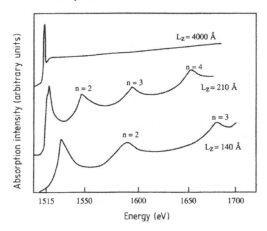

Fig. 4.20. Absorption spectra of 400, 21 and 14 nm thick GaAs quantum wells (after Dingle [1975]).

Photoluminescence can be used for the following: the relative intensity and illumination energy provide knowledge of the relative concentration and type of the luminescence center. The broadening of the peak is related to the quality of the sample. Generally, the higher the quality the narrower the peak. The shape of the peaks reflects the dominant luminescence mechanism. The shift and evolution of the peaks under different physical environments (e.g. pressure, temperature) gives information about the internal structure of the material. Time resolution photoluminescence and excitation photoluminescence spectroscopy provide information about energy transfer inside the material. The above demonstrates that photoluminescence has become one of the main techniques in studying the properties of semiconductors.

4.6. Electromechanical capacitance-voltage and photovoltage spectroscopy

4.6.1. Electrochemical capacitance–voltage measurement

Capacitance–voltage (C–V) measurement is one of the major techniques used for determining the carrier concentration as a function of depth.

The capacitance of a junction is defined as:

Eq. (4.37) $C = dQ / dV$

which gives rise to the explicit formula for the depletion width W_d and carrier concentration N at W_d under the depletion approximation for abrupt junctions.

Eq. (4.38) $W_d = \dfrac{\varepsilon_r \varepsilon_0 A}{C}$

Eq. (4.39) $N = \dfrac{C}{q\varepsilon_0 \varepsilon_r A^2 \left(dC/dV\right)}$

where ε_r is the relative dielectric constant of the semiconductor, ε_0 the vacuum dielectric constant, A is the area of the Schottky junction and q is the charge on the electron.

The distribution of the carrier concentration is profiled by measuring the capacitance of a Schottky junction under different reverse-bias voltages, then the edge of the depletion region and the carrier concentration at that position are calculated using Eq. (4.38) and Eq. (4.39).

The minimum depth and maximum depth of the profile are restricted by the zero-bias voltage and the breakdown voltage, respectively. Outside this range, the depletion approximation fails due to the current flowing through the junction. The lower-range limit is a problem for low-doped semiconductors while the upper-range limit can be especially restrictive in highly doped materials where the depletion width is small. In GaAs, the breakdown field is about 4×10^5 V·cm^{-1} and so the maximum depletion depth for $N_d \simeq 10^{18}$ cm^{-3} is about 0.02 μm [Sze 1981].

To overcome the disadvantages of the conventional C–V method, an alternate chemical etching and profiling with a temporary mercury barrier was originally introduced. However, this is a time consuming and tedious process. Ambridge et al. [1973, 1974, 1975] used an electrolyte to form an electrolyte/semiconductor junction and to remove material electrolytically. Therefore, both etching and measuring processes can be carried out in the same electrochemical cell and controlled electronically to perform the repetitive etching/measuring cycles, generating a profile plot to the desired depth. This method is called electrochemical C–V (ECV).

Fig. 4.21 shows a typical schematic diagram of the electrochemical cell used in ECV. The semiconductor sample is held against a sealing ring, which defines the contact area, by means of spring-loaded back contacts. In certain systems, such as a versatile system manufactured by Polaron Equipment Ltd. [Blood 1986], a front contact is also available, which is very useful for samples with high-resistance substrates. The etching conditions are controlled by the potential across the cell by passing a DC current between the semiconductor sample and the carbon electrode to maintain the required over-potential measured potentiometrically with reference to the saturated calomel electrode (SCE). The AC signals are measured with respect to a platinum electrode located near the semiconductor surface to reduce the series resistance due to the electrolyte.

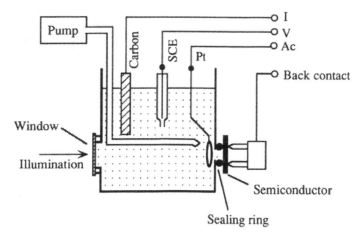

Fig. 4.21. Schematic diagram of the electrochemical cell.

When a semiconductor is in contact with an electrolyte, charge exchange occurs. In an ECV system, charge transfer takes place from the semiconductor to the electrolyte (anodic current), supported by holes in the valence band of the semiconductor. The dissolution reaction of InP, a common III–V semiconductor, is shown below

Eq. (4.40) $InP + 6\oplus \rightarrow In^{3+} + P^{3+}$

The dissolution process results in the separation of electric charge. Equilibrium is achieved when the decrease in chemical free energy, as a result of dissolution, is balanced by the increase in the electrical energy associated with the separation of charge. The separation of charge forms a so-called Helmholtz electrical double layer and the potential drop across the layer is referred to as the equilibrium electrode potential. Provided the electrolyte is fairly concentrated, field penetration into the electrolyte is negligible and the semiconductor/electrolyte interface behaves as a Schottky junction. In this case, the potential drop across the Helmholtz layer is constant and the remaining applied potential is dropped across the depletion layer of width W_d. W_d is determined from Eq. (4.38) and the carrier concentration is determined from Eq. (4.39).

Etching or dissolution is accommodated by the hole transferred from the semiconductor to the electrolyte. With p-type material, the holes required for dissolution are obtained by applying a forward-bias voltage, or the anodic potential in electrochemical notation. The required holes in n-type semiconductors are generated by illuminating the sample with light under a proper anodic potential V_{etch}. Smooth removal of n-type material is achieved when the etching current depends upon the illumination intensity but not upon the anodic potential. So the proper anodic potential should be in the area that has constant and large light current in the I–V curve as in Fig. 4.22. For the sake of consistency with the electrochemical convention, where the anodic potential corresponds to dissolution and the cathodic potential corresponds to deposition, the voltages in ECV are measured with reference to the electrolyte as the zero potential.

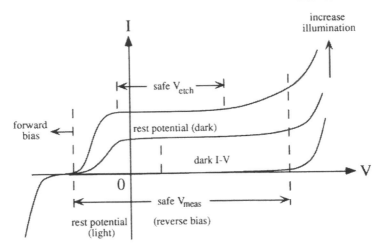

Fig. 4.22. Experimental parameters of the I–V curve of an n-type sample.

The carrier type of the semiconductor is identified by comparing the rest potential under dark and light conditions. For *n*-type material, the rest potential under dark conditions is lower than that at light conditions; *p*-type material is just the opposite. This measurement can be misleading if the structure has electrical junctions, very high doping levels or highly compensated material.

When the sample is etched, Faraday's law of electrolysis is used to calculate the etching depth W_{etch},

Eq. (4.41) $$W_{etch} = \frac{M}{ZFDA} \int I \; dt$$

where M is the molecular weight and D is the density of the semiconductor, F is the Faraday constant (9.64×10^4 C), A is the dissolution area and Z is the charge transferred per molecule dissolved.

The carrier concentration is profiled by repetitive etching and then recording at a fixed measuring voltage. The etching current is integrated and processed to give the etch depth which is added to the depletion depth to give the total depth of the depletion edge with respect to the original surface. Two AC voltages are applied along with a DC bias voltage in the measurement. The AC voltage, called the modulation voltage, of about 3 kHz and 0.14 V peak-to-peak is used to measure the capacitance. Sometimes the modulation frequency has to be decreased to minimize the error from the large series resistance, but a modulation frequency lower than the emission rate of the deep level may ionize the deep center and make the measured carrier concentration imprecise. The AC voltage of about 40 Hz and 0.28 V peak-to-peak, called the carrier voltage, is used to measure dC/dV which is used to calculate the carrier density N from Eq. (4.39). If the leakage current is too large, the carrier voltage has to be decreased to avoid the enclave of the carrier voltage to the large current area where the depletion application fails.

An ECV profiler has the following advantages:

(i)   it is a direct method of measuring the carrier concentrations,

(ii)   no sample preparation is needed except the conventional surface cleaning of the sample,

(iii)   ECV can be used in a wide range of structures and materials provided an adequate electrolyte is chosen,

(iv)   since the electrolyte used is transparent to light of wavelength shorter than 1.2 µm, the optical measurement can be performed in the same chemical cell.

The negative points of the ECV are:

(i)   difficult to measure at very high carrier concentrations ($>5 \times 10^{19}$ cm^{-3}) where the capacitance is outside the measurement range of the system,

(ii)   large surface depletion at very low carrier concentrations ($<1 \times 10^{19}$ cm^{-3}) makes the measurement of carrier concentrations in the top few microns impossible,

(iii)   the use of chemicals destroys the sample and contaminates the unetched part of the sample,

(iv)   the sealing-ring area needs careful calibration to assure reliable results.

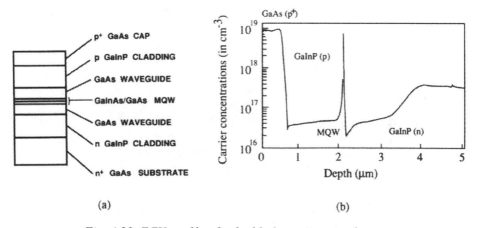

(a)   (b)

Fig. 4.23. ECV profile of a double-heterojunction laser.

Fig. 4.23 shows an ECV profile for a double-heterojunction laser. From the ECV profile, the following information can be acquired:

(i)   the carrier concentration and carrier type of each layer are clearly shown in the profile,

(ii)   the thickness of each layer can be determined to the order of several Debye lengths if uniform etching is achieved.

Recently, *C–V* profiling has been widely used to determine the band discontinuity of the heterojunction since Kroemer and Chien [1981] showed that the Debye averaging process conserves both the charge increment and its total electric moment throughout the profile so the potential drop across the barrier is a precise variable, although the real carrier concentration distribution is different from the profiled one.

However, ECV is not supposed to be a preferred way to get the band offset. Unlike the CV technique where the carrier concentration and the depletion width are both obtained from a capacitance measurement and the potential drop of the junction is independent of the measurement of the area, in the ECV profile the depth scale is obtained predominantly from the integration of dissolution current which gives the potential drop of the junction inversely proportional to the fourth power of the etching area. Even if the etching area can be determined to sufficient accuracy, the measured carrier concentration includes averaging of the Debye tail as well as the etching toughness, which makes the Kroemer and Chien assumption invalid in the ECV case.

The precision of ECV is very sensitive to the appropriate choice of the various measuring conditions. The etching uniformity is the principal instrumental limit to the depth resolution. The area denned limits the absolute accuracy. The Helmholtz capacitance and the electrolyte resistance may limit operation of the instrument on highly doped material where the capacitance of the electrolyte/semiconductor junction is large and the Helmholtz capacitance is comparable to it. The choice of low operating frequencies to minimize the series-resistance effects results in increased sensitivity to deep states in the measurement of capacitance and depletion width.

4.6.2. Photovoltage spectroscopy

An advantage of an electrolyte contact is that it is transparent to radiation at wavelengths shorter than 1.2 μm. With the addition of a variable monochromatic light source, photovoltage spectroscopy (PVS) can be measured in the same ECV system and also at different etching depth.

When light energy $hv \geq E_g$ illuminates a junction (p–n, Schottky, n–n^+ or p–p^+) or within a diffusion length from the junction, the photogenerated electrons and holes are separated by the junction field and create a photogenerated voltage drop across the junction. In ECV, when the electrolyte/semiconductor junction is illuminated from the front side with a continuous fixed wavelength, photovoltage spectroscopy (PVS) is measured. The open-circuit photovoltage V_{oc} is:

Eq. (4.42) $$V_{oc} = \frac{qF\alpha(\lambda)(L+W_d)}{1/R_\ell + (qI_s)/kT}$$

where F is the photon flux density, $\alpha(\lambda)$ is the absorption coefficient at wavelength λ, W_d is the depletion width and L is the diffusion length. R_ℓ is the leakage resistance, I_s is the reverse saturation current, k is the Boltzmann constant, T is the temperature in kelvin. The open-circuit voltage is measured in the ECV profile system to avoid electrochemical etching or deposition at the interface between the sample and electrolyte. PVS, in principle, gives the reflective absorption spectroscopy of the sample. The differences between PVS and absorption spectroscopy are: PVS is undetectable for a too leaky sample (R_ℓ small) while absorption spectroscopy is still available; PVS signals are of opposite sign for n-type and p-type semiconductors while their absorption spectra are the same.

Although analysis of the PVS spectrum might be complicated due to the contribution from various junctions in a material, the combination of etching and PVS provides accurate information on the bandgap at each depth in the sample. The alloy composition of the sample is obtained by exploiting the relationship between bandgap E_g and alloy composition [Casey and Panish 1978] of the known alloy sample. Fig. 4.24 shows a selection of photovoltage spectra recorded at different depths while performing an etching profile of a GaAs/GaAlAs double-heterojunction structure [Webster 1989]. In the case of materials with narrow bandgap ($\lambda > 1.2\ \mu m$), the electrolyte may have large absorption. To avoid this problem, back illumination is used, provided the substrate has a larger bandgap.

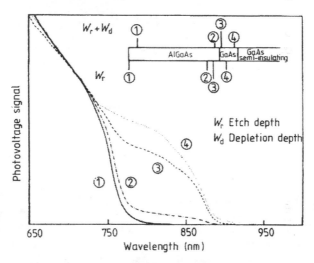

Fig. 4.24. Selection of photovoltage spectra recorded at different depths while performing an etching profile of a GaAs/GaAlAs double-heterojunction structure (after Webster [1989]).

In summary, the ECV technique overcomes the restrictions on profiling depth compared to the case of conventional C–V measurement. Furthermore, due to the additional information that can be obtained from the method combining PVS and controlled etching, an ECV/PVS system is a very powerful characterization instrument in the field of semiconductor research.

4.7. Resistivity and Hall measurement

4.7.1. Resistivity measurement

The conductivity of a semiconductor with electron concentration n, hole concentration p, electron mobility μ_n and hole mobility μ_p is defined by:

Eq. (4.43) $\sigma = q\left(n\mu_n + p\mu_p\right)$

The resistance of a rectangular semiconductor block is given by

Eq. (4.44) $R = \rho L / wt$

where ρ is the resistivity defined as $1/\sigma$, L is the length, w is the width and t is the thickness of the sample. The sheet resistance of a material is defined by

Eq. (4.45) $R_s = \rho / t$

so that the resistance is given by

Eq. (4.46) $R = R_s L / w$

Therefore, if the sheet resistance is known, only the length and width of a sample need to be specified to determine the resistance.

(a) The four-point probe method

The resistivity of a material can be measured by placing two contacts on the sample, sending a fixed current through the probes and measuring the voltage across them, but this method is not accurate, since the resistance measured includes the contact resistance and spreading resistance under the contacts. In order to avoid this problem the four-point probe method is commonly used to measure bulk or sheet resistivity of materials. In this case four contacts are placed on the sample, a fixed current is passed through the outer probes and the potential across the inner pair is measured. The arrangement is shown in Fig. 4.25. Since the current through the voltage probes is very small, the contact potential and spreading resistance at these probes is negligible, leading to a more accurate measurement.

    If the sample is large enough, it can be considered to be semi-infinite in which case resistivity is given by [Valdes 1954]

Eq. (4.47) $\rho = \dfrac{V}{I} \dfrac{2\pi}{\left(1/s_1 + 1/s_3 - 1(s_1 + s_2) - 1/(s_2 + s_3)\right)}$

where V is the voltage across the inner probes, I is the current through the outer probes and s_1, s_2, s_3 are indicated in Fig. 4.25. If the probe spacings are equal $(s_1 = s_2 = s_3 = s)$ Eq. (4.47) simplifies to

Eq. (4.48) $\rho = \dfrac{V}{I} 2\pi s$

A widely used probe spacing is 1.588 mm so that resistivity is given by $\rho = V/I$.

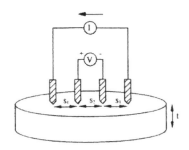

Fig. 4.25. Four-point probe method.

If the geometry of the sample is finite, which is more realistic than the assumption of a semi-infinite volume, the resistivity is defined by

Eq. (4.49) $\rho = \dfrac{V}{I} 2\pi s F$

where F is the correction factor due to several factors including edge, thickness and misalignment in probe placement. For laterally large and thin $(s \gg t)$ samples with uniform probe spacing, the resistivity is given by

Eq. (4.50) $\rho = \dfrac{\pi t}{\ln 2} \dfrac{V}{I} = 4.532 t \dfrac{V}{I}$

Very thin layers are usually characterized by their sheet resistance. Eq. (4.50) gives the sheet resistance as

Eq. (4.51) $R_s = \dfrac{\rho}{t} = 4.532 \dfrac{V}{I}$ $(s \gg t)$

Eq. (4.50) and Eq. (4.51) are valid for a sample of very large lateral extent compared to its thickness. In order to have accurate results by using these expressions, the circular sample must have a diameter much larger than the probe spacing s $(d \geq 40\ s)$ [Schroder 1990].

(b) Van der Pauw's method

Van der Pauw [1958] showed that the sheet resistance of an arbitrarily shaped sample can be measured by placing four contacts on the periphery of the sample (see Fig. 4.26), injecting a fixed current I through one pair of contacts and measuring the voltage V across the other pair of contacts. The following conditions should be satisfied:

(i)   the contacts are at the edge of the sample,
(ii)   the contacts are small enough,
(iii)   the sample is flat and of uniform thickness,
(iv)   the surface of the sample is singly connected (no isolated holes).

Van der Pauw also showed that two of these measurements are related by

Eq. (4.52) $$\exp\left(-\frac{\pi t R_{12,34}}{\rho}\right) + \exp\left(-\frac{\pi t R_{23,41}}{\rho}\right) = 1$$

where t is the thickness of the sample, $R_{12,34} = V_{34}/I_{12}$ and $R_{23,41} = V_{41}/I_{23}$.
    In order to solve for ρ we can write Eq. (4.52) in the following form

Eq. (4.53) $$\rho = \frac{\pi t}{2 \ln 2}\left(R_{12,34} + R_{23,41}\right) F\left(\frac{R_{12,34}}{R_{23,41}}\right)$$

where F is the correction factor for geometrical asymmetry. If the asymmetry is not too large ($R_{12,34}/R_{23,41} < 10$), we can use the following approximation for F:

Eq. (4.54) $$F = 1 - 0.3466\left[\frac{R_{12,34} - R_{23,41}}{R_{12,34} + R_{23,41}}\right] - 0.0924\left[\frac{R_{12,34} - R_{23,41}}{R_{12,34} + R_{23,41}}\right]^{-2}$$

For a symmetrical structure $R_{12,34} = R_{23,41}$ and

Eq. (4.55) $$\rho = \frac{\pi t}{\ln 2}\frac{V_{34}}{I_{12}} = 4.532 t\frac{V_{34}}{I_{12}}$$

which is similar to Eq. (4.50).

    These equations are based on the assumption of infinitesimal ohmic contacts at the edge of the sample. The error introduced by real finite-size contacts which are not exactly on the periphery can be eliminated by the use of the clover-shaped structure shown in Fig. 4.26. This pattern is ideal and simple for making Hall measurements on epilayers in which the layer thickness is much smaller than the lateral dimension of the sample. This structure can be obtained by using Hall pattern makers designed for this purpose.

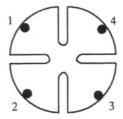

Fig. 4.26. Clover-shaped structure.

4.7.2. Hall measurement

(a) Mobility

The electrical and optical properties of the material used for device fabrication are very important in determining the device performance. Mobility is one of these properties and determines device speed and current handling capabilities. Under low electric field strengths the carrier drift velocity is given by

Eq. (4.56) $v = \mu E$

Since device speed depends strongly on the velocity of the carriers, high mobility is required to obtain a high-frequency response. Due to the low resistivity of high-mobility materials, devices fabricated on these materials can handle high current levels without excessive heating. High current levels are desirable in high-speed applications where the capacitances in the circuits need to be charged rapidly.

The carrier mobility is determined by the combination of scattering processes in the material and depends on the temperature. We can define carrier mobility as the combination of mobilities due to different scattering mechanisms in the form [Jaeger 1990]

Eq. (4.57) $\dfrac{1}{\mu} = \displaystyle\sum_{i=1}^{j} \dfrac{1}{\mu_i}$

where j is the number of scattering mechanisms.

The mobility due to ionized impurity scattering can be expressed by [Sze 1981]

Eq. (4.58) $\mu_i = \dfrac{64\sqrt{\pi}\,\varepsilon_s^2 \left(2kT\right)^{3/2}}{N_1 q^3 m*^{1/2}} \left\{ \ln\left[1 + \left(\dfrac{12\pi\varepsilon_s kT}{q^2 N_1^{1/3}} \right)^2 \right] \right\}^{-1}$

where N_1 is the ionized impurity density, ε_s is the permittivity, T is the temperature, k is the Boltzmann constant and $m*$ is the effective mass. From Eq. (4.58) we can see that mobility due to ionized impurity scattering is proportional to

$(m^*)^{-1/2} N_I^{-1} T^{3/2}$. Therefore ionized impurity scattering is important in determining the mobility at low temperatures (low electron energy) and high ionized impurity densities.

The mobility due to acoustic phonon scattering is given by [Lundstrom 1990]

Eq. (4.59) $$\mu_{AP} = \frac{\sqrt{2\pi} c_\ell q h^4}{D_A^2 m^{*5/2} (kT)^{3/2}}$$

where D_A is the acoustic deformation potential, h is Planck's constant and c_ℓ is the elastic constant. This expression is proportional to $(m^*)^{-5/2} T^{-3/2}$ which shows that this scattering mechanism is important at moderate or high temperatures. In addition to these scattering mechanisms, optical phonon scattering is one of the dominant scattering mechanisms in polar semiconductors such as GaAs and alloy scattering is observed in alloy materials due to the random arrangement of the constituent atoms. Table 4.2 shows the Hall mobilities of epitaxial InP layers grown on semi-insulating substrates.

Sample No	300 K		77 K	
	$N_D - N_A$ (cm^{-3})	μ_H (cm$^2 \cdot$V$^{-1} \cdot$s^{-1})	$N_D - N_A$ (cm^{-3})	μ_H (cm$^2 \cdot$V$^{-1} \cdot$s^{-1})
1	2×10^{15}	5350	1.5×10^{15}	59,800
2	5×10^{15}	5240	3.6×10^{15}	56,700
3	5.7×10^{15}	4950	3.6×10^{15}	53,320
4	1×10^{15}	5500	1×10^{15}	150,000

Table 4.2. The Hall mobilities of epitaxial InP layers grown on semi-insulating substrates.

(b) Hall measurement

Hall measurements yield information about carrier concentration and mobility. Hall data also provide information on the electrically active impurities in a semiconductor. The measurement is based on the Hall effect discovered by Hall in 1879. He found that if a magnetic field is applied perpendicular to the direction of current flow in a conductor, an electric field perpendicular to the current and magnetic field is created.

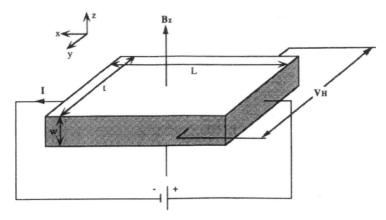

Fig. 4.27. Schematics of the Hall measurement.

Fig. 4.27 shows an *n*-type semiconductor with an electric field, E_x, applied in the *x*-direction and a magnetic field, B_z, applied in the *z*-direction. Due to the electric field in the *x*-direction and the magnetic field in the *z*-direction the Lorentz force on an electron in the *y*-direction is given by

Eq. (4.60) $F = qB_z v_x$

where q is the electronic charge and v_x is the electron velocity.

This force results in piling up electrons in the *y*-direction at one side of the sample and an electric field E_y is produced in this direction. Since there is no current in the *y*-direction, E_y should balance the Lorentz force, i.e.

Eq. (4.61) $qE_y = qB_z v_x$

The current in the *x*-direction due to the applied field is given by

Eq. (4.62) $I_x = qwtnv_x$

Using Eq. (4.62), Eq. (4.61) takes the form

Eq. (4.63) $qE_y = \dfrac{I_x B_z}{nwt}$

The Hall coefficient R_{H} is defined by

Eq. (4.64) $R_{\mathrm{H}} = \dfrac{w V_{\mathrm{H}}}{I_x B_z}$

where V_{H} is the Hall voltage created in the *y*-direction due to E_y. Eq. (4.63) and Eq. (4.64) give

Eq. (4.65) $R_{\mathrm{H}} = -1/qn$

Similarly for *p*-type samples

Eq. (4.66) $R_{\mathrm{H}} = 1/qp$

The sign of the Hall constant tells us the type of doping. In the case of mixed conduction when we have both holes and electrons, R_{H} is given by [Smith 1978]

Eq. (4.67) $$R_{\mathrm{H}} = \frac{\left[\left(p - b^2 n\right) + \left(\mu_n B\right)^2 \left(p - n\right)\right]}{q\left[\left(p - bn\right)^2 + \left(\mu_n B\right)^2 \left(p - n\right)^2\right]}$$

where $b = \mu_n/\mu_p$.

These expressions for the Hall coefficient are derived with the assumption that the relaxation time is energy independent. A numerical factor *r*, between one and two, called the Hall scattering factor, is usually included to account for the energy dependence of scattering mechanisms. The value of *r* depends on the dominant scattering mechanisms in the semiconductor ($r = 1.18$ for acoustic scattering, $r = 1.10$ for piezoelectric scattering and $r = 1.93$ for ionized impurity scattering). In this case the Hall coefficient and mobility are expressed as

Eq. (4.68) $R_{\mathrm{H}} = -r/qn$

and

Eq. (4.69) $\mu = R_{\mathrm{H}}\sigma/r$

We may define a quantity μ_{H} called the Hall mobility which satisfies the relation:

Eq. (4.70) $\mu_{\mathrm{H}} = R_{\mathrm{H}}\sigma$

so that $\mu_{\mathrm{H}} = r\mu$. For thin layers, the sheet Hall coefficient is used which is defined as

Eq. (4.71) $R_{\mathrm{Hs}} = R_{\mathrm{H}}/t$

which gives $\mu_{\mathrm{H}} = R_{\mathrm{Hs}}/\rho_s$.

An arrangement for measuring the Hall effect using a clover-shaped structure is shown in Fig. 4.28. A fixed current *I* is injected at two non-adjacent contacts and the potential across the other pair is measured under a constant magnetic field perpendicular to the sample surface. The Hall coefficient can be calculated using the measured Hall voltage V_{H}, the current *I* and the magnetic field *B* using Eq. (4.64). In order to increase the accuracy of the measurement the voltage across contacts 2 and 3 should be measured first without any magnetic field applied and the voltage drop due to misalignment of the contacts should be subtracted from the Hall voltage under magnetic field. Since the Hall voltage includes voltages generated by other effects, it is recommended to use an average value of V_H based on voltages measured for all permutations of contacts, current and magnetic field directions.

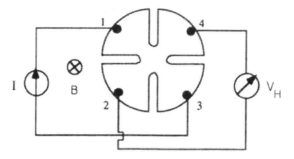

Fig. 4.28. Hall effect setup.

(c) Determination of the compensation ratio

Compensation reduces the free-carrier concentration and mobility in a semiconductor. The reduction in mobility is due to an increase in carrier scattering by ionized impurities. Therefore the compensation ratio is an important parameter to be determined in the characterization of materials.

Variable-temperature Hall measurements yield information about the donor and acceptor concentrations in a crystal. Fig. 4.29 shows a schematic diagram of electron density as a function of temperature in an *n*-type doped semiconductor.

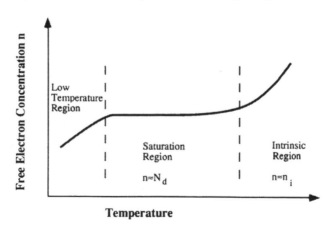

Fig. 4.29. Electron concentration against temperature in an n-type semiconductor.

In the intermediate region, most of the donors are ionized and the free-carrier concentration does not change with temperature. In this region, electron concentration is approximately equal to the donor doping density N_d (assuming the acceptor doping density $N_a = 0$). This region is also called the saturation region. In the low-temperature region, the free-carrier concentration decreases with temperature since the number of ionized donors depends on the temperature. In the intrinsic (high-temperature) region, the number of thermally agitated carriers is much larger than the electrons supplied by the donors and the electron concentration is approximately equal to the intrinsic carrier concentration n_i.

When impurities are added to the material, charge neutrality has to be preserved in the crystal. The equation that satisfies this condition is given by

Eq. (4.72) $n + N_a^- = p + N_d^+$

where N_d^+ is the ionized donor density given by

Eq. (4.73) $N_d^+ = \dfrac{N_d}{1 + 2\exp\left[\left(E_F - E_d\right)/kT\right]}$

and N_a^- is the ionized acceptor density expressed as

Eq. (4.74) $N_a^- = \dfrac{N_a}{1 + 4\exp\left[\left(E_a - E_F\right)/kT\right]}$

E_d is the donor energy and E_a is the acceptor energy. Under non-degenerate conditions electron and hole concentrations are given by

Eq. (4.75) $n = N_c\exp\left[-\left(E_c - E_F\right)/kT\right]$

and

Eq. (4.76) $p = N_v\exp\left[-\left(E_F - E_a\right)/kT\right]$

respectively. In these expressions N_c is the density of conduction band states defined as:

Eq. (4.77) $N_c = 2\left(\dfrac{2\pi m_{dc}kT}{h^2}\right)^{3/2}$

where m_{dc} is the conduction band density of states effective mass. N_v is the density of valence band states given by

Eq. (4.78) $N_v = 2\left(\dfrac{2\pi m_{dh}kT}{h^2}\right)^{3/2}$

where m_{dh} is the valence band density of states effective mass. In the intermediate-temperature region almost all the donors and acceptors are ionized ($N_a^- = N_a, N_d^+ = N_d$). The charge neutrality condition gives

Eq. (4.79) $n - p = N_d - N_a$

Using the relation $np = n_i^2$ with Eq. (4.79) we obtain:

Eq. (4.80) $n = \dfrac{N_d - N_a}{2} + \sqrt{\left(\dfrac{N_d - N_a}{2}\right)^2 + n_i^2}$

and

Eq. (4.81) $p = \dfrac{N_d - N_a}{2} + \sqrt{\left(\dfrac{N_a - N_d}{2}\right)^2 + n_i^2}$

in the intermediate region. If $N_d - N_a$ is much larger than n_i then

Eq. (4.82) $n \approx N_d - N_a$

And

Eq. (4.83) $p = \dfrac{n_i^2}{N_d - N_a}$

Using Eq. (4.75) and Eq. (4.82) we can obtain the Fermi level in the intermediate region as

Eq. (4.84) $E_F = E_c - kT \ln \dfrac{N_c}{N_d - N_a}$

which shows that the Fermi level rises towards the conduction band as the temperature is decreased. Therefore, in the low-temperature region, the Fermi level is close to the conduction band and the assumption $N_d^+ = N_d$ loses its validity ($N_a = N_a^-$ is still valid). In this region the hole density is negligible in an n-type semiconductor. If we write the charge neutrality equation under these conditions, we obtain:

Eq. (4.85) $N_c \exp\left(-\dfrac{E_c - E_F}{kT}\right) + N_a = \dfrac{N_d}{1 + 2\exp\left(E_F - E_d\right)/kT}$

or in terms of n

Eq. (4.86) $\dfrac{2n(n+N_a)}{N_d - N_a - n} = N_c \exp\left(-\dfrac{E_c - E_d}{kT}\right)$

To determine N_a and N_d, Eq. (4.86), which is the theoretical equation for the carrier concentration, must be fitted to the experimental Hall carrier concentration determined from variable-temperature Hall measurements. For semiconductors with deep donor (E_d) levels it is also possible to determine E_d together with N_a and N_d.

Eq. (4.73) is valid for a single donor level. If the impurity levels have excited states, the ionized donor density is given by

Eq. (4.87) $N_d^+ = N_d \left/ \left(1 + \sum_{r=1}^{j} g_r \exp(E_F - E_r)/kT\right)\right.$

where E_r is the energy and g_r is the degeneracy factor of the excited state r, and j is the total number of excited states.

In the high-temperature region, the carrier concentration is approximately equal to the intrinsic concentration which is given by

Eq. (4.88) $n_i = \sqrt{N_c N_v} \exp\left(-E_g / kT\right)$

In this region, n varies theoretically as $T^{3/2} \exp(-E_g/2kT)$ and E_g can be determined by using Eq. (4.88) and the experimental Hall concentration in the high-temperature region.

(d) Depletion correction

In lightly doped thin films it is possible for surface and interface depletion regions to deplete a significant portion or even all of the film. In this case the active film thickness is smaller than the physical film thickness and the bulk resistivity and carrier density must be corrected for surface and interface depletion. We can define the active film thickness by

Eq. (4.89) $t_{act} = t_m - t_i - t_s$

where t_m is the metallurgical thickness, t_i is the thickness of the interface depletion layer and t_s is the surface depletion layer thickness. The thickness of the surface depletion layer is given within the abrupt depletion approximation by

Eq. (4.90) $t_s = \sqrt{\dfrac{2\varepsilon_0 \varepsilon_r}{q(N_d - N_a)}\left(V_{BS} - \dfrac{kT}{q}\right)}$

where V_{BS} is the built-in potential at the surface defined by

Eq. (4.91) $V_{BS} = \Phi_B - \dfrac{kT}{q} \ln \dfrac{N_c}{N_d - N_a}$

In this expression, $\Phi_B = E_{cs} - E_{Fs}$ where E_{cs} and E_{Fs} are the conduction band energy and the Fermi level at the surface, respectively.

The depletion layer thickness at the interface is approximated as

Eq. (4.92) $l_i = \sqrt{\dfrac{2\varepsilon_0 \varepsilon_r}{q(N_d - N_a)}\left(V_{Bi} - \dfrac{kT}{q}\right)}$

where V_{Bi} is the built-in potential at the interface given by

Eq. (4.93) $V_{Bi} = \Phi_i - \dfrac{kT}{q} \ln \dfrac{N_c}{N_d - N_a}$

Φ_i is the interface pinning potential.

Surface and interface depletion region thicknesses depend on the temperature, and unexpected carrier concentration shifts with temperature can be observed especially in thin layers if the necessary depletion corrections are not made [Lepkowski et al. 1987].

4.8. Thickness measurement

Bevel stain measurement is a technique for measuring the thickness of a sample with an accuracy that can measure a few tens of angstroms thick layer with an error of ± 10 Å. It uses a rotating stainless steel ball to mill a hole on a sample and then the hole radius is measured by means of an optical microscope.

4.8.1. Ball polishing

This step mills a spherical hole in a sample by rotating a stainless steel ball as shown in Fig. 4.30. The first step is to choose the right ball for polishing because there are numerous balls of different radius (R) which are used for different types of polishing. The second step is to put the sample on the sample holder before placing a drop of diamond paste on the sample. The diamond paste is added for the purpose of milling. Following this step, the system is turned on for a couple of seconds to let the ball mill the sample.

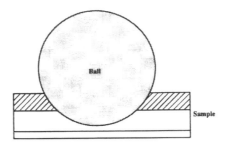

Fig. 4.30. Ball milling of the sample.

4.8.2. Revelation with RCA solution

Following ball polishing, a chemical etchant called RCA solution is used in an etching process to reveal the layer profile. The revelation of layers makes the observation and measurement of diameters easier.

RCA solution is a chemical etchant which consists of the following chemicals:

(1) 250 g of CrO_3 in 300 ml of water
(2) 2 g of $AgNO_3$ in 100 ml of water
(3) 250 ml of HF (48%) (at least 40%) in 100 ml of water.

The solution is prepared by adding (2) to (1) followed by adding (3). For proper use, the reactant must be diluted in 10 volumes of water for one volume of reactant. Using this RCA solution, samples undergo revelation for a few seconds.

4.8.3. Rinsing procedure

Perfect cleaning of the sample and the ball prior to the ball milling procedure is crucial since any dust or impurity on the sample or the ball can damage the sample. This would make the observation of the revealed layers difficult.

After the chemical revelation with RCA solution, the sample has to be rinsed once more to prevent any further attack by the RCA solution. This time it is first rinsed with water and then with methyl alcohol.

4.8.4. Theory behind the thickness measurement

When a hole is etched on the sample, all the different epitaxial layers can be observed. Using an optical microscope of magnification rate G, the thickness of the observed layers can be determined by measuring the corresponding *magnified* diameters D_{ex} and D_{in} shown in Fig. 4.31. Note that the actual diameters (no magnification) are d_{ex} and d_{in}, respectively.

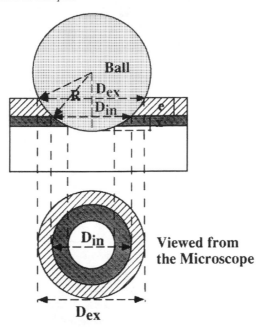

Fig. 4.31. Schematic diagram of the etched hole and diameters.

By simple mathematical relations and manipulation, using the condition d_{ex}, $d_{in} \ll R$, the formula of the thickness e can be derived as shown below

Eq. (4.94) $e = \dfrac{D_{ex}^2 - D_{in}}{8RG^2} \times 1000$

where $D_{ex,in}$ is in mm, G is the magnification of the microscope, R is the radius of the ball and e is in μm.

4.8.5. Other thickness measurement techniques

Besides the bevel stain measurement mentioned above, the chemical bevel revelation discussed in section 4.2 can also be used for thickness measurement of epilayers as illustrated in Fig. 4.3 and Fig. 4.4. Other than chemical etching methods, surface profilers are available for measuring epilayer steps with a resolution as high as 1 Å. As shown in Fig. 4.32, a surface profiler profiles a sample electromechanically. A precision stage, travelling along an optically flat glass block, moves the sample beneath a diamond-tipped stylus. As the sample glides under the diamond-tipped stylus, variations in the sample surface cause vertical translation of the stylus. An LVDT core, mechanically coupled to the stylus, generates an analog signal corresponding to the stylus movement. This signal is then amplified and digitized. Scan data are stored in computer memory for display, manipulation, measurement and readout.

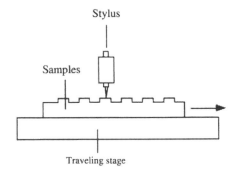

Fig. 4.32. Schematic diagram of a surface profiler.

References

Abrahams, M.S. and Buiocchi, C.J., *J. Appl. Phys.* **36** 2855, 1965.

Ambridge, T., Elliot, C.R., and Faktor, M.M., *J. Appl. Electrochem.* **3** 1, 1973.

Ambridge, T. and Faktor, M.M., *J. Appl. Electrochem.* **4** 135, 1974.

Ambridge, T. and Faktor, M.M., *J. Appl. Electrochem.* **5** 319, 1975.

Bebb, H.B. and Williams, E.W., *Semiconductors and Semimetals* vol 8, ed R.K. Willardson and A.C. Beer (New York: Academic) p 182, 1972.

Blood, P., *Semicond. Sci. Technol.* **1** 7, 1986

Bond, W.L., *Acta. Crystallogr.* **13** 814, 1960.

Bragg, W.H. and Bragg, W.L., *Proc. R. Soc.* **88** 428, 1913a.

Bragg, W.H. and Bragg, W.L., *Proc. R. Soc.* **89** 246, 1913b.

Casey, H.C. and Panish, M.B., *Heterostructure Lasers (Part B)* (New York: Academic), 1978.

Cowley, R.A. and Ryan, T.W., *J. Phys. D: Appl. Phys.* **20** 61, 1987

Dean, P.J., *Prog. Crystal Growth Charact.* **5** 89, 1982

Dingle, R., *Festkörperpröbleme* **15** 21, 1975.

Ewald, P.P., *Ann. Phys.* **49** 117, 1916.

Ewald, P.P., *Ann. Phys.* **54** 519, 557, 1917

Ewald, P.P., *Z. Phys.* **2** 323, 1920

Ewald, P.P., *Phys. Z.* **21** 617, 1921.

Ewald, P.P., *Z. Phys.* **30** 1, 1924.

Ewald, P.P., *Phys. Z.* **26** 29, 1925.

Ewald, P.P., *Acta Crystallogr.* **11** 888, 1958.

Fewster, P.F. and Curling, C.J., *J. Appl. Phys.* **62** 4154, 1987.

Hall, E.H., *Am. J. Math.* **2** 287, 1879.

Huber, A.M., di Persio, J., di Forte-Poisson, M.A., Brylinski, C., Bisaro, R., and Grattepain, C., *Proc. SPIE* **796** 46, 1987.

Huber, A.M., Laurencin, G., and Razeghi, M., *J. Physique Coll.* **4** 409, 1983.

Huber, A.M., Razeghi, M., and Morillot, G., *Gallium Arsenide and Related Compounds 1984 (Inst. Phys. Conf. Ser. 74)* (Bristol: Institute of Physics) p 223, 1984

Jaeger, R.C., *Introduction to Microelectronic Fabrication (Volume V in the Modular Series on Solid State Devices)* (Reading, MA: Addison-Wesley), 1990.

Kroemer, H. and Chien, W.Y., *Solid State Electron.* **24** 655, 1981.

Lang, D.V., *J. Appl. Phys.* **45** 3023, 1974.

Lang, D.V., Grimmeiss, H.G., Meijer, E., and Jaros, M., *Phys. Rev.* B **22** 3917, 1980.

Lepkowski, T.R., *J. Appl. Phys.* **61** 4808, 1987.

Lourenco, J.A., *J. Electrochem. Soc.* **131** 1914, 1984.

Lundstrom, M., *Fundamentals of Carrier Transport* (Reading, MA: Addison-Wesley), 1990.

Miller, G.L., Lang, D.V., and Kimerling, L.C., *Ann. Rev. Mater. Sci.* p 377, 1977.

Pankove, J.I., *Optical Processes in Semiconductors* (New York: Dover), 1975.

Razeghi, M. and Omnes, F., *Appl. Phys. Lett.* **59** 1034, 1991.

Schroder, D.K., *Semiconductor Material and Device Characterization* (New York: Wiley), 1990

Smith, K.K., *Thin Solid Films* **84** 171, 1981

Smith, R.A., *Semiconductors* (Cambridge: Cambridge University Press), 1978.

Sze, S.M., *Physics of Semiconductor Devices* (New York: Wiley), 1981.

Takagi, S., *Acta Crystallogr.* **15** 1311, 1962.

Takagi, S., *J. Phys. Soc. Japan* **26** 1239, 1969

Tapfer, L. and Ploog, K., *Phys. Rev.* B **33** 5565, 1986.

Taupin, D., *Bull. Soc. Fr. Mineral. Crystallogr.* **87** 469, 1964

Thurmond, C.D., *J. Electrochem. Soc.* **122** 1133, 1975

Valdes, L.B., *Proc. IRE* **42** 420, 1954

Van der Pauw, L.J., *Philips Res. Rep.* **13** 1, 1958

Van Vechten, J.A., and Thurmond, C.D., *Phys. Rev.* B **14** 3539, 1976

Webster, G.W., *PVS Instruction Manual* Bio-Rad Microscience Division, 1989.

Weyher, J. and Van de Ven, J., *J. Crystal Growth* **63** 285, 1983.

Weyher, J. and Van de Ven, J., *J. Electrochem. Soc.* **133** 799, 1986.

Weyher, J. and Van de Ven, J., *J. Crystal Growth* **88** 221, 1988.

Williams, E.W. and Bebb, H.B., *Semiconductors and Semimetals* vol 8, ed. R.K. Willardson and A.C. Beer (New York: Academic) p 321, 1972,

Wolfe, J.P. and Mysyrowicz, A., *Sci. Am.* **250** 98, 1984.

5. MOCVD Growth of GaAs Layers

5.1.   Introduction
5.2.   GaAs and related compounds band structure
5.3.   MOCVD growth mechanism of GaAs and related compounds
5.4.   Experimental details
5.5.   Incorporation of impurities in GaAs grown by MOCVD
        5.5.1.   Residual impurities
        5.5.2.   Carbon incorporation in GaAs grown by MOCVD
        5.5.3. n-type GaAs
        5.5.4. p-type GaAs
        5.5.5.   Erbium-doped GaAs

5.1. Introduction

The first half of this book is devoted to GaAs and related compounds, thus we focus on GaAs, $Ga_xAl_{1-x}As$ and $Ga_{0.49}In_{0.51}P$ band structures, and their relevant physical properties. The second half of this book, beginning with chapter 11, is devoted to InP and related compounds.

Over the past decade, metalorganic chemical vapor deposition (MOCVD) has developed into a versatile and reliable crystal growth technique. Very significant improvements have been made in understanding and controlling the purity, perfection and device properties of GaAs-based materials due mainly to the stimulus received from successful device applications in optoelectronics and microwave areas. GaAs has been considered as a possible alternative to silicon in the field of memories and microprocessors. However, technical improvements in silicon have raised requirements, challenging such possibilities. On the other hand, certain fields unexplored by silicon techniques, such as optoelectronics or TV satellite transmission, have turned out to be the major forces in propelling forward the future of GaAs and related compounds.

5.2. GaAs and related compounds band structure

GaAs has a zinc blende structure and consists of two interpenetrating face-centered cubic lattices each of which contains Ga and As, respectively. Each Ga atom has four nearest neighbors of As and vice versa as shown in Fig. 5.1(a). The

Ga and As atoms are tetrahedrally coordinated as shown in Fig. 5.1(*b*). The electronic configurations of a single Ga and As are $3d^{10}4s^24p^1$ and $3d^{10}4s^24p^3$, respectively. The average number of valence electrons per atom is four. When Ga and As atoms form bonds in the crystal, the valence electrons are redistributed in sp^3 hybridized orbitals which are oriented towards the nearest-neighbor atoms in $\langle 1\ 1\ 1 \rangle, \langle \bar{1}\ \bar{1}\ 1 \rangle, \langle 1\ \bar{1}\ \bar{1} \rangle$ and $\langle \bar{1}\ 1\ \bar{1} \rangle$ directions, respectively. The atomic orbitals that are used to form hybridized bonding orbitals are not the same as the ground states of the atoms. Even though this redistribution promotes electrons to a higher-energy state than the ground state, the total energy of the system is lowered by the formation of the bond. The bonding orbital consists of two directed orbitals associated with the two adjacent atoms, and is termed bonding or antibonding orbital depending on whether the two directed orbitals are in phase or out of phase. Such a bond is produced between two unlike atoms is called a heteropolar bond. In the case of GaAs, the valence electrons distribute more in the vicinity of As than Ga due to the larger value of electronegativity of As. Consider N primitive cells in a crystal: due to the hybridization between $2N$ s states and $6N$ p states there are two energy bands; each contains $4N$ hybridized states and thus $8N$ electrons. The $8N$ valence electrons usually fill the lower band corresponding to the bonding orbitals and the upper band is empty.

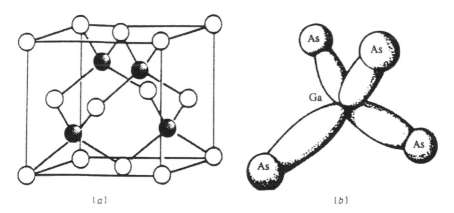

(a)   (b)

Fig. 5.1. (a) Zinc blende lattice: the two interpenetrating cubic face-centered lattices are built by two different types of atom, (b) sp^3 hybrid bond orbitals.

Fig. 5.2 shows the Brillouin zone (see e.g. Kittel [1971] for the theory of band structure in crystals) of the GaAs crystal (zinc blende lattice), and indicates the most important symmetry points and symmetry lines, such as the center of the zone ($\Gamma = (2\pi/a)\ (0,\ 0,\ 0)$), the (1 1 1) axes ($\Lambda$) and their intersections with the zone edge ($L = (2\pi/a)(\frac{1}{2}, \frac{1}{2}, \frac{1}{2})$), the $\langle 1\ 0\ 0 \rangle$ axes (Δ) and their intersections ($X = (2\pi/a)(0,\ 0,\ 1)$) and the $\langle 1\ 1\ 0 \rangle$ axes (Σ) and their intersections ($X = (2\pi/a)(\frac{3}{4}, \frac{3}{4}, 0)$).

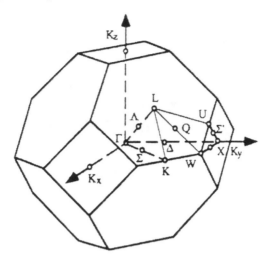

Fig. 5.2. First Brillouin zone for GaAs-based materials (zinc blende lattice),
including important symmetry points and lines.

The currently used method for calculating the band structure around the Γ point
is based on *k·p* perturbation theory [Kane 1966]. The crystal Hamiltonian, which
has translation invariance properties through any lattice vector *R* and its
eigenfunctions, can be written in Bloch's form:

Eq. (5.1) $\Phi_{n,k}(r) = \dfrac{1}{\sqrt{\Omega}} u_{n,k}(r) \exp(ik \cdot r)$

where Ω is the crystal volume, *n* is the index of the energy band and $u_{n,k}$ is a Bloch
function verifying:

Eq. (5.2) $u_{n,k}(r+R) = u_{n,k}(r)$

$u_{n,k}$ can be deduced from the Schrödinger pseudoequation

Eq. (5.3) $H_k u_{n,k} = E_n(k) u_{n,k}$

with

Eq. (5.4) $H_k = \dfrac{\hbar}{2m_0}(-i\nabla + k)^2 + V(r) + H_{SO}$

where m_0 is the electron mass, $V(r)$ is the crystal potential and H_{SO} is the spin–orbit
interaction Hamiltonian.

The *k·p* perturbation method relies on the fact that the Hamiltonian H_k can be
developed around $k = 0$:

Eq. (5.5) $H_k = H_{k=0} + \left(\dfrac{\hbar}{m_0}\right) k \cdot p + \dfrac{\hbar^2 k^2}{2m_0}$

Eq. (5.6) $H_k = H_{k=0} + H_1 + H_2$

By knowing the solutions of the Schrödinger equation around the Γ point ($k = 0$), the band structure can be determined for small k by treating $H_1 + H_2$ as perturbations.

Let L be the orbital kinetic momentum operator and S be the spin operator, with associated quantum numbers (l, m_l) and (s, m_s). For a p-type band, $l = 1$ and $s = \frac{1}{2}$. The total kinetic momentum operator $J = L + S$ with quantum numbers (j, m_j) is introduced, so that due to the theorems relating to the composition of kinetics momentum, electrons verify $j = \frac{1}{2}$ or $j = \frac{3}{2}$.

Spin–orbit coupling allows this degeneracy of a p bonding band to split, so that the valence band in $k = 0$ is separated into two points $\Gamma_8(j = \frac{3}{2})$ and $\Gamma_7(j = \frac{1}{2})$. The spin-orbit energy splitting is currently denoted $\Delta = E_{\Gamma 8} - E_{\Gamma 7}$.

When dealing with a direct bandgap material, the conduction-band minimum is at the Γ point at the zone center (the conduction-band minimum is often called the conduction-band edge, and the valence-band maximum is called the valence-band edge). The minimum energy gap E_g is the energy difference between the conduction-band minimum (E_c) and the valence-band maximum (E_v), so the magnitude E_g of the energy gap in GaAs is given by

Eq. (5.7) $E_g = E_c - E_v$

The non-periodic part $|j, m_j\rangle$ of Bloch's functions in the center of the Brillouin zone is often expressed as a function of the orbitals p_x, p_y, p_z, which are denoted as $|x\rangle$, $|y\rangle$, $|z\rangle$. The results are shown in Table 5.1, the energy origin being taken at the conduction band, in $k = 0$ (Γ_6 point), and the spin sign being denoted $+$ or $-$.

The problem being solved in $k = 0$, a perturbation treatment can be applied to the Hamiltonian $H_1 = \hbar(k \cdot p)/m_0$ in the reduced basis (u_1, ..., u_8). The coupling terms $\langle S|P_\alpha|\beta\rangle$ are null for $\alpha \neq \beta$, due to symmetry considerations. $|S\rangle$ is the wavefunction of an s state electron; $|\beta\rangle$ is the β component of the p state electron and P_α is the α component of the momentum operator. Furthermore, $\langle S|P_\alpha|\alpha\rangle$ is independent of α. Kane [1966] has noted that $1/m_0\langle S|P_\alpha|\alpha\rangle$ is a constant which is denoted as the constant P.

Once this coupling element is known, it is easy to write the matrix elements in the reduced base (u_1, ..., u_8) for the first order in $k \cdot p$ perturbation Hamiltonian [Kane 1966].

$u_i \left| j, m_j \right\rangle$	Φ_j, m_j	$E(k=0)$
$u_1 \left\| \frac{1}{2}, \frac{1}{2} \right\rangle$	$\left\| s, + \right\rangle$	0
$u_3 \left\| \frac{3}{2}, \frac{1}{2} \right\rangle$	$-\sqrt{2}/3 \left\| z, + \right\rangle + 1/\sqrt{6} \left\| x, + iy, - \right\rangle$	$-E_g$
$u_5 \left\| \frac{3}{2}, \frac{3}{2} \right\rangle$	$1/\sqrt{2} \left\| x, + iy, + \right\rangle$	$-E_g$
$u_7 \left\| \frac{1}{2}, \frac{1}{2} \right\rangle$	$1/\sqrt{3} \left\| x + iy, - \right\rangle + 1/\sqrt{3} \left\| z, + \right\rangle$	$-E_g - \Delta$
$u_2 \left\| \frac{1}{2}, -\frac{1}{2} \right\rangle$	$\left\| s, - \right\rangle$	0
$u_4 \left\| \frac{3}{2}, -\frac{1}{2} \right\rangle$	$1/\sqrt{6} \left\| x - iy, + \right\rangle - \sqrt{2}/3 \left\| z, - \right\rangle$	$-E_g$
$u_6 \left\| \frac{3}{2}, -\frac{3}{2} \right\rangle$	$1/\sqrt{2} \left\| x - iy, - \right\rangle$	$-E_g$
$u_8 \left\| \frac{1}{2}, -\frac{1}{2} \right\rangle$	$-1/\sqrt{3} \left\| x - iy, + \right\rangle - 1/\sqrt{3} \left\| z, - \right\rangle$	$-E_g - \Delta$

Table 5.1. Periodic part of Bloch's function.

The eigenvalue equation det $(H - E_1) = 0$ leads to the dispersion relations:

Eq. (5.8) $E = -E_g$

Eq. (5.9) $E\left(E + E_g\right)\left(E + E_g + \Delta\right) = p^2 \hbar^2 k^2 \left(E + E_g + 2\Delta/3\right)$

By developing Eq. (5.9) around $k = 0$, one can find the well known dispersion relation $E(k) = \hbar^2 k^2 / 2m*$ for the conduction band, thus leading to an electronic effective mass

Eq. (5.10) $\dfrac{1}{m*} = \dfrac{2P^2 \left(E_g + 2\Delta/3\right)}{E_g \left(E_g + \Delta\right)}$

From Eq. (5.10), we can deduce that the electron effective mass increases with bandgap. This can be verified for $Ga_{1-x}Al_xAs$, for which the effective mass increases with x from $0.067 m_0$ ($x = 0$) to $0.15 m_0$ ($x = 1$).

Note that the parabolic dispersion relation is only a first approximation. The exact equation deduced from Eq. (5.9) leads to an effective mass increasing with energy (non-parabolicity)

Eq. (5.11) $\dfrac{m*(E)}{m*} = \dfrac{\left(E + E_g\right)\left(E + E_g + \Delta\right)\left(E_g + 2\Delta/3\right)}{E_g \left(E_g + \Delta\right)\left(E + E_g + 2\Delta/3\right)}$

This effect varies with the bandgap approximately as $1/E_g$ so that the non-parabolicity effect is more important for small gaps. Hence, the conduction band can be well defined around the Γ point with a first-order $k \cdot p$ perturbation. This is not the case for the valence band, as seen from Eq. (5.8). This flat dispersion relation leads to an infinite heavy-hole mass. A second-order calculation is necessary to calculate the heavy- and light-hole masses [Dresselhaus et al. 1955, Luttinger 1956]. Fig. 5.3 shows the band structure of GaAs. For an alloy like $Ga_xAl_{1-x}As$, there is a random location of Ga and Al on III sites, so that there is no more translation invariance. To find Bloch's solutions, it must be assumed that "mean" atoms $xGa + (1 - x)Al$ are located on III sites.

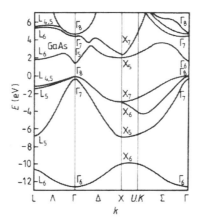

Fig. 5.3. The band structure of GaAs. It has a direct bandgap, with a minimum conduction band and a maximum valence band at the Γ point (after Sze [1981]).

The bandgap of GaAs at room temperature is around 1.42 eV. The main physical parameters of GaAs and of its related compounds are shown in Table 5.2 (after Adachi [1985]).

The bandgap and wavelength of $Ga_xIn_{1-x}As_yP_{1-y}$ compounds lattice matched to GaAs is illustrated in Fig. 5.4 [Adachi 1985]. All of those materials have a direct bandgap. Note that the direct to indirect bandgap crossover for $Ga_xIn_{1-x}P$ compounds is for $x = 0.74$, far away from the lattice-matched GaAs composition ($x = 0.49$).

Fig. 5.5 shows some information about GaInP alloys lattice matched to GaAs substrate. $Ga_xAl_{1-x}As$ compounds show a transition from direct to indirect bandgap, as illustrated in Fig. 5.6. The crossover of Γ and X bands occurs for $x = 0.45$ at 300 K.

For $x < 0.45$: the minimum of the conduction band is at the Γ point, and the bandgap is direct. For $x > 0.45$: the minimum of the conduction band is at the X point, and the bandgap is indirect ($\Gamma \rightarrow X$ transition) [Hilsum 1962].

Fig. 5.7 shows the evolution of the donor binding energy in $Ga_xAl_{1-x}As$, as a function of x, deduced from photoluminescence measurements.

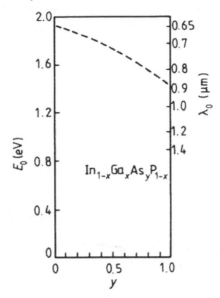

Fig. 5.4. *The bandgap and wavelength of* $Ga_xIn_{1-x}As_yP_{1-y}$, *compounds lattice matched to GaAs (after Adachi [1982]).*

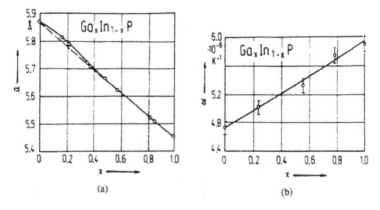

(a)

(b)

Fig. 5.5. *(a)* $Ga_xIn_{1-x}P$. *Composition dependence of the lattice parameter at 300 K. (b)* $Ga_xIn_{1-x}P$. *Composition dependence of the coefficient of linear thermal expansion at 72 K (after Landolt–Bornstein [1982]).*

The donor ionization energy switches from Γ-like $(E_1(\Gamma))$ for $x < 0.3$ to X-like $(E_1(X))$ for $x > 0.6$, as expected from the crossover at $x = 0.45$ (see Fig. 5.7). The energies $E_1(\Gamma)$ and $E_1(X)$ are deduced from the hydrogenoid binding energy, by taking into account the various electron masses in the Γ and X valleys, respectively (see Table 5.2).

Parameter	GaAs	AlAs	$Al_xGa_{1-x}As$
Bandgap energy			
E_g^α (eV)	1.424 $(E_g^\Gamma)^a$	2.168 $(E_g^X)^a$	$1.424 + 1.247x$
			$(0 \leq x \leq 0.45)^a$
			$1.900 + 0.125x + 0.143x^2$
			$(0.45 < x \leq 1.0)^a$
Critical-point energy (eV)			
E_0	1.425^b	3.02^c	$1.425 + 1.155x + 0.37x^2$ [W]
$E_0 + \Delta_0$	1.765^b	3.32^c	$1.765 + 1.115x + 0.37x^2$
$E_1(1)$	2.89^b	3.82^c	$2.89 + 0.94x$
$E_1(2)$	2.96^b	3.96^c	$2.96 + 1.00x$
$E_1(1) + \Delta_1$	3.12^b	4.03^c	$3.12 + 0.91x$
$E_1(2) + \Delta_1$	3.19^b	4.16^c	$3.19 + 0.97x$
E_0'	4.44^b	4.54^c	$4.44 + 0.10x$
$E_0' + \Delta_0'$	4.63^b	4.69^c	$4.63 + 0.06x$
E_2	4.99^b	4.89^{cs}	$4.99 - 0.10x$
Electron			
affinity χ_c (eV)	4.07^d	3.5^d	$4.07 - 1.1x$
			$(0 \leq x \leq 0.45)$
			$3.64 - 0.14x$
			$(0.45 < x \leq 1.0)$
Ionicity f_i	0.310^e	0.274^e	$0.310 - 0.036x$
Pressure coefficient of E_g^α ($\times 10^{-6}$ eV bar^{-1})			
dE_g^Γ / dP	11.5^f	10.2^g	$11.5 - 13x$
dE_g^X / dP	-0.8^h	-0.8^i	-0.8
dE_g^L / dP	2.8^h	2.8^i	2.8
Temperature coefficient of E_g^α ($\times 10^{-4}$ eV K^{-1})			
dE_g^Γ / dT	-3.95^j	-5.1^k	$-3.95 - 1.15x$
dE_g^X / dT	-3.6^l	-3.6^k	-3.6
Conduction-band effective mass			
Γ valley m_c^Γ	0.067^a	0.150^a	...
X valley m_{tX}	0.23^m	0.19^n	...
m_{lX}	1.3^o	1.1^n	...
L valley m_{tL}	0.0754^p	0.0964^q	...
m_{lL}	1.9^r	1.9^s	...
Density-of-states electron mass m_c^α			
Γ valley m_c^Γ	0.067	0.150	$0.067 + 0.083x$
X valley m_c^X	0.85^t	0.71^t	$0.85 - 0.14x$
L valley m_c^L	0.56'	0.66^t	$0.56 + 0.1x$

Parameter	GaAs	AlAs	$Al_xGa_{1-x}As$
Conductivity effective mass m_e^{α}			
Γ valley m_e^{Γ}	0.067	0.150	$0.067 + 0.083x$
X valley m_e^X	0.32^t	0.26^t	$0.32 - 0.06x$
L valley m_e^L	0.11^t	0.14^t	$0.11 + 0.03x$
Valence-band effective mass			
m_{lh}	0.087^u	0.150^v	$0.087 + 0.063x$
m_{hh}	0.62^v	0.76^v	$0.62 + 0.14x$
m_{so}	0.15^v	0.24^v	$0.15 + 0.09x$

Table 5.2. Physical parameters of GaAs and related compounds (after Adachi [1985]).

[a] *H.C. Casey Jr and M.B. Panish Heterostructure Lasers (Academic, New York, 1978) Part A.*

[b] *M. Cardona, K.L. Shaklee and F.H. Pollak Phys. Rev. **154** 696 (1967).*

[c] *A. Onton Proceedings of the 10th International Conference on the Physics of Semiconductors (Cambridge, Mass, 1970) p 107. Note that his symmetry assignment was corrected by W.H. Berninger and R.H. Rediker [Bull. Am. Phys. Soc. **16** 306 (1971)].*

[d] *These values are taken from a tabulation of A.G. Milnes and D.L. Feucht [Heterojunctions and Metal–Semiconductor Junctions (Academic, New York, 1972)].*

[e] *J.C. Phillips Bonds and Bands in Semiconductors (Academic, New York, 1973).*

[f] *R. Zallen and W. Paul Phys. Rev. **155** 703 (1967).*

[g] *Extrapolated from the data of $Al_xGa_{1-x}As$ by N Lifshitz, A Jayaraman, R.A. Logan and R.G. Maines [Phys. Rev. B **20** 2398 (1979)].*

[h] *Calculated by D.L. Camphausen, G.A.N. Connell and W. Paul [Phys. Rev. Lett. **26** 184 (1971)].*

[i] *Assumed similar to GaP [see text].*

[j] *M. Zvara Phys. Status Solidi **27** K157 (1968).*

[k] *B. Monemar Phys. Rev. **B 8** 5711 (1973).*

[l] *Assumed that $dE_g^X / dT = dE_2 / dT$ (see text). The data of dE_2/dT is taken from R.L. Zucca and Y.R. Shen [Phys. Rev. B **1** 2668 (1970)].*

[m] *F.H. Pollak, C.W. Higginbotham, and M. Cardona J. Phys. Soc. Jpn. Suppl. **21** 20 (1966).*

[n] *B. Rheinlander, H. Neumann, P. Fischer, and G. Kuhn Phys. Status Solidi b **49** K167 (1972).*

[o] *E.M. Conwell and M.O. Vassell Phys. Rev. **166** 797 (1968).*

[p] *D.E. Aspnes and A.A. Studna Phys. Rev. B 7 4605 (1973).*

[q] *Calculated from the usual k·p theory.*

[r] *D. Aspnes Phys. Rev. B **14** 5331 (1976).*

[s] *Assumed similar to GaAs.*

[t] *Calculated from $m_e^{\alpha} = N^{2/3}m_{ta}^{2/3}m_{la}^{2/3}$, where N is the number of equivalent α minima ($\alpha = \Gamma$, X or L). Calculated also from $m_e^{\alpha} = (2/m_{ta} + 1/m_{la})^{-1}$.*

[u] *A.L. Mears and R.A. Stradling J. Phys. C: Solid State Phys. **4** L22 (1971).*

[v] *Taken from a tabulation of P. Lawaetz [Phys. Rev. B **4** 3460 (1971)].*

[w] *Taken from H.J. Lee, L.Y. Juravel, J.C. Woolley and A.S. Thorpe Phys. Rev. B **21** 659 (1980).*

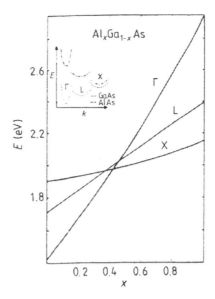

Fig. 5.6. Energetic position of Γ, L and X bands in AlGaAs at T = 300 K (after Landolt–Bornstein [1982]).

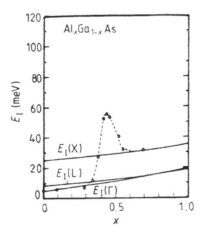

Fig. 5.7. Donor ionization energies, $E_I(\Gamma)$, $E_I(X)$ and $E_I(L)$, and acceptor ionization $E_I(V)$ as a function of composition x for $Al_xGa_{1-x}As$ alloy. The theoretical curves (solid lines) are obtained from the hydrogen approximation using the band-mass parameters. The experimental data (open circles) are taken from Asai and Sugiura [1985].

However, a sharp increase of this binding energy (nearly 50 meV) can be seen around $x = 0.45$, which cannot be explained by a simple Γ-like to X-like switch. From transport measurements, even higher binding energies reaching 150 meV near the $x = 0.45$ crossover point were deduced. All these data agree in establishing that this singularity increases with the composition x of aluminum, reaching its maximum for $x \approx 0.45$, at the crossing of Γ, L and X bands. Hence, the defect centers are called DX centers. The presence of these centers is connected with persistent photoconductivity effects at low temperature: once the centers are optically emptied by irradiation at low temperature, a barrier is introduced to electron recapture (persistent photoconductivity). This barrier can be overcome by thermally heating the sample to higher temperatures where the persistent photoconductivity effect is reduced.

The detailed microscopic structure of DX centers remains unclear [Lang 1974, Bourgoin et al. 1988] although it is clearly connected with the crossover of Γ, L and X bands around $x = 0.45$. Unfortunately, it leads to degradation in the performance of heterostructure lasers and modulation-doped field effect transistors [Stormer et al. 1979]. In this regard, it would be desirable to replace $Ga_xAl_{1-x}As$ by $Ga_{0.51}In_{0.49}P$, composition lattice-matched to GaAs, since the crossover composition is $Ga_{0.74}In_{0.26}P$.

5.3. MOCVD growth mechanism of GaAs and related compounds

The general features of MOCVD growth mechanisms such as flow patterns and chemical reactions have been described in Chapter 2.

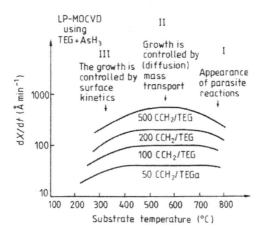

Fig. 5.8. Growth rate of GaAs, as a function of temperature for different TEGa flow rates (after Razeghi et al. [1989]).

Fig. 5.8 shows the growth rate of GaAs layers, as a function of temperature. Three different zones are clearly evident in which:

(i)   growth is limited by the kinetics of cracking of growth species in the low-temperature range,

(ii)   growth is limited by the diffusion of group III elements through the gas phase, and the growth rate is temperature independent,

(iii)   growth is limited by the desorption of species in the high-temperature range.

Van Sark et al. [1990] have first derived the growth rate using an isothermal model by taking into account various flow profiles in a horizontal reactor tube.

Assuming that no diffusion occurs along the side walls (z direction) and along the flow direction (x direction), the diffusion equation can be written in the form [Van Sark et al. 1990]

Eq. (5.12)

$$v(y)\frac{\partial C(x,y)}{\partial x} = \frac{\partial}{\partial y}\left[D(T(y))\left(\frac{\partial C(x,y)}{\partial y} + (\alpha_T + 1)\frac{C(x,y)}{T(y)}\frac{\partial T(y)}{\partial y}\right)\right]$$

where $v(y)$ is the flow velocity, $C(x, y)$ is the concentration of growth species, $D(T(y))$ is the binary diffusion coefficient of the group III elements, α_T is the thermal diffusion factor, $\partial C/\partial y$ represents the diffusion due to the gradient concentration and

Eq. (5.13) $$(\alpha_T + 1)\frac{C(x,y)}{T(y)}\frac{\partial T(y)}{\partial y}$$

represents the diffusion due to the thermal gradient.

Assuming no sidewall diffusion allows the problem to be essentially two dimensional (x, y plane). In an isothermal model, the temperature is assumed to be constant, $T = T_0$, so that there is no diffusion due to the thermal gradient.

The validity of each flow profile model is checked by defining the total deposition parameter which is equal to the ratio of the deposited growth species concentration over the input growth species concentration, and has to be close to unity.

By solving Eq. (5.12), the growth rate is derived:

Eq. (5.14) $$R(x) = D(T_0)\frac{\partial C(x,y)}{\partial y}\bigg|_{y=0}$$

No essential difference is found between the "exact" parabolic profile and the constant "plug" flow $v(y) = v_0$, so that the plug flow is sufficient to calculate the growth rate. The exact solution is found in the form

Eq. (5.15) $$\frac{R(x)}{v_0 C_0} = \frac{D(T)}{v_0 h}\sum_{n=1}^{\infty} A_n \exp\left(-B_n\frac{D(T)}{v_0 h}\frac{x}{h}\right)$$

By retaining only the first-order term, the growth rate can be expressed in the simplified form

Eq. (5.16)
$$\frac{R(x)}{v_0 C_0} = A\frac{D(T)}{v_0 h}\exp\left(-B\frac{D(T)}{v_0 h}\frac{x}{h}\right)$$

The growth rate is found to decrease exponentially along a non-tilted susceptor. This fact has been experimentally evidenced by Van de Ven et al. [1986] and Razeghi [1989].

The influence of surface kinetics is investigated by introducing a reaction rate constant $k(T_0)$, so that Eq. (5.14) is transformed into

Eq. (5.17)
$$D(T_0)\frac{\partial C(x,y)}{\partial y}\bigg|_{y=0} = k(T_0)C(x,0)$$

By defining the dimensionless CVD number: $N_{CVD} = k(T_0)h/D(T_0)$, Van Sark et al. [1990] have found a limit for which:

$N_{CVD} > \pi^2/4$ (fast surface reaction):   growth is diffusion controlled,
$N_{CVD} < \pi^2/4$ (slow surface reaction):   growth is kinetics controlled with slower growth rates.

The diffusion due to thermal gradients has been taken into account in Eq. (5.12). The influence of heating leads to thermal expansion, giving a mean flow velocity higher than in the previous case.

Considering the input temperature $T_0 = 300$ K, and the substrate temperature T_s, the mean flow velocity can be approximated well by

Eq. (5.18)
$$v_T = \frac{T_s/T_0 - 1}{\log(T_s/T_0)}v_0.$$

Heating also has an influence on the diffusion coefficient, which increases with temperature, following the law

Eq. (5.19)
$$D(T) = D_0\left(\frac{T}{T_0}\right)^{\gamma}$$

where γ is a coefficient close to 1.7 in the usual cases.

By solving Eq. (5.12) with thermal diffusion, Van Sark et al. [1990] have expressed the growth rate under the general form (Eq. (5.14)), with an exponential decrease of the growth rate along the susceptor, thus leading to depletion effects. The thermal diffusion factor α_T and the binary diffusion coefficient $D(T)$ have opposite effects on depletion: thermal diffusion toward the cold wall (α_T) weakens the depletion effect while an increase in $D(T)$ increases the mass diffusion, thus leading to an increased depletion effect.

Growth pressure	76 Torr
Growth temperature	510 °C
Total H_2 flow rate	3 l·min^{-1}
AsH$_3$ flow rate	30 cm^3·min^{-1}
H_2 through TEGa bubbler (at 0 °C)	120 cm^3·min^{-1}
Growth rate	150 Å·min^{-1}

Table 5.3. Optimum LP-MOCVD growth conditions for GaAs.

Taking into account the kinetics of a surface reaction with thermal effects leads to identical results without thermal effects. When the N_{CVD} constant is larger than unity, growth is essentially diffusion controlled. Note that depletion effects can be cancelled by tilting the susceptor a few degrees [Razeghi 1989].

Anyway, the modeling and interpretation of the growth mechanism of the MOCVD growth technique without accurate *in situ* diagnosis of the chemical reaction occurring in the gas phase and in the vicinity of the substrate during growth is very difficult.

5.4. Experimental details

The growth apparatus has been described in Chapter 2. GaAs layers can be grown either at atmospheric or low pressure in the temperature range between 500 and 550 °C. Triethylgallium $(C_2H_5)_3$Ga (TEGa) or trimethylgallium (TMGa) $(CH_3)_3$Ga and pure arsine (AsH$_3$) have been used as Ga and As sources. Pure H_2 can be used as a carrier gas. The growth rate depends linearly on the flow rate of group III (Ga) elements and is independent of the AsH$_3$ flow rate, substrate temperature and substrate orientation, suggesting that the epitaxial growth is controlled by the mass transport of the group III species.

Growth of GaAs layers has been carried out on $\langle 1\,0\,0 \rangle$ substrates misorientated up to 2° towards the $\langle 0\,1\,1 \rangle$ plane. The substrates were etched in a 5:1:1 $(H_2SO_4\!:\!H_2O_2\!:\!H_2O)$ solution for 20 s at 40 °C, rinsed with deionized water and dried under pure N$_2$. The substrates were initially heated to the growth temperature under H_2 and AsH$_3$ for 5 min before growth began in order to remove any surface oxides. Table 5.3 lists the optimum growth conditions for the LP-MOCVD growth of GaAs, which were used for this study.

Epitaxial layers are *n*-type for a wide range of V/III ratios and substrate temperatures. Suitable growth conditions such as growth temperature, growth rate, total flow rate, purity of starting materials and reactor design are responsible for the high quality of the epilayers.

Fig. 5.9 shows the SIMS profile for a 10 μm thick GaAs layer grown by MOCVD at 510 °C. The concentration of the impurities in the layer and at the interface is low and homogeneous. This result shows that the preparation of the substrate and the growth conditions were good.

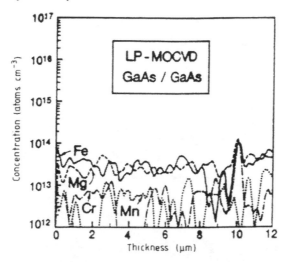

Fig. 5.9. SIMS profile for a 10 μm thick undoped GaAs layer grown by LP-MOCVD
(after Razeghi et al. [1989]).

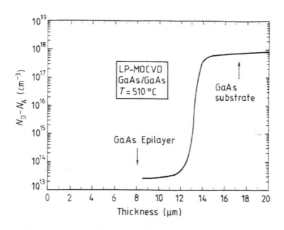

Fig. 5.10. Carrier concentration against depth profile of an ultrapure bulk GaAs
grown by MOCVD (after Razeghi et al. [1989]).

Fig. 5.10 shows the electrochemical $C-V$ profile of a typical undoped GaAs layer. A high-purity epilayer with a residual carrier concentration as low as 10^{13} cm^{-3} for a 12 μm thick layer has been obtained.

Electron Hall mobilities of epitaxial layers grown on semi-insulating ((Cr–O)-doped) GaAs substrates were measured in a magnetic field of 5 kG by a conventional Van der Pauw technique. Table 5.4 indicates the measured mobilities at 300 K, at 77 K and at low temperature (12 K) in the undoped GaAs epilayers.

	A	B	C
$N_d - N_a$ (cm^{-3}) (polaron)	2×10^{13}	2×10^{13}	3×10^{13}
Thickness (μm)	10	10	12
$N_d - N_a$ (cm^{-3}) (Hall)	1.6×10^{13}	1.4×10^{13}	10^{13}
μ(300 K) (cm$^2 \cdot$V$^{-1} \cdot$s^{-1})	8200	9000	9000
μ(77 K) (cm$^2 \cdot$V$^{-1} \cdot$s^{-1})	190,000	200,000	210,000
μ_{max} (cm$^2 \cdot$V$^{-1} \cdot$s^{-1})	290,000	320,000	335,000
	(T = 45 K)	(T = 40 K)	(T = 38 K)
			600,000
			(T = 12 K)

Table 5.4. Measured values of the electron Hall mobility for three samples of undoped GaAs layers grown by LP-MOCVD.

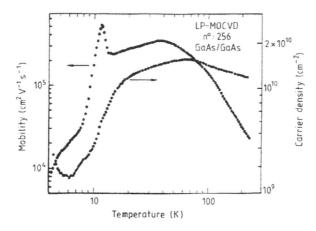

Fig. 5.11. Variation of electron Hall mobility and carrier density as a function of temperature for a high-purity GaAs layer grown by MOCVD (after Razeghi et al. [1989]).

Fig. 5.11 shows the experimental mobility measured at 5 kG as a function of temperature. Electron Hall mobility as high as 335,000 cm$^2 \cdot$V$^{-1} \cdot$s^{-1} at 38 K has been measured. The mobility curve contains an additional peak at 12 K with a mobility of 600,000 cm$^2 \cdot$V$^{-1} \cdot$s^{-1}, which is the highest mobility reported for GaAs layers grown by any technique. This shoulder can result from changes in the ionized impurity concentration due to carrier freezeout [Colter et al. 1983, Stillman and Wolfe 1976]: a phenomenon which has been reported before for VPE GaAs samples [Wolfe and Stillman 1975]. Two factors can contribute to a high low-temperature mobility: the relatively low scattering rate due to the ionized impurities and the relatively high screening effect due to neutral impurities [Colter et al. 1983]. For temperature less than 10 K the resistivity of the sample increases and measurements become very difficult. Fig. 5.12 shows the variation of carrier concentration as a function of temperature for this epilayer. The activation energy measured from this curve is

6 meV, which corresponds to a shallow donor such as silicon or sulfur in the GaAs undoped epilayer (see Fig. 5.12).

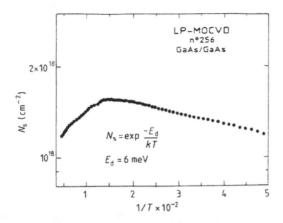

Fig. 5.12. Variation of N_s as a function of 1/T (after Razeghi et al. [1989]).

Ehrenreich [1959] has examined the scattering mechanisms which determine the transport properties of GaAs layers and has shown that a combination of polar optical phonon and ionized impurity scattering yielded qualitative agreement with the temperature and impurity concentration dependence of the electron mobility for the pure bulk GaAs. Later, Stillman and Wolfe [1976] and Wolfe and Stillman [1975] have shown that the combination of polar optical phonons, piezoelectric acoustic phonon, deformation potential acoustic phonon, ionized impurity and neutral impurity scattering in the relaxation time approximation gave results which were in good agreement with the temperature and concentration dependence of the electron mobility in high-purity GaAs grown by vapor phase epitaxy. In their analysis, they have assumed:

each scattering process is described by a relaxation time $\tau(E)$ which may depend on the electron energy (E),

(i)   the scattering mechanisms are independent of each other,

(ii)   the electrons are scattered in a parabolic band. With the above approximation, an average relaxation time $\langle \tau \rangle$ can be calculated from the equation

Eq. (5.20) $\langle \tau \rangle = \frac{4}{3\sqrt{\pi}} x \int_0^\infty \tau(E) \frac{E^{3/2} \exp(-E/k_B T)}{k_B T^{5/2}} dE$

where E is the electron energy and k_B is the Boltzmann constant. The electron mobility is then determined from $\mu = e\langle \tau \rangle / m^*$, where e is the electronic charge and m^* is the electron effective mass.

Satisfying agreement between the experimental results and calculations based on these simplifying assumptions has been obtained, assuming the residual doping level has been lowered down to less than the 10^{14} cm^{-3} range. This is confirmed by Fig. 5.10, which shows the typical electrochemical *C–V* profile of the undoped GaAs layers grown by MOCVD. Carrier concentrations as low as 2×10^{13} cm^{-3} for a 13 μm thick layer have been measured. All layers are *n*-type.

Table 5.4 also shows the measured values of electron Hall mobility for undoped GaAs layers grown by MOCVD, under the growth conditions indicated in Table 5.3.

Hall effect data indicated that the free electrons, of the order of 10^{13} cm^{-3}, are provided by donor impurities which are partially compensated by acceptor impurities. In order to provide additional information on the nature and concentration of the defects contained in these layers, deep-level transient spectroscopy(DLTS) has been used to evaluate the amount of electron compensation they induce and the influence they can have on the electron mobility.

For this experiment, identical layers under conditions described in Table 5.3 and Table 5.4 have been grown on n^{+}-substrates. Ohmic contact is made on the back of the substrate and Au Schottky barriers are evaporated on the top of the layers in order to perform capacitance–voltage (C–V) and DLTS measurements.

The $C^{-2}(V)$ characteristics indicate a uniform carrier profile and a free-electron concentration in the range of 1 to 4×10^{13} cm^{-3} at 300 K, increasing noticeably at 400 K. This increase in electron concentration suggests the presence of a donor defect, which emits electrons between 300 and 400 K. Its concentration can be estimated from the difference between the electron concentrations at 300 and 400 K. This difference varies from 0.3 to 7×10^{13} cm^{-3} depending on the sample studied. From DLTS results (see Fig. 5.13), this defect can be attributed to the so-called EL2 defect [Bourgoin et al. 1988].

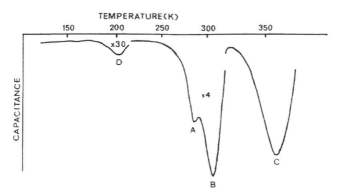

Fig. 5.13. Deep-level transient spectroscopy spectra of GaAs grown by MOCVD (after Feng et al. [1991]).

Capacitance transient measurements have been performed in the range 4–300 K. DLTS spectra show the existence of a large peak, labeled C, around 350 K and two small ones, labeled A and B, in the 250–300 K range (see Fig. 5.13). The signatures—the variation of emission rates with inverse temperature (Fig. 5.14)—give the following thermal ionization energies to the conduction band edge: 0.47 eV (A), 0.65 eV (B) and 0.75 eV (C). The signature of peak C is similar to the one expected for the EL2 defect; in some cases a difference is found which can be accounted for by the fact that the defect concentration is of the order of the free-carrier concentration [Stievenard et al. 1985]. Typically, in a layer containing 1×10^{13} cm^{-3} free electrons at 300 K, the EL2 (C) concentration reaches 4×10^{12} cm^{-3}. The concentrations of the 0.47 and 0.65 eV traps are similar, typically $\sim 10^{12}$ cm^{-3}.

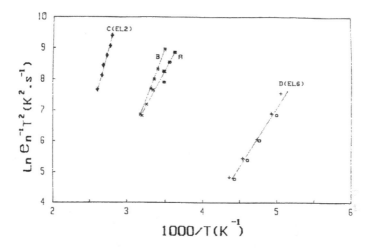

Fig. 5.14. Variation of the emission rates with the inverse of temperature (after Feng et al. [1991]).

Finally, a peak labeled D is also observed at 200 K; its concentration is small $(5 \times 10^{11}$ cm$^{-3})$ and it is associated with an ionization energy of 0.32 eV (see Fig. 5.14). Because it is located only near the surface, and its signature is very similar to the one associated with the so-called EL6 defect [Bourgoin et al. 1988], one can attribute the existence of peak D to the EL6 defect.

Photoluminescence (PL) measurements were made at 4.2 K using a 633 nm He–Ne laser ($P_{exc} = 1$ W·cm^{-2}) as an excitation source. The luminescence was detected using a 1 m double spectrometer equipped with a GaAs photomultiplier tube; the PL spectra are shown in Fig. 5.15. They are very similar to those of the purest VPE samples. The free-exciton (polariton) emission of sample C, which is 12 μm thick, is particularly intense at 1.5149 and 1.5154 eV, with a shoulder at 1.5177 eV associated with the excited state ($n = 2$) of the free exciton. This assignment has been confirmed by reflectivity measurements showing a marked dip at the energy of the ($n = 1$) free exciton between the two polariton branches. The sharp line at 1.5141 eV is due to the recombination of excitons bound to neutral donors (D^0–X). The peaks at 1.5134 and 1.5125 eV originate from the recombination of free holes with neutral donors (D^0–h) and from the recombination of excitons bound to neutral acceptors (A^0–X). The weakness of this (A^0–X) peak is an indication of the low degree of compensation of the samples. Further evidence of the low-acceptor background is given by the small ratio of the band to acceptor luminescence in the 1.49 eV range (not shown in Fig. 5.15) compared to the excitonic luminescence. This ratio amounts to 0.30, 0.23 and 0.16 for samples A, B and C, respectively, the lowest value being obtained for the highest-mobility sample, as expected. The better quality of sample C in comparison to sample A can also be seen in the PL spectra of Fig. 5.15. The relative intensity of the free-exciton emission is smaller for A, whereas the (A^0–X) emission is stronger.

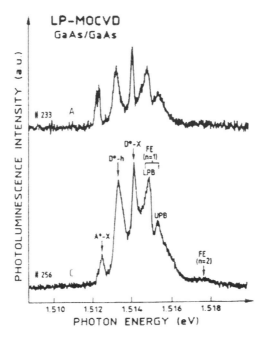

Fig. 5.15. Photoluminescence spectra at T = 4.2 K of two ultrapure thick GaAs epilayers grown by MOCVD (after Razeghi et al. [1989]).

5.5. Incorporation of impurities in GaAs grown by MOCVD

When an impurity atom is introduced in a GaAs lattice, if it occupies an As site, and it provides the crystal with one or more additional electrons than the atom it replaced, then the impurity is called a donor. Thus S, Se and Te on an As site in GaAs are donors. If the impurity atom provides fewer electrons than the atom it replaces, it forms an acceptor, like Zn on a Ga site in GaAs.

Instead of replacing an atom of the GaAs crystal, the impurity may lodge itself in an interstitial position. Then its outer-shell electrons are available for conduction and the interstitial impurity becomes a donor.

A missing atom results in a vacancy and deprives the crystal of one electron per broken band. This makes the vacancy an acceptor. During the MOCVD growth of GaAs, a deviation from stoichiometry (due to the ratio of As/Ga) generates donors or acceptors depending on whether it is the Ga or As which is in excess. The extra electron of the donor is attracted most strongly to the positive charge of the impurity nucleus. Thus it acts as the electron of a hydrogen atom immersed in the high dielectric constant ε of the GaAs crystal. This enables us to calculate the energy binding the electron to the impurity (its ionization energy):

Eq. (5.21) $$E_i = \frac{m*q^4}{2h^2\varepsilon^2 n^2} = \frac{m*}{m\varepsilon^2 n^2}13.6 \text{ eV}$$

where q is the electron charge, m is the mass of an electron in vacuum, n is a quantum number ≥ 1. The ionization energy from the ground state to the conduction band is obtained by making $n = 1$. Since $\varepsilon(\text{GaAs}) = 13.1$ and the effective mass ratio of an electron is 0.067 (and 0.48 for a hole), the ionization energy of a donor and acceptor in GaAs is then 6 meV and 40 meV, respectively.

Some impurities do not agree with the simple hydrogen model and form levels which may lie deep in the energy gap. All the transition elements seem to form deep levels in GaAs. The reasons for which certain impurities form a deep level are not yet completely understood.

Since the impurity is usually either larger or smaller than an atom of the GaAs lattice, a local mechanical strain can be obtained. An interstitial atom evidently induces a deformation potential corresponding to compressional strain, whereas a vacancy will have the opposite effect, since it produces dilational strain. Usually in GaAs layers, both interstitials and vacancies should be present in addition to substitutional impurities. Dislocations are also usually present in GaAs epilayers. They occur at the edge of an extra plane of atoms. The misfit of such an extra plane results in compressional and dilatational strains, with the consequent onset of both lowering and raising of the potentials in the neighborhood of the dislocation.

5.5.1. Residual impurities

The epitaxial layers of GaAs have been grown generally using triethylgallium (TEGa) or trimethylgallium (TMGa), and arsine (AsH_3). The purity of source materials is one of the most important factors to be considered to improve the quality of products. For the MOCVD growth of GaAs, there have been many publications dealing with residual impurities in the epitaxial layers. The quality of epitaxial GaAs layers is known to be affected by the impurities in TEGa or TMGa, for example silicon [Dapkus et al. 1981, Hess et al. 1982, Nakanishi et al. 1981]. Hence, great efforts have been made to improve the purity of TMGa and TEGa. On the other hand, Ge (germanium) as a dominant donor impurity was found in undoped MOCVD-grown GaAs layers and AsH_3 was found to be a possible source of germanium contamination.

5.5.2. Carbon incorporation in GaAs grown by MOCVD

Kuech et al. [1988] studied the influence of hydrocarbons in MOCVD growth of GaAs. Quantitative measurements of carbon concentration in GaAs layers grown by MOCVD is required to elucidate both the growth reactions and the influence of reaction by-products. The influence of methane (CH_4) in MOCVD on both the growth rate and the incorporation of carbon has been studied by several authors [Kuech et al. 1988, El Jani et al. 1982]. In general, the introduction of CH_4 into the growth ambient does not substantially alter the growth rates or result in any additional electrically active carbon [Kuech et al. 1988]. The total additional carbon incorporation from methane, both electrically active and inactive, is also negligible. The low or negligible carbon incorporation has been attributed to both the high reaction efficiency between the hydrogen, which originates on the arsine, and the methyl radicals released on or near the growth surface. The other hydrocarbon sources, such as C_2H_2, C_2H_4 and C_2H_6, should be more reactive due to the presence of the double carbon bond, $-C=C-$. These compounds lead to slight amounts of carbon incorporation when there are high concentrations in the gas phase. Similarly, these compounds may decompose in the gas phase. However, their reaction products must be either non-reactive with respect to the GaAs surface or react rapidly in the gas phase to form other, perhaps saturated, non-reactive hydrocarbons.

Mochizuki et al. [1988] studied the mechanism by which carbon is incorporated into GaAs layers by atomic layer epitaxy using TEGa or TMGa and AsH_3. They showed that the carbon density varied from 1×10^{13} to 8×10^{18} cm^{-3} with TEGa or TMGa and AsH_3 pulse durations and mole fractions. They also observed that carbon incorporation drastically changed at the pulse duration and mole fraction where the growth rate per gas cycle started to saturate to one monolayer (0.283 nm/cycle for a (1 0 0) substrate). They explained the results by the selective adsorption of carbon on surface gallium, the reaction of methylgallium with arsine, and the exchange interaction between arsenic and carbon atoms. They found that even when the TMGa source was used, the epitaxial layers grown under the optimized growth conditions exhibited an electron concentration of 1×10^{14} cm^{-3} and a mobility of 80,000 $cm^2 \cdot V^{-1} \cdot s^{-1}$ at 77 K, with a photoluminescence spectrum with several sharp excitonic lines at the bandgap energy, and an extremely low level of carbon-related peaks.

In conclusion, the TMGa source is the origin of carbon impurities in the GaAs layers grown by MOCVD. The probability of carbon incorporation in the epitaxial layers is increased by increasing the TMGa flow rate. When the ratio of $TMGa/AsH_3 > 1$, the epitaxial layers are *p*-type, and when the ratio of $TMGa/AsH_3 \ll 1$, the epitaxial layers are *n*-type. Arsine plays an important role in the incorporation of carbon in GaAs layers grown by MOCVD.

5.5.3. n-type GaAs

GaAs layers grown by MOCVD can be doped using Si, Se or S as explained in detail below.

(a) Silicon

Silicon is one of the primary dopants in GaAs and related compounds. Controlled carrier concentrations from 10^{15} up to 10^{19} cm^{-3} can be achieved using either silane (SiH_4), or disilane (Si_2H_6). The low diffusion coefficient of silicon in GaAs and the availability of high-purity sources make the silanes attractive doping sources for device applications. Sharp doping profiles can be routinely achieved since neither silane nor disilane exhibits any "memory effect" in the reactor.

Silane exhibits a strong temperature dependence as well as low efficiency for silicon incorporation. As a result, temperature variations across the susceptor lead to doping non-uniformities across the wafer. Kuech et al. [1984] showed that doping from a disilane source has a much higher silicon incorporation efficiency and is relatively temperature independent under most typical growth temperatures and growth conditions. Shimazu et al. [1987] found that for a particular reactor pressure, using Si_2H_6, and at low temperatures, the doping concentration has an Arrhenius dependence on temperature $(K = A \cdot \exp(-k/RT))$, but becomes temperature independent at high temperatures. They showed, by increasing the total reactor pressure, that the overall doping efficiency increases, while the transition temperature at which the doping becomes temperature independent decreases. The reaction mechanism underlying the temperature dependence of Si doping using silane in MOCVD-grown GaAs is the same as in the case of InP [Razeghi 1989].

They proposed a simple model to describe the Si incorporation in GaAs from SiH_4. They assume the following.

(i) The thermal decomposition of SiH_4 is heterogenous, and does not occur in the gas phase. (The Si–H bond strength in SiH_4, 3.2–3.4 eV, is comparable to the As–H bond strength of AsH_3, 2.9–3.2 eV [Saalfeld and Svec 1963]).

(ii) On (1 0 0) GaAs, each chemisorbed species forms two bonds to the surface so they expect $SiH_{2(ads)}$ doping as the initial adsorbed (ads) species leading to incorporation. They proposed that this adsorption reaction is the rate limiter:

Eq. (5.22) $$SiH_4 + V \rightarrow SiH_{2(ads)} + H_2$$

where V is a vacant adsorption site.

(iii) The adsorption rate equals the incorporation rate of Si on GaAs.

(iv) The surface pressure of SiH_4 is approximately the same as the inlet pressure of SiH_4. This is because the adsorption is limited by Eq. (5.22) rather than by diffusion.

Shimazu et al. [1987] explained the variation of Si doping with temperature and reactor pressure using this model, considering also the effects of vacancy coverage and variable adsorption energies. In addition, this model explains the variation of doping with arsine pressure.

Field and Ghandhi [1986] have grown GaAs at a range of temperatures, from 650 to 750 °C, by the MOCVD process at pressures from 0.046 to 1.0 atm, using SiH_4 as a dopant source. They found that the doping is proportional to SiH_4 pressure, inversely proportional to growth rate, and increases with temperature.

Veuhoff et al. [1985] studied the incorporation of silicon in GaAs layers grown by MOCVD using SiH_4 and Si_2H_6 doping sources. The layers were grown on $\langle 1\,0\,0 \rangle$, $\langle 1\,1\,1 \rangle$ A and $\langle 1\,1\,1 \rangle$ B surfaces over a wide range of growth temperatures and gas phase stoichiometries. They found an influence of the substrate orientation on silicon incorporation using SiH_4. They described this behavior by a qualitative model involving both surface-specific adsorption sites and possible surface chemical reactions. In the case of disilane, Si_2H_6, the incorporation process appears to be independent of substrate orientation and growth temperature. This temperature independence results in improved uniformity of the electron concentration over large areas of the substrate when disilane is used as the doping gas.

The incorporation of Si into GaAs epilayers grown by MOCVD using either an SiH_4 or Si_2H_6 source is proportional to its mole fraction in the reactor, inversely proportional to the growth rate and growth temperature. Usually Si substitutes for Ga in the GaAs lattice and acts as a donor: as Si is amphoteric, the donor concentration depends on the III/V ratio.

Omnes et al. [1991] studied silicon incorporation in LP-MOCVD GaAs layers grown at low temperature between 500 and 600 °C. They used 15 ppm silane diluted in pure H_2. The electron concentration varies linearly with the silane flow rate and exponentially with the inverse of growth temperature in the following form:

$$\text{Eq. (5.23)} \qquad N_D - N_A = B\left(P_{SiH_4}, P_{AsH_3} \right) \exp\left(-\frac{E_i}{k_B T} \right)$$

with $E_i = 1.2$ eV, in good agreement with the results of Field and Ghandhi [1986]. A weak dependence of electron-carrier concentration as a function of AsH_3 flow rate of the form $(P_{AsH_3})^{-0.5}$ was observed previously by Field and Ghandi [1986].

Bass and Oliver [1977] found a rise in doping level with increasing TMGa concentration, with a power law of about 0.6. The doping efficiency K is given by:

$$\text{Eq. (5.24)} \qquad K \propto (\text{growth rate})^{-1.6}$$

Luther and di Lorenzo [1975] found a similar law for the trichloride system and attributed it to kinetic effects. Another explanation is that the surface stoichiometry and hence the surface concentration of available arsenic sites is partly determined by the TMGa concentration. They found a considerable increase in doping level and doping efficiency as the growth temperature is decreased.

(b) Selenium

GaAs layers grown by MOCVD can be *n*-type doped using H_2Se, DESe or DMSe as a source of Se. Sakaguchi et al. [1988] showed that the carrier concentration of Se-doped GaAs increases when the V/III ratio decreases. As Se substitutes for As, the carrier concentration of GaAs doped with Se increases when the flow rate of AsH_3 decreases and is independent of group III sources. In the case of using H_2Se as a source of Se, the free-carrier concentration in the GaAs layers decreases when the growth temperature increases. The decomposition of the H_2Se (primary species) is the rate limiting step: after decomposition, most of the secondary species (Se) is incorporated into the growing layer. Contrary to SiH_4, H_2Se shows a strong memory effect.

Asai and Sugiura [1985] proposed a model of the Se incorporation mechanism in MOCVD GaAs layers. Their model is based on the following assumptions.

(i)   Se and As species are competitively adsorbed on a growing surface.
(ii)   Subsequently, Se and As adsorbed at the surface react with gas phase Ga species and as a consequence the Se donor is incorporated into a GaAs epitaxial layer. According to this model, the doping concentration in the GaAs layer depends on the H_2Se/AsH_3 ratio and is independent of the group III Ga sources, which is confirmed experimentally.

(c) Sulfur

GaAs layers grown by MOCVD can be *n*-type doped using H_2S. Bass and Oliver [1977] investigated the variation of the doping level with concentrations of H_2S (hydrogen sulfide), AsH_3 and TEGa, as well as the growth temperature. They found a linear relationship between the H_2S concentration and the doping level up to $>10^{18}$ cm^{-3}. The incorporation of sulfur is given by:

$$H_2S_{(V)} + V_{(As)} \rightarrow S^-_{(As)} + n^- + H_2.$$

Eq. (5.25)
$$\downarrow \qquad\qquad \downarrow$$

Arsenic   Sulfur electron

vacancy   on arsenic

Under equilibrium conditions, the mass action law gives:

Eq. (5.26) $$K=\dfrac{\left[S^-_{As}\right]\left[n^-\right]P_{H_2}}{P_{H_2S}}$$

If $n < n_i$, the intrinsic electron concentration, a linear law would be expected, while if $n < n_i$, a square root law would follow. The doping level is proportional to the inverse of arsenic concentration. One would expect this from the above equation as $V_{(As)}$ should be inversely proportional to the arsenic concentration.

We studied the behavior of H_2S doping in GaAs layers grown by LP-MOCVD. The doping level increases linearly when the growth temperature is increased with $E_i = -1.46$ eV.

When the flow rate of H_2S is kept constant, the free-carrier concentration varies exponentially with $1/T$, as in the case of H_2Se. The free-carriers concentration in the epilayer decreases when the growth temperature increases. S substitutes for As and plays the role of donor in the GaAs lattice. The decomposition of the H_2S (primary species) is the rate limiting step after decomposition. Most of the secondary species (S) are incorporated into the growing layers.

5.5.4. p-type GaAs

(a) Zinc

GaAs layers grown by MOCVD can be p-type doped using pure or diluted DMZn $((CH_3)_2Zn)$ or DEZn $((C_2H_5)_2Zn)$. Bass and Oliver [1977] studied incorporation of Zn in GaAs layers grown by MOCVD. Layers were p-type doped using two different concentrations of DMZn diluted in hydrogen as a doping gas. They found a near-linear relationship between the hole concentration and the DMZn concentration at higher levels. The doping level increases with arsenic concentration which one would expect from a stoichiometric effect on the vacancy concentration. Increasing the TMGa concentration showed an increase in doping level. This is not expected from vacancy theory and hence it must be a kinetic effect. They suggest that if the surface concentration of zinc is higher than the bulk concentration and is only slowly desorbed during growth, a higher doping level might be expected with higher growth rates. When the growth temperature increases, the hole concentration in the epilayer decreases.

Razeghi et al. [1989] studied GaAs layers, doped with DEZn by LP-MOCVD, in the temperature range between 500 and 600 °C. The temperature of DEZn was kept at -15 °C during this study. The H_2 flow through the DEZn bubbler varied between 2.5 and 5 $cm^3 \cdot min^{-1}$. Fig. 5.16 shows that the hole concentration varies linearly with DEZn flow. The influence of growth temperature has been studied by keeping the DEZn, AsH$_3$ and TEGa flows constant. As shown in Fig. 5.17, the hole concentration in the layer decreases when the growth temperature increases. Fig. 5.17 shows that the doping level varies exponentially with the inverse of temperature, with $E_i = -2.6$ eV.

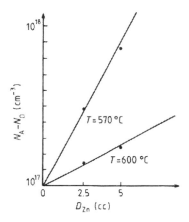

Fig. 5.16. Variation of hole concentration in GaAs as a function of flow rate of DEZn for different growth temperatures (after Omnes et al. [1991]).

Keeping the growth temperature at 510 °C and a DEZn flow of 2.5 $cm^3 \cdot min^{-1}$, the arsine flow has been varied between 60 and 90 $cm^3 \cdot min^{-1}$. The arsine flow has no dependence on the doping level so that both parameters can be considered as independent.

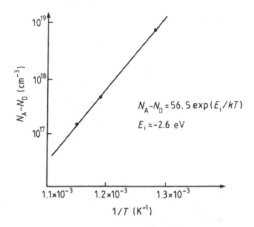

Fig. 5.17. Variation of hole concentration in GaAs as a function of 1/T (after Omnes et al. [1991]).

The acceptor concentration depends on the flow rate of H_2 through the DEZn or DMZn bubbler. When the flow rate of DEZn or DMZn is kept constant, the free-carrier concentration varies exponentially with $1/T$. The incorporation of dopants during MOCVD growth of GaAs using doping species such as DEZn or DMZn can be explained by using the model described in Volume 1 [Razeghi 1989] as in the case of InP layers.

(b) Magnesium

The GaAs layer can be *p*-type doped using $Mg(C_5H_5)_2$ or cyclopentadienyl magnesium (CP_2Mg) as a source of Mg. Lewis et al. [1983] showed that the Mg concentration in the MOCVD growth of GaAs exhibits a squared dependence on the $Mg(C_5H_5)_2$ concentration, $P(Mg) \approx (CP_2Mg)^2$, over the entire range of growth conditions. The doping efficiency is constant with growth temperatures up to ~ 680 °C and drops exponentially at higher temperature with an Arrhenius-type behavior ($\exp(E/k_BT)$, where $E \approx 3$ eV). The reproducibility of doping with $Mg(C_5H_5)_2$ was found to be sensitive to water and oxygen in the reactor, particularly at low growth temperatures [Kuech et al. 1987]. Kuech et al. [1988] observed the formation of a transient center on the GaAs:Mg grown using $Mg(C_5H_5)_2$. The transients associated with the growth of GaAs:Mg layers using $Mg(C_5H_5)_2$ can be influenced by a wide variety of system parameters.

The major cause of dopant transients comes from the interaction of the dopant precursor with the components of the gas panel and the internal surfaces of the reactor. The conditions for minimizing the formation of dopant transients depend largely on the incorporation kinetics of the precursor, as well as reactor design and growth conditions.

The growth of abrupt doping profiles in the case of Mg, Zn and Se is complicated by the interaction of the dopant precursor with the reactor internal surfaces. The adsorption and subsequent desorption of the dopant precursor from these surfaces leads to non-exponential transients in the carrier concentration profile.

These transients can be minimized for the dopant species (Zn, Mg, Se,...) which interact with the reactor components under conditions where the incorporation efficiency in the adjoining regions (reactor tube, lines, valves,...) of the dopant source is minimized.

Wang et al. [1988] studied the deep-level characteristics in MOCVD-grown GaAs layers doped with Zn and Mg over a wide range of doping concentrations and growth temperatures. Four hole traps and a single electron trap were identified in GaAs:Mg while a single hole trap was noted in GaAs:Zn. The presence and characteristics of each deep level in these *p*-type MOCVD GaAs layers depend primarily on the doping concentration in the layer and are relatively insensitive to the growth temperature. Wang et al. [1988] reported that the total trap concentration in GaAs:Mg was consistently higher than in GaAs:Zn.

5.5.5. Erbium-doped GaAs

Erbium-doped GaAs has attracted increasing interest, because of its sharp and temperature-independent emission peaks at 1.54 µm. Uwai et al. [1988] have reported uniform and high (10^{19} cm^{-3}) Er doping of GaAs grown by MOCVD using $Er(C_5H_5)_3$, cyclopentadienyl erbium. Er doping higher than 10^{19} cm^{-3} necessitates source temperatures of about 200 °C, which results in high concentration of impurities and in addition the $Er(C_5H_5)_3$ feedline must be heated to nearly 250 °C to avoid deposition on the tube walls. This severely restricts the system design, making it highly desirable to reduce the source temperature.

Replacing cyclopentadienyl radicals in $Mg(C_5H_5)_3$ (M = rare earth atom) by alkylcyclopentadienyl radicals usually increases the vapor pressure of rare earth metalorganic compounds. For example, tri-isopropylcyclopentadienyl compounds of Nd, Pr and La show vapor pressures more than one order of magnitude higher than those of their cyclopentadienyl counterparts at 250 °C.

GaAs can be doped with Er using $Er(C_5H_5)_3$ or $Er(CH_3C_5H_4)_3$. Er concentrations as high as 10^{20} cm^{-3} can be obtained. The Er concentrations depend linearly on the flow rates through the sources, and do not depend on the growth temperature between 600 and 700 °C. This suggests that $Er(C_5H_5)_3$ and $Er(CH_3C_5H_4)_3$ readily decompose in a simple monomolecular reaction without noticeable desorption from a growing surface at this temperature range. These characteristics guarantee reproducible doping controllability with $Er(CH_3C_5H_4)_3$ or $Er(C_5H_5)_3$ as source materials. The GaAs doped with Er shows a photoluminescence spectrum at 1.54 µm at 77 K. The PL spectrum indicates the coexistence of various kinds of Er center. The Er-related PL peak intensity increases almost linearly with Er concentration below 10^{18} cm^{-3}, above which it decreases abruptly with the highest PL intensity at 5×10^{18} cm^{-3} [Uwai et al. 1988].

References

Adachi, S., *J. Appl. Phys.* **53** 8775, 1982.

Adachi, S., *J. Appl. Phys.* **58** R1, 1985.

Asai, H. and Sugiura, H., *Japan. J. Appl. Phys.* **24** L815, 1985.

Bass, S.J. and Oliver, P.E., *Gallium Arsenide and Related Compounds* (*Inst. Phys. Conf. Ser. 33b*) (Bristol: Institute of Physics) p 1, 1977

Bourgoin, J.C., von Bardeleben, H.J., and Stievenard, D., *J. Appl. Phys.* **64** R65, 1988.

Colter, P.C., Look, D.C., and Reynolds, D.C., *Appl. Phys. Lett.* **43** 282, 1983.

Dapkus, P.D., Manasevit, H.M., Hess, K.L., Low, T.S., and Stillman, G.E., *J. Crystal Growth* **55** 10, 1981.

Dresselhaus, G., Kip, A.F., and Kittel, C., *Phys. Rev.* **98** 368, 1955.

Ehrenreich, H., *J. Phys. Chem. Solids* **8** 130, 1959.

El Jani, B., Leroux, M., Grenet, J.C., and Gibart, P., *J. Physique* **43** 303, 1982.

Feng, S.L., Bourgoin, J.C., Omnes, F., and Razeghi, M., *Appl. Phys. Lett.* **59** 941, 1991.

Field, R.J. and Ghandi, S.K., *J. Crystal Growth* **74** 543, 1986.

Hess, K.L., Dapkus, P.D., Manasevit, H.M., Low, T.S., Skromme, B.J., and Stillman, G.E., *J. Electron. Mater.* **11** 1115, 1982.

Hilsum, C., *Proc. IRE* **5** 185, 1962.

Kane, E.O., *Semiconductors and Semimetals* vol 1, ed R.K. Willardson and A.C. Beer (New York: Academic), 1966.

Kittel, C., *Introduction to Solid State Physics* (New York: Wiley), 1971.

Kuech, T.F., Scilla, G.J., and Cardon, F., *J. Crystal Growth* **93** 550, 1988.

Kuech, T.F., Veuhoff, E., and Meyerson, B.S., *J. Crystal Growth* **68** 48, 1984.

Kuech, T.F., Wolford, D.L., Vauhoff, E., Deline, V., Mooney, P.M., Potemski, R., and Bradley, J., *J. Appl. Phys.* **62** 632, 1987.

Landolt–Bornstein, *Numerical Data and Functional Relationships in Science and Technology* vol 17 (Berlin: Springer), 1982.

Lang, D.V., *J. Appl. Phys.* **45** 3023, 1974.

Lewis, C.R., Ludowise, M.J., and Dietze, F.T., *Electronic Materials Conf.* (*Burlington CA, 1983*), 1983.

Luther, L.C., and Di Lorenzo, J.V., *J. Electrochem. Soc.* **122** 760, 1975.

Luttinger, J.M., *Phys. Rev.* **102** 1030, 1956

Mochizuki, K., Ozeki, M., Kodama, K., and Ohtsuka, N., *J. Crystal Growth* **93** 550, 1988.

Nakanishi, T., Udagawa, T., Tanaka, A., and Kamei, K., *J. Crystal Growth* **55** 255, 1981.

Nelson, C.R., *RCA Rev.* **24** 603, 1963.

Omnes, F., Defour, M., and Razeghi, M., *Rev. Tech. Thomson-CSF* **23** 3, 1991.

Razeghi, M., *The MOCVD Challenge: Volume 1: A Survey of GaInAsP–InP for Photonic and Electronic Applications* (Bristol: Hilger), 1989.

Razeghi, M., Omnes, F., Nagle, J., Defour, M., Acher, D., and Bove, P., *Appl. Phys. Lett.* **55** 1677, 1989.

Saalfeld, F.E., and Svec, H.J., *Inorg. Chem.* **2** 46, 1963.

Sakaguchi, H., Suzuki, R., and Meguro, T., *J. Crystal Growth* **93** 602, 1988.

Shimazu, M., Kamon, K., Kimura, K., Mashita, M., Mihara, M., and Ishii, M., *J. Crystal Growth* **83** 327, 1987.

Stievenard, D., Lannoo, M., and Bourgoin, J.C., *Solid State Electron.* **28** 485, 1985.

Stillman, G.E. and Wolfe, C.M., *Thin Solid Films* **31** 69, 1976.

Stormer, H.L., Dingle, R., Gossard, A.C., Wiegman, W., and Sturge, M.D., *Solid State Commun.* **29** 705, 1979.

Sze, S.M., *Physics of Semiconductor Devices* (New York: Wiley), 1981.

Uwai, K., Nakagome, H., and Takahei, K., *J. Crystal Growth* **93** 583, 1988.

Van de Ven, J., Rutten, G.J., Raaijmakers, M.J., and Giling, L.J., *J. Crystal Growth* **76** 352, 1986.

Van Sark, W.M., Janssen, G., de Croon, M.M., and Giling, L.J., *Semicond. Sci. Technol.* **5** 16, 1990.

Veuhoff, E., Kuech, J.F., and Meyerson, B.S., *J. Electrochem. Soc.* **132** 1958, 1985.

Wang, P.J., Kuech, T.F., Tischler, M.A., Mooney, P.M., Scilla, G.J., and Cardone, F., *J. Crystal Growth* **93** 568, 1988.

Wolfe, C.M. and Stillman, G.E., *Appl. Phys. Lett.* **27** 564, 1975.

6. Growth and Characterization of the GaInP–GaAs System

6.1. Introduction
6.2. Growth details
6.3. Structural order in $Ga_xIn_{1-x}P$ alloys grown by MOCVD
6.4. Defects in GaInP layers grown by MOCVD
6.5. Doping behavior of GaInP
    6.5.1. n-type doping
    6.5.2. p-type doping
    6.5.3. Conclusions
6.6. GaAs–GaInP heterostructures
    6.6.1. Band offset measurements in the $GaAs/Ga_{0.51}In_{0.49}P$ system
    6.6.2. Observation of the two-dimensional properties of the electron gas in $Ga_{0.51}In_{0.49}P/GaAs$ heterojunctions grown by MOCVD
    6.6.3. High-mobility GaInP–GaAs heterostructures
    6.6.4. Electron spin resonance in the two-dimensional electron gas of a $GaAs–Ga_{0.51}In_{0.49}P$ heterostructure
    6.6.5. Magnetotransport measurements in GaInP/GaAs heterostructures using δ-doping
    6.6.6. Persistent photoconductivity in $Ga_{0.51}In_{0.49}P/GaAs$ heterojunctions
    6.6.7. Quantum and classical lifetimes in a $Ga_{0.51}In_{0.49}P/GaAs$ heterojunction
    6.6.8. FIR magnetoemission study of the quantum Hall state and the breakdown of the quantum Hall effect in GaAs–GaInP
6.7. Growth and characterization of GaInP–GaAs multilayers by MOCVD
6.8. Optical and structural investigations of GaAs–GaInP quantum wells and superlattices grown by MOCVD
6.9. Characterization of GaAs–GaInP quantum wells by auger analysis of chemical bevels
6.10. Evaluation of the band offsets of GaAs–GaInP multilayers by electroreflectance
    6.10.1. Study of the GaAs–GaInP superlattice
    6.10.2. The study of the GaAs–GaInP three-quantum-well structure
    6.10.3. III–V intersubband infrared detectors
6.11. Intersubband hole absorption in GaAs–GaInP quantum wells

6.1. Introduction

In the past few years, GaAs–GaAlAs heterostructures have emerged as a promising system for optoelectronic and microwave device applications. However, because of the strong reaction between Al and oxygen, even trace quantities of oxygen have a dramatic effect on the quality of GaAlAs layers due to the effective introduction of deep-level defects. Hence, photonic devices based on GaAs–GaAlAs suffer from the catastrophic dark line defect formation and rapid degradation. This is a major problem for monolithic integrated circuits and the technology of GaAs–GaAlAs heterostructures on Si substrates. One of the solutions to this problem is to replace GaAlAs by $Ga_{0.51}In_{0.49}P$ lattice matched to GaAs which has a direct energy gap of 1.9 eV.

The electrical, optical and structural properties of GaAs–GaInP depend directly on how the system is lattice matched. Concerning ΔE_c, there is a surprise: if we assume that the discontinuity in the conduction band is the difference in the electron affinities (χ) of $\chi(GaAs) = 4.05$ eV, $\chi(InP) = 4.4$ eV and $\chi(GaP) = 4.0$ eV (the electron affinity of GaInP is taken as the average of $\chi(InP)$ and $\chi(GaP)$ and $\chi(Ga_{0.51}In_{0.49}P) = 4.2$ eV), then $\Delta E_c = \chi(GaAs) - \chi(GaInP) = -0.15$ eV. However, the experimental results show that $\Delta E_c = 0.2$ eV and $\Delta E_v = 0.28$ eV.

GaInP lattice matched to GaAs shows a number of unique and interesting features by comparison to GaAlAs. Its large valence band discontinuity makes it very suitable for n–p–n heterojunction bipolar transistors (HBTs) and p-channel FETs [Chen et al. 1988].

Another feature of the $Ga_xIn_{1-x}P$ heterostructure is that the crossover of its direct (Γ) and indirect (X) conduction bands lies at $x = 0.74$ [Casey and Panish 1978] and is therefore far from the lattice-matched composition ($x = 0.51$). In the case of $Ga_{1-x}Al_xAs$, the crossover of direct and indirect conduction bands is around $x = 0.37$. Donor-related deep traps, the so-called DX centers, are particularly important for x exceeding 0.3, i.e. close to the crossover point. Trap activation energies in $Ga_{1-x}Al_xAs$ depend on the composition x in a similar way to the dependence of the indirect conduction band L [Chand et al. 1984]. Since the $Ga_xIn_{1-x}P$ lattice-matched composition corresponds to an x value which is well below the crossover point of the direct and indirect bands, it can be deduced that the presence of DX centers will be very small for this material as explained below.

The DX center is a deep donor level and its activation energies follow one of the indirect conduction bands in III–V compounds. Once the composition of a ternary compound is near or over the crossover point of the bandgap transition, DX centers will start to affect the device's electrical properties. For those compositions well below the crossover point, the trap energy lies above the minimum of the conduction band (Γ) and deep-trap effects are negligible. Since this type of deep trap is quite common in all III–V semiconductors, a large separation between the lattice-matched composition and that of the crossover point helps to eliminate the DX center problems. Therefore, $Ga_{0.51}In_{0.49}P$ will allow operation without significant donor-related deep traps. Strong selective etching between GaAs and GaInP also make this system very promising for device fabrication.

6.2. Growth details

GaInP layers can be grown by MOCVD, either at atmospheric pressure or low pressure and at low temperatures between 500 and 600 °C. One can use different group III alkyls for Ga and In sources, and hydrides or alkyls for group V P sources. Chemical reactions occurring among these sources are as follows

$$0.51 R_3 Ga + 0.49 R'_3 In + EH_3 \xrightarrow{H_2} Ga_{0.51} In_{0.49} P + n C_n H_{2n}$$

where R, R' and E can be methyl, ethyl, alkyl or hydride.
One can also use any of the following as an example

$$0.49 (C_2 H_5)_3 In + 0.51 (C_2 H_5)_3 Ga + PH_3 \xrightarrow{H_2} InGaP + n C_2 H_6$$

$$0.49 (CH_3)_3 In + 0.51 (CH_3)_3 Ga + PH_3 \xrightarrow{H_2} InGaP + n CH_4$$

$$0.49 (CH_3)_3 In + 0.51 (C_2 H_5)_3 Ga + PH_3 \xrightarrow{H_2} InGaP + n C_2 H_6$$

$$0.49 (C_2 H_5)_3 In + 0.51 (CH_3)_3 Ga + PH_3 \xrightarrow{H_2} InGaP + n C_2 H_6$$

The GaInP layers can be grown at low temperature, between 500 and 550 °C, by using triethylgallium (TEGa), trimethylindium (TMIn) and pure phosphine (PH$_3$) in a H$_2$ carrier gas. The optimum growth conditions are given in Table 6.1.

The growth rate (dx/dt) of GaInP depends on the flow rates of TMIn and TEGa (group III element) and is independent of PH$_3$ flow rate (group V element) and growth temperature under the growth conditions listed in Table 6.1. The distribution coefficients of indium and gallium are defined as

Eq. (6.1) $K = X_{Ga}^s / X_{Ga}^v$

and

Eq. (6.2) $K' = X_{In}^s / X_{In}^v$

and are nearly equal to unity. Fig. 6.1 shows the variation of growth rate dx/dt of GaInP lattice matched to GaAs with a growth temperature of $T_G = 540$ °C and growth pressure of 76 Torr. Similar results have been reported by Hsu et al. [1985] at growth temperatures from 600 up to 650 °C. They showed that there was no gas-phase reaction in their reactor leading to premature depletion of In or Ga.

An undoped GaInP layer grown under the conditions of Table 6.1 has a free-electron carrier concentration of 5×10^{14} cm^{-3} with a mobility of 6000 cm$^2 \cdot$V$^{-1} \cdot$s^{-1} at 300 K and 40,000 cm$^2 \cdot$V$^{-1} \cdot$s^{-1} at 77 K. No GaAs buffer layer is grown in this case [Razeghi et al. 1989b].

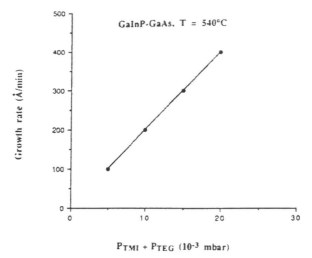

Fig. 6.1. Variation of growth rate of GaInP layers grown by MOCVD at 540 °C as a function of flow rate of group III elements (after Razeghi [1989a]).

	GaAs	GaInP
Growth pressure	76 Torr	76 Torr
Growth temperature	510 °C	510 °C
Total H_2 flow rate	31 min^{-1}	31 min^{-1}
AsH_3 flow rate	30 cc·min^{-1}	—
H_2 through TMIn bubbler at 18 °C	—	200 cc·min^{-1}
H_2 through TEGa bubbler at 0 °C	120 cc·min^{-1}	120 cc·min^{-1}
PH_3 flow rate	—	300 cc·min^{-1}
Growth rate (dx/dt)	150 Å·min^{-1}	200 Å·min^{-1}

Table 6.1 Optimum growth parameters for GaAs and GaInP by MOCVD.

6.3. Structural order in $Ga_xIn_{1-x}P$ alloys grown by MOCVD

It was found that GaInP alloy has two phases, consisting of disordered and ordered structures. The disordered alloy has a bandgap of 1.92 eV at 300 K, while the ordered structure has a lower bandgap varying from 1.83 eV to 1.78 eV [Gomyo et al. 1988, McDermott et al. 1991]. This type of combination of ordered and disordered structures has been found in various alloy systems such as GaInP [Ueda et al. 1987, Bellon et al. 1988, Kondow et al. 1988a, Gomyo et al. 1988], AlInP [Yasuami et al. 1988], AlInAs [Ueda et al. 1988a, b], GaInAs [Nakayama and Fujita 1986], GaAsSb [Jen et al. 1987], GaAlAs [Kuan et al. 1985], etc. Experimental results show that the kinetics of crystal growth plays an important role in the formation of an ordered structure [Kondow et al. 1988a, b, c, Ueda et al. 1988a, b]. If growth kinetics generate ordered alloys, then the CuPt-type ordered structure that has been found in a $Ga_{0.5}In_{0.5}P$ alloy may be also formed in a $Ga_{0.7}In_{0.3}P$ alloy.

Ga_3InP_4 ordered structure exists in either of two variations, "famatinite" or "luzonite" [Landau and Lifshitz 1969]. The famatinite-type ordered structure is a $(GaP)_3$ (InP) superlattice developing along the (2 1 0) direction. If a famatinite-type ordered structure exists, a diffraction pattern with an electron beam incident along the [1 0 0] crystal axis must show extra spots corresponding to the ordered structure [Bellon et al. 1988]. However, the presence of a luzonite-type ordered structure cannot be proved by a diffraction measurement because this ordered structure does not have an equivalent superlattice.

Several studies have been reported on the structural ordering of $Ga_{0.5}In_{0.5}P$ [Kondow et al. 1988a, b, c), $Ga_{0.6}In_{0.4}P$ (Bellon et al. 1988], $Ga_{0.7}In_{0.3}P$ [Kondow et al. 1989], and InGaAlP [Suzuki et al. 1988a, b, c] grown by MOCVD. The structural order has been observed by transmission electron microscopy.

A correlation has been established between the degree of structural order in GaInP and energy-gap measurements by photoluminescence or electroreflectance which shows that the disorder–order transition is reflected in the energy-gap optical transition.

The most interesting structural ordering phenomena are observed in the $Ga_{0.5}In_{0.5}P$ alloy. The structural order of $Ga_{0.5}In_{0.5}P$ has been studied as a function of (i) growth temperature [Kondow et al. 1988b, Gomyo et al. 1987], and (ii) V/III vapor pressure ratio [Suzuki et al. 1988b]. In all the cases, the $Ga_{0.5}In_{0.5}P$ samples were grown on (0 0 1)-oriented GaAs substrates. Transmission electron diffraction (TED) studies have been performed along the [1 1 0], [1 $\bar{1}$ 0] or [0 0 1] directions [Kondow et al. 1988c, Gomyo et al. 1987]. These studies have shown that the disordered phase consists of a cubic face-centered, zinc blende-type crystalline lattice with a random distribution of Ga and In atoms on the group III sublattice. The transmission electron diffraction pattern of the disordered phase shows diffraction spots at $(h\,k\,l)$ positions with h, k and l all being even or odd integers. In this case, no superstructure was observed in the transmission electron diffraction pattern [Suzuki et al. 1988a]. The ordered phase is an ordered distribution of Ga and In atoms on the group III element sublattice of the zinc blende structure.

The most interesting TED studies were made using the [1 1 0] orientation for the incident beam. Using this orientation, the TED pattern of the ordered phase showed $(h-\frac{1}{2}, k+\frac{1}{2}, l+\frac{1}{2})$ spots close to the (h, k, l) spots of the zinc blende structure, corresponding to a double periodicity of the $(\bar{1}\ 1\ 1)$ planes. Therefore, the ordered phase can be described as a regular alternation of GaP and InP monatomic planes in the $[\bar{1}\ 1\ 1]$ direction, as in the CuPt structure [Kondow et al. 1988a, b, c, Suzuki et al. 1988a, b, c].

To confirm this assumption, transmission electron microscopic (TEM) studies have been performed by Suzuki et al. on the ordered $Ga_{0.5}In_{0.5}P$, using a dark-field imaging technique with an aperture including $(-\frac{1}{2}, \frac{1}{2}, \frac{1}{2})$, $(-\frac{1}{2}, \frac{1}{2}, \frac{3}{2})$ and $(\bar{1}\ 1\ 1)$ diffraction points. They observed at atomic resolution a double periodicity of 6.5 Å of the $(\bar{1}\ 1\ 1)$ planes, which can be related to TEM observations. It proves the ordering of GaP and InP monoatomic planes along the $[\bar{1}\ 1\ 1]$ direction.

Suzuki et al. also observed, by TEM imaging, an intermixing of disordered and ordered phases as well as antiphase boundaries on partially ordered phases, with a variation of the relative extensions of the disordered phase, antiphase boundaries, and ordered phases corresponding to a short- or long-range structural order.

The degree of structural order can also be estimated by the TED technique, looking at the intensity and the sharpness of the superstructure spots $(h-\frac{1}{2},k+\frac{1}{2},l+\frac{1}{2})$ of the electron diffraction pattern.

The TED and TEM studies performed on $Ga_{0.5}In_{0.5}P$ alloys grown at different growth temperatures in the range of 550–750 °C showed that the degree of structural order increased with growth temperature, from a very short-range order at 550 °C to a long-range order at 700 °C. At low growth temperatures, the density of antiphase boundaries is high. Intermixing of disordered and ordered phases can also be observed at low growth temperatures [Suzuki et al. 1988a, b, c]. It is also observed that there is a negligible contribution of the V/III vapor pressure ratio on the structural ordering phenomena in $Ga_{0.5}In_{0.5}P$.

Photoluminescence measurements at 300 K have shown that there is a negligible difference in energy gaps for $Ga_{0.5}In_{0.5}P$ alloys grown at the same growth temperature for different V/III ratios, from V/III = 60 to V/III = 440 [Gomyo et al. 1987], although 4 K luminescence would yield more accurate E_g values.

A correlation has been established between energy-gap values and the degree of structural order in $Ga_{0.5}In_{0.5}P$. Photoluminescence and electroreflectance studies have been performed on $Ga_{0.5}In_{0.5}P$ alloys [Kondow et al. 1989, Inoue et al. 1988]. Photoluminescence results showed a minimum energy gap for the $Ga_{0.5}In_{0.5}P$ grown at 650 °C: for this growth temperature, the value of the energy gap of the $Ga_{0.5}In_{0.5}P$ decreased to 1.85 eV at 300 K. As a comparison, the value observed for the energy gap of the disordered $Ga_{0.5}In_{0.5}P$ is 1.90 eV at 300 K. Electroreflectance studies also showed similar results [Inoue et al. 1988].

Gomyo et al. [1987] reported that zinc diffusion through an ordered $Ga_{0.5}In_{0.5}P$ alloy increases its bandgap by 50 meV, from 1.85 eV to 1.90 eV at 300 K; the effect of zinc diffusion is to produce an intermixing of the group III elements, creating structural disorder by randomizing Ga and In positions. This clearly proved the relation between the energy gap of $Ga_{0.5}In_{0.5}P$ and its structural order.

McDermott et al. [1991] have grown $Ga_{0.5}In_{0.5}P$, at temperatures between 480 and 500 °C, using atomic layer epitaxy (ALE). They found that all the samples grown on 2°-misoriented substrates had CuPt structure, with typically two $\frac{1}{2}\langle 111\rangle$ variants. For samples grown on the $(1\ 0\ 0)$ nominal substrates, no CuPt ordering was observed. However, the accurate origin of this anomaly in bandgap energies is not well understood.

6.4. Defects in GaInP layers grown by MOCVD

Most of the studies performed on GaInP layers have been limited to the conditions of growth [Horng et al. 1988, Hoshino et al. 1986a, b, c, Kondow et al. 1988a, b, c], in particular to investigate the influence of the growth temperature on structural ordering [Ueda et al. 1989], and to the evaluation of a few optical devices [Hsieh et al. 1984, Ishikawa et al. 1986, Ikeda et al. 1987].

Using deep-level transient spectroscopy (DLTS), the defects present in GaInP layers grown by MOCVD were characterized by Feng et al. [1991]. The layers were grown on n^+-doped GaAs substrates in a horizontal cold-wall reactor using trimethylindium, triethylgallium and PH_3 as sources [Razeghi et al. 1990]. The layers, ~1 μm thick, have the following composition: Ga (0.51), In (0.49) for lattice matching on GaAs; they are not intentionally doped. After an ohmic contact has been deposited on the back of the wafer (Au–Ge alloy annealed at 450 °C for a few minutes), Au Schottky barriers (area: 0.07 mm^2) are deposited by evaporation. Capacitance–voltage (C–V) measurements show that the layers are n-type with a uniform concentration throughout the whole layer of 2.4–3.5×10^{15} cm^{-3}, depending on the location of the Schottky diode, at room temperature. This concentration increases slightly above room temperature from 2.8×10^{15} cm^{-3} at 300 K to 3×10^{15} cm^{-3} at 400 K. It also decreases with decreasing temperature from 2.4×10^{15} cm^{-3} at 300 K to 2.2×10^{15} cm^{-3} at 30 K, for another sample. This suggests that defects are present, which trap $\sim 2 \times 10^{14}$ cm^{-3} electrons in the 300–400 K range as well as below 300 K.

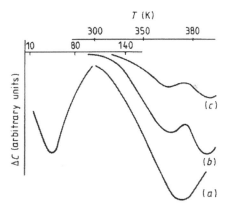

Fig. 6.2. DLTS spectra of GaInP grown on GaAs obtained in the temperature range 4–400 K. The low-temperature peak (60 K) was measured at bias -1 V, bias pulse 1 V, emission rate 60 s^{-1}, pulse duration 1ms. High-temperature peaks were measured with an emission rate 30 s^{-1}, pulse duration 1 ms, bias pulse 1 V, and bias: (a) -3 V, (b) -2 V and (c) -1 V (after Feng et al. [1991]).

DLTS detects electron traps which account for the changes observed in the electron concentration above room temperature, but not below. As shown in Fig. 6.2, a first peak is observed around 60 K corresponding to an ionization energy of 75 meV (Fig. 6.3) in a concentration of 3×10^{13} cm^{-3}. This spectrum cannot be attributed to a possible emission over band discontinuities induced by the existence of ordered domains (having a larger bandgap). Indeed, in this case the emission rate should be strongly sensitive to the electric field and the width of the spectrum on the low-temperature side should increase then with this field, i.e. with the applied reverse bias. Thus, there should be no correlation between this spectrum and the large photoluminescence shift versus excitation intensity commonly observed in similar layers [Delong et al. 1990]. A second peak, rather wide and exhibiting a double structure, appears between 350 and 450 K. The shape of this structure varies with the duration of the filling pulse, suggesting that one of the traps is not completely filled (see Fig. 6.2). The activation energy associated with the first maximum of this peak is 0.92 eV (see Fig. 6.4). Only the order of magnitude of concentration of this trap, 15% of the free-carrier concentration (i.e. 4.5×10^{14} cm^{-3}), can be evaluated because it cannot be completely filled (see Fig. 6.5). The dependence of the high-temperature spectrum on the reverse bias, at constant pulse amplitude, indicates a non-uniform distribution of the associated defect: its concentration increases near the epi–substrate interface.

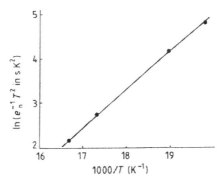

Fig. 6.3. Variation of the emission rate e_n versus temperature of the peak appearing around 60 K (after Feng et al. [1991]).

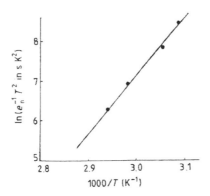

Fig. 6.4. Variation of the emission rate e_n versus temperature of the high-temperature peak under the bias conditions of Fig. 6.2(a) (after Feng et al. [1991]).

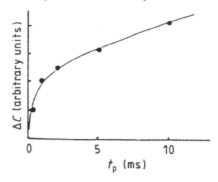

Fig. 6.5. Amplitude ΔC of the high-temperature peak versus the filling pulse duration (after Feng et al. [1991]).

The GaInP layer, grown by MOCVD at optimum growth conditions, contains only one defect species emitting below room temperature (60 K) with a concentration of the order of 1% of the concentration of residual uncompensated donors. This defect cannot be attributed to a DX-like center since its concentration is saturated with short filling times (~50 μs). If it existed, a DX center should have an energy level resonant in the conduction band. According to Bourgoin [1991], donor impurities are expected to introduce a DX-like energy level in the forbidden gap when the X or L band lies at an energy not larger than typically 0.2 eV from the bottom of the conduction band. In $Ga_{1-x}In_xP$ the condition is fulfilled only when $x \geq 0.7$.

Consequently, the persistent photoconductivity which was observed by Ben Amor et al. [1989] should be ascribed to electron emission from deep traps emitting above room temperature. These traps, because of their relatively high concentration ($\sim 10^{14}$ cm^{-3}), should dominate the electrical properties of these layers for low doping densities.

Consequently, the unintentionally doped $Ga_{1-x}In_xP$ ($x = 0.51$) layers grown by MOCVD exhibit an uncompensated electron concentration of a few 10^{15} cm^{-3}. They contain one dominant defect, which is ionized just above room temperature, in a concentration of the order of 10^{14}–10^{15} cm^{-3}. For the alloy composition considered, the DX center, if present, is not localized in the gap.

6.5. Doping behavior of GaInP

6.5.1. n-type doping

GaInP layers grown by MOCVD can be *n*-type doped using group VI or group IV elements such as: hydrogen sulfide, H_2S; diethyl selenium, $(C_2H_5)_2Se$; dimethyl selenium, $(CH_3)_2Se$; hydrogen selenide, H_2Se; diethyltellurium, $(C_2H_5)_2Te$; dimethyltellurium, $(CH_3)_2Te$; silane, SiH_4; disilane, Si_2H_6; germane, GeH_4.

H₂S

GaInP layers grown by MOCVD can be *n*-type doped using H₂S. The probability of incorporation of sulfur (S) in GaInP is very high. At the same growth temperature, growth conditions and H₂S flow rate, a GaInP layer is doped 100 times higher than a GaAs layer.

Fig. 6.6 shows the electrochemical polaron profile of a GaInP–GaAs layer grown by MOCVD using the growth conditions of Table 6.1. It is clear from this experimental result that for the same H₂S flow rate, the donor concentration in GaInP is much higher than in GaAs with the different curves corresponding to different H₂S flow rates. Fig. 6.7 confirms this result. For different growth temperatures and constant H₂S flow rate, the donor concentration in the layer increases when the growth temperature decreases [Razeghi 1989c].

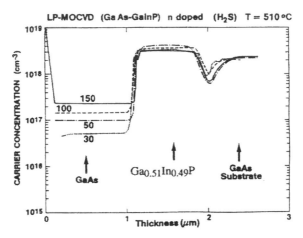

Fig. 6.6. Electrochemical polaron profile of a GaInP layer grown by MOCVD using H₂S as doping source. Different curves are related to different H₂S flow rates into the reactor (after Razeghi [1989a]).

Fig. 6.6 shows that the doping level in $Ga_{0.51}In_{0.49}P$ saturates at 4×10^{18} cm^{-3}. The H₂S flow has been first varied between 3 and 15 cc·min^{-1}, keeping the growth temperature equal to 510 °C and the PH₃ flow equal to 450 cc·min^{-1}. The H₂S flow was even lowered to 1 cc·min^{-1} with no change. This means that much more diluted H₂S is necessary to reach the linear zone where the doping level is proportional to the H₂S flow. This saturation effect is in good agreement with work done on a chloride VPE system [Kitahara et al. 1986].

The effect of growth temperature change is shown in Fig. 6.7. The concentration of donor decreases as growth temperature increases. It can be fitted using the relation

Eq. (6.3) $N_d - N_a = D_{PH_3} \exp(-E_i / k_B T)$

assuming that by using 1000 ppm H_2S and doping flows higher than 1 cc·min^{-1}, the doping level is independent of H_2S flow. $E_i = -2.16$ eV. The phosphine flow has been varied between 300 and 750 cc·min^{-1} by keeping the growth temperature at 510 °C. A relation between the doping level and the phosphine flow is found in the form: $(D_{PH_3})^{-07}$.

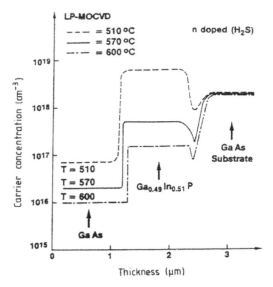

Fig. 6.7. Electrochemical polaron profile of a GaInP layer grown by MOCVD using H_2S as the doping source. Different curves are related to different growth temperatures, keeping other growth parameters constant (after Razeghi [1989a]).

In conclusion, the doping level of H_2S 1000 ppm in $Ga_{0.51}In_{0.49}P$ grown with a fixed III element flow (i.e. a growth rate of 200 Å·min^{-1}) can be written in the form

Eq. (6.4) $N_d - N_a = 7 \times 10^5 \exp(2.16 / k_B T)(D_{PH_3})^{-0.7}$

where $k_B T$ is in eV and D_{PH_3}, is in cc·min^{-1}. In the flow range used for this study $N_d - N_a$ saturates with D_{H_2S} so that it does not depend on that parameter [Razeghi and Omnes 1991a].

SiH4

GaInP layers grown by MOCVD can be n-type doped using SiH_4 or Si_2H_6. In the GaInP layer, Si substitutes Ga or In and plays the role of a donor. The concentration of Si in GaInP depends on different growth parameters such as growth temperature, growth pressure, growth rate, the ratio of III/V and the flow rate of SiH_4.

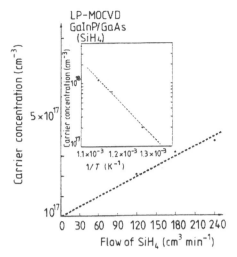

Fig. 6.8. Variation of electron carrier concentration in a GaInP layer grown by MOCVD at 540 °C, using SiH₄ as a doping source, as a function of SiH₄ flow rate; the inset shows the variation of carrier concentration as a function of 1/T (after Razeghi [1989b]).

In contrast to H_2S doping of $Ga_{0.51}In_{0.49}P$, the doping level using SiH_4 varies linearly with silane flow (Fig. 6.8). The growth temperature is maintained at 540 °C while the phosphine flow is kept at 450 °C. By changing the growth temperature, an exponential dependence is found (Fig. 6.8 and Fig. 6.9). The doping level can be written in the form

Eq. (6.5) $N_d - N_a \approx A \exp\left(-E_i / kT\right)$

with $E_i = 1.26$ eV.

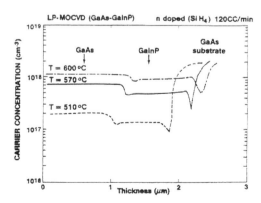

Fig. 6.9. Electrochemical polaron profile of GaInP layers grown by MOCVD using SiH₄ as a doping source for different growth temperatures (after Razeghi [1989b]).

Omnes et al. [1991] have studied the influence of PH_3 flow rate on doping concentration in GaInP layers grown by MOCVD using SiH_4 as n-type dopant. By keeping the growth temperature equal to 540 °C and the SiH_4 flow at 120 cc·min^{-1}, the PH_3 flow has been varied from 300 to 750 cc·min^{-1}.

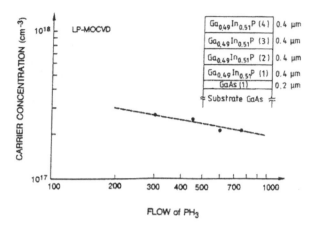

Fig. 6.10. *Variation of electron carrier concentration in GaInP layers grown by MOCVD, using SiH_4 as a doping source, as a function of PH_3 flow rate. Inset is the structure of the epilayer (after Razeghi [1989b]).*

Fig. 6.10 shows that the doping level varies very slowly with PH_3 flow rate as $(D_{PH_3})^{-0.2}$. Similar results have been reported by Hsu et al. [1985] and Kawamura and Asahi [1983]. Due to the low diffusion coefficient and low vapor pressure, a sharp doping profile can be obtained with silicon. Ladany [1971] reported that Si-doped GaAs has a very high electroluminescence quality.

Gomyo et al. [1989] studied the effects of Si doping on the E_g behavior and on the $\langle\frac{1}{2},\frac{1}{2},\frac{1}{2}\rangle$ sublattice ordering in GaInP, grown at 700 °C by MOCVD under the conditions in which crystals show anomalously low E_g values unless dopants are introduced. An anomalously low E_g value, which was 1.836 eV for an undoped crystal, increased to a normal E_g value of 1.92 eV when the Si doping level increased at least up to $n = 3.8 \times 10^{17}$ cm^{-3}. At this doping condition, the TEM study revealed substantial disordering of the $\langle\frac{1}{2},\frac{1}{2},\frac{1}{2}\rangle$ superlattice (SL). In a range from 3.8×10^{17} cm^{-3} to 2×10^{18} cm^{-3}, the normal E_g value remained virtually unchanged. They observed a Burstein shift (i.e. the energy shifts towards a higher value due to a band-filling effect by free electrons) for $n \geq 2 \times 10^{18}$ cm^{-3}. Anomalous E_g values, which were around 1.87 eV for undoped crystals, started to increase at $n = 1 \times 10^{18}$ cm^{-3}. After the increase due to disordering of the $\langle\frac{1}{2},\frac{1}{2},\frac{1}{2}\rangle$ SL, the Burstein shift occurred consecutively when n exceeded $\approx 2 \times 10^{18}$ cm^{-3} in contrast to the "two-stage" E_g behavior for the Si doping due to the lower threshold doping concentration $\lesssim 3.8 \times 10^{17}$ cm^{-3} for the disordering [Gomyo et al. 1989].

Anyway, such behavior was not observed in GaInP layers doped with Si up to 5×10^{18} cm^{-3} at low growth temperature between 510 and 550 °C [Razeghi et al. 1989b]. The E_g change appears to depend more on the alloy composition than on the doping level.

Se

Gomyo et al. [1989] studied GaInP layers doped with Se under the same conditions as Si. They found that the threshold for the SL disordering by Si doping ($n = 3.8 \times 10^{17}$ cm^{-3}) is higher than that for Se doping ($n \approx 2 \times 10^{18}$ cm^{-3}). They gave two reasons for this difference. The first one is the difference in the sublattice occupation between Si and Se. Si atoms substitute the group III sublattice and Se atoms substitute the group V sublattice. This difference in sublattice occupation affinity between Si and Se could be a cause for the $n_{\text{th}}^{\text{disord}}$ difference. The second reason was the difference between the growth conditions. The V/III ratios of the growth for Si and Se doping were different. However, further studies will be needed to understand the reason behind ordering and disordering mechanisms in this system.

Te

GaInP layers grown by MOCVD can be *n*-type doped using Te atoms as a doping element. In GaInP Te usually occupies the P site. Hsu et al. [1985] used diethyltelluride (DETe) diluted to 5.45 ppm in H$_2$ as the source for the *n*-type dopant. They found a linear relation between the dopant mole fraction in the gas phase and carrier concentrations in the range 10^{17} cm^{-3} to 10^{19} cm^{-3} at a growth temperature of 625 °C. They reported a high distribution coefficient for Te with

Eq. (6.6) $K_{\text{Te}} = X_{\text{Te}}^{\text{s}} / X_{\text{Te}}^{\text{v}} = 54$

where

Eq. (6.7) $X_{\text{Te}}^{\text{s}} = \dfrac{\text{concentration of electron sites}}{\text{concentration of group-V sites}}$

Eq. (6.8) $X_{\text{Te}}^{\text{v}} = \dfrac{\text{concentration of group-VI dopant in the vapor}}{\text{concentration of P in the vapor}}.$

Since the concentration of group VI dopant is much smaller than the concentration of P in both the solid and the vapor, they neglect the group VI concentration in the denominator in calculating both X^s and X^v. Hsu et al. [1985] reported that the PL intensity of Te-doped GaInP increases with carrier concentration until $n = 2 \times 10^{18}$ cm^{-3} and the PL half-width increases with increasing carrier concentration. They found that the electron mobility decreases from 1020 to 500 cm^2 V^{-1}·s^{-1} as carrier concentration increases from 10^{16} to 10^{19} cm^{-3}.

6.5.2. p-type doping

GaInP layers grown by MOCVD can be *p*-type doped using Mg, Zn and Cd.

DEZn

The same type of study can be carried out on $Ga_{0.51}In_{0.49}P$ for *p*-type doping. The growth temperature is kept at 510 °C and the phosphine flow rate at 450 cc·min^{-1}. The doping level is found to vary linearly with DEZn flow as for GaAs (Fig. 6.11).

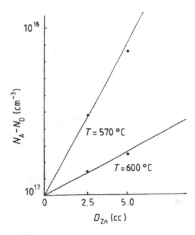

Fig. 6.11. Variation of hole concentration in GaInP as a function of flow rate of DEZn for different growth temperatures (after Razeghi et al. [1990]).

The effect of growth temperature has been investigated between 500 and 600 °C by keeping the DEZn flow constant and the phosphine flow equal to 450 cc·min^{-1}. Fig. 6.12 shows the evolution of the doping level as a function of $1/T$. It varies exponentially and decreases as the growth temperature increases

Eq. (6.9) $$N_d - N_a = 1.6 \times 10^2 \, \text{cm}^{-3} \exp\left(-E_i / kT\right)$$

with $E_i = -2.5$ eV.

The influence of the phosphine flow has been investigated without significant results. No clear influence of group V element flow has been found on *p*-type doping of GaInP with DEZn [Razeghi et al. 1989b]. The acceptor doping level of DEZn in GaInP can be written in the general form

Eq. (6.10) $$N_d - N_a = 10^2 \, D_{\text{DEZn}} \exp\left(2.5 / k_B T\right)$$

with DEZn in cc·min^{-1} and $k_B T$ in eV.

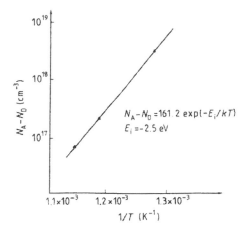

Fig. 6.12. Variation of the hole concentration in GaInP as function of 1/T (T is the growth temperature) (after Razeghi et al. [1990]).

Hsu et al. [1985] used dimethylzinc (DMZn) as a p-type dopant source with a concentration of 474 ppm in H_2. By varying the DMZn mole fraction in the gas phase, they obtained p-carrier concentrations in the range 10^{17} cm^{-3} to 10^{19} cm^{-3}. They calculated the Zn distribution coefficient as

Eq. (6.11) $K_{Zn} = X_{Zn}^s / X_{Zn}^v$

with

Eq. (6.12) $X_{Zn} = [DMZn] / [TMIn + TMGa] = 0.0038$

at $p = 10^{18}$ cm^{-3} and a growth temperature of 625 °C. They attributed this low doping efficiency to the fact that Zn has a vapor pressure of 15 Torr at the growth temperature [Honig 1962] and most of the zinc is evaporated from the crystal surface before incorporation. They reported GaInP hole mobilities in the range of 20–40 cm$^2 \cdot$V$^{-1} \cdot$s^{-1}. They did not observe any deep level in the PL spectrum of Zn-doped GaInP. They observed that at room temperature the PL intensity increases with carrier concentration, the maximum PL intensity for p-type GaInP being about the same as that of n-type samples, and the PL half-width increased at higher concentrations.

Mg

Suzuki et al. [1988a, b, c] have used bis-magnesium [(C$_2$H$_5$)$_2$Mg] as a p-type dopant for GaInP layers grown by MOCVD. They reported similar behavior for Mg as Zn in GaInP layers. They studied the photoluminescence properties of Mg-doped GaInP layers as a function of hole concentration. The bandgap energy (E_g) value for Mg-doped GaInP, grown under conditions in which undoped GaInP shows an anomalously low E_g, showed a steep increase for $p > 1 \times 10^{18}$ cm^{-3}. This anomalous behavior was attributed to the Mg (or Zn) diffusion-enhanced randomization of the naturally formed monolayer $(\frac{1}{2},\frac{1}{2},\frac{1}{2})$ superlattices on the column III sublattice.

Doping source	Materials	Doping flow rate	Group V flow rate	E_i (eV)	Group III flow rate
DEZn	GaAs	D_{DEZn}	$D_{AsH_3}^{0.3}$	-2.6	
	GaInP	D_{DEZn}	$D_{PH_3}^{0.1}$	-2.5	
	GaInAs	D_{DEZn}		-2.2	
	InP	D_{DEZn}		-1.2	
H_2S	GaAs	D_{H_2S}	$D_{AsH_3}^{-1}$	-1.4	
	GaInP	D_{H_2S}	$D_{PH_3}^{-0.7}$	-2.1	
	GaInAs	D_{H_2S}		-1	
	InP	D_{H_2S}		-2.1	
H_2Se	GaAs	D_{H_2Se}	$D_{AsH_3}^{-1}$	-2.8	D_{TEGa}
SiH_4	GaAs	D_{SiH_4}	$D_{AsH_3}^{-0.5}$	$+1.2$	D_{TEGa}^{-1}
	GaInP	D_{SiH_4}	$D_{PH_3}^{-0.2}$	$+1.2$	
	GaInAs	D_{SiH_4}			
	InP	D_{SiH_4}		$+1.2$	
TMSn	GaAs	D_{TMSn}	$D_{AsH_3}^{0.8}$	$+1$	
GeH_4	GaAs	D_{GeH_4}		$+1.5$	D_{TEGa}^{-1}

Table 6.2 Summary of n- and p-type behavior in III–V alloys.

6.5.3. Conclusions

n- and p-type doping of GaAs and $Ga_{0.51}In_{0.49}P$ suggest an analytical expression giving the doping level as a function of dopant flow rate, growth temperature and the ratio of V/III. It can be written in the general form

Eq. (6.13) $\qquad n = \alpha D_{dopant}^{\alpha} D \exp\left(-E_i / k_B T\right)$

With the exception of the case of $Ga_{0.51}In_{0.49}P$ doping with H_2S where saturation occurs ($\alpha = 0$), all the doping levels have been found to vary linearly with the amount of dopant introduced: $\alpha = 1$.

The doping level is always found to decrease slightly while increasing the V/III ratio with $-1 < a \leq 0$ (DEZn doping). The temperature dependence allows us to separate the dopants into two groups.

- For group IV elements (Sn, Ge, Si), the energy E_i, is positive so that the incorporation efficiency of the dopant increases while increasing the temperature. E_i is weakly dependent on the doping element and is close to 1 eV. Here, the main limiting step for incorporation of the dopant is thermal cracking.
- For group II and VI elements (Zn, S, Se), the energy E_i is negative so that the incorporation efficiency decreases while increasing the temperature. Its value is found to vary between -1 and -2 eV without being a characteristic of the dopant or of the host material. Here, the main limiting step for incorporation of the dopant is its desorption from the crystal surface (Table 6.2).

6.6. GaAs–GaInP heterostructures

Originally, it was assumed that the shortcomings of AlGaAs could be overcome by replacing the AlGaAs by $Ga_{1-x}In_xP$, which for $x = 0.51$ is lattice matched to GaAs. GaInP is comparable to AlGaAs in that it acts as a barrier material for both electrons and holes. There are reports of the growth of corresponding heterostructures and the formation of two-dimensional electron gases (2DEGs) at the heterointerface. However, these heterostructures also reveal the persistent photoconductivity effect (PPC) upon illumination, which is probably due to a sort of DX center in doped GaInP. Hence, besides the absence of Al, it is the general interest in an alternative to the GaAs–AlGaAs system that motivates the investigation of GaAs–GaInP. However, the study of two-dimensional electrons in GaAs–GaInP heterostructures suffers from extremely low mobilities compared to MBE-grown GaAs–AlGaAs heterostructures due to the interface quality of GaAs–GaInP–GaAs and the difficulty in growing high-quality multilayers.

6.6.1. Band offset measurements in the GaAs/Ga$_{0.51}$In$_{0.49}$P system

The band offsets of GaAs–GaInP were measured using different techniques as detailed below.

(a) Capacitance–voltage (C–V) profiling [Rao et al. 1987]
δE_v *measurements:* δE_v was determined from C–V profiling through the p-GaInP/p-GaAs heterojunction by reverse biasing the n^+–p junction between the substrate and the p-GaAs epilayer. The measurement frequency was 1 MHz and the magnitude of the applied differential voltage was 30 mV. The apparent majority-carrier concentration profile $p^*(x)$ was derived from the C–V profile according to the standard relation:

Eq. (6.14) $$p^*(x) = \frac{2}{q\varepsilon}\left(\frac{\mathrm{d}}{\mathrm{d}V}\frac{1}{C^2}\right)^{-1}$$

where

Eq. (6.15) $x = \varepsilon / C$

is the width of the depletion layer (the distance of the depletion layer edge from the p–n junction), $p^*(x)$ is the apparent majority-carrier (hole) concentration at position x, V is the reverse bias voltage, C is the capacitance per unit area, ε is the dielectric permittivity of the semiconductor.

The value of δE_v is obtained from the electrostatic dipole moment associated with the charge imbalance between a presumably known doping distribution $p(x)$ and the experimental $p^*(x)$ curve, using the relation given by Kroemer et al. [1980], adapted to the p–p case:

Eq. (6.16) $\delta E_v = \dfrac{q^2}{\varepsilon} \displaystyle\int_{\varepsilon}^{+\infty} \left[p(x) - p^*(x) \right] (x - x_i) \ dx - kT \ \ln\left(\dfrac{p_2 N_{v1}}{p_1 N_{v2}} \right)$

where $p(x)$ is the ionized acceptor distribution, assumed to be known and to level out far away from the heterojunction, $p_{1,2}$ are the asymptotic values of the doping levels in the GaAs (1) and the GaInP (2), x_i is the distance of the GaAs–GaInP interface from the n^+-GaAs substrate, $N_{v1,2}$ are the valence band densities of states in GaAs and GaInP regions.

To evaluate δE_v, Kroemer et al. [1980] assumed:

(i)   uniform doping on both sides of the interface at $x = x_i$,
(ii)   an abrupt step in the doping *at* $x = x_i$,
(iii)   the existence of a localized interface defect charge σ_i at $x = x_i$. For a given value of x_i, the value of σ_i is obtained from the requirement of overall charge neutrality:

Eq. (6.17) $\sigma_i = \displaystyle\int_0^{+\infty} \left[p(x) - p^*(x) \right] dx$

In such a model an accurate knowledge of the interface position x_i, the thickness of the epilayers and the doping level in epilayers are essential. They found $\delta E_v = 0.238$ eV.

They also performed an additional check by performing a computer reconstruction of what the $p^*(x)$ profile should have been for an abrupt interface, with given values of the band offset and interface charge, using a computer program that basically solves Poisson's equation for incremental voltage steps applied to the heterojunction. They obtained a satisfactory agreement between experimental and theoretical results.

δE_c *measurements:* Rao et al. [1987] used the following structure for δE_c measurements: first a 0.4 μm thick GaAs layer with an *n*-type doping level of 9×10^{16} cm^{-3} was grown on a (1 0 0)-oriented, n^+-doped GaAs substrate; the next layer was a 0.5 μm thick GaInP layer with an *n*-type doping level of 3×10^{16} cm^{-3}.

An In ohmic contact was realized on the back side of the substrate. Finally, an Al Schottky contact was deposited on the GaInP epilayer [Rao et al. 1987].

The Al Schottky barrier was reverse biased to obtain a $C - V$ profile through the GaAs–GaInP heterojunction. The analysis of the data was analogous to that for the p–p structure. The value of $\delta E_c = 0.22$ eV was obtained.

Possible sources of errors: The largest source of uncertainty in this measurement is the quality of the samples. Another source of error is the presence of deep levels near or at the interface, the ionization energy of which changes with changing bias. Errors in the capacitance measurements, such as errors in the diode area, can occur that may yield an incorrect apparent charge concentration.

(b) Evaluation of the band offsets using DLTS [Biswas et al. 1990]

δE_c *and* δE_v *measurements:* Deep-level transient spectroscopy (DLTS) measurements were carried out on GaInP/Au Schottky diodes having a GaAs single quantum well (SQW) in the depletion region. The emission energies of electrons and holes obtained from DLTS measurements are related to the band offsets δE_c and δE_v. The emission rate of electrons from a QW can be derived from the thermionic current due to electrons emitted from the well to the barrier region. This emission rate has been calculated by Martin et al. [1983, 1988] and is given by:

$$\text{Eq. (6.18)} \qquad e_n = \left(\frac{kT}{2\pi n_w^*} \right)^{1/2} \frac{1}{L_w} \exp\left(-\frac{\delta E_c}{kT} \right)$$

It is also possible, by drawing an analogy between an electron in a QW and a deep-level trap, to formulate a detailed balance between thermal capture and emission of electrons from the QW. From such a detailed balance, the emission rate of electrons is given by

$$\text{Eq. (6.19)} \qquad e_n = \frac{16\pi^{3/2}}{3h^3} m_w^* X \left(kT\right)^{1/2} \left(\delta E_e\right)^{3/2} \exp\left(-\frac{\delta E_e}{kT} \right)$$

where δE_e is the electron emission energy from the conduction band well, m_w^* is the effective mass of electrons in the well material and X is a parameter related to the capture of carriers by the wells.

This is valid for hole emission from a valence band well from which δE_v can be measured. The thermal emission energy of carriers from a QW is related to the appropriate band offset.

Next, consider a SQW in the depletion region of a Schottky barrier. The existence of confined electrons in the well changes the depletion width W. Solution of Poisson's equation in the well and barrier regions with appropriate boundary conditions gives

Eq. (6.20) $\quad W^2 = W_0^2 \left(1 + \dfrac{2 n_w L_w}{N_D W_0^2} \right)$

where

Eq. (6.21) $\quad W_0^2 = \left(2\varepsilon / q N_D \right) V$

is the depletion region width in the absence of the well, N_D is the net donor density in the barrier and $V = V_{app} + V_{b1}$, where V_{b1}, is the built-in potential of the junction. The transient capacitance δC is then given by

Eq. (6.22) $\quad \dfrac{\delta C}{C(W)} = \dfrac{n_w L_w}{N_D W_0^2}$

The DLTS signal for rate windows t_1 and t_2 is then given by

Eq. (6.23) $\quad S(t) = C(t_2) - C(t_1) \dfrac{C_0 n_{w0} L_w}{N_D W_0^2} \left[\exp(-e_n t_1) - \exp(-e_n t_2) \right]$

and

Eq. (6.24) $\quad \delta E_c = \delta E_e + E_{e1} + L_w F$

Eq. (6.25) $\quad \delta E_v = \delta E_h + E_{h1} + L_w F$

where E_{e1} and E_{h1} are the electron and hole ground-state subband energies, L_w is the well width and F is the electric field across the well region due to the applied reverse bias. The materials were grown by LP-MOCVD at 510 °C. The structure for δE_c measurements was as follows. First, a 0.2 μm thick n^+-doped GaAs buffer layer was grown on a n^+-GaAs substrate, followed by a 0.5 μm n-doped GaInP layer with a doping level of 2×10^{16} cm^{-3}. A 120 Å thick GaAs n-doped quantum well was grown, followed by a 0.5 μm thick n-doped GaInP layer with doping levels of 2×10^{16} cm^{-3}. Finally, an Au Schottky contact was deposited on the GaInP cap layer.

The structure for δE_v measurements was obtained by growing p-type doped materials on a p^+-doped GaAs substrate, with the same structure and doping levels used for δE_c measurements.

The values of δE_c and δE_v derived from this experiment are 0.198 eV and 0.285 eV, respectively. Also $\delta E_c + \delta E_v = \delta E_g = 0.48$ eV which agrees with the measured value from photoluminescence. The values of δE_c and δE_v estimated in this study agree reasonably well with the measured values reported by Miyoko et al. [1987], using MOCVD materials grown at 700 °C. They also agree with the measured values reported by Rao et al. [1987] using MBE materials.

Possible sources of errors: (i) the quality of the samples, the QW thickness, the carrier concentration in the QW and barrier, the interface quality, etc, (ii) the spread in the subband energies and the excess energy of carriers above the barrier during emission, (iii) reduced energy due to tunneling, (iv) the values of E_{e1} and E_{h1} which are obtained from theoretical analysis.

(c) δE_c evaluation from the I–V curve of HEMTs [Chan et al. 1990]

GaInP–GaAs HEMT structures have been grown by LP-MOCVD at 510 °C. Devices with a 1 μm × 150 μm gate had an extrinsic transconductance of 163 mS·mm^{-1} at 300 K and 213 mS·mm^{-1} at 77 K with $V_{ds} = 3$ V.

An extremely low gate inverse bias current ($I_{rev} < 200$ nA at $V_g = -4$ V) was achieved, and a 0.87 eV Schottky barrier height was evaluated from C–V measurements. The turn-on voltage was 0.75 V and the value of the ideality factor was 1.18, indicating that thermionic emission and diffusion currents are present in the carrier transport. These results demonstrate the high quality of the sample.

According to the model proposed by Chen et al. [1988], the I–V characteristics of HEMTs can be described by the combination of two diodes in series: D_1 and D_2 (see Chan et al. [1990]). D_1 represents the metal–semiconductor Schottky diode and D_2 is the heterojunction diode. Since the Schottky barrier (0.87 eV in this case) is much higher than the conduction band discontinuity, the gate current at low forward bias is determined by D_1. As the bias increases, especially when the potential drop across D_1 is larger than the Schottky barrier height, the change of gate current is determined by D_2. Two slopes are observed in the gate current–voltage characteristics. By extrapolating the I–V curve under D_2 operation it is possible to determine the value of the reverse saturation current of D_2 (I_{ss})

Eq. (6.26) $$I_{ss} = AA^* T^2 \exp\left(-q\Phi(0)/kT\right)$$

where A is the diode area, A^* is the Richardson constant ($A^* = 4\pi \cdot q \cdot k^2 \cdot m_0/h^3$) and $\Phi(0)$ is the potential difference between the channel and the top of the conduction band discontinuity at zero gate bias. In order to reduce the influence of the series resistance, a large-gate (200 × 50 μm^2) diode was measured. The value of $\Phi(0)$ evaluated from I_{SS} is 196 meV.

Using a self-consistent solution of the Schrodinger and Poisson equations, the potential difference between the Fermi level in the channel and the bottom of the conduction band discontinuity is found to be 35 meV. Therefore the conduction band discontinuity of GaInP–GaAs is estimated to be $\delta E_c = 0.231$ eV, which is quite close to the previous results.

(d) δE_c measured by the temperature dependence of the collector current of HBTs [Kobayashi et al. 1989]

The band offset in the emitter–base heterojunction can be evaluated from the temperature dependence of the collector current of HBTs.

Thermionic electrons from the emitter are injected into the base and diffuse towards the collector, recombining slightly with the majority holes. The collector current can be represented by

Eq. (6.27) $I_c = A * T^2 \exp(-E_a / kT)$

where $A*$ is the effective Richardson constant and E_a is the activation energy for injected electrons. E_a can be equated with the energy difference between the conduction band in the emitter edge adjacent to the base and the Fermi energy δ_1 in the neutral emitter, and is given by

Eq. (6.28) $E_a = (V_D - V_{be}) \Gamma / (1 + \Gamma) + \delta_1$

V_D is the diffusion potential at equilibrium

Eq. (6.29) $V_D = E_g (GaAs) + \delta E_c - \delta_1 - \delta_2$

where δ_2 is the Fermi level in the base, V_{be} is the emitter–base bias voltage. From these equations one gets:

Eq. (6.30) $\ln(I_c / A * T^2) = -[E_g (GaAs) + \delta E_c - V_{be}] / nkT$

where

Eq. (6.31) $n = (\Gamma + 1) / \Gamma$

Γ is the partition ratio of the diffusion potential under biasing

Eq. (6.32) $\Gamma = V_1 / V_2 = \varepsilon_2 N_A / \varepsilon_1 N_D$

ε_1 and ε_2 are the dielectric constant in the emitter and the base and N_D and N_A are the dopant densities in the emitter and the base.

Kobayashi et al. [1989] used MOCVD to grow the *npn* HBT structure at 700 °C. They obtained a current gain as high as 200 in the current range of about 10 Å·cm^{-2}. The base doping was 8×10^{18} cm^{-3} (using Zn as a *p*-type dopant), and the base thickness was 1500 Å.

They used Eq. (6.30) to determine the band offset in GaAs–GaInP and they deduced a value of $\delta E_c = 30$ meV at 300 K.

Possible sources of errors: (i) An increase in *n* of 0.01 resulted in an increase in δE_c of 6 meV, (ii) temperature measurement resulted in an error in δE_c of 2.3 meV/°C, (iii) the quality of the sample, (iv) if zinc diffusion caused the *p–n* junction to be in the GaInP emitter, the evaluated E_a and hence δE_c in Eq. (6.30) are apparently larger than the true E_a and δE_c, (v) if the alloy composition of GaInP is not matched to GaAs, or if GaInP is contaminated with As at the heterointerface,

δE_c decreases, (vi) in this sample, the bandgap energy of GaInP grown for HBT was 60 meV smaller than the energy gap of the intrinsic material grown by LPE. The band lineup must then be arranged so as to compensate for this 60 meV gap shrinkage, which can further reduce δE_c. (vii) In order to confirm the validity of these measurements, it is necessary to evaluate δE_v by using a *pnp* HBT structure, as was the case for other measurements.

(e) Conclusions

It was generally assumed that the energy bands on both sides of the junction in the GaInP–GaAs system were aligned so that approximately 40% of the total, direct bandgap difference (δE_g) appeared as the conduction band discontinuity (δE_c) and the remaining 60% made up the valence band discontinuity (δE_v).

6.6.2. Observation of the two-dimensional properties of the electron gas in $Ga_{0.51}In_{0.49}P/GaAs$ heterojunctions grown by MOCVD [Razeghi et al. 1986a]

The optimum conditions for the growth of GaAs and $Ga_{0.51}In_{0.49}P$, as determined during these investigations, are presented in Table 6.1. The heterojunctions studied here were grown on a semi-insulating (1 0 0) GaAs substrate. A nominally undoped, 3000 Å thick GaAs layer with electron concentration of 2×10^{16} cm^{-3} was covered by a sulfur- or silicon-doped, 1000 Å thick $Ga_{0.51}In_{0.49}P$ layer with an electron concentration of 2×10^{17} cm^{-3}.

From Auger profiling of phosphorus in a multiquantum well of GaAs–$Ga_{0.51}In_{0.49}P$ grown under identical conditions, the interface was estimated to be abrupt within two atomic layers.

Magnetotransport measurements were performed on long bar-shaped samples with current and potential contacts. Standard Hall bridges were photo-lithographically defined. Shubnikov–de Haas (SDH) and quantum Hall effects (QHE) were measured in magnetic fields up to 18 T.

The low-field measurements performed on these heterojunctions gave electron mobilities of 49,000 cm$^2 \cdot$V$^{-1} \cdot$s^{-1} at 4 K, 38,000 cm$^2 \cdot$V$^{-1} \cdot$s^{-1} at 77 K and 6000 cm$^2 \cdot$V$^{-1} \cdot$s^{-1} at 300 K. SDH measurements were performed in normal and tilted-field configurations.

Fig. 6.13 shows SDH oscillations for various angles between the field direction and the normal to the interface. Extrema of resistivity are observed at constant values of the perpendicular component of the magnetic field $B_\perp = B \cos \theta$ (Fig. 6.14), giving evidence of the two-dimensionality of electrons at the interface. Fig. 6.15 shows the diagonal resistivity ρ_{xx} and the Hall resistivity ρ_{xy} as functions of magnetic field at 4.2 K. In the vicinity of 5.2, 7.8 and 16 T, ρ_{xx} approaches the zero-resistance state. At the same field positions ρ_{xy} develops plateaus at h/ie^2 for $i = 6, 4, 2$ characteristic of the QHE, hence giving further proof of the ideal 2D behavior of this system.

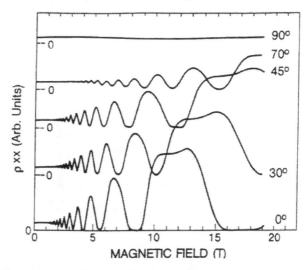

Fig. 6.13. Shubnikov–de Haas oscillations in a GaInP–GaAs heterostructure as a function of magnetic field for various tilt angles measured at 4.2 K (after Razeghi et al. [1986a]).

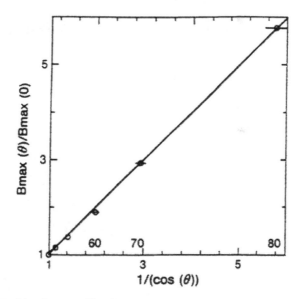

Fig. 6.14. Field value $B_{max}(\theta)$ of magnetoresistance extrema at angle θ divided by $B_{max}(\theta = 0)$ of corresponding extrema at $\theta = 0$ as a function of $[cos(\theta)]^{-1}$ (after Razeghi et al. [1986a]).

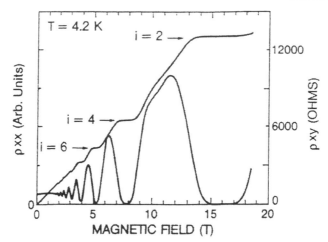

Fig. 6.15. Shubnikov–de Haas and quantum Hall effect as a function of magnetic field (after Razeghi et al. [1986a]).

Another important feature of this system is the persistent increase of the 2D electron concentration which was observed after illuminating the sample at low temperature (persistent photoconductivity effect), similar to that observed in GaAs/GaAlAs heterostructures [Stormer et al. 1981, Kastalsky and Hwang 1984].

Fig. 6.16 shows the SDH oscillations before and after illuminating the sample. After illumination, a very clear shift towards higher fields is observed. The fundamental field $N \times B_n$ increases from 31.6 to 39.0 T giving a change of the total concentration from 7.64×10^{11} to 9.43×10^{11} cm^{-2}. The oscillations become more pronounced with increasing temperature because of the increased population of the first excited subband.

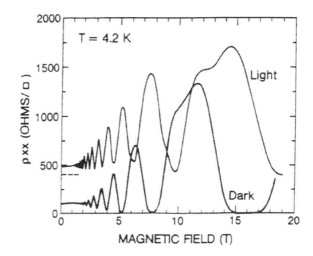

Fig. 6.16. Shubnikov–de Haas oscillations before and after illuminating the samples as a function of magnetic field (after Razeghi et al. [1986a]).

The population of the ground subband derived from the low-field SDH periodicity changes from 7.51×10^{11} to 8.44×10^{11} cm^{-2}, leaving the difference $n_1 = n_{tot} - n_0$ to be 0.14×10^{11} and 0.98×10^{11} cm^{-2}, respectively, for the first excited subband's population.

Using the persistent photoconductivity effect to tune the 2D electron concentration, one can measure the population of each subband as a function of the total population (Fig. 6.17). The results are fitted well by linear dependences and are very similar to the predictions of numerical calculations [Ando 1982] for GaAs/GaAlAs heterojunctions. Extrapolation of this plot indicates that the second subband starts to be populated at about 7.3×10^{11} cm^{-2}.

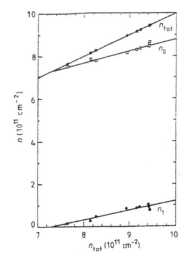

Fig. 6.17. Plot of electron populations in the first two electric subbands n_0 and n_1 as a function of the total electron concentration, n_{tot} (after Razeghi et al. [1986a]).

It is worth adding that the low-field Hall concentration underestimated the total 2D electron concentration n_{tot}. The difference $\Delta n_H = n_{tot} - n_H$ clearly exceeded the experimental error and was obviously connected with a two-subband conduction, each subband having a different scattering rate. The Hall mobility decreased slightly as the second subband was populated, probably due to intersubband scattering.

6.6.3. High-mobility GaInP–GaAs heterostructures

Electron mobilities as high as 780,000 cm$^2 \cdot$V$^{-1} \cdot$s^{-1} at a two-dimensional electron concentration $N_s = 4.1 \times 10^{11}$ cm^{-2} have been measured in GaAs–GaInP heterostructures grown by low-pressure metalorganic chemical vapor deposition (LP-MOCVD) [Razeghi et al. 1989b].

The sample reported was grown on a semi-insulating GaAs substrate, oriented 2° off the (0 0 1) plane towards the (1 1 0) plane. On top of the substrate, a 1 μm thick unintentionally doped GaAs layer followed by a 2000 Å thick unintentionally doped layer of GaInP lattice matched to GaAs were grown. The GaInP layer was grown at a low pressure ($p = 76$ Torr) at a substrate temperature in the range 500–550 °C using triethylgallium, trimethylindium and pure phosphine with hydrogen as the carrier gas. The optimum growth conditions are summarized in Table 6.1. For the growth conditions of the GaAs layer, refer to Razeghi et al. [1989b]. Corresponding layers of GaAs and GaInP with thicknesses of 2–3 µm, grown under identical conditions as the much thinner layers in the heterostructure, revealed $N_D - N_A \leq 5 \times 10^{14}$ cm^{-3} for GaAs and $N_D - N_A \leq 5 \times 10^{15}$ cm^{-3} for GaInP ($N_D - N_A$: donor and acceptor density), respectively, determined by Hall measurements. The sample was structured into a standard Hall bar geometry, Au–Ge and Ni metallizations were evaporated and then alloyed for 2 min at 420 °C in a hydrogen atmosphere to act as ohmic contacts.

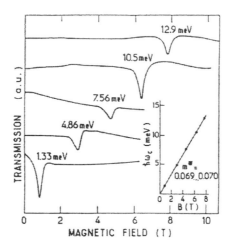

Fig. 6.18. Cyclotron resonance absorption in GaInP–GaAs for different photon energies (at $T = 1.6$ K). The far-infrared radiation originated from a CO_2-pumped molecular laser and a carcinotron. The inset shows the cyclotron energy, derived from the absorption peak position, as a function of magnetic field (after Razeghi [1989a]).

Surprisingly, this nominally undoped heterostructure contains, depending slightly on the cooling procedure, about 1.7×10^{11} cm^{-2} electrons, determined from the periodicity of the Shubnikov–de Haas oscillations. The fact that there are electrons and that they are on the GaAs side of the interface is clear from cyclotron resonance experiments (Fig. 6.18), as the dependence of the cyclotron energy $\hbar\omega_c = \hbar eB / m^*$ (see inset in Fig. 6.18) yields the effective mass expected for GaAs heterostructures $m^*/m_0 = 0.069$–0.070. (Here, $\hbar$ is Planck's constant, ω_c the cyclotron frequency, e the elementary charge, B the magnetic field and m^* and m_0 the effective mass and the free-electron mass, respectively.) This m^* is clearly distinguishable from the estimated value $m^*/m_0 \approx 0.1$ in GaInP [Watanabe and Ohba 1987]. The question of where the electrons come from remains open. If they originate from donors in the barrier material, their concentration has to be at least of the order of $N_D \approx 2 \times 10^{16}$ cm^{-3}, as a simple calculation (disregarding surface depletion) shows. It is possible that the 2000 Å thick GaInP layer in the heterostructure has properties different from the 3 μm thick reference layer ($N_D - N_A \leq 10^{15}$ cm^{-3}) although these were grown under identical conditions. Such a big difference is rather surprising.

Successive illumination of the sample with light pulses from a red light emitting diode (LED) at liquid-helium temperatures persistently increases the electron concentration. The photon energy of the LED is centered around $E = 1.87$ eV, comparable to the bandgap of $Ga_{0.51}In_{0.49}P$ (1.9 eV) [Madelung 1991] and above that of GaAs. The PPC is well known for GaAs–AlGaAs heterostructures [Stormer et al. 1979] and was also recently reported for GaAs–GaInP heterostructures with intentionally n-doped barrier material. The mechanism for the PPC in the latter material combination, however, is not yet understood in detail, but it involves the ionization of a DX-center-like deep trap [Ben Amor et al. 1989]. Using the PPC, the electron concentration could be increased up to $N_s = 3.9 \times 10^{11}$ cm^{-2} (N_s was determined from the periodicity of the Shubnikov–de Haas oscillations; a typical trace is shown in Fig. 6.19). Steady illumination increased N_s further on up to 4.1×10^{11} cm^{-2} (transient photoconductivity [Schubert et al. 1985a, b]). One cannot exclude that a stronger illumination might increase N_s even more since, in the intensity range explored, there were no signs of saturation.

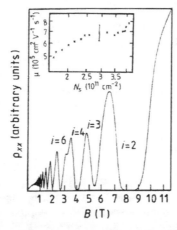

Fig. 6.19. Magnetoresistivity ρ_{xx} of a GaAs–GaInP heterostructure for an electron density of $N_s = 3.9 \times 10^{11}$ cm^{-2} at a temperature $T = 1.4$ K. This N_s has been reached by illumination with red light using the persistent photoconductivity effect. Some filling factors $i = N_s h/(eB)$ are indicated. The inset shows the mobility as a function of the electron density varied by light pulses of a red light emitting diode using the persistent photoconductivity effect. For clarity, the error bar is only shown on one of the points (after Razeghi [1989a]).

Razeghi et al. [1986b] do not see any evidence for parallel conduction through the barrier material after strong illumination. A current bypass would manifest itself in a non-zero magnetoresistivity ρ_{xx} in the plateau regions of the quantum Hall effect [von Klitzing and Ebert 1984], especially also at filling factor $i = 2$ (see Fig. 6.19). The lack of parallel conduction might find its explanation in the low doping level in the GaInP which is not intentionally doped.

After each successive illumination, the mobility μ was determined from the electron concentration N_s and the resistivity ρ at $B = 0$, according to the classical relation $\mu = (e\rho N_s)^{-1}$, as shown in the inset of Fig. 6.19. The sheet resistivity ρ deduced from the resistance R by using $\rho = (w/l)R$ (where w is the width of the current carrying channel, l is the distance of the potential probes) involves an uncertainty of $\pm10\%$ due to the finite extent of the potential probes. The mobility increases with N_s up to a concentration of about 2.5×10^{11} cm^{-2} and then remains constant upon further increasing N_s until $\simeq 3.8 \times 10^{11}$ cm^{-2} when μ again increases with N_s, leading in this sample to a final value of $\mu = 780{,}000$ cm$^2 \cdot$V$^{-1} \cdot$s^{-1} at $N_s = 4.1 \times 10^{11}$ cm^{-2}. The half width at half amplitude $\Delta B_{1/2}$ of the cyclotron resonance line confirmed the high mobility if one used the semi-empirical expression $\Delta B_{1/2} = 0.63 \cdot \mu^{-1/2} \cdot B^{1/2}$ established in the case of GaAs–AlGaAs heterojunctions [Voisin et al. 1981]. The value $\Delta B_{1/2} = 0.18$ T measured for a resonance magnetic field $B = 6.5$ T (a photon energy of 10.5 meV in Fig. 6.18 would lead to $\mu = 800{,}000$ cm$^2 \cdot$V$^{-1} \cdot$s^{-1}) is in good agreement with the value deduced from transport experiments. This extremely high mobility proves the very high purity of the GaAs. For GaAs–AlGaAs heterostructures, such high mobilities are only reached by the growth of an undoped spacer layer, which separates the two-dimensional electrons from the ionized impurities in the barrier material. If we assume that the electrons are coming from the barrier material (residual donors in GaInP), there is nothing equivalent to a spacer layer, provided the growth of the barrier layer is uniform. The reduction of ionized impurity scattering in this heterostructure may be due to the fact that MOCVD-grown GaAs samples are slightly n-type, yielding an accumulation layer, contrary to the usual case for MBE-grown GaAs heterostructures having quasi-two-dimensional electron inversion layers. For accumulation layers, the penetration of the electronic wavefunction into the barrier is weaker, thus reducing ionized impurity scattering in the barrier material. In addition, the density of ionized impurities is small, as the barrier is not intentionally doped. Therefore, the whole barrier might be assumed to act as a spacer layer.

The plateau-like behavior of the mobility beginning at $N_s \approx 2.5 \times 10^{11}$ cm^{-2} (cf. inset in Fig. 6.19) is due to the onset of intersubband scattering, and indicates that the first excited electric subband is starting to be populated. A comparison of this critical N_s with corresponding calculations of the confining interface potential and the electronic wavefunctions leads to a depletion charge of the order of 10^9 cm^{-2} [Bastard 1988]. These calculations have been carried out for the GaAs–AlGaAs system. The results may be slightly different for GaAs–GaInP (e.g., due to a different conduction band offset), but in any case, the order of magnitude of the depletion charge would be far too small for an inversion layer. This confirms again the well known fact that unintentionally doped MOCVD-grown GaAs is usually slightly n-type. In ρ_{xx} (see Fig. 6.19), a slight occupation of the second subband can be inferred from the positive magnetoresistance. For some electron concentrations, additional structure in the Shubnikov–de Haas oscillations is observed, which is typical of second-subband occupation.

Magnetotransport measurements showed that LP-MOCVD-grown GaAs–GaInP heterostructures can reach extremely high electron mobilities, which up to now have been exclusively attained by MBE-grown GaAs–GaAlAs heterostructures. One can interpret this very high mobility as an indication of reduced ionized impurity scattering due to the high purity of the materials in the nominally undoped heterostructures.

6.6.4. Electron spin resonance in the two-dimensional electron gas of a GaAs–Ga$_{0.51}$In$_{0.49}$P heterostructure [Dobers et al. 1989a, b]

One possibility for the magneto-optic investigation of two-dimensional electron gases (2DEGs) is electron spin resonance (ESR). The most frequently used method, cyclotron resonance [Merkt 1987], yields the effective mass m^*, but ESR yields the g factor. Both of these band structure parameters enter into the energy spectrum of a 2DEG in a perpendicular magnetic field [Ando et al. 1982]

$$\text{Eq. (6.33)} \qquad E_{Nm_s} = E_0 + \left(N + \tfrac{1}{2}\right)\left(e\hbar / m^*\right)B + m_s g \mu_B B$$

where E_0 is the energy of the lowest electric subband (higher ones are not considered), e is the elementary charge, $\hbar$ Planck's constant, μ_B Bohr's magneton and B the magnetic field. $N = 0, 1, 2,\dots$ is the Landau level index and m_s is the magnetic spin quantum number, which takes the values $m_s = \tfrac{1}{2}$ ("spin up") and $m_s = -\tfrac{1}{2}$ ("spin down"). Strictly speaking, this energy spectrum is only valid in the case of parabolic bands, where m^* and g are really constants, independent of the magnetic field and the Landau level. In reality, however, the non-parabolicity of the conduction bands of the semiconductors involved makes both of them vary with the magnetic field and the Landau quantum number [Rossler et al. 1989].

The dependence of the g factor on B and N has been studied systematically in the 2DEG of GaAs–AlGaAs heterostructures [Dobers et al. 1988a, b]. These experiments were done by looking at the change of the magnetoresistivity ρ_{xx}. Using the samples themselves as a detector for ESR, the sensitivity was high enough to see ESR. A basic requirement for this, however, turned out to be the following: the spin splitting must be resolved in ρ_{xx}, i.e. minima in the Shubnikov–de Haas oscillations corresponding to odd filling factors $i = N_s h / (eB)$ have to be visible (N_s: concentration of 2D electrons). This limits the measurability of ESR to high-mobility heterostructures.

The availability of very high-mobility GaAs–GaInP heterostructures grown by low-pressure metalorganic vapor deposition (LP-MOCVD) [Razeghi et al. 1989b] also permits the study of the spin splitting of the Landau levels in the 2DEG of GaAs–GaInP heterostructures.

Consider a GaAs–GaInP heterostructure, grown by LP-MOCVD, which consists of a 1 µm thick undoped GaAs layer on top of a semi-insulating GaAs substrate followed by a 200 nm thick, unintentionally doped layer of Ga$_x$In$_{1-x}$P lattice matched to GaAs ($x = 0.51$). The concentration of 2D electrons was about $N_s = 1.7 \times 10^{11}$ cm^{-2}, depending slightly on the cooling procedure. By illumination with short light pulses of a red LED, N_s could be increased using the persistent photoconductivity effect [Razeghi et al. 1989a]. In order to see two resolved spin minima in the SDH oscillations within the available magnetic field range

$(B \leq 12 \text{ T})$, N_s was increased to $N_s = 2.7 \times 10^{11} \text{ cm}^{-2}$. The mobility at this concentration was $\mu = 650{,}000 \text{ cm}^2 \cdot \text{V}^{-1} \cdot \text{s}^{-1}$. At this concentration, which remained constant for all the ESR experiments to be discussed, not only the $i = 1$ minimum but also the $i = 3$ minimum of the SDH oscillations became resolved (at $T = 1.3 \text{ K}$), thus permitting the study of ESR in the two lowest Landau levels. At about $N_s = 2.5 \times 10^{11} \text{ cm}^{-2}$ the second subband may begin to be populated [Razeghi et al. 1989a]. This might also be the case during the ESR study ($N_s = 2.7 \times 10^{11} \text{ cm}^{-2}$). However, there is neither an additional oscillation in ρ_{xx} typical for second subband occupation nor a positive magnetoresistance at small magnetic fields (see Fig. 6.20). Therefore, the possibility of second-subband occupation must be extremely small, and it will not be considered further.

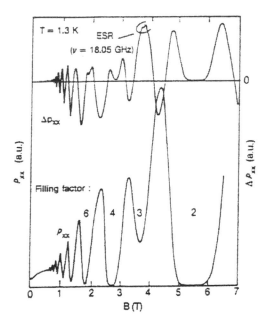

Fig. 6.20. Magnetoresistivity ρ_{xx} of a GaAs–GaInP heterostructure (lower trace) and its change due to microwave radiation $\Delta\rho_{xx}$ (upper trace) as a function of the magnetic field. ESR occurs at $B \approx 3.7$ T. Note that the scale for $\Delta\rho_{xx}$ is greatly magnified compared with that of ρ_{xx} (after Dobers et al. [1989b]).

The sample had a standard Hall bar geometry, and was located inside an oversized waveguide, immersed in liquid helium. In magnetic fields perpendicular to the plane of the 2DEG, ρ_{xx}, was measured by a standard lock-in detection utilizing frequencies of the order of 10 Hz. Changes of the sample resistivity due to the chopped microwave radiation (of the order of 1 kHz) were detected using a second lock-in amplifier. Different klystrons and backward-wave oscillators were used to cover the frequency range up to 60 GHz, with magnetic fields up to 12 T. The output power of these microwave sources lay between a few mW and several hundred mW.

In Fig. 6.20, both ρ_{xx} and its change due to microwave radiation $\Delta\rho_{xx}$ are shown. The lower trace, ρ_{xx}, reveals pronounced SDH oscillations already beginning at $B \approx 0.5$ T. There are broad regions of vanishing ρ_{xx}, typical for the quantum Hall effect. At $B \approx 3.7$ T, the spin minimum corresponding to $i = 3$ is visible, whereas the next higher one ($i = 5$) is not resolved, because the magnetic field is too low. The upper trace shows $\Delta\rho_{xx}$. The broad spectrum reflects the periodicity of the SDH oscillation. It is due to a non-resonant heating of the sample [Stein et al. 1984, Guldner et al. 1987], and is related to the derivative of ρ_{xx} with respect to the temperature. On top of this broad spectrum, there is a sharp resonance structure, in this case ($v = 18.05$ GHz) at $B \approx 3.7$ T. Whereas the broad non-resonant spectrum does not depend on the microwave frequency, the position of the ESR structure shifts with the applied microwave frequency on top of the non-resonant background. This can be seen in Fig. 6.21 which shows $\Delta\rho_{xx}$ for several microwave frequencies yielding ESR in the vicinity of the filling factor $i = 1$. Here, the background changes because of a continuous variation of the bath temperature between $T \approx 2$ K and $T \approx 3$ K, and also due to different microwave powers at different frequencies; the width of the ESR structures is of the order of $\Delta B = 100$ mT. This is a factor of two to three larger than has, up to now, been seen in the of case of GaAs–AlGaAs [Dobers et al. 1988a, b].

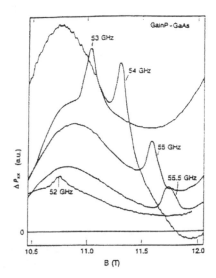

Fig. 6.21. Typical ESR structures in the change of magnetoresistivity due to microwave radiation $\Delta\rho_{xx}$ for different microwave frequencies. These ESR structures occur around filling factor i = 1 (after Dobers et al. [1989b]).

The ESR structures shown in Fig. 6.21 consist of an increase of ρ_{xx} due to ESR. Approaching the even filling factor $i = 2$, the effect of ESR on ρ_{xx} (and $\Delta\rho_{xx}$) turns into a decrease. The same observation has been made in the $N = 1$ Landau level: around $i = 3$, ESR results in an increase; approaching the neighboring even filling factors, say around $i = 2.5$ and $i = 3.5$, ESR decreases ρ_{xx}. This is the same finding as in GaAs–AlGaAs heterostructures [Dobers et al. 1989a, b], which might give some evidence to the assumption that this is an intrinsic property of a 2DEG.

A very weak hysteresis of the ESR structure, found during a decrease of the magnetic field, is an indication of an Overhauser shift of the ESR [Dobers et al. 1988a, b] which is due to a dynamic nuclear spin polarization of the lattice nuclei via hyperfine interaction. This effect is only visible during very slow magnetic field sweeps (≈ 0.1 T·min^{-1}), and even then it is very weak. Thus in practice, there was no complication for a determination of the electronic g factors. Although one is not able to quantify properly the "strength" of a dynamic nuclear spin polarization, it became obvious that it is much weaker in the present case of a GaAs–GaInP heterostructure compared with the GaAs–AlGaAs heterostructures [Dobers et al. 1988a, b], as all experiments were done with the same experimental setup. Basically, there are two possibilities for the weakness of the Overhauser shift: either the nuclear spin relaxation times are very short, so that a nuclear spin polarization, once created, immediately relaxes, or more likely the electrons relax directly, without polarizing the nuclei, thus leaving the nuclear spin polarization unchanged from the beginning. The broader ESR structures in GaAs–GaInP indicate smaller electronic spin relaxation times. Thus, the electrons feel more spin-flip scattering and the second interpretation seems to be preferable: among the different electronic spin relaxation channels, the dynamic nuclear spin polarization is less important, because all the others are more effective here.

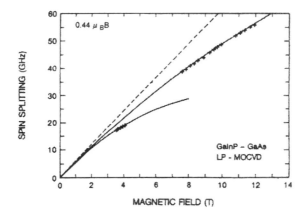

Fig. 6.22. Measured spin splittings of the two lowest Landau levels. The full curves are least-squares fits to the experimental data leading to the two coefficients in Eq. (6.34). For comparison, the spin splitting due to the bulk GaAs conduction band edge g factor, g = 0.44, is indicated as a broken line (after Dobers et al. [1989b]).

From a series of experiments with different microwave frequencies, as shown in Fig. 6.21, the magnetic field dependence of the spin splitting can be deduced. This is shown in Fig. 6.22. The experimental results for a certain Landau level cover only a limited magnetic field range, as does the Fermi level. The spin splittings are smaller than the spin splitting in bulk GaAs (broken line). Both of the two sets of data ($N = 0$ and $N = 1$) follow the quadratic magnetic field dependence ($1/T$ = inverse tesla)

Eq. (6.34) $\Delta E = 0.418 \mu_B B - 0.0133 (1/T)(N + \tfrac{1}{2}) \mu_B B^2$

the two coefficients being determined by a common least-squares fit of all the data. From Eq. (6.34), a magnetic-field- and Landau-level-dependent g factor can be deduced

Eq. (6.35) $g(B, N) = 0.418 - 0.0133(1/T)(N + \tfrac{1}{2}) B$

This qualitative magnetic field dependence is predicted by theory [Lommer et al. 1985], and confirmed by experiments [Dobers et al. 1988a, b, 1989a, b], in the case of GaAs–AlGaAs. It turns out to be valid also in GaAs–GaInP heterostructures. Furthermore, the two coefficients are also found to be in the same range [Dobers et al. 1989a, b]. The similarity of the spin splitting in GaAs–GaInP heterostructures and those consisting of GaAs–AlGaAs is because the 2DEG is in both cases on the GaAs side of the interfaces. The extent of the electronic wavefunctions into the barrier material is only very small, in either case. Therefore, an exchange of the barrier material does not alter the properties of the 2DEG very much.

MOCVD-grown GaAs is known to be slightly *n*-type. Thus, the 2DEG in MOCVD-grown GaAs heterostructures forms an accumulation layer, in contrast with MBE-grown heterostructures. The 2DEG, therefore, is confined at a slightly larger distance to the interface and enters to a lesser extent into the barrier material, an effect that is still supported by the high conduction band offset in GaAs–GaInP. The spin splitting, therefore, should be more "GaAs-like." There is slight evidence for this: according to Lommer et al. [1985], the magnetic field dependence of the g factor

Eq. (6.36) $g(B, N) = g_0 - c(N + \tfrac{1}{2}) B$

is simply an average of two corresponding quantities, one for each of the two materials, weighted with the probability of finding the electrons on either of the two sides of the interface. The coefficient c is equal to 0.0133 T^{-1} (inverse tesla). If the electrons are more on the GaAs side, the coefficient c is bigger [Lommer et al. 1985]. In fact, for the actual GaAs–GaInP heterostructure, the magnetic field dependence of the g factor is slightly stronger than found in GaAs–AlGaAs heterostructures [Dobers et al. 1989a, b], but, unfortunately, not strong enough to verify the significance of this difference. Anyway, the experimental findings are consistent with the 2DEG being an accumulation layer at a higher conduction band offset, compared with that in GaAs–AlGaAs.

From these investigations, one can obtain the dependence of the g factor on the magnetic field and the Landau level which turned out to be quantitatively comparable to its dependence in GaAs–AlGaAs heterostructures.

6.6.5. *Magnetotransport measurements in GaInP/GaAs heterostructures using δ-doping*

The electrical properties of the two-dimensional electron gas in selectively S-and Si-doped GaInP/GaAs heterostructures grown by low-pressure MOCVD have been studied by Ranz et al. [1990]. The influence of the spacer thickness on the sheet carrier density derived from Shubnikov–de Haas oscillations is analyzed. Persistent photoconductivity is studied through the temperature dependence of the Hall effect and is used to increase the electronic density. At each step of illumination the classical to quantum lifetimes ratio is derived and an increase of this ratio is observed. Comparison with AlGaAs/GaAs heterostructures shows that DX-like centers are not necessarily responsible for the observed persistent photoconductivity effect [Ranz et al. 1990].

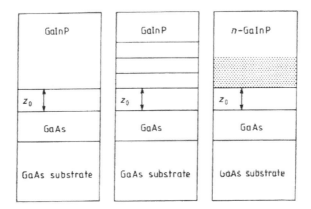

Fig. 6.23. *Schematic diagram of GaInP/GaAs heterostructures with single δ-doping, multi-δ-doping and uniform doping (after Ranz et al. [1990]).*

The samples were single-interface GaInP/GaAs heterostructures grown by low-pressure MOCVD [Razeghi et al. 1986b] schematically represented in Fig. 6.23. Several types of doping process were made use of in the GaInP layer: using δ-doping with variable spacers, multi-δ-doping and uniform doping. Both the transverse magnetoresistance and the Hall effect were measured under magnetic fields up to 20 T. Shubnikov–de Haas (SDH) oscillations and Hall plateaus (with $i = 2$, 4, 6) were observed at 4.2 K indicating the formation of a two-dimensional electron gas (2DEG) at the interface. The temperature dependence of the Hall effect at low magnetic field was also used before and after optical excitation at 77 K using a LED emitting at 1.8 eV.

For the uniform and multi-δ-doped samples, the shape of the transverse magnetoresistance curves are very similar and indicate that parallel conduction is present in the doped GaInP layer. Strong oscillations of the Hall effect traces were also detected and can be explained by assuming the existence of a Hall current in the interface region opposite to the one for the parallel conduction region.

Parallel conduction was completely eliminated by using δ-doped GaInP layers with spacer thickness less than 500 Å. Mobilities of the order of 10^5 cm^2·V^{-1}·s^{-1} and evidence of only one occupied electrical subband were then obtained. The carrier density of this subband (N_s) was derived from the period of the SDH oscillations. The carrier density decrease observed with increasing spacer thickness (z_0), as shown in Fig. 6.24, can be explained by the increase of the electrostatic potential in the spacer layer, the Fermi level being fixed at the donor-level positions.

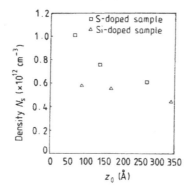

Fig. 6.24. Electron density N_s as a function of the spacer thickness (z_0) in δ-doped GaInP/GaAs heterostructures with S and Si (after Ranz et al. [1990]).

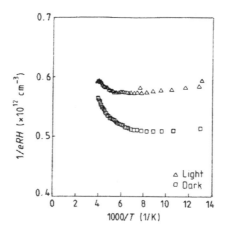

Fig. 6.25. Hall concentration versus 1/T in the dark and after illuminations (after Ranz et al. [1990]).

Temperature-dependent Hall measurements have been performed on the highest-mobility S- and Si-doped samples. The data before and after illumination shown in Fig. 6.25 indicate a small persistent photoconductivity (PPC) effect already described for similar heterostructures [Razeghi et al. 1986b]. However, the behavior of the photoexcited Hall density is very different from the one reported by Ben Amor et al. [1989], where a continuous decrease of the Hall density after illumination is observed with increasing temperature. This decrease, which was attributed to a capture process through a lattice-relaxation energy barrier, is not observed in this sample (Fig. 6.25). The small enhancement of the Hall mobility obtained after illumination can be attributed to an increased screening of the impurity potential by the 2DEG, involving an increase of the small-angle scattering process. Such a conclusion is also supported by the following analysis of the classical to quantum lifetimes ratio.

In order to study the change of the scattering processes when the PPC effect is used to increase the Fermi wavevector, low-field SDH oscillations and Hall effect data have been analyzed to obtain the quantum and classical lifetimes. The scattering (or classical) lifetime τ_s is related to the mobility through the relation $\mu = e\tau_s/m^*$ where m^* is the effective mass. The single-particle (or quantum) lifetime τ_q is related to the lifetime of the unperturbed one-electron Hamiltonian eigenstate in the presence of scatterers. This characteristic time is partly responsible for the damping of the SDH oscillations at low magnetic field. Using the Ando formula, the amplitude A of the SDH oscillations can be expressed as [Ando et al. 1982]

$$\text{Eq. (6.37)} \qquad A \; \alpha \; T\omega_c^{-3} \exp\left(-\frac{\pi}{\omega_c \tau_{\text{eff}}}\right)$$

where ω_c is the cyclotron frequency and τ_{eff} is denned by $\tau_{\text{eff}}^{-1} = \tau_T^{-1} + \tau_q^{-1}$ with $\tau_T = h/2\pi kT$. It is worth noting that this formula is valid for $\omega_c \tau_q \gg 1$ and for the sinusoidal oscillation regime.

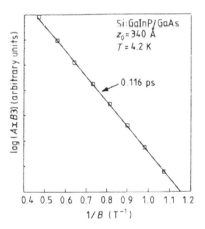

Fig. 6.26. Analysis of the SDH amplitude; the effective lifetime τ_{eff} derived from the slope is indicated (after Ranz et al. [1990]).

The magnetic field limit for the analysis was chosen to be below the field at which quantization of the Hall plateaus becomes observable (for $B \leq 2$ T). The determination of τ_q is illustrated in Fig. 6.26. At each step of illumination, the Hall mobility and τ_q were derived in order to obtain the classical to quantum lifetimes ratio ($R = \tau_s/\tau_q$). At the same time the Fermi wavevector k_F was derived from the SDH carrier density $\left(k_F = \sqrt{2\pi N_s}\right)$. The plot of R as a function of the normalized

Fermi wavevector k_F/q_{TF} (where $q_{TF} = 2 \times 10^6$ cm^{-1} is the Thomas–Fermi screening parameter in GaAs) is shown in Fig. 6.27. This ratio is always greater than one, indicating that small-angle scattering is the predominant scattering process. This ratio increases after each illumination dose.

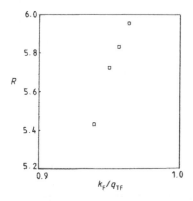

Fig. 6.27. Classical to quantum lifetimes ratio $R = \tau_s/\tau_q$ versus normalized Fermi wavevector in GaInP/GaAs. Saturation corresponds to an increase of 6% of the initial density (after Ranz et al. [1990]).

It is interesting to compare these results with those reported in AlGaAs/GaAs heterostructures. In AlGaAs/GaAs a decrease of R with increasing k_F was found after illumination. This decrease was attributed to the effect of DX centers by assuming a negative-U character of the defect.

In this case, such a conclusion is no longer valid and the predominance of remote impurity scattering is sufficient to explain the behavior of R with a light-induced increase of k_F. Hence, the PPC effect observed here does not have the same origin as in AlGaAs/GaAs heterostructures.

6.6.6. Persistent photoconductivity in $Ga_{0.51}In_{0.49}P/GaAs$ heterojunctions [Ben Amor et al. 1989]

It was originally predicted that several problems associated with the use of AlGaAs (oxidation, presence of DX center) could be avoided by using GaInP. Razeghi et al. [1986b] reported the observation of a strong persistent photoconductivity effect (PPC) in these structures and PPC was recently reported to occur in S-doped bulk GaInP layers. Ben Amor et al. [1989] focus on the study of PPC in single-heterojunction systems. They observe an important and persistent increase of the electron concentration which remains at surprisingly high temperatures, as high as 330 K for one sample. Temperature-, as well as time-dependent Hall measurements were also performed. They have investigated the effect of hydrostatic pressure ($P < 12$ kbar) on quantum transport (quantum Hall effect and Shubnikov–de Haas (SDH) oscillations) at 4.2 K, as well as on the number of photoexcited carriers and their recombination rate at higher temperature. Thermal cycling of the samples decreased the number of free 2D electrons (in the dark) until the sample became semi-insulating. However, illumination restored the samples to their original precycling state. The critical temperature at which the PPC is suppressed is also strongly affected by the thermal cycles. Finally, optical experiments were performed which reveal an optical threshold energy of 1.15 eV for the PPC and a partial infrared quenching after saturation of the PPC [Ben Amor et al. 1989].

	A	B
GaInP layer	1000 Å	3000 Å
GaInP doping	$3 \times 10^{17} \, \text{cm}^{-3}$	$1 \times 10^{17} \, \text{cm}^{-3}$
Carrier density (dark)	$6 \times 10^{11} \, \text{cm}^{-2}$	$1 \times 10^{11} \, \text{cm}^{-2}$
Mobility (4.2 K)	$42{,}000 \, \text{cm}^2 \cdot \text{V}^{-1} \cdot \text{s}^{-1}$	$27{,}000 \, \text{cm}^2 \cdot \text{V}^{-1} \cdot \text{s}^{-1}$
Quenching temperature	330 K	210 K
Growth temperature	550 °C	650 °C

Table 6.3 Characteristics of the studied samples.

The samples studied were grown by low-pressure MOCVD [Omnes and Razeghi 1991]. They consist of a semi-insulating (1 0 0) GaAs substrate followed by an undoped GaAs layer and finally by a S- and Si-doped $Ga_{0.51}In_{0.49}P$ layer. The parameters of the samples studied are summarized in Table 6.3. The temperature- and time-dependent Hall measurements were performed in a small 0.25 T electromagnet. The excitation wavelength dependence of the PPC was studied by illuminating the sample, cooled down in the dark, with monochromatic light and waiting for any change in the conductivity to saturate. However, the transport measurements were performed using a pulsed red electroluminescent diode (LED) to control the PPC. For the temperature-dependent Hall effect the sample was illuminated at low temperature and then slowly heated (about 1 K·min^{-1}). The time-dependent conductivity and Hall measurements were performed at a stabilized temperature (± 0.3 K) using the internal clock of the monitoring computer. The pressure apparatus was a liquid helium cell allowing us to reach 12 kbar at 4.2 K. A small LED was inserted in the cell and allowed us to illuminate the sample under pressure. The light-induced concentration increase is in the 2D layer over the whole range of temperatures (4.2–330 K). This was demonstrated by angle-dependent SDH oscillations at 4.2 K [Razeghi et al. 1986b]. At higher temperatures, the Hall coefficient is found to be unaffected by a strong magnetic field (19 T), thus proving that only one type of carrier is present (i.e. only 2D carriers).

(a) Temperature- and pressure-dependent transport in GaInP–GaAs

Typical temperature-dependent Hall measurements are shown in Fig. 6.28 before and after illumination. Several traces were observed in the dark, of which only two are represented here. The Hall density slightly decreases with increasing temperature. After illumination, the photoexcited Hall concentration drops smoothly with increasing temperature and finally joins the "dark" trace at a critical temperature T_c. The most striking point is the very high critical temperature at which the PPC is suppressed. Due to this high quenching temperature, cycling the sample to room temperature is not sufficient to bring the sample to an equilibrium state, hence the different temperature dependences in the dark. The temperature-independent Hall density is characteristic of a predominantly shallow-donor behavior in the barrier (as is also reported in Kitahara et al. [1988]). The slight decrease of the 2D density for increasing temperatures is due to a less efficient electron transfer from GaInP to GaAs as the temperature is raised. The smooth temperature decrease of the photoexcited Hall density is in sharp contrast to the behavior of photoexcited DX centers in $Al_xGa_{1-x}As$ ($x = 0.2$–0.4) which usually exhibits an abrupt drop in density at around 150 K, but is consistent with the smooth decay of PPC observed in bulk S-doped GaInP of comparable compositions by Kitahara et al. At a given temperature, the photoexcited Hall concentration measured immediately after the end of the illumination is always the same regardless of the history of the sample. The metastability is, therefore, confirmed to be associated with the PPC donor centers.

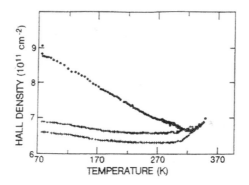

Fig. 6.28. Hall density as a function of temperature. Circles stand for the density after illumination. Two dependences in the dark are shown (crosses), as the sample did not reach thermal equilibrium during the temperature cycle (after Ben Amor et al. [1989]).

The number of photoexcited carriers (immediately after saturation of the PPC) diminishes strongly with increasing pressure at high temperatures, close to T_c, as can be seen in Fig. 6.29. It has been shown by Chand [1984] that if a DX-like center is associated with a non-gamma minimum of the conduction band, then applying pressure to the sample will increase the recombination barrier energy in the case of a small relaxation center and decrease it in the case of a large lattice relaxation center (LLR). Kitahara et al. [1988] have reported that the center responsible for PPC in S-doped GaInP is associated with a non-gamma minimum. Therefore, the observations of this work are compatible with a LLR center in which the recombination barrier is smaller at higher pressure and more easily overcome at higher temperatures (the large optical activation energy (1.15 eV) and the infrared quenching of the PPC (as shown below) are further indications of the LLR character of the center).

Pressure measurements at 4 K (after saturation of the PPC to keep the system in a quasistable state) are also shown in Fig. 6.29. One can observe a weak dependence of the density on applied pressure. Possible reasons for such a pressure dependence are a change of the band offset (as the bandgaps on each side of the interface have rather different pressure dependences) or some other non-gamma character impurity in the barrier, as found in GaAlAs/GaAs [Mercy et al. 1984]. In this case, however, the latter effect should be small because the donor level is emptied by the PPC and the electrons cannot easily recombine with their parent center.

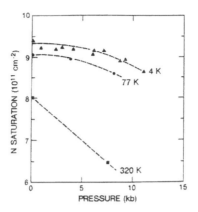

Fig. 6.29. Hall density after saturation of the PPC as a function of pressure for three temperatures. The dashed lines are guides for the eye (after Ben Amor et al. [1989]).

As can be seen in Table 6.3, the second sample studied (B) showed a very different temperature dependence of the PPC quenching. The higher growth temperature resulted in a much lower quenching temperature of 210 K (still much higher than the quenching temperature of a DX center in AlGaAs). This behavior was simulated in the first sample by a short (2 min) annealing at 700 K. This produced the striking result that the sample became semi-insulating in the dark, while PPC was no longer observed at room temperature.

These observations show that the PPC is correlated with extrinsic defects. In this system, the interface stress is very sensitive to the composition and the growth temperature. If defects were due to interface stress, annealing would then change the defect density at the interface, and, as for InAs/GaSb structures [Beerens et al. 1987], the interface defects would influence the number of electrons in the well.

Quantum transport was performed at 4.2 K, and from SDH oscillations and the quantum Hall effect, one can obtain the illumination-dependent concentration in each populated electric subband, as well as the classical and quantum mobility of the electrons. After each temperature cycle (a few minutes annealing at 350 K), i.e. after a change in both the 2D concentration and the ionized defect density, a different electronic density prior to illumination is observed (Fig. 6.30) but all the carriers appear to be in the lowest electric subband. After each cycle, illumination of the sample will populate the second electric subband with a different density threshold (Fig. 6.30), suggesting that the depletion charge in the GaAs has somewhat changed. After saturation of the PPC, the total 2D density is always the same as is the mobility (Fig. 6.31). The density of photoexcited donors thus appears to control simultaneously the electron density, the electron mobility and the population of electrons between subbands. The classical mobility is proportional to the density in the dark, as measured after each annealing cycle, as expected for a sample with no spacer layer where the dominant scattering centers are ionized impurities [Ando 1982]. On illumination, the drop in mobility corresponds to the population of the second electric subband which causes increased scattering [Stormer et al. 1983], [Harris et al. 1987]. Analysis of the amplitude of the SDH oscillations shows that the single-particle lifetime is of the order of 0.1 of the classical lifetime for the fundamental subband (prior to illumination), which is further proof that the dominant scattering center is by remote ionized impurity scattering (small angle) [Harrang et al. 1985].

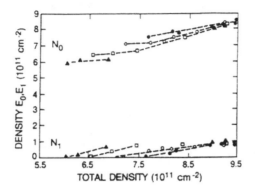

Fig. 6.30. Division of the electrons (as obtained from SDH measurements at 4.2 K) between the different subbands as a function of the total two-dimensional density for several temperature cycles. The longer the sample is annealed at 350 K, the lower the electron density before illumination. After saturation, the electron densities are always the same whatever the history of the sample. The dashed lines have no physical meaning (after Ben Amor et al. [1989]).

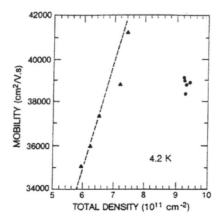

Fig. 6.31. Classical mobility as a function of the total density. After each annealing cycle, a new density (mobility) is obtained in the dark (triangles); after saturation of the PPC, the electron density and mobility always return to the same values. The dashed line represents the calculated ionized-impurity-limited mobility in a sample with no spacer (after Ando [1982]).

(b) Time-dependent transport in GaInP–GaAs grown by MOCVD

In order to investigate further the recombination of the photoexcited donors in GaInP–GaAs Ben Amor et al. [1989] performed time-dependent Hall effect and conductivity measurements. The Hall mobility was almost independent of the illumination at temperatures over 77 K. Typical normalized conductivity results are displayed in Fig. 6.32. They observe a very slow relaxation tail at room temperature. At room temperature and ambient pressure, the relaxation of all the photoexcited carriers takes several days. However, a much quicker relaxation process is also present in these samples, as can be observed in the first seconds of the measurement, by focusing on the slower recombination process. The decay of the photoexcited concentration is strongly non-exponential. The relaxation rate increases only slightly with increasing temperature, in contrast with DX centers in AlGaAs. Both the non-exponential behavior and the weak thermal sensitivity made the determination of a recombination barrier meaningless; however, the observation of a room-temperature PPC several thousands of seconds after the end of illumination suggests the presence of a high barrier. Moreover, the application of a hydrostatic pressure strongly increases the relaxation rate as expected in the LLR description. Such a non-exponential decay was observed in AlGaAs/GaAs structures and was attributed to the spread in the DX center's energies [Mooney et al. 1985] or to tunneling through the barrier to the impurity site [Schubert 1985a, b].

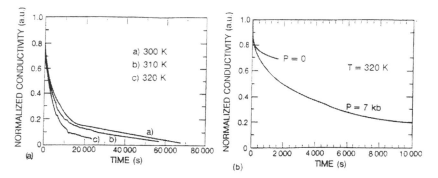

Fig. 6.32. Normalized conductivity as a function of time spent since the end of illumination, (a) Temperature effect at constant pressure (7 ± 0.3 kbar) and (b) pressure effect at constant temperature (320 K) (after Ben Amor et al. [1989]).

The latter process can be excluded, as it depends strongly on temperature (through the temperature increase of the capture cross-section of the impurity). Queisser and Theodorou [1986] observed a logarithmic decay with time which they associated with the recombination of spatially separated electron–hole pairs. In the present case, the decay is not logarithmic and one can exclude this mechanism for the observed PPC. The observed non-exponential behavior is therefore due to the spread of the recombination barrier energies of the centers. This is confirmed by the broad (300 meV) spread of the optical activation energies shown in Fig. 6.33.

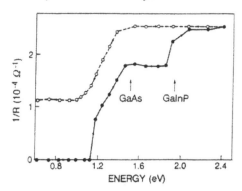

Fig. 6.33. Effect of illumination on a semi-insulating sample as a function of photon energy. The dark circles (and continuous line) represent the absolute changes in the conductivity. The open circles (and dashed line) represent the relative recovery when the photon energy is decreased after PPC saturation (after Ben Amor et al. [1989]).

Ben Amor et al. [1989] performed photoconductivity experiments at 77 K in GaAs–GaInP layers grown by MOCVD. Fig. 6.33 shows the variation of photoconductivity as a function of the excitation wavelength for sample A after a high-temperature anneal (semi-insulating sample). The sample was cooled in the dark and monochromatic light was applied of progressively increasing energy, having waited for the response to saturate. A large PPC occurred (four orders of magnitude) above a threshold of 1.15 eV, rather similar to that found in AlGaAs [Nathan et al. 1983]. The response saturated just below the GaAs bandgap, confirming that the electron–hole spatial separation process is not of major importance. There is a further significant doubling of the PPC at energies corresponding to the GaInP bandgap. This may be due to interband excitation and subsequent capture of free holes at impurities in the GaInP or at the interface.

Following illumination at high energies, the photon energy was then reduced, and it was found that below 1.4 eV a partial quenching of the PPC occurs. This effect peaks at around 1 eV with the removal of over 50% of the PPC at an energy below that where any photoresponse was seen before the PPC. This is again reminiscent of infrared quenching (IRQ) measurements in AlGaAs [Nathan et al. 1983, Kastalsky and Hwang 1984]. Both activation and quenching energies are spread over 300 meV. This shows that the defects are broadly distributed in energy and explains the non-exponential decays.

The optical measurements on the annealed sample support the idea of a LLR DX-like center, where the optical ionization energy E_I (see Fig. 6.34) is much greater than the trap energy (E_T) or thermal barrier energy (E_B). Similarly, the optical capture energy E_q is large but significantly lower than E_I to favor quenching at low energies.

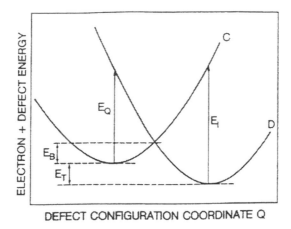

Fig. 6.34. A configuration coordinate model for a donor defect center in GaInP–GaAs heterostructures (after Ben Amor et al. [1989]).

Most of these observations are very similar to the effect of DX centers in the barrier of AlGaAs/GaAs heterostructures. A donor and defect-related DX-like center accounts for the PPC and the IRQ, as well as for the low-temperature behavior of the electron density and mobility. The strong pressure dependence of the PPC and the optical measurements suggest a large lattice relaxation center. The non-exponential decays are presumably related to the dispersion of the center's energies due to inhomogeneities or alloy disorder. However, the high quenching temperature corresponds to recombination barriers in the eV range which are much larger than typical recombination barriers in DX centers (0.2 eV).

6.6.7. Quantum and classical lifetimes in a $Ga_{0.51}In_{0.49}P/GaAs$ heterojunction

The subband structure of $Ga_{0.51}In_{0.49}P/GaAs$ heterostructures has been studied by magnetotransport under hydrostatic pressure [Ben Amor et al. 1990]. The samples were single-interface $Ga_{0.51}In_{0.49}P/GaAs$ heterojunctions grown by low-pressure metalorganic chemical vapor deposition (MOCVD) [Razeghi et al. 1990]. The 1000 Å wide $Ga_{0.51}In_{0.49}P$ layer was S and Si doped (3×10^{17} cm^{-3}) and no undoped spacer separated the doped layer from the two-dimensional electron gas (2DEG). The GaAs layer was n-type residual in contrast with molecular-beam-epitaxy-grown GaAs. Hydrostatic pressure was applied at room temperature and the samples were cooled in the dark using a liquid-pressure cell that permitted up to 12 kbar at 4.2 K to be applied [Ben Amor et al. 1989]. A small light-emitting diode (LED) was inserted in the cell allowing variation of N_{2D} under pressure by controlled light exposure.

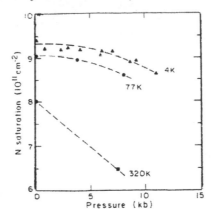

Fig. 6.35. Hall density after saturation of the PPC as a function of pressure for three temperatures. The broken curves are guides for the eye (after Ben Amor et al. [1989]).

The effect of pressure on N_{2D} is shown in Fig. 6.35. In AlGaAs/GaAs heterostructures [Mercy et al. 1984], the pressure deepening of the DX center energy level drives the Fermi level deeper in the gap, thus decreasing N_{2D} and the ionized donor density. However, illumination damps the pressure effects [Maude et al. 1987] on DX centers and reionizes the donors. These results were obtained after saturation of PPC at each pressure so N_{2D} is expected to have a weak pressure dependence. The striking point of the observations is that there is a slight increase of the 2DEG density at low pressures, while N_{2D} decreases for $P > 3$ kbar.

The only mechanism to give an overall increase in N_{2D} is the pressure-related increase of the effective mass (m^*) (around 1% kbar), which increases the density of states and in turn N_{2D}. Since there is a turnover and decrease in N_{2D} for $P > 3$ kbar, it is also necessary to account for a mechanism that decreases N_{2D} as pressure is applied. The bandgaps on both sides of the interface have different pressure dependences, so the decrease in N_{2D} for $P > 3$ kbar could arise from a pressure-induced decrease in the band discontinuity. At ambient pressure it was shown [Rao et al. 1987] that the bandgap discontinuity (ΔE_g) was equally divided between conduction and valence bands. Assuming this partition of band discontinuities to be pressure independent (a reasonable assumption in the present case of hydrostatic pressure), the conduction band discontinuity should decrease at a rate of 2 ± 1 meV·kbar^{-1}. Using a simple model of the band structure [Gregoris et al. 1987] that accounts for the pressure decrease of ΔE_g and the pressure increase of m^*, good agreement was obtained with the observed behavior in N_{2D} (including a predicted turnover at ≈ 5 kbar).

One can obtain the classical scattering time τ_0 from the DC conductivity and the low-field Hall effect (the classical mobility is about 40,000 cm^2·V^{-1}·s^{-1}). The broadening of the Landau levels is reflected through the single-particle (i.e. quantum) lifetime τ_q. The quantum lifetime was obtained from the magnetic field dependence of the SDH oscillations by fitting Ando's expression for the conductivity [Ando 1982]

Eq. (6.38) $$\delta\sigma \propto \left(\omega_c\tau_q\right)^2 \frac{\exp\left[-\pi/\left(\omega_c\tau_q\right)\right]\Phi\cosh(\Phi)}{\left[1+\left(\omega_c\tau_q\right)^2\right]^2}$$

where

Eq. (6.39) $\Phi = 2\pi^2 kT / \hbar\omega_c$

Eq. (6.40) $\omega_c = eB / m^*$

and T is the temperature. The amplitude A of the SDH oscillations can be expressed as [Fang et al. 1988]

Eq. (6.41) $$A = T\omega_c^{-3}\exp\left(-\frac{\pi}{\omega_c\tau_q}\right)\exp(-\Phi)$$

A typical fit is displayed in Fig. 6.36. It is important to note that in this approximation, τ_q is determined at a given temperature without any fitting parameters. Spatial fluctuations of N_{2D} along the sample can alter the shape of the SDH oscillations (i.e. τ_q). In the present case, however, no evidence of amplitude anomalies or beating structures was seen in the SDH envelopes.

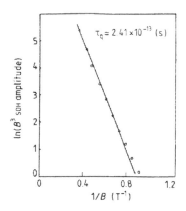

Fig. 6.36. Typical fit of the SDH amplitudes at 4.2 K. The quantum lifetime is directly deduced from the slope (after Ben Amor et al. [1990]).

Fig. 6.37 displays the dependence of the ratio $R_a = \tau_c/\tau_q$ on the PPC-induced increase in N_{2D}. The quantum lifetime is almost an order of magnitude smaller than the scattering time. This confirms the dominance of small-angle scattering mechanisms in heterostructures [Harrang et al. 1985, Stern and Sarma 1985] (i.e. scattering by remote ionized impurities). The surprising point is the 2DEG density dependence of R_a, which decreases with increasing N_{2D}, although theory predicts the opposite even if the scattering centers are at the interface [Stern and Sarma 1985]. A similar discrepancy was reported by Mani and Anderson [1988] in an AlGaAs/GaAs heterostructure and was attributed to the dominant effect of DX centers rather than shallow impurities. In the sample studied here, N_{2D} has been shown to be high enough so that the first two subbands are occupied; hence it is likely that intersubband scattering causes the PPC-induced decrease in R_a.

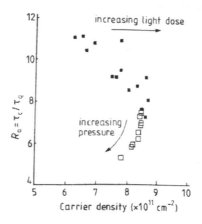

Fig. 6.37. Ratio of the classical to quantum lifetimes, R_a, as a function of PPC-induced increases in N_{2D} (■). and the pressure-induced changes in R_a plotted against the corresponding pressure-induced values of N_{2D} (□) (after Ben Amor et al. [1990]).

It is obvious that the misfit strain at the interface of heterostructures of GaAs and III–V alloys increases with increasing difference in the lattice constants of the binary members. From this point of view the GaAs–$Ga_xAl_{1-x}As$ system, where the lattice constants, a_0, of GaAs and AlAs are 5.6533 and 5.661 Å [Pierron et al. 1967], presents fewer problems than that of GaAs and $Ga_xIn_{1-x}P$, where $a_{Gap} = 5.451$ Å and $a_{Inp} = 5.8694$ Å [Alferov et al. 1974]. It was shown that the quality of $Ga_xIn_{1-x}P$ layers grown on GaAs substrates, and consequently the quality of the $Ga_xIn_{1-x}P$ interface itself, are strongly affected by small deviations of the composition from $x \approx 0.5$, at which the lattice constant of $Ga_xIn_{1-x}P$ is equal to the lattice constant of the GaAs substrate. Because of the large difference in compressibility of GaAs and GaInP, application of hydrostatic pressure (similar to a change in alloy composition) creates a misfit strain at the interface, even in a perfectly lattice-matched system at ambient pressure. It was shown in photoluminescence studies of GaAs/GaInP heterostructure grown by vapor phase epitaxy that outside some range of misfit strain, arrays of misfit dislocations are generated and the interface recombination velocity between GaAs and GaInP increases approximately linearly with misfit strain [Ettenberg and Olsen 1977]. The strong decrease of R_a with pressure is due to the increasing role of interface-related scattering mechanisms (i.e. some interface states induced by increasing misfit strain or interface dislocations that relaxed the pressure-induced misfit strain).

6.6.8. FIR magnetoemission study of the quantum Hall state and the breakdown of the quantum Hall effect in GaAs–GaInP [Raymond et al. 1990]

The Landau emission has been suggested as a very useful tool to investigate two-dimensional systems. An analysis of the emission spectra allows us to determine, for example, the cyclotron mass of the 2D electrons [Chaubet et al. 1991], the amount and the position of impurities in the quantum well [Gornik et al. 1987, Robert et al. 1988, Knap et al. 1990], and to study the polaron effects [Seidenbush 1987]. The Landau emission experiments are performed in a quantizing magnetic field and coincide, for a degenerated 2D electron gas, with the quantum Hall effect (QHE) regime. In the emission technique, Raymond et al. [1990] studied both the FIR emission spectra and the components ρ_{xx} and ρ_{xy} of the resistivity tensor in the QHE regime of GaAs–GaInP heterojunctions. By varying the heating electric field, Raymond et al. [1990] analyzed the evolution of the emission spectra and the breakdown of the QHE when the electric field increases.

Using cyclotron emission, they performed simultaneously optical far-infrared (FIR) and transport measurements on quasi-2DEGs in GaInP/GaAs heterojunctions. In low-electric-field conditions, the FIR spectrum showed, beside the cyclotron resonance line, supplementary lines in coincidence with minima of ρ_{xx} (Hall plateaus). These supplementary lines were interpreted in terms of a non-homogeneous distribution of potential in the quantum Hall state.

As a result of a non-equilibrium carrier distribution, in high-electric-field conditions ($E > 3$ V·cm^{-1}), the width of the Hall plateaus is reduced and the "zeros" of ρ_{xx} are less pronounced. This phenomenon is due to the breakdown of the equation $N_s = i(eB/h)$ and leads to the breakdown of the QHE when the electric field increases further.

Sample No II.31	Growth time (s)	Thickness (Å)
GaAs buffer layer	600	1000
GaAs wells	120	$d_1 = 200$
10×		
GaInP barriers	40	$d_2 = 135$
GaInP cap layer	210	700

Table 6.4. Structural details of a GaInP–GaAs superlattice grown by LP-MOCVD.

For Landau emission experiments, the electric field is used to heat the electrons. To spectroscopically investigate impurities in quantum wells by Landau emission, the heating electric field must be very low (≈ 1 V·cm^{-1}).

6.7. Growth and characterization of GaInP–GaAs multilayers by MOCVD

Extensive experimental studies of low-dimensional structures of GaInP–GaAs have been performed [Razeghi et al. 1990, 1991a, b]. The quantum size effect and excitonic absorption in GaAs–GaInP multiquantum wells (MQWs) and superlattices have attracted a great deal of interest because of their novel physical properties and their potential use in optoelectronic [Guanapala et al. 1990] and electronic devices [Chan et al. 1988].

Sample No. II.31 is a typical ten-period GaAs–$Ga_{0.51}In_{0.49}P$ superlattice. Its structure, deduced from growth rates of GaAs and $Ga_{0.51}In_{0.49}P$, is briefly described in Table 6.4.

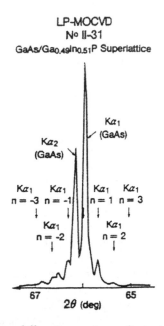

Fig. 6.38. Simple x-ray diffraction pattern of a typical GaAs–$Ga_{0.51}In_{0.49}P$ superlattice (after Razeghi [1989a]).

Fig. 6.38 shows the simple diffraction pattern of the sample, performed using $K\alpha$ of Cu as x-ray source, in a classical (θ, 2θ) configuration. $K\alpha_1$ and $K\alpha_2$ components of the satellites, due to the superperiod $D = d_1 + d_2$ artificially introduced by the growth, can be seen up to order $n = \pm 3$. This fact is remarkable, especially using simple diffraction, and suggests that the thicknesses of the wells and barriers are perfectly controlled. The corresponding data are summarized in Table 6.5.

Order n of the $K\alpha_1$ component of the satellite	2θ (deg)
−3	66.97
−2	66.66
−1	66.35
0	66.02
+1	65.73
+2	65.43
+3	65.12

Table 6.5. X-ray data of GaInP–GaAs superlattices grown by LP-MOCVD.

A mean value of $2\delta\theta = 0.31°$, corresponding to the spacing between two satellites, is found. The superperiodicity $D = d_1 + d_2$ can then be deduced by the well known relation

Eq. (6.42) $$D = \frac{\lambda}{2\delta\theta \, \cos\theta_0}$$

where θ_0 represents the Bragg angle at which the (4 0 0) reflection of the GaAs occurs ($\theta_0 = 33.01°$). λ is the x-ray wavelength, which is about 1.54 Å for the Cu $K\alpha_1$ ray.

One finds $D = 340$ Å, which is in good agreement with the value obtained from the growth times and deposition rates (see Table 6.4).

The morphology of this GaInP–GaAs superlattice sample was examined by cross-sectional transmission electron microscopy (TEM). A dark-field image of the complete superlattice is shown in Fig. 6.39(*a*). This image was taken using the (2 0 0) reflection, with GaInP appearing brighter than GaAs. The ten periods of GaInP/GaAs are clearly visible with each GaInP layer measuring 140 Å and each GaAs layer 205 Å in thickness. The layers are extremely regular and no local layer thickness variations were seen in the area examined. The GaAs buffer layer can be distinguished from the GaAs substrate by a dark contamination line, most clearly visible in Fig. 6.39(*b*), whose presence has not affected the quality of the buffer layer. One defect in the superlattice was found and is shown in Fig. 6.39(*b*). It originates at the boundary between the buffer layer and the first GaInP layer.

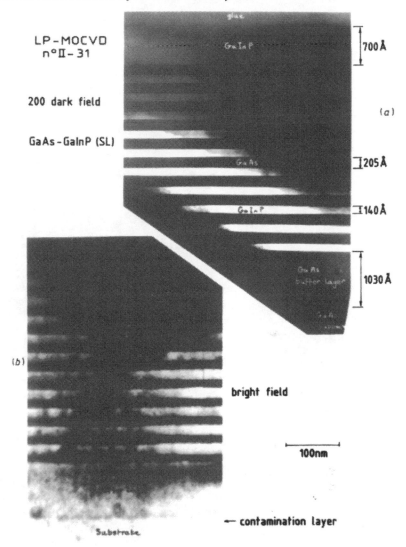

Fig. 6.39. Transmission electron micrograph of a typical GaAs–$Ga_{0.51}In_{0.49}P$ superlattice (after Razeghi [1989a]).

The SIMS profile of the sample is shown in Fig. 6.40. The signals relative to the majority species (Ga, In, As and P) have been plotted. The ten periods of the superlattice are clearly evidenced by following the oscillations of the different signals. The $Ga_{0.51}In_{0.49}P$ top layer and GaAs substrate appear on each side.

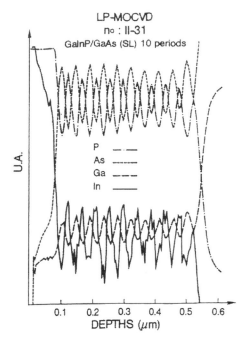

Fig. 6.40. SIMS profile relative to the majority species (In, Ga, As and P) of a typical GaAs–Ga$_{0.51}$In$_{0.49}$P superlattice (after Razeghi [1989a]).

Fig. 6.41 shows the electrochemical (polaron) profiles of Ga$_{0.51}$In$_{0.49}$P layers. The residual doping level is 8×10^{14} cm^{-3}. This explains why a concentration of carriers of $n = 2.9 \times 10^{11}$ cm^{-2} is measured at $T = 4$ K by the classical Hall method, in the unintentionally doped multiquantum well structure.

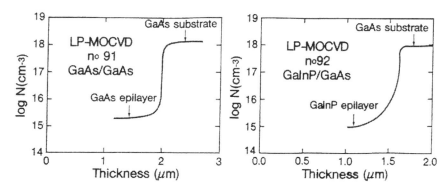

Fig. 6.41. Polaron profile of bulk GaAs and GaAs–Ga$_{0.51}$In$_{0.49}$P (after Razeghi [1989a]).

Carriers arising from the donor impurities are confined in the GaAs quantum wells, where they form a two-dimensional electron gas. This fact is evidenced by the mobility curve plotted against temperature (Fig. 6.42). The mobility reaches a constant value of 50,000 cm^2·V^{-1}·s^{-1} at temperatures below 50 K. This is clearly characteristic of the transport properties of a two-dimensional electron gas. Because of the spatial separation between carriers and donors, the ionized impurity scattering becomes much less efficient than in three-dimensional systems, so the mobility does not decrease at low temperatures under this effect.

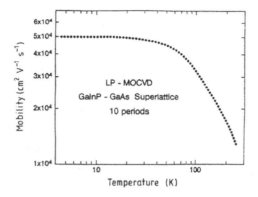

Fig. 6.42. Mobility curve versus temperature of a typical GaAs–Ga$_{0.51}$In$_{0.49}$P superlattice (after Razeghi [1989a]).

An Auger spectrum has been taken from a GaAs–Ga$_{0.51}$In$_{0.49}$P multiquantum well structure, with GaAs wells of 15, 50 and 100 Å thickness and GaInP barriers of 1000 Å (Fig. 6.43). Resolution good enough to identify wells of thickness down to 15 Å was obtained by scanning the surface of a chemical bevel and correcting the arsenic signal.

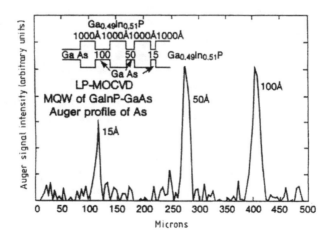

Fig. 6.43. Auger spectrum of As species of a typical GaAs–Ga$_{0.51}$In$_{0.49}$P multiquantum well (after Razeghi [1989a]).

6.8. Optical and structural investigations of GaAs–GaInP quantum wells and superlattices grown by MOCVD

High-quality GaInP–GaAs multiquantum wells and superlattices were grown by LP-MOCVD [Omnes and Razeghi 1991]. Growth was carried out at 76 Torr and at a substrate temperature of 510 °C. Trimethylindium (TMIn) and triethylgallium (TEGa) were used as sources of In and Ga, while pure arsine (AsH_3) and (AsH_3) and phosphine (PH_3) provided As and P, respectively. Hydrogen (H_2) was used as a carrier gas.

The GaAs–GaInP interfaces were realized by turning off the AsH_3 flow and turning on both the TMIn and PH_3 flows. The reverse procedure was used to obtain GaInP–GaAs interfaces. The GaInP–GaAs interfaces are more critical due to the memory effect of In, so purging of In at the GaInP–GaAs interface improves the optical confinement in the GaAs quantum well.

The growth rate was small and the gas flow stabilized to its new steady-state value in less than a few seconds after switching. The optimum growth conditions of GaAs–GaInP as determined during these investigations are given in Table 6.6.

	GaAs	GaInP
Growth temperature	510 °C	510 °C
Growth pressure	76 Torr	76 Torr
Total H_2 flow rate	$3\ l{\cdot}min^{-1}$	$3\ l{\cdot}min^{-1}$
AsH_3 flow rate	$20\ cm^3{\cdot}min^{-1}$	—
PH_3 flow rate	—	$200\ cm^3{\cdot}min^{-1}$
H_2 through TMIn	—	$220\ cm^3{\cdot}min^{-1}$
H_2 through TEGa	$180\ cm^3{\cdot}min^{-1}$	—
Growth rate	$180\ Å{\cdot}min^{-1}$	$360\ Å{\cdot}min^{-1}$

Table 6.6. Optimum growth parameters.

The quantum well and superlattice samples grown for this study were unintentionally doped. Residual impurity concentrations are assumed to be of the order of those determined for bulk layers. Corresponding layers of GaAs and GaInP with thicknesses of 3 μm, grown under identical conditions, revealed $N_D - N_A \leq 10^{14}\ cm^{-3}$ for GaAs [Razeghi et al. 1989a] and $N_D - N_A \leq 5 \times 10^{14}\ cm^{-3}$ for GaInP [Razeghi et al. 1989b], and a uniform distribution of impurities in the direction perpendicular to the layers.

Fig. 6.44 shows a typical PL spectrum of a three-well structure of GaAs–GaInP. The GaAs (well) and GaInP (barrier) indicated "nominal" thicknesses are those deduced from the measured growth rate of thick layers. The substrates were GaAs (1 0 0) misoriented 2° off axis towards $\langle 0\ 1\ \bar{1} \rangle$.

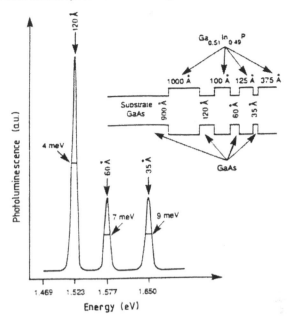

Fig. 6.44. Photoluminescence spectrum of GaAs–Ga$_{0.51}$In$_{0.49}$P sample (No 1219) measured at 4 K using a 5435 Å argon laser operating at 2 mW. Inset: schematic representation of sample No 1219 grown by LP-MOCVD (after Omnes and Razeghi [1991]).

The PL measurements were done at 4 K using a 5435 Å argon laser operating at 2 mW, and dispersed on a HRS Jobin-Yvon monochromator equipped with a photomultiplier tube. The full widths at half maximum of the PL spectra are 4, 7 and 9 meV for the nominal QW thicknesses of 120, 60 and 35 Å.

The quality of PL peaks of GaAs wells is proved by PLE spectroscopy, using a tunable pyridine dye laser pumped with an argon laser. The dye laser power is locked at 2 mW in the whole wavelength range. Fig. 6.45 shows the PLE spectrum of a GaAs–GaInP quantum well of 120 Å thickness at 4 K. The PLE spectrum exhibits a series of peaks, which can be attributed to the heavy-hole exciton and light-hole exciton as shown in Fig. 6.45. These results show that high-quality GaAs–GaInP MQWs have been achieved. The intensity of the PLE peaks indicates that the GaAs–GaInP system is of type 1, i.e. the GaAs layers are simultaneously QWs for both the conduction and valence band states.

By comparing the PL and PLE spectra taken at 4 K in Fig. 6.45, we find that the energy of the $E_1 \rightarrow HH_1$ subband transition, as determined by PLE, is 1524 meV and the peak energy of the (e, h) emission in the PL spectrum is at 1522 meV, indicating the high quality of the interfaces of GaInP–GaAs MQWs grown by LP-MOCVD. The variation of the PL intensity as a function of excitation power at low temperature is linear. X-ray diffraction measurements carried out on this sample showed a $\Delta a/a$ of about 10^{-4}.

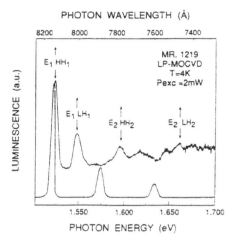

Fig. 6.45. Photoluminescence and photoluminescence excitation spectrum of the quantum well of 120 Å thickness at 4 K (sample No 1219) (after Omnes and Razeghi [1991]).

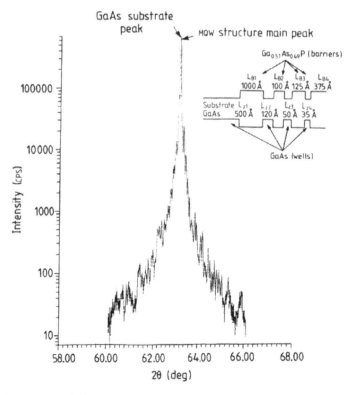

Fig. 6.46. X-ray diffraction pattern of a three-well GaAs–$Ga_{0.51}In_{0.49}P$ MQW structure. Inset: schematic representation of this MQW sample grown by LP-MOCVD (after Papalus et al. [1991]).

Fig. 6.46 shows the x-ray (4 0 0) diffraction pattern of a three-well structure of GaAs/Ga$_{0.51}$In$_{0.49}$P (shown inset), with GaInP barriers of about $L_{B1} = 1000$ Å, $L_{B2} = 100$ Å, $L_{B3} = 125$ Å and $L_{B4} = 375$ Å and GaAs wells of about $L_{Z1} = 900$ Å, $L_{Z2} = 120$ Å, $L_{Z3} = 60$ Å and $L_{Z4} = 35$ Å. A synchrotron beam ($\lambda = 1.47596$ Å) is used as the x-ray source. The GaAs (well) and GaInP (barrier) indicated "nominal" thicknesses are those deduced from the measured growth rate of thick layers. This multiquantum well structure is shown in Fig. 6.46 (inset); the substrate was GaAs (1 0 0) misoriented 2° off axis towards $\langle 0\ 1\ \bar{1} \rangle$ [Razeghi and Omnes 1991].

The fringe spacing is inversely related to layer thickness. This means that the larger the fringe spacing, the thinner the epilayer. Thus, due to the different fringe spacings seen in Fig. 6.46, different layer thicknesses can be determined. From the angular separations $\Delta\theta_1 = 120"$, $\Delta\theta_2 = 720"$, $\Delta\theta_3 = 1080"$ and $\Delta\theta_4 = 540"$, we deduce, using the classical expression

Eq. (6.43) $\qquad L'_B + L'_Z = \dfrac{\lambda}{2\Delta\theta \cos\theta}$

that the thicknesses of the quantum well and barrier layer are L_1 $(L'_{B1} + L'_{Z1}) = 1490$ Å, $L_2 = 248$ Å, $L_3 = 166$ Å and $L_4 = 331$ Å; these values are consistent with those expected from growth parameters.

Although there is no hyperfine structure observed near the substrate Bragg peak in Fig. 6.46, the angular separation arising from Pendellosung oscillations can also be determined by means of the classical equation

Eq. (6.44) $\qquad \Delta\theta = \dfrac{\lambda}{2L \cos\theta}$

with $L = 2750$ Å; we find $\Delta\theta = 65"$. Comparing this value to the x-axis scale shown in Fig. 6.46, it is obvious that this kind of angular separation is too small to be observed.

Because of the cubic structure of this sample and the differential equation of the Bragg law with respect to θ, we obtain

Eq. (6.45) $\qquad \dfrac{\Delta a}{a} = -\dfrac{\Delta\theta}{\tan\theta}$

where a is the lattice parameter, θ the average diffraction angle and $\Delta\theta$ is half the angular separation of layer and substrate Bragg peaks. From the angle $\Delta\theta = 90"$ between the quantum well diffraction peak and the GaAs substrate Bragg peak, Papalus et al. [1991] concluded, using Eq. (6.45), that x-ray diffraction measurements exhibited a $\Delta a/a$ up to 7.1×10^{-4}. This result shows the high quality of GaAs–GaInP MQWs grown by LP-MOCVD [Omnes and Razeghi 1991].

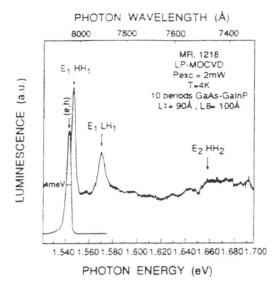

Fig. 6.47. Photoluminescence and photoluminescence excitation spectrum of a GaAs–GaInP superlattice consisting of ten periods of 90 Å thick GaAs quantum wells and 100 Å thick $Ga_{0.51}In_{0.49}P$ barriers (after Omnes and Razeghi [1991]).

Fig. 6.47 shows the low-temperature PL and PLE spectra of a GaAs–GaInP superlattice grown by LP-MOCVD under the conditions indicated in Table 6.6. The superlattice consists of ten periods of approximately 100 Å thick GaInP barriers and 90 Å thick GaAs wells. The 4 K PL spectrum has a full width at half maximum of 4 meV evidencing a good layer to layer reproducibility. The PLE spectrum of the 90 Å QWs exhibits a series of sharp peaks, which can be attributed to the electron to light hole and electron to heavy hole transitions as indicated in Fig. 6.47. The energy of the $E_1 \rightarrow HH_1$ subband transition, as determined by PLE, is 1545 meV and the peak energy of the (e, h) emission in the PL spectrum is at 1542 meV, providing evidence of the high quality of interfaces and layer to layer homogeneity.

Using a standard quantum well calculation of energy levels, where we introduce, in a simple iterative form, the non-parabolicity of the energy bands, Omnes and Razeghi [1991] have calculated the transition energies for the nominal QW thicknesses reported here. The parameters used for this study are the following where all values are taken at a temperature of 4 K [Landolt–Börnstein 1982]

GaAs: $E_g = 1518$ meV, $m_0^* = 0.0665m_0$

$m_{hh} = 0.475m_0$, $m_{1h} = 0.087m_0$

GaInP: $E_g = 1984$ meV, $m_0^* = 0.1175m_0$

$m_{hh} = 0.660m_0$, $m_{1h} = 0.145m_0$

GaAs/GaInP: $\Delta E_c / \Delta E_v = 0.4$.

Table 6.7 indicates the experimental and theoretical values of the energy levels of GaAs–GaInP quantum wells of 90, 120, 60 and 35 Å thickness. A satisfying agreement between theoretical and experimental values is observed for quantum wells of 90, 60 and 35 Å thickness. There is a noticeable discrepancy between the measured and calculated heavy-hole–light-hole splitting for the widest quantum well which could be attributed to compressive strain within this quantum well. However, the fundamental transition should increase in this case, which is contrary to the observed trend. Such behavior is not clear at present but could be related to In segregation and subsequent localized strain. This could also be a source of uncertainty on the quantum well thickness. One could expect the widest quantum well to be the least sensitive to interface effects, but one should note that this widest well is the first of the whole structure: we carried out a mechanochemical bevel on this structure which showed that the first quantum well interface was not as good as the following ones.

Samples	L_z (Å)	Energy level	Calculation (meV)	Exp. (meV)	PL (e, h) (meV)
Superlattice					
No 1218	90	$E_1 \rightarrow HH_1$	1555	1546	1544
		$E_1 \rightarrow LH_1$	1576	1575	
		$E_2 \rightarrow HH_2$	1661	1664	
MQW					
No 1219	120	$E_1 \rightarrow HH_1$	1541	1524	1522
		$E_1 \rightarrow LH_1$	1554	1550	
		$E_2 \rightarrow HH_2$	1608	1597	
		$E_2 \rightarrow LH_2$	1659	1661	
	60	$E_1 \rightarrow HH_1$	1582	1580	1574
		$E_1 \rightarrow LH_1$	1617	1618	
	35	$E_1 \rightarrow HH_1$	1637	1635	1634

Table 6.7. Experimental and theoretical values of energy levels for GaAs–GaInP MQWs.

Fig. 6.48 shows the x-ray (4 0 0) diffraction pattern of a 10-period GaAs–$Ga_{0.51}In_{0.49}P$ superlattice, with approximately 100 Å thick (L_B) GaInP barriers and 90 Å thick (L_Z) GaAs wells [Razeghi et al. 1991b]. A synchrotron beam of wavelength $\lambda = 1.47596$ Å is used as the x-ray source. Satellites corresponding to the artificial crystalline period $L_B + L_Z$ introduced during growth have been resolved up to $n = \pm 7$. This rarely observed result demonstrates the excellent structural quality of the sample under study. The $n = 0$ satellite, attributed to the (4 0 0) Bragg diffraction of the mean parameter of the superlattice, appears close to the position of the GaAs substrate Bragg peak. The x-ray measurements carried out on this sample showed a $\Delta a/a$ of about 2.8×10^{-3} using Eq. (6.45).

From the angular separation $\Delta\theta = 1080''$ between two adjacent satellites, one can deduce, using Eq. (6.43), that the superlattice period is $L'_B + L'_Z = 166$ Å ; this corresponds well to the value expected from the growth parameters.

The hyperfine structure observed in Fig. 6.48 near the substrate Bragg peak may be attributed to the interaction between incident and reflected waves through the whole epitaxial layer (Pendellosung oscillations). According to the angle difference value $\Delta\theta = 90''$ between two peaks in the hyperfine structure, if we use Eq. (6.43), one can find $L = 1985$ Å, in good agreement with the growth parameters. The behavior observed in Fig. 6.48 has rarely been observed in III–V superlattices.

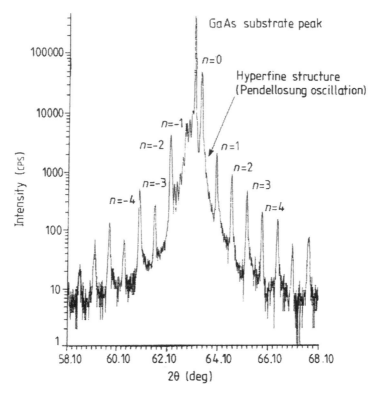

Fig. 6.48. *X-ray diffraction pattern of a 10-period GaAs–Ga$_{0.51}$In$_{0.49}$P superlattice grown by LP-MOCVD. L$_B$(GaInP) = 100 Å and L$_Z$(GaAs) = 90 Å (after Papalus et al. [1991]).*

6.9. Characterization of GaAs–GaInP quantum wells by auger analysis of chemical bevels

Understanding the initial stage of epitaxial growth is essential for multilayer semiconductor materials, where interfaces between layers play a prominent part in their optical and electrical properties. The physical study of the chemical species concentrations at the interfaces can give valuable information on the initial steps of epitaxial growth.

The theoretical calculations of Auger currents given by Auger line scan measurements on chemically beveled heterostructures were performed by Olivier et al. [1987], assuming exponentially varying concentrations at the interface and compared with experimental results.

Combined with ion etching, Auger electron spectroscopy (AES) can be used to get elemental concentration profiles within overlayers. However, several factors affect the resolution depth as described below. There are different mechanisms which broaden the concentration profiles:

(i)   Auger electron escape depth and ion bombardment effects (ion knock-on mixing, preferential sputtering and ion-induced roughness).

(ii)   In order to avoid the disadvantages of ion milling, Bisaro et al. [1982] developed a method of chemical beveling coupled with line scan Auger measurements to check interfaces of epitaxial III–V compounds.

The chemical bevels are produced by a technique in which the liquid–liquid interface between pure methanol and a bromine–methanol solution is raised progressively over the sample. The bevel angle α can be controlled by the speed of the etching solution flowing up the sample, and can be as small as 0.05 or less giving a magnification coefficient of $M = (\sin \alpha)^{-1} > 10^3$. In contrast to sputtering, the absolute depth resolution R of the bevel Auger profiling method is independent of the film thickness to be examined and is of the order of the electron beam size s plus the beam deflection resolution d divided by the lateral magnification of the bevel angle [Tarng and Fisher 1978]

Eq. (6.46) $R = (s + d)/M$

A schematic diagram of the bevel is shown in Fig. 6.49.

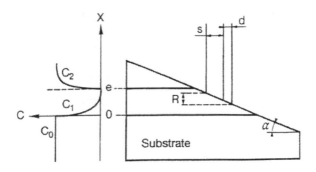

Fig. 6.49. Schematic diagram of the bevel. The mean parameters are the bevel angle α, the electron beam size, the beam deflection resolution d and the depth resolution R. The phosphorus exponential composition profiles taken into account are $C_1(x) = C_0 \exp(-x/\Lambda_1)$ and $C_2(x) = C_0[1 - \exp(-(x - e)/\Lambda_2)]$ (after Olivier et al. [1987]).

In the present experiment, the beam diameter was 2000 to 7500 Å (primary energy and current: 10 keV, 1 nA and 5 keV, 20 nA respectively) and the deflection resolution ≤500 Å. With a magnification coefficient $M = 1000$, the absolute resolution R is 8 Å in the worst case.

The Auger current produced at x' by dx' is partially absorbed by the thickness $(x - x')$ before leaving the sample and entering the analyzer with an angle α

Eq. (6.47) $$J(x')\,dx' = kn(x')\exp\left(-\frac{x - x'}{L_A \cos\alpha}\right)dx'$$

where k is proportional to the product of the cross-section for ionization of a level, of the Auger transition probability and of the secondary-electron coefficient $(1 + r)$ (Fig. 6.50).

Eq. (6.48) $$n(x')\,dx' = N\exp\left(-\frac{x - x'}{L_p}\right)$$

is the attenuated primary electron beam. $C(x') = C_B\,f(x')$ is the material atomic concentration. L_A is the mean free path of the Auger electron. Thus the Auger bulk current of a substrate of concentration C_0 is

Eq. (6.49) $$I_B = kNC_B L$$

with

Eq. (6.50) $$\left(L\right)^{-1} = \left(L_p\right)^{-1} + \left(L_A \cos\alpha\right)^{-1}$$

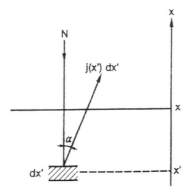

Fig. 6.50. Schematic diagram of the Auger current produced at x' by dx', which is partially absorbed by the thickness (x − x') before leaving the sample and entering the analyzer an angle α (after Olivier et al. [1987]).

The elementary Auger signal emerging from the surface x, normalized by the bulk value, is expressed as:

Eq. (6.51) $\quad J(x')dx' = \dfrac{f(x')}{L}\exp\left(-\dfrac{x-x'}{L}\right)dx'$

In the case of a QW with exponentially varying concentrations at the interfaces, the normalized Auger currents are expressed by

$x < 0$:

Eq. (6.52) $\quad I(x) = 1$

$0 < x < e$:

Eq. (6.53) $\quad I(x) = \dfrac{\Lambda_1}{\Lambda_1 - L}\exp\left[-\dfrac{x}{\Lambda_1}\right] - \dfrac{L}{\Lambda_1 - L}\exp\left[-\dfrac{x}{L}\right]$

$x > e$:

Eq. (6.54)
$$I(x) = 1 + \left\{ \dfrac{L}{\Lambda_2 - L} - \dfrac{L}{\Lambda_1 - L}\exp(e - L) \right.$$
$$\left. -\left[\dfrac{L(\Lambda_1 - \Lambda_2)}{(\Lambda_2 - L)(\Lambda_1 - L)}\right]\exp\left(-\dfrac{e}{\Lambda_1}\right)\right\}\exp\left[-\dfrac{x - e}{L}\right]$$
$$-\dfrac{\Lambda}{(\Lambda_2 - L)}\left[1 - \exp\left(-\dfrac{e}{\Lambda_1}\right)\right]\exp\left[-\dfrac{x - e}{\Lambda_2}\right]$$

where e is the escape depth.
   For abrupt interfaces: $\Lambda_1 = \Lambda_2 \approx 0$

Eq. (6.55) $\quad 0 < x < e \quad I(x) = \exp(-x/L)$

Eq. (6.56) $\quad x > e \quad I(x) = 1 - \exp\left[-(x - e)/L\right] \quad (e \gg L)$.

A log scale for $I(x)$ or $1 - I(x)$ gives straight lines, the slope of which allows the escape depth of the Auger electron to be deduced.
   The exact interface position does not correspond to a relative intensity of about 50% as usually assumed: it corresponds either to the beginning of the decrease from unity ($0 < x < e$) or the beginning of the increase from zero ($x > e$). The QW profiles are asymmetric.
   For gradual interfaces: Λ_1 and $\Lambda_2 \gg L$

Eq. (6.57) $\quad 0 < x < e \quad I(x) = \exp(-x/\Lambda_1)$

Eq. (6.58) $x > e$ $I(x) = 1 - \exp\left[-(x-e)/\Lambda_2\right]$ $(e \gg \Lambda_1)$

Thus in a log scale, the characteristic widths Λ_1 and Λ_2 govern the slope of the straight lines, when Λ_1 and $\Lambda_2 \gg L$.

Although one now has two interfaces of finite thickness, here one can also precisely determine the point that corresponds to the start or interruption of the gas flow in the growth reactor. The distance between these two points corresponds to the same thickness as deduced from the steady-state growth speed determined by measurements on thicker layers.

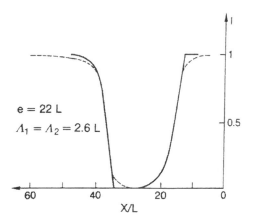

Fig. 6.51. Theoretical phosphorus Auger line scan of a GaInP/GaAs/GaInP QW. The dashed lines are the experimental curves (after Olivier et al. [1987]).

Fig. 6.51 shows the experimental phosphorus Auger line scan of the MOCVD growth of a QW of GaInP/GaAs/GaInP. The good agreement between experimental and theoretical curves *a posteriori* justifies the reasonable hypothesis of exponentially varying concentrations at the interfaces.

The experimental increase of the Auger signal that is less steep than the theoretical one at the two extremities of the well can be explained by a slight atomic mixing resulting from a short ion cleaning before the Auger line scan was taken. The two characteristic lengths are very different. The explanation is likely to come from the dynamics of low-pressure MOCVD growth: in the neighborhood of the growth surface, there is a gas layer whose characteristics change gradually from those of the convective gas phase, to those of the growth–surface-gas-phase interface [Razeghi 1985]. The evolution of that boundary layer by stopping or setting up again the phosphine flux in the reactor is the key to the problem.

A GaAs QW of intended width $e = 100$ Å is fitted with a theoretical curve calculated using the following parameters: $L = 4.75$ Å, $e = 105$ Å, $\Lambda_1 = \Lambda_2 = 12.5$ Å. The characteristic lengths Λ are equal, showing sharp interfaces and that both interfaces are equivalent.

6.10. Evaluation of the band offsets of GaAs–GaInP multilayers by electroreflectance [Razeghi et al. 1992]

The band offsets of GaAs/GaInP have already previously been measured by different techniques. The results are summarized in Table 6.8.

Technique	ΔE_c (eV)	ΔE_v (eV)	Reference
Capacitance–voltage profiling	0.238		Rao et al. [1987]
DLTS	0.198	0.285	Biswas et al. [1990]
I–V curves of HEMTs	0.231		Chan et al. [1990]
Temperature dependence of the collector current of HBTs	0.03		Kobayashi et al. [1989]

Table 6.8. Previously measured values for the band offsets of GaAs/GaInP.

Electroreflectance (ER) is the best and most direct technique for the determination of band profiles [Kassel et al. 1990, 1991a, b] in general and band offsets in particular. This is because ER produces especially sharp optical spectra and yields enhanced signals from interfacial surfaces. Thus, ER allows the determination of many more transition energies than does photoluminescence. In particular, ER allows one to observe and measure the energies of transitions between states allowed only on one side of an interface and the tails of states allowed only on the other side; such transitions are called crossover transitions. For a type-I quantum well or superlattice, these crossover transitions are between well states and the barrier band edges. Because of the ability to obtain a better signal to noise ratio, to control better the modulating electric field and to apply bias voltages, electrolyte electroreflectance (EER) also gives many more transition energies than does photoreflectance. In particular, the ability to apply a bias voltage allows one to observe and identify many transitions forbidden by symmetry in the absence of an electric field, as well as crossover transitions. This allows one to measure directly intersubband, intraband transition energies without the use of band calculations. It also allows one to measure almost directly the band offset in both samples, with the only theoretical input being the energy of the first heavy-hole level relative to the valence band edge in the well.

Razeghi et al. [1992] obtained approximately 20 EER spectra, at different bias voltages. This result is the first reported direct optical measurement of either of these types of quantity for type-I superlattices or quantum wells. Two GaAs–Ga$_{0.51}$In$_{0.49}$P-based samples (a superlattice and a multiple quantum well) have been used for this study. The EER spectra were fitted using the generalized ER lineshape [Garland and Raccah 1986] to obtain the transition energies and linewidths, as well as the coefficients of the first-, second- and third-derivative terms in the generalized lineshape. Well state to well state transitions were assumed to have only a first-derivative term and crossover transitions were assumed to be primarily first derivative in nature. This study of the spectra as a function of the bias voltage helped to confirm the identification of the types of transition observed. For example, for quantum wells and superlattices, the linewidth of the crossover transitions increases much more rapidly with internal electric field than does that of any other type of transition.

For the superlattice, 35 transitions were identified. This large number of observed transitions gives a multiplicity of self-consistency conditions and thus gives many checks both on the accuracy of our values for the transition energies and the identification of the transitions. All of the self-consistency conditions were well satisfied. For the multiple quantum well, which contained wells of three different widths, 30 transitions were identified. For both samples, the observed transition energies were fitted with a parameterized envelope function band calculation in which the energy dependence of the effective masses was taken into account. The well widths obtained from these fittings were in excellent agreement with those obtained by calibration of the growth rate.

The GaAs–GaInP multiquantum wells and superlattices used for this study were grown by low-pressure metalorganic chemical vapor deposition. The growth temperature was 510 °C and the growth pressure was 76 Torr. Triethylgallium (TEG), trimethylindium (TMI), pure arsine (AsH_3) and pure phosphine (PH_3) were used as Ga, In, As and P sources, respectively. Pure H_2 was used as the carrier gas. The quantum well and superlattice samples grown for this study were unintentionally doped. Residual impurity concentrations are assumed to be of the order of those determined for bulk layers. $N_d = N_a \approx 10^{14}$ cm^{-3} for GaAs [Razeghi et al. 1989a] and 5×10^{14} cm^{-3} for GaInP layers [Omnes and Razeghi 1991].

For both samples studied (a superlattice (SL) and a three-quantum-well (3QW) structure) the linewidths of the well-state to well-state transitions were very narrow, from 3 meV for the lowest-energy transitions up to 8 meV for the highest-energy well-state to well-state transitions. These linewidths are narrower than we have seen before at room temperature, even on other quantum-well systems. These narrow linewidths are evidence of the unusually high quality of these samples. More evidence of their exceptional quality was our ability to see many more transitions than is usual for such samples.

6.10.1. Study of the GaAs–GaInP superlattice

The GaAs–GaInP SL consists of 10 periods of 90 Å thick GaAs wells between $Ga_{0.51}In_{0.49}P$ barriers approximately 100 Å thick. Because of the thickness of the barriers, the well subbands are very narrow, of the order of 1 meV or less in width. For this reason, one could not clearly distinguish transitions associated with the bottom of a given subband from those associated with the top. Thus, the observed spectrum is equivalent in form to that which would be observed for a single quantum well.

Fig. 6.52 shows the conduction band offset and the heavy-hole and conduction band levels for a single well, along with examples of the different types of transition observed. For simplicity, the light-hole and split-off states and the transitions to and from these states are not shown. The energies of the well states are determined by the following quantities:

E(e*m*)   the energy from the bottom of the conduction band well to the *m*th conduction band well level

E(hh*m*)   the energy from the *m*th heavy-hole level up to the top of the valence band well

E(lh*m*)   the energy from the *m*th light-hole level up to the top of the valence band well

E(so*m*)   the energy from the *m*th split-off well level up to the top of the split-off well

Δ_w the energy from the split-off well edge to the valence band well edge

Δ_b the energy from the split-off barrier edge to the valence band barrier edge

E_w the well material bandgap

E_b the barrier bandgap

ΔE_c the conduction band discontinuity.

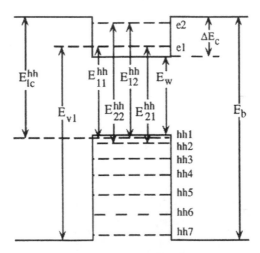

Fig. 6.52. Schematic diagram of a single GaAs–Ga$_{0.51}$In$_{0.49}$P quantum well (after Razeghi et al. [1992]).

The EER spectrum determines the optical transition energies from which all of these quantities can be directly calculated, given only E(hh1) and either E_w or E(e1). The quantity E(hh1) can be calculated easily to an accuracy of better than 1 meV; E(e1) can be calculated to an accuracy of 1–2 meV. Thus, the energies of all the well states are accurately determined from the EER data, as are the band discontinuities ΔE_c and $\Delta E_v = E_b - E_w - \Delta E_c$. The observed optical transition energies are of the following types:

E_b the barrier bandgap

$E_{mn}^{hh}, (E_{mn}^{lh}, E_{mn}^{so})$ the transition energy from the *m*th heavy-hole (light-hole, split-off) level to the *n*th conduction band level in the well

$E_{mc}^{hh}, (E_{mc}^{lh}, E_{mc}^{so})$ the transition energy from the mth heavy-hole (light-hole, split-off) level to the conduction band edge in the barriers

$E_{vm} (E_{sm})$ the transition energy from the valence band (split-off) edge in the barriers to the mth conduction band level in the well.

Twenty-two experimental EER spectra were obtained and analyzed, at different bias voltages and modulating voltages. Fig. 6.53(*a*) shows a typical EER spectrum at room temperature (RT) and the fit to that spectrum. The details of the spectrum and its fit from 1.48 eV to 1.88 eV are shown in Fig. 6.53(*b*). The spectrum has a rich structure due to the large number of transitions observed. Table 6.9 lists the transitions observed and gives the transition energies measured by EER at RT and by photoluminescence (PL) at low-temperature (LT). Only transitions observed in all of the spectra are listed. The uncertainties listed for the energies represent maximum deviations from the values listed, considering all 22 spectra and different fitting methods; the root-mean-square deviations are smaller by factors of two to five.

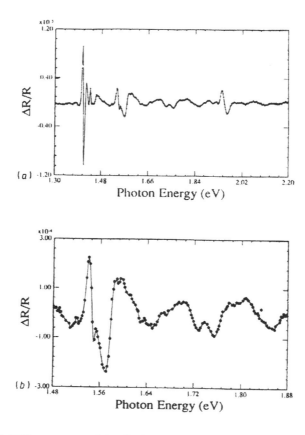

Fig. 6.53. (a) EER spectrum of the GaAs–GaInP superlattice and its fit. (b) Detailed structure of the EER spectrum and of its fit. The dots show the experimental data points and the solid lines show the fit (after Razeghi et al. [1992]).

Transition	RT EER	LT PL
Band to band transitions		
E_0 (bulk GaAs)	1412 ± 3	1518
$E_0 + \Delta_0$ (bulk GaAs)	1757 ± 3	
E_b	1952 ± 8	1984
Well state to well state transitions		
E_{11}^{hh}	1437 ± 1	1546
E_{21}^{hh} and E_{11}^{lh}	1455 ± 2	1575
E_{31}^{hh}	1512 ± 2	
E_{12}^{hh}	1520 ± 4	
E_{22}^{hh} and E_{12}^{lh}	1540 ± 2	1661
E_{41}^{hh} and E_{21}^{lh}	1550 ± 2	
E_{32}^{hh}	1576 ± 3	
E_{51}^{hh}	1614 ± 4	
E_{42}^{hh} and E_{22}^{lh}	1628 ± 2	
E_{52}^{hh}, E_{61}^{hh} and E_{31}^{lh}	1704 ± 6	
E_{11}^{so}	1773 ± 3	
E_{62}^{hh}	1785 ± 3	
E_{32}^{lh}	1796 ± 3	
E_{12}^{so}	1865 ± 3	
E_{22}^{so}	1908 ± 3	
Transitions between well states and band edges		
E_{1c}^{hh}	1572 ± 2	
E_{2c}^{hh} and E_{1c}^{lh}	1598 ± 2	
E_{3c}^{hh}	1643 ± 2	
E_{4c}^{hh} and E_{2c}^{lh}	1683 ± 2	
E_{5c}^{hh}	1760 ± 2	
E_{v1}	1817 ± 2	
E_{6c}^{hh} and E_{3c}^{lh}	1828 ± 2	
E_{v2}	1888 ± 2	
E_{1c}^{so} and E_{s1}	1917 ± 2	
E_{1c}^{hh}	1937 ± 2	
E_{2c}^{so}	1989 ± 2	

Table 6.9. Observed optical transition energies for the GaAs–GaInP superlattice.

From the measured optical transition energies, the conduction band discontinuity can be deduced from any of the following relations

$$\Delta E_c = E_{nc}^{hh} - E_{nm}^{hh} + E(em)$$
$$= E_{nc}^{lh} - E_{nm}^{lh} + E(em)$$

Eq. (6.59)
$$= E_b - E_{vm} + E(em)$$
$$= E_{nc}^{hh} - E_w + E(hhm)$$
$$= E_{nc}^{lh} - E_w + E(lhm)$$

The first of these relations is the most precise: with $n = m = 1$ it yields a value of ΔE_c correct to within 4 meV; using other values of n and m and the other relations yields 15 other determinations of ΔE_c, all of which are consistent with the first, most precise, determination. The conduction band offset was found to be 159 ± 4 meV and the valence band offset was found to be 388 ± 6 meV.

The energy difference between any two valence band levels is given directly in terms of the measured optical transition energies by the equations

Eq. (6.60) $E(\alpha m) - E(\beta n) = E_{mr}^{\alpha} - E_{nr}^{\beta}$

or

Eq. (6.61) $E(\alpha m) - E(\beta n) = E_{mc}^{\alpha} - E_{nc}^{\beta}$

where α and β can be hh, lh or so, and r can be 1 or 2. These energy differences are important to know for the conception of III–V IR detectors and for other applications. Eq. (6.60) and Eq. (6.61) provide two direct self-consistency conditions for each pair $(\alpha m, \beta n)$; each of these conditions is well satisfied. For the low-lying subbands, the energy differences $E(\alpha m) - E(\beta n)$ are determined within ± 2 meV.

Knowing $E(hh1)$ from theory or from a theoretical fit to the measured transition energies, one can find $E(hhm)$, $E(lhm)$ and $E(som)$ from the relations

Eq. (6.62) $E(hhm) = E(hh1) + E_{mn}^{hh} - E_{1n}^{hh}$

Eq. (6.63) $E(lhm) = E(hh1) + E_{mn}^{lh} - E_{1n}^{hh}$

Eq. (6.64) $E(som) = E(hh1) + E_{mn}^{so} - E_{1n}^{hh} - \Delta_w$

or from the relations

Eq. (6.65) $E(hhm) = E(hh1) + E_{mc}^{hh} - E_{1c}^{hh}$

Eq. (6.66) $E(\text{lh}m) = E(\text{hh}1) + E_{mc}^{\text{lh}} - E_{1c}^{\text{hh}}$

Eq. (6.67) $E(\text{so}m) = E(\text{hh}1) + E_{mn}^{\text{so}} - E_{1n}^{\text{hh}} - \Delta_{\text{w}}$

These relations provide further, more direct, self-consistency checks on the data and on the interpretation of the spectra. All of these self-consistency checks are well satisfied. Similarly, knowing $E(e1)$, one can find $E(e2)$ from any of six equations analogous to Eq. (6.62) – Eq. (6.67), each valid for different values of m. The resultant self-consistency conditions are also well satisfied.

The bandgap, E_{w}, and spin splitting energy, Δ_{w}, for the wells is not exactly the same as for bulk GaAs, presumably because of strain. From the relations

Eq. (6.68) $E_{\text{w}} = E_{11}^{\text{hh}} - E(\text{hh}1) - E(e1)$

and

Eq. (6.69) $\Delta_{\text{w}} = E_{11}^{\text{so}} + E(\text{hh}1) - E_{11}^{\text{hh}} - E(\text{so}1)$

or from a fitting of all the observed transition energies to a parameterized band calculation, one can find $E_{\text{w}} = 1.405$ eV and $\Delta_{\text{w}} = 0.33$ eV (0.009 and 0.01 eV less than the values for bulk GaAs). The fitting yields a 91 Å well width, in excellent agreement with the value of 90 Å found by calibration of the rate of epitaxial growth. The values found for the conduction band, heavy-hole and light-hole masses in the barriers and the wells and for the split-off well mass are in excellent agreement with the literature.

6.10.2. The study of the GaAs/GaInP three-quantum-well structure

A 3QW structure was grown by LP-MOCVD under the same conditions as the superlattice (SL). The quantum wells were 120, 60 and 35 Å in width, as shown in Fig. 6.54. The barriers separating the wells were sufficient to decouple the wells. All of the analysis discussed for the SL is equally applicable to the 3QW structure, but with each of the three wells contributing its own spectrum. However, the weaker transitions seen for the SL are not seen for the 3QW structure because in the 3QW structure each transition occurs in only one well.

Nineteen experimental EER spectra were obtained and analyzed for the 3QW structure. Fig. 6.54 also shows a typical EER spectrum at RT and our fit to that spectrum. Table 6.10 lists some of the transitions observed and gives the transition energies measured by EER at room temperature (RT) and by PL at low temperature (LT). The conduction and valence band offsets found for the 3QW structure were the same as those found for the SL within experimental error, as were the band masses.

Transition	RT EER	LT PL
Band to band transitions		
E_0 (bulk GaAs)	1419 ± 1	1522
$E_0 + \Delta_0$ (bulk GaAs)	1760 ± 2	
E_b	1912 ± 1	1984
Well state to well state transitions (120 Å QW)		
E_{11}^{hh}	1424 ± 1	1524
E_{11}^{lh} and E_{21}^{hh}	1439 ± 2	1550
E_{31}^{hh}	1460 ± 2	
E_{22}^{hh}	1487 ± 2	1597
E_{51}^{hh}	1532 ± 1	
E_{22}^{lh} and E_{42}^{hh}	1555 ± 1	1661
E_{31}^{lh}	1600 ± 2	
E_{71}^{hh}	1641 ± 2	
E_{62}^{hh} and E_{71}^{hh}	1641 ± 2	
E_{82}^{hh}	1769 ± 2	
E_{11}^{so}	1772 ± 1	
E_{42}^{lh}	1792 ± 1	
E_{22}^{so}	1866 ± 1	
Well state to well state transitions (60 Å QW)		
E_{11}^{hh}	1464 ± 2	1580
E_{11}^{lh} and E_{21}^{hh}	1504 ± 2	1618
E_{22}^{hh}	1515 ± 1	
E_{31}^{hh}	1585 ± 4	
E_{22}^{lh}	1677 ± 4	
Well state to well state transitions (35 Å QW)		
E_{11}^{hh}	1515 ± 1	1635
E_{11}^{lh} and E_{21}^{hh}	1585 ± 4	

Table 6.10. Transition energies identified for the GaAs/GaAs/Ga$_{0.51}$In$_{0.49}$P multiple quantum well.

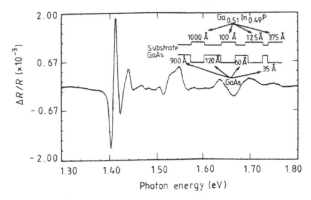

Fig. 6.54. EER spectrum of a GaAs/GaInP multiple QW along with a schematic diagram of the band structure (after Razeghi et al. [1992]).

6.10.3. III–V intersubband infrared detectors

There has been considerable research in the area of III–V intersubband quantum well infrared photodetectors (QWIP) for the 8–12 μm wavelength region as an alternative to HgCdTe (MCT) detectors [Levine et al. 1991, 1992, Rosencher et al. 1992]. Of all the III–V quantum well intersubband detectors, AlGaAs–GaAs quantum wells and superlattices have been extensively studied and they show great promise for focal plane array applications. However, the performance of these detectors remains inferior, with a lower detectivity and quantum efficiency, higher dark current and lower operating temperature, compared to MCT detectors. The major disadvantages of MCT technology are non-uniform composition, lack of high-quality lattice-matched substrates, mechanical softness, small wafer size, high cost and low yield. On the other hand, the advantages of III–V QWIPs are uniform composition, large excellent-quality substrates, mature processing technology, higher possibility of monolithic integration, multispectral detectors, ultrahigh speed and radiation hardness. At present there are many factors which are limiting the performance of AlGaAs–GaAs QWIPs and most of the problems arise from the AlGaAs barrier, the poor AlGaAs/GaAs interface due to higher growth temperature, oxygen-induced defects in the AlGaAs layer and low barrier height due to the limitation of AlGaAs composition. Hence, an improvement in the quality of the barrier material or an alternative barrier material is necessary for further improvement in device performance. GaInP has been considered as the best alternative for AlGaAs because of its superior material properties such as its insensitivity to oxygen, higher carrier mobility and carrier mobility insensitivity to DX centers. Also, it was recently found that the GaInP/GaAs interface is better due to its lower growth temperature [Razeghi 1985], and the interface recombination velocity is 10 times lower than AlGaAs/GaAs [Olsen et al. 1989]. All these properties clearly suggest that by replacing AlGaAs with GaInP, it is possible to improve the performance of III–V QWIPs.

The first step is to obtain high-quality GaInP/GaAs multiquantum wells (MQWs) and superlattices by LP-MOCVD on GaAs, InP and Si substrates. In order to achieve the above objective, growth conditions were varied for GaInP/GaAs heterostructures. Both multiquantum well structures and superlattices were grown by LP-MOCVD [Razeghi et al. 1989a]. Heterostructures were characterized by several techniques such as x-ray diffraction, photoluminescence and photovoltage spectroscopy for interface quality and subband energies.

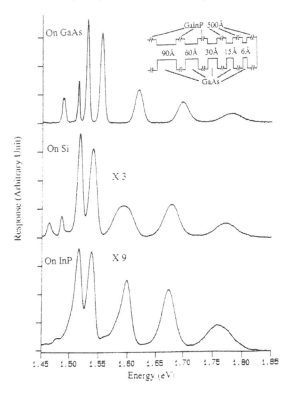

Fig. 6.55. 4 K PL spectra of the GaAs–GaInP MQWs grown on GaAs, Si and InP substrates. The inset shows the schematic of the multiquantum well structures (after He and Razeghi [1992]).

Several high-quality GaInP–GaAs quantum wells and superlattices were grown by LP-MOCVD. In order to demonstrate the quality of GaInP–GaAs quantum well structures grown by LP-MOCVD, five quantum wells with different well thicknesses were grown simultaneously on GaAs, Si and InP substrates and the structure is shown in Fig. 6.55. First photoluminescence measurements were performed to assess the quality of the samples [He and Razeghi 1992]. Sharp photoluminescence peaks (FWHM = 5 meV at 4 K) are observed on all three samples, indicating good interfaces and uniform layer thicknesses. Also, another important result observed was that the optical quality of the MQW grown on the Si substrate is better than that grown on the InP substrate. Theoretical calculations were performed to determine the subband energies and it was found that we were able to observe the subband due to the thinnest well (6 Å), which further confirms the high quality of the interface. The photoluminescence spectra of the samples grown on Si and InP substrates showed a shift of 15 meV from the corresponding peaks of the sample grown on the GaAs substrate. There are two possible explanations for this behavior: diffusion of In along the dislocation threads from GaInP to GaAs or localized strain induced by defects and In segregation [He and Razeghi 1992].

Room-temperature photovoltage measurements were performed on a 10-period GaInP–GaAs superlattice with a nominal well width of 90 Å and barrier width of 100 Å. Fig. 6.56 shows the room-temperature photovoltage spectroscopy (PVS) of a GaInP–GaAs superlattice. The well resolved peaks correspond to transitions from several sublevels. Also, the sharp peak, rather than a step-like profile, of the photovoltage spectrum indicates a pronounced exciton absorption behavior at room temperature. These results clearly demonstrate that photovoltage spectroscopy is a simple and powerful tool to study quantum confinement structures which will be of use in the mass production of these devices [He and Razeghi 1993a].

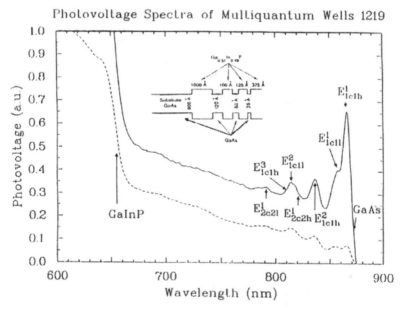

Fig. 6.56. The structure of a three-well MQW sample and its PVS. The marked positions are the calculated peak positions. The broken line displays the raw PVS. The solid line displays the calibrated PVS (after He and Razeghi [1993a]).

High-resolution x-ray diffraction measurements were performed to investigate the interface properties of GaInP/GaAs structures. Fig. 6.57 shows the x-ray diffraction pattern obtained in one of the samples. The superlattice satellites were observed over a 4 degree angle indicating that the interface roughness and disorder in the sample are low. In summary, we have shown that good-quality GaInP–GaAs multiple quantum wells and superlattices can be obtained by LP-MOCVD [He and Razeghi 1993b]. The fabrication of QWIPs using GaInP/GaAs will be the next step towards the realization of focal plane arrays.

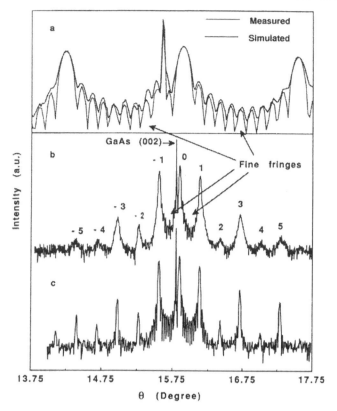

Fig. 6.57. (0 0 2) X-ray diffraction pattern of GaInP–GaAs superlattice. (a) Magnified view of both the measured and simulated patterns for θ between 15.75 and 16.2°. (b) Measured pattern, and (c) simulated pattern for the layer model (after He and Razeghi [1993b]).

6.11. Intersubband hole absorption in GaAs–GaInP quantum wells

In some semiconductor systems (GaAs/GaInP, GaAs/AlGaAs, GaInAs/AlGaAs), band offsets are such that they induce potential wells for electrons and/or holes in the different bands. Progress in growth technology such as molecular beam epitaxy and metalorganic chemical vapor deposition allow us to obtain extremely thin films of semiconductors with a precision of one atomic layer. These wells are so thin that the electron and hole motion is quantized in the direction normal to the layers. The carrier energy is then distributed in different subbands, the bottoms of which are the one-dimensional quantized states E_i in the wells. Light can then induce transitions between these subbands. The resulting absorption is resonant between these quantized states (or bound states), with values as high as several 10^3 cm^{-1}. This resonant absorption between quantized states is termed intersubband absorption.

Intersubband hole absorption has been observed in *p*-type GaAs–GaInP multiquantum wells. The multiquantum well consists of 50 zinc-doped GaAs quantum wells, nominally 45 Å thick, separated by undoped lattice-matched GaInP barriers nominally 500 Å thick. This multiple quantum well is grown between zinc-doped GaAs contact layers of thickness 1 μm (bottom) and 0.5 μm (top). All *p*-type layers have been doped with $N_a - N_d = 1 \times 10^{18}$ cm^{-3}. Only the middle 35 Å of each well has been doped in order to prevent dopant migration into the barrier region.

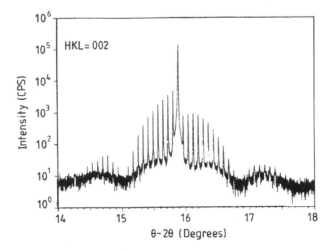

Fig. 6.58. (0 0 2) x-ray diffraction pattern of a GaAs–GaInP quantum well.

Fig. 6.58 shows the (0 0 2) x-ray diffraction pattern for a GaAs–GaInP quantum well. Satellites are clearly observed up to 20th order, evidencing the high structural quality of the superlattice. A superlattice period of 560 Å is deduced from the satellite spacing. A modulation of the superlattice satellite peaks is clearly observed every other 11 satellites. This modulation period is consistent with the nominal barrier to well width ratio. The slight asymmetry of the diffraction pattern indicates a difference between the two quantum well interfaces [He and Razeghi 1993a, b]. The angular separation between the GaAs peak and the zero-order superlattice satellite yields a lattice mismatch of 0.024% for the barrier, indicating good composition control.

Fig. 6.59 shows a low-temperature (4 K) photoluminescence spectrum of the sample obtained using a pump Ar laser, after selectively etching away the top contact layer. We attribute the sharp peak at 1.57 eV to the e1–hh1 transition, which is consistent with the nominal superlattice parameters. The full width at half maximum (FWHM) is 11 meV, which is quite small for doped quantum wells. The doping level in the contact layers was ascertained to be close to the nominal value by electrochemical *C–V* profiling. Since the wells were doped under the same conditions, we expect the doping level of the wells to lie in the 10^{18} cm^{-3} range as well. We attribute this narrow FWHM to the good superlattice quality. The broader peak at 1.55 eV originates from an acceptor-related transition (e1–A0) because its relative magnitude decreases with increasing pump intensity [Bastard 1988]. The weak peak at 1.49 eV corresponds to bulk GaAs. At room temperature, an additional peak is found at 1.87 eV corresponding to the GaInP barriers.

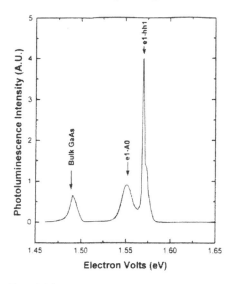

Fig. 6.59. Low-temperature PL spectrum.

After selectively etching away the top contact layer, the sample was processed into an optical waveguide with polished 45° facets. The waveguide transmission was measured in a Fourier transform infrared (FTIR) spectrometer. The infrared beam was focused onto the polished facets using KBr lenses, and its polarization was controlled with a grid wire polarizer. Fig. 6.60 shows the transmission spectra for two perpendicular polarizations labeled s and p (where for s polarization, the infrared light electric field lies in the plane of the quantum wells, and p polarization is perpendicular to the s polarization). A broad minimum in the transmission is observed around 9 μm for the two polarizations, which is attributed to intersubband transitions. The doublet around 13–14 μm and the complete fallout of the transmission beyond 15 μm are due to multiphonon absorption processes in the GaAs substrate. These same features are observed in a simple GaAs substrate waveguide as well. The slowly decreasing transmission background with increasing wavelength originates from free-carrier absorption in the bottom GaAs doped layer, as was checked by selectively etching away all the quantum wells. The intersubband absorption of s-polarized light is roughly twice as strong as that of p-polarized light. Since the p-polarization can be decomposed into equal TE and TM polarizations, $p = \frac{1}{2}(TE + TM)$, and since the s-polarization is purely TE, $s = TE$, this means that the TM absorption is almost negligible compared to the TE absorption. Here, TE polarization (TM polarization, respectively) corresponds to the light electric field in the plane (perpendicular to the plane, respectively) of the quantum wells. This polarization dependence corresponds to an enhanced normal incidence absorption, which is in qualitative agreement with results on *p*-doped GaAs–GaAlAs QWIPs [Levine 1991], as well as with theoretical calculations [Chang and James 1989,

Teng et al. 1992]. Because of the large valence band discontinuity in excess of 200 meV [Omnes and Razeghi 1991], the observed absorption around 9 μm originates from transitions between bound states. The broad character of the absorption peak suggests that several of such bound to bound transitions contribute to it. A Kronig-Penney calculation yields a 180 meV energy difference between the heavy-hole ground state and heavy-hole second excited state corresponding to an expected transition at 7 μm wavelength. Calculated energy differences between the heavy-hole ground state and light-hole states do not fall within the wavelength range of interest, but it should be pointed out that these calculated transitions are sensitive to slight discrepancies between the nominal and actual band diagram. Taking into account the waveguide dimensions (length 6 mm × thickness 0.4 mm), the observed intersubband absorption is estimated to peak at 10^{-3} per well (i.e. 2.2×10^3 cm^{-1} normalized to the 45 Å well width), in reasonable agreement with values calculated for GaAs–GaAlAs by Man and Pan [1992].

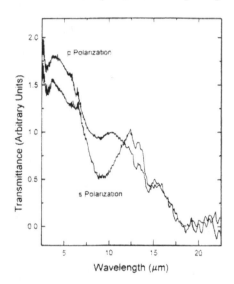

Fig. 6.60. Transmission spectra for two perpendicular polarizations.

References

Alferov, Z.I., Garbuzov, D.Z., Mishurnyi, V.A., Rumyantsev, V.D., and Tret'yakov, D.N., *Sov. Phys.–Semicond.* **7** 1534, 1974.

Ando, T., *J. Phys. Soc. Japan* **51** 3900, 1982.

Ando, T., Fowler, A.B., and Stern, F., *Rev. Mod. Phys.* **54** 437, 1982.

Bastard, G., *Wave Mechanics Applied to Semiconductor Heterostructures* (Les Ulis: Les Editions de Physique) p 175, 1988.

Beerens, J., Gregoris, G., Portal, J.C., Mendez, E.E., Chang, L.L., and Esaki, L., *Phys. Rev.* B **36** 4742, 1987.

Bellon, P., Chevalier, J.P., Martin, G.P., Dupont-Nivel, E., Theibant, C., and Andre, J.A., *Appl. Phys. Lett.* **52** 567, 1988.

Ben, A.S., Dmowski, L., Portal, J.C., Martin, K.P., Higgins, R.J., and Razeghi, M., *Appl. Phys. Lett.* **57** 2925, 1990.

Ben, A.S., Dmowski, L., Portal, J.C., Pulsford, N.J., Singleton, J., Nicholas, R.J., and Razeghi, M., *J. Appl Phys.* **65** 2756, 1989.

Bisaro, R., Laurincin, G., Friederich, A., and Razeghi, M., *Appl. Phys. Lett.* **40** 978, 1982.

Biswas, D., Debbar, N., Bhattacharya, P., Razeghi, M., and Omnes, F., *Appl. Phys. Lett.* **56** 833, 1990.

Bourgoin, J.C., *Appl. Phys. Lett.* **59** 941, 1991.

Casey, H.C.Jr and Panish, M.B., *Heterostructure Lasers Part B* (New York: Academic) p 20, 1978.

Chan, Y.J., Pavilidis, D., Razeghi, M., Jaffe, M., and Singh, J., *Gallium Arsenide and Related Compounds* (*Inst. Phys. Conf. Ser. 96*) ed J S Harris (Bristol: Institute of Physics) p 459, 1988.

Chan, Y.J., Pavilidis, D., Razeghi, M., and Omnes, F., *IEEE Trans. Electron. Dev.* **ED-37** 2141, 1990.

Chand, N., Henderson, T., Klem, J., Masselink, W.T., Fischer, R., Chang, Y.C., and Morkoc, H., *Phys. Rev.* **B 30** 4481, 1984.

Chaubet, C., Raymond, A., Knap, W., Mulot, J.Y., Baj, M., and Andre, J.P., *Semicond. Sci. Technol.* **6** 160, 1991.

Chen, C.H., Baier, S.M., Arch, D.K., and Shur, M., *IEEE Trans. Electron. Dev.* **ED-35** 570, 1988.

Chang, Y.C. and James, R.B., *Phys. Rev.* B **39** 12 672, 1989.

Delong, M.C., Taylor, P.C., and Olson, J.M., *Appl. Phys. Lett.* **57** 620, 1990.

Dobers, M., Malcher, F., von Klitzing, K., Rossler, U., Ploog, K., and Weimann, G., *Semiconductors in High Magnetic Fields* (*Springer Series in Solid State Sciences*) ed G Landwehr (Berlin: Springer), 1989a.

Dobers, M., Vieren, J.P., Razeghi, M., Defour, M., and Omnes, F., *Semicond. Sci. Technol.* **4** 687, 1989b.

Dobers, M., von Klitzing, K., Schneider, J., Weimann, G., and Ploog, K., *Phys. Rev. Lett.* **31** 1650, 1988a.

Dobers, M., von Klitzing, K., and Weimann, G., *Phys. Rev.* B **38** 5453, 1988b.

Ettenberg, M. and Olsen, G.H., *J. Appl. Phys.* **48** 4275, 1977.

Fang, F., Smith, T.P., and Wright, S.L., *Surf. Sci.* **196** 310, 1988.

Feng, S.L., Bourgoin, J.C., Omnes, F., and Razeghi, M., *Appl. Phys. Lett.* **59** 941, 1991

Garland, J.W. and Raccah, P.M., *SPIE* **659** 32, 1986.

Gomyo, A., Hotto, H., and Hino, I., *Japan. J. Appl. Phys.* **28** L1330, 1989.

Gomyo, A., Suzuki, T., and Iijima, S., *Phys. Rev. Lett.* **60** 2645, 1988.

Gomyo, A., Suzuki, T., Kohayashi, K., Kawata, S., Hino, I., and Yuara, T., *Appl. Phys. Lett.* **50** 673, 1987.

Gornik, E., Seidenbush, W., Christanell, R., Lasnig, R., and Pidgeon, C.R., *Surf. Sci.* **196** 339, 1987.

Gregoris, G., Beerens, J., Ben, A.S., Dmowski, L., Portal, J.C., Sivco, D.L., and Cho, A.Y., *J. Phys. C: Solid State Phys.* **20** 425, 1987.

Guldner, Y., Voos, M., Vieren, J.P., Hirtz, J.P., and Heiblum, M., *Phys. Rev.* B **36** 1266, 1987.

Gunapala, S.D., Levine, B.F., Logan, R.A., Tanbun-Ek, T., and Humphrey, D.A., *Appl. Phys. Lett.* **57** 1802, 1990.

Harrang, J.P., Higgins, R.J., Goodall, R.K., Jay, P.R., Laviron, M., and Deslecluse, P., *Phys. Rev.* B **32** 8126, 1985.

Harris, J.J., Lacklison, D.E., Foxon, C.T., Selten, F.M., Suckling, A.M., Nicholas, R.J., and Barnham, K.J., *Semicond. Sci. Technol.* **2** 783, 1987.

He, X. and Razeghi, M., *Appl. Phys. Lett.* **61** 1703, 1992.

He, X. and Razeghi, M., *Appl. Phys. Lett.* **62** 618, 1993a.

He, X. and Razeghi, M., *J. Appl. Phys.* **73** 3284, 1993b.

Honig, R.E., *RCA Rev.* **23** 657, 1962.

Horng, R.H., Wuu, D.S., and Lee, M.K., *Appl. Phys. Lett.* **53** 2614, 1988.

Hoshino, M., Kodama, K., Kitahara, K., Komeno, J., and Ozeki, M., *Appl. Phys. Lett.* **48** 983, 1986a.

Hoshino, M., Kodama, K., Kitahara, K., Komeno, J., and Ozeki, M., *Japan. J. Appl. Phys.* **25** L534, 1986b.

Hoshino, M., Kodama, K., Kitahara, K., and Oseki, M., *Appl. Phys. Lett.* **48** 770, 1986c.

Hsieh, S.J., Patten, E.A., and Wolfe, C.M., *Appl. Phys. Lett.* **45** 1125, 1984.

Hsu, C.C., Yuan, J.S., Cohen, R.M., and Stringfellow, G.B., *J. Appl. Phys.* **59** 395, 1985.

Ikeda, M., Toda, A., Nakano, K., Mori, Y., and Watanabe, N., *Appl. Phys. Lett.* **50** 1033, 1987.

Inoue, Y., Nishino, T., Hamakawa, Y., Kondow, M., and Minagawa, S., *Optoelectron. Dev. Technol.* **3** 61, 1988.

Ishikawa, M., Ohla, Y., Sugawara, H., Yamamoto, M., and Nakanisi, T., *Appl. Phys. Lett.* **48** 207, 1986.

Jen, H.R., Jou, M.J., Cerng, Y.T., and Stringfellow, G.B., *J. Crystal Growth* **85** 175, 1987.

Kassel, L., Garland, J.W., Abad, H., Raccah, P.M., Potts, J.E., Haase, M.A., and Cheng, H., *Appl. Phys. Lett.* **56** 42, 1990.

Kassel, L., Garland, J.W., Raccah, P.M., Haase, M.A., and Chang, H., *Semicond. Sci. Technol.* **6** A146, 1991a.

Kassel, L., Garland, J.W., Raccah, P.M., Tamargo, M.C., and Farrell, H., *Semicond. Sci. Technol.* **6** A152, 1991b.

Kastalsky, A. and Hwang, J.M., *Solid State Commun.* **51** 317, 1984.

Kawamura, Y. and Asahi, H., *Appl. Phys. Lett.* **43** 780, 1983.

Kitahara, K., Hoshino, M., Kodama, K., and Ozeki, M., *Japan. J. Appl. Phys.* **25** L534, 1986.

Kitahara, K., Hoshino, M., and Ozeki, M., *Japan. J. Appl. Phys.* **27** L110, 1988.

Knap, W., Huant, S., Chaubet, C., and Etienne, B., *Proc. 5th Int. Conf. on the Physics of Electro-optic Microstructures and Microdevices* (*Heraklion, 1990*), 1990.

Kobayashi, T., Taira, K., Nakamura, F., and Kawai, H., *J. Appl. Phys.* **65** 4898, 1989.

Kondow, M., Kakibayashi, H., and Minagawa, S., *J. Crystal Growth* **88** 291, 1988a.

Kondow, M., Kakibayashi, H., Minagawa, S., Inoue, Y., Nishino, T., and Hamakawa, Y., *Appl. Phys. Lett.* **53** 2053, 1988b.

Kondow, M., Kakibayashi, H., Minagawa, S., Inoue, Y., Nishino, T., and Hamakawa, Y., *J. Crystal Growth* **93** 412, 1988c.

Kondow, M., Kakibayashi, H., Tanaka, T., and Minagawa, S., *Phys. Rev. Lett.* **63** 884, 1989.

Kroemer, H., Chien, W.Y., Harris, J.S., and Edwall, D.D., *Appl. Phys. Lett.* **36** 295, 1980.

Kuan, T.S., Kuech, T.F., Wang, W.I., and Wilkie, E.L., *Phys. Rev. Lett.* **54** 201, 1985.

Kudman, I. and Poff, R.J., *J. Appl. Phys.* **43** 3760, 1972.

Ladany, I., *J. Appl. Phys.* **42** 654, 1971.

Landau, L.D. and Lifshitz, E.M., *Statistical Physics* (Oxford: Pergamon) ch 14, 1969.

Landolt–Börnstein, *New Series Group III*, vol **17a** (Berlin: Springer), 1982.

Levine, B.F., Gunapala, S.D., Kuo, J.M., Pei, S.S., and Hui, S., *Appl. Phys. Lett.* **59** 1864, 1991.

Levine, B.F., Zussman, A., Gunapala, S.D., Asom, M.T., Kuo, J.M., and Hobson, W.S., *J. Appl. Phys.* **72** 4429, 1992.

Lommer, G., Malcher, F.F., and Rossler, U., *Phys. Rev.* B **32** 6995, 1985.

Madelung, O., *Semiconductors (Data in Science and Technology)* (Berlin: Springer), 1991.

Man, P. and Pan, D.S., *Appl. Phys. Lett.* **61** 2799, 1992.

Mani, R.G. and Anderson, J.R., *Phys. Rev.* B **37** 4299, 1988.

Martin, K.P., Higgins, R.J., Rascol, J.L., Yoo, H.M., and Arthur, J.R., *Surf. Sci.* **196** 323, 1988.

Martin, P.A., Meehan, K., Gavrilovic, P., Hess, K., Holonyak, N., and Coleman, J.J., *J. Appl. Phys.* **54** 4689, 1983.

Maude, D.K., Portal, J.C., Dmowski, L., Foster, T., Eaves, L., Nathan, M., Heiblum, M., Harris, J., and Beall, R.B., *Phys. Rev. Lett.* **59** 815, 1987.

McDermott, B.T., El-Marry, N.A., Jiang, B.L., Hyuga, F., and Bedair, S.M., *J. Crystal Growth* **107** 96, 1991.

Mercy, J.M., Bousquet, C., Robert, J.L., Raymond, A., Gregoris, G., Beerens, J., Portal, J.C., Frijlink, P.M., Delescluses, P., Chevrier, J., and Linh, N.T., *Surf. Sci.* **142** 298, 1984.

Merkt, U., *Festkorperproblem* **27** 109, 1987.

Miyoko, O., Watanabe, N., and Ohba, Y., *Appl. Phys. Lett.* **50** 906, 1987.

Mooney, P., Fisher, R., and Morkoc, H., *J. Appl. Phys.* **57** 1928, 1985.

Nakayama, H. and Fujita, H., *GaAs and Related Compounds 1985 (Inst. Phys. Conf. Ser. 79)* ed M. Fujimoto (Bristol: Institute of Physics) p 289, 1986.

Nathan, M.I., Jackson, T.N., Kiirchner, P.D., Mendez, E.E., Petit, G.D., and Worlock, J.W., *J. Electron. Mater.* **12** 719, 1983.

Olivier, J., Etienne, P., Razeghi, M., and Alnot, P., *Nat. Res. Soc. Symp. Proc.* **91** 497, 1987.

Olsen, J.M., Ahrenkiel, R.K., Dunlavy, D.J., Keys, B., and Kibbler, A.E., *Appl. Phys. Lett.* **55** 1208, 1989.

Omnes, F., Defour, M., and Razeghi, M., *Rev. Tech. Thomson-CSF* **23** No 3 September, 1991.

Omnes, F. and Razeghi, M., *Appl. Phys. Lett.* **59** 1034, 1991.

Papalus, G., Lin, S.Z., and Razeghi, M., unpublished results, 1991.

Pierron, E.D., Parker, D.L., and McNeely, J.B., *J. Appl. Phys.* **38** 4669, 1967.

Queisser, H.J. and Theodorou, D.E. *Phys. Rev.* B **33** 4027, 1986.

Ranz, E., Lavielle, D., Cury, L.A., Portal, J.C., Razeghi, M., and Omnes, F., *Superlatt. Microstruct.* **8** 245, 1990.

Rao, M.A., Caine, E.J., Kroemer, H., Long, S.J., and Babic, D., *J. Appl. Phys.* **61** 643, 1987.

Raymond, A., Chaulet, C., and Razeghi, M., *Proc. SPIE* **1362** 275, 1990.

Razeghi, M., *Semiconductors and Semimetals* vol 22 Part A, ed R.K. Willardson and A.C. Beer (New York: Academic) chapter 5 p 305, 1985.

Razeghi, M., *EIT Workshop on Photonics* (*Vienna, 6–8 November 1989*) (Banca Popolare di Verona) p 49, 1989a.

Razeghi, M., *The MOCVD Challenge Volume 1: A Survey of GaInAsP–InP for Photonic and Electronic Applications* (Bristol: Hilger), 1989b.

Razeghi, M., unpublished Razeghi M 1989c

Razeghi, M., Defour, M., Omnes, F., Dobers, M., Vieren, J.P., and Guldner, Y., *Appl. Phys. Lett.* **55** 457, 1989a.

Razeghi, M., Duchemin, J.P., Portal, J.C., Dmowski, L., Remeni, G., Nicholas, R.J., and Briggs, A., *Appl. Phys. Lett.* **48** 712, 1986a.

Razeghi, M., Machado, A., Koch, S.M., Acher, O., Omnes, F., and Defour, M., *SPIE* **1283** 64, 1990.

Razeghi, M., Maurel, P., Omnes, F., Ben, A.S., Dmowski, L., and Portal, J.C., *Appl. Phys. Lett.* **48** 1267, 1986b.

Razeghi, M. and Omnes, F., *Appl. Phys. Lett.* **59** 1034, 1991a.

Razeghi, M., Omnes, F., Maurel, P., Chan, J., and Pavlidis, D., *Semicond. Sci. Technol.* **6** 103, 1991b.

Razeghi, M., Omnes, F., Nagle, J., Defour, M., Acher, O., and Bove, P., *Appl. Phys. Lett.* **55** 1677, 1989b.

Razeghi, M., Yang, D., Garland, J.W., Zhang, Z., and Xue, D., *Proc. SPIE* **1676** 130, 1992.

Robert, J.L., Raymond, A., Mulot, J.Y., Bousquet, C., Andre, J.P., Knap, W., Kubiza, M., Zawadski, W., Gornik, E., Seidenbush, W., and Vitzani, M., *Proc. 19th ICPS* ed W Zawadski (Warsaw: Institute of the Polish Academy of Science), 1988.

Rosencher, E., Vinter, B., and Levine, B., *Intersubband Transitions in Quantum Wells* (New York: Plenum), 1992.

Rossler, U., Malcher, F., and Lommer, G., *Semiconductors in High Magnetic Fields* (*Springer Series in Solid State Sciences*) ed G. Landwehr (Berlin: Springer), 1989.

Schubert, E., Fisher, A., and Ploog, K., *Phys. Rev.* B **31** 7937, 1985a.

Schubert, E., Knecht, J., and Ploog, K., *J. Phys. C: Solid State Phys.* **18** L215, 1985b.

Seidenbush, W., *Phys. Rev.* B **36** 1877, 1987.

Smith, J.S., Chiu, L.C., Margalit, S., and Yariv, A., *J. Vac. Sci. Technol.* B **1** 376, 1983.

Stein, D., Ebert, G., von Klitzing, K., and Weimann, G., *Surf. Sci.* **142** 406, 1984.

Stern, F. and Sarma, S.D., *Phys. Rev.* B **32** 8442, 1985.

Stormer, H.L., Dingle, R., Gossard, A.C., Wiegman, W., and Sturge, M.D., *Solid State Commun.* **29** 705, 1979.

Stormer, H.L., Gossard, A.C., and Wiegmann, W., *Solid State Commun.* **41** 707, 1983.

Stormer, H.L., Gossard, A.C., Wiegmann, W., and Baldwin, K., *Appl. Phys. Lett.* **39** 912, 1981.

Suzuki, T., Gomyo, A., Hino, I., Kobayashi, K., Kawata, S., and Lijina, S., *Japan. J. Appl. Phys.* **27** L1549, 1988a.

Suzuki, T., Gomyo, A., and Iijima, S., *J. Crystal Growth* **93** 389, 1988b.

Suzuki, T., Gomyo, A., Ijima, S., Kobayashi, K., Kawata, S., Hino, I., and Yuasa, T., *Japan. J. Appl. Phys.* **27** 2098, 1988c.

Tarng, M.L., and Fisher, D.G., *J. Vac. Sci. Technol.* **15** 50, 1978.

Teng, D., Lee, C., and Eastman, L.F., *J. Appl. Phys.* **72** 1539, 1992.

Ueda, O., Fujii, T., Nakada, Y., Yamada, H., and Umebu, I., *Workbook 5th Int. Conf. on Molecular Beam Epitaxy (Sapporo, 1988)* unpublished p 45, 1988a.

Ueda, O., Takechi, M., and Komeno, J., *Appl. Phys. Lett.* **54** 2312, 1989.

Ueda, O., Takikawa, M., Komeno, J., and Umebu, I., *Japan. J. Appl. Phys* (Part 2) **26** L1824, 1987.

Ueda, O., Takikawa, M., Takechi, M., Komeno, J., and Umebu, I., *J. Crystal Growth* **93** 418, 1988b.

Voisin, P., Guldner, Y., Vieren, J.P., Voos, M., Deslecluse, P., and Lihh, N.T., *Appl. Phys. Lett.* **39** 982, 1981.

von Klitzing, K.V. and Ebert, G., *Two-Dimensional Systems, Heterostructures and Superlattices (Springer Series in Solid State Sciences 53)* ed G. Bauer, F. Kuchar and H. Heinrich (Berlin: Springer) p 242, 1984.

Watanabe, M.O. and Ohba, Y., *Appl. Phys. Lett.* **50** 906, 1987.

Yasuami, S., Nozaki, C., and Ohba, Y., *Appl. Phys. Lett.* **52** 2031, 1988.

7. Optical Devices

7.1.   **Electro-optical modulators**
    7.1.1.   Introduction
    7.1.2.   Multiple-quantum-well modulators based on the quantum-confined Stark effect
    7.1.3.   Superlattice modulators based on Wannier–Stark localization
    7.1.4.   Perspectives
7.2.   **GaAs-based infrared photodetectors grown by MOCVD**
    7.2.1.   Basic physics of photodetectors
    7.2.2.   II–VI material-based photodetectors
    7.2.3. $InAs_{1-x}Sb_x$ materials
    7.2.4.   Intersubband GaAs quantum well photodetectors
7.3.   **Solar cells and GaAs solar cells**

7.1. Electro-optical modulators

7.1.1. Introduction

Electro-optical modulators are key components that enable electrical control of light. This can refer to both phase modulation and intensity modulation. Phase modulation is obtained directly through refractive index modulation. Historically, the variation of the refractive index under application of an electric field is known as the Pockels linear electro-optic effect. It finds its origin in the displacement of the bond charges and in a possible slight deformation of the ion lattice [Yariv and Yeh 1984]. The Pockels effect displays negligible dispersion which eases device design and operation specifications. The counterpart of this negligible dispersion is that this effect is weak. This is the reason why guided-wave devices have to be used in order to increase the interaction length between light and the electro-optic material. Phase modulators, as well as directional couplers have thus been realized in GaAs- and InP-based materials [Carenco 1987]. Intensity modulation has also been made possible using the Pockels effect through the use of Mach–Zehnder interferometers.

By the end of the 1950s, the effect of an electric field on the absorption spectrum of a semiconductor had been investigated by Franz [1958] and Keldysh [1958]. The net result is a red-shift of the absorption edge. Because of this shift, absorption can become significant under the application of an electric field at a wavelength where the semiconductor is normally transparent. These absorption variations (electroabsorption) have been exploited to realize intensity modulators. However, these absorption variations are only moderate and relatively long waveguide devices have to be used in order to obtain a significant contrast ratio.

Semiconductor microstructures, such as multiquantum wells and superlattices, whose dimensions are comparable to atomic dimensions and whose interfaces are atomically smooth, exhibit novel optical properties not encountered in the parent compounds. Advances in material growth and processing have made it possible to fabricate these thin semiconductor layers with unprecedented precision, making it possible to fabricate practical devices based on novel quantum mechanical effects. Most of these new phenomena rely on the original electroabsorption properties of these new structures. They have led to the improvement of previously existing devices as well as to the development of new functionalities.

When compared to previous devices made out of bulk III–V semiconductor or dielectric materials such as $LiNbO_3$, electro-optic devices made out of quantum structures offer the advantages of improved performance, reduced dimensions and possible integration with other semiconductor devices such as semiconductor lasers. In the following, we present the two major electroabsorption phenomena that have led to considerable research activity over the past few years: the quantum-confined Stark effect in multiple quantum wells and Wannier–Stark localization in superlattices, along with the corresponding device applications. This chapter will conclude with a discussion of new trends in this research area.

7.1.2. Multiple-quantum-well modulators based on the quantum-confined Stark effect

(a) The absorption spectrum of a multiple quantum well

The absorption edge (absorption spectrum near the bandgap) of a quantum well structure displays several interesting features that originate from the carrier quantum confinement [Bastard 1988].

First of all, the carrier motion in the direction perpendicular to the layers is quantized, which results in a two-dimensional density of states. Therefore, the absorption edge displays a step-like behavior which reflects the step-like two-dimensional density of states. This is in contrast with the typical parabolic three-dimensional density of states of a bulk semiconductor.

The second interesting feature originates from the Coulombic interaction between confined electrons and holes, also called excitonic interaction. Excitons are electron–hole (e–h) pairs forming a bound state analogous to the hydrogen atom. They produce very sharp resonance peaks just below the bandgap where a large oscillator strength is concentrated in a narrow spectral domain. This excitonic interaction is also present in bulk semiconductors, but here the excitonic orbital dimension is large and the interaction is weak. Consequently, excitonic transitions are usually observed in bulk semiconductors only at low temperature: as the temperature is increased, the resonances broaden due to phonon collisions and by room temperature they are usually only marginally resolvable from the interband absorption. By reducing the well width below the typical excitonic orbital dimensions, the excitonic orbit is squeezed in the growth direction as illustrated in Fig. 7.1. As a result, the interaction is strengthened and sharp exciton absorption peaks have been observed in quantum wells at room temperature [Miller et al. 1982].

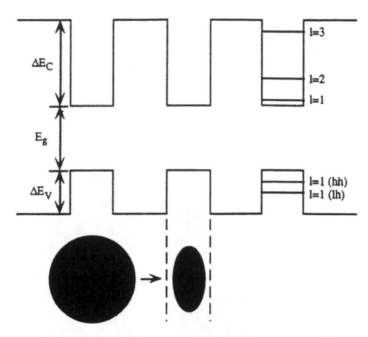

Fig. 7.1. Schematic of the band structure of a multiple-quantum-well structure. The shaded circle and ellipse illustrate how the exciton is compressed by the confinement.

Another distinction between bulk and quantum well semiconductors is the change in symmetry because of the layered structure. The degeneracy between light and heavy holes at the zone center is removed by introducing splitting between the heavy and light holes. The quantum confinement energies of the two holes are different and the exciton resonances associated with the two holes are now distinct, giving two clear peaks in the spectrum near the band edge as seen in Fig. 7.2(a).

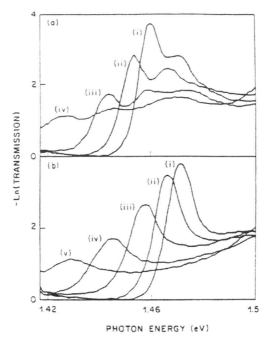

Fig. 7.2. Absorption spectra of a quantum well waveguide as a function of electric field applied perpendicular to the layers [Weiner et al. 1985]. (a) Incident optical polarization parallel to the plane of the layers for fields of (i) 1.6×10^4 V·cm^{-1}, (ii) 10^3 V·cm^{-1}, (iii) 1.3×10^5 V·cm^{-1} and (iv) 1.8×10^5 V·cm^{-1}. (b) Incident optical polarization perpendicular to the layers for fields of (i) 1.6×10^4 V·cm^{-1}, (ii) 10^5 V·cm^{-1}, (iii) 1.3×10^5 V·cm^{-1}, (iv) 1.8×10^5 V·cm^{-1} and (v) 2.2×10^5 V·cm^{-1} (after Weiner et al. [1985]).

(b) Quantum-confined Stark effect [Miller et al. 1985]

The application of an electric field perpendicular to the layers of a quantum well structure modifies the potential well profile as schematized in Fig. 7.3. Wavefunctions and energy levels are modified, and an easy way to guess the final result is to consider the classical counterpart: if classical particles were located in such distorted potential wells, they would be pulled against the potential barriers, and their potential energies would be reduced by half the potential drop over the quantum well (Fig. 7.4). The trends are the same for the present case: electrons and holes are pulled toward opposite sides of the well and their confinement energies are reduced. Therefore the fundamental transition energy of the quantum well is reduced which leads to a red-shift of the absorption edge. In contrast to bulk semiconductors where excitons are ionized very rapidly by the electric field, strong excitonic transitions are maintained in that case. The reasons for this behavior are the following: firstly, the well is narrow compared to the bulk exciton diameter so that the Coulomb interaction is still strong even though electrons and holes are pulled towards opposite sides of the well, and secondly, potential barriers inhibit the field ionization of the exciton.

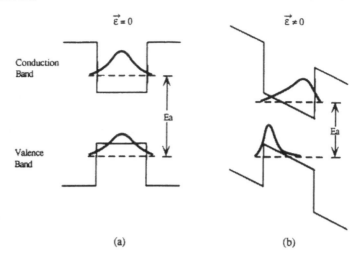

Fig. 7.3. Schematic view of the electroabsorption effect in MQWs. The conduction band minimum and valence band maximum and the energy levels of the ground states for electrons and holes along with the wavefunction envelopes are shown. These items are shown in the absence of an electric field in (a) and with an electric field perpendicular to the quantum well layers in (b).

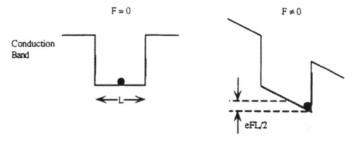

Fig. 7.4. Effect of a distorted potential well on a classical particle.

This red-shift of the absorption edge together with the persistence of strong excitonic interaction, is called the quantum-confined Stark effect. One should be aware that the excitonic oscillator strength undergoes a decrease under the application of the electric field because of the reduction in the overlap between the electron and hole wavefunctions which reduces the probability of finding the electron and the hole in the same unit cell.

A typical set of absorption spectra with increasing field is shown in Fig. 7.2(b). The exciton peaks remain resolved up to very high fields of 2×10^5 V·cm^{-1} and show very large shifts of 40 meV. This corresponds to a shift of about four times the zero-field binding energy at a field of about 100 times the classical ionization field of an unconfined exciton of comparable binding energy.

(c) Device application

Large absorption variations are obtained for wavelengths corresponding to the
normally transparent region. These absorption variations enable light intensity
modulation as was first demonstrated by Wood et al. [1984]. In the initial
demonstrations, light was propagating perpendicularly to the layers. Such
modulators are particularly well suited for optical signal processing and computing
because they can be integrated in two-dimensional arrays.

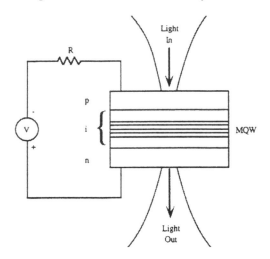

Fig. 7.5. Schematic of the quantum well SEED.

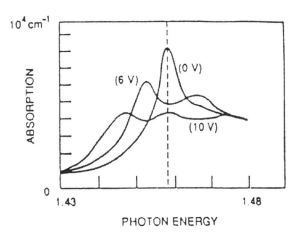

Fig. 7.6. Schematic of SEED operation.

Starting from this elementary modulator, a new bistable device was invented by
Miller et al. [1984]: this device, called the self-electro-optical device (SEED),
consists of a reverse-biased modulator connected with a series resistance as shown
in Fig. 7.5. The modulator is illuminated with a wavelength corresponding to the
zero-field exciton position. This is a normally absorbing wavelength region as

indicated by the dashed line in Fig. 7.6. When weak light is shining, there is a negligible current flow and all of the supply voltage appears across the diode, resulting in a relatively small absorption. As the illumination power increases, a voltage drop occurs across the resistor due to the photocurrent thus reducing the voltage across the diode. Consequently, the absorption of the diode increases as the exciton recovers its original position. This demonstrates the feedback mechanism that is necessary to reach bistability. The basic operation of a SEED is represented in Fig. 7.6.

Other types of modulator with light propagating perpendicularly to the layers have been developed based on the quantum-confined Stark effect. For example, devices consisting of Fabry–Pérot interferometers made out of epitaxially grown Bragg reflectors, that include quantum wells in between, have been realized. Such devices usually operate in a reflection mode and are based on the vanishing of a Fabry–Pérot resonance because of increased absorption.

Most of these vertical cavity devices have been realized out of GaAs-based materials because higher-quality quantum wells are more readily obtained in this material system. Guided-wave devices have been proposed as well. Although the first demonstrations were realized in the same material system [Wood et al. 1985], a lot of recent work has concentrated on InP-based materials because the use of external intensity modulators is an attractive way to reduce wavelength chirping occurring in directly modulated semiconductor lasers [Koch and Bowers 1984]. This chirping is detrimental to high-bit-rate and long-distance optical fiber transmission systems operating in the long-wavelength range [Koyama and Iga 1988]. Modulation bandwidths of over 20 GHz have been achieved in both MBE-grown InGaAlAs quantum wells [Wakita et al. 1990, 1992] and MOCVD-grown InGaAsP quantum wells [Devaux et al. 1992, Razeghi et al. 1986, 1989]. Several groups are now concentrating their efforts towards the monolithic integration of such modulators with DFB lasers [Wakita et al. 1992, Kato et al. 1992, Aoki et al. 1992].

Although large absorption variations can be obtained using the quantum-confined Stark effect, the required electric fields are large because of its quadratic relationship [Bastard et al. 1983]. Large electric fields usually imply large drive voltages which is a major drawback for any application. The need for smaller drive voltages has been one of the stimuli for research in new electroabsorption phenomena [Rao et al. 1987].

7.1.3. Superlattice modulators based on Wannier–Stark localization

(a) Wannier–Stark localization [Bleuse et al. 1988a, Mendez et al. 1988]

When quantum wells are separated by very thin barriers in a periodic structure called a superlattice, wells are coupled to each other through the resonant tunneling process. Due to this resonant coupling, carriers are delocalized and the original QW discrete energy levels broaden into minibands of widths A_c and Δ_v [Esaki and Tsu 1970]. This carrier derealization and miniband formation is quite similar to the band formation in a crystal lattice, and the superlattice exhibits, to some extent, a three-dimensional behavior. Because of the miniband broadening, the bandgap of the superlattice E_g^{SL} is smaller than that of an isolated quantum well E_g^{QW} by about $\frac{1}{2}(\Delta_c + \Delta_v)$. The absorption edge of a superlattice reflects the miniband density of states that is given by an arccosine function, in the tight-binding approach [Bastard 1988]. This broadened absorption edge is represented schematically in Fig. 7.7 (solid line) in the case of an ideal superlattice without any source of broadening and without taking any excitonic contribution into account.

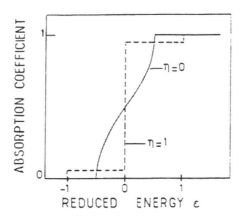

Fig. 7.7. Absorption spectra of an ideal superlattice without applied electric field (solid line) and in the high-field limit: $eFd/(\Delta_c + \Delta_v) = 1$ (dashed line).

The application of an electric field parallel to the growth axis of the SL shows interesting behavior [Bleuse et al. 1988a]. As schematized in Fig. 7.8, with the application of an electric field F, the genuine energy levels in adjacent quantum wells become misaligned by eFd, where d is the SL period and e the electron charge. Due to this potential drop, the tunneling probability decreases drastically and the resonant tunneling process is turned off. Carriers localize into the wells and the superlattice tends to recover its original isolated quantum-well behavior. This phenomenon is called Wannier–Stark localization. The scaling factor is the ratio of the energy misalignment to the total miniband width.

The SL broadened absorption spectrum tends to recover its QW step-like absorption spectrum. This should result in a blue-shift of the absorption edge of the order of $\frac{1}{2}(\Delta_c + \Delta_v)$. However, a blue-shift of the absorption edge is only effective because carrier localization is incomplete as the wavefunctions exhibit additional lobes in adjacent quantum wells. As a consequence the main step at E_g^{QW} is accompanied by smaller ones at

Eq. (7.1) $E_g^{QW} + peFd \quad (p = \pm 1, \pm 2, \ldots)$

corresponding to "oblique" transitions in real space, connecting an electron localized near the nth quantum well with a hole in the $(n + p)$th quantum well. In the "high-field" limit, $eFd/(\Delta_c + \Delta_v) = 1$, the SL behaves more as a series of uncoupled quantum wells with the wavefunction localized over each quantum well and only the "oblique" transitions connecting adjacent wells ($p = \pm 1$) are significant. The corresponding absorption spectra for an ideal SL at zero field and in the high-field limit are shown in Fig. 7.7.

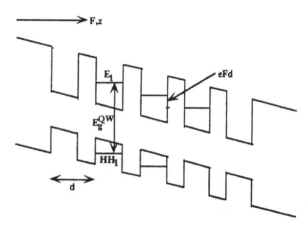

Fig. 7.8. Band structure in real space of a superlattice when an electric field F is applied in the z-direction. E_1 and HH_1 are the ground conduction and valence subbands.

These theoretical predictions have been confirmed experimentally. However, excitonic features play an important role in determining the actual electroabsorption spectra. Since the structure evolves from a three-dimensional behavior to a two-dimensional behavior, an enhancement of the excitonic oscillator strength is expected. This is observed experimentally and is in contrast with the usual electric-field-induced exciton ionization in bulk materials or multiple quantum wells. Fig. 7.9 shows experimental electroabsorption spectra for a InGaAs–InAlAs superlattice and it can be seen that the excitonic enhancement as well as the low-energy oblique transition tend to hide the effective blue-shift of the absorption edge. Nevertheless, large absorption variations can be obtained both below and above the superlattice bandgap using the effective blue-shift of the absorption edge and the low-energy oblique transitions, respectively.

SLs are unique systems where the conduction and valence miniband widths can be tailored with a judicious choice of the host materials and layer thicknesses, in the range of a few tens of meV while the corresponding periods fall in the range of a few tens of Å. This means that the high-field limit can be attained with moderate electric fields in the range of a few tens of kV·cm^{-1}. These electric fields are significantly smaller than those required for observation of the quantum-confined Stark effect.

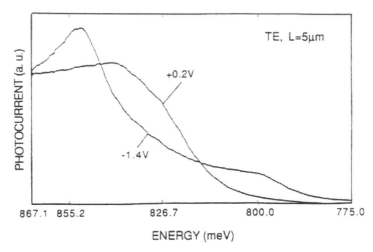

Fig. 7.9. Electroabsorption spectra of a InGaAs–InAlAs superlattice (after Bigan et al. [1990]).

(b) Device application [Bleuse et al. 1988b]

Most of the various modulators first introduced using the quantum-confined Stark effect have been reproduced using Wannier–Stark localization, taking advantage of both the effective blue-shift of the absorption edge and the low electric fields required. Vertical cavity devices such as Fabry–Pérot reflection modulators and SEEDs [Bar-Joseph et al. 1989] have been implemented using GaAs–GaAlAs superlattices. SEEDs based on Wannier–Stark localization strongly benefit from the large negative differential photoconductivity occurring in the blue-shifted spectral region. Such negative differential photoconductivity is necessary in order to achieve bistability in a SEED device.

There have been fewer reports of waveguided modulators based on Wannier–Stark localization. These devices operate in the normally transparent spectral region and use the low-energy oblique transitions to circumvent the problems of large on-state attenuation encountered in the blue-shifted spectral region [Bigan et al. 1990]. They benefit from the low electric fields required and subvolt drive voltage intensity modulators operating in the 1.5 μm wavelength range have been demonstrated and optimized [Bigan et al. 1992a]. The advantage of superlattice waveguide modulators has been demonstrated using a figure of merit that allows comparison of different electroabsorptive waveguide modulators. As this subject closely follows the work on multiple quantum wells, work towards monolithic integration of these modulators with DFB lasers is under progress [Bigan et al. 1992b].

7.1.4. Perspectives

One of the main problems encountered in absorptive devices is saturation at high optical powers. These saturation effects originate from electric-field screening through carrier accumulation [Wood et al. 1990]. These problems stimulate the use of refractive index changes associated with electroabsorption effects; here only virtual transitions are involved and there is no carrier generation. As a result, saturation effects are expected to be eliminated. This stimulates the development of intensity modulators based on Mach–Zehnder interferometers [Zucker et al. 1989]. Such interferometers operating in the 1.5 μm wavelength range have thus been realized.

New structures based on carrier transfer from a delta-doped layer into a quantum well located nearby have also been proposed to increase refractive index variations. These structures named BRAQWETs (barrier reservoir and asymmetric quantum well electron transfer), might prove to be interesting alternatives to conventional modulators [Zucker et al. 1990, 1991] although their integration with DFB lasers might be more difficult.

7.2. GaAs-based infrared photodetectors grown by MOCVD

7.2.1. Basic physics of photodetectors

Photodetectors are semiconductor devices that can convert optical signals into electrical signals. A general photodetector has basically three processes: (i) carrier generation by incident light, (ii) carrier transport and/or multiplication by whatever current-gain mechanism may be present and (iii) interaction of current with the external circuit to provide the output signal. Also, photodetectors must satisfy stringent requirements such as high sensitivity at operating wavelengths, high response speed and minimum noise.

The performance of a photodetector in general and a photoconductor in particular is measured in terms of three parameters: the quantum efficiency or gain, the response time and the sensitivity (detectivity).

(a) Quantum efficiency (η) and gain

The quantum efficiency η is the number of electron–hole pairs generated for each incident photon. For a photoconductor with an illuminated area $A = W \cdot L$ and thickness T, the photocurrent flowing between the electrodes is

Eq. (7.2)
$$I_p = (\Delta \sigma E)WT = (q\mu_n \Delta nE)WT = (q\Delta n \upsilon_d)WT = q(\eta P_{out} / h\nu)(\mu_n \tau E / L)$$

where n is the number of carriers per unit volume (carrier density), Δn is the excess carrier density, E is the electric field inside the photoconductor, μ_n is the mobility, υ_d is the drift velocity, τ is the carrier lifetime, P_{out} is the incident optical power and $h\nu$ is the photon energy. If the primary photocurrent is defined as

Eq. (7.3) $I_{ph} = q(\eta P_{out} / h\nu)$

then the photocurrent gain can be expressed as

Eq. (7.4) $\text{gain} = I_p / I_{ph} = \mu_n \tau E / L = \tau / t_r$

where $t_r = L/v_d$ is the carrier transit time. The gain which depends on the ratio of carrier lifetime to the transit time is a critical parameter in photoconductors.

(b) Figures of merit

The infrared system is often designed around the characteristics of the infrared detector. They are defined by certain figures of merit describing the performance of the detector under specified operating conditions.

Many infrared detectors exhibit a signal which is a function of the wavelength of the radiation, being sensitive only to radiation in a given spectral interval and exhibiting a response which is dependent on wavelength within that interval. Also, like all other types of electromagnetic radiation detector, the ultimate performance of infrared detectors is limited by noise. Because most detectors are based upon photoeffects in semiconductors, the mechanisms encountered most frequently are those which pertain to semiconductors: current noise, thermal noise, generation–recombination noise and photon noise. The measured noise voltages for all noise mechanisms is proportional to the square root of the electrical bandwidth, and for some mechanisms is a function of frequency.

The most important figures of merit which relate to the problem of characterizing the signal-to-noise ratio to be expected from a detector for a given intensity of radiant power can be described as follows.

(i) One of the most commonly used figures of merit relating the radiation power capable of producing a signal voltage equal to the noise voltage, that is, a signal-to-noise ratio of unity, is the noise-equivalent power, NEP. The NEP is defined as the RMS value of the sinusoidally modulated radiant power falling upon a detector which will give rise to an RMS signal voltage equal to the RMS noise voltage from the detector. The NEP is determined experimentally by measuring the signal-to-noise ratio in a specified narrow electrical bandwidth, called the measurement bandwidth. The value obtained for a known amount of radiant power is extrapolated linearly to the power required to give a signal-to-noise ratio of unity. Thus the noise-equivalent power is given by

Eq. (7.5) $\text{NEP} = HA_D \left(V_n / V_s \right) \left[1 / \left(\Delta f \right)^{1/2} \right]$

where H is the RMS value of the irradiance falling on the detector of area A_D, and V_n/V_s is the ratio of the RMS noise voltage in the bandwidth Δf to the RMS signal voltage. Here it is assumed that Δf is sufficiently small so that the noise voltage per cycle within Δf is independent of frequency. The units of NEP are $W \cdot Hz^{-1/2}$. The noise-equivalent power for a 500 K black-body source, 900 Hz chopping frequency and 1 Hz bandwidth is written as NEP(500 K, 900, 1). Also, the detecting capability of the detector improves as the NEP decreases.

(ii) Another common figure of merit is the noise-equivalent incidence NEI. This is the radiant power per unit area of detector required to give rise to a signal-to-noise ratio of unity. The conditions of measurement are the same as those for the NEP. Thus the NEI is simply the NEP divided by the detector area. The NEI can be expressed as follows

Eq. (7.6) $NEI = NEP/A_D$

The units of NEI are $W \cdot Hz^{-1/2} \cdot cm^{-2}$.

(iii) Most detectors exhibit a noise-equivalent power which is directly proportional to the square root of the area of the detector. An area-independent figure of merit can be obtained by dividing the NEP by the square root of the area. The reciprocal of this quantity is known as D^*. Thus D^* is defined as

Eq. (7.7) $$D^* = \frac{A_D^{1/2}}{NEP} = \frac{1}{A_D^{1/2}(NEI)}$$

The units of D^* are $cm \cdot Hz^{1/2} \cdot W^{-1}$. The reference bandwidth is always 1 Hz and the measured value of D^* is written in a manner analogous to that of the NEP, such as $D^*(500 \text{ K}, 900, 1)$. Therefore, D^* is a widely used figure of merit to be referred to as detectivity. Wherever the term detectivity occurs herein, it refers to D^*.

The figures of merit with which we have been concerned refer to the response of the detector to radiation from a black-body source of a specified temperature. Since the responses of many types of detector are dependent upon the wavelength of the incident radiation, the values of NEP, NEI and D^* can be expressed in terms of their responses to monochromatic radiation. In this case, the wavelength λ is specified and the figures of merit are written as NEP_λ, NEI_λ and D_λ^*. If the performance at only a single wavelength is given, it is usually that at which the detector performs best.

The other thing we have to consider is the problem of characterizing the signal voltage per unit radiant power. Because we are only interested in the amplitude of the signal, the noise can be neglected. However, knowledge is still required of the black-body temperature, the temperature of the detector and the area of the detector. Knowing these, the responsivity R can be defined to be the RMS signal voltage V_s per unit RMS radiant power P incident upon the detector. Thus

Eq. (7.8) $R = V_s / P = V_s / HA_D$

Also, we can see that the responsivity is related to the NEP and D^* by

Eq. (7.9) $R = V_n / NEP(\Delta f)^{1/2} = D^* V_n / (A_D \Delta f)^{1/2}$

We also specify a spectral responsivity R_λ, the responsivity to monochromatic radiation of wavelength λ.

7.2.2. *II–VI material-based photodetectors*

The most important II–VI semiconductors used for thermal imaging are mercury cadmium telluride ($Hg_{1-x}Cd_xTe$) bulk material and HgTe–CdTe superlattices.

(a) $Hg_{1-x}Cd_x$ *Te-based photodetectors*

In $Hg_{1-x}Cd_x$ Te, the electronic structure is dramatically changed by varying the Hg composition x while the crystalline structure is only slightly perturbed (Fig. 7.10).

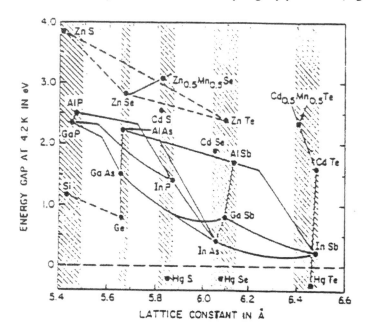

Fig. 7.10. *Bandgap energy of the main semiconductors and compounds as a function of lattice parameter (after Esaki [1986]).*

For $x = 0$, the compound is a semimetal and for $x = 1$, it is a wide-gap (1.5 eV) semiconductor. It is therefore possible to tune the bandgap of the semiconductor to a desired photon energy by adapting the Hg content. The optical detection is obtained by interband transitions (from valence to conduction band). Absorption coefficients of 10^4 cm^{-1} are obtained in the 8–12 μm window which leads to quantum yields higher than 90% in the *p–n* junctions. However, this technology has a certain number of disadvantages:

- Small uncertainties in the Hg composition lead to large variations in bandgap energy. For example, a relative non-uniformity of 3% in Hg content yields a variation of cutoff wavelength from 12 μm to 14 μm. Very good uniformity is extremely difficult to obtain in the growth of HgCdTe compounds.

- The Hg diffusion coefficient is so high that the detectors must not be exposed even to moderate temperatures.
- Large numbers of Shockley–Read–Hall trap centers throughout the bandgap produce substantial and detrimental tunneling currents.
- For large arrays, high yields are difficult to obtain.

Furthermore, high dislocation density, lack of low-cost and high-quality lattice-matched substrate, fabrication difficulties, $1/f$ noise and sensitivity to elevated temperatures are also problems associated with this technology.

Some of these difficulties have been overcome by the use of MOCVD to grow HgCdTe layers. This technique has been used for the fabrication of high-quality HgCdTe photoconductors [Specht et al. 1986, Bethea et al. 1988, Doll et al. 1990, Druilhe et al. 1990], and photovoltaic detectors [Bubulac et al. 1990, Irvine 1992, Summers et al. 1994], operated in the near-, middle- and long-wavelength ranges of the infrared spectrum. Direct alloy growth or the interdiffused multilayer process, in which thin layers of HgTe and CdTe are alternatively grown and subsequently interdiffused during a post-grown annealing are used by various groups. Near room temperature double-heterojunction long-wavelength devices have also been reported by Elliott et al. [1994]. MOCVD seems to be the most promising for future large-scale and low-cost production of HgCdTe epilayers. The flexibility of MOCVD largely enables us to integrate the growth and device processing in one run. Arsenic is the preferred dopant for *p*-type layers, while indium and iodine are preferable for *n*-type layers. Close to 100% doping efficiency with control of dopants over a wide range of concentrations was obtained. The devices are usually passivated with CdTe or ZnS. Lattice and thermal mismatch, spurious growth and impurity doping from the substrates are the most serious remaining problems which would require considerable development to meet the stringent requirements of high-performance devices.

Despite the direct fabrication of sensitive elements, MOCVD is also frequently used for preparation of the hybrid substrates GaAs- and CdZnTe-coated silicon or sapphire, which are then used for deposition of HgCdTe by other methods. High-quality focal plane arrays (FPAs) have been fabricated with this approach with a number of elements exceeding 640×480 [Tung et al. 1992, Johnson et al. 1993].

(b) HgTe–CdTe-superlattice-based photodetectors

The HgTe–CdTe superlattice was originally proposed as an infrared (IR) material by Schulman and McGill [1979]. Smith et al. [1983] proposed that this superlattice should exhibit properties superior to those of the HgCdTe alloy as an infrared detector material. Their calculations showed that the superlattice tunneling length is shorter than that of the alloy with the same gap. For a given cutoff wavelength tolerance, it has been found that less fractional precision is needed in the superlattice control parameter (layer thickness) than in the alloy control parameter (composition). They have also stated that *p*-side diffusion currents should be reduced due to the larger superlattice electron effective mass. Later, Arch et al. [1986] proposed that interdiffusion of the layers of HgTe–CdTe superlattices at temperatures near the growth temperature ($< 200\,°C$) would be large due to the large equilibrium vacancy concentration normally present in the HgCdTe system. They have investigated the extent of intermixing of the HgTe and CdTe layers by making high-temperature x-ray diffraction measurements on HgTe–CdTe superlattices. The interdiffusion coefficients they obtain (4.23×10^{-18} cm^2·s^{-1} at 185 °C) show that the interface intermixes rapidly at this temperature, and abrupt interfaces are difficult to achieve with this system. They have stated that the interface would intermix about 12 Å for a 1 h growth run and they have concluded that an unstable HgTe–CdTe interface might pose serious problems in using this material for optoelectronic devices.

7.2.3. InAs$_{1-x}$Sb$_x$ materials

(a) InAs$_{1-x}$Sb$_x$-based photodetectors
InAs$_{1-x}$Sb$_x$ has received increasing attention over the past few years for applications as infrared sources and detectors. The theoretical results indicate that the bandgaps of these InAsSb strained-layer superlattices (SLS) vary relatively slowly with composition. Thus, InAsSb SLS detectors can possess a broad spectral response. As a result, small, lateral compositional variations across the superlattice wafer will not result in larger variations in wavelength response. Furthermore, it has recently been proposed that, in low-bandgap cases, band-to-band tunneling in superlattice *p–n* junctions may be reduced compared to such tunneling in bulk materials [Smith et al. 1983]. Junctions in InAsSb strained-layer superlattices would also benefit from this effect. All these results indicate that InAsSb strained-layer superlattice materials offer potential advantages for 8–12 µm applications and are a better alternative to HgCdTe-based detectors.

Another important advantage of InAsSb-based detectors is the monolithic integration of devices. Silicon is currently the best semiconductor adapted to logical circuitry. InAsSb can be epitaxially grown on Si substrate and therefore high-speed optoelectronic integrated circuits are possible. Hence, InAsSb is a natural choice of material for infrared imaging, both for 3–5 µm and the 8–12 µm applications. Kurtz et al. demonstrated a photodiode consisting of a *p–n* junction embedded in an InAs$_{0.09}$Sb$_{0.91}$–InSb strained-layer superlattice. The non-optimized device exhibited a photoresponse out to a wavelength of 8.7 µm at 77 K [Kurtz et al. 1988]. Later, Kurtz et al. fabricated long-wavelength infrared photodiodes using InAs$_{0.15}$Sb$_{0.85}$–InSb strained-layer superlattices. The detectors displayed broad spectral responses with a detectivity $\geq 10^{10}$ cm·Hz$^{1/2}$·W^{-1} at wavelengths of ≤ 10 µm [Kurtz et al. 1990].

However, for InAsSb materials, the valence-to-conduction-band transition is the most efficient way to detect light. We can obtain small energy gaps by extending the energy gap of small-gap semiconductors towards smaller values using strained-layer structures such as InSb/InAsSb or InAs/GaSb. However, in these structures, type-II superlattices are obtained. In type-II superlattices optical transitions between the valence and conduction bands are indirect in space and this results in less absorption and a small quantum yield. Other disadvantages of this structure are: firstly, InAsSb materials can only be lattice matched to some suitable substrate, secondly, due to strain, a high density of Shockley–Read generation centers is created and this leads to a high leakage current density.

(b) Strained-layer superlattices (SLSs) and multiple quantum wells

An attractive way to tailor semiconductor bandgaps by alloying is the use of multiquantum wells (MQW) and superlattices (SL), which are usually periodic structures made from two compounds with different bandgaps. The use of multiquantum wells also opens up the possibilities of using quantum size effects and special exciton-related properties in these materials for switching and coupling applications. The exciton binding energies are greatly enhanced, giving rise to a large non-linearity in the refractive index and absorption [Frank and van der Merwe 1949]. The use of strained-layer superlattices (SLS) and strained MQW eliminates the need for lattice matching and, in principle, high-quality defect-free samples can be grown by MBE or MOCVD [Matthews 1975]. The dominant effect of strain is to lift certain degeneracies in the Brillouin zone and alter the density of states at band edges and to affect bandgaps. These effects can have a significant influence on transport and optical properties of devices.

The strain-induced band structure changes may lead to increased charged-carrier mobility within the pseudomorphic layers. Pseudomorphic layers are layers that have a slight lattice mismatch between them so that it is possible to accommodate this lattice mismatch through elastic strain rather than dislocation formation. Strain induces modifications of the energy band structure which result in significant modifications of the bulk optical and electronic properties of pseudomorphic layers.

The strain-induced bandgap shift is given by

Eq. (7.10) $\Delta E = -\alpha \varepsilon$

where α is the constant including all elastic hydrostatic and shear deformation potentials of the material and ε is the strain parameter. Its value is negative for compressive strain and positive for tensile strain. As a result, ΔE is positive for negative strains, which results in a net bandgap increase with biaxial compression (Fig. 7.11) [Kato et al. 1986].

Fig. 7.11 shows the bandgap shifts due to strain. These bandgap shifts (ΔE) are experimentally determined by directly comparing the photoluminescence peak energies of strained layers with those of unstrained bulk-like films.

From photoluminescence experiments it has been established that quantum wells of different thicknesses give rise to peaks at different wavelengths. Each wavelength corresponds to a certain bandgap energy which is different from the bandgap of the bulk layer by an amount ΔE. If we reduce the thickness of the quantum well, the strain associated with it becomes higher and ΔE becomes larger. This is a useful mechanism to tailor the wavelength of absorbed light without a change in the material composition.

Bandgaps of strained-layer superlattices vary relatively slowly with composition, so that small compositional variations across the wafer will not result in large variations in wavelength response.

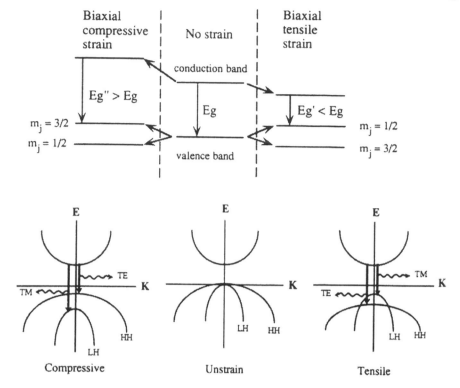

Fig. 7.11. Changes in the energy diagram due to strain.

7.2.4. Intersubband GaAs quantum well photodetectors

Another important technology of thermal infrared detection is the intersubband GaAs quantum well (QW) detectors. They have a number of potential advantages with respect to HgCdTe and InAsSb devices. For example, growth, processing, and passivation technologies of GaAs are more mature than those of HgCdTe. With monolithic integration of these detectors with GaAs field-effect transistors (FETs), charge-coupled devices and high-speed signal processing electronics are possible. In addition, GaAs substrates are larger, cheaper and of higher quality than CdTe substrates. Also, GaAs is more thermally stable than HgCdTe. Furthermore, compared to InAsSb-based detectors, GaAs QW devices can achieve higher quantum yields and reduce the tunneling dark current. Finally, for GaAs QW technology, we also can vary the peak absorption wavelength over the atmospheric window region (8–14 μm) by varying the GaAs well dimensions and composition, and the thickness of the wider-gap materials.

In some semiconductor systems (such as GaAs–GaInP, GaAs–AlGaAs, GaInAs–AlGaAs), band offsets are such that they induce potential wells for electrons and/or holes in the different bands. Progress in growth technology such as molecular beam epitaxy and metalorganic chemical vapor deposition allow us to obtain extremely thin films of semiconductors with a precision of one atomic layer. These wells are so thin that the electron and hole motion is quantized in the direction normal to the layers. The carrier energy is then distributed in different subbands, the bottoms of which are the one-dimensional quantized states E_i in the wells. Light can then induce transitions between these subbands. The resulting absorption is resonant between those quantized states (or bound states), with values as high as several 10^3 cm^{-1} [West and Eglash 1985]. In the GaAs–AlGaAs system, perfect lattice matching allows us to vary independently the thickness of the GaAs well and the potential height of the AlGaAs barrier in order to tune the transition towards the desired wavelength.

(a) Photoconductive detectors

Levine et al. at Bell Laboratories have constructed infrared detectors using intersubband transitions. In the first version, optical transitions occurred between the bound states in the well (E_1 and E_2) [Levine et al. 1987]. The structure is biased so that the electron on E_2 has a much higher probability to tunnel out of the well than on E_1 (Fig. 7.12(a)). However, because of the high electric field necessary to obtain this tunneling mechanism, the dark current is very important in these photoconductors. In a more recent version, Levine and his coworkers have optimized their structure so that the level E_2 is in the conduction band continuum of the AlGaAs barrier (extended state) and the transition is thus bound to extended [Levine et al. 1988] (Fig. 7.12(b)). The dark current has sharply decreased. They have demonstrated a high-detectivity (1×10^{10} cm·Hz$^{1/2}$·W^{-1} at 77 K) GaAs/GaAlAs quantum well infrared photodetector with a sensitivity peak at 8 μm. Later, by lowering the quantum well barriers, they extended the responsivity peak out to 10 μm with a sensitivity covering the spectral region from 8–19 μm [Levine et al. 1989, 1992, Levine 1993]. The detectivity for $\lambda = 19\ \mu$m increased from 10^9 cm·Hz$^{1/2}$·W^{-1} at 50 K up to 2×10^{12} cm·Hz$^{1/2}$·W^{-1} at 20 K.

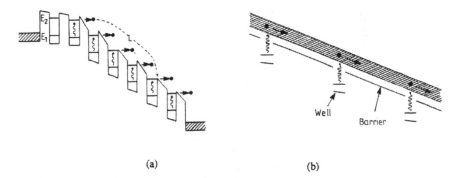

(a)   (b)

Fig. 7.12. Energy band diagrams of quantum well photodetectors. Bound-to-bound transitions (a) and bound-to-extended transitions (b).

Levine et al. have also showed that thermionic-assisted tunneling is a major source of dark current [Levine et al. 1990] and, by increasing the quantum well barrier width from 300 Å to 500 Å, they have reduced the tunneling dark current by an order of magnitude and increased the black-body detectivity. For a GaAs quantum well infrared detector having a cutoff wavelength of 10.7 μm, they have achieved a black-body detectivity of 1×10^{10} cm·Hz$^{1/2}$·W^{-1} at 68 K. Hasnain et al. demonstrated a GaAs/AlGaAs quantum well infrared photodetector with an optical gain of 8.1, showing that bound-to-continuum-state intersubband transitions can have a photoconductive gain much greater than unity similar to extrinsic photoconductors [Hasnain et al. 1990]. Hasnain's work also shows that the photocarrier lifetime is not limited by the transit time (i.e. photocurrent gain is not limited to $\frac{1}{2}$). Thus, in general, carrier lifetimes or optical coupling can be increased with proper device optimization and design. In other words, quantum well infrared photodetectors (QWIPs) simultaneously having large quantum efficiencies and large gains can be expected and therefore lead to significantly improved detectivity.

In spite of achievements in GaAs/AlGaAs QWIPs technology, direct bandgap intrinsic semiconductors, as typified by HgCdTe, exhibit superior performance to the GaAs/AlGaAs superlattice at the same cutoff wavelength and temperature. Kinch and Yariv [1989] have presented an investigation of the fundamental physical limitations of individual GaAs/AlGaAs multiple quantum well IR detectors as compared to ideal HgCdTe photoconductors with cutoff wavelengths $\lambda_c = 8.3$ and 10 μm. It appears that in the range 40 to 100 K the thermal generation rate is approximately five orders of magnitude smaller than the corresponding GaAs/AlGaAs superlattice (see Fig. 7.13). The dominant factor favoring HgCdTe in this comparison is the excess carrier lifetime, which for n-type HgCdTe is greater than 10^{-6} s at 80 K, compared to about 10^{-11} s for the GaAs–AlGaAs superlattice. In the superlattice the confined carriers are free to move within the plane (there is no energy gap separating confined from unconfined states), so the net carrier recombination rate is very high.

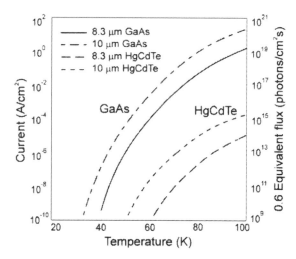

Fig. 7.13. Thermal generation current versus temperature for GaAs/AlGaAs multiple quantum wells and HgCdTe alloys at $\lambda_c = 8.3$ μm and 10 μm. The assumed effective quantum efficiencies are $\eta = 12.5\%$ and 70% for AlGaAs/GaAs and HgCdTe, respectively (after Kinch and Yariv [1989]).

(b) Focal plane arrays

It is well known that the detectivity D^* is not the relevant figure of merit for large arrays. One measure of performance of a thermal image system is the noise-equivalent temperature difference (NEDT), the temperature change of a scene required to produce a signal equal to the root mean square (RMS) noise. An array having $D^* = 10^{10}$ cm·Hz$^{1/2}$·W^{-1} would achieve a very sensitive NEΔT = 0.01 K. This would produce an excellent image, comparable to that of present arrays which are uniformity limited. Despite large research and development efforts, large photovoltaic HgCdTe focal plane arrays (FPAs) remain expensive, primarily because of the low yield of operable arrays. The low yield is due to the high incidence of electrically active defects and to the sensitivity of long-wavelength infrared (LWIR) HgCdTe devices to defects and surface leakage, which is a consequence of basic material properties. With respect to HgCdTe detectors, GaAs/AlGaAs quantum well devices have a number of potential advantages, including the use of standard manufacturing techniques based on mature GaAs growth and processing technologies (monolithic integration of these detectors with GaAs field effect transistors, CCDs and high-speed signal processing electronics is possible), highly uniform and well-controlled MBE and MOCVD growth on greater than 4 inch GaAs wafers, high yield and thus low cost, more thermal stability, and intrinsic radiation hardness.

At present, large 128×128 and 256×256 focal plane arrays (FPAs) with long-wavelength infrared imaging performance comparable to state of the art HgCdTe have been fabricated. Levine et al. [1991] have presented thermal imaging data of hybrid 128×128 GaAs/AlGaAs FPAs consisting of 50×50 μm^2 photoconductors having peak response at $\lambda_p = 9$ μm. To improve the optical coupling, different types of grating are used [Levine 1993]. The 99% yield of this array technology is a result of the excellent MBE growth uniformity (1%) in thickness and the mature processing technology. After correction, the measured non-uniformity of the array was better than 0.1%, and a NEΔT of 0.01 K was observed at 60 K. Recently, Kozlowski et al. [1994] have reported a background limited 128×128 GaAs/AlGaAs QWIP FPA with $\lambda_p = 8.8$ μm and > 99.4% pixel operability at 35–40 K operating temperature. Typical NEATs of ≈0.008 K and detectivity >4.0 × 10^{10} cm·Hz$^{1/2}$·W^{-1} were measured at conventional imaging backgrounds at operating temperatures up to 65 K. For comparison, at 77 K, the operability of HgCdTe photodiodes is ≥95%, but at 40 K the sport population limits the operability from 70 to 80%. The uniformity of QWIPs is higher and this characteristic is superior to HgCdTe photodiode arrays operated at 40 K.

GaAs QWIPs combine the advantages of PtSi Schottky barrier arrays (high uniformity, high yield, radiation hardness, large arrays with monolithically integrated electronics, low cost) with the advantages of HgCdTe (high quantum efficiency and long-wavelength response).

(c) Advantages of intersubband GaAs quantum well detectors

(i)   *Wavelength range*: The intersubband spacing and thus the cutoff wavelength of the detectors can be tuned. Moreover, multicolor detectors may be realized in the same structure by growing multiquantum wells with different growth parameters.

(ii) *Spectral bandwidth*: The spectral bandwidth may also be tuned by choosing a bound-to-bound or a bound-to-extended transition. The spectral width of the bound-to-bound absorption peak is determined by the electron dephasing time in the subbands. This bandwidth is extended to larger values by choosing a bound-to-extended transition.

(iii) *Temporal bandwidth*: The intrinsic transient behavior of the intersubband QW detectors is determined by the electron lifetime in the excited state (E_2 for bound to bound, the conduction band of the wider-gap material for bound to extended). This lifetime, though not accurately known, is considerably faster than in HgCdTe detectors. The spectral bandwidth is thus controlled by the reading electronics and a time response smaller than 5 ns has been measured in a QW photoconductor.

(iv) *Integration capabilities*: The industrial skills developed in GaAs integrated circuits could allow the complete integration of detectors with the reading electronics. This would also enhance the reliability of the arrays. Finally, localized GaAs epitaxy on patterned silicon is developing very fast. GaAs materials on Si could allow the use of Si circuitry for reading electronics.

(d) GaAs/GaInP p-type quantum well infrared photodetector

The presence of aluminum in the GaAlAs barrier layers suggests reliability and fabrication limitations due to oxidation. Therefore, in an effort to further the development of QWIPs, researchers have turned to other material systems such as GaAs–InGaAs, GaInAs–InP and GaAs–GaInP [Levine 1993]. Compared to AlGaAs, GaInP presents several advantages which are expected to lead ultimately to better device performance [Omnes and Razeghi 1991]. Dislocation and impurity motion is much lower than in AlGaAs because of the large difference in atomic size between indium and gallium. GaAs–GaInP should ultimately lead to better performance because of the better transport properties in direct bandgap semiconductors. Finally, surface and recombination velocities are lower than for AlGaAs, and selective etchants are available.

The first n-type GaAs/GaInP QWIPs have been fabricated by Gunapala et al. [1990]. They have determined the conduction band and valence band offsets to be 221 meV and 262 meV respectively by comparing the theoretical absorption spectrum and the measured responsivity spectrum. An absorption coefficient of 1900 cm^{-1} and quantum efficiency of 6.1% at 8 μm have been obtained. More recent work on n-type GaAs/GaInP QWIPs by Jelen et al. [1997] have been able to achive cutoff wavelenghts as long as 10 μm.

However, for n-type QWIPs, intersubband transitions are limited by dipole selection rules to couple only to radiation with a polarization component perpendicular to the quantum wells (i.e. normal incidence is forbidden). Since the pioneering work of Chang and James [1989], it has been known that, unlike *n*-type QWIPs, *p*-type QWIPs allow absorption of infrared radiation at normal incidence. Special attention has to be paid to the spin-orbit mixing of the valence bands in addition to the QW induced admixtures as shown by Hoff et al. [1996]. The spin orbit mixing calculated in $k \cdot p$ theory can be used to explain the normal incidence absorption. [Hoff et al. 1996]

Hoff et al. [1995] have fabricated *p*-type QWIPs from the GaAs–GaInP material system. The structures consist of 50 GaAs quantum wells doped with zinc with nominal thicknesses of 55, 32 and 22 Å, respectively. The barrier is an undoped lattice-matched $Ga_{0.51}In_{0.49}P$ nominally 280 Å thick. This MQW structure is cladded between zinc-doped GaAs contact layers with thicknesses of 1 μm (bottom) and 0.5 μm (top). All the GaAs layers were doped with zinc to a net acceptor concentration of 3.0×10^{18} cm^{-3}.

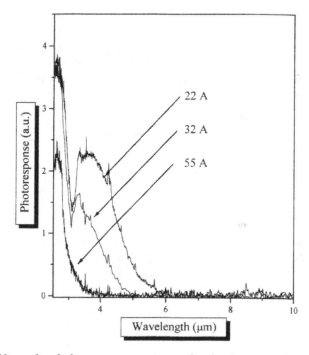

Fig. 7.14. Normalized photoresponse spectra for the three samples, measured under normal incidence illumination at 77 K (after Hoff et al. [1995]).

400 μm × 400 μm mesa detectors were fabricated using photolithography and wet chemical etching, and 100 μm × 100 μm square Au/AuZn electrodes were evaporated and alloyed. Fig. 7.14 shows the normalized photoconductive response spectra for the three samples measured using a Fourier transform infrared spectrometer. The spectral response shifts toward longer wavelength as the well width is reduced, which is expected as the hole ground state is pushed up towards the top of the barrier. At 77 K, the current responsivity of the 22 Å wide well QWIP

at 2.5 µm wavelength was 0.5 mA·W^{-1} at 5 V bias and increased linearly to 2 mA·W^{-1} at 20 V bias. The current responsivity spectral shape and magnitude were found to be independent of temperature from 77 K up to 200 K. The photoconductive gain was rather small, about 0.18, which combined with the current responsivity of 0.5 mA·W^{-1} at 2.5 µm yields an absorption quantum efficient of 1.4×10^{-3}. The estimated detectivities for this preliminary device at 2.5 µm were 9×10^6 cm·Hz$^{1/2}$·W^{-1} and 8×10^8 cm·Hz$^{1/2}$·W^{-1} at 200 K and 77 K, respectively.

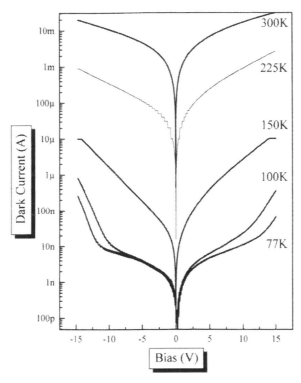

Fig. 7.15. Dark current versus bias for different temperatures between 77 K and 300 K for the 22 Å wide well sample (after Hoff et al. [1995]).

The dark current was measured versus bias voltage at temperatures between 77 K and 300 K. Fig. 7.15 shows a set of measurements for the 22 Å wide well sample. At low bias, tunneling is negligible and the dark current originates from thermionic emission above the top of the barrier. Arrhenius plots of the dark current measured at a fixed low bias of 0.1 V are shown in Fig. 7.16 for each of the three samples, which reveal respective activation energies of 235 meV, 300 meV, and 330 meV for the 22 Å, 32 Å and 55 Å wide well QWIPs. These activation energies correspond to wavelengths of 5.28 µm, 4.13 µm, and 3.75 µm, respectively. These values closely correspond to the observed cutoff wavelengths in Fig. 7.14, which is expected because the Fermi level lies within a few meV of the heavy hole ground state.

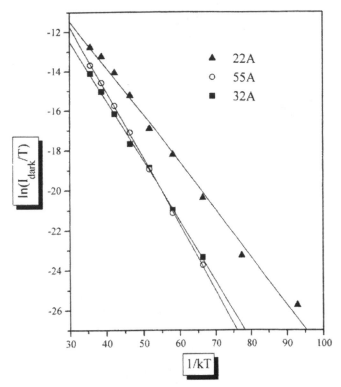

Fig. 7.16. Arrhenius plot of the normalized dark current versus inverse thermal energy for the three samples. The dark current values have been taken at a fixed bias value of 0.09 V (after Hoff et al. [1995]).

Hoff et al. [1995] have also performed theoretical calculations of the cutoff wavelength of GaAs/Ga$_{0.51}$ In$_{0.49}$ P and GaAs/Ga$_x$In$_{1-x}$P$_{1-y}$ QWIPs using a modified Kronig–Penney model and taking into account the effective mass discontinuity and the well/barrier interface, and reported parameters for GaAs and Ga$_{0.51}$In$_{0.49}$P (see Omnes and Razeghi [1991]). The cutoff wavelength corresponds to the energy difference between the top of the barrier and the ground-state heavy-hole energy level. Fig. 7.17 shows measured and calculated cutoff wavelengths as a function of well width for a 280 Å barrier. Theoretical predictions closely match experimental observations, and further show that extremely narrow GaAs quantum wells must be used in order to reach the 8–12 μm atmospheric window. Such narrow wells can be difficult to manufacture with high yield. It can be seen that GaInAsP alloys should be used in order to reduce the valence band discontinuity, thereby permitting operation at longer wavelength.

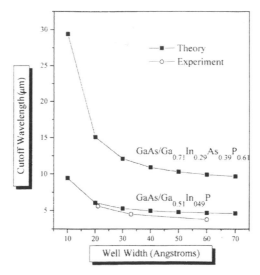

Fig. 7.17. Calculated cutoff wavelength of GaAs/Ga$_{0.51}$In$_{0.49}$P and GaAs/Ga$_x$In$_{1-x}$As$_y$P$_{1-y}$ QWIPs versus quantum well width. Circles represent experimental data for the three GaAs/Ga$_{0.51}$In$_{0.49}$P QWIPs (after Hoff et al. [1995]).

7.3. Solar cells and GaAs solar cells

The invention of the *p–n* junction in 1949 formed the basis of the discovery of the crystalline Si Solar cell. Since then, solar cells have been developed and produced with polysilicon, CdTe, and GaAs. Over 95% of solar cells in production are silicon based. Solar cells continue to be a critical technology for overcoming global environmental and energy problems.

The radiation reaching Earth is scattered and absorbed in the atmosphere and the intensity is dependent on the angle of incidence. Depending on this angle, the intensity can vary between 500 and 1000 W·m^{-2}. The power level of the solar spectrum in outer space, where there is no absorption of the radiation, is about 140 mW·cm^{-2}. This is commonly termed the air-mass-zero (AMO) spectrum. On Earth at sea level, with the sun at zenith, the power level is reduced to nearly 100 mW·cm^{-2}. This is the AMI spectrum. At an angle of incidence that results in twice the path length through the atmosphere, the power level drops to approximately 80 mW·cm^{-2} and the corresponding spectrum is termed AM2.

A solar cell is normally a *p–n* diode. When sunlight falls on the surface of a semiconductor, electron-hole pairs are generated which are then separated by the potential barrier across the *p–n* junction. Since the solar cell is a diode, the *I–V* characteristics of a solar cell are given by

Eq. (7.11) $I_d = I_s \left(\exp(qV / kT) - 1 \right)$

where I_d is the diode current, I_s is the reverse saturation current and V is the applied voltage. Upon illumination, the diode equation is modified as

Eq. (7.12) $I = I_s\left(\exp\left(qV / kT\right)-1\right)-I_L$

where I_L is the photocurrent which opposes the dark current of the diode and I is the current through the diode. The equivalent circuit of a solar cell is shown in Fig. 7.18 and the corresponding I–V characteristic is shown in Fig. 7.19. Upon illumination, the I–V curve shifts down to the fourth quadrant and, hence, power can be extracted from the device. The most important parameter of a solar cell is the efficiency. The efficiency is defined as the ratio of the power output (P_m) to the power input (P_{in})

Eq. (7.13)   efficiency $= P_m / P_{in}$

The output power of the solar cell can be written as $P_m = I_{mp}V_{mp}$ where I_{mp} and V_{mp} are the current and voltage corresponding to the maximum power point. P_m can also be defined as $P_m = I_{sc}V_{oc}FF$ where I_{sc} is the short-($V = 0$), and V_{oc} is the open-circuit voltage ($I = 0$), and FF is the fill factor which is defined as the measure of squareness. Then, the efficiency can be written as

Eq. (7.14)   efficiency $= V_{oc}I_{sc}FF / P_{in}$

The solar cell parameters such as open-circuit voltage (V_{oc}). short-circuit current (I_{sc}), fill factor (FF) and efficiency can be calculated from Fig. 7.19. The open-circuit voltage (V_{oc}) can be calculated by substituting $I = 0$ and $V = V_{oc}$ in Eq. (7.12) and rearranging terms

Eq. (7.15) $V_{oc} = \left(kT / q\right)\ln\left\{\left(I_{sc} / I_s\right)\right\}$

The fill factor (FF) is defined as the measure of squareness, which is defined as

Eq. (7.16) $FF = \left(V_{mp}I_{mp}\right) / \left(V_{oc}I_{sc}\right)$

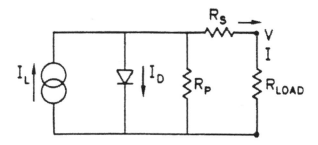

Fig. 7.18. Equivalent circuit of the solar cell.

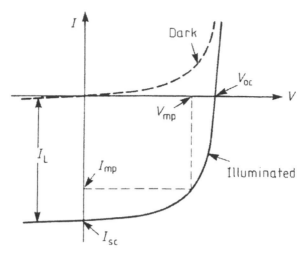

Fig. 7.19. I–V characteristics of the circuit shown in Fig. 7.18 equivalent to a solar cell.

In solar cell conversion efficiency, it is important to note that photons which have an energy hv smaller than the semiconductor bandgap will not produce any electron–hole pairs. Also, photons with energy greater than the bandgap will produce electrons and holes with the same energy (E_g) regardless of how large ($hv - E_g$) is. The excess energy $hv - E_g$ is simply dissipated as heat. Thus the solar cell efficiency depends quite critically on how the semiconductor bandgap matches with the solar energy spectra.

Two major paths have been identified from the past two decades of photovoltaic research and development to bring the cost of photovoltaic modules down to target values. The first is to achieve very high efficiencies using a variety of materials, such as Si, GaAs and related alloys. However, due to the high cost of material growth and fabrication, an alternative approach was developed. The alternative approach is to develop low-cost thin-film polycrystalline solar cells grown on an inexpensive substrate such as glass with moderate efficiencies. For the first approach, the metalorganic chemical vapor deposition technique has been widely used because of its advantages such as compositional control over a large area, suitability for large-scale production, reproducibility and versatility.

The properties of GaAs are well suited for photovoltaic applications. As a single-junction cell, its bandgap is almost ideally suited for the solar spectrum. GaAs is also more radiation resistant than Si, making it more attractive for space applications. The largest component of the production cost of high-efficiency compound semiconductor solar cells today is the cost of epitaxy. Almost all the III–V solar cells with record efficiencies to date have been produced by MOCVD as shown in Table 7.1. The first GaAs cell with an efficiency of 22% was grown by LPE [Woodall and Hovel 1977], while early MOCVD-grown GaAs single-junction cell efficiencies were only in the range of 18–20% [Tobin et al. 1988]. More recently, MOCVD has surpassed other techniques and a record efficiency of 24.8% (AM 1.5) was achieved for a single-junction GaAs cell with the structure shown in

Fig. 7.20 [Tobin and Vernon 1989]. This enhanced performance was primarily attributed to an improved MOCVD process where an increased growth temperature from 720 °C to 790 °C resulted in reduced recombination in the Si-doped GaAs base layer as well as at the GaAs/AlGaAs interface. Even though the active epitaxial region is thin in the above cell design, the use of crystalline GaAs wafer makes it expensive for terrestrial applications. The challenge of growing high-quality crystalline GaAs films without consuming a GaAs wafer is being addressed by the cleavage of lateral epitaxial films for transfer (CLEFT) technique [McClelland et al. 1980] and by heteroepitaxy of GaAs on Si and Ge substrates. In the CLEFT process, the MOCVD-grown single-junction epitaxial cell is cleaved from the substrate which is reused for epitaxial growth [Okamoto et al. 1988, Yeh et al. 1990]. Efficiencies over 23% (AM 1.5) for a 4 cm^2 total area have been obtained using the CLEFT process [McClelland et al. 1990]. An MOCVD-grown single-junction GaAs/Ge cell with an AlGaAs window has resulted in a 4 cm^2 cell efficiency of 24.3% (AM 1.5) [Benner 1991] and a 16–36 cm^2 GaAs/Ge cell efficiency of 18% (AMO) [Chu et al. 1991].

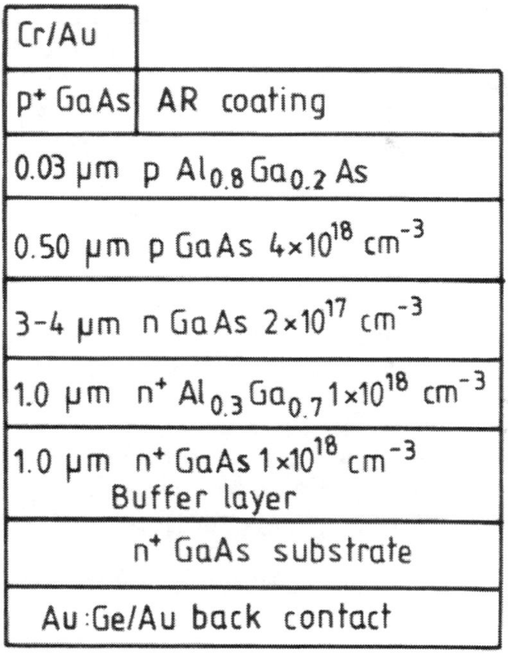

Cr/Au
p$^+$ GaAs │ AR coating
0.03 μm p Al$_{0.8}$Ga$_{0.2}$As
0.50 μm p GaAs 4×10^{18} cm^{-3}
3–4 μm n GaAs 2×10^{17} cm^{-3}
1.0 μm n$^+$ Al$_{0.3}$Ga$_{0.7}$ 1×10^{18} cm^{-3}
1.0 μm n$^+$ GaAs 1×10^{18} cm^{-3} Buffer layer
n$^+$ GaAs substrate
Au:Ge/Au back contact

Fig. 7.20. *Structure of a high-efficiency (24.8%) GaAs solar cell (after Tobin and Vernon [1989]).*

Cell type	V_{oc} (mV)	J_{sc} (mA·cm^{-2})	FF (%)	Efficiency (%)
Single junction				
GaAs/AlGaAs	1022	28.2	87.1	25.1
GaAs/Ge	1035	27.6	85.3	24.3
GaAs/Si	891	25.5	77.7	17.6
GaAs/GaInP	1038	28.7	86.4	25.7
Multijunction/Concentrator				
GaAs concentrator				28.7
GaInAsP concentrator				27.5
Monolithic InP/GaInAs (3 terminal)				31.8
Monolithic GaInP/GaAs				27.5
Monolithic AlGaAs/GaAs				27.6
Stacked GaAs on GaSb				34.2
Stacked GaAs on Si				31.0
Stacked GaAs on GaInAsP				30.2

Table 7.1. High-efficiency, compound semiconductor solar cells.

One of the first terrestrial products using III–V MOCVD is likely to be high-efficiency (> 30%) concentrator cells. In order to improve the efficiency beyond 25%, two- or four-terminal multijunction or tandem cells are being fabricated using MOCVD. Fig. 7.21 illustrates the different configurations and concepts of tandem cells [Kazmerski 1989]. MOCVD is best suited for these structures because of its controlled multilayer growth.

It has been shown that for *a* >30% efficient multijunction cell, the top cell should have a bandgap of ~1.75 eV and the bottom cell bandgap should be ~1.1 eV [Fan and Palm 1984]. The first cell with efficiency greater than 30% was reported for a mechanically stacked MOCVD-grown GaAs cell on top of an Si cell. This four-terminal multijunction cell gave an efficiency of 31% at AM 1.5 under a concentration of 347 suns [Gee and Virshap 1988]. This record efficiency was surpassed recently by another multijunction structure with an efficiency of 34% at 100 suns concentration. This cell consists of an MOCVD-grown GaAs cell on a GaSb cell [Frass et al. 1990].

The best two monolithic multijunction solar cells to date were also grown by MOCVD. A two-terminal AlGaAs cell on a GaAs cell was the first monolithic tandem cell to produce one-sun efficiency of 27.6% under AM 1.5 conditions [Macmillan et al. 1988]. A GaInP/GaAs monolithic tandem cell was suggested as an alternative for AlGaAs/GaAs, and it produced a one-sun efficiency of 27.3% under AM 1.5 conditions [Olson et al. 1988]. This structure is very attractive because the GaInP cell efficiency can be improved further by optimizing the growth conditions. More recently, three-terminal monolithic multifunction cells, based on InP and GaInAs materials, gave an efficiency of 31.8% at a concentration of 50 suns [Wanlass et al. 1991]. To reduce the cost of concentrator cells further, GaAs/Ge tandem cells are being investigated because of the low cost of lattice-matched Ge substrate [Wojtczuk et al. 1990].

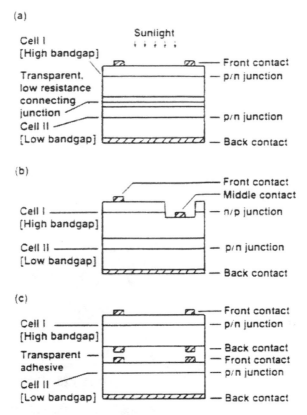

(a)

Cell I
[High bandgap]

Sunlight

Transparent.
low resistance
connecting
junction
Cell II
[Low bandgap]

Front contact
p/n junction

p/n junction

Back contact

(b)

Front contact
Middle contact

Cell I
[High bandgap]

n/p junction

Cell II
[Low bandgap]

p/n junction

Back contact

(c)

Cell I
[High bandgap]

Transparent
adhesive

Cell II
[Low bandgap]

Front contact
p/n junction

Back contact
Front contact
p/n junction

Back contact

Fig. 7.21. Structure of the tandem cell. (a) Two-terminal, monolithic; (b) three-terminal, monolithic and (c) four-terminal, mechanically stacked (after Kazmerski [1989]).

References

Aoki, T., Takahashi, M., Suzuki, M., Suno, H., Uomi, K., Kawano, T., and Takai, A., *IEEE Photon. Technol. Lett.* **4** 580, 1992.

Arch, D.K., Staudenmann, J.L., and Faurie, J.P., *Appl. Phys. Lett.* **48** 1588, 1986.

Bar-Joseph, I., Goossen, K.W., Kuo, J.M., Kopf, R.F., Miller, D.B., and Chemla, D.S., *Appl. Phys. Lett.* **55** 340, 1989.

Bastard, G., *Wave Mechanics Applied to Semiconductor Heterostructures* (Paris: Editions de Physique), 1988.

Bastard, G., Mendez, E.E., Chang, L.L., and Esaki, L., *Phys. Rev.* B **28** 3241, 1983.

Benner, J.P., *Proc. 22nd IEEE Photovoltaic Specialists Conf. (Las Vegas, 7–11 October),* 1991.

Bethea, C.G., Levine, B.F., Lu, P.Y., Williams, L.M., and Ross, M.H., *Appl. Phys. Lett.* **53** 1629, 1988.

Bigan, E., Allovon, M., Carré, M., Carenco, A., and Voisin, P., *IEEEJ. Quantum Electron.* **QE-28** 214, 1992a.

Bigan, E., Allovon, M., Carré, M., and Voisin, P., *Appl. Phys. Lett.* **57** 327, 1990.

Bigan, E., Harmand, J.C., Allovon, M., Carré, M., Carenco, A., and Voisin, P *Appl. Phys. Lett.* **60** 1936, 1992b.

Bleuse, J., Bastard, G., and Voisin, P., *Phys. Rev. Lett.* **60** 220, 1988a.

Bleuse, J., Voisin, P., Allovon, M., and Quillec, M., *Appl. Phys. Lett.* **53** 2632, 1988b.

Bubulac, L.O., Edwall, D.D., McConnel, D., DeWames, R.E., Blazejewski, E.R., and Gertner, E.R., *Semicond. Sci. Technol.* **5** S45, 1990.

Carenco, A., Semiconductor waveguides in III–V materials for integrated optics *Proc. 4th Eur. Conf. on Integrated Optics (ECIO 87) (Glengow, 1987)*, 1987.

Chang, Y.C. and James, R.B., *Phys. Rev.* B **39** 12 672, 1989.

Chu, C., Iles, P., Yoo, H., Reed, B., and Krogen, J., *Proc. 22nd IEEE Photovoltaic Specialists Conf. (Las Vegas, 7–11 October)*, 1991.

Devaux, F., Bigan, E., Ovgazzaden, A., Pierre, B., Hvet, F., Carré, M., and Carenco, A., *IEEE Photon. Technol. Lett.* **4** 720, 1992.

Doll, B., Bruder, M., Wendler, J., Ziegler, J., and Maier, H., *4th Int. Conf. Advanced Infrared Detectors and Systems* (London: IEE) p 120, 1990.

Druilhe, R., Katty, A., and Triboulet, R., *4th Int. Conf. Advanced Infrared Detectors and Systems* (London: IEE) p 20, 1990.

Elliott, T., Gordon, N.T., Hall, R.S., Philips, T.J., Jones, C.L., Matthews, B.E., Maxey, C.D., and Metcalfe, N.E., *Proc. SPIE* **2269** 618, 1994.

Esaki, L., *IEEE J. Quantum Electron.* **QE-22** 1611, 1986.

Esaki, L. and Tsu, R., *IBM J. Res. Dev.* **14** 61, 1970.

Fan, J.C. and Palm, B.J., *Solar Cells* **12** 401, 1984.

Frank, F.C. and van der Merwe, J.H., *Proc. R. Soc.* A **198** 205, 1949.

Franz, W., *Naturf. Z.* **13** 484, 1958.

Frass, L.M., Avery, J.E., Martin, J., Sundaram, V.S., Givard, G., Dinh, V.T., Davenport, T.M., Yerkes, J.W., and O'Neill, J., *IEEE Trans. Electron Devices* **ED-37** 443, 1990.

Gee, J.M. and Virshup, G.F., *Proc. 20th IEEE Photovoltaic Specialists Conf.* p 754, 1988.

Gunapala, S.D., Levine, B.F., Logan, R.A., Tanbun-Ek, T., and Humphrey, D.A., *Appl. Phys. Lett.* **57** 1802, 1990.

Hasnain, G., Levine, B.F., Gunapala, S., and Chand, N., *Appl. Phys. Lett.* **57** 608, 1990.

Hecht, E., *Optics* (New York: Addison-Wesley), 1987.

Hoff, J., Bigan, E., Brown, G.J., and Razeghi, M., *Proc. SPIE* **2397** 445, 1995.

Hoff, J.R., Razeghi, M., and Brown, G., Physical Rev. B **54** 10773, 1996.

Irvine, J.C., *Proc. SPIE* **1735** 92, 1992.

Irvine, S.C., Gertner, E.R., Bubulac, L.O., Gill, R.V., and Edwall, D.D., *Semicond. Sci. Technol.* **6** C15, 1991.

Jelen, C., Slivken, S., Hoff, J., Razeghi, M., and Brown, G., *App. Phys. Lett.* **70** 360, 1997.

Johnson, S.M., Vigil, J.A., James, J.B., Cockrum, C.A., Konkel, W.H., Kalisher, M.H., Risser, R.F., Tung, T., Hamilton, W.J., Ahlgren, W.L., and Myrosznyk, J.M., *J. Electron. Mater.* **22** 835, 1993.

Kato, H., Iguchi, N., Chika, S., Nakayama, M., and Sano, N., *J. Appl. Phys.* **59** 588, 1986.

Kato, T., Sasoki, T., Komatsu, K., and Mito, I., *Electron. Lett.* **28** 253, 1992.

Kazmerski, L.L., *Int. Mater. Rev.* **34** 185, 1989.

Keldysh, L.V., *Sov. Phys.–JETP* **7** 788, 1958.

Kinch, M.A., and Yariv, A., *Appl. Phys. Lett.* **55** 2093, 1989.

Koch, T.L. and Bowers, J.E., *Electron. Lett.* **20** 1038, 1984.

Koyama, F. and Iga, K., *IEEE J. Lightwave Technol.* **6** 87, 1988.

Kozlowski, L.J., Arias, J.M., Williams, G.M., Vural, K., Cooper, D.E., Cabelli, S.A., and Bruce, C. *Proc. SPIE* **2274** 93, 1994.

Kurtz, S.R., Dawson, L.R., Zipperian, T.E., and Lee, S.R., *Appl. Phys. Lett.* **52** 1581, 1988.

Kurtz, S.R., Dawson, L.R., Zipperian, T.E., and Whaley, R.D., *IEEE Electron Devices Lett.* **11** 54, 1990.

Levine, B.F., *J. Appl. Phys.* **74** R1, 1993.

Levine, B.F., Bethea, C.G., Glogovsky, K.G., Stay, J.W., and Leibenguth, R.E., *Semicond. Sci. Technol.* **6** C114, 1991.

Levine, B.F., Bethea, C.G., Hasnai, G., Shen, V.O., Pelve, E., Abbot, R.R., and Hsieh, S.J., *Appl. Phys. Lett.* **56** 851, 1990.

Levine, B.F., Bethea, C.G., Hasnai, G., Walker, J., and Malik, R.J., *Electron Lett.* **24** 747, 1988.

Levine, B.F., Choi, K.K., Bethea, C.G., Walker, J., and Malik, R.J., *Appl. Phys. Lett* **50** 1092, 1987.

Levine, B.F., Hasnain, G., Bethea, C.G., and Chand, N., *Appl. Phys. Lett.* **54** 2704, 1989.

Levine, B.F., Zussman, A., Kuo, J.M., and de Jong, J., *J. Appl. Phys.* **71** 5130, 1992.

Macmillan, H.F., Hamaker, H.C., Virshup, G.F., and Werthen, J.E., *Proc. 20th IEEE Photovoltaic Specialists Conf.* p 48, 1988.

Matthews, J.W., *Epitaxial Growth* (New York: Academic) ch 8 p 559, 1975.

McClelland, R.W., Bozler, C.O., and Fan, J.C., *Appl. Phys. Lett.* **37** 560, 1980.

McClelland, R.W., Dingle, B.D., Gale, R.P., and Fan, J.C., *Proc. 21st IEEE Photovoltaic Specialists Conf.* p 168, 1990.

Mendez, E.E., Agullo-Rueda, F., and Hong, J.M., *Phys. Rev. Lett.* **60** 2426, 1988.

Miller, D.B., Chemla, D.S., Damen, T.C., Gossard, A.C., Wiegmann, T.H., Wood, C.A., and Burrus, C.A., *Appl. Phys. Lett.* **45** 13, 1984.

Miller, D.B., Chemla, D.S., Damen, T.C., Gossard, A.C., Wiegmann, T.H., Wood, C.A., and Burrus, C.A., *Phys. Rev.* B **32** 1043, 1985.

Miller, D.B., Chemla, D.S., Eilenberger, D.J., Smith, P.W., Gossard, A.C., and Tsang, W.T., *Appl. Phys. Lett.* **41** 679, 1982.

Mullin, J.B., Irvine, S.C., Gies, J., and Ryple, A., *J. Crystal Growth* **88** 161, 1988.

Okamoto, H., Kadota, Y., Watanabe, Y., Fukuda, Y., Oh'Hara, T., and Ohmachi, Y., *Proc. 20th IEEE Photovoltaic Specialists Conf.* p 475, 1988.

Olson, J.M., Kurtz, S.R., and Kibber, A.E., *Proc. 20th IEEE Photovoltaic Specialists Conf.* p 777, 1988.

Olson, J.M., Kurtz, S.R., Kibber, A.E., and Faine, P., *Appl. Phys. Lett.* **55** 1208, 1989.

Omnes, F., and Razeghi, M., *Appl. Phys. Lett.* **59** 1034, 1991.

Rao, M., Caine, E., Kroemer, H., Long, S., and Babic, D., *J. Appl. Phys.* **61** 643, 1987.

Razeghi, M., Defour, M., Omnes, F., Dobers, M., Vieren, J.P., and Guldner, Y., *Appl Phys Lett.* **55** 457, 1989.

Razeghi, M., Maurel, P., Omnes, F., Amor, S.B., Dmowski, L., and Portal, J.C., *Appl. Phys. Lett.* **48** 1267, 1986.

Schulman, J.N., and McGill, T.C., *Appl. Phys. Lett.* **34** 663, 1979.

Smith, D.L., McGill, T.C., and Schulman, J.N., *Appl. Phys. Lett.* **43** 180, 1983.

Specht, L.T., Hoke, W.E., Oguz, S., Lemonias, P.J., Kreismanis, V.G., and Korenstein, R., *Appl. Phys. Lett.* **48** 417, 1986.

Summers, C.J., Wagner, B.K., Benz, R.G., and Conte Matos, A., *Proc. SPIE* **2021** 56, 1994.

Tobin, S.P., and Vernon, S.M., *4th Int. Photovoltaic Science Eng. Conf. (Sydney, Australia)* p 47, 1989.

Tobin, S.P., Vernon, S.M., Bajgar, C., Geoffory, L.M., Keavney, C.J., Sanfacon, M.M., and Haven, V.E., *Solar Cells.* **24** 103, 1988.

Tung, T., DeArmond, L.V., Herald, R.F., Herning, P.E., Kalisher, M.H., Olson, D.A., Risser, R.F., Stevens, A.P., and Tighe, S.J., *Proc. SPIE* **1735** 109, 1992.

Wakita, K., Kotaka, I., Asai, H., Okamoto, M., Kondo. Y., and Naganuma, M., *IEEE Photon. Technol. Lett.* **4** 16, 1992.

Wakita, K., Kotaka, I., Mitomi, O., Asai, H., Kawamura, Y., and Naganuma, M., *IEEE J. Lightwave Technol.* **LT-8** 1027, 1990.

Wanlass, M.W., Coutts, T.J., Ward, J.S., Emery, K.A., Gessert, T.A., and Osterwald, C.R., *Proc. 22nd IEEE Photovoltaic Specialists Conf. (Las Vegas, 7–11 October),* 1991.

Watanabe, M.O., and Ohba, Y., *J. Appl. Phys.* **60** 1032, 1986.

Weiner, J.S., Miller, D.B., Chemla, D.S., Damen, T.C., Burrus, C.A., Wood, T.H., Gossard, A.C., and Wiegmann, W., *Appl. Phys. Lett.* **47** 1148, 1985.

West, L.C. and Eglash, S.J., *Appl. Phys. Lett.* **46** 1156, 1985.

Wojtczuk, S.J., Tobin, S.P., Keavney, C.J., Bajgar, C.J., Sanfancon, M.M., Geoffroy, L.M., Dixon, T.M., Vernon, S.M., Scofield, J.D., and Ruby, D.S., *IEEE Trans. Electron Devices* **ED-37** 455, 1990.

Wood, T.H., Burrus, C.A., Miller, D.B., Chemla, D.S., Damen, T.C., Gossard, A.C., and Wiegmann, W., *Appl. Phys. Lett.* **44** 16, 1984.

Wood, T.H., Burrus, C.A., Tucker, R.S., Weiner, J.S., Miller, D.B, Chemla, D.S., Damen, T.C., Gossard, A.C., and Wiegmann, W., *Electron. Lett.* **21** 693, 1985.

Wood, T.H., Pastalan, J.Z., Burrus, C.A.Jr, Johnson, B.C., Miller, B.I., de Miguel, J.L., Koren, U., and Young, M.G., *Appl. Phys. Lett.* **57** 1081, 1990.

Woodall, J.M. and Hovel, H.J., *Appl. Phys. Lett.* **30** 492, 1977.

Yariv, A. and Yeh, P., *Optical Waves in Crystals* (New York: Wiley), 1984.

Yeh, Y.M., Cheng, C., Ho, F. and Yoo, H., *Proc. 21st IEEE Photovoltaic Specialists Conf.* p 79, 1990.

Zucker, J.E., Jones, K.L., Toung, M.G., Miller, B.I., and Koren, V., *Appl. Phys. Lett.* **55** 2280, 1989.

Zucker, J.E., Jones, K.L., Wegener, M., Chang, T.Y., Sauer, N.J., Divino, M.D., and Chemla, D.S., *Appl. Phys. Lett.* **59** 201, 1991.

Zucker, J.E., Wegener, M., Jones, K.L., Chang, T.Y., Sauer, N., and Chemla, D.S., *Appl. Phys. Lett.* **56** 1951, 1990.

8. GaAs-Based Lasers

8.1. **Introduction**
8.2. **Basic physical concepts**
    8.2.1.   Optical gain and feedback
    8.2.2.   Threshold current
    8.2.3.   Spectrum of longitudinal modes
    8.2.4.   Carrier and light confinement
    8.2.5.   High-speed modulation
    8.2.6.   Quantum size effects
8.3. **Laser structures**
    8.3.1.   Threshold current density and transverse waveguiding
    8.3.2.   Threshold current and lateral mode control
8.4. **New GaAs-based materials for lasers**
    8.4.1.   Strained InGaAs layer laser structures
    8.4.2.   AlGaInP visible lasers
    8.4.3.   GaInAsP/GaAs lasers

8.1. Introduction

Semiconductor lasers based on GaAs and related materials are the subject of a vast part of the research in this field. The first suggestion to use semiconductors as laser materials was expressed by Aigrain as early as 1958. Electroluminescence spectrum narrowing in GaAs at 77 K observed by Nasledov et al. in 1962 as experimental evidence of high-efficiency radiative recombination gave a strong impulse to the laser-oriented activity. The first homojunction semiconductor lasers were created using GaAs in 1962 almost simultaneously by three groups of American scientists [Hall et al. 1962, Nathan et al. 1962 and Quist et al. 1962]. In the same period, Holonyak and Bevacqua [1962] reported lasing in GaAsP. Those early devices were only able to operate at low temperatures due to the very high threshold current density. The solution to this problem was proposed in 1963 by Alferov and Kazarinov [1963] and soon after that by Kroemer [1963]. The idea was to use double heterostructures to ensure electrical and optical confinement and reduce the threshold current. At that time the possibility of efficient carrier injection through a heterojunction still seemed to be very vague because of the extremely high carrier recombination rate on the interface of polycrystalline heterojunctions. However, by 1969 Kressel and Nelson [1969], Panish et al. [1969] and Hayashi et al. [1969]

successfully applied liquid phase epitaxy to obtain laser-quality heterojunctions in GaAs–AlGaAs with a decreased threshold current density. A breakthrough was made in 1970 when Alferov's group achieved continuous wave operation of a GaAs–AlGaAs double-heterostructure laser at room temperature [Alferov et al. 1971], again in GaAs-based material. A breathtaking competition took place at that time as witnessed by Casey and Panish [1978], impressive by its scientific significance as well as by emotional strength.

Studies of heterostructure lasers in the very attractive GaAs–AlGaAs system of solid solutions and other related materials stimulated the development of new growth techniques for producing ultrathin layers and multilayer structures: molecular beam epitaxy (MBE) and metalorganic chemical vapor deposition (MOCVD). Extensive research resulted in obtaining a miscellany of new laser materials and in designing even more efficient types of laser structure. Major advances have been made in achieving such parameters as low threshold current density and absolute threshold current, high output power, stable single longitudinal and transverse mode operation and a broad-bandwidth frequency response. A wide range of applications arose for these new devices: high-power lasers for optical pumping of solid-state laser crystals, fast-switching and high-coherence single-mode lasers for high-data-rate optical communications, surface-emitting and ultralow-threshold lasers for optical interconnects in integrated circuits and many others.

Since the first attempts to use MOCVD for the growth of laser heterostructures this technique has demonstrated an unimpaired ability to produce high-quality materials for all kinds of application. MOCVD-grown GaAs-based lasers cover wavelengths ranging from mid-infrared in Sb-containing solid solutions to visible AlGaInP devices. Its flexibility and potential to yield a broad range of growth rates resulted in layers featuring thicknesses from tens of microns down to several nanometers. Planar structures containing quantum wells with atomically flat interfaces, superlattices, strained or graded-index layers were successfully grown by MOCVD. Furthermore, MOCVD proved its efficiency in producing a variety of laser devices by overgrowth and epitaxy on patterned substrates. One aspect of the importance of MOCVD is strongly enhanced by the possibility of large-scale production by simultaneous growth on several substrates in one process. Parameter variation as low as 2% has been observed across wafers 3 inches in diameter.

8.2. Basic physical concepts

The fundamental principles of semiconductor laser operation are briefly discussed in this chapter. Some useful references are given where a more detailed discussion of the subject may be of interest. We also indicate the most important problems arising when specific applications impose requirements on the basic characteristics of laser devices.

8.2.1. *Optical gain and feedback*

A semiconductor laser is a source of highly coherent radiation generated by recombining carriers, excited electrons and holes, in the semiconductor medium with optical gain and feedback.

Free electrons pumped from valence to conduction bands by either electric current or external irradiation may recombine with a hole spontaneously, emitting a random photon (Fig. 8.1(a)) or undergo stimulated emission. In this latter process an electron interacts with a stimulating photon and recombines giving birth to another photon of identical energy and phase (Fig. 8.1(b)). If pumping makes the electron concentration in the conduction band high enough to equalize the probabilities of valence band absorption and stimulated emission, the medium becomes transparent. In this situation light amplification due to stimulated emission neutralizes absorption loss. The transparency condition corresponds to zero optical gain. Under stronger pumping more and more electrons are excited to the conduction band, resulting in a population inversion, when photons stimulate recombination from the conduction band rather than become absorbed by valence-band electrons. A positive optical gain is thus achieved [Sze 1981].

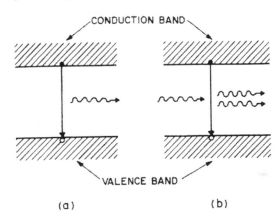

Fig. 8.1. *Schematic illustration of (a) spontaneous emission, and (b) stimulated emission processes wherein an electron–hole pair recombines to generate a photon. In the case of stimulated emission the two outgoing photons match in their frequency and direction of propagation.*

In order for the gain medium to lase it should be confined in a resonator cavity providing optical feedback, that is restricting the range of amplified wavelengths to a limited number of resonant modes. A semiconductor crystal with mirror-like facets perpendicular to propagating beam is a Fabry–Pérot resonator by itself (Fig. 8.2(a)). Otherwise it may be integrated with Bragg mirror gratings (Fig. 8.2(b)) or equipped with an external mirror (Fig. 8.2(c)) [Buus 1991].

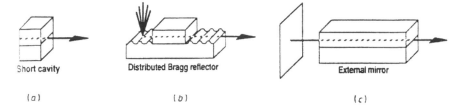

Fig. 8.2. Examples of single-mode lasers.

8.2.2. Threshold current

Let us consider a semiconductor crystal with a *p–n* junction plane and cleaved mirror facets (Fig. 8.3). Under forward bias voltage, electrons are injected into the *p*-region and holes into the *n*-region thus providing the population inversion necessary for achieving optical gain. However, lightwaves resonating in the Fabry–Pérot cavity are absorbed by free carriers and scattered on defects and optical inhomogeneities. Due to the limited reflectivity of the mirrors some part of the radiation comes out of the resonator or becomes absorbed. All this results in internal and mirror losses. The optical field is just partly confined within the excited region, while some fraction of the light propagates outside the gain layer adjacent to the *p–n* junction plane. The threshold condition for laser light emission to occur is that the gain should be high enough to compensate for all losses through the length of the cavity during one period of oscillation. A fundamental equation for the threshold gain can be derived from this consideration:

Eq. (8.1) $g_{th} = 1/G \left(\alpha_i + (1/L) \ln(1/R) \right)$

where G is the optical confinement factor, α_i is the internal loss, L is the cavity length and R is the reflectivity of the end mirrors [Casey and Panish 1978].

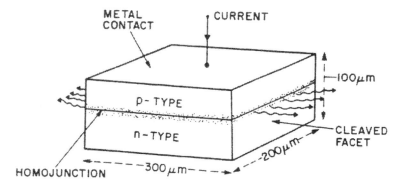

Fig. 8.3. Schematic illustration of a homostructure semiconductor laser with its typical physical dimensions.

The pumping electric current providing the threshold gain is a threshold current, characterizing the efficiency of the entire laser device. Minimization of the threshold current is one of the key issues to achieve high-speed modulation of laser output as long as this value limits the current rise and drop necessary to switch the device on and off.

It is convenient to describe the quality of a planar laser structure in terms of its threshold current density rather than its absolute threshold current. Since a laser bar is long and broad enough to neglect all mirror and lateral effects, the threshold current per unit area depends exclusively on the specific quality of the wafer. A minimal threshold current density is highly desirable in most cases. It results in low Joule heating of the laser device and therefore is crucial for high-power applications.

A useful expression for the threshold current density for a uniformly excited GaAs layer at room temperature is given by Casey and Panish [1978]

Eq. (8.2) $J_{th}\left(A\,cm^{-2}\right)=d\,/\,\eta\left(4.5\times10^{3}+20g_{th}\right)$

Here d is the layer thickness in microns and η is the internal quantum efficiency given by the ratio of the radiative recombination rate to the total recombination rate. The quantity in the bracket represents the so called nominal current density or current density required to excite 1 μm thick layer with unit quantum efficiency. This important parameter is the number of photons emitted in the medium per electron injected. It summarizes the influence of intrinsic features of the material (Auger recombination and spontaneous emission probabilities), extrinsic characteristics (concentration of defect centers where non-radiative recombination occurs) and pumping level (carrier concentration) [Agrawal and Dutta 1986]. An internal quantum efficiency close to 100% may be achieved in laser-quality GaAs.

Another useful quantity describing laser operation is the differential (or external) quantum efficiency taking into account all the losses in the laser structure. It is given by the ratio of the surplus in coherently emitted power to the rise of pumping power required to give this surplus. In other words, it is a fraction of the number of electrons that may be converted into stimulated emission if injected by increasing the driving current

Eq. (8.3) $\eta_{d}=e\Delta P_{out}\,/\,hv\Delta I$

At threshold η_{d} is related to the internal quantum efficiency by

Eq. (8.4) $1/\eta_{d}=\left(1/\eta\right)\left(1+\alpha_{i}L\,/\ln\left(1/R\right)\right)$

MOCVD-grown semiconductor lasers may exhibit an external quantum efficiency above 80% [Wagner et al. 1988].

8.2.3. Spectrum of longitudinal modes

A sequence of equidistant wavelengths (longitudinal modes) satisfies a resonant condition in the laser structure shown in Fig. 8.4. For a typical resonator cavity length of several hundred microns the wavelength separation is small enough with respect to the range of photon energies where gain is higher than loss. This permits a number of modes to be amplified and results in a multimode lasing spectrum. At the same time, single-longitudinal-mode operation is required for many important applications, primarily for optical fiber communication. Different modes travel with different velocities through the fiber and the optical pulse broadens thus limiting the data transmission rate [Buus 1991].

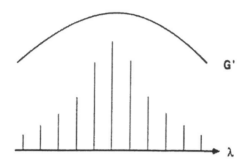

Fig. 8.4. Schematic of a laser spectrum.

8.2.4. Carrier and light confinement

In order to achieve the high gain necessary for lasing, both excited carriers and resonant lightwaves should be spatially confined in some limited volume (active region). Poor optical confinement results in a low value of G and high loss, while excited carriers leaking from the gain area can prevent the build-up of a population inversion. These effects are the reason for the high threshold current densities observed in homostructure lasers like the one shown in Fig. 8.3. Efficient optical and electrical confinement may be ensured by placing the narrow-bandgap active region between wider-bandgap cladding regions having lower refractive index. In this case the bandgap difference provides a potential barrier to confine injected carriers as illustrated in Fig. 8.5. At the same time, light traveling through the high-refractive-index active layer is guided by the cladding layers due to total internal reflection. Fig. 8.6 presents this kind of sandwich structure, called a dielectric waveguide. The optical confinement factor G in Eq. (8.1) refers to transverse waveguiding in the plane normal to the p–n junction. Carrier confinement between p- and n-cladding layers in Fig. 8.5 also prevents them from leakage in the direction normal to the junction plane. In this way a low threshold current density may be achieved. For a low total threshold current to be obtained, carriers and light should also be confined in the lateral direction parallel to the junction plane. This requires formation of a stripe contact locating the current flow to the finite recombination (active) region (see Fig. 8.7). A well defined gain region may provide a refractive index step sufficient for optical confinement (gain guiding). Otherwise a stripe geometry active region may be surrounded by higher-index cladding layers to form a dielectric waveguide similar to that in Fig. 8.6 (index guiding). Near-and far-field patterns of the laser beam may be tailored by varying the stripe width and waveguiding refractive index step as shown in Fig. 8.8(*a*) and Fig. 8.8(*b*) [Yonezu et al. 1973]. A variety of laser device structures were designed on the way to obtaining the most efficient transverse and lateral confinement. A profound discussion of the physical aspects of this problem and basic device concepts can be found in Casey and Panish [1978].

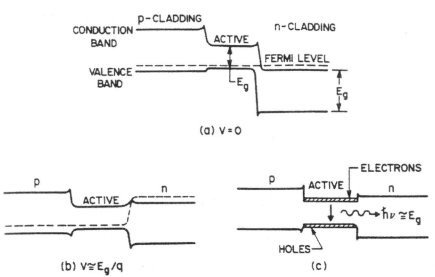

Fig. 8.5. Energy band diagram of a double-heterostructure semiconductor laser at
(a) zero bias and (b) forward bias, (c) The bandgap discontinuities at the two
heterojunctions help to confine electrons and holes inside the active region, where
they recombine to produce light.

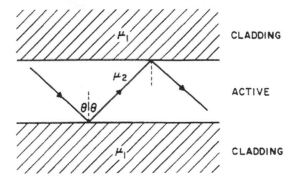

Fig. 8.6. Dielectric waveguiding in a heterostructure semiconductor laser. The
relatively higher refractive index $\mu_2 > \mu_1$ of the active layer allows total internal
reflection to occur at the two interfaces for angles such that $\sin \theta > \mu_1/\mu_2$.

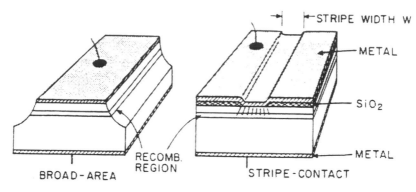

Fig. 8.7. Stripe contact: a simple way to confine the recombination region in the lateral direction (after Kressel [1980]).

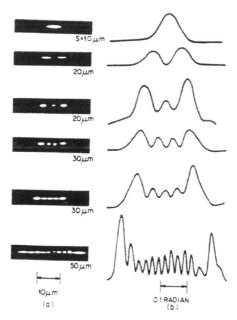

Fig. 8.8. Modes along the junction planes as a function of stripe width S for planar stripe DH lasers, (a) Near-field patterns, (b) Far-field patterns (after Yonezu et al. [1973]).

8.2.5. High-speed modulation

Since high-data-rate transmission is one of the fields where semiconductor lasers are most extensively applied, modulation of laser output is of great interest. A semiconductor laser can be directly modulated by applying a periodical or pulsing driving current. The modulation efficiency is characterized by the frequency-dependent transfer function $H(\omega)$, representing a change in external quantum efficiency due to impaired device response

Eq. (8.5) $\eta_d(\omega) = H(\omega)\eta_d(0)$

A typical transfer function is shown in Fig. 8.9 in units of dB, that is $10 \cdot \log(H(\omega))$. The range of frequencies where the response function is higher than −3 dB is usually referred to as the modulation bandwidth. It is limited by the delay time between the pumping current rise and light pulse onset and by a transitional period when relaxation oscillations occur. Parasitic electrical effects of the device structure as well as of the driving circuit also decrease the modulation bandwidth. For further details see Buus [1991] and also a review by Arnold et al. [1982].

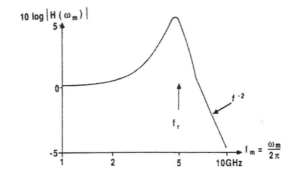

Fig. 8.9. Amplitude modulation transfer function as a function of the modulation frequency (after Agrawal and Dutta [1986]).

8.2.6. Quantum size effects

As soon as the active layer of a semiconductor laser is made very thin (of the order of the de Broglie wavelength for the carrier), it acts like a two-dimensional potential well for carriers, and quantum size effects are manifested in carrier motion in the z-direction perpendicular to the layer plane. That is, instead of a continuum of conduction band states available for electrons in bulk material, these are confined to a series of discrete levels separated from the band edge by the energies $E_n = n^2 h^2 / 8 m^* L_z^2$ for the ideal case of an infinitely deep well. Bound-state energies depend on the well depth (width) (see Fig. 8.10). Quantization effects resulted in a number of consequences which proved to be very advantageous for laser applications. First of all, with of the volume of the active layer being small in comparison with the DH laser, the transparency current necessary to reach population inversion is proportionally lower. Free-carrier absorption contributes less to the internal losses for exactly the same reason. Therefore, quantum well lasers operate at low threshold current densities, high differential quantum efficiencies and narrow spectral linewidth. The density of states in a quantum well demonstrates a step-like energy dependence as shown in Fig. 8.11. It rises rapidly when reaching the first energy level and so does the concentration of injected carriers taking part in stimulated emission. This consideration implies a high differential optical gain which yields a more abrupt optical pulse front and permits improved modulation properties. The last, but not least, advantage is a possibility to tailor the lasing wavelength precisely by changing the quantum well width and shifting energy levels in the well. In Yariv's book [1989b], an excellent introduction to the physics of quantum well lasers is provided.

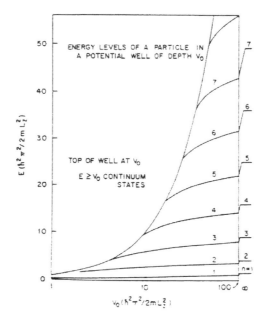

Fig. 8.10. Dependence of bound-state energies (reduced energy units $\hbar^2\pi^2 / 2mL_z^2$) on the well depth of a rectangular potential well (after Adams [1973]).

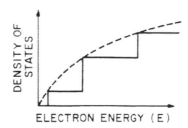

Fig. 8.11. Schematic representation of the density of states in a quantum well.

8.3. Laser structures

Since 1962 when the first semiconductor laser appeared, scientific interest and application requirements have brought to life miscellaneous structures and designs elaborated as a result of extensive research efforts. Whole new fields of application have arisen with the continuous improvement of basic device parameters: threshold current density and total threshold current, output power, frequency stability and tuning, modal pattern and modulation bandwidth. In this section we shall present laser structures which provided major steps and breakthroughs for this progress in semiconductor laser design.

8.3.1. Threshold current density and transverse waveguiding

(a) Double-heterostructure lasers

The transition from early homostructure lasers to heterostructures was probably the most impressive innovation affecting the very attitude of the scientific community and related industries to the future of this new class of semiconductor devices. Continuous wave operation at room temperature has become possible due to the two orders of magnitude decrease in the threshold current density from $\sim 50,000$ A·cm^{-2} for homostructure lasers to less than 500 A·cm^{-2} for heterostructure devices.

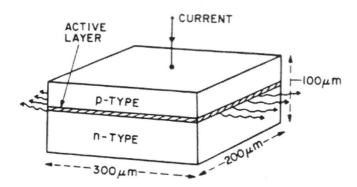

Fig. 8.12. Schematic illustration of a double-heterostructure laser with its typical physical dimensions.

Fig. 8.12 shows a common broad-area double-heterostructure (DH) laser with thin GaAs active layer confined between n-type and p-type layers of Al$_x$Ga$_{1-x}$As solid solution. A typical active layer thickness is 0.1–0.5 micron while values of x are usually chosen to be 0.3–0.4. Under forward electric bias, majority carriers are injected from Al$_x$Ga$_{1-x}$As cladding layers to the active region giving rise to population inversion and thus providing conditions for light amplification. Since the solid solution has a broader bandgap than GaAs, the gap discontinuity leads to the formation of a potential barrier confining injected carriers within the active layer. Refractive index dependence on x is given by $n = 3.6 - 0.7x$ and the structure described behaves as a dielectric waveguide concentrating lightwaves in the active region.

The operating properties of the GaAs–Al$_x$Ga$_{1-x}$As DH laser are dependent primarily upon the structure geometry and cladding-layer composition which is critical for the optical and electrical confinement. Fig. 8.13 presents a plot of the room-temperature threshold current density dependence on active layer thickness. The minimal threshold current density is limited by the rise at small thickness due to the low optical confinement factor G, as can be seen from Eq. (8.1) and Eq. (8.2) of the previous section. The lowest threshold current densities obtained for double heterostructure lasers with $d = 0.06$ micron are about 390 Å·cm^{-2} which is close to the theoretical limit [Tsang 1984]. Low threshold current densities for DH lasers grown by MOCVD were observed by Dupuis et al. [1978].

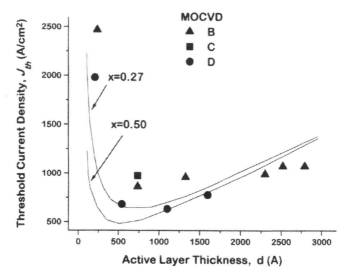

Fig. 8.13. Experimental dependence of J_{th} on d and x for MOCVD broad-area DH lasers (closed data points). For comparison, the results of theoretical calculations (curves) are also shown (after Dupuis et al. [1978]).

Transverse mode behavior is governed by the ability of the laser waveguide to support specific modes. Strong optical confinement resulting from large refractive index steps at the heterojunction permits higher-order transverse modes to propagate in a thick waveguide. However, a sufficiently small active layer thickness leads these modes to a cutoff and stable fundamental transverse mode operation is observed for lasers with an active layer thickness of less than 0.2 micron. At the same time a thin active layer in a DH laser increases the optical power density on the laser mirror and, subsequently, the risk of mirror damage. Laser beam divergence as high as 47.5° is observed for 0.1 micron thick active layers [Casey et al. 1973] which may prevent the efficient coupling of emission into an optical fiber.

The above-mentioned limitations imposed on the performance of double-heterostructure lasers by their very nature stimulated a search for a more efficient laser structure design. An important step on the way was provided by the introduction of multilayer structures of which separate confinement heterostructures proved to be the most promising.

(b) Separate confinement heterostructure lasers

A separate confinement heterostructure (SCH) laser, as proposed in 1973 by Thompson and Kirkby [1973] and as shown in Fig. 8.14, is basically a four- or five-layer dielectric waveguide where, first, a smaller composition step gives an energy-band discontinuity necessary to confine injected carriers within the active layer, while a larger step in the refractive index between the waveguide and cladding layers provides light confinement within the optical cavity. Separate optical and electrical confinement ensures moderate beam divergence and low optical power density on the laser mirror while preserving fundamental transverse mode operation and low threshold current density for the lasers with thin active layers. Though SCH lasers did not immediately demonstrate threshold current densities lower than that of contemporary double heterostructures, they readily helped to reduce the beam half width to 10° and improve the uniformity of near-field emission and external quantum efficiency [Thompson et al. 1976].

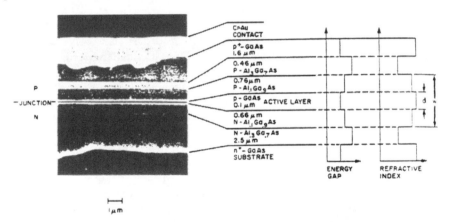

Fig. 8.14. Scanning electron photomicrograph of the selectively etched surface of the SCH laser. The layer thickness, Al composition and the conductivity type are indicated for each layer. A representative variation in the energy gap and refractive index is also shown. The active layer thickness is d and the optical waveguide thickness is ω (after Casey et al. [1974]).

Fig. 8.14 illustrates the sequence of epitaxial layers grown on a GaAs substrate by liquid phase epitaxy to form a SCH laser with thick waveguide. Bandgap and refractive index profiles are also shown. Structure geometry can be optimized easily to minimal threshold current density by selecting the waveguide and active layer thickness and composition corresponding to the maximal optical confinement factor G for the active region [Casey et al. 1974]. With a thin active layer $(d < 0.2 \ \mu m)$, fundamental transverse mode emission is retained for thick optical cavities which would give rise to higher-order modes in DH lasers. A typical waveguide thickness for SCH lasers is 0.2–1.0 micron depending on the application supposed, while an active layer thickness is usually 0.1 micron or less. Four-layer heterostructures and asymmetric composition profiling were used for different purposes some of which will be discussed later.

Versatile and efficient by itself, the SCH laser has unavoidably become dominant after progress in crystal growth technology resulted in the introduction of laser structures with quantum well size active layers and graded-index heterostructure materials.

(c) Graded-index separate confinement heterostructure lasers (GRIN-SCH)

In 1976, Kazarinov and Tsarenkov [1976] theoretically predicted that a graded composition region surrounding a DH laser active region would efficiently guide the lightwave and collect electrons in the active layer, thus helping to achieve a lower threshold current density. As soon as molecular beam epitaxy developed into an efficient tool to control the growing layer composition, Tsang [1981a] modified a separate confinement heterostructure by gradually changing the waveguide layer composition rather than by making an abrupt step to the active layer. This enabled the matching of a far-field profile to a particular type of optical fiber, further improving higher-order mode discrimination and increasing the cutoff thickness for the first-order mode in asymmetrical waveguide structures. The threshold current density, however, remained within the range of that for DH lasers. The sequence of layers shown in Fig. 8.15 is referred to as a GRIN-SCH, standing for graded-index separate confinement heterostructure.

GRIN-SCH demonstrated its full potential after extremely thin ($d < 0.05$ micron) active layers had become possible to grow by modern epitaxial methods, MBE and MOCVD. Conventional double-heterostructure thin active layers resulted in a negligible optical confinement factor and the threshold current densities observed in these early experiments varied from 2 to 3 kA·cm^{-2} [Dupuis et al. 1978, Tsang et al. 1979]. The new-born GRIN-SCH technology helped to change the situation dramatically, immediately yielding 250 A·cm^{-2} for the same device area and as low as 160 A·cm^{-2} for a longer optical cavity [Tsang 1982]. The latest world records for low threshold current density have been made by employing basically the same generic GRIN-SCH structure [Chen et al. 1987, Chand et al. 1991]. Its efficiency is rooted in its different $G(d)$ dependence when compared with double heterostructures, as well as in a better carrier confinement due to the "funneling" effect [Kazarinov and Tsarenkov 1976, Hersee et al. 1984]. Some contradictory evidence for this latter suggestion shows that a convincing analysis has not yet been performed. The former can be easily seen from Fig. 8.16 where the relative contributions of three terms to the value of J_{th} are shown (compare this to equations Eq. (8.1) and Eq. (8.2) of the previous section). For a three-layer waveguide, G is approximately proportional to d^2 and losses increase with decreasing d, while for a separate confinement case the optical intensity distribution within the waveguide is independent of the thin active layer size making G directly proportional to d. This, in turn, results in d-independent losses and the threshold current density drops together with the transparency current densities for thin layers. Further progress in achieving low threshold current densities was associated with ultrathin (quantum well) active layers.

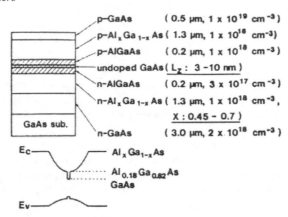

Fig. 8.15. Schematic diagram of a GaAs–AlGaAs GRIN-SCH laser grown by MBE
and the corresponding energy band diagram of the laser (after Fujii et al. [1984]).

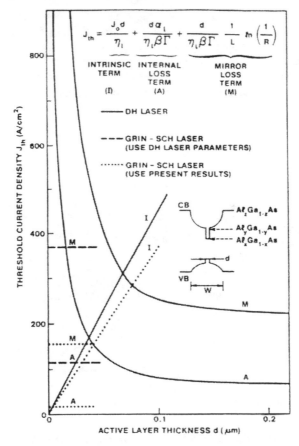

Fig. 8.16. Relative importance of the contributions to J_{th} of intrinsic, internal loss
and mirror loss terms. The solid curves were calculated for regular DH lasers,
while the dashed curves were calculated for GRIN-SCH lasers (after Tsang [1982]).

(d) Quantum well lasers

The concept of quantized carrier behavior in extremely thin layers of semiconductor material originated as early as 1970 [Esaki and Tsu 1970] and was later recognized by the Nobel Prize. Experimental realization required a growth technique to produce high-quality heterostructures consisting of layers of alternating composition of thickness less than 50 nm. Laser oscillation from MBE-grown multiquantum well (MQW) GaAs–AlGaAs heterostructures under optical pumping was first reported in 1975 [van der Ziel et al. 1975]. Soon after that, progress made in MOCVD crystal growth resulted in the first room-temperature GaAs–AlGaAs single-quantum-well (SQW) injection laser [Dupuis et al. 1978] and initiated the overwhelming domination of this type of semiconductor laser as predicted in a paper by Yariv [1989a]. A combination of high gain, typical for QW lasers, and tight optical confinement in a GRIN-SCH structure made it possible to fabricate a laser with a threshold current density of 232 A·cm^{-2} for a reasonable cavity length of 413 μm, while low intrinsic losses in the quantum well and excellent heterointerface abruptness within two monolayers performed by MOCVD helped to achieve the value of 121 A·cm^{-2} for an extremely long 1788 μm cavity [Hersee et al. 1982]. Quantum well lasers demonstrate high-temperature stability of the threshold current density thanks to the relative independence of carrier temperature in the well on the lattice temperature [Chinn et al. 1980]. SQW-GRIN-SCH structures, and separately confined modified multiquantum well (MMQW) structures [Tsang 1981b] were employed in numerous device designs for low-threshold, single-frequency and high-power applications. Fig. 8.17 gives a schematic view of the structures discussed above.

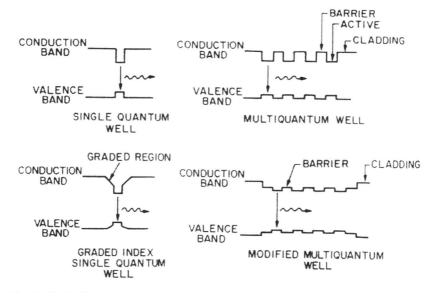

Fig. 8.17. Different single-quantum-well and multiquantum-well laser structures shown schematically.

8.3.2. Threshold current and lateral mode control

While the threshold current density is a representative characteristic of laser-structure quality, it is the absolute threshold current which has to be as low as possible in an actual device. That is why the search operates in parallel to find an optimal way to define a stripe or array of stripes thus limiting device area to the extent dictated by a specific application. Issues of concern here are the minimal current spreading outside the stripe, preventing lateral carrier diffusion, optical confinement providing single-lobe, i.e. fundamental-mode, operation, and the absence of fabrication-induced defects leading to a high irradiative recombination rate and shorter device lifetime. Simplicity of the fabrication process is always a plus, albeit often sacrificed to the efficiency of the eventual laser.

(a) Broad-area lasers

Although broad-area lasers are of small practical importance because of their high threshold currents, the data obtained for them are often used as reference values or for comparison purposes and therefore some discussion of these could be useful. Early semiconductor lasers were just bars sawn or cleaved from an initial wafer and had a large area due to the easy process of their fabrication. Later on, broad-area lasers were prepared by depositing metal contact through the mask, leaving a wide window open. These devices show a non-uniform optical intensity distribution along the junction plane in the form of separate filaments. Current spreading to as far as 50 μm from the stripe may occur, strongly affecting the actual width of the pumped region. Finally, internally circulating modes may be excited in the chips with four cleaved facets. All these effects should be taken into account when fabricating and characterizing broad-area or wide-stripe samples.

(b) Stripe geometry lasers

A miscellany of techniques has been applied to the fabrication of lasers with stripe geometry active regions in order to ensure low threshold current, single lateral mode and high output power. Most of them are discussed in detail elsewhere [Casey and Panish 1978, Sze 1981, Agrawal and Dutta 1986] and so we shall only dwell upon some of the more most recent innovations and achievements.

The simplest contact stripe geometry laser as first introduced [d'Asaro 1973] is shown in Fig. 8.18. The structure was extensively employed using SiO_2 or Si_3N_4 as an insulating layer to block the electric current outside the stripe and no major changes could be imagined. However, the possibility of forming a high-quality native oxide on the AlGaAs surface was recently reported and a 20 mA threshold current was observed for a 10 μm wide stripe [Dallesasse and Holonyak 1991]. The lasers demonstrated high performance capability and a number of promising features, of which simplicity of fabrication is the most important advantage. This innovation opens a major pathway to circuit technology for AlGaAs–GaAs similar to that existing for silicon.

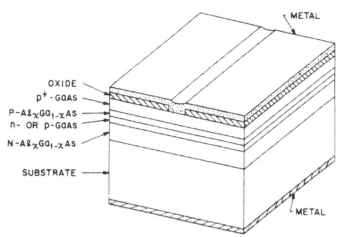

Fig. 8.18. Stripe geometry DH laser with four heteroepitaxial layers on the n⁺ GaAs substrate. The oxide layer isolates all but the stripe contact. The rectangular bar is typically 380 μm long, 250 μm wide and 120 μm thick. The drawing is not to scale in order to show the various layers.

In terms of both optical and electrical confinement, buried heterostructures (BH) described by Tsukada [1974] proved to be the most efficient. In this structure the active layer is totally surrounded by a wider-bandgap material giving tight confinement and, therefore, permitting extremely narrow stripe widths (Fig. 8.19(*a*)). Recently, submilliamp threshold currents were achieved for AlGaAs–GaAs BH lasers by several groups using MBE-grown lasers with a high-reflectivity coating [Lau et al. 1988] and MOCVD on a non-planar substrate [Narui et al. 1992]. The latter technique is of special interest since it combines such advantages as an extremely low continuous wave threshold current of 0.88 mA without facet coating, high external quantum efficiency of 70%, nearly round beam shape suitable for coupling to optical fiber and relatively simple one-step MOCVD growth. The stripe width in this case is as narrow as 0.6 μm. A schematic view of this unique structure is presented in Fig. 8.19(*b*).

Since a thin active layer taken to its ultimate conclusion results in quantum well phenomena, an extremely narrow stripe at some point exhibits carrier quantization in the lateral direction. A one-dimensional structure of this kind is a quantum wire (QWR) promising further progress in perfecting the static and dynamic qualities of lasers. MOCVD growth on V-grooved substrates was successfully used to yield both single-quantum-wire and multiquantum-wire lasers (see Fig. 8.20) [Simhony et al. 1991]. Low threshold currents were observed for both uncoated (2.5 mA) and coated (0.6 mA) devices. However, even lower threshold current should be achievable in order to guarantee lasing from the fundamental subband and take full advantage of the attractive properties inherent to the new generation of low-dimensional laser structures.

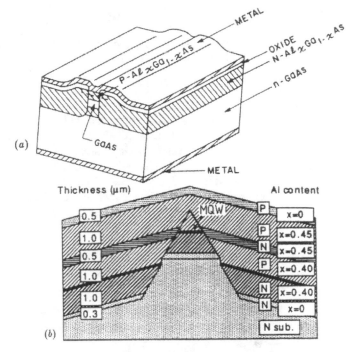

Fig. 8.19. (a) Buried heterostructure laser. The GaAs active region is completely
surrounded by $Al_xGa_{1-x}As$ [Tsukada 1971]. (b) Schematic representation of a
separated double-heterostructure (SDH) laser (after Narui et al. [1992]).

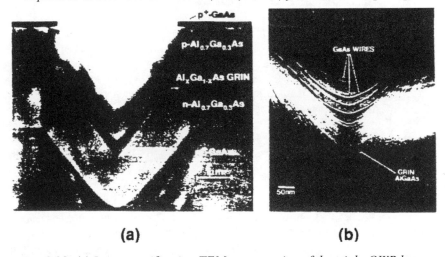

Fig. 8.20. (a) Low-magnification TEM cross-section of the triple-QWR laser,
showing the two-dimensional, tapered optical waveguide formed by the growth of
the graded-index (GRIN) layers in the V groove (x linearly graded between 0.2 and
0.7). (b) High-magnification TEM cross-section of the waveguide core, showing the
three crescent-shaped QWRs (after Simhony et al. [1991]).

8.4. New GaAs-based materials for lasers

Most of the results cited earlier in this chapter were obtained in the AlGaAs–GaAs system. This range of solid solutions is attractive due to its perfect lattice match to GaAs substrate for all Al compositions. High-quality material can be readily grown by any known growth technique: LPE, VPE, MBE and MOCVD. Some reports on LPE and MOCVD growth of InGaP and InGaAsP lattice matched to GaAs also appeared [Alferov et al. 1984, Brunemeir et al. 1985, Razeghi 1993] and efficient laser devices have been fabricated by this method [Garbuzov et al. 1990]. In recent years, however, new materials have appeared, which essentially require the non-equilibrium growth conditions provided by MBE or MOCVD for controllable layer formation. This includes InGaAs strained layers for the 1 μm wavelength range and AlGaInP for visible range emission. In the following section we shall cover both the major considerations favoring these materials and some new device developments.

8.4.1. Strained InGaAs layer laser structures

Semiconductor lasers emitting in the range 0.8 μm to 1.0 μm are important sources of light for a variety of applications such as pumping erbium-doped optical fiber, pumping solid state lasers and direct frequency doubling for the generation of blue and other visible light lasers. They have a number of attractive features, including the possibility of fine bandgap adjustment, low threshold current density [Chand et al. 1991] and large gain permitting high-frequency modulation. Of particular importance is the use of 0.98 μm laser diodes in communication systems for optical fiber pumping.

GaInAs/GaAs strained-layer quantum well structures have been shown to be suitable for the fabrication of laser diodes used for these purposes. In most of the previous works, AlGaAs served for the cladding layers. However, in some recent work GaInP was used in place of GaAlAs [Chen et al. 1991, Ohkubo et al. 1992]. It was shown that GaInP has advantages over AlGaAs. These advantages include selective chemical etching between $Ga_{0.51}In_{0.49}P$ and GaAs in addition to less surface oxidation during the fabrication process and device operation due to the absence of Al. Also the growth temperature of GaInP is 510 °C, compared with 700 °C or higher for AlGaAs, which makes this material compatible with monolithic integration for optoelectronic integrated circuits. Finally due to the absence of Al and presence of In, it is expected that these devices will not be affected by dark-line defects causing catastrophic degradation in AlGaAs-based lasers emitting in the same wavelength range.

Extremely low threshold current densities of 45 A·cm^{-2} have been reported for MBE-grown InGaAs quantum well diodes [Chand et al. 1991]. An example of recent progress in the area of 0.98 μm lasers is the high-quality $Ga_{0.8}In_{0.2}As/GaAs/Ga_{0.51}In_{0.49}P$ multiple-quantum-well laser emitting at 0.98 μm that has been grown by low-pressure MOCVD (LP-MOCVD) [Mobarhan et al. 1992a]. Continuous wave operation with an output power of 500 mW per facet was achieved at room temperature for a broad-area laser with 130 μm width and 300 μm cavity length. The external quantum efficiency exceeded 75% with excellent homogeneity and uniformity.

The epitaxial layers were grown on a n^+ silicon-doped GaAs substrate (1 0 0) with a 2° misorientation. The process of growth was carried out at a pressure of 76 Torr in a horizontal reactor at a substrate temperature of 510 °C. Trimethylindium (TMI) and triethylgallium (TEG) were used as sources of In and Ga, while pure arsine (ASH_3) and phosphine (PH_3) provided As and P, respectively. Hydrogen (H_2) was used as a carrier gas. SiH_4 and diethylzinc (DEZn) were used as n-type and p-type dopants, respectively. The laser structure consists of (i) 1 μm thick $Ga_{0.51}In_{0.49}P$, silicon-doped ($N_D - N_A \approx 10^{18}$ cm^{-3}) cladding layer, (ii) 500 Å GaAs waveguide layer, (iii) three 40 Å $Ga_{0.8}In_{0.2}As$ strained-layer quantum wells separated by 100 Å thick GaAs barriers, (iv) a second 500 Å GaAs waveguide layer, (v) 1 μm thick $Ga_{0.51}In_{0.49}P$, Zn-doped ($N_A - N_D \approx 5 \times 10^{17}$ cm^{-3}) cladding layer and (vi) 0.5 μm thick GaAs, Zn-doped ($N_A - N_D \approx 10^{19}$ cm^{-3}) contact layer. Fig. 8.21 shows a typical electrochemical doping profile of the laser structure.

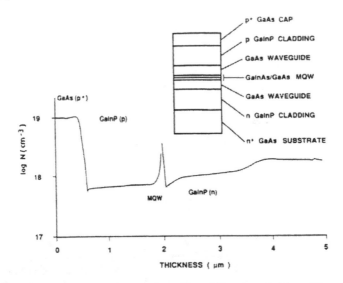

Fig. 8.21. Typical electrochemical profile of $Ga_{0.8}In_{0.2}As/GaAs/Ga_{0.51}In_{0.49}P$ strained layer multiple-quantum-well structure grown by LP-MOCVD for 0.98 μm lasers. SiH_4 and DEZn were used as n-type and p-type dopants, respectively (after Mobarhan et al. [1992a]).

The sample was then thinned down to about 100 μm thickness. The contacts consist, on the p-side, of Pt(1500 Å)/Au(3000 Å) and, on the n-side, AuGe(n)(900 Å)/Mo(2000 Å)/Au(6000 Å). After cleaving individual diodes were mounted p-side down on Au-plated copper heat sinks using In preforms. No antireflection or high-reflectivity coatings of any kind were applied on the cleaved mirror facets. Excellent homogeneity of the device was established through a very uniform near-field emission pattern at an injection current just above threshold. The wavelength of emission was found to be at 0.98 μm and the device operated in a multilongitudinal mode.

Long term reliability is one of the most critical issues for laser diodes. It is more crucial for the lasers for communication systems such as 0.98 μm lasers due to the high cost of replacement and/or repair. Operation of InGaAs/InGaP/GaAs lasers for over 2000 h at 50 °C with emitting output power of 1 W in CW mode has been reported [Razeghi 1994]. After 2000 h of aging, the increase of the operating current had not exceeded 15% (Fig. 8.22). Operation at high temperature is a good indicator as to how robust the device is. Operation of Al-free InGaAs/GaAs lasers at 100 °C with emitting power of 800 mW was achieved (Fig. 8.23(a)). The characteristic temperature T_0 was measured up to 350 K in temperature range of 15–41 °C while $T_0 \sim 250$ K overall temperature range (Fig. 8.23(b)). This is favorably compared with a recent report on an Al-containing counterpart where $T_0 = 160$ K was achieved [Chand et al. 1994].

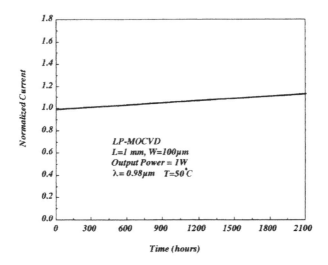

Fig. 8.22. 1 W constant-power CW aging test at 50 °C for
$In_{0.20}Ga_{0.80}As/GaAs/In_{0.49}Ga_{0.51}P$ SCH-QW lasers emitting at 0.98 μm (T = 50 °C, L = 1 mm, stripe width W = 100 μm).

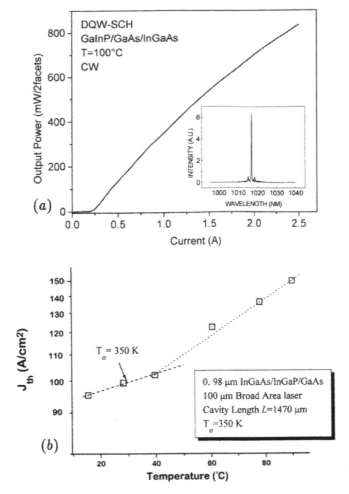

Fig. 8.23. (a) Output power for $In_{0.20}Ga_{0.80}As/GaAs/In_{0.49}Ga_{0.51}P$ ($\lambda = 0.98 \mu m$) double QW-SCH lasers emitting at $T = 100$ °C. (b) Temperature dependence of the threshold current density (J_{th}) for a typical $In_{0.20}Ga_{0.80}As/GaAs/In_{0.49}Ga_{0.51}P$ ($\lambda = 0.98 \mu m$) 100 μm broad-area laser ($L = 1470$ μm). The characteristic temperature T_0 is 350 K in the temperature range 15–41 °C while $T_0 \sim 250$ K over the whole temperature range.

A substantial reduction of the threshold current for this kind of laser has been recently achieved by using buried-ridge structures (BRS) very similar to the buried heterostructures described above [Mobarhan et al. 1992b]. The laser structure was grown by two-step LP-MOCVD. Photolithographic and selective chemical etching techniques were used to prepare the ridge. Lateral current localization was obtained by shallow ion implantation. The laser diodes were mounted *p*-side down on copper heat sinks using In bonding. With this laser, a continuous wave output power of 40 mW has been obtained without any facet coating. The threshold current was

30 mA for a laser with a 2 μm wide ridge. The laser diodes processed from a 10 cm^2 area of the same wafer showed excellent homogeneity and uniformity. The characteristic temperature T_0 was measured to be about 95 K. Fig. 8.24 shows the schematic diagram of this laser.

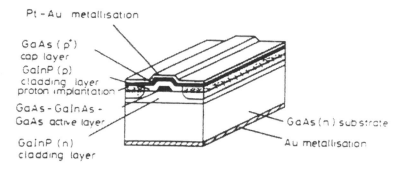

Fig. 8.24. Schematic diagram of a buried ridge structure (BRS), $Ga_{0.8}In_{0.2}AsZGaAs/Ga_{0.51}In_{0.49}P$ single-quantum-well laser, grown by LP-MOCVD with proton implantation (after Mobarhan et al. [1992b]).

Submilliamp threshold InGaAs/GaAs/AlGaAs laser array elements grown by single-step MOCVD growth on non-planar substrates have been reported [Zhao et al. 1994]. The epitaxial growths were performed in an atmospheric-pressure MOCVD reactor. The substrate non-planarity is introduced by chemical etching of mesas with (1 0 0) tops and (1 1 1)A sidewalls. The growth properties of InGaAs/GaAs/AlGaAs on the mesa top and the mesa sidewall are strongly influenced by the sidewall orientation, growth temperature and V/III ratio. By controlling these variables during the lower cladding layer growth, the mesa top width is reduced by 1 to 2 μm depending on the growth conditions. The InGaAs quantum well growth rate on the (1 1 1)A mesa sidewall is four times slower than on the (1 0 0) mesa top, resulting in a cathodoluminescence (CL) emission wavelength on the mesa sidewall that is about 100 nm shorter than on the mesa top. The width of this active region is controlled by the starting mesa width and the growth of the $Al_{0.6}Ga_{0.4}As$ lower cladding layer. Current confinement in this laser is provided by the orientation-selective doping properties of the undoped $Al_{0.6}Ga_{0.4}As$ cladding layer which results in p-type doping on the (1 0 0) mesa top and n-type doping on the sidewalls. The L–I curve of a SQW laser with 0.6 μm wide active region is shown in Fig. 8.25. The room-temperature threshold current for uncoated SQW lasers is 0.7 mA and for uncoated DQW lasers it is 0.5 mA. These lasers have an excellent external quantum efficiency of 80–85%. High reflectance/low reflectance (HR/LR) coated 120 μm long SQW lasers with 0.9 μm wide active regions showed a RT CW threshold current of 0.28 mA. They also exhibit excellent threshold and quantum efficiency uniformity and high yield because of the simple growth and processing procedures. Fig. 8.26 shows the threshold currents of 1 mA laser array elements on a single bar.

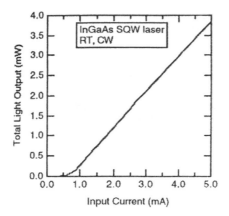

Fig. 8.25. The L–I curve of a SQW laser with a 0.6 μm wide active region (after Zhao et al. [1994]).

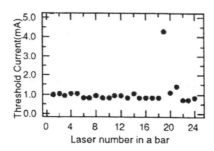

Fig. 8.26. Threshold currents of a laser array (after Zhao et al. [1994]).

The use of a GaInP–GaAs superlattice as an optical confinement layer (SL-OCL) for a laser has recently been reported for 0.98 μm InGaAs/GaInP lasers [Usami and Matsushima 1994]. Fig. 8.27 shows the band diagrams for three types of SCH structure for comparison. Fig. 8.28, Fig. 8.29, and Fig. 8.30 show the inverse differential quantum efficiency versus cavity length, the threshold current density versus inverse cavity length and temperature dependence of threshold current, respectively. The beneficial effect of SL-OCL is obvious from the figures. An internal quantum efficiency as high as 90% and the threshold current density as low as 280 A·cm^{-2} are obtained in a laser with a superlattice optical confinement layer. The SL-OLC structure, in addition to the ordinary graded index (GRIN) effect, is also expected to have the multiple-quantum-barrier (MQB) effect [Iga et al. 1986], which enhances the confinement of carriers.

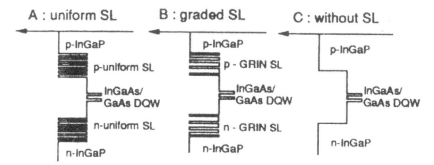

Fig. 8.27. *Band diagrams of LDs with SL-OCL (type A, type B) and without SL-OCL (type C) (after Usami and Matsushima [1994]).*

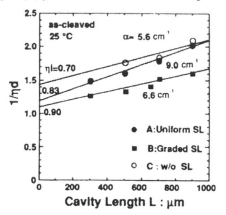

Fig. 8.28. *Inverse differential quantum efficiency versus cavity length (after Usami et al. [1994]).*

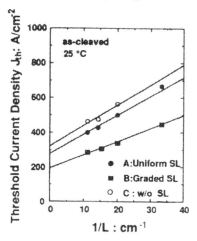

Fig. 8.29. *Threshold current density versus inverse cavity length (after Usami et al. [1994]).*

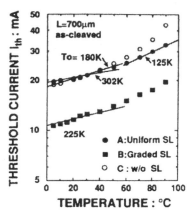

Fig. 8.30. Temperature dependence of threshold current (after Usami et al. [1994]).

8.4.2. AlGaInP visible lasers

Visible light laser diodes are extremely important light sources having a wide range of applications, including plastic fiber optics, optical recording, high-resolution television, pumping of Cr-doped solid-state lasers, information processing systems, laser printers and bar-code scanners. Shorter-wavelength laser operation is better for obtaining small focused spots and highly luminous light sources for the human eye. It is expected that visible light laser diodes will gradually replace HeNe gas lasers in these fields.

Although the AlGaAs system is the most intensively investigated, the shortest continuous wave lasing wavelength achieved is in the neighborhood of 0.68 μm, which is very near to the limit for this material system. The AlGaInP system has the potential for the shortest-wavelength emission (0.580 μm) of any III–V semiconductor, with the advantage of complete lattice matching to GaAs substrate. However, the AlGaInP single epitaxial layers are extremely difficult to grow in a controlled manner by any near-equilibrium growth method such as LPE or VPE. This is due to the large Al segregation coefficient. For example, in the growth of the quaternary alloy AlGaInP the Al distribution coefficient from vapor phase to the solid phase is of the order of unity. As a result, MBE and MOCVD (not being equilibrium growth techniques) are the best methods for growing these compounds. The theoretically predicted wavelength region for the AlGaInP double-heterostructure system is 0.580 μm to 0.680 μm which, as mentioned, includes the lasing wavelength of HeNe gas lasers (0.6328 μm). However, even for completely lattice-matched GaInP, longer emission wavelengths than theoretically predicted usually result. This can be due to unintentional dopant incorporation into the GaInP causing radiative emission via impurity levels, which have a longer wavelength than that of band to band emission. A slightly excessive In composition in GaInP can also give rise to a smaller bandgap energy. In either case the result is a longer-wavelength emission.

There are different mechanisms available to shorten the wavelength and tailor the bandgap to a specific desired value. One method is by increasing the Al content

x of the $(\text{Al}_x\text{Ga}_{1-x})_{0.5}\text{In}_{0.5}\text{P}$ active layer. Another way is to use a GaInP quantum well active layer and AlGaInP barrier layers. The application of quantum well structures to laser diodes makes it possible to achieve laser operation at a low threshold current density and high T_0 value, as well as to tailor the bandgap. A reduction in the threshold current density is expected, for example, by optimizing barrier height, barrier layer thickness and well layer numbers as well as the growth conditions. Also, the application of strained-layer quantum well structures to laser diodes can make it possible to achieve laser operation at shorter wavelengths without increasing the Al content, which can give rise to surface oxidation problems and dark-line-induced degradation effects.

High-power continuous wave operation of broad-area InGaAlP [Kobayashi et al. 1985] and high-power operation of InGaAlP visible lasers using buried ridge waveguide structures have been reported [Fujii et al. 1987]. The performance of these visible lasers is limited by the temperature dependence of the threshold current which is more severe than in AlGaAs lasers. However, 340 mW continuous wave and approximately 1 W pulse output has recently been made possible by reducing the threshold current density to 425 A·cm^{-2} in the GaInP strained-layer single-quantum-well GRIN separate confinement heterostructure (SQW-GRIN-SCH) shown in Fig. 8.31 [Serreze et al. 1991]. The emission wavelength was 665 nm, which is useful for practical applications. Fundamental transverse-mode operation of a buried heterostructure laser emitting at 600 nm has been achieved with a relatively thick (400 Å) GaInP active layer.

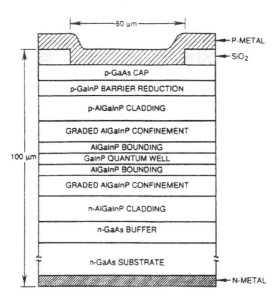

Fig. 8.31. Schematic cross-section (not to scale) of the 60 μm wide, oxide stripe, strained layer (SQW-GRIN-SCH) visible light laser (after Serreze et al. [1991]).

AlGaInP lasers are very susceptible to facet degradation at high-power operation. To overcome this problem, several window structures [Ueno et al. 1990, Arimoto et al. 1993], which reduce optical absorption near the facets, have been reported. Among these window structures, the window grown on a facet structure is rather simple and effective [Sasaki et al. 1991]. The kink-free maximum output of 295 mW was obtained at room temperature, which was about twice as much as that of the conventional one (no window on the facets). The threshold current and the slope efficiency under 150 mW were the same as those of the conventional one, which indicates that the window layers do not affect the laser properties.

8.4.3. GaInAsP/GaAs lasers

(a) 0.808 μm GaInAsP/GaAs high-power lasers

Up to now, commercially available diode pumped Nd-YAG solid-state lasers used diode lasers emitting at 808 nm fabricated from materials in the $Al_{1-x}Ga_xAs$ system. However, the lifetime of AlGaAs-based laser diodes limits the reliability of a whole system due to the problem connected with the presence of Al in the diode materials. High interaction of Al with oxygen leads to oxide formation at the mirror facet and enhances the non-radiative recombination of injected carriers near the mirror facet which creates the overheating of the mirror, and decreases the lifetime of the device. Another lifetime limiting factor for AlGaAs-based laser diodes is the formation of dark-line defects, as a consequence of the spreading of the dislocations in the active region during high-power operations.

Both of these problems can be solved by the replacement of $Al_{1-x}Ga_x$ As with the Al-free $Ga_xIn_{1-x}As_yP_{1-y}$ system [Garbuzov et al. 1991, Razeghi 1993]. It has been demonstrated that the non-radiative recombination velocity near the mirror facet for the GaInAsP system is at least two orders of magnitude less than for the AlGaAs system. On the other hand, due to the presence of big atoms such as In, the dislocation mobility of GaInAsP alloys is much less than for the AlGaAs system. The other advantages of GaInAsP systems include the possibility of selective etching and regrowth which is essential for laser diode processing. Table 8.1 lists the advantages of the GaInAsP–GaAs system over the AlGaAs system based on cost efficiency.

$Ga_{0.51}In_{0.49}P$–$Ga_{0.87}In_{0.13}As_{0.75}P_{0.25}$–$Ga_{0.51}In_{0.49}P$–GaAs double-heterostructure (DH) lasers (Fig. 8.32(a)) emitting at 0.808 μm have been fabricated by LP-MOCVD [Razeghi 1991]. Broad-area contact laser diodes with a stripe width of 100 μm were prepared by a photolithographic technique. Measurements of the absolute values of external efficiency for spontaneous emission as well as the temperature dependence of the emission efficiency have shown that the internal efficiency of radiative recombination in the GaInAsP active layer for these laser diodes is close to 100% [Diaz et al. 1994b]. This means that all carriers injected in the $Ga_xIn_{1-x}As_yP_{1-y}$ active region recombine radiatively at room temperature. In accordance with these results, the threshold current density for these diodes was close to the theoretical limit and better than the data known for the best double-heterostructure AlGaAs/GaAs lasers [Dyment et al. 1974].

AlGaAs	InGaAsP
High growth temperature (~700 °) necessary to suppress oxidation of Al during epitaxial growth	Lower growth temperature (~500 °C) entails lower consumption of source materials
Ion implantation necessary to achieve high doping level of the p-GaAs contact layer	Lower growth temperature facilitates obtaining high p-doped GaAs ($N_a \sim 10^{20}$ cm^{-3}) with excellent surface morphology
Fabrication of mesa structures and gratings for DFB lasers requires plasma or ion beam etching to avoid the oxidation of AlGaAs	Selective chemical etching is possible for InGaP, GaAs and InGaAsP materials without oxidation
Non-absorbing mirror fabrication necessary to avoid mirror facet overheating and catastrophic damage	Higher (~10 times) COD limit eliminates the necessity of non-absorbing mirror preparation
Pre-selection of diode chips required to eliminate occasionally damaged or defective diodes	Highly uniform defect-free MOCVD material and ruggedness of In-containing compounds provide for the yield of good chips close to 100%
Extreme caution required when handling laser wafers and chips because of material sensitivity to dislocation spreading	Dislocation movement is inhibited by large In atoms and material is insensitive to minor impacts
Burn-in testing and selection required to eliminate diodes demonstrating rapid degradation	Material resistance to dark-line and dark-spot defect formation provides for the yield of good lasers close to 100%

Table 8.1. The advantages of Al-free InGaAsP high-power diode allow lower production costs in comparison with AlGaAs lasers.

Separate confinement heterostructure quantum well (SCH-QW) laser diodes have been fabricated. The band diagram and details of the structure are shown in Fig. 8.32(b). By changing the thickness and composition of the GaInAsP active layer, one can vary the emitting wavelength of the lasers between 0.7 and 1 µm. Due to the very thin active region (L_z in Fig. 8.32(b)) and quantum size effects, the energy distribution of the carriers is narrower in SCH-QW lasers than for DHs with a thick active layer, which leads to the improvement of the main laser parameters such as the threshold current density (J_{th}), differential efficiency (η_d), temperature dependence of the threshold current density (T_0), etc.

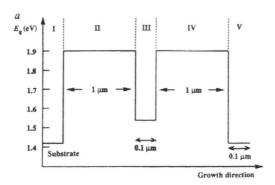

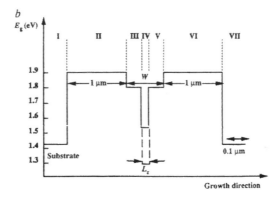

Fig. 8.32. Schematic diagram of the band structure of (a) a double-heterostructure (DH) and (b) separate confinement heterostructure (SCH) single-quantum-well (SQW) GaInAsP lasers emitting at 0.808 μm (after Razeghi [1994]).

As in the case of DHs, broad-area lasers have been fabricated, and diodes were In bonded onto copper heat sinks. The value of the series resistance measured under the continuous wave (CW) regime for bonded laser diodes with 1 mm cavity length was as low as 0.04 Ω which is three times less than for similar GaAlAs/GaAs lasers.

An 808 nm GaInAsP laser diode with a facet coating gave out as high as 4.6 W of power from the antireflective- (AR) coated facet alone (Fig. 8.33(*a*)). One laser facet was high reflection (HR) coated to 93% and the other facet was AR-coated to 7%. The differential quantum efficiency from the AR-coated facet was 82%. A slope efficiency of 1.2 $W \cdot A^{-1}$ was obtained from Fig. 8.33(*a*).

In fact, a record low of 4.5 A of input current for 4.6 W output power was observed. The quantum differential efficiency and output power were even more improved by using more sophisticated ohmic contact formation. When Ti/Pt/Au was used for *p*-ohmic contact material instead of Au/Zn as used in previous cases, the differential efficiency reached 1.3 $W \cdot A^{-1}$ (87%) without mirror coating (Fig. 8.33(*b*)). It is almost impossible to obtain such high power and efficiency from uncoated AlGaAs/GaAs lasers due to higher non-radiative recombination and degradation related to surface effects.

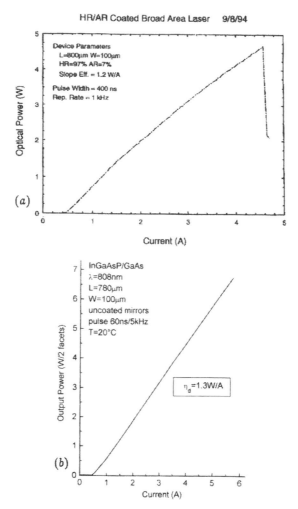

Fig. 8.33. Output power as a function of injection current for GaInP/GaInAsP/GaAs 0.808 µm laser diodes with (a) coated and (b) uncoated facets.

The laser diodes were tested in the CW regime as well as for the quasi-CW regime (pulse width 200 µs, repetition rate 20 Hz), which is typical for a Nd:YAG laser pumping application. In quasi-CW operation the differential efficiency obtained was 1.2 W·A^{-1} (Fig. 8.34(a)) at high-power operation of 27 W, while the maximum output could reach up to 67 W (Fig. 8.34(b)) when a laser bar was made up of 32 diodes. In CW operation, the output power from a single laser diode was up to 6 W, limited by the maximum driving current available in our setup (Fig. 8.34(c)). All of the lasers used in the quasi-CW and CW operation tests were without mirror coating. The distribution of the radiation intensity along 100 µm of diode aperture was highly uniform in a wide range of driving current densities.

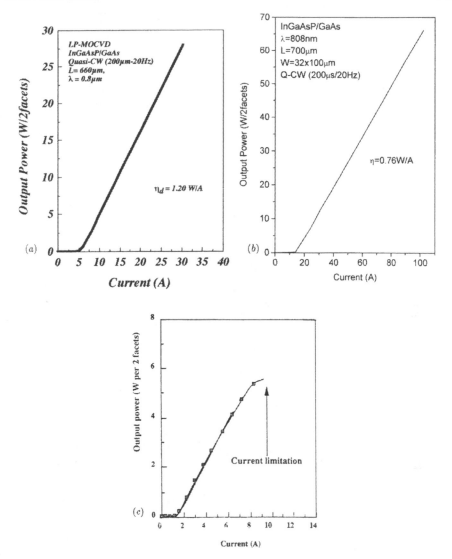

Fig. 8.34. Output power versus injection current for GaInP/GaInAsP/GaAs 0.808 μm laser diodes: (a) quasi-CW, 5 mm wide laser bar; (b) quasi-CW, 1 cm wide laser bar; and (c) CW single laser diode.

Typical threshold current densities (J_{th}) were as low as 220 A·cm^{-1}, with a differential efficiency as high as 1.2 W·A^{-1} for 1 mm cavity length diodes, with an emitting wavelength of 808 nm. For comparison, the best commercially available AlGaAs-based diodes emitting at 0.808 μm wavelength have J_{th} = 300–400 A·cm^{-2}, and η_d = 0.5–0.8 W·A^{-1}. T_0 reached 155 °C (Fig. 8.35) which is comparable with the best results obtained for AlGaAs (130–160 °C).

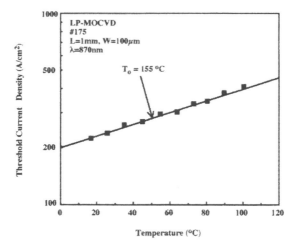

Fig. 8.35. Determination of T_0 from $J(T)$–T relation (after Razeghi et al. [1994]).

Optical pumping applications impose additional requirements to high-power diodes—primarily the possibility of obtaining a specific emission wavelength. The lasing wavelength may be adjusted precisely to a desired position by varying the cavity length and/or heatsink temperature of the laser diodes. A spectrum half width as narrow as 2 nm was maintained up to 3–4 I_{th} under pulse testing.

	AlGaAs	InGaAsP (results of CQD)
λ (nm)	800	800
J_{th} (A·cm^{-2})	230[a]–400[b]	240–500
ηd (W·A^{-1})	1.3[a]–0.8[b]	1.3
T0 (°C)	130[a]–160[b]	155–175
R (Ω)	0.25[b]	0.1
Thermal resistance (K·W^{-1})	10[b]	1
Transverse beam divergence	32–48[o][b]	26°
COD limit for uncoated facets (MW·cm^{-2})	0.5–1[c]	6
Lifetime under 1 MW·cm^{-2} (h)	200 (uncoated facets)d	> 6000 (uncoated facets)

Table 8.2. Comparison of AlGaAs versus aluminum-free InGaAsP high-power laser diodes (aperture 100 µm).
[a] Yariv, A. et al. 1986 Electron. Lett. **22** 79.
[b] Spectra Diode Labs, 1994 Product Catalog.
[c] Fukuda, M. 1991 Reliability and Degradation of Semiconductor Lasers and LEDs (Boston: Artech House).
[d] Tang, W.C., Altendorf, E.H., Rosen, H.J., Webb, D.J., and Vettiger, P. 1994 Electron. Lett. **30** 143.

The half width of the optical beam divergence in the direction perpendicular to the active layer is also another important factor for pumping diodes. The value of the half width for these lasers is 27° (Fig. 8.36) which is much less than for pumping AlGaAs lasers with a 40–48° divergence for a similar structure. The small difference in refractive indices of the waveguide and cladding layers in the InGaAsP system is the reason for such a narrow half width of beam divergence.

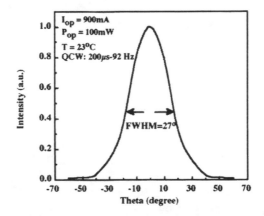

Fig. 8.36. Half width of optical beam divergence in the direction perpendicular to the active layer (after Razeghi et al. [1994]).

Preliminary lifetesting of GaInAsP (SCH-QW) lasers emitting at 808 nm for uncoated diodes at room temperature for more than 4000 h shows less than 1% decrease of output power (Fig. 8.37(*a*)). This device is currently running continuously. Selected laser diodes were tested in a constant-output-power condition (Fig. 8.37(*b*)). These devices also did not show any noticeable increase of driving current due to degradation after over 600 h of operation. Equivalent results cannot be obtained with uncoated AlGaAs/GaAs lasers.

The major laser characteristic parameters of GaInAsP/GaAs-based high-power laser diodes, such as J_{th}, T_0 and η_d, as listed in Table 8.2, are comparable with the best results obtained with AlGaAs-based laser diodes. On the other hand, the series resistance, beam divergence and resistance to degradation under high-power operation of GaInAsP-based lasers are superior to AlGaAs system. The advantages of the GaInAsP system mentioned above are sufficient for replacing the AlGaAs-based lasers in present-day high-power laser applications.

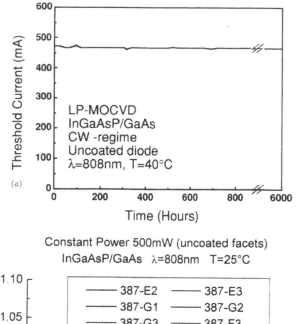

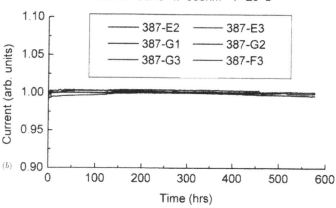

Fig. 8.37. Aging histories of GaInP/GaInAsP/GaAs 0.808 μm uncoated laser diodes: (a) single laser diode, air ambient; (b) 6 identical laser diodes, N_2 ambient.

(b) Temperature dependence of the Threshold currentdensity J_{th} and differential efficiency η_d of high-power GaInAsP/GaAs ($\lambda = 0.8$ μm) lasers

Separate confinement heterostructure (SCH) InGaAsP/GaAs with 300 Å active layer thickness grown on (1 0 0)-oriented Si-doped GaAs substrates by low-pressure MOCVD are used in this study. Undoped quaternary active region $In_{0.13}Ga_{0.87}As_{0.74}P_{0.26}$ and waveguide $In_{0.37}Ga_{0.63}As_{0.25}P_{0.75}$ were formed between Si-doped $(N_d \sim 5 \times 10^{17}\ cm^{-3})$ and Zn-doped $(N_a \sim 6 \times 10^{17}\ cm^{-3})$ InGaP cladding layers. Broad-area contact laser diodes with a stripe width of 100 μm were fabricated using standard processing methods. The diodes were mounted on copper heat sinks by indium bonding without a mirror coating.

The dependences of J_{th} and η_d on temperature were measured under pulse operation (pulse width 100–400 ns, repetition rate 5–2.5 kHz) using an integrating sphere with an Si photodiode for temperatures ranging from 15 °C to 120 °C. The measured data are shown in Fig. 8.38 and Fig. 8.39 by full squares.

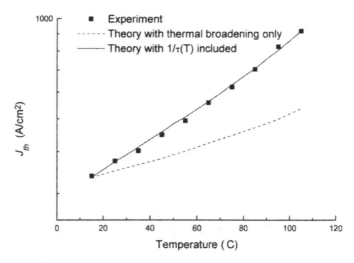

Fig. 8.38. The threshold current density (J_{th}) of lasers with cavity length L = 1090 mm for different operating temperatures (T) (after Yi et al. [1995]).

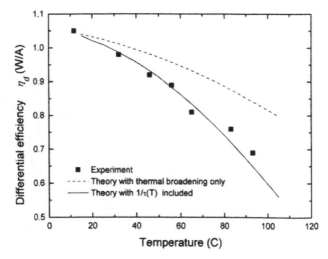

Fig. 8.39. The differential efficiency (η_d ($W \cdot A^{-1}$)) of lasers with cavity length L = 1090 mm for different operating temperatures (T) (after Yi et al. [1995]).

The variation of J_{th} and η_d as a function of cavity length, considering the thermal broadening of the gain spectrum with momentum relaxation rate ($\hbar/\tau_0$, $t_0 = 4 \times 10^{-14}$ s) have been calculated [Yi et al. 1995]. The gain thermal broadening effect originates from the microscopic gain expression [Chinn et al. 1988].

Eq. (8.6) $$g(\hbar\omega) = \frac{q^2 |M|^2}{\hbar\omega\varepsilon_0 m^2 c_0 \hbar n L_z} \sum_{ij} m_{r,ij} A_{ij} C_{ij} [\tilde{f}_c - (1 - \tilde{f}_v)]\theta(\hbar\omega - E_{ij})$$

where $|M|^2$ is the bulk momentum transition matrix element, ε_0 is the permittivity, m is the free-electron mass, c_0 is the vacuum speed of light, n is the effective refractive index, i, j are conduction-electron valence-heavy-hole quantum numbers, $m_{r,ij}$ is the reduced mass for transition i, j, C_{ij} is the spatial overlap factor between states i and j, $\theta(x)$ is the Heaviside step function and E_{ij} is the transition energy between stages i and j. A_{ij} is the anisotropy factor for transition i, j given by

Eq. (8.7)
$$A_{ij} = (3/4)(1 + \cos^2 \Theta_{ij}) \quad (\text{for TE})$$
$$A_{ij} = (3/4)\sin^2 \Theta_{ij} \quad (\text{for TM})$$

The angular factor Θ_{ij} is given by $\cos^2 \Theta_{ij} = E_{ij} / E$. $\tilde{f}_c(E)$ and $\tilde{f}_v(E)$ are convoluted quasi-Fermi distribution functions, e.g.

Eq. (8.8) $$\tilde{f}_c(E) = \int_{E_g}^\infty d\xi \left[\frac{1}{\pi} \frac{\hbar/\tau}{(\xi - E)^2 + (\hbar/\tau)^2} \right] \left[\exp\left(\frac{\xi - E_{fc}}{kT} \right) + 1 \right]^{-1}$$

where E_{fc} is the quasi-Fermi level of electron carriers in an active medium, determined by the injection carrier density, and $\hbar/\tau$ is the momentum relaxation rate (=scattering rate of carriers). $\tilde{f}_c(E)$ and $\tilde{f}_v(E)$ have broader distributions over energy levels at high temperature, causing a broader gain spectrum.

Numerical calculation of J_{th} and η_d was carried out, taking into account Auger recombination in the active region, free-carrier absorption in the waveguide and current leakage to the cladding barriers. The total current through a diode includes the following components

Eq. (8.9) $$J_{total} = J_{active} + J_{wg} + J_{Auger} + J_{leakage} \equiv J_{majority} + J_{leakage}$$

where J_{active} and J_{wg} represent radiative recombination currents in the active region and waveguide, respectively. J_{Auger} is the non-radiative Auger recombination current in the active region and $J_{leakage}$ is the current of minority carriers overflowing from the waveguide to the cladding layers, $J_{leakage}$ is essentially a current of electrons leaking to the p-cladding layer and eventually recombining on the boundary with the highly doped p-contact layer. The higher effective mass of holes results in a quasi-Fermi level position close to the valence band edge of the active layer so that the hole leakage current may be effectively neglected [Zory 1993]. The first three terms in Eq. (8.9) were calculated according to standard theory [Chinn et al. 1988] to give $J_{majority}$ Closer attention was paid to the calculation of the leakage current in order to explain the experimental results. In a realistic model, the electric field $E(x)$ inside the cladding layer should be computed self-consistently with excess electron and hole distributions ($dn(x)$ and $dp(x)$). In order to perform this analysis, the boundary condition for the electric field E_0 at the interface of the cladding and contact layers was determined from the majority carrier current

Eq. (8.10) $J_{\text{majority}} = qm_p P_0 E_0 - qD_p \nabla dp$

which gives

Eq. (8.11) $E_0 \cong \dfrac{1}{qm_p P_0}\left(J_{\text{majority}} + qD_p \nabla dn\right) = \dfrac{1}{\sigma_p}\left(J_{\text{majority}} + \dfrac{D_p}{D_n} J_{\text{leakage}}\right)$

Here, P_0, m_p and σ_p are the hole doping concentration, mobility and conductivity in the cladding layer, and $dn(x) \approx dp(x)$ was used in Eq. (8.10). dn at the interface of the cladding layer and waveguide layer is obtained from the thermal equilibrium distribution: $\delta n \approx (N_c N_v / P_0)\exp[-\beta(\Delta E_{\text{g,cladding}} - (E_{\text{fc}} + E_{\text{fv}}))]$. At the interface of the cladding layer and the contact layer, $dn = 0$ is satisfied because of surface recombination [Casey 1978]. In the above expression, N_c and N_v are the density of states for the conduction band and the heavy-hole valence band, and $\Delta E_{\text{g,cladding}}$ is the difference between the energy gap of the cladding layer and the waveguide layer. Once all the boundary conditions are given, $E(x)$ and $dn(x)$ can be determined by numerically solving Poisson's equation and the diffusion–drift equation. With the obtained $dn(x)$, J_{leakage} is calculated by $J_{\text{leakage}} = qD_n \nabla \delta n(x)$. With reference to parameters such as the internal optical loss α_0 and optical confinement factor G, the values used by Diaz et al. [1994c] were employed ($\alpha_0 = 4$ cm^{-1} G = 0.050). The theoretical results are shown in Fig. 8.38 and Fig. 8.39, showing that the experimental data have a much stronger temperature dependence than this effect alone could explain. This led us to consider the temperature dependence of the momentum relaxation rate $\hbar/\tau$ and the way it affects J_{th} and η_{d}.

Theoretical calculations [Asada et al. 1984, Fortini et al. 1978, Masu et al. 1983] show that the major mechanisms contributing to the momentum relaxation rate $\hbar/\tau$ are (i) carrier–carrier scattering [Asada et al. 1984], (ii) carrier–phonon scattering [Fortini et al. 1978] and (iii) alloy scattering [Masu et al. 1983] and their temperature dependences are well known. Since the three major sources of the momentum relaxation have comparable magnitudes in the temperature range of this investigation, it is difficult to theoretically predict the temperature dependence of the relaxation width $\hbar/\tau$ when the mechanisms are combined. Yi et al. [1995] have obtained the temperature dependence of $\hbar/\tau$ by measuring photoluminescence spectra and comparing them with the theoretically calculated spectrum, as done in Yamada et al. [1980]. The jagged curves in Fig. 8.40 show the experimental photoluminescence spectrum obtained from the studied laser structure in the temperature range from 15 to 105 °C. The detailed setup for this measurement was described elsewhere [Diaz et al. 1994c]. The solid curves in Fig. 8.40 represent the theoretical calculation of the spontaneous emission spectrum $R(E)$ obtained as follows [Chinn et al. 1988]

Eq. (8.12) $R(E) = \dfrac{16p^2 N q^2 E |M|^2}{e_0 m^2 c_0^3 h^4 L} \sum_{i,j} m_{r,ij} c_{ij} [\tilde{f}_c \tilde{f}_v] q(E - E_{ij})$

At several temperatures between 15 and 105 °C, the theoretical spectrum is fitted to experimental data by adjusting two parameters, E_g and $\hbar/\tau$, as was done in Fig. 8.40. Roughly speaking, E_g determines the peak position, and $\hbar/\tau$ gives the half width of the spectrum. In this way, we obtained E_g and $\hbar/\tau$ of the active layer material at each temperature and the results are shown in Fig. 8.41.

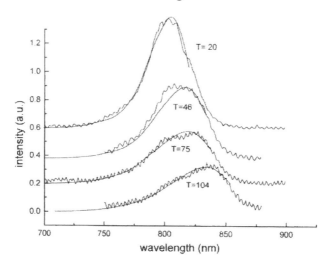

Fig. 8.40. Measured photoluminescence spectra of the studied laser structure at different temperatures (jagged curve). The spectra were reproduced by calculation with best-fit parameters for (i) momentum relaxation rate and (ii) energy gap, at each temperature (full curve) (after Yi et al. [1995]).

Both E_g and τ decrease from 1.51 eV to 1.45 eV and from 4×10^{-14} s to 1.8×10^{-14} s, respectively, as the temperature varies from 20 to 105 °C. Fig. 8.41 shows that $E_g(T)$ is an almost linear function of temperature, while $\tau(T)$ has a sublinear dependence approximated by $T^{-1/2}$. It is well known that the alloy scattering time (= momentum relaxation time t) has a $T^{-1/2}$ dependence from mobility measurements and first-principles calculation [Masu et al. 1983, Look et al. 1992], while the polar optical phonon scattering has an approximately T^{-2} dependence [Fortini et al. 1978] and carrier–carrier scattering (mainly from hole–hole scattering) has a very small temperature dependence [Asada et al. 1984]. Therefore the photoluminescence experiment data imply that alloy scattering may be the dominant mechanism behind momentum relaxation for the studied active layer material.

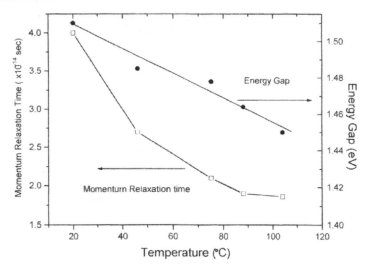

Fig. 8.41. Temperature dependence of the momentum relaxation time and energy gap, obtained from best fitting of the calculated photoluminescence spectrum to the experimental spectrum, as was done in Fig. 8.40 (after Yi et al. [1995]).

The experimental temperature dependence of the momentum relaxation time $\tau(T)$ given in Fig. 8.41 is best fitted with the following formula

Eq. (8.13) $$\tau(T) = \tau_0 \left(1 + \beta \left(\frac{T - T_0}{T_1 - T_0}\right)\right)^{-1}$$

where $T_0 = 15\ ^\circ C$, $T_1 = 105\ ^\circ C$, $\beta \approx 1.1$ and $\tau_0 = 4 \times 10^{-14}$ s. With this $\tau(T)$, the gain $g(E)$ was recalculated using Eq. (8.6) and Eq. (8.7). The decrease of $\tau(\Gamma)$ at high temperature (increase of relaxation width $\hbar/\tau$) makes $\tilde{f}_{c(v)}(E)$ and $g(E)$ broader and reduces their peak heights. Besides the additional gain broadening, $1/\tau(T)$ affects our theoretical model in two more ways. Firstly, the hole mobility decreases at high temperature because it is inversely proportional to the scattering rate [Look et al. 1992] (electron mobility does not play a crucial role in the calculation of J_{th} and η_d as discussed in Diaz et al. [1994c])

Eq. (8.14) $$\mu(T) = \mu_0 \left(\frac{\tau_0}{\tau(T)}\right)$$

where $\mu_0 = 300\ cm^2 \cdot V^{-1} \cdot s^{-1}$ was used for the InGaP cladding layer. The mobility of the cladding layer is an important parameter affecting the minority-carrier leakage current which was shown previously to be anomalously large in non-optimized structures [Diaz et al. 1994a, b]. The other effect influences the free-carrier absorption in the waveguide layer. This is because the free-carrier absorption originates from the scattering of photoexcited free carriers, and therefore the absorption rate is linearly proportional to the scattering rate. Thus the free-carrier absorption loss $\alpha_{fc}(T)$ is given by

Eq. (8.15) $\alpha_{fc}(T) = \alpha_{fc,0}\left(\dfrac{\tau(T)}{\tau_0}\right) \cong \left[3N_{wg} + 7P_{wg}\right] \times 10^{-18}\left[\dfrac{\tau(T)}{\tau_0}\right](cm^{-1})$

where $\alpha_{fc,0}$, the room-temperature free-carrier absorption loss, was obtained from Casey and Panish [1978], and N_{wg}, P_{wg} are non-equilibrium electron and hole concentrations in the waveguide layer. All the temperature dependences of $g(T$, $\tau(T))$, $m(T)$ and $\alpha_{fc}(T)$ which originated from $\tau(T)$ require J_{th}, higher and η_d lower than in the case when only thermal broadening $g(T)$ was considered. The results calculated using Eq. (8.6), Eq. (8.7), Eq. (8.9)–Eq. (8.11) are shown in Fig. 8.38 and Fig. 8.39, showing an excellent quantitative agreement.

(c) Theoretical investigation of minority-carrier leakage of high-power 0.8 μm GaInAP/GaAs laser diodes

Laser diodes with cavity lengths varying from 200 to 1700 μm were prepared without mirror coating and light–current characteristics were recorded under short-pulse operation (pulse width 400 ns, repetition rate 1250 Hz). Fig. 8.42 shows the reciprocal differential efficiency as a function of cavity length (L) for the diodes studied. According to conventional theory, assuming constant internal efficiency and internal loss, η_d increases with decreasing L in the range of cavity lengths longer than 1 mm. However, η_d saturates at the level ~0.85 W·A^{-1} in the range $L = 600$–1200 μm. A sharp drop in η_d occurs for cavity lengths shorter than 500 μm. Fig. 8.43 shows data on threshold current density J_{th} as a function of output losses ($\alpha_m \approx 1/L$). Similar to the differential efficiency discussed above, J_{th} for long-cavity lasers ($\alpha_m < 20$ cm^{-1}) increases linearly with output losses in agreement with conventional theory [Casey and Panish 1978]. For shorter cavities, the J_{th} dependence on output loss deviates from the linearity predicted by conventional theory. Characteristically, the deviation from linear dependence of J_{th} occurs at the same cavity length as the sharp drop in differential efficiency.

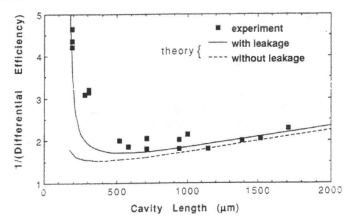

Fig. 8.42. Reciprocal differential efficiency versus cavity length for 0.8 μm
InGaAsP/GaAs laser diodes measured in the pulse regime (pulse width 400 ns,
repetition rate 1250 Hz). The solid curve is the result of theoretical calculation
taking current leakage to the p-cladding layer into account. The dashed curve was
calculated neglecting the current leakage (after Diaz et al. [1994b]).

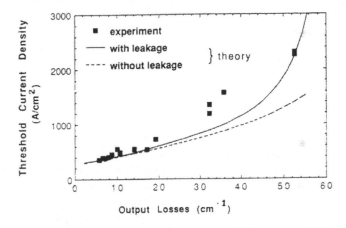

Fig. 8.43. Threshold current density versus output losses ($\alpha_m \approx 1/L$) for 0.8 μm
InGaAsP/GaAs laser diodes measured in the pulse regime (pulse width 400 ns,
repetition rate 1250 Hz). The solid curve shows the result of calculation taking
current leakage to the p-cladding layer into account. The dashed curve was
calculated neglecting current leakage (after Diaz et al. [1994b]).

(d) Comparison of gain and threshold current density for diodes with different QW thicknesses—quantum-size effect of QW lasers

Here we will demonstrate how the alloy-scattering-induced high relaxation rate
discussed in section 8.4.3(*b*) affects the quantum-size effect to gain and threshold
current density. We illustrate this work rather heuristically so that readers without a
background in the theoretical frame for the calculation of gain and other laser

characteristics easily understand one of the most interesting and sophisticated topics in semiconductor lasers: the quantum-size effect in laser diodes. During the discussion below, for the band-structure calculation based on the Luttinger–Kohn Hamiltonian, and the gain calculation based on their mathematical derivation, see Appendices C and D.

Through experiment and the theoretical model for the leakage current (due to minority-carrier leakage to the p-cladding layer) discussed in section 8.4.3(c), Yi et al. [1995] revealed that for InGaAsP/GaAs ($\lambda = 808$ nm) laser diodes, 300 Å is the optimal thickness for the active layer. It was found that laser diodes with active layers thinner than 300 Å had a significantly large minority-carrier leakage to the cladding layer, degrading the threshold condition and internal efficiency. This was primarily due to too small an optical confinement.

Given that 300 Å is the optimal thickness in this case, the question naturally arises as to whether a multiple quantum well with total thickness 300 Å is superior to a single quantum well 300 A thick. This investigation was motivated by the fact that in AlGaAs/GaAs lasers, narrow QWs are known to have better performance (less temperature dependence and wavelength stability) than thicker active layer lasers (the advantages of quantum well lasers were discussed in section 8.3.1). Arakawa and Yariv [1985] attributed the lower threshold current density of thinner QWs to the lower transparency current (the current at which the gain is zero). They calculated the gains of QW lasers, assuming that the electrons and holes are free particles with masses determined by crystal potential (this approximation is called the effective-mass approximation) as described in detail in Appendix A.3.

However, their argument (that thinner quantum wells give lower threshold current densities) may not be an accurate conclusion when other factors are considered. Firstly, the effective-mass energy band approximation used in their work fails to describe the mixing of heavy holes and light holes, which has a significant effect on optical transition properties. Since this mixing becomes more prominent in narrow quantum wells, a comparison of optical properties for QWs with different well thickness without considering the mixing effect would not give a valid conclusion. Secondly, a relatively high intraband relaxation rate usually found in the InGaAsP quaternary material due to alloy scattering may deform the gain spectra significantly. Even though lasers with higher relaxation rate are favored for the single longitudinal mode operation due to the homogeneous broadening of the gain spectrum [Zee 1978], the high relaxation rate may lower the difference of the gain for the lasers with different QW thickness by smoothing the gain spectra.

The above question, i.e., whether lasers composed of narrow QWs demonstrate better performance compared with lasers with thicker QWs when the same optical confinement applies (same total active layer thickness), is addressed in this work. Yi et al. [1995] compared the characteristics of three different QW structures with the same total QW active layer thickness of 300 Å (i) triple 100 Å QWs, (ii) double 150 Å QWs, and (iii) a single 300 Å QW for the active layer. A schematic diagram for these structures is shown in Fig. 8.44. Details of the sample preparation and experimental procedure for measuring J_{th} and η_d were described in section 8.4.3(a)– (c). The experimental results (symbols) for J_{th} as a function of the cavity length L show no significant difference of J_{th} between the laser structures with one, two and three quantum wells (Fig. 8.45). The differential efficiency η_d was also found to be independent of the inner structure of the active region.

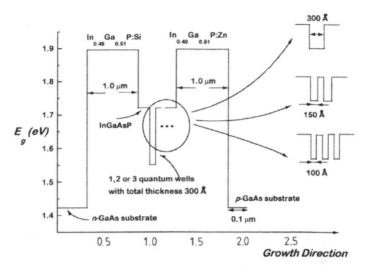

Fig. 8.44. Schematic diagram of the three kinds of laser (a triple 100 Å QW, a double 150 Å QW and a single 300 Å QW) that were used for this experiment.

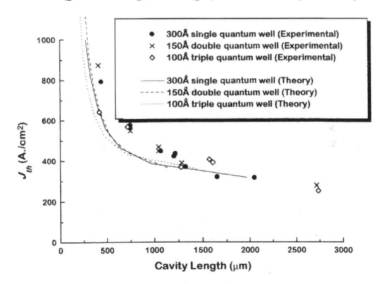

Fig. 8.45. The threshold current densities of the three lasers shown in Fig. 8.44. The symbols represent experimental results while lines represent theoretical results.

To determine whether this result comes from intrinsic effects such as valence-band mixing or a high relaxation rate as mentioned earlier, or from extrinsic effects such as interface-induced degradation, we compared experiment to theory by taking the valence mixing and high relaxation rate into account. The energy levels of the

valence bands should be calculated by employing the Luttinger–Kohn Hamiltonian
as explained in detail in Appendix A.4.

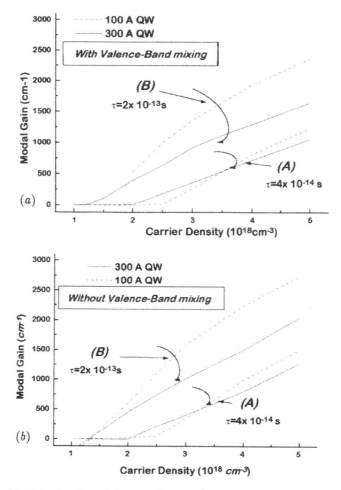

*Fig. 8.46. (a) Calculated modal gains for 100 Å QW lasers (dashed lines) and 300 Å
QW lasers (solid lines) when valence-band mixing was considered. Curve A is
related to the relaxation time τ = 4 × 10^{-14} s, while curve B is for τ = 2 × 10^{-13} s.
(b) Same as (a) except that valence-band mixing was not taken into account for the
calculation.*

Material parameters such as the Luttinger parameter and the energy gap are
linearly interpolated from the values of binary materials as in Table 8.3. In
Eq. (A.38) in Appendix A.4, the heavy hole (upper row) and light hole (lower row)
are mixed due to the non-zero off-diagonal elements of the Hamiltonian. The space-
dependent Hamiltonian was solved using the eigenfunction expansion method
[Gershoni et al. 1993]. The energy band and the envelope functions obtained from
the Hamiltonian (Eq. (A.38)) of Appendix A.4 are used in the calculation of the

optical gain. In the gain calculation, we used $\tau = 0.04$ ps for the relaxation time as determined from photoluminescence measurements as in section 8.4.3(b). The calculated gains for the 100 Å QW and 300 Å QW versus injection current density are shown in Fig. 8.46(a) (curve A). The two curves (in curve A) show a remarkably similar dependence on the injection carrier density. As discussed above, this is quite contrary to the work of Arakawa and Yariv [1985] where a significantly lower transparency carrier density and higher gain were obtained from a thinner QW. The difference may originate from either of two effects considered here: valence-band mixing, or high intraband relaxation rate. To clarify the origin, we recalculated the gain without considering either effect. Curve B in Fig. 8.46(a) is the gain versus carrier density without considering a high relaxation rate. Instead of $\tau = 0.04$ ps, $\tau = 0.2$ ps was assumed. In this case, the quantum-size effect of the QW is evident: the 100 Å QW lasers would have >30% higher gain than the 300 Å QW laser. This difference was almost absent for the $\tau = 0.04$ ps case.

To estimate the effect of valence-band mixing, the effective-mass approximation (non-mixing) was applied to the gain calculation. For a systematic comparison, we made the following change to the Hamiltonian, while keeping the form of Eq. (A.38) in Appendix A.4

Eq. (8.16)

$$H_v \rightarrow H_v^{\text{non-mixing}} = \frac{1}{2}\left(\frac{\hbar^2}{m_0}\right)\begin{pmatrix} (\gamma_1 - 2\gamma_2)(k_z^2 + k_\rho^2) & 0 \\ 0 & (\gamma_1 - 2\gamma_2)(k_z^2 + k_\rho^2) \end{pmatrix} + V(z)$$

In this way, we could see the difference between mixing and non-mixing directly without being affected by changes in other parameters. Note that Eq. (A.38) and Eq. (8.16) would give the same eigenenergies when $V(z) = 0$ (i.e. for bulk material). Fig. 8.46(b) shows the calculated gains when the energies and wavefunction for the holes are calculated with the effective-mass Hamiltonian (Eq. (8.16)) for $\tau = 0.04$ ps (curve A) and $\tau = 0.2$ ps (curve B). This calculation also indicates the absence of a difference in the gains for the 100 Å and 300 Å QW lasers when the relaxation rate is high (curve A). Thus it is not the valence-band mixing that makes the gains for the two QW lasers less different but the high relaxation rate.

In order to directly verify whether the calculated gain correctly models experiment, we performed a calculation of J_{th} using the model discussed in section 8.4.3(c). The calculation reveals almost the same threshold current densities for all three laser structures. This result was expected since the three lasers have the same gain, optical confinement, active layer volume, and loss. Fig. 8.45 shows the comparison between calculation (lines) and experiment (symbols). The agreement is excellent for the entire range of cavity length investigated.

The excellent match between experiment and theory indicates that the interfaces between well layers and barrier layers do not contribute to loss or non-radiative recombination to a detectable level. If this were not so, the triple 100 Å QW lasers, having 3 times more hetero-interfaces than single 300 Å QW lasers, would have shown a higher threshold current due to the higher loss and non-radiative recombination. The absence of additional loss due to the interface is consistent with

Materials constants

	GaAs	Al$_{0.50}$Ga$_{0.50}$As	Ga$_{0.49}$In$_{0.51}$P	In$_{0.13}$Ga$_{0.87}$As$_{0.74}$P$_{0.26}$	In$_{0.37}$Ga$_{0.63}$As$_{0.25}$P$_{0.75}$	Ga$_{0.9}$In$_{0.1}$As
Energy gap E_g (eV)	1.42	1.79	1.90	1.52	1.80	1.31
Refractive index n	3.655	3.41	3.450	3.58	3.48	3.64
Effect electron mass m_c (m_0 unit)	0.067	0.108	0.0794	0.0664	0.072	0.0626
Luttinger parameter γ_1	6.790	–	5.29	7.428	6.635	8.078
γ_2	1.924	–	1.541	2.335	2.117	2.568
γ_3	2.782	–	2.221	3.153	2.847	3.432
Electron affinity χ (eV)	4.5	–	4.204	4.37	4.231	4.495
Lattice constant a (Å)	5.641	5.65	5.663	5.646	5.652	5.683

Table 8.3. *Material constant for GaAs, ternary and quaternary materials commonly used for lasers. The effective mass of electrons and Luttinger parameters for quaternary materials are interpolated from the related binary materials (i.e. GaAs, InAs, InP, GaP), using the following equation:*

$$A_{(In_{1-x}Ga_xAs_yP_{1-y})} = A_{GaAs}xy + A_{InAs}(1-x)y + A_{GaP}x(1-y) + A_{InP}(1-x)(1-y)$$

Eq. (8.17)

In bulk material, the effective mass of heavy holes (m_{hh}) and light holes (m_{lh}) along (1 0 0) and (1 1 1) directions are related to the Luttinger parameters as:

Eq. (8.18)
$$\frac{m_0}{m_{hh}(100)} = \gamma_1 - 2\gamma_2 \qquad \frac{m_0}{m_{lh}(100)} = \gamma_1 + 2\gamma_2$$

Eq. (8.19)
$$\frac{m_0}{m_{hh}(111)} = \gamma_1 - 2\gamma_3 \qquad \frac{m_0}{m_{lh}(111)} = \gamma_1 + 2\gamma_3$$

recent work on Hall mobility measurements [Mitchel et al. 1994]. In that work, it was shown that the interface roughness should be less than two monolayers. This small fluctuation does not cause any detectable optical loss according to light-scattering calculations [Henry et al. 1981].

Numerical calculations of the $\eta_d(L)$ and $J_{th}(1/L)$ dependences were performed as described in section 8.4.3(*b*). Calculated curves for $\eta_d(L)$ and $J_{th}(1/L)$ dependences are shown in Fig. 8.42 and Fig. 8.43 and exhibit good quantitative agreement with experimental data. For comparison, the dashed lines in Fig. 8.42 and Fig. 8.43 demonstrate values of η_d and J_{th} calculated without taking the current leakage into account. For short-cavity lasers the leakage current is the single most important factor responsible for the increase in J_{th} and drop in η_d for the investigated laser diodes. The leakage current contribution to J_{th} increases with output losses because higher threshold gains shift the quasi-Fermi level position to higher energies and thus enhance excess carrier spillover to the waveguide and cladding layers. Leaking excess carriers recombine non-radiatively in the cladding and contact layers and decrease the internal efficiency above the threshold. The comparison of theoretical and experimental data clearly demonstrates that current leakage to the *p*-cladding layer affects the laser parameters significantly only in the range of cavity lengths below 500 µm. For cavity lengths longer than 500 µm the current leakage results in no more than a 10–15% change in the values of efficiency and threshold. Long-cavity 0.8 µm InGaAsP/GaAs lasers have already been successfully used for high-power operation in the quasi-CW and CW regimes.

References

Adams, M.J., *Opto-Electron.* **5** 201, 1973.

Agrawal, G.P. and Dutta, N.K., *Long-Wavelength Semiconductor Lasers* (New York: Van Nostrand Reinhold), 1986.

Aigrain P., *Proc. Conf. on Quantum Electronics (Paris, 1963)* p 1762, 1958.

Alferov, Z.I., Andreev, V.M., Garbuzov, D.Z., Zhilyaev, Y.V., Morozov, E.P., Portnoi, E.L., and Trofim, V.G., *Sov. Phys.–Semicond.* **4** 1573, 1971.

Alferov, Z.I., Arsentyev, I.N., Vavilova, L.S., Garbuzov, D.Z., and Krasovskyi, V.V., *Sov. Phys.–Semicond.* **18** 1035, 1984.

Alferov, Z.I. and Kazarinov, R.F., *Author's Certificate, USSR No 181737* with priority of 30 March, 1963

Arakawa, Y. and Yariv, A., *IEEE J. Quantum Electron.* **QE-21** 1666, 1985.

Arimoto, S., Yasuda, M., Shima, A., Kadoiwa, K., Kamizato, T., Watanabe, H., Omura, E., Aiga, M., Ikeda, K., and Mitsui, S. *IEEE J. Quantum Electron.* **QE-29** 1874, 1993.

Arnold, G., Pusser, P., and Petermann, K. *Semiconductor Devices for Optical Communications (Springer Topics in Applied Physics 39)* 2nd edn (Berlin: Springer), 1982.

Asada, M., Kameyama, A., and Suematsu, Y. *IEEE J. Quantum Electron.* **QE-20** 745, 1984.

Brunemeir, P.E., Hsieh, K.C., Deppe, D.G., Brown, J.M., and Holonyak, N., *J. Crystal Growth* **71** 705, 1985.

Buus, J. *Single Frequency Semiconductor Lasers* (Bellingham, WA: SPIE Optical Engineering), 1991.

Casey, H.C. Jr, *J. Appl. Phys.* **49** 3684, 1978.

Casey, H.C. Jr and Panish, M.B., *Heterostructure Lasers* (New York: Academic), 1978.

Casey, H.C. Jr, Panish, M.B., and Merz, J.L., *J. Appl. Phys.* **44** 5470, 1973.

Casey, H.C Jr, Panish, M.B., Schlosser, W.O., and Paoli, T.L., *J. Appl. Phys.* **45** 322, 1974.

Chand, N., Becker, E.E., van der Ziel, J.P., Chu, S.G., and Dutta, N.K., *Appl. Phys. Lett.* **58** 1704, 1991.

Chand, N., Chu, S., Dutta, N., Geva, J., Syrbu, A., Mereutza, A., and Takolev, V. *IEEE J. Quantum Electron.* **QE-30** 424, 1994.

Chen, H.Z., Ghaffari, A., Morkoc, H., and Yariv, A., *Appl. Phys. Lett.* **51** 2094, 1987.

Chen, Y.K., Wu, M.C., Kuo, J.M., Chin, M.A., and Sergent, A.M., *Appl. Phys. Lett.* **59** 2, 1991.

Chin, R., Holonyak, N. Jr, Vojak, B.A., Hess, K., Dupuis, R.D., and Dapkus, P.D., *Appl. Phys. Lett.* **36** 19, 1980.

Chinn, K., Zory, P., and Reisinger, A., *IEEE J. Quantum Electron.* **QE-24** 2191, 1988.

Dallesasse, J.M., and Holonyak, N. *Appl. Phys. Lett.* **58** 394, 1991.

d'Asaro, L.A. *J. Lumin.* **7** 310, 1973.

Diaz, J., Eliashevich, I., He, X., Yi, H., Kolev, E., Garbuzov, D., and Razeghi, M. *Appl. Phys. Lett.* **65** 1004, 1994a.

Diaz, J., Eliashevich, I., Yi, H., He, X., Stanton, M., Erdtmann, M., Wang, L., and Razeghi, M., *Appl. Phys. Lett.* **65** 2260, 1994b.

Diaz, J., Yi, H., Erdtmann, M., He, X., Kolev, E., Garbuzov, D., Bigan, E., and Razeghi, M., *J. Appl. Phys.* **76** 700, 1994c.

Dupuis, R.D. and Dapkus, P.D. *IEEE J. Quantum Electron.* **QE-15** 128, 1979.

Dupuis, R.D., Dapkus, P.D., Holonyak, N. Jr, Rezek, E.A., and Chin, R., *Appl. Phys. Lett.* **32** 295, 1978.

Dupuis, R.D., Dapkus, P.D., and Moudy, L.A., *Tech. Digest 1977 Int. Electron Devices Meeting* p 575, 1977.

Dyment, J.C., Nash, F.R., Hwang, C.J., Rozgonyi, G.A., Hartman, R.L., Marcos, H.M., and Haszko, S.E., *Appl. Phys. Lett.* **24** 481, 1974.

Esaki, L. and Tsu, R. *IBM J. Res.* **14** 61, 1970.

Fortini, A., Diguet, D., and Lugand, J. *J. Appl. Phys.* **41** 3121, 1978.

Fujii, H., Kobayashi, K., Kawata, S., Gomyo, A., Nino, I., Hotta, H., and Suzuki, T., *Electron. Lett.* **23** 938, 1987.

Fujii, T., Yamakoshi, S., Nanbu, K., Wada, O., and Hiyamizu, S., *J. Vac. Sci Technol.* B **2** 259, 1984.

Garbuzov, D.Z., Antonishkis, N.J., Bondarev, A.D., Gulakov, A.B., Zhigulin, S.N., Kotsavets, N.I., Kochergin, A.V., and Rafailov, E.V., *IEEE J. Quantum Electron.* **QE-27** 1531, 1991.

Garbuzov, D.Z., Gulakov, A.B., Kochergin, A.V., Shkurko, A.P., Strugov, N.A., Ter-Martirosyan, A.L., and Chalyi, V.P., *Tech. Digest Conf. Lasers Electron-Optics (Optical Society of America, Washington, DC)* p 468, 1990.

Gershoni, D., Henry, C.H., and Baraff, G.A., *IEEE J. Quantum Electron.* **QE-29** 2433, 1993.

Hall, D.N., Fenner, G.E., Kingsley, J.D., Soltys, T.J., and Carlson, R.O., *Phys. Rev. Lett.* **9** 366, 1962.

Hayashi, I., Panish, M.B., and Foy, P.W., *IEEE J. Quantum Electron.* **QE-5** 211, 1969.

Henry, C.H., Logan, R., and Merritt, F., *IEEE J. Quantum Electron.* **QE-17** 2196, 1981.

Hersee, S.D., Baldy, M., Assenat, P., Cremoux, B., and Duchemin, J.P., *Electron. Lett.* **18** 870, 1982.

Hersee, S.D., Cremoux, B., and Duchemin, J.P., *Appl. Phys. Lett.* **44** 476, 1984.

Holonyak, N. Jr and Bevacqua, S.F., *Appl. Phys. Lett.* **1** 82, 1962.

Iga, K., Uenohara, H., and Koyama, F. *Electron. Lett.* **22** 1008, 1986.

Itaya, K., Watanabe, Y., Ishikawa, M., Hatakoshi, G., and Uematsu, Y., *Appl. Phys. Lett.* **58** 1718, 1991.

Kazarinov, R.F. and Tsarenkov, G.V., *Sov. Phys.–Semicond.* **10** 178, 1976.

Kobayashi, K., Kawata, S., Gomyo, A., Nino, I., and Suzuki, T., *Electron. Lett.* **21** 931, 1985.

Kressel, H., *Semiconductor Devices for Optical Communication* (Berlin: Springer), 1980.

Kressel, H. and Nelson, H. *RCA Rev.* **30** 106, 1969.

Kroemer, H., *Proc. IEEE* **51** 1782, 1963.

Lau, K.Y., Derry, P.L., and Yariv, A., *Appl. Phys. Lett.* **52** 88, 1988.

Look, D., Lorance, D., Sizelov, J., Stutz, C., Evans, K., and Whitson, D., *J. Appl. Phys.* **71** 260, 1992.

Luttinger, J.M., *Phys. Rev.* **102** 1030, 1956.

Masu, K., Tokumitsu, E., Konagai, M., and Takahashi, K., *J. Appl. Phys.* **54** 5785, 1983.

Mitchel, W.C., Brown, G.J., Lo, I., Elhamri, S., Ahoujja, M., Ravindran, K., Newrock, R.S., Razeghi, M., and He, X., *Appl. Phys. Lett.* **65** 1578, 1995.

Mobarhan, K., Razeghi, M., and Blondeau, R., *Electron. Lett.* **28** 1510, 1992a.

Mobarhan, K., Razeghi, M., Marquebielle, G., and Vassilaki, E., *J. Appl. Phys.* **72** 4447, 1992b.

Narui, H., Hirata, S., and Mori, Y., *IEEE J. Quantum Electron.* **QE-28** 4, 1992.

Nasledov, D.N., Rogachev, A.A., Ryvkin, S.M., and Tsarenkov, B.V., *Sov. Phys.– Solid State* **4** 782, 1962.

Nathan, M.I., Dumke, W.P., Burns, G., Dill, F.H.Jr, and Lasher, G., *Appl. Phys. Lett.* **1** 62, 1962.

Ohkubo, M., Ijichi, T., Iketani, A., and Kikata, T., 1992 *Appl. Phys. Lett.* **60** 23, 1992.

Panish, M.B., Hayashi, I., and Sumski, S., *IEEE J. Quantum Electron.* **QE-5** 210, 1969.

Quist, T.M., Rediker, R.H., Keyes, R.J., Krag, W.E., Lax, B., McWhorter, A.L., and Zeiger, H.J., *Appl. Phys. Lett.* **1** 91, 1962.

Razeghi, M., *The MOCVD Challenge: Volume 1: A Survey of GaInAsP–InP for Photonic and Electronic Applications* (Bristol: Hilger), 1989.

Razeghi, M., *Materials for Photonic Devices, 15* (New Jersey: World Scientific), 1991.

Razeghi, M., *InGaAsP Diodes, US Air Force Philips Laboratory, Diode Laser Technology Program Conf. (Albuquerque, 20–22 April),* 1993.

Razeghi, M., *Nature* **369** 631, 1994.

Razeghi, M., Diaz, J., Eliashevich, I., He, X., Yi, H., Erdtmann, M., Kolev, E., Wang, L., and Garbuzov, D., *Conf. Digest IEEE Int. Semiconductor Laser Conf. '94 (Maui, Hawaii, 19–23 September 1994)* p 159, 1994.

Sasaki, K., Matsumoto, M., Kondo, M., Ishizumi, T., Takeoka, T., Yamanoto, S., and Hijikata, T., *Japan. J. Appl. Phys.* **30** L904, 1991.

Serreze, H.B., Chen, Y.C., and Waters, R.G., *Appl. Phys. Lett.* **58** 3, 1991.

Simhony, S., Kapon, E., Colas, E., Hwang, D.M., Stoffel, N.G., and Worland, P., *Appl. Phys. Lett.* **59** 2225, 1991.

Sze, S.M., *Physics of Semiconductor Devices* 2nd edn (New York: Wiley), 1981.

Thompson, G.B., Henshall, G.D., Whiteaway, J.A., and Kirkby, P.A., *J. Appl. Phys.* **47** 1501, 1976.

Thompson, G.B. and Kirkby, P.A., *IEEE J. Quantum Electron.* **QE-9** 311, 1973.

Tsang, W.T., *Appl. Phys. Lett.* **39** 134, 1981a.

Tsang, W.T., *Appl. Phys. Lett.* **39** 786, 1981b.

Tsang, W.T., *Appl. Phys. Lett.* **40** 217, 1982.

Tsang, W.T., *IEEE J. Quantum Electron.* **QE-20** 1119, 1984.

Tsang, W.T., Weisbuch, C., Miller, R.C., and Dingle, R., *Appl. Phys. Lett.* **35** 673, 1979.

Tsukada, T., *J. Appl. Phys.* **45** 4899, 1974.

Ueno, Y., Fujii, H., Kobayashi, K., Endo, K., Gomyo, A., Hara, K., Kawata, S., Yuasa, T., and Suzuki, T., *Japan. J. Appl. Phys.* **29** L1666, 1990.

Usami, U. and Matsushima, Y., *Proc. 14th IEEE Int. Semiconductor Laser Conf. '94 (Maui, Hawaii, 19–23 September 1994)*, 1994.

van der Ziel, J.P., Dingle, R., Miller, R.C., Wiegmann, W., and Nordland, W.A. Jr, *Appl. Phys. Lett.* **26** 463, 1975.

Wagner, D.K., Waters, R.G., Tihanyi, P.L., Hill, D.S., Roza, A.J. Jr, Vollmer, H.J., and Leopold, M.M., *IEEE J. Quantum Electron.* **QE-24** p 1258, 1988.

Yamada, M., Ishiguro, H., and Nagato, H., *Japan. J. Appl. Phys.* **19** 135, 1980.

Yariv, A., *IEEE Circuits Devices Mag.* **5**(6) 25, 1989a.

Yariv, A., *Quantum Electron.* 3rd edn (New York: Wiley), 1989b.

Yi, H.Y., Diaz, J., Eliashevich, I., Stanton, M., Erdtmann, M., He, X., Wang, L.J., and Razeghi, M., *Appl. Phys. Lett.* **66** 253, 1995.

Yi, H.Y., Diaz, J., Wang, L.J., Eliashevich, I., Kim, S., William, R., Erdtmann, M., He, X., Kolev, E., and Razeghi, M., *Appl. Phys. Lett.* **66** 3251, 1995.

Yonezu, H., Sakuma, I., Kobayashi, K., Kamejima, T., Ueno, M., and Nannichi, Y., *Japan. J. Appl Phys.* **12** 1585, 1973.

Zee, B., 1978 *IEEE J. Quantum Electron.* **QE**-14 727, 1978.

Zhao, H., MacDougal, M.H., Uppal, K., and Dapkus, P.D., *Proc. 14th IEEE Int. Semiconductor Laser Conf. '94 (Maui, Hawaii, 19–23 September 1994)*, 1994.

Zory, P., (ed) *Quantum Well Lasers* (San Diego, CA: Academic) p 315, 1993.

9. GaAs-Based Heterojunction Electron Devices Grown by MOCVD

9.1.   **Introduction**
     9.1.1.   Heterojunctions
9.2.   **Heterostructure filed-effect transistors (HFETs)**
     9.2.1.   AlGaAs/GaAs MODFETs
     9.2.2.   AlGaAs/InGaAs MODFETs
     9.2.3.   GaInP/GaAs MODFETs
     9.2.4.   MODFETs on silicon substrate
     9.2.5.   GaInP/GaAs heterostructure insulated gate field effect transistors
9.3.   **Heterojunction bipolar transistors (HBTs)**
     9.3.1.   Introduction
     9.3.2.   Frequency response and design considerations
     9.3.3.   AlGaAs/GaAs HBTs
     9.3.4.   GaInP/GaAs HBTs

9.1. Introduction

In high-speed electronic device applications GaAs and its related compounds are superior to silicon due to their higher carrier mobilities which enhance device speed and current handling capability. Molecular beam epitaxy (MBE) and metalorganic chemical vapor deposition (MOCVD) have been the most important growth techniques in the production of high-quality epilayers for device fabrication. An important problem with MBE is the difficulty in controlling the phosphorus vapor pressure during MBE growth. Therefore phosphorus-based materials are not easy to grow with this technique. Another problem is the low throughput of MBE growth. MOCVD, with its batch capability, is more suitable for volume manufacturing. This chapter will concentrate on the growth, fabrication and characteristics of MOCVD-grown, GaAs-based heterostructure field effect and bipolar transistors which have been the most important devices in digital and microwave applications. It is assumed that the reader is familiar with the basic principles involving the operation of homojunction field effect and bipolar transistors at the level of an undergraduate course on solid-state devices.

9.1.1. Heterojunctions

A heterojunction can be defined as a junction formed between two materials with different electrical properties. Fig. 9.1 shows the construction of the band diagram for an *n–p* heterojunction.

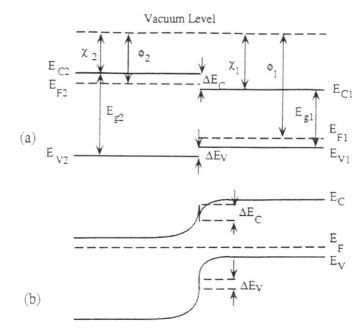

Fig. 9.1. Band diagram of a p–n heterojunction (a) before and (b) after the contact is formed.

The difference in energy between the vacuum level and the conduction band is defined as the electron affinity χ which is the energy necessary to promote an electron from the conduction band to the vacuum level. The work function ϕ is the energy required to take an electron from the Fermi level to the vacuum level. Hence the work function varies with the doping level. When the junction is formed, discontinuities in the conduction and valence bands (ΔE_c and ΔE_v) appear due to the difference in the electron affinities. The conduction band discontinuity is given by [Anderson 1962]:

Eq. (9.1) $\Delta E_c = \chi_1 - \chi_2$

From Fig. 9.1 it can be seen that the valence band discontinuity can be expressed as

Eq. (9.2) $\Delta E_v = \Delta E_g - \Delta E_c$

where ΔE_g is the difference in bandgaps of two semiconductors.

Due to the initial difference in the Fermi levels in the two semiconductors, band bending is observed at the interface as shown in Fig. 9.1(*b*) (similar to homojunctions). The built-in potential is

Eq. (9.3) $qV_{\mathrm{D}} = E_{\mathrm{F1}} - E_{\mathrm{F2}}$

The ratio of the two components of V_{D} that drop on p (V_{D1}) and n (V_{D2}) regions is given by

Eq. (9.4) $\dfrac{V_{\mathrm{D1}}}{V_{\mathrm{D2}}} = \dfrac{\varepsilon_2 N_2}{\varepsilon_1 N_1}$

where N_1, N_2 are the doping concentrations and ε_1, ε_2 are the permittivities in *p*- and *n*-type semiconductors, respectively.

9.2. Heterostructure field-effect transistors (HFETs)

One of the promising heterojunction devices in digital and microwave applications is the modulation-doped field effect transistor (MODFET). The transistor has several other names such as HEMT (high-electron-mobility transistor), TEGFET (two-dimensional electron gas field effect transistor) and SDHT (selectively doped heterojunction transistor).

A modulation-doped heterostructure in its simplest form can be defined by a heterojunction between two semiconductors with different bandgaps where only the semiconductor with the larger bandgap is doped. The energy band diagram for a modulation-doped heterostructure is shown in Fig. 9.2. In order to minimize their energy, electrons are transferred into the lower-bandgap material by diffusion and they form a two-dimensional electron gas confined at the heterojunction interface in a quasitriangular potential well. Since the electrons are separated from the ionized donors, ionized impurity scattering is significantly reduced. This scattering process is the dominant mechanism that scatters electrons at low temperatures where phonon scattering is not significant. The absence of ionized impurity scattering enhances electron mobility and therefore device speed. A combination of the elimination of ionized impurity scattering with the ability to confine a large number of electrons in a thin layer makes it possible to switch large currents rapidly. This is especially important in digital circuits where high current levels are necessary for fast switching of device and interconnection capacitances. The most commonly used material systems are AlGaAs–GaAs, AlGaAs–InGaAs, GaInP–GaAs on GaAs substrate and InAlAs–InGaAs on InP substrate. In the following sections the basics of MODFET operation and the factors affecting its characteristics will be discussed and the performance of devices based on different material systems will be compared.

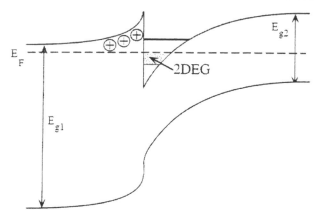

Fig. 9.2. Energy band diagram for a modulation-doped heterostructure.

9.2.1. AlGaAs/GaAs MODFETs

AlGaAs/GaAs has been the most extensively studied material system for modulation-doped field effect transistors. $Al_xGa_{1-x}As$ is lattice matched to GaAs over the entire compositional range. This permits the growth of high-quality heterojunctions without misfit dislocations. However, at cryogenic temperatures where a MODFET's superiority over a homojunction field effect transistor becomes significant, abnormal behavior is observed in AlGaAs/GaAs MODFETs as will be discussed later. Since this system has been the most widely used material system in heterostructures, this section is devoted to the discussion of AlGaAs/GaAs MODFETs.

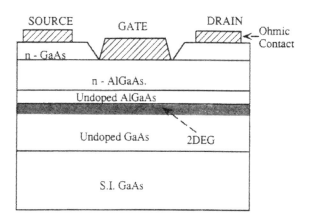

Fig. 9.3. Typical cross-section of an AlGaAs/GaAs MODFET.

The cross-sectional structure of an AlGaAs/GaAs MODFET is shown in Fig. 9.3. The undoped AlGaAs layer (spacer) is used to eliminate remote ionized impurity scattering of the electrons in the GaAs layer by screening the potential of the ionized impurities in the doped AlGaAs layer. The top *n*-GaAs layer (cap layer) is designed for good ohmic contact formation and for the protection of AlGaAs against oxidation. A widely used metal alloy for ohmic contacts is AuGeNi. During metallization Ge makes contact with the 2DEG channel by diffusing to the AlGaAs/GaAs interface. Due to the excellent lattice match between Ge and GaAs good ohmic contact with a low interface state density is formed. AlTi is commonly used for Schottky barrier contacts.

Since source and drain resistances affect the device performance, these resistances should be minimized. This can be done by ion implantation using a T-bar gate structure. However ion implantation requires subsequent annealing at elevated temperatures for the activation of the implant. An increase in the 2DEG density and a decrease in the mobility have been observed in MBE-grown modulation-doped AlGaAs/GaAs structures during annealing at around 800 °C [Ishikawa et al. 1981]. This led to the conclusion that Si impurities in the AlGaAs layer were diffusing into the undoped GaAs layer during annealing.

Later it was reported that rapid thermal annealing, which requires only a few seconds, can reduce Si diffusion [Pearah et al. 1984].

The operation of a MODFET is similar to that of a homogeneous field effect transistor. The Schottky barrier (gate) on the doped AlGaAs layer modulates the two-dimensional electron gas and the drain–source current by depleting the region under the gate. The doped AlGaAs layer is also depleted at the AlGaAs/GaAs interface due to electron transfer into GaAs. Since the conduction in the doped AlGaAs layer is inferior to the transport in undoped GaAs, it is important that the AlGaAs layer is totally depleted (the two depletion regions overlap) in order to avoid conduction through this layer. Therefore, in a MODFET, the maximum gate voltage is limited. Above a certain value, conduction occurs in the doped AlGaAs layer and a degradation in performance is observed. The limitation in the forward gate voltage also limits the logic swing.

If the AlGaAs layer is sufficiently thin or a large negative bias is applied, the junction depletion region (AlGaAs/GaAs) and gate depletion region (metal/AlGaAs) overlap. Under these conditions, the sheet carrier concentration at the interface can be expressed approximately as [Delagebeaudeuf and Ling 1982 and Drummond et al. 1983]

Eq. (9.5) $$n_s = \frac{\varepsilon}{q(d + \Delta d)}(V_g - V_T)$$

where q is the electron charge, and d is the total thickness of the doped and undoped AlGaAs layers. V_g is the gate voltage and V_T is the threshold voltage given by

Eq. (9.6) $$V_T = \Phi_b - \Delta E_c - \frac{qN_d d_d^2}{2\varepsilon} + \Delta E_{FO}$$

Here Φ_b is the Schottky barrier height, N_d is the doping density and d_d is the thickness of the doped AlGaAs layer. Δd, ΔE_{FO} are the fitting parameters

($\Delta d = 80$ Å and $\Delta E_{FO} = 0$ and 25 mV at 300 K and 77 K respectively for the AlGaAs/GaAs system).

The threshold voltage is an important parameter in determining the noise margin of a logic circuit. Large variations in the threshold voltage result in poor logic margins. It is necessary that the sensitivity of the threshold voltage to fabrication process variations be minimized in order to achieve an acceptable yield in the integrated circuits.

A very important characteristic of a field effect transistor is the intrinsic transconductance which is a measure of its gain and is given by

Eq. (9.7) $$g_m = \left. \frac{\partial I_{DS}}{\partial V_{GS}} \right|_{V_{DS}} = \frac{C_G}{\tau}$$

where I_{DS} is the drain–source current, V_{GS} is the gate–source voltage, V_{DS} is the drain–source voltage, C_G is the gate capacitance and τ is the transit time of the carriers through the channel. The maximum intrinsic transconductance of a MODFET is given approximately by [Lee et al. 1983]

Eq. (9.8) $$g_{m_{max}} = \frac{q \mu n_s}{L \sqrt{1 + \left[q \mu n_s (d + \Delta d) / \varepsilon v_s L \right]^2}}$$

where L is the gate length and v_S is the saturation velocity. For short-gate MODFETs Eq. (9.8) can be expressed approximately in the form

Eq. (9.9) $$g_{m_{max}} = \frac{\varepsilon v_s}{(d + \Delta d)}$$

Eq. (9.9) shows that the maximum transconductance becomes independent of gate length. However, for short-gate devices, electron velocity overshoot effects arc important and the carrier velocity depends on the gate length.

The frequency response of a field effect transistor is governed by the transit time of the carriers under the gate (τ). The cutoff frequency of a field effect transistor is given by

Eq. (9.10) $$f_T = \frac{1}{2 \pi \tau}$$

or if velocity saturation occurs in the channel

Eq. (9.11) $$f_T = \frac{v_s}{2 \pi L}$$

Using Eq. (9.7) and Eq. (9.10) we obtain

Eq. (9.12) $$f_{\mathrm{T}} = \frac{g_{\mathrm{m}}}{2\pi C_{\mathrm{G}}}$$

Where f_{T} sets an upper limit on the operation frequency of a field effect transistor.

(a) Design considerations for AlGaAs/GaAs MODFETs

Spacer thickness, Al mole fraction and doping concentration of AlGaAs are the most important parameters in determining the carrier concentration and mobility of a two-dimensional electron gas. The background impurity concentration in GaAs is also an important parameter in determining the mobility of a 2DEG. By decreasing the spacer layer thickness, both the transconductance and drain current can be increased but this lowers the mobility of the 2DEG due to an increase in ionized impurity scattering. Hiyamuzu et al. [1983 and 1990] investigated the mobility of a two-dimensional electron gas in MBE-grown AlGaAs/GaAs heterostructures as a function of the spacer layer thickness (d_0). They found that the mobility increases with increasing spacer layer thickness for thicknesses less than about 200 Å and decreases with d_0 for spacer layer thicknesses larger than 800 Å. The decrease in mobility for large spacer thicknesses is due to the reduction in the 2DEG density.

Another way to increase the transconductance is to reduce the AlGaAs layer thickness but this calls for an increase in the doping level of this layer in order to obtain an acceptable threshold voltage. However, very high doping levels result in a leaky Schottky barrier which is not desirable for field effect transistors.

Since the conduction-band discontinuity in the AlGaAs–GaAs system, the ionization energy of Si impurities and the trap concentration in AlGaAs depend on the Al mole fraction in AlGaAs, this parameter has to be carefully chosen. We can increase both the conduction-band discontinuity and the Schottky barrier height by increasing this parameter. A large conduction-band discontinuity is desirable, since it reduces carrier injection from GaAs into AlGaAs and increases carrier density by providing good carrier confinement at the interface. A large Schottky barrier permits high forward gate voltages and therefore increases logic swing. However, for Al mole fractions larger than or around 0.3, DX centers (donor-related traps) become significant and device performance is degraded, as will be discussed in the next section. The maximum 2DEG mobility and electron concentration are obtained at Al mole fractions in the range $0.25 < x < 0.28$.

(b) Effect of traps on the performance of AlGaAs/GaAs MODFETs

Since the diffusion constant of Si in AlGaAs is small, abrupt doping profiles can be obtained by doping AlGaAs with Si; however, Si impurities form deep donor levels called DX centers in AlGaAs. There are three kinds of Si donor in AlGaAs: one shallow and two deep-level donors [Tachikawa et al. 1985]. For Al mole fractions less than about 0.2, almost all the electrons populate shallow donors. For larger x values, the deep-level donor population increases with x and a persistent photoconductivity effect is observed due to the presence of DX centers. The trap density in the AlGaAs layer at Al mole fractions of about 0.3 is comparable to the desired shallow-dopant density and the *I–V* characteristics of MODFETs with this composition are shown to be a strong function of trap occupancy [Chi et al. 1984 and Drummond et al. 1983]. Under sufficiently high drain–source bias, electrons accelerating through the channel of the device become hot near the drain end and are injected into AlGaAs where traps are present. In dark conditions and at low temperatures, depopulation of the traps by thermal emission progresses at a slow rate and, due to the decrease in the 2DEG density, a collapse in the $I-V$ characteristics and a degradation of g_m are observed. Fig. 9.4 shows the $I-V$ characteristics and transconductance versus gate voltage of AlGaAs/GaAs MODFETs before and after bias stress at 77 K in the dark demonstrating a collapse in the drain current and a reduction in the transconductance [Chan et al. 1990].

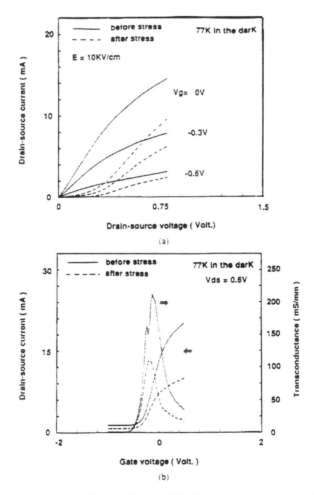

Fig. 9.4. Characteristics of AlGaAs/GaAs MODFETs before and after bias stress at 77 K in the dark (after Chan et al. [1990]).

Another abnormal behavior observed in AlGaAs/GaAs MODFETs is a positive shift in the threshold voltage with temperature (0.2 V between 77 K and 300 K)

[Valois et al. 1983]. If the gate voltage is low enough, the trap energy level will be above the Fermi level. At sufficiently high temperatures the traps are depopulated by thermal emission. If a positive gate voltage is applied at room temperature, the carriers in the channel are attracted into the doped AlGaAs and neutralize the traps. If the gate voltage is subsequently reduced, the trapped carriers return to the channel. If the device is cooled down to 77 K under a positive gate bias, the carriers remain trapped after the gate voltage is lowered. This is due to having insufficient thermal energy to overcome the potential barrier. An increase in the trapped charge density modifies the potential distribution in the AlGaAs layer and therefore the drain current and threshold voltage.

The disadvantages of the AlGaAs–GaAs system can be summarized as follows.

(i)   The sheet carrier concentration at the interface is limited to about $10^{12}\,cm^{-2}$. The carrier concentration can be increased by using larger Al mole fractions, in which case the defect concentration is increased and performance at low temperatures is degraded due to DX centers.

(ii) Diffusion of Si impurities from AlGaAs into GaAs during annealing after ion implantation.

(iii) High reactivity with oxygen causing degradation of the device [Stormer et al. 1979].

(iv) High-quality AlGaAs is grown at temperatures around 700 °C; this leads to a slight decrease in the semi-insulating properties of the substrate.

9.2.2. AlGaAs/InGaAs MODFETs

Because of the disadvantages mentioned in the preceding section, the AlGaAs–GaAs system, although widely used, is not an ideal material system for heterostructure field effect transistors. Therefore researchers have been looking for alternative systems to replace AlGaAs–GaAs. A solution consists of using InGaAs as the undoped narrow-bandgap material and GaAs as the doped large-bandgap material. This approach seems to be promising due to the larger conduction-band discontinuity of this system when compared with low-Al-mole-fraction AlGaAs–GaAs, and because of the good electrical properties of InGaAs (high electron mobility and velocity). If the InGaAs layer is sufficiently thin the mismatch is accommodated as an elastic strain without the formation of misfit dislocations and the structure is called pseudomorphic. Rosenberg et al. [1985] fabricated pseudomorphic InGaAs/GaAs MODFETs that demonstrated good performance. $In_{0.15}Ga_{0.85}As$ (200 Å thick) was grown pseudomorphically on a GaAs substrate. An extrinsic transconductance of $175\,mS{\cdot}mm^{-1}$ for 1 μm gate length devices was obtained at 300 K. Ketterson et al. [1985] replaced the doped GaAs layer with a low-Al-mole-fraction AlGaAs layer ($Al_{0.15}Ga_{0.85}As$) in order to increase the conduction-band discontinuity and improve carrier confinement at the interface. Since x was sufficiently low, DX center effects were not observed. DC transconductances of $270\,mS{\cdot}mm^{-1}$ at 300 K and $360\,mS{\cdot}mm^{-1}$ at 77 K were obtained for 1 μm gate devices with a 300 K current gain cutoff frequency of 20 GHz. Later, Nguyen et al. [1988] reported an f_T of 120 GHz for 0.2 μm gate length pseudomorphic $Al_{0.3}Ga_{0.7}As/In_{0.25}Ga_{0.75}As$ MODFETs with high transconductance (peak transconductance of 550 mS·mm^{-1}). A large indium mole fraction results in a large conduction-band discontinuity and a high two-dimensional electron gas density.

Kikkawa et al. [1991] fabricated MOCVD-grown AlGaAs/GaAs and pseudomorphic AlGaAs/InGaAs/GaAs MODFETs using tertiarybutylarsine (TBAs). Transconductances of 324 mS·mm^{-1} and 350 mS·mm^{-1} were obtained for 0.5 μm gate *n*-AlGaAs MODFETs and pseudomorphic AlGaAs/InGaAs/GaAs MODFETs, respectively, verifying that layers grown using TBAs are of device quality.

9.2.3. GaInP/GaAs MODFETs

n-channel devices.

Using the AlGaAs/InGaAs/GaAs structure instead of AlGaAs/GaAs has permitted the use of low Al mole fractions and avoided DX center problems. The Al mole fraction in this structure should be less than about 0.15 to completely eliminate the DX center problem. However, low Al mole fractions result in a small conduction-band discontinuity and poor carrier confinement. Sufficiently large conduction-band discontinuities can be obtained by increasing the indium mole fraction but this results in an increase in the lattice mismatch between AlGaAs and InGaAs.

The DX center density becomes significant at compositions close to the crossover of direct and indirect conduction bands (at $x = 0.37$ for Al$_x$Ga$_{1-x}$As and $x = 0.74$ for Ga$_x$In$_{1-x}$P). Since Ga$_x$In$_{1-x}$P is lattice matched to GaAs at $x = 0.51$, which is far from the crossover point, DX centers are expected to be negligible for this material. DLTS measurements have shown that there are no detectable deep levels in *n*-doped Ga$_{0.51}$In$_{0.49}$P layers [Watanabe and Ohba 1986 and Tanaka et al. 1987] and the concentration of electron traps is lower than 1×10^{13} cm^{-3}.

Initially, the study of two-dimensional electrons in GaInP/GaAs heterostructures suffered from extremely low mobilities when compared to MBE-grown AlGaAs/GaAs heterostructures. Later, Razeghi et al. [1989] have shown that extremely high mobilities can be obtained with this material system. They have reported a mobility of 780,000 cm^2·V^{-1}·s^{-1} at an electron density of 4.1×10^{11} cm^{-2}.

Razeghi et al. [1991] fabricated lattice-matched (Ga$_{0.51}$In$_{0.49}$P/GaAs) and strained (Ga$_{0.51}$In$_{0.49}$P/Ga$_{0.85}$In$_{0.15}$As/GaAs) modulation-doped field effect transistors by using LP-MOCVD. Growth was carried out at 76 Torr, at a substrate temperature of 510 °C. The growth conditions of GaAs and GaInP are listed in Table 9.1. The residual doping of both materials was *n*-type. The undoped GaAs grown under these conditions had an electron Hall mobility of 9000 cm^2·V^{-1}·s^{-1} at 300 K and 335,000 cm^2·V^{-1}·s^{-1} at 40 K for 12 μm thick layers. 3 μm thick GaInP layers grown on GaAs substrates had an electron carrier concentration of 8×10^{14} cm^{-3} and a Hall mobility of 6500 cm^2·V^{-1}·s^{-1} at 300 K and 50,000 cm^2·V^{-1}·s^{-1} at 77 K.

Fig. 9.5 shows the structures of lattice-matched (*a*) and strained (*b*) MODFETs after the layers are grown. A 1 μm thick GaAs buffer layer was grown followed by a 70 Å GaInP spacer and a doped GaInP layer 200 Å thick. A 250 Å thick undoped GaInP layer was grown for the purpose of reducing the gate leakage. Finally a

200 Å n^+-GaAs cap layer was grown to reduce the contact resistance. Fig. 9.6 shows the cross-section of the lattice-matched MODFET after the device is fabricated.

n + GaAs > 3×10^{18} cm^{-3}	200 Å
undoped Ga$_{0.49}$In$_{0.51}$P	250 Å
n + Ga$_{0.49}$In$_{0.51}$P 2×10^{18} cm^{-3}	200 Å
undoped Ga$_{0.49}$In$_{0.51}$P	70 Å
undoped GaAs	1.0 μm
SI-GaAs substrate	

(a)

n + GaAs > 3×10^{18} cm^{-3}	200 Å
undoped Ga$_{0.49}$In$_{0.51}$P	250 Å
n + Ga$_{0.49}$In$_{0.51}$P 2×10^{18} cm^{-3}	200 Å
undoped Ga$_{0.49}$In$_{0.51}$P	70 Å
undoped In$_{0.15}$Ga$_{0.85}$As	150 Å
undoped GaAs	1.0 μm
SI-GaAs substrate	

(b)

Fig. 9.5. MODFET structures after the layers are grown: (a) lattice-matched n-GaInP/GaAs HEMT, and (b) 15% In strained n-GaInP/GaAs HEMT (after Razeghi et al. [1991]).

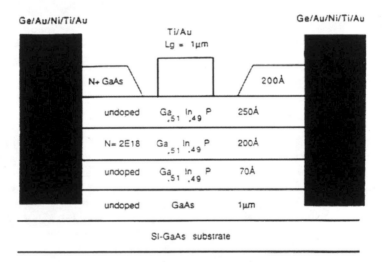

Fig. 9.6. Cross-section of the lattice-matched GaInP/GaAs MODFET (after Razeghi et al. [1991]).

Sulfur or silicon are used as *n*-type dopants for GaInP. Using H_2S for *n*-type doping, with the same growth conditions and the same H_2S flow rate, the probability of sulfur incorporation in the GaInP lattice is an order of magnitude higher than in GaAs. Therefore sulfur doping of GaInP is useful in modulation and delta doping. However, sulfur-doped GaInP shows persistent photoconductivity, which is not observed in Si doping. Recently it has been reported [Ginoudi et al. 1992] that considerable degradation in the g_m and the drain current of MOMBE-grown S-doped GaInP/GaAs MODFETs is observed at low temperatures while the performance of Si-doped devices remains good. In order to avoid this problem, Si was used as the *n*-type dopant in the fabrication of the MODFETs [Razeghi et al. 1991].

MODFETs were fabricated using conventional 1 μm long gate optical lithography. $NH_4OH:H_2O_2:H_2O$ (10:4:500) and $H_3PO_4:HCl$ (1:1) were used to etch GaAs and GaInP respectively to define the active regions of the device. $NH_4OH:H_2O_2:H_2O$ is selective for GaInP and attacks only GaAs, therefore uncontrolled gate recess variation is avoided. Ge/Au/Ni/Ti/Au (700/1400/500/200/1000 Å) was deposited for ohmic contact formation and a rapid thermal annealing at 440 °C for 10 s was carried out. Before depositing the gate metal (Ti/Au 500/3000 Å) a gate recess was made to remove the n^+-GaAs (200 Å) layer by $NH_4OH:H_2O_2:H_2O$ (10:4:500) selective etching.

Device type	Temperature of measurement			Carrier density increases at 77 K
	300 K	77 K(dark)	77 K (light)	
A	$\mu = 3500$ $n_s = 1.89 \times 10^{12}$	$\mu = 21300$ $n_s = 1.26 \times 10^{12}$	$\mu = 21900$ $n_s = 1.42 \times 10^{12}$	13%
B	$\mu = 5300$ $n_s = 2.45 \times 10^{12}$	$\mu = 78,500$ $n_s = 9.79 \times 10^{-11}$	$\mu = 76600$ $n_s = 1.13 \times 10^{12}$	16%

Table 9.1. Mobility μ ($cm^2 V^{-1} s^{-1}$) and sheet carrier density n_s (cm^{-2}) of lattice-matched (A) and strained (B) Ga(In)As/GaInP heterostructures (after Razeghi et al. [1991]).

Table 9.1 shows the Hall measurement results for lattice-matched (A) and strained (B) GaInP/GaAs heterostructures. Note that at 300 K, the mobility and carrier concentration change considerably by adding 15% excess In in the channel. It can also be seen that the heterostructures are not light sensitive at low temperatures. The carrier density has a slight dependence on the illumination. The mobilities obtained were high, despite the high doping of GaInP.

Fig. 9.7 shows the *I–V* characteristics of a typical metal-GaInP Schottky diode. The reverse-bias leakage current is very low ($I_{rev} < 200$ nA at $V_{GS} = -4$ V) demonstrating the high quality of the undoped GaAs layers. The Schottky barrier height evaluated from *C–V* measurements was 0.87 eV.

Fig. 9.8 shows $I_{DS}–V_{DS}$ and g_m, *Ids–V*$_{GS}$ transfer characteristics for the lattice-matched device. The extrinsic transconductance was 163 mS·mm^{-1} at 300 K and 213 mS mm^{-1} at 77 K. Fig. 9.9 shows the measured short-circuit current (H_{21}) and power (G) gain for lattice-matched MODFETs at $V_{DS} = 3.5$ V and $V_{GS} = -1.85$ V which revealed cutoff frequencies $f_T = 17.8$ GHz and $f_{max} = 23.5$ GHz.

I–V characteristics and g_m, $I_{DS}–V_{GS}$ transfer characteristics for strained devices are shown in Fig. 9.10. The extrinsic transconductance was 133 mS·mm^{-1} at 300 K which was lower than for the lattice-matched MODFET. The devices showed very good channel pinch-off behavior and extremely low output conductance ($g_{DS} < 2$ mS·mm^{-1}). The cutoff frequency was 11 GHz and f_{max} was 23.5 GHz.

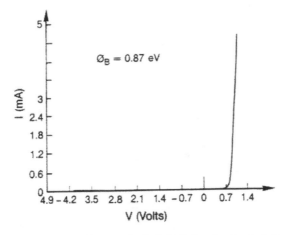

Fig. 9.7. *I–V characteristics of a metal–GaInP Schottky diode (after Razeghi et al.* [1991]).

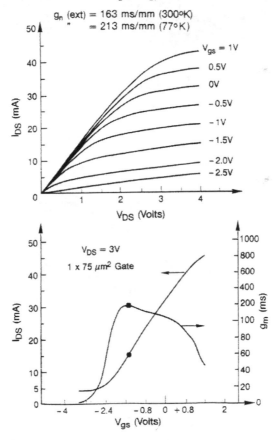

Fig. 9.8. *Characteristics of a lattice-matched GaInP/GaAs MODFET (after Razeghi et al. [1991]).*

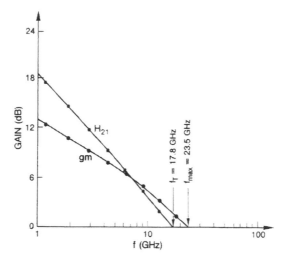

Fig. 9.9. Short-circuit current (H_{21}) and power (g_m) gain for lattice-matched
MODFETs (after Razeghi et al. [1991]).

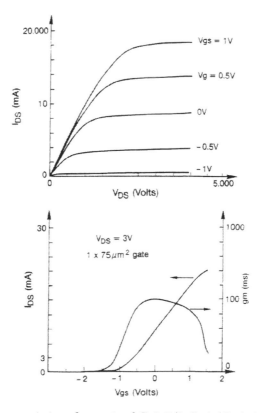

Fig. 9.10. Characteristics of a strained GaInP/InGaAs/GaAs MODFET (after
Razeghi et al. [1991]).

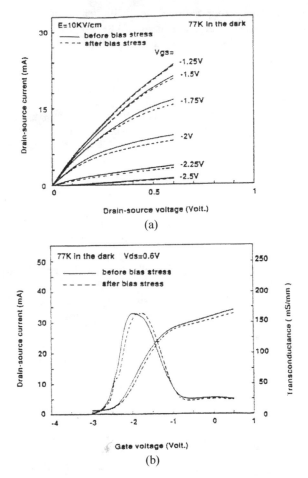

Fig. 9.11. Characteristics of GaInP/GaAs MODFETs at 77 K in the dark before and after bias stress (after Chan et al. [1990]).

Fig. 9.11 shows the *I–V* characteristics and I_{DS}–g_m versus gate voltage at 77 K before and after bias stress. As shown in the figure only a slight decrease in the current and transconductance was observed. The threshold voltage of the devices biased in the dark at $V_{DS} = 3$ V was −3.34 V at 300 K and −3.38 V at 77 K. These results demonstrate that traps in GaInP are negligible and good device characteristics can be maintained at low temperatures.

Takikawa et al. [1991] reported high-performance MOCVD-grown strained layer GaInP/InGaAs/GaAs MODFETs. They obtained good uniformity for doping concentration, thickness and composition over a 3 inch wafer. The 2DEG concentration and mobility for an *n*-InGaP–InGaAs–GaAs structure with a 25 Å thick spacer were 1.5×10^{12} cm^{-2} and 7000 cm^2·V^{-1}·s^{-1} at 300 K and 1.4×10^{12} cm^{-2} and 45,000 cm^2·V^{-1}·s^{-1} at 77 K. They have observed that the sheet carrier concentration increases monotonically with InAs mole fraction but the mobility shows a peak at an InAs mole fraction of about 0.15. The mobility and concentration at this composition were higher than those obtained for AlGaAs/InGaAs/GaAs by Ketterson et al. [1986]. They have also observed that short-channel effects are less significant in GaInP/InGaAs/GaAs MODFETs when compared with AlGaAs/GaAs devices.

p-channel device.

The valence-band discontinuity in the GaInP–GaAs system is larger than that of AlGaAs–GaAs resulting in better carrier confinement when *p*-channel MODFETs are considered. In comparison with *n*-channel devices, the transconductance of *p*-channel lattice-matched MODFETs is inferior due to the lower mobility of holes. For lattice-matched AlGaAs/GaAs MODFETs the best experimentally demonstrated ratio of transconductance between *n*- and *p*-channel lattice-matched devices is $g_{mn}/g_{mp} = 8$. The low transconductance of *p*-channel devices can be increased by increasing the width of the device but this is not desirable due to the increase in the chip dimensions. It has been shown that the performance of *p*-channel devices can be increased by introducing strain into the structure [Drummond et al. 1986 and Lee et al. 1987]. Strained *p*-channel MODFETs have been attracting considerable interest in digital applications [Kiehl et al. 1987]. The reduced hole mass and improved transport properties make them useful for complementary logic circuits with small power-delay products.

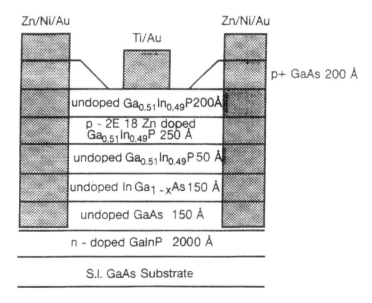

Fig. 9.12. Cross-section of the lattice-matched (x = 0) and strained (x = 0.1) p-channel GaInP/InGaAs/GaAs MODFETs (after Chan et al. [1988]).

To study the effect of strained channels, Chan et al. [1988] carried out a theoretical and experimental study of device performance in both lattice-matched and strained GaInP/GaAs heterostructures. Their simulations have shown that the average density of state mass at room temperature decreases by 30% due to the strain (from $0.57m_0$ to $0.36m_0$) and the reduction is more dramatic at lower temperatures. The cross-sections of the designed lattice-matched ($x = 0$) and strained ($x = 0.1$) GaInP/InGaAs/GaAs MODFETs are shown in Fig. 9.12. The channel consists of an undoped GaAs and a strained $In_xGa_{1-x}As$ ($x = 0.1$) layer. A 50 Å undoped $Ga_{0.51}In_{0.49}P$ spacer was grown between the strained channel and the Zn-doped (2×10^{18} cm^{-3}) p-$Ga_{0.51}In_{0.49}P$ layer. A 200 Å thick undoped $Ga_{0.51}In_{0.49}P$ layer was grown to reduce the gate leakage current. Finally a p^+-GaAs layer was grown to improve the ohmic contacts.

The devices were grown by low-pressure MOCVD. A $NH_4OH/H_2O_2/H_2O$ solution was used for etching the GaAs cap layer. The etching of GaInP for the mesas was achieved using a $HCl:H_3PO_4$ solution. ZnNiAu ohmic contacts were deposited by lift off and followed by conventional annealing in N_2 at 425 °C for 5 min. Ti/Au was used for gate metallization.

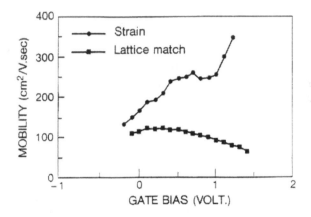

Fig. 9.13. Gate bias dependence of mobility in p-channel GaInP/InGaAs/GaAs MODFETs (after Chan et al. [1988]).

The layers were characterized to confirm the hole mobility enhancement by strain. Fig. 9.13 shows the gate bias dependence of the mobility for both lattice-matched and strained structures. In the lattice-matched case the reduction in mobility with gate bias is due to a decrease in the sheet carrier density (and therefore a decrease in the screening) and an increase of heavy-hole band occupation. In the strained structure, the HH and LH bands are split and an increase in the gate bias increases the LH population resulting in mobility enhancement. The g_m–V_G and I_D–V_G characteristics of a 1 μm × 100 μm MODFET are shown in Fig. 9.14. It was found that the velocity improvement by strain was 68%.

Significant threshold voltage shifts with temperature have been reported for p-AlGaAs/GaAs MODFETs grown by MBE [Hirano et al. 1986]. Chan et al. [1988] have also found that threshold voltage shifts with temperature are negligible in p-GaInP/InGaAs/GaAs MODFETs making this system very promising for high-speed digital applications.

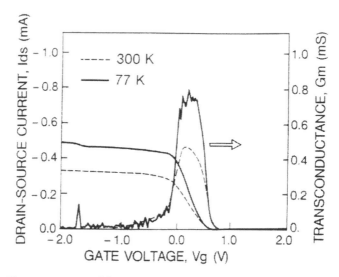

Fig. 9.14. Characteristics of the strained p-channel GaInP/InGaAs/GaAs MODFET (after Chan et al. [1990]).

9.2.4. MODFETs on silicon substrate

There has been considerable interest in the integration of III–V semiconductors with Si technology using the best aspects of both technologies to produce better performance and cheaper devices. Si technology is mature and Si has a higher thermal conductivity and mechanical strength compared to III–V materials. Also, large-area Si substrates are available. Therefore the growth of III–V-based devices on Si substrates is an important approach for ultralarge-scale integration.

A significant amount of research has been done on GaAs-based materials and devices on Si substrates [Razeghi et al. 1989, Houdre and Morkoc 1990, and Fang et al. 1990].

Studies on the fabrication of AlGaAs/GaAs MODFETs on Si substrate have shown that their performance is comparable to MODFETs on GaAs substrates [Fischer et al. 1986, Ren et al. 1992, and Ohari et al. 1988]. Omnes and Razeghi [1991] used low-pressure MOCVD to grow multiquantum-well structures of GaAs/GaInP on GaAs, InP and Si substrates. It was found from x-ray diffraction and photoluminescence measurements that the films grown on Si substrate are of comparable quality to the films grown on GaAs substrate. Suehiro et al. [1991] fabricated MOCVD-grown 0.6 μm gate enhancement- and depletion-mode GaInP/InGaAs/GaAs pseudomorphic MODFETs on Si substrate. The performance of the devices on Si substrate ($g_m = 242$ mS·mm^{-1}, $K = 324$ mS·V^{-1}·mm^{-1} for e-mode, $g_m = 211$ mS·mm^{-1}, $K = 184$ mS·V^{-1}·mm^{-1} for d-mode) was comparable to devices built on GaAs substrates under the same conditions ($g_m = 236$ mS·mm^{-1}, $K = 371$ mS·V·mm^{-1} for e-mode, $g_m = 224$ mS·mm^{-1}, $K = 265$ mS·V·mm^{-1} for d-mode). The threshold voltage standard deviation for the devices on Si substrate (22 mV) was higher than that for the devices on GaAs substrate (6 mV) but it was still at an acceptable level. These results verify that GaInP/InGaAs/GaAs MODFETs on Si substrates are potential devices for high-speed, low-power ultralarge-scale integration.

9.2.5. GaInP/GaAs heterostructure insulated gate field effect transistors

Heterostructure insulated gate field effect transistors (HIGFETs) have demonstrated promising characteristics for high-speed digital circuit applications [Solomon et al. 1984 and Feuer et al. 1989]. This device is similar to a conventional MODFET except that the higher-bandgap material is left undoped. Under flat-band conditions, charge neutrality is preserved throughout the device. If a sufficiently large positive gate voltage is applied, a negative charge (consisting of electrons in a two-dimensional well) is induced in the semiconductor (Fig. 9.15). Similarly, a two-dimensional hole gas can be formed by applying a negative gate bias. Since the threshold voltage of the device is determined by the material independent of the doping density and thickness, an improved threshold voltage uniformity is achieved. Furthermore, since the active channel is induced under the gate only by the application of a bias, both n- and p-type devices can be fabricated on the same wafer by changing the type of the implant in the source and drain contacts. This allows the realization of complementary logic circuits with low power dissipation.

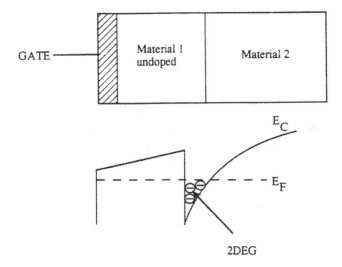

Fig. 9.15. A heterostructure with an undoped semiconductor and its band diagram under a sufficiently large, positive gate bias.

To date, the most commonly used material system in heterostructure FET applications has been AlGaAs/GaAs, which has major drawbacks as was discussed in the previous sections. Razeghi et al. [1990a] fabricated MOCVD-grown n- and p-type GaInP/InGaAs/GaAs lattice-matched and strained HIGFETs in order to investigate the potential of this system in HIGFET applications as an alternative to the AlGaAs–GaAs system. The device cross-section is shown in Fig. 9.16. A 1 μm undoped GaAs layer was used as a buffer followed by 250 Å undoped GaInP used as an insulator. The growth conditions of GaInP were optimized in order to obtain a high-purity material (necessary for good insulating properties) lattice matched to GaAs. The growth parameters are the same as those listed in Table 6.1. The background doping levels in the GaInP and GaAs layers were $1 \times 10^{15}\,\mathrm{cm}^{-3}$ and $1 \times 10^{14}\,\mathrm{cm}^{-3}$ respectively.

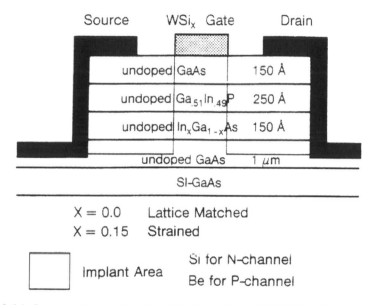

Fig. 9.16. Cross-section of the GaInP/InGaAs/GaAs HIGFET (after Razeghi et al. [1990a]).

A 1.55 μm long WSi$_x$ gate was defined by optical lithography and lift off. Self-aligned Si and Be implantations were used to access *n*- and *p*-channels, respectively. In order to activate the implant, the samples were treated by rapid thermal annealing at 800 °C for 5 s. Studies on the implant activation for Si in GaInP showed that activation was 5% at 800 °C, and 36% at 900 °C for 5 s rapid thermal annealing. Very high temperatures necessary for good activation are not compatible with the conditions required to maintain material integrity: phosphorus, for example, evaporates and degrades material stoichiometry.

Following mesa etching, source and drain contacts were defined by depositing and annealing Ge/Au/Ni/Ti/Au (700/1400/500/200/1000 Å) at 460 °C for 10 s for *n*-type and Zn/Ni/Au (300/500/2500 Å) at 425 °C for 2 min for *p*-type devices.

Fig. 9.17 shows the I_{DS}–V_{DS} and g_m–I_{DS}–V_{GS} characteristics of the *n*-type lattice-matched HIGFETs with 1.5 × 75 μm² gates. The devices showed excellent pinch-off characteristics with enhancement-mode operation (V_{th} = 0.19 V). The corresponding characteristics for the strained *n*-channel HIGFETs are shown in Fig. 9.18. The improvement in device performance with the introduction of In in the channel is apparent. The improvement is reflected by electron mobility and sheet carrier density enhancement. The full advantages of GaInP/InGaAs/GaAs heterostructures can be obtained by fully optimizing the post-implantation capless rapid-thermal-annealing conditions. This allows the optimization of access regions and therefore larger extrinsic g_m values. By using sufficiently thick gate metal and high-energy implantation, the access resistances can be minimized.

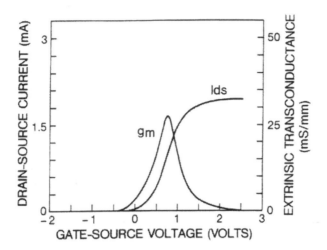

Fig. 9.17. Characteristics of the n-type lattice-matched HIGFET (after Razeghi et al. [1990a]).

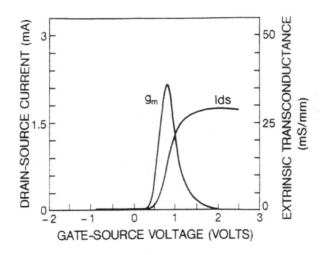

Fig. 9.18. Characteristics of the n-type strained HIGFET (after Razeghi et al. [1990a]).

The threshold voltages for the strained structure at 300 K and 77 K were 0.56 V and 0.62 V respectively, demonstrating the absence of carrier trapping at low temperatures. The high-bandgap GaInP allows the application of large gate voltages before leakage by thermionic emission through the barrier becomes significant. The gate leakage current at $V_{GS} = 2$ V was 250 mA·mm^{-2} on 520 Å thick GaInP devices. This is small compared with AlGaAs/GaAs or InGaAs/AlInAs where the maximum gate voltage is limited to about 1.5 V.

The I_{DS}–V_{DS} and g_m–I_{DS}–V_{GS} characteristics of p-type lattice-matched devices are shown in Fig. 9.19. The intrinsic transconductance was 74 mS·mm^{-1} which is considered to be high for p-type HIGFETs. The high channel leakage current at $V_{GS} = 0$ V is associated with the penetration of Be ions through the thin WSi$_x$ gate.

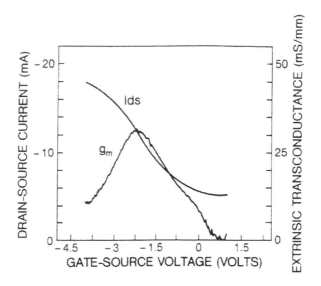

Fig. 9.19. *Characteristics of the p-type lattice-matched HIGFET (after Razeghi et al. [1990a]).*

9.3. Heterojunction bipolar transistors (HBTs)

9.3.1. Introduction

The bipolar junction transistor, since its invention in 1948 by Bardeen, Brattain and Shockley, has played the most important role in the development of semiconductor electronics. The idea of a heterojunction bipolar transistor (HBT), or, in other words, using wider-bandgap material in the emitter is almost as old as the transistor itself [Kroemer 1957]. However, the difficulties in the material processing technologies delayed successful implementation of this idea until the 1970s. Starting in the early 1970s, with the breakthroughs in material technology, especially with the development of MBE and MOCVD, high-performance HBTs have been fabricated and the gain, frequency and power capabilities of the bipolar transistor have increased significantly.

In comparison with field effect transistors, the HBT has the following important advantages. Since it is a vertical device, which means that the electrons travel perpendicular to the wafer surface, the transit time of the carriers is determined by the layer thickness which can be controlled precisely by epitaxy. Therefore, a high-frequency response can be obtained without using sophisticated lithographic techniques required for submicron design. HBTs can handle higher current levels when compared with FETs and the switch-on voltage of the HBT is relatively insensitive to process variations resulting in good uniformity of this parameter over the wafer. In this section, after a basic introduction to the heterojunction bipolar transistor and its advantages over the homojunction bipolar transistor, we will discuss growth, fabrication and performance features of MOCVD-grown, state of the art HBTs based on AlGaAs–GaAs and GaInP–GaAs material systems.

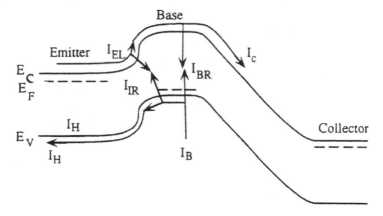

Fig. 9.20. Energy band diagram and current components in an n–p–n homojunction bipolar transistor.

Fig. 9.20 shows the energy band diagram and the current components in an n–p–n homojunction transistor biased in the active region. The forward bias at the emitter–base junction reduces the energy barrier for electron flow from the emitter into the base. Electrons that overcome the barrier travel through the base by drift and diffusion and are swept into the collector due to the high electric field in the base–collector depletion region.

Some of the injected electrons from the emitter into the base and some of the injected holes from the base into the emitter are lost due to recombination in the emitter–base space charge region (I_{IR}). Depending on the base thickness and the lifetime of the minority carriers in the base, part of the injected electron current (I_{EL}) is lost due to recombination with the holes in the quasineutral base (I_{BR}). The holes lost due to injection into the emitter (I_H) and recombination (in the emitter–base depletion layer (I_{IR}) and in the base (I_{BR})) are supplied through the base contact which results in the base current I_B. We have neglected the generation current in the collector–base depletion region. The emitter current is given by

Eq. (9.13) $\qquad I_E = I_{EL} + I_{IR} + I_H$

and the collector current is

Eq. (9.14) $I_{c} = I_{EL} - I_{BR}$

The base current, which is the difference between the emitter and the collector currents, can be expressed as

Eq. (9.15) $I_{B} = I_{BR} + I_{IR} + I_{H}$

The common emitter current gain β is a figure of merit for a bipolar transistor and is given by

Eq. (9.16) $\beta = \dfrac{I_{c}}{I_{B}} = \dfrac{I_{EL} - I_{BR}}{I_{H} + I_{IR} + I_{BR}}$

The current transfer ratio of a bipolar transistor α is the ratio of the collector current to the emitter current which is related to β by

Eq. (9.17) $\beta = \dfrac{\alpha}{1 - \alpha}$

Therefore, in an efficient bipolar transistor with high current gain, α should be very close to unity; α is determined by two factors: the ratio of the injected electron current to the total emitter current, which is the emitter injection efficiency γ, and the fraction of the injected electrons which reach the collector without being recombined in the base, which is the base transport factor B. In order to obtain an α close to unity and therefore high β, both γ and B should be near unity, or in other words, the emitter current should be due mostly to electrons (for n–p–n transistors) and the fraction of injected electrons that recombine in the base should be as small as possible.

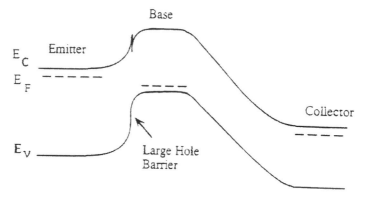

Fig. 9.21. Energy band diagram for a heterojunction bipolar transistor.

If we assume that the current gain is limited by the emitter injection efficiency (that is, the recombination currents are negligible), the common emitter current gain

β_{max} is given by the ratio of the injected electron current from the emitter into the base to the injected hole current from the base into the emitter. If the base is doped lightly with respect to the emitter doping level, the hole current injected into the emitter will be small in comparison to the electron current injected into the base from the emitter, resulting in a large current gain. However, device performance is very sensitive to the doping level in the base. Low base doping results in high base resistance, which degrades device performance by causing parasitic effects such as emitter crowding and reach-through of the base depletion region from the collector to the emitter. A high base resistance also degrades the speed performance of the device, as will be discussed later. The main purpose of using a wide-bandgap emitter or, in other words, a heterostructure emitter–base junction, is to improve β_{max} which is an upper limit for the gain. Fig. 9.21 shows the energy band diagram of a heterojunction bipolar transistor. Due to the extra barrier for the holes, the hole injection current from the base to the emitter is significantly reduced and a high current gain is achieved even if the base is very highly doped. In this case β_{max} is given approximately by [Kroemer 1982]

Eq. (9.18) $\qquad \beta_{\text{max}} = \dfrac{N_E \upsilon_{\text{nb}}}{P_B \upsilon_{\text{pe}}} \exp\left(\dfrac{\Delta E_g}{kT} \right)$

where N_E, P_B are the doping densities in the emitter and base, respectively, υ_{pe} and υ_{nb} are the effective velocities of holes in the base end of the emitter and electrons in the emitter of the base. Noting that ΔE_g is zero for a homojunction, the improvement factor in the current gain is $\exp(\Delta E_g/kT)$. Since this factor is quite large, very high values of current gain can be obtained regardless of the ratio of the doping levels in the emitter and base.

9.3.2. Frequency response and design considerations

One of the figures of merit for bipolar transistors is the current gain cutoff frequency f_T where the transistor incremental current gain drops to unity, f_T is given by

Eq. (9.19) $\qquad f_T = \dfrac{1}{2\pi\tau_{\text{ec}}}$

where τ_{ec} is the emitter to collector transit time expressed in the form

Eq. (9.20) $\qquad \tau_{\text{ec}} = \tau_e + \tau_b + \tau_{\text{cdl}} + \tau_c$

In this expression, τ_e is the emitter charging time which is $r_e C_e$ where r_e is the emitter resistance and C_e is the emitter capacitance. τ_b is the base transit time determined by the base thickness, diffusivity of electrons in the base and the drift fields in this region. τ_{cdl} is the collector depletion layer transit time given by $W_c/2\upsilon_s$ where W_c is the collector depletion layer width and υ_s is the saturation-limited electron velocity in the collector. τ_c is the collector charging time which is r_cC_c where r_c is the collector bulk resistance and C_c is the collector capacitance.

Another figure of merit is the frequency at which the maximum power gain of the transistor drops to unity, which is called the maximum frequency of oscillation f_{max} and is given by

Eq. (9.21) $$f_{max} = \left(\frac{f_T}{8\pi r_b C_{bc}} \right)^{1/2}$$

where r_b is the base resistance and C_{bc} is the base collector capacitance. Due to the r_bC_{bc} time constant, the base resistance is an important factor in the frequency response of a bipolar transistor. In heterojunction bipolar transistors, the base can be doped very heavily, resulting in a very low base resistance without sacrificing the emitter injection efficiency. Since high base doping causes low lifetimes in the base for the injected electrons, it is important that the base region be thin to keep the recombination current I_{BR} small and the base transport factor close to unity.

Assuming that the base is heavily doped, the emitter capacitance is determined by the doping level in the emitter. In order to keep the emitter depletion layer charging time τ_e low, the emitter can be doped lightly to provide a small emitter capacitance and the emitter resistance can be kept low by decreasing the total emitter thickness. A low doping level in the emitter also provides a high emitter reverse breakdown voltage.

The conduction-band discontinuity at the emitter–base junction establishes a barrier for electron flow from the emitter into the base. The emitter injection efficiency can be further increased by compositionally grading the emitter–base junction which eliminates the conduction-band discontinuity. However, the conduction-band discontinuity lowers the base transit time by injecting the electrons into the base with high velocities [Ankri et al. 1982]. The base transit time can be further reduced by compositionally grading the base. The gradual change in the conduction-band energy establishes a quasielectric field in the base. In this case the injected electrons are driven by both diffusion and drift. Fig. 9.22 shows the energy band diagrams for graded-junction and graded-base HBTs. The collector depletion layer transit time τ_{cdl} plays an important role in determining the total transit time and therefore the frequency response. Using a heavy doping level in the collector results in a low collector depletion layer transit time due to the small depletion layer width, but this results in a low collector–base breakdown voltage and high collector capacitance. Monte Carlo simulations for AlGaAs/GaAs HBTs have shown that the electron velocity may exceed the saturation velocity in the collector depletion layer [Asbeck et al. 1982]. Maziar et al. [1986] have demonstrated that by using an inverted field structure, velocity overshoot can be extended over a longer portion of the collector space-charge region reducing the transit time through this layer.

In double-heterojunction transistors, wide-bandgap material is used for both emitter and collector. By using a wide-bandgap collector, the breakdown voltage of the collector junction can be increased, and carrier injection from the collector to the base can be suppressed, which results in faster operation in digital applications.

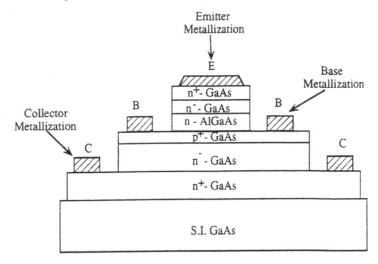

Fig. 9.22. Cross-sectional structure of an AlGaAs/GaAs HBT.

9.3.3. AlGaAs/GaAs HBTs

Due to the excellent lattice match between $Al_xGa_{1-x}As$ and GaAs over the entire compositional range, the AlGaAs–GaAs system has been the most widely used system for heterojunction bipolar transistors. Fig. 9.22 shows the cross-sectional structure of an AlGaAs/GaAs HBT. In order to avoid DX center problems, the Al mole fraction x in $Al_xGa_{1-x}As$ is usually kept around 0.25 which results in a conduction-band discontinuity of 0.2 eV and a valence-band discontinuity of 0.1 eV. Due to the large conduction-band discontinuity, the emitter–base junction is usually compositionally graded.

Device isolation is performed by deep ion implantation to make the layers outside the device semi-insulating or by using a mesa structure. Using composition-selective etches, vias are etched to the base and collector layers to make the corresponding contacts.

In order to obtain good device performance, the contact resistances of the ohmic contacts should be minimized. One of the ways that has been used to reduce the emitter contact resistance is to use lattice-mismatched InGaAs cap layers grown on the emitter. Due to the small metal–semiconductor barrier height good ohmic contacts can be achieved. Commonly used contacts are AuGe/Ni and Ge/Au/Cr. Minimization of the parasitic resistances is also very important in obtaining good device performance. The base–emitter separation should be a few tenths of a micron. This can be accomplished by using self-aligned techniques [Nagata et al. 1987, Hayama et al. 1987, and Chang et al. 1987]. By using a shallow proton implant in the collector region under the base contacts, extrinsic collector doping and therefore the base-collector capacitance can be reduced [Ginoudi et al. 1992].

It is important that the base–emitter *p–n* junction coincide with the heterojunction between AlGaAs and GaAs. Therefore good doping profile and material composition control is required in the growth of the epilayers. Most of the

AlGaAs/GaAs HBT research has been done on MBE-grown devices. Since the early 1980s the performance of MOCVD-grown AlGaAs/GaAs HBTs has increased significantly. An f_{max} of 94 GHz and an f_T of 45 GHz have been obtained by Enquist and Hutchby [1989] using a self-aligned structure.

One of the difficulties in HBT fabrication is the diffusion of impurities from the heavily doped GaAs base into the AlGaAs emitter at high temperatures during or subsequent to growth. This causes the p–n junction to move into the AlGaAs layer and the current gain of the device is reduced due to the reduction in the barrier to hole injection. This problem can be avoided by introducing a thin undoped GaAs spacer layer between the base and emitter or by reduction of the growth temperature before the AlGaAs layer is grown. Common p-type dopants in MOCVD are magnesium, zinc and carbon. Mg doping shows abnormal memory effects which requires growth interruptions in order to obtain an abrupt doping profile [Kuech et al. 1988, Landgren et al. 1988]. Zn has a large diffusion coefficient and carbon doping needs a low growth temperature, both of which are incompatible with the growth of high-quality AlGaAs which requires high temperatures. However, very high base doping levels are possible with carbon doping due to the very low diffusion coefficient of carbon [Ashizawa et al. 1991]. Using carbon doping in the base ($p = 4 \times 10^{19}$ cm^{-3}), Twynam et al. [1991] have reported MOCVD-grown AlGaAs/GaAs microwave HBTs with an f_T of 42 GHz, f_{max} of 117 GHz and a current gain of 50.

9.3.4. GaInP/GaAs HBTs

As mentioned in the previous chapters, the $Ga_{0.51}In_{0.49}P$–GaAs system has some major advantages over AlGaAs–GaAs. For n–p heterojunction bipolar transistors, the GaInP–GaAs system has an additional advantage when compared with the widely used AlGaAs–GaAs structure. The valence-band discontinuity in the $Ga_{0.51}In_{0.49}P$–GaAs system is about 0.28 eV and conduction-band discontinuity is 0.2 eV [Biswas et al. 1990]. A large valence-band discontinuity is an exciting property for n–p–n HBTs. In the AlGaAs–GaAs system, the same amount of valence-band discontinuity requires that the Al mole fraction be about 0.6, in which case there would be a very large conduction-band spike at the emitter–base junction together with an indirect-gap emitter, neither being acceptable. In the AlGaAs–GaAs system, about 60% of the energy gap difference occurs in the conduction band and the emitter–base junction of the device is usually graded to eliminate the conduction-band spike which decreases the emitter injection efficiency and increases the emitter switch-on voltage. However, theoretical investigations [Das and Lundstrom 1988] have shown that grading of the emitter–base junction increases the recombination in the emitter–base junction and therefore the current gain may not be increased considerably by junction grading. Because of the relatively small conduction-band discontinuity and large valence-band discontinuity of $Ga_{0.51}In_{0.49}P$/GaAs, it can be estimated that the current gain of n–p–n HBTs based on this material system will be significantly higher than that of AlGaAs/GaAs HBTs.

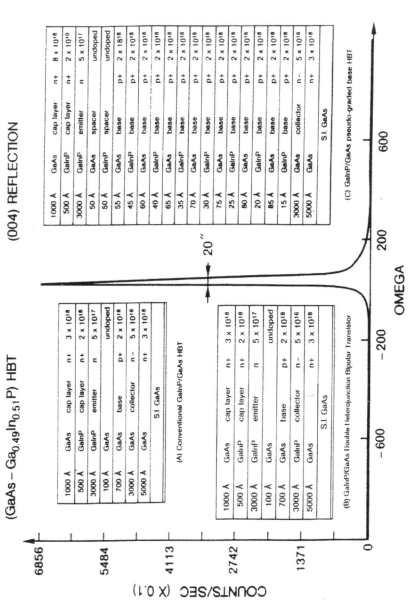

Fig. 9.23. Typical x-ray double-diffraction pattern of a GaAs–GaInP HBT: (A) conventional, (B) double heterojunction, (C) pseudograded base (after Razeghi et al. [1990b]).

Modry and Kroemer [1985] have reported a GaInP/GaAs HBT grown by MBE. The current gain was low at small current densities suggesting a high recombination rate at the emitter–base junction due to a large number of defects at the heterojunction interface. A maximum current gain of 30 was obtained at 3000 A·cm^{-2}. Later, MOCVD- and chemical-beam-epitaxy-grown GaInP/GaAs HBTs with better performance were reported [Kobayashi et al. 1989, Razeghi et al. 1990b, Alexandre et al. 1990, and Bachem et al. 1992]. Razeghi et al. [1990b] reported a current gain of 400 for a low-pressure MOCVD-grown GaInP/GaAs HBT. Three different HBT structures were grown: (i) conventional, (ii) double heterojunction, (iii) pseudo-graded base. Optimized growth parameters are given in Table 6.1. The details of collector, base and emitter thicknesses, carrier concentrations and a typical x-ray diffraction pattern for the structures around the (4 0 0) reflection peak are shown in Fig. 9.23. The signal is very intense and has a full width at half maximum (FWHM) of 20 arcsec, demonstrating that GaInP is perfectly lattice matched to GaAs and the pseudograded base has excellent crystallographic properties, which is necessary to allow optimal transport properties of the injected minority carriers. Fig. 9.24 shows the doping profile demonstrating an abrupt and perfectly controlled transition from emitter to base and from base to collector. The device structure was a conventional mesa type. NH_4:H_2O_2:H_2O (10:4:500) and HCl:H_3PO_4 (1:1) were used to etch GaAs and GaInP respectively. The emitter and collector contacts were defined by depositing and annealing Ge/Au/Ni/Au. The base contact was defined by the deposition and annealing of Zn/Au. Fig. 9.25 shows the emitter-grounded *I–V* characteristics of the device which exhibited a current gain of 400 at 20 mA.

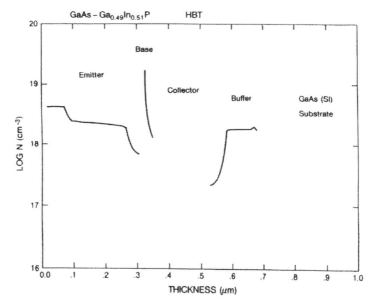

Fig. 9.24. Typical doping profile of the HBTs (after Razeghi et al. [1990b]).

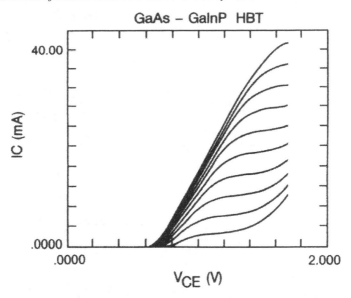

Fig. 9.25. Emitter-grounded I–V characteristics of structure (A) Fig. 9.23 (after Razeghi et al. [1990b]).

References

Alexandra, F., Benchimol, J.L., Dangla, J., Dubon-Chevallier, C., and Amarger, V. *Electron. Lett.* **26** 1753, 1990.

Anderson, R.L. *Solid-State Electron.* **5** 341, 1962.

Ankri, D., Schaff, W., Smith, P., Wood, C.E., and Eastman, L.F. *IEDM Tech. Digest* p 788, 1982.

Asbeck, P.M., Miller, D.L., Asatourian, R., and Kirkpatrick, C.G. *IEEE Electron Devices Lett.* **EDL-3** 403, 1982.

Ashizawa, Y., Noda, T., Morizuka, K., Asaka, M., and Obara, M. *J. Crystal Growth* **107** 903, 1991.

Bachem, K.H., Lauterbach, T.H., Maier, M., Pletschen, W., and Winkler, K. *Gallium Arsenide and Related Compounds 1991* (*Inst. Phys. Conf. Ser. 120*) ed G.B. Stringfellow (Bristol: Institute of Physics) ch 6, p 293, 1992.

Biswas, D., Debhar, N., Bhattacharya, P., Razeghi, M., Defour, M., and Omnes, F. *Appl. Phys. Lett.* **56** 833, 1990.

Chan, Y.J., Pavlidis, D., Razeghi, M., Jaffe, M., and Singh, J. *Gallium Arsenide and Related Compounds* (*Inst. Phys. Conf. Ser. 96*) ed J.S. Harris (Bristol: Institute of Physics) ch 7, p 459, 1988.

Chan, Y.J., Pavlidis, D., Razeghi, M., and Omnes, F. *IEEE Trans. Electron Devices* **ED-37** 2141, 1990.

Chang, M.F., Asbeck, P.M., Wang, K.C., Sullivan, C.J., Sheng, N.H., Higgins, J.A., and Miller, D.L. *IEEE Electron Devices Lett.* **EDL-8** 7, 1987.

Chi, J.Y., Holmstrom, R.P., and Salerno, J.P. *IEEE Electron Devices Lett.* **EDL-5** 381, 1984.

Das, A. and Lundstrom, M.S. *IEEE Trans. Electron Devices* **ED-35** 863, 1988.

Delagebeaudeuf, D. and Ling, N.T. *IEEE Trans. Electron Devices* **ED-29** 955, 1982.

Drummond, T.J., Fischer, R., Kopp, W., Morkoc, H., Lee, K., and Shur, M.S. *IEEE Trans. Electron Devices* **ED-30** 1806, 1983.

Drummond, T.J., Masselink, W.T., and Morkoc, H. *IEEE Proc.* **74** 773, 1986.

Enquist, P.M. and Hutchby, J.A. *Electron. Lett.* **25** 1124, 1989.

Fang, S.F., Adomi, K., Iyer, S., Morkoc, H., and Zabel, H. *J. Appl. Phys.* **68** R31, 1990.

Feuer, M.D., Tennant, D.M., Kuo, J.M., Shunk, S.C., Tell, B., and Chang, T.Y. *IEEE Electron Devices Lett.* **10** 70, 1989.

Fischer, R., Morkoc, H., Neumann, D.A., Zabel, H., Choi, C., Otsuko, N., Longerbone, M., and Erickson, L.R. *J. Appl. Phys.* **60** 1640, 1986.

Ginoudi, A., Paloura, E.C., Kostandinidis, G., Kiriakidis, G., Maurel, P.H., Garcia, J.C., and Christou, A. *Appl. Phys. Lett.* **60** 3162, 1992.

Hayama, N., Okamoto, A., Madihian, M., and Hanjo, K. *IEEE Electron Devices Lett.* **EDL-8** 246, 1987.

Hirano, M., Oe, K., and Yanagawa, F. *IEEE Trans. Electron Devices* **ED-33** 620, 1986.

Hiyamuzu, S. *Semicond. Semimet.* **30** 53, 1990.

Hiyamuzu, S., Saito, J., Nanbu, K., and Ishikawa, T. *Japan. J. Appl. Phys.* **22** L609, 1983.

Houdre, R. and Morkoc, H. *CRC Crit. Rev. Solid State Mater. Sci* **16** 91, 1990.

Ishikawa, T., Hiyamuzu, S., Mimura, T., Saito, J., and Hashimoto, H. *Japan. J. Appl. Phys.* **20** L814, 1981.

Ketterson, A.A., Masselink, W.T., Gedymin, J.S., Klem, J., Peng, C.K., Kopp, W.F., Morkoc, H., and Gleason, K.R. *IEEE Trans. Electron Devices* **ED-33** 564, 1986.

Ketterson, A.A., Moloney, M., Masselink, W.T., Klem, J., Fischer, R., Kopp, W., and Morkoc, H. *IEEE Electron Devices Lett.* **EDL-6** 628, 1985.

Kiehl, R.A., Scontras, M.A., Widiger, D.J., and Kwapien, W.M. *IEEE Trans. Electron Devices* **ED-34** 2412, 1987.

Kikkawa, T., Ohori, T., Tanaka, H., Kasai, K., and Komeno, J. *J. Crystal Growth* **115** 448, 1991.

Kobayashi, T., Taira, K., Nakamura, F., and Kawai, H. *J. Appl. Phys.* **65** 4898, 1989.

Kroemer, H. *Proc. IRE* **45** 1535, 1957.

Kroemer, H. *Proc. IEEE* **70** 13, 1982.

Kuech, T.F., Wang, P.J., Tischler, M.A., Potenski, R., Scilla, G.J., and Cardone, F., *J. Crystal Growth* **93** 624, 1988.

Landgren, G., Rask, M., Anderson, S.G., and Lundberg, A. *J. Crystal Growth* **93** 646, 1988.

Lee, J.S., Iwasa, Y., and Muera, N. *Semicond. Sci. Technol.* **2** 675, 1987.

Lee, K., Shur, M., Drummond, T.J., Su, S.L., Lyons, W.G., Fisher, R., and Morkoc, H. *J. Vac. Sci. Technol.* **B 1** 186, 1983.

Maziar, C.M., Klausmeier-Brown, M.E., and Lundstrom, M.S. *IEEE Electron Devices Lett.* **EDL-7** 483, 1986.

Modry, M.J. and Kroemer, H. *IEEE Electron Devices Lett.* **EDL-6** 175, 1985.
Morkoc, H. and Unlu, H. *Semicond. Semimet.* **24** 135, 1987.
Nagata, K., Nakajima, O., Nittono, T., Ito, H., and Ishibashi, T. *Electron. Lett.* **23** 64, 1987.
Nguyen, L.D., Radulescu, D.C., Tasker, P.J., Schaft, W.J., and Eastman, L.F. *IEEE Electron Devices Lett.* **EDL-9** 374, 1988.
Ohari, T., Takechi, M., Suzuki, M., Takikawa, M., and Komeno, J. *J. Crystal Growth* **93** 905, 1988.
Omnes, F. and Razeghi, M. *Appl. Phys. Lett.* **59** 1034, 1991.
Pearah, P., Henderson, T., Klem, J., Morkoc, H., Nilsson, B., Wu, O., Swanson, A.W., and Ch'en, D.R. *J. Appl. Phys.* **56** 1851, 1984.
Razeghi, M., Defour, M., Omnes, F., Dobers, M., Vieren, J.P., and Guldner, Y. *Appl. Phys. Lett.* **55** 457, 1989.
Razeghi, M., Omnes, F., Defour, M., Maurel, P.H., Bove, P.H., Chan, Y.J., and Pavlidis, D. *Semicond. Sci. Technol.* **5** 274, 1990a.
Razeghi, M., Omnes, F., Defour, M., Maurel, P.H., Hu, J., Woik, E., and Pavlidis, D. *Semicond. Sci. Technol.* **5** 278, 1990b.
Razeghi, M., Omnes, F., Maurel, P.H., Chan, Y.J., and Pavlidis, D. *Semicond. Sci. Technol.* **6** 103, 1991.
Ren, F., Abernathy, C.R., Peatron, S.J., Lothian, J.R., Chu, S.N., Wisk, P.W., Fullowan, T.R., Tseng, B., and Chen, Y.K. *Electron. Lett.* **28** 2250, 1992.
Rosenberg, J.J., Benlamri, M., Kirchner, P.D., Woodall, J.M., and Pettit, J.P. *IEEE Electron Devices Lett.* **EDL-6** 491, 1985.
Solomon, P.M., Knoedler, C.M., and Wright, S.L. *IEEE Electron Devices Lett* **EDL-5** 379, 1984.
Stormer, H.L., Dingle, R., Gossard, A.C., Wiegmann, W., and Sturge, M.D. *Solid State Commun.* **29** 705, 1979.
Suehiro, M., Hirata, T., Maeda, M., Hihara, M., Yamada, N., and Hosomatsu, H. *Japan. J. Appl. Phys.* **30** 3410, 1991.
Tachikawa, M., Mizuta, M., Kukimoto, H., and Minomura, S. *Japan. J. Appl. Phys.* **24** L821, 1985.
Takikawa, M., Ohori, T., Takechi, M., Suzuki, M., and Komeno, J. *J. Crystal Growth* **107** 942, 1991.
Tanaka, H., Kawamura, Y., Nojima, S., Wakita, K., and Asahi, H. *J. Appl. Phys.* **61** 1713, 1987.
Twynam, J.K., Sato, H., and Kinosada, T. *Electron. Lett.* **27** 141, 1991.
Valois, A.J., Robinson, G.Y., Lee, K., and Shur, M.S. *J. Vac. Sci. Technol.* **B 1** 190, 1983.
Watanabe, M.O. and Ohba, Y. *J. Appl. Phys.* **60** 1032, 1986.

10.   Optoelectronic Integrated Circuits (OEICs)

10.1. Introduction
10.2. Material considerations
10.3. OEICs on silicon substrates
10.4. The role of optoelectronic integration in computing
10.5. Examples of optoelectronic integration by MOCVD
     10.5.1. Monolithic integration of photodetectors with transistors
     10.5.2. Monolithic integration of photodetectors with optical waveguides
     10.5.3. Monolithic integration of optical modulator with lasers
     10.5.4. Surface emitting laser diodes

10.1. Introduction

With the advancements in epitaxial growth techniques such as MOCVD, optoelectronics has been attracting increasing attention in a wide area of applications such as communication systems, optical discs and computers. Most of the optoelectronics research has been focused on improving the speed and reliability of optical communication. Optical fiber communication systems with bit rates in multigigabits·s^{-1} have been demonstrated and a significant amount of research is in progress to achieve faster, more reliable and lower-cost optical communication. Recently, optoelectronic integration has also received attention in building the optical interconnects in computing systems as will be discussed later.

An optoelectronic integrated circuit (OEIC) includes both optical devices, such as lasers and photodetectors, and electronic devices, such as transistors (which are necessary for driving the lasers), amplification and signal processing [Hirano 1987]. Most of the optoelectronic circuits which are commercially available are still fabricated using a hybrid assembly technology, and the interconnections between optical and electronic discrete devices are performed using conventional techniques such as wire bonding. However, as the bit rate is increased, the parasitic capacitances and inductances associated with discrete packaging and bonding wires lead to significant degradations in the performance of the circuit. Therefore, monolithic integration of optical and electronic devices on the same substrate (preferably a semi-insulating one) has received increasing attention since the first demonstration of an optoelectronic integrated circuit by Lee et al. [1978] which consisted of the integration of a laser diode with a Gunn oscillator. However,

399

present integration techniques are still complicated due to the different structural and material requirements of optical and electronic devices. It looks like the future of optoelectronics relies not only on improvements in the performance of electronic and optical devices, but also on advancements in monolithic integration techniques, which have the advantage of compactness, higher performance and lower cost over conventional hybrid technology.

10.2. Material considerations

OEICs can be separated into two major groups: transmitters and receivers. A transmitter basically consists of a laser and the necessary electronics to drive the laser. A receiver includes a detector and amplifying circuits. During initial OEIC research, most of the circuits reported contained one or two optical devices with several electronic devices. However, with the improvements in integration technologies, the number of components in integrated circuits is increasing in parallel with significant improvements in performance.

AlGaAs–GaAs has been one of the most widely studied III–V material systems due to the desirable electronic and optical properties of GaAs and the perfect lattice match between GaAs and AlGaAs. Therefore, GaAs technology is more mature compared with other III–V technologies. However, unfortunately, in the wavelength offered by GaAs (0.8 μm) the losses incurred using the optical fibers are large, limiting the transmission distance. InGaAsP–InP has become an important material system offering longer wavelengths (0.9–1.6 μm) which cover the range where the losses from silica-based fibers are minimal (1.3 μm). There have been numerous reports on the fabrication of transmitters and receivers based on AlGaAs–GaAs and InGaAsP–InP systems. In most of the transmitters and receivers based on the AlGaAs–GaAs system, MESFETs have been used in the driving and amplifying electronics due to their easy fabrication. However, the Schottky barrier heights for InP and InGaAsP lattice matched to InP (≈ 0.2 eV) are too small for reliable operation of a MESFET. Therefore, MISFETs, JFETs or HBTs are generally used as the electronic components in OEICs based on the InGaAsP–InP system.

PIN photodiodes are usually employed in the receivers based on both material systems due to the desirable properties of PINs such as high speed, low dark current and low bias voltage [Kolbas et al. 1983 and Kasahara et al. 1984]. Fig. 10.1 shows the performance of InP-based receiver OEICs reported in the last few years [Wada 1990]. For receivers on GaAs substrate, metal–semiconductor–metal (MSM) photodiodes have also received attention due to the compatibility between their fabrication process and that of MESFETs [Ito et al. 1984] and MODFETs.

There have been two methods of integration used: horizontal and vertical. In the vertical integration technique [Fukuzawa et al. 1980], electronic devices are fabricated above optical ones in a two-storied manner on conductive substrates. However, due to the difficulties in large-area fabrication, this method has not received much interest. In the horizontal technique which employs semi-insulating substrates, the difference in height between optical and electronic devices is eliminated by embedding optical devices into the substrate. This can be achieved by etching the substrate to a depth equal to the thickness of the optical device before the epilayers are grown. For more detailed information on transmitter and receiver OEICs on GaAs and InP substrates, the interested reader is referred to the comprehensive review presented by Matsueda et al. [1990].

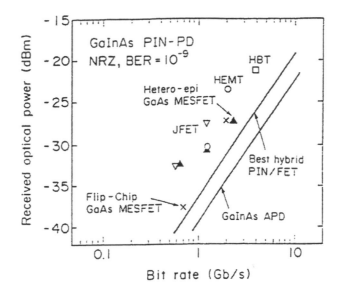

Fig. 10.1. *Performance improvement of* **OEICs** *based on* **InP** *receivers (after Wada [1990]).*

10.3. OEICs on silicon substrates

Recently, heteroepitaxy of GaAs- and InP-based materials on Si substrates and the integration of III–V devices with mature Si technology have become one of the major research areas in order to benefit from the advantages offered by Si [Choi 1987]. These advantages include high mechanical strength and thermal conductivity and the possibility of using large-diameter substrates. However, the growth of III–V semiconductors on Si substrates is a difficult task. The major problems are a large lattice mismatch, a significant difference in the thermal expansion coefficients and a difference in lattice symmetry. The large lattice mismatch creates a high density of dislocations at the interface which can also propagate upward into the active region leading to performance degradation. However, research on GaAs/Si heteroepitaxy is making rapid progress and dislocation densities as low as $10^5 \, \text{cm}^{-2}$ have been reported [Shimomura et al.1992]. Compared to GaAs-based materials, not much research has been performed on InP-based materials on Si, since the lattice mismatch between InP and Si (8%) is higher than that between GaAs and Si (4%). Table 10.1 and Table 10.2 list the first GaAs- and InP-based devices fabricated on Si which have been major steps in the realization of OEICs on Si substrates.

Device	Material	Growth	Reference
LED	GaAs/Ge/Si	MBE	Shinoda et al. [1983]
DH laser	GaAs/AlGaAs/Si	MBE	Windhorn et al. [1984]
SLS laser	InGaAs/AlGaAs	MOCVD	Choi et al. [1991]
PIN photodiode	GaAs/Si	MBE	Paslaski et al. [1988]
MESFET	GaAs/Ge/Si	MBE	Choi et al. [1984]
MODFET	GaAs/AlGaAs/Si	MBE	Fischer et al. [1984]
Bipolar transistor	GaAs/Si	MBE	Fischer et al. [1986]
Solar cell	GaAs/Ge/Si	MOCVD	Gale et al. [1981]
Modulator	AlGaAs/GaAs/Si	MBE	Dobbelaere et al. [1988]
Waveguide	GaAs/AlGaAs/Si	MBE	Kim et al. [1990]

Table 10.1. First GaAs-based electronic and optical devices fabricated on Si substrate.

Device	Material	Reference
LED at RT	InGaAs/InP/Si (1.15 μm)	Razeghi et al. [1988a]
DH laser	GaIn AsP/InP/Si(1.3 μm)	Razeghi et al. [1988c]
BRS CW laser	GaInAsP/InP/Si (1.3 μm)	Razeghi et al. [1988b]
PIN photodiode	GaInAs/InP/Si	Razeghi et al. [1989]
Waveguide	MQW GaInAs/InP/Si (1.5 μm)	Razeghi et al. [1989]
Solar cell	InP/Si	Yamaguchi et al. [1987]
Bipolar transistor	InGaAs/InP/Si	Makimoto et al. [1991]
Heterojunction Phototransistor	InGaAs/InP/Si	Aina et al. [1991]
MQW laser at RT	InGaAs/InGaAsP/Si	Sugo et al. [1991]

Table 10.2. First InP-based electronic and optical devices fabricated on Si substrate.

Compared to the extensive research on OEIC fabrication on GaAs and InP substrates, there has been very limited work performed toward the realization of OEICs on Si substrates. The fabrication of a monolithic photoreceiver on Si using selective growth has been reported [Aboulhouda et al. 1990]. The photoreceiver consists of a GaAs MESFET (0.5 μm wavelength), a long-wavelength (1.3–1.55 μm) MSM photodetector and an inductor in order to achieve the resonance effect. Fig. 10.2 shows the circuit diagram of the receiver. The inductor couples the photodetector and the preamplifier. The load can be either active or passive as indicated on the figure. Fig. 10.3 presents a schematic view of the device. The epitaxial layers have been grown by LP–MOCVD on high-resistivity Si substrate. The epilayer structure consists of a 3 μm thick undoped buffer, a 0.2 μm thick n-type layer doped to 3×10^{17} cm^{-3} which serves as the channel and a 0.2 μm thick n^+ top layer for ohmic contact formation.

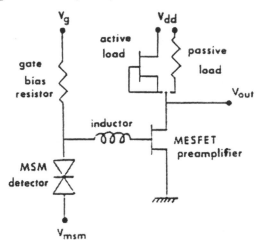

Fig. 10.2. Circuit diagram of the monolithic resonant photoreceiver (after
Aboulhouda et al. [1990]).

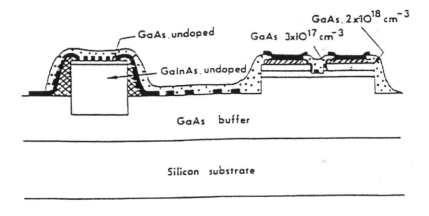

Fig. 10.3. Schematic view of the monolithic resonant receiver (after Aboulhouda et
al. [1990]).

The epilayers for MSM have been deposited with the MOCVD growth
technique by using a dielectric mask. The structure includes a 1 μm thick
$In_{0.53}Ga_{0.47}As$ layer for long-wavelength operation and a 100 Å thick GaAs cap layer
for Schottky contact formation. In order to achieve a planar structure, the

InGaAs/GaAs heteroepitaxy has been embedded into a groove previously etched in the GaAs layer. The first step of the local epitaxy is SiO_2 deposition (1200 Å). After the SiO_2 is etched where the photodetector will be grown, the groove is made using GaAs wet etching. Following the growth of InGaAs/GaAs layers, the SiO_2 layer is removed.

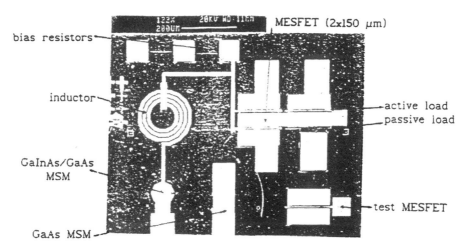

Fig. 10.4. SEM photograph of the integrated resonant receiver (after Aboulhouda et al. [1990]).

A scanning electron microscopy (SEM) photograph of the device is presented in Fig. 10.4. The spiral inductor has been designed to have an inductance of 3 nH. A GaAs MSM (80 μm–80 μm) has been fabricated for comparison. The GaAs MESFET exhibited transconductance and a cutoff frequency close to 110 mS·mm^{-1} and 14 GHz respectively. The dark current of the InGaAs/GaAs MSM has been found to be lower than 20 μA at 2 V and a capacitance of 0.4 pF has been measured at 500 MHz. Fig. 10.5 shows the measured and calculated frequency response of the photoreceiver. Results obtained on a PIN photodiode loaded on a 50 Ω bias resistor are also shown in the figure for comparison. An improvement of 10 dB at 7.25 GHz has been achieved compared to the response of the PIN photodiode. The results clearly show that selective epitaxy can be a useful technique to combine GaAs electronics and long-wavelength optoelectronics on the same wafer.

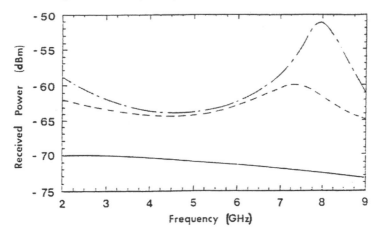

Fig. 10.5. Measured (broken curve) and calculated (chain curve) frequency
response of the photoreceiver compared to that of a PIN loaded on a 50 Ω resistor
(solid curve) (after Aboulhouda et al. [1990]).

10.4. The role of optoelectronic integration in computing

A computing system consists of central processing units (CPUs), data storage systems and input/output (I/O) interfaces. Mature Si technology offers low-cost, high-density integration and the complexity of Si VLSI chips used in each subsystem is steadily increasing due to the need for faster and higher-performance computing. Conventional data transfer between Si VLSI chips and different subsystems is performed through electrical transmission lines, which suffer from limited bandwidth, and packaging problems as the scale of integration is increased. The time constant RC of the wire interconnect increases strongly with decreasing design rule leading to a significant increase in the interconnect delay. This introduces a major obstacle to the improvement of the speed performance of the computing system. Fig. 10.6 shows the improvement in delay time in Si MOS large-scale-integration circuits [Hirose et al. 1990]. With the rapid progress in device fabrication, gate delay has been reduced significantly. However, as the length of the interconnects increases with increasing chip size, an increase in the interconnect delay is seen. It is expected that the interconnection delay will be the major speed degrading mechanism in the future for ultralarge-scale-integration (ULSI) circuits using conventional electrical wiring and sequential data transfer.

Photons have major advantages in signal transfer. Optical interconnects offer high speed, need relatively low power and their large bandwidth and immunity to crosstalk provide a way to decrease the number of interconnects by multiplexing/demultiplexing. Fig. 10.7 compares the delay time of electrical and optical interconnects [Hayashi 1993]. The speed advantage of optical interconnects becomes more significant as the transmission distance is extended. This is especially important in the clock and bus distribution in ULSI microprocessors. A schematic view of a proposed clock distribution and optical bus are shown in Fig. 10.8 [Hayashi 1993]. Besides the speed advantage, high fan-in and fan-out problems are eliminated since the performances of laser diodes and photodiodes are not considerably affected by fan-out and fan-in.

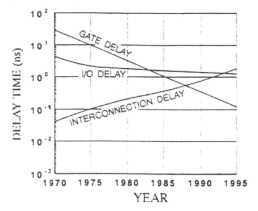

Fig. 10.6. Improvement in the delay time in Si MOS LSI circuits (after Hirose et al. [1990]).

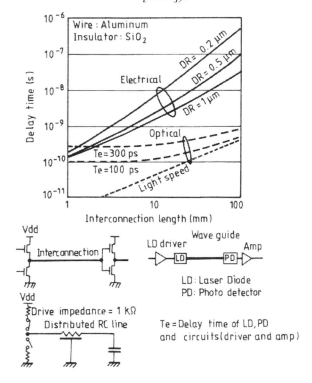

Fig. 10.7. Comparison of the delay time of electrical and optical interconnects (after Hayashi [1993]).

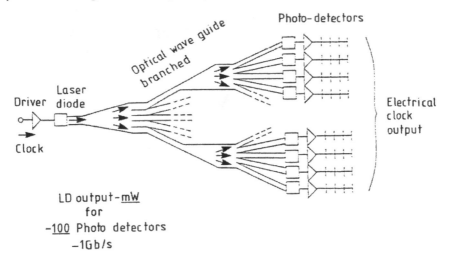

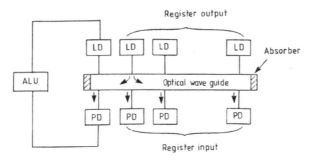

Fig. 10.8. Proposed clock distribution and optical bus line (after Hayashi [1993]).

The advantages offered by the optical interconnects over conventional electrical wiring open an important application field for optoelectronic integration in the rapidly improving computer industry. It is reasonable to assume that optical interconnection in computers will soon reach the board to board, chip to chip and finally intrachip level. A schematic drawing of a proposed optical interconnection between microprocessor boards is presented in Fig. 10.9 [Wada 1990]. Surface emitting laser and photodiode arrays can be used for parallel interconnection between boards. This makes the two-dimensional integration of lasers, photodiodes and switching circuits an important research field.

A three-dimensional common memory offering ultrafast data transfer has been suggested [Hirose et al. 1990] as shown schematically in Fig. 10.10. The two-dimensional silicon memory (static RAM) cell arrays with monolithically integrated light emitting diodes and photoconductors are optically coupled.

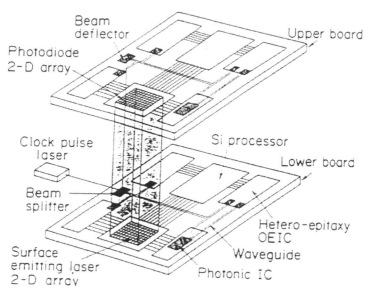

Fig. 10.9. Schematic drawing of a proposed optical interconnection system between processor boards (after Wada [1990]).

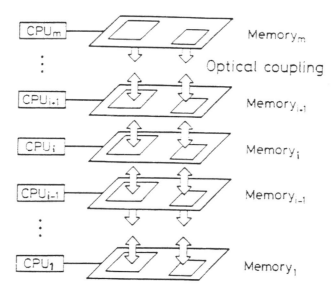

Fig. 10.10. Three-dimensional optically coupled common memory (after Hirose et al. [1990]).

Optical data transfer between two memory layers can be performed between individual cells or cell blocks. Data transfer occurs sequentially through the memory layers, and after the transfer is completed, each CPU can execute its individual process by using the common data on the three-dimensional memory. Fig. 10.11

presents a diagram of the memory cell circuit. The circuit has a data storage area which is similar to the conventional static RAM cell and a data transfer part. The high resistive load (photoconductor) detects the light signal sent by the following or preceding level memory cell. The flip-flop node with the photoconductor receiving light is set to high. Optical data transfer is started by selecting the LED control line which makes the LED connected to the "high" node of the flip-flop emit light that is received by the photoconductors in the upper and lower memory cells. The LEDs and photoconductors in memory cells of any layer must face the photoconductors and LEDs of the upper and lower level memory cells as shown in Fig. 10.12. It is clear that the Si substrate must be thin enough (a few microns) to minimize the optical attenuation of the light signal through the substrate. It has been suggested that each of the memory layers be fabricated by selective epitaxial growth of GaAs on a patterned Si window at low temperatures following the conventional silicon LSI process.

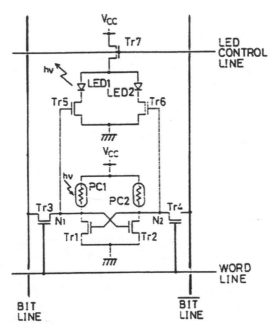

Fig. 10.11. Circuit diagram of the memory cell circuit (after Hirose et al. [1990]).

In conclusion, with the developments in heteroepitaxy and monolithic integration techniques, a new era in the semiconductor industry which allows close collaboration of electrons and photons is expected to take place in the near future. The developments in the heteroepitaxy of GaAs- and InP-based materials on Si substrates will facilitate the integration of III–V devices with Si technology using the best aspects of both technologies. Heteroepitaxy on Si substrate will also have a major impact on the computer industry by permitting the fabrication of optical interconnects inside Si ULSI chips. However, there are still many problems to be solved. First of all, to facilitate high-density integration, micron-sized laser diodes with low threshold currents (as low as microamperes) are necessary. Considering the fact that thousands of laser diodes and photodetectors will be needed for optical coupling, very high yield and highly reliable fabrication techniques are necessary which also calls for defectless III–V heteroepitaxy on Si.

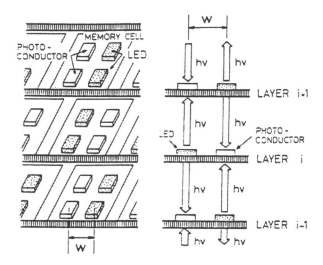

Fig. 10.12. Layout of LEDs and photodetectors in a three-dimensional optically coupled common memory (after Hirose et al. [1990]).

10.5. Examples of optoelectronic integration by MOCVD

MOCVD growth technology of GaAs–GaInP, GaAs–GaInAsP, and GaAs–GaInAlP for photonic and electronic devices, as well as on patterned substrates, makes OEIC possible. MOCVD is perfectly suitable for OEIC fabrication since using the local and (in some cases) mismatched epitaxy, it is possible to deposit what we want exactly where we want it, independent of the integration functions and of the type of substrate. This allows separate optimization of each component independently of the others on an OEIC. Since epilayers suitable for lasers, for example, are not optimized for photodetectors, waveguides, modulators or transistors, such a method would solve the difficult problem of technological compatibility among all the components of an OEIC, which is the key for the development of such devices. This will open the door to mass production, low-cost optoelectronic devices. In this chapter, silicon has been suggested as the common substrate for any monolithic integration for two reasons. First, large-scale electronic integration is now well developed on this material. Second, this material displays a large thermal conductivity which makes it attractive for any application requiring high power dissipation.

In this section we describe four different examples of monolithic integration of optoelectronic devices with different functions.

10.5.1. Monolithic integration of photodetectors with transistors [Hosserni et al. 1988]

This can significantly improve the signal to noise ratio by reducing parasitic elements compared to hybrid systems. A GaAs-based photoreceiver has been demonstrated by the monolithic association of a GaAs Schottky photodiode (PD) and a GaAs metal–semiconductor field effect transistor (MESFET). In the first step, the Schottky contact of the photodiode was deposited on an *n*-type GaAs layer for the MESFET gate leading to a large PD capacitance, and to a low sensitivity. In a second step, in order to optimize separately the transistor and the PD, a special undoped layer for the fabrication of the Schottky PD was grown by molecular beam epitaxy (MBE) on the epilayers suitable for the fabrication of the MESFET. The capacitance of the PD could be reduced and the performance of the photoreceiver increased. In the third step, the MESFET epilayers were replaced by epilayers suitable for high electron mobility transistor (HEMT) fabrication. With the same structure as for the previous device, an improvement of 3 dB·m on the sensitivity has been recorded. It should be noted that precise etching processes are needed for such a technology.

A promising technique for the fabrication of OEICs is the use of mismatch epitaxy to fabricate FETs on GaAs-based materials and PDs on InP-based materials. In the first demonstration, a GaInAs photoconductive detector was monolithically associated with a GaAs MESFET. Good performance (sensitivity of 28 dB·m at 250 Mbit·s^{-1}) has been recorded with this integrated circuit.

10.5.2. Monolithic integration of photodetectors with optical waveguides [Mallecot et al. 1988]

This is suitable for coherent applications since it could reduce the difficult problem of achieving alignment between several optical and optoelectronic components.

Several structures have been proposed, mainly on InP substrate, based on evanescent field coupling or the butt-coupling technique. The evanescent field coupling technique seems more convenient from a technological point of view, since epilayers for photodetection can be deposited on the epilayers of the optical waveguide in a one-step growth. Because of the difference in refractive indices between the absorbing layer and the waveguide layer, part of the light propagating inside the waveguide is detected by the photodetector (PD). However, in general, the coupling length needed to detect the totality of the light is large and thus increases the capacitance of the PD. A thin and short metal–semiconductor–metal (MSM) PD was integrated with a GaAlAs optical rib waveguide on GaAs semi-insulating substrate by MBE. Two undoped GaAlAs epilayers were deposited to optically isolate the GaAlAs guiding layer from the GaAs semi-insulating substrate. An inverted rib was etched before depositing the guiding layer. A very thin (0.2 μm) undoped GaAs epilayer was grown to form the MSM photodetector. This quasi-planar structure is completely compatible with the fabrication of a MESFET. Experimental results performed at 0.85 μm wavelength show that only a 100 μm long GaAs absorbing layer is needed to detect 90% of the light intensity, and that the cutoff frequency is in excess of 15 GHz.

This technique was also applied to long-wavelength applications. In the first device, a GaInAs photoconductor was monolithically integrated with a n^-/n^+ GaAs rib waveguide. In the improved device, a short-length GaInAs photoconductor was monolithically integrated with a GaAs/GaAlAs optical waveguide on a GaAs semi-insulating substrate. As for the previous device, a GaInAs length of only 100 μm is needed to detect 90% of the light.

10.5.3. *Monolithic integration of optical modulator with lasers [Oe 1991]*

This technique offers a solution to overcome the high-speed modulation limit of direction modulation imposed by such phenomena as electron–photon resonance, and increases the modulation frequency up to hundreds of GHz. Among various modulation schemes, an intensity modulator based on electro-absorption effects using the quantum-confined Stark effect in multiple-quantum-well structures provides very efficient modulation because of its higher extinction ratio and smaller driving voltage and power consumption. A new integration scheme for efficient modulation involves depositing two separate electrodes on the same MQW epilayers so that one of them can act as the electrode for the laser while the other acts as the modulator electrode. Then the laser part is biased by the forward voltage and the modulator part is biased by the reverse voltage to provide the electric field needed to change the absorption coefficient. The advantages of this scheme for the modulator are the short modulator length arising thanks to the strong absorption effect, and the very low modulation power consumption since it is under reverse bias. When this device is fabricated on a silicon substrate, heating due to the CW current of the laser part and the light absorption in the modulation part can be easily dissipated.

10.5.4. *Surface emitting laser diodes*

These are key components for the development of optoelectronic systems, offering two-dimensional arrays that are suitable for optical signal processing. So far most of the results have been obtained in GaAs/AlGaAs or InGaAs/GaAs heterostructures that are known to suffer from reliability problems related to aluminum-induced oxidation. In this respect, GaInP/GaAs heterostructures which provide much lower surface recombination are an attractive alternative to surface-emitting lasers. Power dissipation problems resulting from the very small active volume can be solved also by employing Si as the substrate.

References

Aboulhouda, S., Razeghi, M., Vilcot, J.P., Decoster, D., Francois, M., and Maricot, S. *Proc. SPIE* **1362** 494, 1990.
Aina, O., Serio, M., Mattingly, M., O'Connor, J., Shastry, S.K., Hill, D.S., Salerno, J.P., and Ferm, P. *Appl. Phys. Lett.* **59** 268, 1991.

Choi, H.K. *Optoelectronics* **2** 265, 1987.

Choi, H.K., Tsaur, B.Y., Metze, G.M., Turner, G.W., and Fan, C.C. *IEEE Electron Devices Lett.* **EDL-5** 207, 1984.

Choi, H.K., Wang, C.A., and Karam, N.H. *Appl. Phys. Lett.* **59** 2634, 1991.

Dobbelaere, W., Huang, D., Unlu, M.S., and Morkoc, H. *Appl. Phys. Lett.* **53** 94, 1988.

Fischer, R., Henderson, T., Klem, J., Masselink, W.T., Kopp, W., Morkoc, H., and Litton, C.W. *Electron. Lett.* **20** 945, 1984.

Fischer, R., Peng, C.K., Klem, J., Henderson, T., and Morkoc, H. *Solid-State Electron.* **29** 269, 1986.

Fukuzawa, T., Nakamura, M., Hirao, M., Kuroda, T., and Umeda, J. *Appl. Phys. Lett.* **36** 181, 1980.

Gale, R.P., Fan, C.C., Tsaur, B.Y., Turner, G.W., and Davis, G.M. *IEEE Electron Devices Lett.* **EDL-2** 169, 1981.

Hayashi, I. Critical Reviews in Optical Science and Technology *Integrated Optics and Optoelectronics* ed K K Wong and M Razeghi, **CR45**, 3, 1993.

Hirano, M. *Optoelectronics* **2** 137, 1987.

Hirose, M., Takata, H., and Koyanagi, M. *Proc. SPIE* **1362** 316, 1990.

Hosserni, T.A., Decoster, D., Vilcot, J.P., and Razeghi, M. *J. Appl. Phys.* **64** 2215, 1988.

Ito, M., Wada, O., Nakai, K., and Sakurai, T. *IEEE Electron Devices Lett.* **EDL-5** 531, 1984.

Kasahara, K., Hayashi, J., Makita, K., Taguchi, K., Suzuki, A., Nomura, H., and Matsushita, S. *Electron. Lett.* **20** 314, 1984.

Kim, Y.S., Lee, S.S., Ramaswamy, R.V., Sakai, S., Kao, Y.C., and Shichijo, H. *Appl. Phys. Lett.* **56** 802, 1990.

Kolbas, R.M., Abrokwah, J., Carney, J.K., Bradshaw, D.H., Elmer, B.R., and Biard, J.R. *Appl. Phys. Lett.* **43** 821, 1983.

Lee, C.P., Margalit, S., Ury, I., and Yariv, A. *Appl. Phys. Lett.* **32** 806, 1978.

Makimoto, T., Kurishima, K., Kobayashi, T., and Ishibashi, T. *Japan. J. Appl. Phys.* **30** 3815, 1991.

Mallecot, F., Vinchant, I.E., Razeghi, M., and Vandermoere, D. *Appl. Phys. Lett.* **53** 2522, 1988.

Matsueda, H., Tanaka, T.P., and Nakamura, M. *Semiconductors and Semimetals* vol 30, ed Ikoma, T. (Boston: Academic) p 231, 1990.

Oe, K. *Optoelectronic Materials and Device Concepts* ed M. Razeghi (Bellingham: SPIE Optical Engineering) p 222, 1991.

Paslaski, J., Chen, H.Z., Morkoc, H., and Yariv, A. *Appl. Phys. Lett.* **52** 1410, 1988.

Razeghi, M. *Prog. Crystal Growth Charact.* **19** 21, 1989.

Razeghi, M., Blondeau, R., Defour, M., Omnes, F., Maurel, P.H., and Brillouet, F. *Appl. Phys. Lett.* **53** 854, 1988a.

Razeghi, M., Defour, M., Blondeau, R., Omnes, F., Maurel, P.H., Ocher, O., Brillouet, F., Fan, C.C., and Salerno, J. *Appl. Phys. Lett.* **53** 2389, 1988b.

Razeghi, M., Defour, M., Omnes, F., Maurel, P.H., Chazelas, J., and Brillouet, F. *Appl. Phys. Lett.* **53** 725, 1988c.

Shimomura, H., Okada, Y., and Kawabe, M. *Japan. J. Appl. Phys.* **31** L628, 1992.

Shinoda, Y., Nishioka, T., and Ohmachi, Y. *Japan. J. Appl. Phys.* **22** L450, 1983.

Sugo, M., Mori, H., Itoh, Y., Sakai, Y., and Tachikawa, M. *Japan. J. Appl. Phys.* **30** 3876, 1991.

Wada, O. *Proc. SPIE* **1362** 598, 1990.

Windhorn, T.H., Metze, G.M., Tsaur, B.Y., and Fan, C.C. *Appl. Phys. Lett.* **45** 309, 1984.

Yamaguchi, M., Yamamoto, A., Itoh, Y., and Nishioka, T. *Proc. 19th IEEE Photovoltaic Specialists Conf.* p 267, 1987.

11.   InP–InP System: MOCVD Growth, Characterization, and Applications

11.1.   Introduction
11.2.   Energy band structure of InP
11.3.   Growth and characterization of InP using TEIn
   11.3.1.   Preparation of substrates
   11.3.2.   Orientation effects
   11.3.3.   Source-purity effects
   11.3.4.   Material characterization
   11.3.5.   Interfaces
11.4.   Growth and characterization of InP using TMIn
11.5.   Incorporation of dopants
   11.5.1.   p-type
   11.5.2.   n-type
11.6.   Applications of InP epitaxial layers
   11.6.1.   Gunn diodes

11.1. Introduction

The early work on InP was stimulated by the need for long-wavelength lasers as light sources adapted to lower loss and lower dispersion at $\lambda = 1.3–1.55$ μm in silica fibers. Future telecommunications systems will require narrow-spectrum 1.55 μm lasers and very sensitive detectors, and this is driving significant research on the technology of InP and related compounds. For microwave devices, with the exception of mobility, which is higher in GaAs, the other characteristics favor InP over GaAs in terms of superior Gunn device performance. The development of these devices has been the direct result of basic research on InP and related compounds. Knowledge of the band structures, electrical properties, optical properties and carrier recombination mechanisms has led a number of physicists to conclude that multilayers of InP and related compounds are the best candidates for a number of future optoelectronic and microwave devices.

11.2. Energy band structure of InP

InP (indium phosphide) has the zinc blende crystal structure. Its lattice may be considered as two interpenetrating face-centered-cubic sublattices, one made up of indium atoms and the other of phosphorus atoms.

Fig. 11.1 shows the Brillouin zone (see e.g. Kittel [1971] for the theory of band structure in crystals) of the InP crystal (zinc blende lattice), and indicates the most important symmetry points and symmetry lines, such as the center of the zone ($\Gamma = (2\pi/a)(0,0,0)$), the $\langle 1\,1\,1 \rangle$ axes (Λ) and their intersections with the zone edge $\left(L = (2\pi/a)\left(\frac{1}{2},\frac{1}{2},\frac{1}{2}\right) \right)$, the $\langle 1\,0\,0 \rangle$ axes (Δ) and their intersections ($X = (2\pi/a)(0,0,1)$), and the $\langle 1\,1\,0 \rangle$ axes (Σ) and their intersections $\left(K = (2\pi/a)\left(\frac{3}{4},\frac{3}{4},0\right) \right)$.

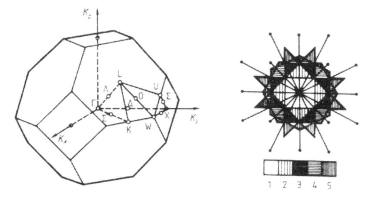

Fig. 11.1. First Brillouin zone for InP-based materials (zinc blende lattice), including important symmetry points and lines.

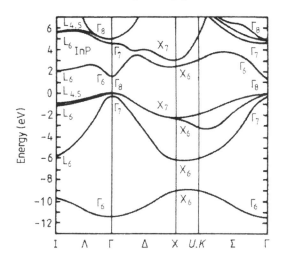

Fig. 11.2. The energy band structure of InP.

The E–k diagram of InP is shown in Fig. 11.2. Around an extremum, the energy E varies with k–k_0 parabolically, where k_0 is the value of k at the extremum. The expression relating E has the simplest form for the Γ point minimum. The relation is

Eq. (11.1) $E - E_\Gamma = \hbar^2 k^2 / 2m_e^*$

The constant m_e^* is the effective mass of an electron [Kane 1966]. The E–k relation for the minima lying on the Λ or Δ directions is of the form

Eq. (11.2) $E - E_L = \dfrac{\hbar^2}{2}\left(\dfrac{\Delta k_\parallel^2}{m_\parallel^*} + \dfrac{\Delta k_\perp^2}{m_\perp^*} \right)$

where $\Delta k = k - k_L$, $m_\parallel$ is the effective mass for the $\langle 1\,1\,1 \rangle$ direction and $m_\perp$ that for a perpendicular direction.

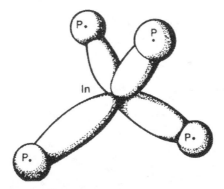

Fig. 11.3. sp³ hybrid bond orbitals.

InP and related compounds have a covalent bond structure. When a covalent bond is produced between two unlike atoms, such as In and P, it is called a heteropolar bond. In InP semiconductor crystals, the bond orbitals are constructed from sp^3 hybrids (see Fig. 11.3) (a linear combination of p orbitals and s orbitals is called an sp hybrid; the 3 indicates that there is three times the probability of finding an electron in a ρ state than in an s state). Indium has the electronic configuration $4d^{10}5s^25p^1$ and P has the electronic configuration $3s^23p^3$. After hybridization there are two atoms, and hence eight electrons, per primitive unit cell in the InP crystal. There are one 5s level and three 5p levels in the In atom and one 3s level and three 3p levels in the P atom, each giving rise to a bond in the crystal. This gives a total of four bonds arising from the atomic energy levels containing the valence electrons of InP. These four bonds are called the valence bonds of the band structure of InP. We now consider the allocation of the eight valence electrons of InP among the four valence bonds, since there are two primitive unit cells of the InP lattice. There are a total of eight electrons to put into four valence bonds. Considering, for example, the

(1 0 0) and (1 1 1) directions near the center of the Brillouin zone, the point in *K* space (reciprocal lattice) at which a valence electron has the highest energy is the Γ point at the center of the Brillouin zone. We therefore say that the valence-band maximum occurs at the center of the Brillouin zone. At $T = 0$ K, all the valence-band states are occupied by electrons, and all the higher bands of the InP band structure are empty. At temperatures above absolute zero, a few electrons will be thermally excited from the highest valence band into the next vacant higher band.

The next empty band above the highest valence band is called the conduction band, because electrons thermally excited into it will find empty states available for the electrical conduction process. The conduction-band minimum is at the Γ point at the zone center (the conduction-band minimum is often called the conduction-band edge, and the valence-band maximum is called the valence-band edge). The minimum energy gap E_g is the energy difference between the conduction-band minimum (E_c) and the valence-band maximum (E_v), so the magnitude of E_g of the energy gap in InP is given by

Eq. (11.3) $E_g = E_c - E_v$

The valence bands are characteristically threefold near the edge, with the heavy hole (hh) and light hole (lh) bands degenerate at the center, and a band spin–orbit hole (soh) split off by the spin–orbit splitting Δ;

Eq. (11.4) $E_v(\text{hh}) = -\hbar^2 k^2 / 2m^*_{\text{hh}}$

Eq. (11.5) $E_v(\text{lh}) = -\hbar^2 k^2 / 2m^*_{\text{lh}}$

Eq. (11.6) $E_v(\text{soh}) = -\Delta - \hbar^2 k^2 / 2m^*_{\text{soh}}$

The values of the mass parameters of InP are given in Table 11.1.

Electron effective mass	m^*_e / m_0	0.073
Heavy hole	m^*_{hh} / m_0	0.40
Light hole	m^*_{lh} / m_0	0.078
Split-off hole	m^*_{soh} / m_0	0.15
Spin-orbit splitting	Δ (eV)	0.11

Table 11.1. Effective masses of electrons and holes in InP crystal.

11.3. Growth and characterization of InP using TEIn

High-quality InP layers have been grown by MOCVD. TEIn and PH_3 are used for In and P sources, respectively. A mixture of H_2 and N_2 is used as carrier gas. The presence of H_2 is necessary to avoid the deposition of carbon, and the presence of N_2 is necessary to avoid the parasitic reaction between TEIn and PH_3.

The InP layers can be grown at 76 Torr and low temperature, between 500 and 650 °C, by using TEIn and phosphine (PH_3) in a H_2+N_2 carrier gas. The growth rate depends linearly upon the TEIn flow-rate and is independent of the flow-rate of PH_3 within the range 200–800 $cm^3 \cdot min^{-1}$ (Fig. 11.4), of the substrate temperature and of the substrate orientation, which suggests that the epitaxial growth is controlled by the mass transport of the group III species.

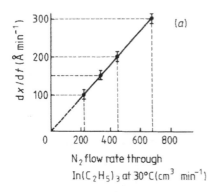 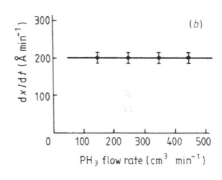

Fig. 11.4. Variation of the growth rate of InP with (a) TEIn flow (T_G = 550 °C, PH_3 flow 300 $cm^3 \cdot min^{-1}$, total flow 6 $1 \cdot min^{-1}$) and (b) PH_3 flow (T_G = 550 °C, N_2 flow-rate through TEIn 450 $cm^3 \cdot min^{-1}$).

Razeghi and Duchemin [1983] have studied the growth of InP layers by using 100% H_2 and H_2+N_2 mixtures as the carrier gas. The best morphology and the highest PL intensity were obtained by using 50% H_2 and 50% N_2.

Growth temperature (°C)	N_2–TEIn bubbler flow ($cm^3 \cdot min^{-1}$)	PH_3 flow ($cm^3 \cdot min^{-1}$)	Total flow ($1 \cdot min^{-1}$)	Growth rate ($Å \cdot min^{-1}$)
550	450	260	6	200 ±10
	225	200	6	100 ±10
650	450	520	6	220 ±10
	225	400	6	110 ±10

Table 11.2. Optimum growth conditions.

Using Ar instead of N_2 gives InP layers with the same surface quality. Table 11.2 lists the optimum growth conditions for MOCVD growth of InP at 550 and 650 °C that were used for this study. The InP layers grown by MOCVD are less compensating at lower growth temperature. Fig. 11.5 shows that the carrier concentration of InP is $N_D - N_A = 6 \times 10^{14}$ cm^{-3} when grown at 520 °C and 6×10^{15} cm^{-3} at 650 °C if all other growth conditions remain the same.

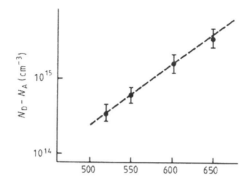

Fig. 11.5. Variation of residual carrier concentration of InP layers grown by MOCVD as a function of growth temperature.

11.3.1. Preparation of substrates

Semi-insulating or n^+ substrates are supplied by MCP Crystacomm Metals Research or Sumitomo in the form of ingots or polished wafers. Wafers are sliced 2° off $(1\bar{1}0)$ and chemically etched for 10 s at room temperature in a 15% bromine–methanol solution in order to remove 20 μm from each side. The wafers are then mechanochemically polished in a solution of 1.5% bromine in methanol, removing 80–100 μm more. The substrates are finally cleaned in methanol and rinsed in isopropyl alcohol. The substrates are etched again, just before use, by dipping in 1% bromine in methanol at room temperature for 1 min, rinsing in warm isopropyl alcohol, and drying. An n^+ Sn-doped substrate and a semi-insulating Fe-doped substrate are generally used for each experiment.

Pretreatment of the substrates prior to epitaxial growth is critical. The optimum pretreatment procedure is as follows:

(1)   Dipping in H_2SO_4 for 3 min with ultrasonic agitation.
(2)   Rinsing in deionized H_2O.
(3)   Rinsing in hot methanol.
(4)   Dipping in 3% Br in methanol at room temperature for 3 min (ultrasonic bath).
(5)   Rinsing in hot methanol.
(6)   Dipping in H_2SO_4 for 3 min.
(7)   Rinsing in deionized H_2O.
(8)   Rinsing in hot methanol.

After this pretreatment, it is possible to keep the substrate for one or two weeks without repeating this treatment prior to growth.

11.3.2. Orientation effects

Growth of InP layers has been carried out on (1 0 0) substrates misoriented up to 4° towards (1 1 0). At the low growth temperature (550 °C), the surfaces of the grown layers are generally mirror-smooth to the naked eye whether the substrates were accurately oriented or not, but at high growth temperature the best-quality InP layers were obtained by using (1 0 0) substrates 2° off towards $(1\,\overline{1}\,0)$. The density of surface defects is higher on (1 0 0)-oriented substrates compared to 2° off material. Saxena et al. [1981] reported a similar investigation on GaAs substrates.

Most epitaxial growth of III–V semiconductors has been performed on either (1 0 0) or (1 1 1) crystal faces. Shaw [1968] has studied the chloride VPE growth of GaAs on (1 0 0), (1 1 1), (1 1 2), (1 1 3) and (1 1 5) GaAs substrates and concluded that the best morphologies in the 700–800 °C temperature range are obtained on (1 1 3)A substrates. Olsen et al. [1982] reported the chloride VPE growth of InP on InP substrates at 700 °C on (1 0 0), (1 1 1), (3 1 1) and (5 1 1) both exact and 2° off towards (1 1 0) orientation. They found the best morphology and highest PL intensity for the (3 1 1)B 2° off substrate orientation.

Razeghi et al. [1984] have performed a study of MOCVD simultaneous growth on nine InP substrates with orientations of (1 0 0), (1 1 1) and (1 1 5) placed adjacent to each other within the reactor for a growth temperature of 650 °C. Table 11.3 indicates relative PL intensity I, PL halfwidth Δhv, growth rate dx/dt, net carrier concentration evaluated by C–V measurements and relative surface appearance of MOCVD InP grown on InP substrates of various orientations at 650 °C. The relative PL intensity is a measure of the radiative recombination efficiency within the material, whereas the PL half-width is a measure of crystalline quality and impurity incorporation (see section 11.4).

Orientation	$N_D - N_A$ (cm^{-3})	Surface quality[b]	I	Δhv at 300 K (MeV)
InP(Sn)(1 0 0) 2° $1\,\overline{1}\,0$	3×10^{17}	E	16	60
InP(Fe)(1 0 0) exact	3×10^{17}	G	9	60
InP(Fe)(1 0 0) 2° $1\,\overline{1}\,0$	3×10^{17}	VG	16	60
InP(Fe)(1 0 0) 3° $1\,\overline{1}\,0$	3×10^{17}	VG	16	60
InP(Sn)(1 0 0) 4° $1\,\overline{1}\,0$	3×10^{17}	VG	16	60
InP(S)(1 1 1B) 2°	2×10^{17}	VG	45	60
InP(S)(1 1 5B) 2°	2×10^{17}	E	180	95
InP(Fe)(1 1 1B) exact	3×10^{17}	G	10	60
InP(Sn)(1 0 0) exact	3×10^{17}	VG	8	60

Table 11.3. Relative PL intensity I, relative surface appearance and carrier concentrations of the InP grown on InP substrates by MOCVD. (Growth rate 220 ± 10 Å·min$^-$; E = excellent, VG = very good, G = good.)

Table 11.4 presents S, Si, Cr, Fe, Mg and Mn distributions measured by secondary-ion mass spectroscopy (SIMS) in these layers. All of the layers were doped by H$_2$S, and C–V measurements give $N_D - N_A = 2 \times 10^{17}$ cm^{-3} for all of them.

The SIMS analyses were performed using a Cameca IMS3F instrument [Huber et al. 1982]. The surface of the sample was scanned with a focused oxygen primary-ion beam. The scanned area was 250×250 μm for the working conditions, and the analyzed region was 60 μm in diameter.

The precision of measurements is 50% for each impurity element. Analysis of a large number of epilayers is necessary for any given set of growth conditions in order to specify accurately the contribution of each impurity source and its relation to the substrate in MOCVD layer growth.

Reference Layer	S	Si	Cr	Fe	Mg	Mn
InP(S)						
(1 1 1)	3×10^{17}	5×10^{16}	4×10^{14}	2×10^{14}	2×10^{14}	4×10^{14}
(1 1 5)	6×10^{17}	5×10^{16}	4×10^{14}	2×10^{15}	1×10^{14}	5×10^{14}
(1 0 0) 2°	3×10^{17}	5×10^{16}	4×10^{15}	1×10^{16}	2×10^{14}	5×10^{14}
InP(Fe)						
(1 0 0) 2°	6×10^{16}	3.5×10^{16}	3×10^{15}	2×10^{14}	4×10^{14}	7×10^{14}
(1 0 0) 3°	8×10^{16}	3×10^{16}	5×10^{14}	2×10^{15}	1×10^{15}	1×10^{15}
(1 0 0) 4°	1.5×10^{16}	5×10^{16}	5×10^{14}	2×10^{14}	2×10^{14}	5×10^{15}
InP(Sn)						
(1 0 0) 2°	2×10^{17}	7×10^{16}	5×00^{15}	2×10^{14}	2×10^{16}	5×10^{14}

Table 11.4. S, Si, Cr, Fe, Mg and Mn distributions measured by SIMS in the InP layers grown by MOCVD at 650°C. (All layers doped by H_2S ($N_D - N_A = 2 \times 10^{17}$ cm^{-3}).

11.3.3. Source-purity effects

The importance of source purity is illustrated in Table 11.5, where we have listed the range of 300 K and 77 K mobilities and carrier concentrations with a variety of source materials under optimum growth conditions, as denned in Table 11.2. The purity of undoped InP grown by MOCVD depends very strongly on the particular source of TEIn and PH_3 used in the growth. Dapkus et al. [1981] reported a similar investigation for GaAs layers grown by MOCVD using TMGa and AsH_3 and showed a similar dependence on source purity.

Sources		$N_D - N_A$ (cm^{-3})	μ(300 K) (cm$^2\cdot$V$^{-1}\cdot$s^{-1})	μ (77 K) (cm$^2\cdot$V$^{-1}\cdot$s^{-1})
TEIn	PH$_3$			
α-Ventron (I)	Matheson (I)	6×10^{14}	5300	85,000
α-Ventron (I)	Air Liquide (I)	3×10^{16}	3400	42,000
Sidercom	Matheson (I)	1.3×10^{16}	3000	25,000
α-Ventron (II)	Matheson (I)	7×10^{16}	3990	58,000
α-Ventron (II)	Air Liquide (I)	7×10^{16}	3550	35,000
Texas-Alkyle	Phoenix	3×10^{15}	4500	65,000
α-Ventron (III)	Matheson (II)	5×10^{15}	5500	95,000
SMI	Matheson (II)	10^{14}	5500	150,000

Table 11.5. Effect of source purity on carrier concentration and 300 K and 77 K mobility on InP layers. Different batches are indicated by producer and batch number.

11.3.4. Material characterization

(a) Photoluminescence

Photoluminescence is the optical radiation emitted by a physical system resulting from excitation to a non-equilibrium state by irradiation with light. Three processes can be distinguished: creation of electron–hole pairs by absorption of the exciting light, radiative recombination of electron–hole pairs, and escape of the recombination radiation from the sample. Photoluminescence (PL) is very useful for the assessment of semiconductor materials. Impurities and native defects that are present in concentrations as low as about 1×10^{15} cm^{-3} can be detected without destruction of the sample, and any surface irregularities are unimportant. By calibrating with known impurity concentrations, it is possible to calculate any unknown carrier concentration from the halfwidth of the spectral line associated with a particular impurity. From the lineshapes and halfwidths as a function of temperature, it is possible to distinguish between simple and complex centers, and between simple donors and acceptors if the effective masses of the electrons and holes are different. By simple donors and acceptors are meant those centers which give rise to luminescence lines whose lineshapes can be fitted by the effective-mass theory for hydrogenic levels. A simple center is defined as an impurity that sits on the In or P lattice site and which contributes only one additional carrier to the system. The activation energy of a single carrier bound to the substitutional impurity is small and close to that calculated from the hydrogenic model.

Fig. 11.6 shows the PL spectrum measured at 6.6 K of an undoped InP layer grown by MOCVD using TEIn α-Ventron and PH$_3$ from Matheson. The PL was excited by using the 5145 Å line of an Ar$^+$ laser and was analyzed with a Ge photodiode.

Similar to GaAs, several elementary recombination mechanisms may occur and cause near-band-gap emission lines: free excitons (X: 1.4181 eV), excitons and shallow impurities (D^0X: 1.4169 eV; A^0X: 1.4147 eV). Silicon is the dominant donor and Zn the acceptor in the undoped InP samples grown by MOCVD, using TEIn.

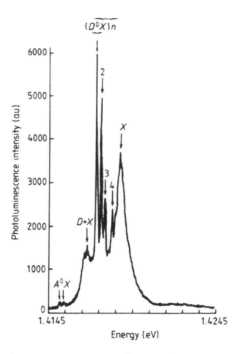

Fig. 11.6. Photoluminescence spectrum of InP undoped layer grown by MOCVD.

(b) Study of impurity redistribution in the InP layer by SIMS

In SIMS analysis the surface of the sample under study is bombarded by a beam of energetic ions (O_2^+, Ar^+ or Cs^+ generally). A fraction of the sputtered species are ions. The secondary ions provide information about the sample composition. A mass spectrometer selects the mass of the ions of interest. The secondary-ion intensity measurements are carried out by an electron multiplier.

The analyses at Thomson are performed with a Cameca IMS3F [Lepareux 1980 and Huber 1986], which has the essential characteristics required for semiconductor analysis, such as high trace sensitivity (parts-per-billion range), high mass resolution ($M/\Delta M = 10,000$), high depth resolution (nanometer range), wide dynamic range in depth profiling (5 in current use, 7 in special conditions), mass-separated primary-ion beam, simultaneous depth profiling of several elements and cesium gun equipment. The use of Cs^+ ion bombardment in conjunction with negative secondary-ion detection in an ultra-high-vacuum environment has dramatically increased the usefulness of SIMS, particularly for III–V compound technology [Magee 1979]. Therefore SIMS, with its excellent sensitivity and depth resolution (30–100 Å), provides doping profiles and positions.

The replacement in the primary- and secondary-ion optics of the stainless-steel apertures by tantalum ones was one of the modifications. As a result of this the background of Cr diminished by a factor of 100 and a lower limit of detection of about 5×10^{12} atoms·cm^{-3} was obtained in InP for these elements. This sensitivity is necessary for a better evaluation of InP, because Fe and Cr are deep-acceptor-level elements in InP. In the analysis of InP, the surface is scanned with a focused mass-filtered oxygen-ion beam ($I_p \sim 1.5$ µA at 10 keV). The scanned area is 250×250 µm and the analyzed region is 150 µm in diameter [Huber et al. 1984]. An offset target voltage of -50 V was applied to avoid possible interference with ionized hydrocarbons. Ion-implanted samples were the standard for quantitative calibration of the instrument. The statistical results of various experiments show that the quantitative results of SIMIS are given with an accuracy of ±20% at a concentration level of 1×10^{16} atoms·cm^{-3}. Below this level, results are less accurate: ±50% at 1×10^{14} atoms·cm^{-3}. Depth measured on a bevel has a precision estimated at 10%. The detection limits of the impurities that were measured are Mg, Cr, Mn: 5×10^{12} atoms·cm^{-3}; Fe: 1×10^{13} atoms·cm^{-3}; and Si: 7×10^{13} atoms·cm^{-3}.

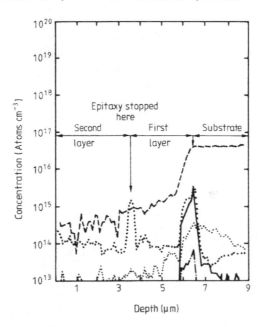

Fig. 11.7. Depth profiles of Mg (—), Si (....), Cr (....), Fe (— — —) and Mn (— —) in an MOCVD InP layer grown on an InP(Fe) substrate. Epitaxy was stopped, the wafer was taken out of the reactor and a second layer was grown on the first.

Fig. 11.7 shows typical SIMS depth profiles of Mg, Si, Cr, Fe and Mn in an InP layer grown on semi-insulating Fe-doped InP substrate by MOCVD under standard growth conditions. Each sample was analyzed in two different areas about 10 mm apart. Generally, analysis of two clean areas gives reproducible and representative results for the material.

Accumulation of impurities such as Mg, Fe, Cr, Mn and Si was detected by SIMIS at the interface between substrate and epilayer. A similar phenomenon was observed in GaAs epilayers grown by MOCVD [Huber et al. 1982]. The accumulation of these impurities may have different origins such as:

(1)   Incorporation of impurity from the ambient gas phase during preheating.
(2)   Incorporation of impurity from exposure to air.
(3)   Contamination of the substrate by chemical etching before growth.
(4)   Incorporation of impurity during *in situ* HCl etching.
(5)   Out-diffusion of impurities from substrate to epilayer.
(6)   The effect of substrate orientation on impurity redistribution.

To determine the origin of these impurities, Razeghi and Duchemin [1983] performed the following experiment. An InP epilayer 3 μm thick was grown on an InP substrate under the conventional procedure. The wafer was removed from the reactor. After 30 min exposure to air, the wafer was transferred again into the reactor and a second epilayer 3.5 μm thick was grown under the same growth conditions.

Fig. 11.7 shows the "pile-up" of impurities at the epilayer–substrate interface where peaks in the concentration of Mg, Cr, Mn, Fe and Si were analyzed. At the epilayer–epilayer interface, the only concentration change was some increase in the Si level, which may be due to the heating of the susceptor. These experiments indicate that exposure to air and heating under $H_2+N_2+PH_3$ pressure under conventional growth conditions do not account for the accumulation of impurities.

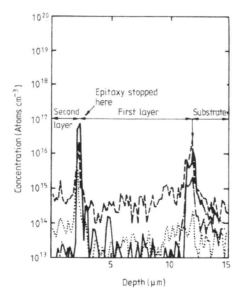

Fig. 11.8. Depth profiles of Mg (—), Cr (....), Fe (— — —) and Mn (— —) in an MOCVD InP layer grown on an InP(S) substrate. The first layer was partly removed by controlled chemical etching. A second layer was grown on the chemically treated first epilayer.

In order to examine the possibility of contamination of the substrate by pregrowth chemical treatment, Razeghi [1983] grew an InP epilayer 10 μm thick under conventional growth conditions. She cut the sample into two parts. On one part, Razeghi did the pregrowth chemical etching to remove 2 μm of epilayer. Then the sample was placed into the reactor and another 2 μm InP was grown under the conventional growth conditions. Fig. 11.8 shows the impurity profile of the sample after regrowth. One can observe similar accumulations of impurities at the epilayer–substrate interface and epilayer–epilayer interface. These results show clearly that the majority of impurities result from the chemical etching, even though very pure reactants were used (e.g. 'suprapure' Merck H_2SO_4, Fe (10^{-6}%), Cr (10^{-7}%) and Mg (10^{-6}%)).

In order to evaluate the effect of some growth parameters such as *in situ* HCl etching prior to layer growth on impurity redistribution, the second part of the 10 μm thick InP epilayer was again placed in the reactor and Razeghi et al. did *in situ* HCl etching for 1 min at 550 °C. After that, they grew 2 μm InP on the sample. Fig. 11.9 shows the SIMS impurity profiles of the sample. The considerable peaks of Si, Fe, Mg, Mn and Cr at the epilayer–epilayer interface show the impurities incorporated from the HCl.

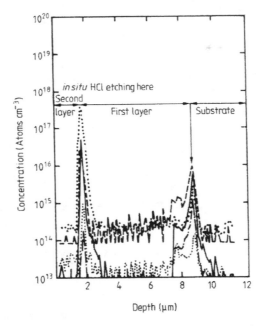

Fig. 11.9. Depth profiles of Mg (—), Si (....), Cr (— — —), Fe(....) and Mn (— —) in an MOCVD InP layer grown on an InP(S) substrate. The first layer was partly removed by in situ *HCl etching in PH$_3$/H$_2$ flow.*

In situ HCl etching without PH$_3$ flow created the same interface contamination with an enhanced impurity redistribution in the layer for Fe and Cr which occurred as shown in Fig. 11.10. It is possible that in this case a surface with a large degree of non-stoichiometry and a large density of defects (phosphorus vacancies) accelerates impurity diffusion.

To evaluate the effect of doping elements of substrates, Razeghi and Duchemin [1983] have performed a study of MOCVD growth simultaneously on three InP substrates doped respectively with Fe, Sn and S placed adjacent to each other within the reactor for growth under conventional growth conditions.

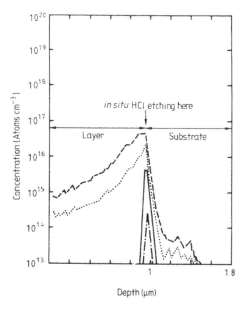

Depth (μm)

Fig. 11.10. Depth profiles of Mg (—), Cr (....), Fe (— — —) and Mn (— —) in an MOCVD InP layer grown on an InP(S) substrate. The substrate was in situ *HCl etched without PH₃ in H₂.*

Fig. 11.11 shows the Fe profiles. One can observe that the Fe concentration and the Fe accumulation at the Fe-doped substrate–epilayer interface and in the layer are about 10 times lower than in the case of Sn- or S-doped substrates. The results are similar for Mg, Mn and Si. The different levels of the same elements in the layers grown in the same run are most probably due to the substrate quality. The degree of non-stoichiometry on the surface and in the substrates could influence chemisorption, the crystal quality of layers, the formation of vacancies, and consequently, the impurity redistribution. Layers grown on Fe-doped substrates often present lower impurity peaks at the interface and lower impurity density in epitaxial layers. Nevertheless, similar results have also been obtained on S- and Sn-doped substrates from different growths and different suppliers [Huber et al. 1984].

In order to estimate the effect of substrate orientation on the impurity redistribution, growth of InP layers has been carried out on Fe-doped substrates oriented 2°, 3° and 4° off (1 0 0) towards (1 1 0) (from the same InP ingot) [Razeghi and Duchemin 1983].

The three samples were prepared simultaneously and layers were grown in the same run under conventional growth conditions. Fig. 11.12 shows the Fe depth profiles of layers. No significant difference in Fe redistribution due to misorientation of InP substrate can be observed. Similar results were found concerning Mg, Cr, Si and Mn.

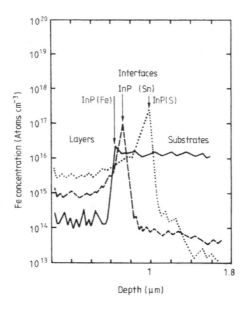

Fig. 11.11. Depth profiles of Fe in MOCVD on InP(Fe) (—), InP(S) (...) and InP(Sn) (— — —) substrates (same run).

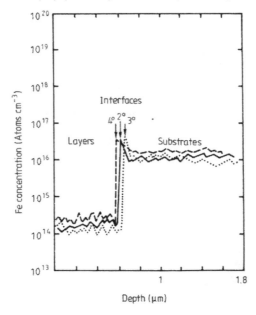

Fig. 11.12. Depth profiles of Fe in MOCVD InP on InP(Fe) semi-insulating substrates oriented (1 0 0) off 2° (—), 3° (....) and 4° (— — —) towards (1 1 0).

(c) Carrier concentration measurements

When a metal is brought into intimate contact with a semiconductor, the difference between the conduction band of the semiconductor at the interface and the Fermi level in the metal is established. The energy difference is called the built-in voltage eV_{bi} and serves as a boundary condition on the solution of the Poisson equation in the semiconductor, which proceeds in exactly the same manner as in p–n junctions. Energy band diagrams of a Schottky barrier on an n-type semiconductor under different biasing conditions, viz. (a) thermal equilibrium, (b) forward bias and (c) reverse bias, are shown in Fig. 11.13 [Sze 1982].

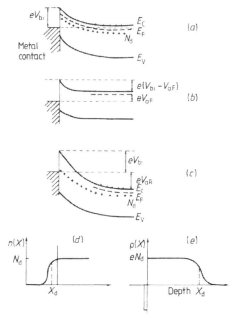

Fig. 11.13. Energy band diagram of a Schottky barrier on an n-type semiconductor under different biasing conditions: (a) thermal equilibrium; (b) forward bias; (c) reverse bias; (d) electron density distribution n(X) at the edge of the depletion region; (e) space-charge density plotted as a function of depth.

A depletion region is established by the combined effect of the built-in voltage (V_{bi}) and applied reverse bias (V_{aR}) or applied forward bias (V_{aF}), and the electron density distribution $n(X)$ at the edge of the depletion region is as illustrated in Fig. 11.13(d). The plot of space-charge density $\rho(X)$ as a function of depth for uniform material illustrates the presence of positive fixed ionized donor charge eN_d in the semiconductor and an accumulation of electrons in the metal (Fig. 11.13(e)). Under the abrupt approximation that $\rho \simeq qN_d$ for $X \le X_d$ and $\rho \simeq 0$ and $dV/dX \simeq 0$ for $X > X_d$, where X_d is the depletion width, N_d can be given by

Eq. (11.7) $\qquad N_d = \dfrac{2}{q\varepsilon_s}\left(-\dfrac{1}{d\left(1/C^2\right)/dV}\right)$

If N_d is constant throughout the depletion region, one should obtain a straight line by plotting $1/C^2$ versus V.

For determining the carrier concentration in InP and related compounds, the conventional non-destructive, capacitance–voltage (C–V) method is usually used. The main disadvantage of the *C–V* method is that the maximum depth that can be profiled is limited by electrical breakdown at high reverse bias, and this can be restrictive in highly doped materials where the depletion depths are small. Originally this was overcome by alternate chemical etching and profiling with a temporary mercury barrier, but this process is time-consuming and tedious. Ambridge et al. [1973, 1974, and 1975] used an electrolyte to make the barrier and to remove material electrolytically so both processes can be carried out in the same electrochemical cell and controlled electronically, using automatic equipment to perform the repetitive etch/measure cycle and generate a profile plot. The etch depth can be measured continuously by integrating the etch current and applying Faraday's law. The basic principles of electrochemical *C–V* profiling are documented in a number of original papers by Ambridge et al. [1975] and a review paper by Blood [1985]. This method is destructive, but the profile can, in principle, be measured to unlimited depth. The requirements for the electrolyte are rather demanding, calling for satisfactory barrier and dissolution properties on *n*- and *p*-type material. Actually, a versatile instrument manufactured by Polaron Equipment Ltd, Watford, UK (sometimes called the "Post Office Plotter" or POP) is available commercially.

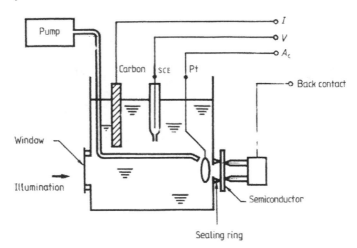

Fig. 11.14. Schematic diagram of the electrochemical cell used in the profiles.

Fig. 11.14 shows a schematic diagram of the electrochemical cell used in the profiles. The semiconductor sample is held against a sealing ring, which defines the contact area, by means of spring-loaded back-contacts. The etching and measuring conditions are controlled by the potential across the cell and this is established by passing a DC current between the semiconductor and the carbon electrode to maintain the required overpotential measured potentiometrically with reference to the saturated calomel electrode (SCE). The AC signals are measured with respect to a Pt electrode located near the semiconductor surface to reduce the series resistance of the electrolyte. The capacity associated with the charged depletion zone in the semiconductor material is then determined. When the contact is illuminated with photons of energy greater than the band gap of the semiconductor, the reverse current is increased, due to the flow of holes or electrons from the semiconductor into the electrolyte in *n*- or *p*-type material, respectively. This causes a change in the voltage for zero current which is of opposite sign for *n*- and *p*-type material and can therefore be used to indicate the material type. Material is dissolved when an anodic current is drawn by a flow of holes from the InP, whereas a cathodic current causes deposition of material from the electrolyte (which is HCl for InP) onto the semiconductor surface. In order to etch *n*-type material the holes are generated by illumination under reverse bias. Smooth removal of *n*-type material is achieved when the anodic current depends upon the illumination intensity but not upon the potential. For a reverse-bias *C–V* measurement on *p*-type material, the potential is switched from an anodic to a cathodic value, and, to avoid contamination of the sample surface by the cathodic reaction promoted by electrons from the conduction band, it is important that the cathodic potential is such that the reverse current during the measurement is very small.

To calculate the depth etched (X_e), one can use Faraday's law of electrolysis [Blood 1985] for the total charge transferred by integrating the etch current *I*:

Eq. (11.8) $X_e = (M/qFDS) \int I dt$

where *M* is the molecular weight and *D* the density of the semiconductor, *F* the Faraday constant, *S* the dissolution surface and *q* the charge transferred per molecule dissolved. For InP, $q = 6$. The depletion depth (X_d) can be obtained from

Eq. (11.9) $X_d = \varepsilon \varepsilon_0 S / C$

by measurement of *C* at 3 kHz (≈ 0.14 V peak-to-peak), and $N(X_d)$ is derived from

Eq. (11.10) $N(X_d) = -\dfrac{C^3}{\varepsilon \varepsilon_0 S^2} \left(\dfrac{\Delta C}{\Delta V}\right)^{-1}$

where $\Delta C / \Delta V$ is measured by modulation at 30 Hz (≈ 0.28 V peak-to-peak) at a low fixed reverse bias.

Fig. 11.15 shows a typical electrochemical profile of an *n*-type InP grown by MOCVD using the 3 mm sealing ring and HCl electrolyte. Fig. 11.15(*a*) shows the electrochemical (polaron) profile for InP over an Sn-doped InP substrate (N_D-

$N_A = 10^{18}$ cm^{-3}), establishing uniform doping through the layer thickness and an abrupt change in carrier concentration.

Fig. 11.15(*b*) shows the carrier concentration profile evaluated from *C–V* measurements for InP over a semi-insulating InP substrate, exhibiting uniform doping through the layer thickness and an abrupt change in carrier concentration. The *C–V* measurements and polaron profile give the concentration of shallow dopants (N_D–N_A), which agrees with the carrier concentration deduced from Hall effect measurements.

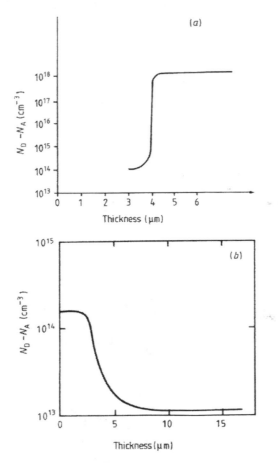

Fig. 11.15. *Carrier concentration profiles from (a) polaron profile and (b) C–V measurements for InP (TEIn) layer on semi-insulating InP substrate.*

(d) Hall mobility measurements

All of the devices using InP-based materials require an exact specification in terms of carrier concentration and thickness, and top performance often depends on material quality. Use of the Hall effect is extremely common to determine carrier density and mobility between room temperature and 2 K. In many cases mobility is used as a simple figure of merit for quality of starting materials such as hydride (PH$_3$ or AsH$_3$) and alkyl (TEIn or TMIn). Hall data can yield excellent quantitative information on the electrically active impurities in a semiconductor.

The total mobility in InP material results from the combination of different scattering processes such as:

(1)   defect scattering—impurity (neutral or ionized) scattering and crystal defects;

(2)   lattice scattering—inter-valley scattering, intra-valley optic (polar or non-polar) scattering, intra-valley acoustic (piezoelectric or deformation potential) scattering; and

(3)   carrier–carrier scattering.

The lack of a center of symmetry in InP-based materials leads to piezoelectric behavior and through this the phonons can scatter the electrons. The most important defect scattering is charged-impurity scattering, in which the electron scatters off the long-range Coulomb potential. Neutral centers can also interact in a complex manner but this is generally only important in the highest-purity samples at low temperatures. The dependence of mobility from ionized impurities (μ_i) on temperature and impurity density can be given by [Sze 1982]

Eq. (11.11) $\mu_i \simeq \left(m*\right)^{-1/2} N_I^{-1} T^{3/2}$

where N_I is the ionized-impurity density.

The lattice is in constant motion even at absolute zero, the quantized normal modes of its vibration being known as phonons. The contractions and dilatations of the lattice in the presence of phonons create potential fluctuations of various origins which interact with and scatter the electron. The interaction may be through local band structure perturbations, known as deformation potential scattering, involving either acoustic phonons or non-polar optic phonons.

The temperature dependence of the mobility due to deformation potential acoustic phonon scattering is

Eq. (11.12) $\mu_A \simeq \left(m*\right)^{-5/2} T^{-3/2}$

In polar semiconductors such as InP, the lattice vibrations give rise to an electrostatic potential on which the carriers can scatter. For acoustic phonons, this mechanism is known as piezoelectric scattering. Above 200 K polar optical phonon scattering dominates the mobility.

For making Hall measurements the Van der Pauw clover-leaf [Van der Paw 1958] geometry is most commonly used. This pattern is ideal and simple for making Hall measurements on epitaxial layers in which the layer thickness is orders of magnitude less than the lateral dimension of the sample.

Fig. 11.16 shows the typical Hall mobility as a function of temperature for InP layers grown by MOCVD. From these data and the measured carrier density, one can usually estimate the net donor density N_D–N_A, the compensation ratio $K = N_A/N_D$ and the donor ionization energy E_d.

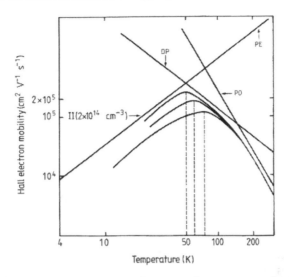

Fig. 11.16. Electron Hall mobility as a function of temperature, for MOCVD InP on InP substrate.

If we assume that all common donor species in InP and related compounds are hydrogenic with binding energy E_d given by

Eq. (11.13) $$E_d = 13.6 \times \frac{m^*}{m_0} \frac{1}{\varepsilon_r^2} (eV)$$

with $m_e^* = 0.08 m_0$, $m_p = 0.5 m_0$ and $\varepsilon_r = 12.4$ (ε_r is the relative dielectric constant), we obtain for donor impurities E_d (donor) ≈ 7 meV and for acceptor impurities E_d (acceptor) ≈ 40 meV. This shows that for InP and related compounds the donors are all shallow and the acceptors rather deep. This means that for temperatures between, say, 4 and 400 K, the Fermi-energy movement is not sufficient to change the charge state of the deeper compensating acceptors.

Hall mobilities of epitaxial InP layers grown on semi-insulating substrates were measured in a magnetic field of 4000 G by a conventional Van der Pauw technique. In Table 11.6 we give the measured mobility at 300 and 77 K in undoped InP, using pure starting materials (TEIn from α-Ventron and PH$_3$ from Matheson).

Sample no.	Thickness (μm)	300K		77 K	
		N_D-N_A (cm^{-3})	μ_H $(cm^2 \cdot V^{-1} \cdot s^{-1})$	N_D-N_A (cm^{-3})	μ_H $(cm^2 \cdot V^{-1} \cdot s^{-1})$
119	2.5	2×10^{15}	$5350 \pm 2\%$	1.5×10^{15}	59,800
151	2.5	5×10^{15}	$5240 \pm 2\%$	3.6×10^{15}	56,700
127	3.3	5.7×10^{15}	$4950 \pm 2\%$	5.7×10^{15}	53,320
43	4	10^{14}	$5500 \pm 2\%$	10^{14}	150,000

Table 11.6. Average measured values of mobility for InP epilayer.

(e) Deep-level transient spectroscopy

This is a method for obtaining information about defects giving rise to electrically active deep energy levels. Conventional deep-level transient spectroscopy (DLTS)) can, in a Schottky-barrier junction, only detect majority-carrier traps. It is easy, however, to resolve different traps and derive their concentrations. The DLTS measurements show that there are two electron traps in InP layers grown by MOCVD (see Table 11.7) [Lim et al. 1982].

Activation energy, E (meV)	Capture cross section, S (cm^{-2})
E_5, 433	3×10^{14}
E_6, 661	8×10^{14}

Table 11.7. Electron traps in InP epilayer.

11.3.5. Interfaces

The surface of InP layers and the epitaxial layer–substrate interface have been studied by Auger electron spectroscopy. Fig. 11.17 shows the Auger spectrum for such an interface eroded by sputtering. Here it can be seen that there are no impurities at the interface. Fig. 11.17 indicates the Auger spectrum for the surface of the InP epilayer and shows the presence of O and C as well as P and In. The number of dislocations in the epilayer is the same as in the substrate, and it is not possible to see the interface between the substrate and the epilayer after normal chemical etching.

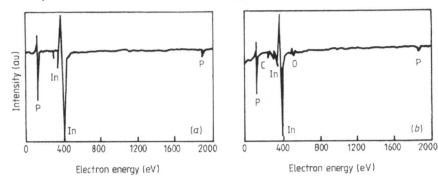

Fig. 11.17. Auger spectrum for (a) the InP epilayer–substrate interface and (b) the surface of the epilayer.

11.4. Growth and characterization of InP using TMIn

TEIn is a liquid and its vapor pressure is low (see appendix A.6). In MOCVD at low pressure, one has to be careful, and a small change of pressure inside or outside the TEIn container produces secondary sources of TEIn in the gas panel. In this case it is very difficult to control the growth rate and quality of the epitaxial layer.

Trimethylindium is solid under 80 °C, and its vapor pressure is higher than that of TEIn. Therefore one can maintain the temperature of TMIn sources lower than room temperature, avoiding the condensation of TMIn in the gas panel, and also eliminate the problem of secondary sources of TMIn in the gas panel. So Razeghi et al. [1988] decided to use TMIn instead of TEIn for the growth of InP and related compounds. In that case one can use pure H_2 as a carrier gas (see Chapter 2). Essentially for these reasons the results of growth using TMIn are generally better than those from TEIn.

High-quality InP epilayers have been grown by MOCVD using trimethylindium (TMIn) (Alfa product) and PH_3 (Matheson) for In and P sources, respectively. Pure hydrogen is used as the carrier gas. The growth temperature is 550 °C and growth pressure is 100 mbar. Details on optimized growth conditions determined during these investigations are indicated in Table 11.8.

Growth temperature	(°C)	550
Total H_2 flow	$(1 \cdot min^{-1})$	6
H_2 through TMIn bubbler	$(cm^3 \cdot min^{-1})$	50
PH_3 flow	$(cm^3 \cdot min^{-1})$	100
Growth rate	$(\text{Å} \cdot min^{-1})$	300

Table 11.8. Optimized conditions of InP growth using TMIn.

All the InP layers have been grown on (1 0 0)-oriented Sn- and Fe-doped InP substrates. The main characteristics of the different samples are summarized in Table 11.9.

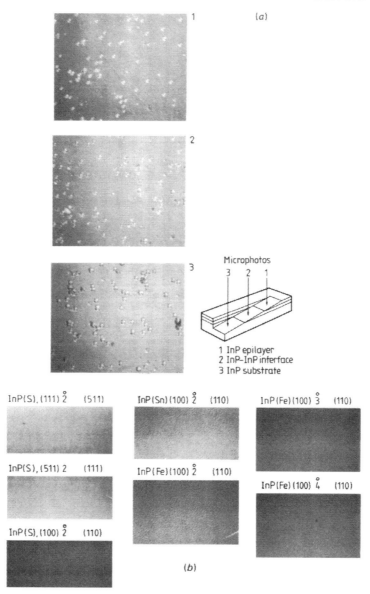

Fig. 11.18. (a) Transition region after chemical etching of an InP–InP epilayer
grown by MOCVD. (b) Optical surface photographs of InP layers on InP substrates
of various orientations as indicated.

Fig. 11.18(*a*) shows the transition region after chemical etching of an InP
epilayer grown by MOCVD using TMIn. The etch-pit density (EPD) was the same
in the substrate as in the confinement InP layers, and the interfaces are defect-free.
The surface photographs of InP layers on InP substrates of various orientations are
shown in Fig. 11.18(*b*).

Sample no.	A	B	C	D	E
Thickness (µm)	5	8	10	5	3.5
N_D-N_A (cm$_{-3}$)	7×10^{13}	2×10^{14}	3×10^{13}	6×10^{13}	10^{14}
μ(300 K)(cm$^2 \cdot$V$^{-1} \cdot$s^{-1})	5500	4500	6000	5500	5000
μ(50 K)(cm$^2 \cdot$V$^{-1} \cdot$s^{-1})	150,000	100,000	200,000	145,000	110,000

Table 11.9. Some of the characteristics of InP epilayers using TMIn.

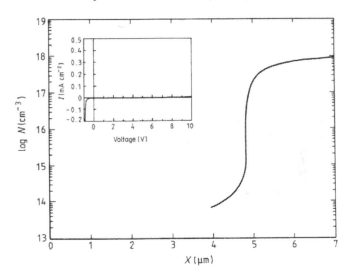

Fig. 11.19. Typical electrochemical polaron profile of an InP epilayer grown by MOCVD: TMIn+PH$_3$, μ(77 K) $\square$ 150,000 cm$^2 \cdot$V$^{-1} \cdot$s^{-1}, dx/dt ~ 300 Å·min^{-1}.

Fig. 11.19 indicates the electrochemical polaron profile of an InP layer using TMIn. A carrier concentration as low as 3×10^{13} cm^{-3} has been measured, which is the purest InP epilayer yet grown by any growth technique.

Electrical measurements were performed, using the classical Van der Pauw technique, with ohmic contacts produced by evaporation and annealing of Au–Ge. The maximum mobility was generally found between 50 and 55 K, and exceeded 100,000 cm$^2 \cdot$V$^{-1} \cdot$s^{-1}, proving the high quality of the samples under study. At the same time, the residual doping level was found to be very low, generally less than 10^{14} cm^{-3}.

Photoluminescence measurements were performed at 2 K, using a dye laser as the source of excitation. The excitation energy was tuned just above the band gap of InP, at about 1430 meV, with a slit width of 50 µm.

Fig. 11.20 to Fig. 11.22 show the near-gap photoluminescence (NGPL) spectra of samples A, B and C. All of them exhibit different lines identified as free exciton (X), exciton bound to neutral donor (D^{0}X) and charged donor (D$^+$ X), donor–valence band recombination (D^0–h) and exciton bound to acceptor (A^{0}X). The positions in energy of all of the peaks and their identification are summarized in Table 11.10.

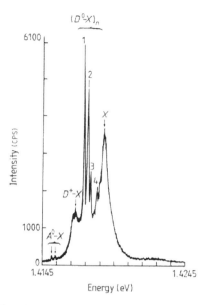

Fig. 11.20. Near-gap photoluminescence (NGPL) at 2 K of an InP (TMIn) epilayer grown by LP-MOCVD (sample A). $E_{exc} = 1.426$ eV, $I_{exc} = 4$ mW.

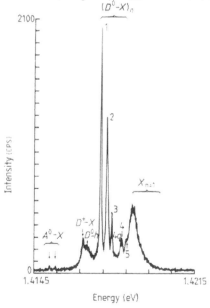

Fig. 11.21. Low-temperature NGPL spectrum at 2 K of an InP (TMIn) epilayer grown by MOCVD (sample B). $E_{exc} = 1.426$ eV, $I_{exc} = 8$ mW.

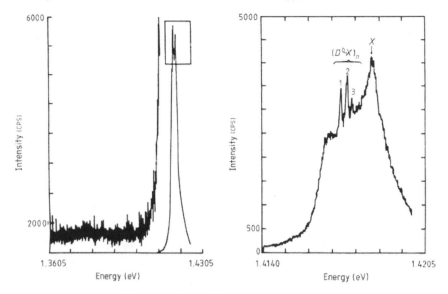

Fig. 11.22. Low-temperature photoluminescence spectrum at 2 K of an InP (TMIn) epilayer grown by MOCVD (sample C). $I_{exc} = 6.6$ mW. Inset: detail of NGPL spectrum at 2 K. $E_{exc} = 1.43$ eV, $I_{exc} = 4$ mW.

The free-exciton line is located at about 1418.8 meV, in good agreement with previous work on the subject. Various components of the bound-exciton (D^0X) recombination, identified as $(D^0X)_n$ with $n = 1,...,5$, have been observed. They are more complicated, since an additional electron is present, making the donor neutral before binding the exciton. The different components $(D^0X)_n$ result from different angular momentum states of the $j = \frac{3}{2}$ hole [Dean 1983]. The states of the complex are generally described by those of the hole in a central potential, with a pair of electrons in the singlet state. Their linewidths are less than 0.1 meV, thus necessitating a very sensitive optical system.

In two of them (A and B), two very weak lines due to an exciton bound to the acceptor (A^0X), located at 1414.2 and 1414.4 meV, appear. We have summarized in Table 11.11 the intensity ratio of X/D^0X and A^0X/D^0X lines for the three samples. In all cases, the acceptor-bound exciton line appears very weak compared to the other recombination processes. The sample therefore should be very little compensated.

For sample C (Fig. 11.22), the sensitivity of detection was multiplied by 100 in the region of the A^0X peak, with no trace of an acceptor. This further demonstrates the purity of the sample, and can explain the exceptionally high mobility measured ($\mu(50$ K$) = 200,000$ cm$^2 \cdot$V$^{-1} \cdot$s^{-1}).

On the other hand, the free-exciton peak (X) appears more intense than the donor-bound exciton peak (D^0X). This is the first time that such a fact has been observed, in any InP crystal, thus confirming the very low concentration of donors (3×10^{13} cm^{-3}) determined by a polaron profile.

Type of recombination	Energy (meV)
X	1418.8
D^0X, $n = 1$	1417.3
D^0X, $n = 2$	1417.6
D^0X, $n = 3$	1417.8
D^0X, $n = 4$	1418.3
D^0X, $n = 5$	1418.5
D^+X	1416.6
D^0–h	1416.8
A^0X	1414.2
A^0X	1414.4

Table 11.10. Positions in energy of all of the PL recombination lines in an InP epilayer

Sample no.	A	B	C
X/D^0X	5.8×10^{-1}	2.8×10^{-1}	1.25
A^0X/D^0X	2.0×10^{-2}	3.0×10^{-2}	No acceptor

Table 11.11. The ratio of PL intensity of X/D^0X and A^0X/D^0X in InP epilayers grown by MOCVD.

Residual donors in a high-purity InP layer grown by MOCVD were identified in 1987 by magnetophotoluminescence (MPL) and photothermal ionization spectroscopy (PTIS) measurements at the University of Illinois by Professor Stillman's group. The donors identified in two samples from MPL and PTIS are 88% Si and 12% S. (For the experimental details, the reader should consult the original reference by Bose et al. [1987].)

11.5. Incorporation of dopants

11.5.1. p-type

InP layers grown by MOCVD can be doped *p*-type by using diethylzinc (DEZn). Fig. 11.23 shows the variation of net carrier concentration as a function of H_2 flow through the DEZn bubbler ($-15°C$) (growth temperature 650 °C). When the flow rate of DEZn is kept constant, the free-carrier concentration varies exponentially with $1/T$, as shown in Fig. 11.24, where T is growth temperature.

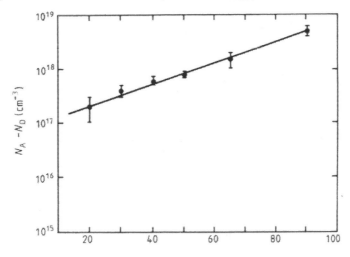

Fig. 11.23. Variation of acceptor level N_A-N_D in InP with DEZn flow-rate ($T_G = 650\ °C$, total flow 7 $1·min^{-1}$).

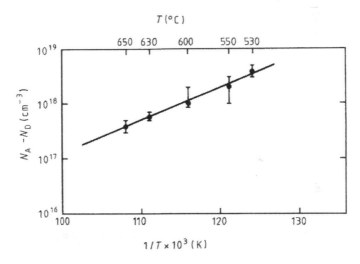

Fig. 11.24. Variation of acceptor level in InP with growth temperature: N_A-N_D
$$\propto e^{\left(\frac{-E_1}{kT}\right)}\ (E_1 \simeq 10\ meV,\ 30\ cm^3·min^{-1}\ H_2\ flow\ through\ DEZn\ bubbler).$$

The incorporation of dopants during growth on InP using doping species such as DEZn can be explained by using a model described by Duchemin [1977]. If we assume that all the DEZn arriving at the hot surface is decomposed, then the Zn concentration becomes limited by the diffusion of the DEZn through the boundary layer to the hot surface. After decomposition there are two possible limiting cases:

(1)   In the simplest case, all of the decomposed material is incorporated into the growing layer. Thus here the impurity concentration is independent

of temperature and inversely proportional to the growth rate. This behavior is observed for the doping of silicon by germanium using germane (GeH$_4$) [Duchemin 1977].

(2)   The second case is that in which only a small fraction of the secondary form (Zn) of the dopant is incorporated into the growing layer. Here the major part of the dopant is vaporized and is then lost by diffusion away from the substrate. As the temperature is raised, more of the dopant is vaporized, and the doping concentration decreases. Here, then, the doping concentration decreases as the temperature increases but is independent of the growth rate. This behavior is typical of the doping of InP by zinc (Fig. 11.25).

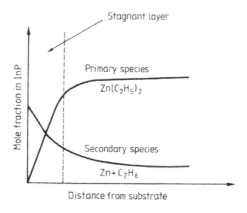

Fig. 11.25. Total decomposition of DEZn (primary species) with partial incorporation of Zn (secondary species).

11.5.2. n-type

Epilayers of InP and related compounds grown by MOCVD can be doped η-type using H$_2$S or SiH$_4$. When the flow-rate of H$_2$S is kept constant, the free-carrier concentration varies exponentially with $1/T$ as in the case of DEZn. The free-carrier concentration in the epilayer decreases when the growth temperature increases. Fig. 11.26 shows the variation of N_D-N_A as a function of H$_2$S flow rate.

The decomposition of the H$_2$S (primary species) is the rate-limiting step; after decomposition, most of the secondary species (S) is incorporated into the growing layer (Fig. 11.27).

A typical polaron electrochemical profile (using a 0.1 cm^2 area and 0.5 M HCl) of an InP Gunn-diode structure (n^+-n-n^+) grown by MOCVD is shown in Fig. 11.28. Using H$_2$S for n$^-$ doping, the interfaces between the layers are sharp. Fig. 11.29 presents a SIMIS profile for a similar Gunn-diode structure of InP.

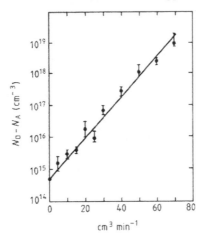

Fig. 11.26. *Variation of donor level N_D-N_A in InP layers with H_2S in H_2*
($T_G = 550$ °C, dx/dt = 200 Å·min⁻¹).

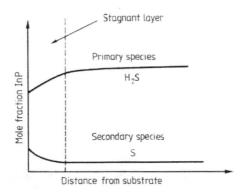

Fig. 11.27. *Partial decomposition of H_2S (primary species) with incorporation of S*
(secondary species).

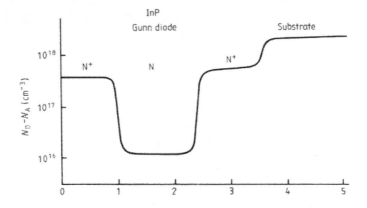

Fig. 11.28. *Electrochemical profile of an InP Gunn-diode structure.*

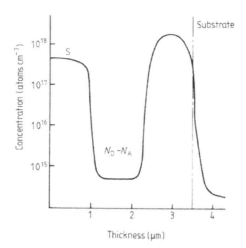

Fig. 11.29. SIMS profile for Gunn-diode structure of InP by MOCVD.

Sulfur has been widely used as an *n*-type dopant for DH layers grown by MOCVD for laser applications. There have been indications in the literature [Benz et al. 1982, Giles et al. 1984, and Stoermer et al. 1979] that, at high growth temperature, significant sulfur diffusion can occur, which will seriously degrade doping interfaces.

In the case of SiH_4, when the flow rate is kept constant, the free-carrier concentration increases with increasing growth temperature. At higher temperature, the decomposition of the SiH_4 is more efficient. We found that for high doping levels of about 10^{18} cm^{-3} (such as used in lasers) it is better to use H_2S. Using H_2S the epitaxial layers are less compensated. Silicon in III–V compounds is amphoteric, and incorporation in InP depends on the ratio of III/V elements. But the diffusion coefficient of Si is less than that of S, so for modulation doping it is better to use Si.

In order to examine these topics and to identify a satisfactory *n*-type dopant for Gunn-diode InP and modulation-doped GaInAs–InP layers grown by MOCVD, Razeghi [1985] examined the electrical properties of epitaxial layers doped with SiH_4. Fig. 11.30 shows electrochemical profiles of typical doping interfaces for InP doped with S and Si, indicating that the interfaces of S-doped layers do not differ significantly from those doped with Si.

For the same carrier concentration, the measured Hall mobility is lower in Si-doped InP layers than in the S-doped one (Fig. 11.31). The results show that the Si-doped InP layers are compensated. Autocompensation was not present in the S-doped samples. Also, with S it is generally simpler to obtain highly doped InP layers. There is a satisfying agreement between the results of Razeghi [1985] and those obtained by Giles et al. [1984] using chloride-process VPE growth of InP layers.

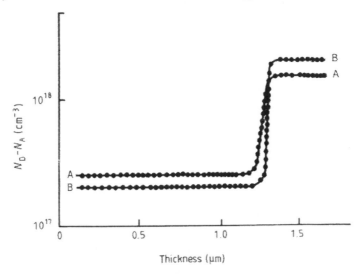

Fig. 11.30. Electrochemical profiles of typical doping interfaces for InP doped with S (A) and Si (B): LP-MOCVD; full circles indicate values with light; full curves indicate values without light.

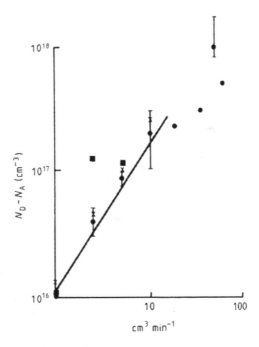

Fig. 11.31. Variation of donor level in an InP layer with SiH_4 flow-rate: MOCVD InP; (X) C–V, (•) electrochemical profile, (■) Hall mobility (250 PPM SiH_4 in H_2) ($T_G = 550$ °C, dx/dt = 200 Å·min^{-1}, N_D–N_A = 66, 57 and 47% for μ = 2600, 1400 and 2700 cm^2·V^{-1}·s^{-1}, respectively).

11.6. Applications of InP epitaxial layers

InP is an excellent candidate for microwave diodes. Microwave diodes can be made with operating frequencies covering the range from about 0.1 to 200 GHz with corresponding wavelengths from 300 cm to 1.5 mm.

The applications in this frequency range are numerous, including radar for both military and civil applications. For example, communication via satellites is made at 12 GHz for broadcasting and at 20–30 GHz for communication. But communication flow is increasing so rapidly that new channels are necessary, and the use of frequencies up to 100 GHz is anticipated in the near future.

11.6.1. Gunn diodes

The Gunn diode [Gunn 1963] is a transferred-electron device. Coherent microwave output is generated when a DC electric field in excess of a critical threshold value of several thousand volts per centimeter is applied across an *n*-type sample of GaAs or InP. Hilsum [1962] and Ridley and Watkins [1961] predicted that such oscillations should be observed because of an intrinsic negative differential resistance. To understand this effect, consider the energy–momentum diagrams for GaAs and InP in Fig. 11.32, the two most important semiconductors for transferred-electron devices. Fig. 11.32 shows that the band structures of GaAs and InP are similar. The energy separation between the two valleys is ΔE, which is about 0.31 eV for GaAs and 0.53 eV for InP. An applied electric field induces transfer of conduction-band electrons from the low-energy, high-mobility valley to higher-energy, low-mobility satellite valleys.

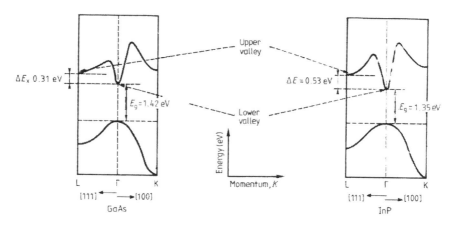

Fig. 11.32. Band structures of InP and GaAs.

As a consequence the electrons are slowed down and accumulate in domains which are propagated along the sample. The propagation time, which is proportional to the sample length, determines the frequency.

Fig. 11.33 shows the measured room-temperature velocity–field characteristics for GaAs and InP. The threshold field ε_T defining the onset of negative differential resistivity is 3.2 kV·cm^{-1} for GaAs and 10.5 kV·cm^{-1} for InP. The peak velocity V_P is about 2.2×10^7 cm·s^{-1} for high-purity GaAs and 2.5×10^7 cm·s^{-1} for high-purity InP. In order to have the electron-transfer mechanism, the following conditions are necessary:

(1) In the absence of a bias electric field, most electrons are in the lower conduction-band minimum, or $kT \leq \Delta E$.

(2) In the lower conduction-band minimum, the electrons must have high mobility, small effective mass and low density of states; whereas in the upper satellite valleys, the electrons must have low mobility, large effective mass and high density of states.

(3) The energy separation between the two valleys must be smaller than the semiconductor band gap so that avalanche breakdown does not set in before electrons are transferred into the upper valleys.

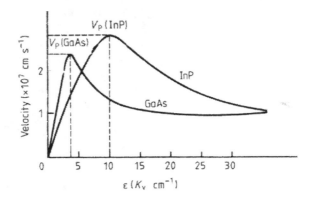

Fig. 11.33. Room-temperature velocity–field characteristics for GaAs and InP.

High-power and high-efficiency InP Gunn diodes made from layers grown by MOCVD have been developed in the millimeter-wave range. Gunn diodes processed using the integral heatsink technique have delivered up to 100 mW continuous-wave (CW) output power with 2.5% efficiency at 94 GHz, while average power levels in excess of 90 mW were obtained at 94 GHz [Poisson 1984].

Gunn diodes, which are well known as low-noise devices, are better adapted for communications receivers than are avalanche diodes. Until recently, gallium arsenide had been the only material for such applications. But since its introduction a few years ago, indium phosphide has been recognized to be a very attractive material for making millimeter-wave oscillators [Corlett 1975, Crowley 1980, and Hamilton 1976].

The cutoff frequency is approximately 200 GHz at 3 dB for InP compared with 100 GHz for GaAs. The key characteristic necessary to obtain high efficiency is a high peak-to-valley ratio V_p/V_v of the electron velocity as a function of applied electric field. The peak-to-valley ratio is significantly higher in InP than in GaAs (4 for InP versus 2.4 for GaAs).

Other InP properties compared to those in GaAs which aid in obtaining higher power output are higher threshold field, higher electric breakdown field and higher thermal conductivity. For these reasons, indium phosphide offers significant advantages over gallium arsenide for high power and efficiency at millimeter wavelengths.

Until recently, the output power and efficiency obtained using InP represented only a modest improvement over GaAs. Recent progress in 94 GHz InP Gunn-device development at Thomson's laboratory [Bose et al. 1987] showed that higher output power and efficiency can be obtained from InP in the millimeter range.

Thomson has developed a microwave diode structure whose performance is claimed to outclass any other device reliable enough to be used in professional equipment. In the CW mode output power up to 103 mW was measured at 94 GHz with 2.5% efficiency. This device is based on a three-layer n^+–n–n^+ structure where Si (from silane) was used as the dopant [Razeghi et al. 1987]. This dopant allows precise control of doping profiles. High-quality non-intentionally doped and Si-doped epitaxial InP material was reproducibly obtained by low-pressure metalorganic chemical vapor deposition.

The device structure consists of an n^+ buffer layer with carrier densities of 10^{18} cm^{-3} followed by an active layer doped in the 10^{16} to 1.5×10^{16} cm^{-3} range and a contact layer in the 10^{18} cm^{-3} range.

Excellent results have been obtained from wafers ranging in active layer thickness from 0.9 to 1 μm. A typical 94 GHz InP Gunn-device doping profile is shown in Fig. 11.34.

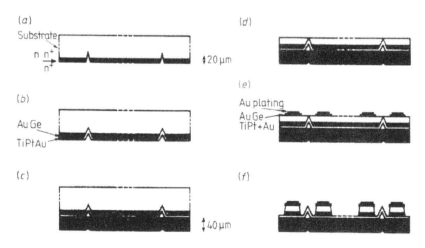

Fig. 11.34. The various sequences of Gunn-diode technology, (a) Etched channels in the epilayer side; (b) Au-Ge, Ti, Pt, Au metallization; (c) Au heatsink; (d) substrate removed to expose channels; (e) metalizations (30–60 μm diameter, Au-plated); (f) etched mesas.

The grown wafers are 2.5 × 4 cm and are processed using the integral heatsink technique. This technique consists of various epitaxial layers, alloyed Au–Ge, Ti, Pt and Au contacts and an electroplated 40 μm gold heatsink to reduce thermal resistance and facilitate chip bonding and packaging. On the epilayer side of the as-grown wafers, channels are etched 10–20 µm deep, spaced 3 mm apart to provide a depth gauge when the substrate side is thinned. The entire epilayer side is then metalized with Au–Ge, Ti, Pt and Au. The Au heatsink is electroplated 40 µm thick. The wafer is then turned over and uniformly thinned to 10 µm with a lap and chemical polish technique until the channels on the other side are exposed (see Fig. 11.34).

A total substrate thickness less than 10 µm is required at 94 GHz in order to minimize significant resistive loss created by the substrate. Au–Ge, Ti, Pt and Au contacts are formed on this new thinned substrate surface. Mesas of 30–60 µm diameter are defined using a light-sensitive $FeCl_3$ etch.

A low-inductance bond was obtained by using two 100 µm gold ribbons stitch-bonded from the metalized ceramic across the mesa. The best result, however, was obtained using a star pattern top bond. In both cases the reduced parasitic inductance allowed the diode to operate at a higher frequency with no power loss.

The thermal resistance of these 94 GHz InP Gunn devices with 60 µm diameter mesa ranges between 40 and 45 °C·W^{-1}. The working temperature of the active layer is less than 200 °C.

Both CW power level and efficiency of these devices were studied over a temperature range of −40 to +60 °C. A power shift of 0.005 dB·°C^{-1} and a frequency shift of 3.5 MHz·°C^{-1} have been observed.

Amplitude-modulated (AM) noise (−140 dB$_c$·Hz^{-1} single-sideband (SSB) at 10 kHz from carrier) and frequency-modulated (FM) noise (−60 dB$_c$·Hz^{-1} SSB at 10 kHz from carrier) at 94 GHz were found to be similar to those obtained with gallium arsenide [Poisson 1984].

References

Ambridge, T., Elliott, C.R., and Faktor, M.M. *J. Appl. Electrochem.* **3** 1–15, 1973.
Ambridge, T. and Faktor, M.M. *J. Appl. Electrochem.* **4** 135–42, 1974.
Ambridge, T. and Faktor, M.M. *J. Appl. Electrochem.* **5** 319–28, 1975.
Benz, K., Hagleko, H., and Bosch, R. 1982 *J. Physique* **43** C5, 1975.
Blood, P. 1985 *Semicond. Sci. Technol.* **1** 7–27, 1975.
Bose, S.S., Lee, B., Kim, N.H., and Stillman, G.E. *Appl. Phys. Lett.* **51** 937, 1987.
Corlett, R. *Indium Phosphide CW Transferred Electron Amplifiers* (Inst. Phys. Conf. Ser. 24) ch 2, 1975.
Crowley, J.D. *Electron. Lett.* **16** 705, 1980.
Dapkus, P., Manasevit, H., and Hess, K. *J. Cryst. Growth* **55** 10, 1981.
Dean, P.J. and Skolnick, M.S. *J. Appl. Phys.* **54** 346, 1983.
Duchemin, J.P. *Rev. Res. Thomson-CSF* **9** 1, 1977.
Duchemin, J.P. *Rev. Res. Thomson-CSF* **9** 2, 1977.
Giles, P., Davies, P., and Hardell, N. *J. Cryst. Growth* **65** 351, 1984.
Gunn, J.B. *Solid State Commun.* **1** 88, 1963.
Hamilton, R.J. *IEEE Trans. Microwave Theor. Tech.* **MTT-24** 775, 1976.
Hilsum, C. *Proc. IRE* **50** 185, 1962.

Huber, A., Morillot, G., Bonnet, M., Merenda, P., and Bessonneau, G. *Appl. Phys. Lett.* **41** 638, 1982.

Huber, A. *Rev. Tech. Thomson-CSF* **18** 1, 1986.

Huber, A.M., Morillot, G., Hersee, S.D., and Kazmierski, K. *Proc. Third Conf. on Semi-insulating III–V Materials* Oregon (Nantwich: Shiva) p 466, 1984.

Huber, A.M., Morillot, G., Bonnet, M., Meranda, P., and Bessonneau, G. *Appl. Phys. Lett.* **41** 638–9, 1982.

Huber, A.M., Razeghi, M., and Morillot, G. *Gallium Arsenide and Related Compounds, 1984* (Inst. Phys. Conf. Ser. 74) p 223, 1984.

Kane, E.O. *Semicond. Semimet.* **1** 75, 1966.

Kittel, C. *Introduction to Solid State Physics* (New York: Wiley), 1971.

Lepareux, M. *Rev. Tech. Thomson-CSF* **12** 225, 1980.

Lim, H., Sagres, G., Bastide, G., and Gouskov, L. *J. Appl. Phys.* **53** 3085, 1982.

Magee, C. *J. Electrochem. Soc.* **126** 660, 1979.

Olsen, G., Zamerowski, J., and Hawrylo, F. *J. Cryst. Growth* **59** 654, 1982.

Poisson, M.A. *Proc. Int. Conf. on GaAs and Related Compounds,* Biarritz, 1984.

Razeghi, M. and Duchemin, J.P. *J. Cryst. Growth* **64** 76, 1983.

Razeghi, M., Blondeau, R., Bouley, J.C., Decremoux, B., and Duchemin, J.P. *Proc. 9th IEEE Int. Laser Conf.*, 1984.

Razeghi, M., Maurel, P., Defour, M., Omnes, F., Neu, G., and Kozacki, A. *Appl. Phys. Lett.* **52** 117, 1988.

Razeghi, M. and Duchemin, J.P. *J. Vac. Sci.* B **1** 262, 1983.

Razeghi, M. *Rev. Tech. Thomson-CSF* 1983 **15** 1, 1983.

Razeghi, M. *Semiconductors and Semimetals* ed W.T. Tsang (New York: Academic) ch 5, 1985.

Razeghi, M., Blondeau, R., Decremoux, B., and Duchemin, J.P. *Appl. Phys. Lett.* **46** 131, 1985.

Razeghi, M., Poisson, M.A., Larivain, J.P., and Duchemin, J.P. *J. Electron. Mater.* **12** 371, 1983.

Ridley, B.K. and Watkins, T.B. *Proc. Phys. Soc. Lond.* **78** 293, 1961.

Saxena, R., Cooper, C., Ludowise, M., Hikido, S., and Borden, P.G. J. *Cryst. Growth* **55** 58, 1981.

Shaw, D. *J. Electrochem. Soc.* **115** 405, 1968.

Störmer, H.J., Dingle, R., Gossard, A., Weigmann, W., and Sturge, M. *Solid State Commun.* **29** 705, 1979.

Sze, S.M. *Physics of Semiconductor Devices* (New York: Wiley), 1982.

Van der Pauw, J. *Philips Res. J.*, 1958.

12.   GaInAs–InP System: MOCVD Growth, Characterization, and Applications

12.1. Introduction
12.2   Growth conditions
12.3   Optical and crystallographic properties, and impurity incorporation in
       GaInAs grown by MOCVD
       12.3.1. Sample preparation
       12.3.2. Electrical, crystallographic and optical experiments
       12.3.3. Results and discussion
       12.3.4. Exciton line
       12.3.5. Determination of Eg (x_{Ga}) near $x_{Ga} = 47\%$
       12.3.6. Donor–acceptor pair recombination
       12.3.7. Zn-doped samples
       12.3.8. Deep Fe and intrinsic defect levels in $Ga_{0.47}In_{0.53}As/InP$
       12.3.9. Comparison of transport, optical and crystallographic properties
12.4. Shallow p^+ layers in GaInAs grown by MOCVD by mercury
       implantation
12.5   GaInAs–InP heterojunctions. Multiquantum wells and superlattices
       grown by MOCVD
       12.5.1. Growth technique
       12.5.2. Structural characterization of GaInAs–InP quantum wells grown by
                MOCVD
       12.5.3. Optical properties of GaInAs–InP quantum wells
       12.5.4. Room-temperature excitons in GaInAs–InP superlattices grown by
                MOCVD
       12.5.5. Negative differential resistance at room temperature from resonant
                tunneling in GaInAs/InP double-barrier heterostructures
12.6. Magnetotransport in GaInAs–InP heterojunctions grown by MOCVD
       12.6.1. Shubnikov–de Haas and quantum Hall effects
       12.6.2. Observation of a two-dimensional hole gas in a $Ga_{0.47}In_{0.53}As/InP$
                heterojunction grown by MOCVD
       12.6.3. Precise quantized Hall resistance measurements in $In_xGa_{1-x}As/InP$
                heterostructures
       12.6.4. Persistent photoconductivity and the quantized Hall effect in
                $In_{0.53}Ga_{0.47}As/InP$ heterostructures grown by MOCVD
       12.6.5. The effect of hydrostatic pressure on a $Ga_{0.47}In_{0.53}As/InP$
                heterojunction with three electric subbands
       12.6.6. Cyclotron resonance

12.6.7.   Shallow-donor spectroscopy and polaron coupling in $Ga_{0.47}In_{0.53}As$–
          InP grown by MOCVD
12.7.   Applications of GaInAs–InP system grown by MOCVD
12.7.1.   PIN photodetector
12.7.2.   Field-effect transistor
12.7.3.   GaInAs–InP optical waveguides

12.1. Introduction

A useful material for long-wavelength optical communication devices is $Ga_{0.47}In_{0.53}As$, lattice-matched to InP. Moreover, its high mobility and large drift velocity make this material very promising for use in high-frequency field-effect transistors (FET) [Stoermer et al. 1979], optical-fiber communication, satellite transmission, radar linkage, direct satellite broadcasting, use in supercomputers and high-speed signal processing [Hilsum and Reese 1970 and Linh 1983]. This chapter describes MOCVD growth, characterization and applications of $Ga_{0.47}In_{0.53}As$–InP heterojunctions (HJ), multiquantum wells (MQW) and superlattices (SL), which show mobility values comparable with the best reported values obtained on epitaxial growth by other techniques.

12.2. Growth conditions

Layers of GaInAs can be grown at 76 Torr and low temperature, between 500 and 650 °C, using TEIn or TMIn, TEGa and AsH_3 in an $H_2 + N_2$, or H_2 in the case of TEIn, carrier gas. Fig. 12.1 shows that the growth rate depends linearly upon the TEIn + TEGa flow-rate and is independent of that of AsH_3 within the range 60–90 $cm^3 \cdot min^{-1}$, which suggests, as in the case of InP, that epitaxial growth is controlled by mass transport of the group III species. Uniform-composition $Ga_{0.47}In_{0.53}As$ over a large area of 10 cm^2 of InP substrate has been obtained (Fig. 12.2). The epitaxial-layer quality is sensitive to the alloy composition, as in the case of GaInAs grown by other techniques.

The surface morphology of GaInAs–InP layers depends on the pretreatment of the substrate and is independent of the lattice mismatch, even in the case of a GaAs layer on an InP substrate with a lattice mismatch of 10^{-2} or more. $Ga_{0.47}In_{0.53}As$/InP layers with an excellent surface morphology and state-of-the-art electron mobility can be grown under the following reactor conditions:

Growth temperature	550 °C
Total flow	7 $l \cdot mm^{-1}$
N_2–TEIn bubbler flow	450 $cm^3 \cdot min^{-1}$
H_2–TEGa bubbler flow	180 $cm^3 \cdot min^{-1}$
AsH_3 flow	90 $cm^3 \cdot min^{-1}$
Growth pressure	76 Torr.

Table 12.1. Reactor conditions.

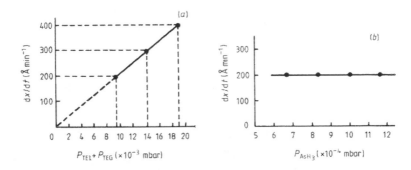

Fig. 12.1. Variation of the growth rate dx/dt of $Ga_{0.47}In_{0.53}As$–InP with (a) TEIn + TEGa flow-rate and (b) AsH_3 flow-rate ($T_G = 550$ °C; P_{TEIn}+ $P_{TEGa} = 9.4 \times 10^{-3}$ mbar; total flow-rate – 6 $l \cdot min^{-1}$).

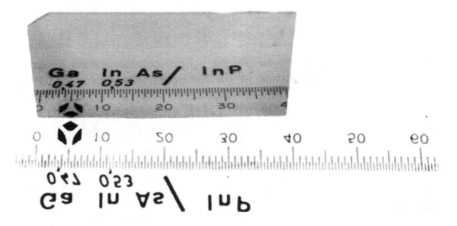

Fig. 12.2. Surface quality of typical GaInAs–InP layers grown by MOCVD.

Under these conditions, the growth rate is 270 $Å \cdot min^{-1}$. InP–GaInAs interfaces can be obtained by turning off the phosphine flow and turning on both the TEGa and AsH_3. GaInAs–InP interfaces can be obtained by turning off the AsH_3 and TEGa flow and turning on the PH_3 flow. The growth rate is small (1 to 5 $Å \cdot s^{-1}$), and it takes less than 1 s for a gas flow to reach its new steady state.

The thickness of epilayer is measured by a bevel stain technique (solution) and the composition calculated either from the PL wavelength or from the value of the lattice parameter as measured by single-crystal x-ray diffraction. The x-ray measurement yields the strained (s) value of the lattice mismatch

Eq. (12.1) $(\Delta a / a)_s = (a_{InP} - a_{GaInAs}) / a_{InP}$

by means of

Eq. (12.2) $(\Delta a / a)_s = \Delta\theta \cos\theta$

where a is the lattice parameter, θ the average diffraction angle and $\Delta\theta$ is half the angular difference between the $K_{\alpha 1}$ x-ray peak for the substrate and $K_{\alpha 1}$ for the epitaxial layer.

The unstrained (un) value of lattice mismatch $(\Delta a/a)_{un}$ that is necessary to calculate the true composition of the epitaxial layer is related to the strained value by

Eq. (12.3) $(\Delta a / a)_s = (1 + 2c_{12} / c_{11})(\Delta a / a)_{un}$

where c_{11} and c_{12} are the elastic constants of $Ga_x In_{1-x} As$ [Bisaro et al. 1979]. The variation of growth rate dx/dt, lattice mismatch $\Delta a/a$ and PL wavelength of GaInAs layers over an area of 10 cm^2 of an InP substrate is less than 5% (Fig. 12.3).

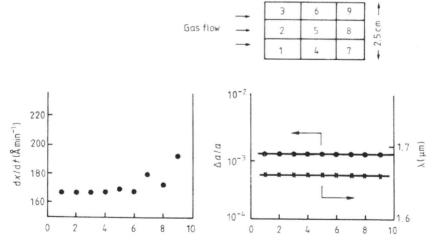

Fig. 12.3. Variation of PL wavelength (X), lattice mismatch Δ a/a (•) and growth rate dx/dt over an area of 10 cm^2 for a GaInAs–InP epitaxial layer; the layer thickness is 5 µm.

Razeghi and Duchemin [Razeghi et al. 1983] have studied the growth of GaInAs layers by using 100% H_2 and mixtures of $H_2 + N_2$ as the carrier gas. The highest PL intensity and the best mobility were obtained by using 60% H_2 and 40%

N$_2$. Razeghi and Duchemin [Razeghi et al. 1983] have performed a study of MOCVD growth of GaInAs simultaneously on six InP substrates with orientations of (1 0 0), (1 1 1) and (1 1 5) placed adjacent to one another within the reactor at a growth temperature of 550 °C, similar to the study on InP described earlier in Chapter 11.

Table 12.2 indicates relative PL intensity, PL halfwidth $\Delta h\nu$, growth rate dx/dt and electron carrier concentration measured by a *C–V* method. Measurements by DLTS show that there are no detectable electron traps in GaInAs layers grown by MOCVD [Goetz et al. 1984].

The photoluminescence (PL) intensity and PL halfwidth of GaInAs layers over InP substrates grown by MOCVD depend directly on the purity of the starting material.

Orientation	N_D-N_A (cm^{-3})	dx/dt (Å·min^{-1})	$\Delta a/a$	I (au)	$\Delta h\nu$ (meV) at 300K
InP(Sn)(1 0 0) 2°	5×10^{14}	260 ± 10	$+1 \times 10^{-3}$	1	45
InP(S)(1 0 0) 2°	6×10^{14}	260 ± 10	$+1 \times 10^{-3}$	0.5	60
InP(Fe)(1 0 0) 2°	3×10^{14}	260 ± 10	$+1 \times 10^{-3}$	2	40
InP(Sn)(1 0 0)	6×10^{14}	260 ± 10	$+1 \times 10^{-3}$	0.2	60
InP(S)(1 1 1) 2°	5×10^{14}	260 ± 10	$+1 \times 10^{-3}$	3	50
InP(S)(1 1 5) 2°	5×10^{14}	260 ± 10	$+1 \times 10^{-3}$	10	80

Table 12.2 PL intensity I, PL halfwidth $\Delta h\nu$ and C–V carrier concentrations of GaInAs grown on InP substrates by MOCVD.

Fig. 12.4 presents a typical absorption spectrum [Voisin et al. 1981] at 1.5 K, the light beam being perpendicular to the interface. The absorption edge occurs around 807 meV and is characteristic of free-exciton absorption. The corresponding luminescence spectrum exhibits two lines around 805 and 788 meV, which are likely to be due, as in bulk GaInAs grown by LPE [Chen et al. 1981 and Marzin et al. 1989], to donor–valence band and donor–acceptor recombination processes, respectively. These optical results show that the quality of the bulk material is good.

The Hall mobilities were measured in a magnetic field of 4000 G on cloverleaf samples cut from epitaxial wafers. Ohmic contacts were formed by evaporating approximately 2500 Å of 12% Ge in Au, then annealing for 4 min at 460 ° C under nitrogen. This procedure yielded contacts that were ohmic at 300 and 77 K.

Typical values as a function of the thickness of the layer, composition, electron carrier concentration and temperature are summarized in Table 12.3. The electron carrier concentration remained roughly constant between 300 and 77 K. The mobilities and doping of nominally undoped Ga$_{0.47}$In$_{0.53}$As grown at a substrate temperature between 550 and 650 °C are also shown in Table 12.3. The mobilities increase and the backgrounds (all *n*-type) decrease at lower growth temperature. The increased mobility in Ga$_{0.47}$In$_{0.53}$As is comparable to the best reported mobility achieved by LPE, VPE, and MBE for this composition. In Table 12.3, the change of mobility at 77 K with temperature shows that the material is apparently less compensated at 550 than at 650 °C.

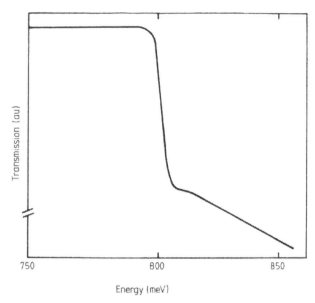

Fig. 12.4. Transmission spectrum measured for a Ga$_{0.47}$In$_{0.53}$As–InP heterojunction at 1.5 K (after Razeghi [1985]).

GaInAs layer	T_G (°C)	Thickness (μm)	$N_D–N_A$ (cm^{-3})	μ (cm^2·V^{-1}·s^{-1}) 300 K	μ (cm^2·V^{-1}·s^{-1}) 77 K	$\Delta a/a$ strained
1	550	6	1.8×10^{15}	11,800	48,000	$+5 \times 10^{-3}$
2	550	3.4	2.7×10^{15}	9,732	32,150	-4.9×10^{-3}
3	550	1	4.8×10^{15}	8,700	41,000	$+2.6 \times 10^{-3}$
4	650	0.78	1.2×10^{16}	8,100	17,600	$+4 \times 10^{-4}$
5	650	1.21	1.1×10^{16}	9,020	21,000	$+10^{-4}$
6	550	0.9	4.2×10^{15}	9,000	37,300	-10^{-3}
7	550	0.66	2.5×10^{15}	11,900	60,000	$+10^{-4}$
8	550	0.6	2×10^{14}	12,000	100,000	$+10^{-4}$
9	530	1	2×10^{14}	13,000	130,000	$+10^{-4}$

Table 12.3. Typical mobility for GaInAs/InP grown by MOCVD as a function of layer thickness, composition, electron carrier concentration and growth temperature.

Compensation and mismatch effects would reduce the mobility for a particular electron concentration. The effect of mismatch on mobility is pronounced. The variation of Hall mobility with lattice mismatch for a particular electron concentration $\left(N_D - N_A \simeq 5\times10^{15}\right)$ is plotted in Fig. 12.5. The mobility is independent of the lattice mismatch in the range $\left(\Delta\, a/a\right)_s \leq \pm2\times10^{-3}$, although Razeghi and Duchemin [Razeghi et al. 1983] have found a serious decrease for values of mismatch in the range $(\Delta\,a\,/\,a)_s \geq \pm5\times10^{-3}$. In general, layers that are in tension (positive mismatch) have lower mobilities than those in compression with the same, but negative, mismatch [Oliver and Eastman 1980].

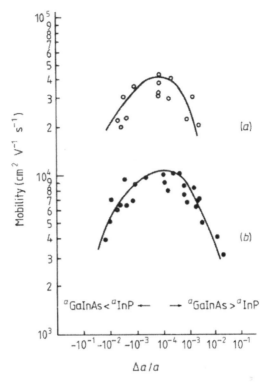

Fig. 12.5. *Variation of the measured Hall mobility with lattice mismatch for an electron concentration $N_D - N_A \simeq 2.5\times10^{15}$ at (a) 77 and (b) 300 K (after Razeghi et al. [1982]).*

The variation of Hall mobility as a function of temperature is shown in Fig. 12.6. The best mobility obtained was $12{,}000~cm^2V^{-1}s^{-1}$ at 300 K and $700{,}000~cm^2V^{-1}s^{-1}$ at 2 K, with electron carrier concentration of $N_D - N_A \simeq 2.5\times10^{14}~cm^{-3}$, for a thickness of 0.5 μm.

The carrier concentration profile evaluated from *C–V* measurements and the electrochemical profile for GaInAs layers over an InP substrate exhibit uniform doping along the layer thickness and an abrupt change in the carrier concentration (Fig. 12.7). GaInAs layers grown by LP-MOCVD can be doped *n*-type by using H_2S or SiH_4, and *p*-type by using DEZn.

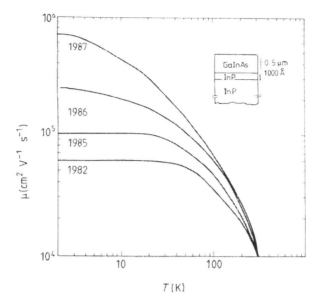

Fig. 12.6. Variation of the measured Hall mobility as a function of temperature.

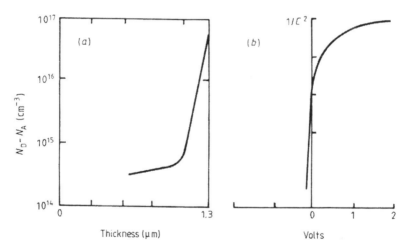

Fig. 12.7. (a) Carrier concentration profile evaluated by C–V measurements and (b) electrochemical profile (shown for a $Ga_{0.47}In_{0.53}As$ layer on an Sn-doped substrate) (after Razeghi et al. [1983]).

For the same growth temperature and the same flow-rate of DEZn, the hole carrier concentration of GaInAs layers is higher than that of InP. Fig. 12.8 shows the electrochemical profile of a GaInAs–InP DH layer for FET applications. The growth conditions and DEZn flow-rate for InP and GaInAs layers are the same, and we can see that the interfaces are sharp and that N_A–N_D is higher for GaInAs than for InP.

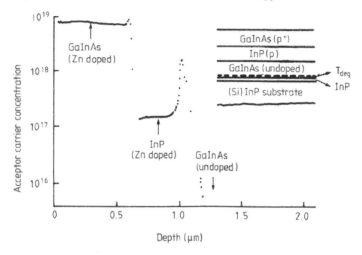

Fig. 12.8. Electrochemical profiles of typical doping interfaces for GaInAs–InP doped with DEZn for FET applications.

12.3. Optical and crystallographic properties, and impurity incorporation in GaInAs grown by MOCVD

Goetz et al. [1983] have performed a comparative study of some crystallographic and optical properties of *n*- and *p*-type $Ga_xIn_{1-x}As$/InP, $0.44 < x < 0.49$, grown by liquid-phase epitaxy (LPE), vapor-phase epitaxy (VPE) and metalorganic chemical vapor deposition (MOCVD). Three different substitutional acceptors were chemically identified as C, Zn and Si, and their binding energies determined. The presence and concentration of these acceptors in nominally undoped material depend on the crystal growth process and its parameters. X-ray measurements and photoluminescence spectra reveal clearly and in mutual agreement inhomogeneities of the composition and deviations of the $Ga_xIn_{1-x}As$ composition from the composition at lattice match. The energy of the direct band gap as a function of the Ga concentration and the bowing parameter (see Chapter 11) were derived with great precision and yielded values different from earlier experimental work. Details of the photoluminescence and absorption spectra such as free- and bound-exciton lines and pair bands were identified and provided relevant exciton and acceptor binding energies and the LO-phonon energy.

12.3.1. Sample preparation

The LPE layers were grown in a conventional horizontal sliding system within a lamp-heated furnace, which permits fast changes in temperature. After baking the In melt for about 20 h at 800 °C and the In + GaAs + InAs melt for about 10 h at 660 °C, growth was carried out at temperatures of about 640 °C by a step cooling technique with a temperature step of 3 °C. PH_3 was added to the H_2 ambient to avoid substrate degradation at elevated temperatures [Beneking et al. 1981]. Nominally undoped n-type layers with a thickness of up to 8 µm exhibited carrier concentrations in the low 10^{15} cm^{-3} range and mobilities of up to 11,000 cm$^2 \cdot$V$^{-1} \cdot$s^{-1} at 300 K and 43,000 cm$^2 \cdot$V$^{-1} \cdot$s^{-1} at 77 K (Table 12.4). The p-type Ga$_{0.47}$In$_{0.53}$As layers were grown from a Zn-doped solution with Zn mole fractions equal to 5×10^{-7} or more, leading to p doping levels lower than 10^{16} cm^{-3}.

The VPE material was grown using the In/Ga–AsH$_3$–HCl–H2 system [Kordos et al. 1981]. An In/Ga alloy source (Ga concentration 6%, source temperature 700 °C) was used for the generation of metal chlorides; it was shown earlier that this method has the advantage of improved control of film composition compared to the commonly used two-source system. High-purity HCl was generated by decomposition of AsCl$_3$ in a separate furnace (730 °C) and freeze-out of the arsenic. The deposition temperature was between 680 and 715 °C. Partial pressures of the chlorides were adjusted at 7.5×10^{-3} bar and of AsH$_3$ at 1.2×10^{-2} bar in the deposition zone. Undoped n-type layers with a thickness of up to 5.6 µm exhibited carrier concentrations in the low 10^{15} cm^{-3} range and mobilities of up to 9700 cm$^2 \cdot$V$^{-1} \cdot$s^{-1} at 300 K and 29,000 cm$^2 \cdot$V$^{-1} \cdot$s^{-1} at 77 K (Table 12.4).

Razeghi et al. [1982] have grown MOCVD undoped n-type layers at 550 °C with a thickness of 0.7 µm exhibiting carrier concentrations of 2×10^{14} cm^3 and mobilities of 12,000 cm$^2 \cdot$V$^{-1} \cdot$s^{-1} at 300 K, 155,000 cm$^2 \cdot$V$^{-1} \cdot$s^{-1} at 77 K and 700,000 cm$^2 \cdot$V$^{-1} \cdot$s^{-1} at 2 K. The details are indicated in Table 12.4.

12.3.2. Electrical, crystallographic, and optical experiments

Carrier concentrations and mobilities were measured by the conventional Van der Pauw technique (Table 12.4). Double-crystal x-ray diffraction was used to assess the lattice mismatch and the Ga concentration x_{Ga} of the epitaxial films. Fig. 12.9 shows diffraction spectra of VPE, LPE, and MOCVD samples.

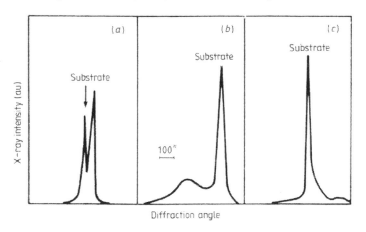

Fig. 12.9. X-ray double-crystal diffraction pattern of (a) LPE, (b) VPE, and (c) MOCVD Ga$_x$In$_{1-x}$As on an InP substrate. The LPE layer is better matched to the substrate lattice and shows less compositional fluctuations (smaller width of layer x-ray peak). Layer thicknesses: (a) 3.7 µm; (b) 1.4 µm; (c) 0.7 µm.

ID	Method	μm	x_{Ga}	$\Delta\theta$ arcsec	I_{lum} (au)	Doping	n,p (10^{15} cm^{-3}) 300 K	77 K	μ (cm^2·V^{-1}·s^{-1}) 300 K	77 K
8.81	LPE	6.8	0.485	61	25	und	n = 80	–	1,800	–
69.81	LPE	3.7	0.472	39	40	und	n = 1.7	–	11,000	43,000
86.81	LPE	4.1	0.462	61	–	Zn	p = 14	–	170	450
105.81	LPE	2.7	0.475	65	33	und	n = 1.8	–	10,000	36,000
107.81	LPE	8.2	0.473	44	5	und	n = 6.0	–	8,000	–
108.81	LPE	5.1	0.473	119	–	Zn	p = 4.4	–	169	–
113.81	LPE	2.5	0.481	33	–	Zn	n = 3.8	–	4,800	–
128.81	LPE	1.2	0.486	58	–	Zn	n = 16	–	5,200	–
129.81	LPE	0.9	0.464	54	13	und	n = 40	–	6,800	–
8.82	LPE	1.1	–	–	15	und	n = 2.4	3.1	9,600	29,000
9.82	LPE	6.1	0.471	46	24	und	n = 5.9	4.4	7,200	23,000
16.82	LPE	0.5	–	–	–	und	p = 4.0	–	110	–
17.82	LPE	0.6	0.475	43	24	und	n = 4.0	–	5,800	–
34.82	LPE	0.6	0.475	72	2.4	und	–	–	–	–
M 23	LPE	11.0	–	–	57	und	n = 5.0	–	10,000	31,000
M 65	LPE	5.5	0.473	33	–	Zn	p = 1.6	–	137	–
J 32	VPE	2.0	0.472	65	3.8	und	n = 2.6	2.9	9,800	27,000
J 37	VPE	3.0	0.457	144	1.8	und	n = 4.6	4.1	9,500	26,000
J 39	VPE	5.0	0.458	147	2.9	und	n = 4.7	4.2	9,700	29,000
JQ 49	VPE	0.8	0.441	272	1.4	und	n = 2.8	–	9,300	–
JQ 51	VPE	1.1	0.447	188	1.3	und	n = 2.4	–	8,500	–
JQ 55	VPE	0.7	0.457	251	0.8	und	n = 2.8	–	8,200	–
JQ 62	VPE	1.4	0.456	232	0.1	und	n = .67	–	9000	–
P 79	VPE	5.6	0.457	586	1.2	und	n = 17	13	6,200	15,000
PS 82	VPE	3.3	0.451	337	0.4	und	n = 3.4	2.9	9,400	27,000
58	MOCVD	0.9	0.477	–	1.0	und	n = 4.2	–	9,000	37,000
59	MOCVD	0.9	–	–	0.2	und	n = 5.1	–	8,600	33,000
85	MOCVD	6.0	0.431	344	5.8	und	–	–	12,000	48,500
235	MOCVD	0.7	–	–	39	und	n = 2.0	–	12,000	55,000
365	MOCVD	1.0	0.462	70	7.0	und	n =200	–	7,500	–
131	MOCVD	0.6	0.47	30	–	und	n = 0.2	0.2	12,600	15,5000

Table 12.4. Sample parameters (x_{Ga} is the Ga concentration, $\Delta\theta$ is the width of the x-ray peaks and I_{lum} is the luminescence intensity in relative units of the exciton line at 2 K and 30 mW cm^{-2} excitation density).

The strain due to the lattice mismatch between layer and substrate leads to a tetragonal distortion of the layer lattice with two different lattice constants $a_\parallel$ (parallel to the interface) and $a_\perp$ (in the growth direction) [Ishida et al. 1975 and Bortels et al. 1978] and eventually to an inclined growth of the layer relative to the substrate surface with a misorientation angle α [Nagi 1974 and Kawamura et al. 1974]. Goetz et al. [1983] determined the three quantities $(\Delta a_\parallel/a)_s$, $(\Delta a_\perp/a)_s$ and α from the x-ray double-crystal diffractometer measurements in the conventional way (described in Ishida et al. [1975], Bartels et al. [1978], Nagi [1974], and Kawamura et al. [1979]) by using the (1 1 3) and (1 1 5) Cu K_α Bragg reflections in different geometries (InP substrate: (0 0 1) oriented). They found $(\Delta a/a)_s$ for MOCVD samples in the range -2.6 to $+3.7 \times 10^{-3}$. In agreement with Kawamura and Okamoto's results [Kawamura et al. 1984], $(\Delta a_\parallel/a)_s$, was at least one order of magnitude smaller (coherent growth). Within experimental error (± 10 arcsec) they did not find any misorientation α. The Ga concentration x_{Ga} (Table 12.4) was determined from a_f, the lattice constant of the strain-free epilayer, by applying Vegard's law [Nahory et al. 1978]; a_f was estimated from $(\Delta a/a)_s$ using the relation $(\Delta a_f/a)_s \simeq 0.5 \times (\Delta a_\perp/a)_s$. This relation follows from Eq. (12.3), assuming $c_{11} \simeq c_{12}$ in $Ga_{0.47}In_{0.53}As$ [Marzin et al. 1984].

12.3.3. Results and discussion

Fig. 12.10 shows typical photoluminescence spectra of nominally undoped n-type LPE, VPE, and MOCVD samples at $T = 2$ K. Part of the information gained from these spectra is similar and in agreement with that gained from x-ray experiments. The width of the edge luminescence bands and their intensities are governed largely by the quality of the crystal. Clearly the MOCVD material has a much more homogeneous composition, and a more perfect surface, leading to a lower surface recombination velocity: in the LPE and MOCVD (no. 235) spectra the dominating near-band-gap line is considerably more intense and has a linewidth of about 2–3 meV ($P_{laser} = 30$ mW·cm^{-2}), much narrower than in VPE material (4–5 meV) (for details see Table 12.4).

In section 12.3.4, we shall show that the dominant high-energy line (Fig. 12.10) is due to bound excitons. Fig. 12.10 reveals that already small variations of the Ga concentration lead to a considerable shift of the band gap and thus of the luminescence lines. In section 12.3.5, we use this sensitive feature to derive the dependence of the gap energy at $T = 2$ K from the composition in the range $x_{Ga} = 45$–49%. Down to 25 meV below the exciton line, further bands can be seen. They will be identified as donor–acceptor pair recombinations (see section 12.3.6).

The mobility of sample 235 increases to 90,000 cm^2·V^{-1}·s^{-1} at 2 K, indicating the presence of a two-dimensional electron gas at the GaInAs/InP interface. The luminescence of such samples differs strongly from the luminescence of the MOCVD layer shown in Fig. 12.10 which has a 77 K mobility of 33,000 cm^2·V^{-1}·s^{-1}. The intensity is higher by more than one order of magnitude, and the near-gap spectrum does not show any indication of recombination of donor-bound electrons with holes bound to Si, Zn and C acceptors. An example of such a remarkable spectrum is given in Fig. 12.11.

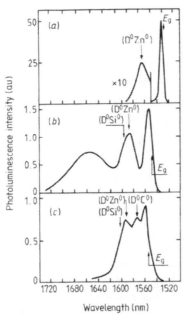

Fig. 12.10. Typical photoluminescence spectra of nominally undoped
$(n = 2$ to 5×10^{15} cm$^{-3})$ LPE (a), VPE (b) and MOCVD (c) $Ga_xIn_{1-x}As$. Small
changes in the Ga concentration x cause considerable shifts of the emission lines.
$T = 2$ K, $P = 30$ mW·cm^{-2}.

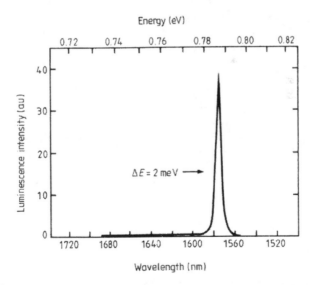

Fig. 12.11. Photoluminescence spectrum of a very-high-mobility MOCVD
$Ga_xIn_{1-x}As$ layer. The emission intensity and linewidth are comparable to the
spectra of the LPE layers (Fig. 12.10). No impurity bands occur. $T = 2$ K,
$P = 30$ mW·cm^{-2}, $x = 0.447$.

12.3.4. Exciton line

Similar to GaAs, several elementary recombination mechanisms may occur and cause near-band-gap emission lines: free excitons (X), excitons bound to shallow impurities (D ^{0}X, D $^+$X, A ^{0}X) and donor–valence band recombinations (D ^{0}h). Since only one rather broad near-band-gap line was observed under any circumstances, we estimated roughly the relative spectral positions at which these recombinations should be found in $Ga_{0.47}In_{0.53}As$. Table 12.5 shows the resulting values. The donor and free-exciton binding energies E_D and E_X used in this calculation were determined from the usual hydrogenic model:

Eq. (12.4) $E_{D,X} = [(\mu / m_0) / \varepsilon_0^2] \times 13.6 \text{ eV}$

where $\mu = m_e$ (conduction-band effective mass) or $1/\mu = 1/m_e + \frac{1}{2}(1/m_{hh} + 1/m_{lh})$ (exciton reduced mass), respectively. Using a dielectric constant $\varepsilon_0 = 13.7$ (linear interpolation between $\varepsilon_0(GaAs) = 12.56$ and $\varepsilon_0(InAs) = 14.74$) and the effective masses $m_e/m_0 = 0.041$, $m_{hh}/m_0 = 0.47$ and $m_{lh}/m_0 = 0.050$ (derived by Alavi et al. [1978] from magnetoabsorption measurements) the effective-mass donor and free-exciton binding energies of three shallow acceptors (identified as C, Zn and Si) are determined as $E_C = 13 \text{ meV}$, $E_{Zn} = 22 \text{ meV}$ and $E_{Si} = 25 \text{ meV}$. According to Haynes' rule the coefficients a, a', β, β' and β'', which are a measure of the localization energy of the exciton, have a value of about 0.1. The resulting relative spectral positions of the different recombination lines (Table 12.5) are displayed in Fig. 12.12. Two typical examples of near-gap lines (2 and 5 meV halfwidth) are superimposed. It is obvious from this comparison that the observed peaks are caused by a superposition of several of these recombination processes. Rather broad linewidths are typical of ternary and quaternary semiconductors and are attributed to compositional inhomogeneities and to alloy clustering [Pikhtin 1977].

Recombination process	Theoretical emission energy using $a, a', \beta, \beta', \beta'' \approx 0.1$	
X	$E_g - E_X$	$= E_g - 2.0 \text{ meV}$
D^{0}X	$E_g - (E_X + aE_D)$	$= E_g - 2.3 \text{ meV}$
D$^+$X	$E_g - (E_D + a'E_D)$	$= E_g - 3.3 \text{ meV}$
A$_1^0$X	$E_g - (E_X + \beta E_{A_1})$	$= E_g - 3.3 \text{ meV}$
A$_2^0$X	$E_g - (E_X + \beta' E_{A_2})$	$= E_g - 4.2 \text{ meV}$
A$_3^0$X	$E_g - (E_X + \beta'' E_{A_3})$	$= E_g - 4.5 \text{ meV}$
D^{0}h	$E_g - E_D$	$= E_g - 3.0 \text{ meV}$

Table 12.5. Predicted spectral positions of several near-band-gap emission lines in $Ga_{0.47}In_{0.53}As$. E_g is the energy of the fundamental gap; E_X, E_D and E_A are the free-exciton, donor and acceptor binding energies, respectively.

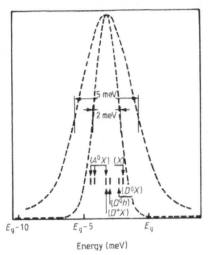

E_g-10 E_g-5 E_g

Energy (meV)

Fig. 12.12. Theoretically derived positions of several near-band-gap emission processes relative to E_g, (compare Table 12.5). Typical near-band-gap emission lines with widths of 2 and 5 meV are superimposed.

The excitonic nature of the dominant high-energy peak is proved by its excitation and temperature dependence. Whereas the impurity band shows a sublinear increase due to saturation effects at higher laser powers, typical of pair bands, the near-band-gap peak increases linearly with excitation over four orders of magnitude (Fig. 12.13). Detailed balance calculations show that this is the expected excitation dependence of bound-exciton recombination processes [Bimberg et al. 1970].

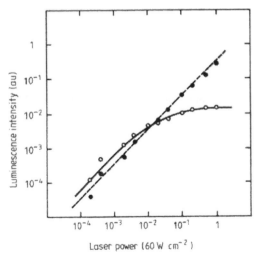

Fig. 12.13. Integrated intensity of the pair emission band (○) and the exciton line (•) as a function of the laser power density for an undoped $Ga_{0.47}In_{0.53}As$ sample at $T = 1.6$ K.

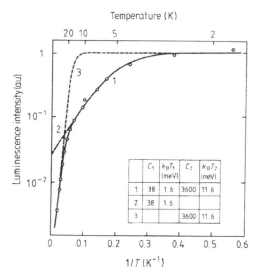

Fig. 12.14. Integrated intensity of the exciton line for Ga$_x$In$_{1-x}$As as a function of 1/T. The open circles are experimental points. The broken and full curves are theoretical fits with Eq. (12.5) assuming either one (broken) or two (full) activation energies $k_BT_{1,2}$ = 1.6, 11.6 meV, respectively.

The thermal activation energies of the exciton line are determined from the integrated intensity shown in Fig. 12.14 versus the reciprocal temperature. Fig. 12.14 shows that there are two competing ionization mechanisms with different activation energies. The measured temperature dependence is fitted with a theoretical expression derived by Bimberg et al. [1971]:

Eq. (12.5) $I_T / I_0 = \left[1 + C_1 \exp\left(T_1 / T\right) + C_2 \exp\left(T_2 / T\right)\right]^{-1}$

Here I_T and I_0 are the emission intensities at the temperatures T and $T = 0$ K, respectively, c_1 and c_2 are ratios of degeneracies, and k_BT_1 and k_BT_2 are activation energies. Excellent agreement between the measured points and Eq. (12.5) is achieved with the parameters given in the Fig. 12.12. The resulting activation energies are k_BT_1 = 1.6 ±0.1 meV and k_BT_2 = 11.6 ±2 meV. In a hydrogenic model the thermal activation is equal to three-quarters of the binding energy (the ratio between the $n = 1$ and $n = 2$ binding energies of a Rydberg series). Thus the activation energy of k_BT_1 = 1.6 meV yields a binding energy of 2.1 meV, in excellent agreement with the theoretical binding energy of the free exciton (2.0 meV) derived above. The second activation energy in Fig. 12.14 (although less precise) is almost one order of magnitude larger and can be unambiguously identified with the excitation of a bound hole.

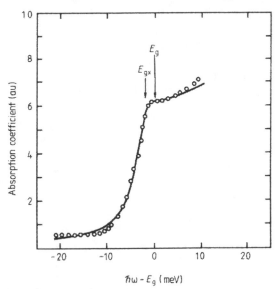

Fig. 12.15. Absorption spectrum of an undoped LPE Ga$_x$In$_{1-x}$As layer at T = 77 K. The open circles are experimental points. The full curve is a theoretical one from Eq. (12.6) (parameters given in the text). E$_{gx}$ is the position of the exciton energy gap derived from the fit. X = 0.472.

Absorption measurements confirm the excitonic nature of the near-band-gap fundamental absorption edge. In Fig. 12.15 an absorption spectrum (circles) is fitted with a theoretical expression of the absorption coefficient (full curve) based on Elliott's [1957] theory of allowed, direct transitions between non-degenerate parabolic energy bands, including exciton effects in the approximation of the hydrogenic model:

Eq. (12.6)
$$\alpha(\hbar\omega) = A\left[\sum_{n=1}^{\infty} 2\frac{R}{n^3}\delta\left(\hbar\omega - E + \frac{R}{n^2}\right) + \frac{U(\hbar\omega - E_g)}{1 - \exp\left\{-2\pi\left[R/(\hbar\omega - E_g)\right]^{1/2}\right\}}\right]$$

where A contains the transition-matrix element, R is the excitonic Rydberg and $U(x) = 1$ for $x \geq 0$ and $U(x) = 0$ for $x < 0$. The first term in Eq. (12.6) describes transitions into the discrete states of the exciton and the second term transitions into the continuum. The temperature broadening of the lineshape has been taken into account by convoluting Eq. (12.6) with a Lorentzian.

The best agreement between theory and the experimental spectrum is found for the width $\Gamma = 2.7$ meV, $E_g = 0.813$ eV and $E_{gx} = 0.811$ eV (Fig. 12.15). So again we find for the free-exciton binding energy a value of $E_X = 2$ meV in agreement with the photoluminescence results and the hydrogenic model.

Careful examination of the absorption in Fig. 12.15 reveals a faint shoulder in the ascent of the absorption coefficient, 2–3 meV below the main edge. This shoulder might be caused by a bound exciton.

At high excitation intensity several additional lines appear in the photoluminescence spectrum (Fig. 12.16) separated from each other and the exciton line by 32 meV, respectively. The donor–acceptor pair emission band at 0.793 eV is completely saturated at these laser intensities (compare Fig. 12.13). The additional bands can be attributed to LO-phonon replicas of the exciton line. The LO-phonon energy of 32 ± 0.5 meV is in good agreement with earlier values by Shah et al. [1980], who fitted spectra of hot carriers with a value of 34 meV, by Pinczuk et al. [1978], who found 33.8 meV (Raman effect measurements), and by Brodsky and Lucovsky [1968], who found two LO-phonons with the energies of 32.9 and 29.1 meV (infrared reflectivity).

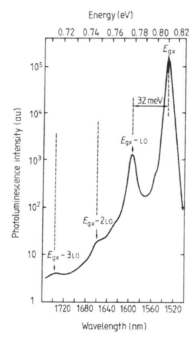

Fig. 12.16. *Low-temperature photoluminescence spectrum of the LPE sample of Fig. 12.15 at high excitation density: several LO-phonon replicas of the exciton line appear; the donor–acceptor band at 0.793 eV is completely saturated and hardly visible.* $Ga_xIn_{1-x}As$ *(undoped),* $T = 2$ *K,* $P = 60$ *W·cm*$^{-2}$.

12.3.5. Determination of E_g (x_{Ga}) near $x_{Ga} = 47\%$

A small mismatch between the lattice constants of the layer and the InP substrate ($\Delta a/a$) or corresponding small deviations of the Ga concentration (x_{Ga}) from the optimum value for complete lattice match (46.8%) lead to considerable change in the band gap. For a series of LPE and VPE samples, Goetz et al. [1983] derived the band gap at $T = 2$ K from the exciton emission-line position using Fig. 12.12 by adding the binding energy of 2.1 meV. In Fig. 12.17 the resulting values are displayed as a function of the Ga concentration x_{Ga} which was determined from the x-ray measurements described in section 12.3.4.

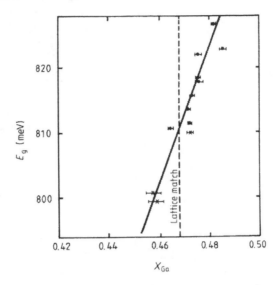

Fig. 12.17. *Band-gap energy at $T = 2$ K for $Ga_xIn_{1-x}As$ as a function of the Ga concentration X_{Ga} of several LPE (•) and VPE (X) samples. The error bars are due to the compositional inhomogeneities. The full line is a fit with* Eq. (12.7).

The varying error bars for x_{Ga} are due to varying compositional fluctuations in the layers and were extracted from the width of the x-ray layer peaks. The straight line in Fig. 12.17 is a fit to the experiment with the expression

Eq. (12.7) $$E_g(x) = E_g^{InAs} + \left(E_g^{GaAs} - E_g^{InAs} \right) x - Cx(1-x)$$

with $E_g^{InAs} = 0.4105$ eV and $E_g^{GaAs} = 1.5192$ eV ($T = 2$ K) [Landolt-Boerstein 1982]. The bowing parameter $C = 0.475$ describes the deviation of $E_g(x)$ from linearity between the two extremes E_g^{InAs} and E_g^{GaAs}. Thus Eq. (12.7) can be rewritten

Eq. (12.8) $$E_g(x)\big|_{2K} = 0.4105 + 0.6337x + 0.475x^2$$

For $x_{Ga} = 0.468$, the composition of optimum lattice match, Eq. (12.8) yields a gap energy of $E_g = 0.811$ eV ($T = 2$ K), which is probably the most precise value reported. This value is smaller than the value given by Towe [1982]: $E_g = 0.821$ eV.

12.3.6. Donor–acceptor pair recombination

Three bands are observed (8, 17 and 20 meV) below the near-band-gap line and are identified as donor–acceptor pair transitions (Fig. 12.18). Fig. 12.18 shows a spectrum of an MOCVD sample in which all three acceptors occur simultaneously. In order to determine the binding energies of the acceptors, Goetz et al. [1983] tried to fit the pair bands with a theoretical lineshape as derived by Lorentz et al. [1968] for GaP:

Eq. (12.9) $I(R) \sim R^4 \exp\left[-4\pi / 3 N_D R^3\right] \exp\left(-2R / a_D\right)$

This expression describes the intensity of saturated distant pair emission as a product of the number of "isolated" pairs and the transition probability for radiative recombination of the electrons and holes. R is the distance between donor and acceptor and is connected with the emission energy $h\omega$, the band-gap energy E_g and the donor and acceptor binding energies E_D and E_a by

Eq. (12.10) $h\omega = E_g - \left(E_A + E_D\right) + e^2 / \left(\varepsilon_0 R\right)$

Here the last term (Coulomb interaction term) takes into account the electrostatic interaction between the ionized donor and acceptor in the final state of the transition. The factor R^4 in Eq. (12.9) accounts for the number of donor sites per unit energy range; the first exponential (nearest-neighbor factor) describes the probability that an acceptor has no donor nearer than R, and the second exponential describes the overlap of the hole and electron, assuming that this overlap is predominantly determined by the effective Bohr radius a_D of the donor-bound electron. Further parameters are the donor concentration N_D and the static dielectric constant ε_0.

It turned out, however, that Eq. (12.9) yields lineshapes that are much narrower than the observed pair bands. As in the case of the exciton line, considerable broadening, familiar in solid solutions, has to be taken into account: microscopic fluctuations of the composition lead to a variation of the impurity energy levels. Fig. 12.19 shows an experimental pair spectrum (circles) and a theoretical lineshape according to Eq. (12.9) (full curve). The broadening effect is taken into account by convoluting Eq. (12.9) with a Gaussian distribution of width σ (broken curve). The donor concentration N_D is set equal to the carrier concentration at room temperature. For the donor Bohr radius a_D and the donor binding energy E_D we assume hydrogenic values with $a_D = (\varepsilon_0 \, m_0/m_e) \times 0.53$ Å ($\varepsilon_0 = 13.7$, $m_0/m_e = 0.041$, see section 12.3.4) and $E_D = 3.0$ meV (according to Eq. (12.1)), respectively. For the gap energies E_g we use the values determined in section 12.3.5. E_A and σ serve as fitting parameters. The acceptor binding energies derived from fits on 25 samples are $E_{A1} = 13 \pm 1$ meV, $E_{A2} = 22 \pm 1$ meV and $E_{A3} = 25 \pm 1$ meV. Goetz et al. [1983] identify these acceptors as $A_1 = C$, $A_2 = Zn$ and $A_3 = Si$. The shallowest acceptor C occurs nearly exclusively in MOCVD samples (Fig. 12.10). Furthermore secondary-ion mass spectrometry (SIMS) analysis shows that the LPE and VPE samples contain Zn and Si. So the two other bands must be attributed to these impurities. Comparison with GaAs (Fig. 12.20) suggests that Zn is the shallower one.

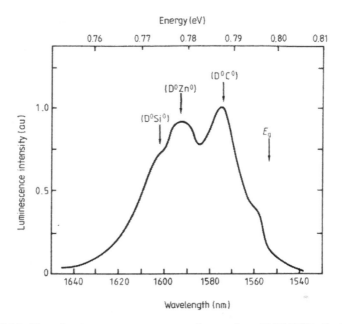

Fig. 12.18. *Photoluminescence spectrum of an undoped MOCVD $Ga_xIn_{1-x}As$ sample at $T = 1.8$ K. Three pair bands due to C, Zn and Si acceptors are resolved. The near-band-gap exciton line appears only weakly, $x = 0.457$.*

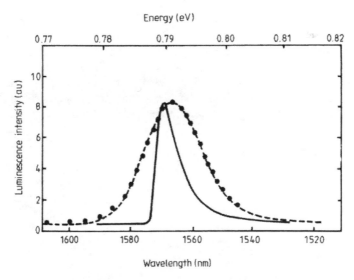

Fig. 12.19. *Pair band of an undoped $Ga_xIn_{1-x}As$ layer at $T = 2$ K (circles). Good agreement with the theoretical lineshape (Eq. (12.9) is achieved by convoluting the theoretical shape (full curve) with a Gaussian of width 7.6 meV (broken curve).*

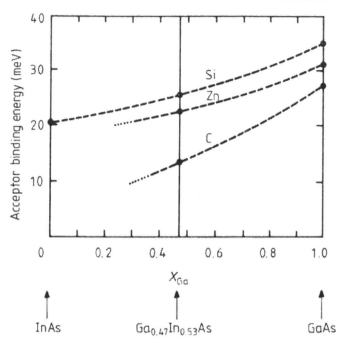

*Fig. 12.20. Si, Zn and C acceptor binding energies in GaAs and InAs (after
Beneking et al. [1981]) and their respective values for $Ga_{0.47}In_{0.53}As$.*

12.3.7. Zn-doped samples

Zn-doped samples (Fig. 12.21) show one broad intense band in the spectral region
where Goetz et al. [1983] found donor–Zn acceptor pair emission in undoped
samples, in agreement with the earlier LPE results [Chen and Kim 1981]. The near-
band-gap exciton line appears only very weakly. They found another weak emission
32 meV below the intense pair band, which can be identified as a LO phonon
replica of this band. The spectra suggest that these samples are strongly
compensated and have impurity concentrations orders of magnitude larger than the
free-carrier concentration.

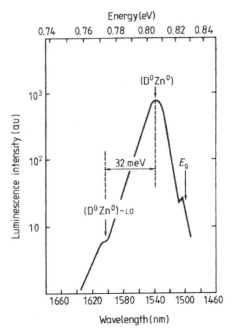

Fig. 12.21. *Photoluminescence spectrum of a lightly Zn-doped p-type $Ga_xIn_{1-x}As$ layer at $T = 2$ K. A strong donor–Zn acceptor pair emission with its LO-phonon replica can be seen.*

12.3.8. Deep Fe and intrinsic defect levels in $Ga_{0.47}In_{0.53}As/InP$

Only a few reports exist on deep traps in $Ga_{0.47}In_{0.53}As$ [Forrest and Kim 1982, Yagi et al. 1983, Rao et al. 1983, and Tromer and Albrecht 1983]. Among these only Yagi et al. [1983] and Rao and Bhattacharya [1983] report on layers grown on semi-insulating InP:Fe substrates. Rao and Bhattacharya [1983] tentatively proposed the existence of a trap 0.56 eV above the valence band, in contrast to our results. Yagi et al. [1983] observed an intense luminescence band at 0.66 eV for liquid-phase epitaxy layers grown on both Sn- and Fe-doped substrates. The luminescence intensity was found to depend strongly on lattice mismatch.

Goetz et al. [1985] have used low-temperature photoluminescence and deep-level transient spectroscopy (DLTS) to identify and characterize a deep Fe trap in high-purity undoped n-type and slightly Sn- and Zn-doped $(n \leq 10^{17}\ \mathrm{cm^{-3}})$ $Ga_xIn_{1-x}As$ $(x \simeq 0.47)$, epitaxially grown on InP. The InP substrate material was either semi-insulating Fe-doped or n-type S-doped. Results on samples grown by LPE, VPE and MOCVD (see previous section) are compared with each other and related to the different growth processes.

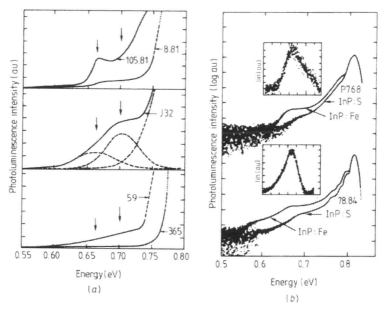

Fig. 12.22. (a) Typical deep-level photoluminescence spectra of pure LPE, VPE and MOCVD GaInAs–InP samples. The arrows denote the positions of the 0.66 and 0.70 eV bands. The broken curves are Gaussians. (b) Near-gap and deep-level photoluminescence spectra of LPE samples from different laboratories grown on InP:Fe and InP:S, respectively. The emission band at 0.66 eV is absent in the layers grown on InP:S. The insets show the 0.66 eV band alone, extracted from the spectra by subtraction of the $Ga_{0.47}In_{0.53}As/InP:S$ spectra (multiplied by a scaling factor) from the $Ga_{0.47}In_{0.53}As/InP:Fe$ spectra, respectively. For both (a) and (b), $T = 2$ K, $P = 60$ $W \cdot cm^{-2}$.

Fig. 12.22(a) shows typical deep-level photoluminescence spectra taken at 2 K and an excitation density of 60 $W \cdot cm^{-2}$ of pure LPE, VPE and MOCVD samples all grown on InP:Fe substrates. The intensity of the bands shown in Fig. 12.22(a) is 3–5 orders of magnitude weaker than the near-gap emission band (not shown in Fig. 12.22(a); see previous section) [Chemla et al. 1984]. The spectra exhibit two bands at 0.66 eV and at about 0.70 eV. The intensity of these bands from the 2.7 μm thick sample no. 105.81 is much larger than the intensity from the 6.8 μm thick sample no. 8.81. The precise position of the 0.70 eV band cannot be derived exactly due to the superposition with phonon sidebands of the near-gap emission. The 0.66 and 0.70 eV bands are absent in the luminescence spectra of InP:Fe substrates. More than 40 layers grown on Fe-doped InP substrates have been investigated and almost every one shows these bands. One can conclude that the luminescence definitely originates from the $Ga_{0.47}In_{0.53}As$ layers. No dependence of the luminescence intensity on doping or lattice mismatch is observed. Thus the new bands are not shallow-impurity- or interface-related.

The 0.66 eV band is completely absent in layers grown on S-doped InP substrates (Fig. 12.22(b)) whereas the 0.70 eV band is also present in the spectra of these layers. The spectra of the two samples (P768 and 78.84) in Fig. 12.22(*b*) can be compared quantitatively with each other. In both cases two layers were grown under identical conditions from the same melt, but on different substrates, as indicated. The insets of Fig. 12.22(*b*) show nicely the 0.66 eV band, which is extracted by subtraction of the spectra of the layers grown on different substrates.

The deep trap emission bands of the MOCVD layers are always extremely weak, broad and structureless; MOCVD layers of larger thickness do not show the bands at all (sample no. 59 in Fig. 12.22(*a*) has a thickness of 0.7 µm, whereas no. 365 has a thickness of 1.1 µm).

Obviously there exists one Fe-related trap with an energy depth of 150 meV and another center that is not related to Fe at a depth of 110 meV. The Fe center is definitely introduced by out-diffusion from the substrate during the growth process. This conclusion is based on a comparison of a series of samples grown (a) at identical temperatures but with increasing thickness and (b) at increasing temperature but with similar thickness. With decreasing growth temperature and/or increasing layer thickness the 0.66 eV luminescence intensity drops. The 0.70 eV band is most likely due to an intrinsic defect of $Ga_{0.47}In_{0.53}As$. It is very weak in layers of high crystallographic perfection (monitored by x-ray experiments) showing high mobilities. In LPE layers grown at a temperature as low as 500 °C and VPE layers the 0.70 eV luminescence intensity is comparatively large.

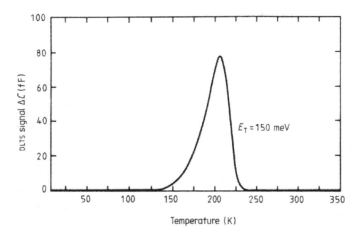

Fig. 12.23. DLTS spectrum of a MIS structure on a p-type LPE $Ga_{0.47}In_{0.53}As$:Zn layer grown on an InP:Fe substrate. $P = 1.4 \times 10^{17}$ cm^{-3}.

DLTS measurements were performed on metal–insulator–semiconductor (MIS) structures fabricated on *p*-type LPE samples. SiO_2 is deposited by chemical vapor deposition at 350 °C. Aluminum is used as insulator contact. The carrier concentration of the $Ga_{0.47}In_{0.53}As$ layer is $p = 1.4 \times 10^{17}$ cm^{-3}. The DLTS measurements reveal one trap only (Fig. 12.23) due to the inherently lower resolution of this method. The sign of the DLTS signal shows that this trap is correlated with the valence band. An Arrhenius plot yields an activation energy of 150 meV, in excellent agreement with the position of the 0.66 eV band found in photoluminescence. Depth profiles of this level agree with the above interpretation. Similar results were obtained on MOCVD-grown GaInAs layers.

The dependence of the integrated luminescence intensity of the 0.66 eV band on laser power and temperature has been studied in detail. The intensity increases linearly with increasing excitation power over three orders of magnitude up to 100 $W \cdot cm^{-2}$. No saturation effects can be observed at higher laser powers. With increasing temperature the 0.66 eV band disappears at 40–50 K and the 0.70 eV band remains visible. The temperature dependence of the luminescence intensity of the 0.66 eV band discloses two competing ionization mechanisms with different activation energies. The measured temperature dependence is fitted using Eq. (12.5). The resulting activation energies are $k_B T_1 = 2.5 \pm 0.5$ meV and $k_B T_2 = 100 \pm 20$ meV. In a hydrogenic model the thermal activation is equal to three-quarters of the binding energy (the ratio between the $n = 1$ and $n = 2$ binding energies of a Rydberg series). Thus the activation energy of $k_B T_1 = 2.5$ meV yields a binding energy of 3.3 meV, which is in reasonable agreement with the theoretical binding energy $E_D = 3.0$ meV of a hydrogenic shallow donor [Goetz et al. 1983]. The second activation energy can be unambiguously identified with the excitation of a bound hole of the 150 meV deep trap. From these results we conclude that at the lowest temperatures the 0.66 eV band is due to a shallow donor–Fe acceptor transition:

Eq. (12.11) $D^0 + Fe^{3+} + \rightarrow D^+ + Fe^{2+} + \hbar\omega_{photon}$

is a model of the underlying process. At higher temperatures the band is due to a free electron–Fe acceptor transition. The emission energies of these two transitions differ by less than the already small shallow donor binding energy of 3 meV. This does not lead to a perceptible difference in the spectral position of the rather broad ($\simeq$ 40 meV) and featureless 0.66 eV band.

So, a deep Fe-related trap at E_v+150 meV in $Ga_{0.47}In_{0.53}As/InP$:Fe layers and an intrinsic trap at a depth of 110 meV in $Ga_{0.47}In_{0.53}As$ layers grown on both InP:Fe and InP:S substrates are observed and characterized using photoluminescence and DLTS. The Fe trap is absent in layers grown on Fe-free InP substrates and is obviously caused by Fe diffusing out of the substrate during the growth process. The lower growth temperature of the MOCVD samples (550 °C) leads to a lower out-diffusion compared to LPE (640 °C) and VPE ($\simeq$ 700 °C) samples.

12.3.9. Comparison of transport, optical, and crystallographic properties

High mobility of the electrons is advantageous for a number of devices like the two-dimensional electron-gas field-effect transistor (TEGFET). This study shows that decreasing acceptor-related luminescence—monitored by the intensity of acceptor-related luminescence—causes increasing mobilities, in particular at 77 K and lower, where ionized-impurity scattering is most important. The link between decreasing acceptor concentration and increasing mobility is most striking for the MOCVD layers. Layers that no longer contain C as monitored by the luminescence spectra show a really dramatic increase of mobility at temperatures below 77 K, indicating two-dimensional (2D) electron conduction at the interface. Layers grown with Air Liquide starting material, showing lower mobilities and strong impurity lines in the luminescence spectra, still exhibit indications of 2D mobility. The same layers show a luminescence yield improved by almost two orders of magnitude, indicating more perfect control of morphology and interfaces by MOCVD.

Double-crystal diffractometry, transport, low-temperature photoluminescence and absorption measurements of high-purity n-type and lightly Zn-doped p-type $Ga_xIn_{1-x}As/InP$ layers $(x \simeq 0.47)$ grown with different epitaxial techniques (LPE, VPE, MOCVD) are presented. The MOCVD samples have properties superior to the LPE samples. Peaks related to free and bound excitons are found in the photoluminescence and absorption spectra of the undoped samples. The binding energy of the exciton is determined as 2.1 ± 0.1 meV, in agreement with a hydrogenic theory. LO-phonon replicas of the exciton line establish a LO-phonon energy of 32 ± 0.5 meV. The exact dependence of the energy gap at $T = 2$ K from the solid-solution composition in the range $x_{Ga} = 45–49\%$ is determined and yields a bowing parameter $C = 0.475$. A gap value of $E_g = 0.811$ eV at optimum lattice match is found. Data on donor–acceptor pair transitions observed in the photoluminescence spectra are combined with secondary-ion mass spectrometry data to identify different acceptors: C, Zn and Si. Their binding energies are 13 ± 1.22, 22 ± 1 and 25 ± 1 meV, respectively. C is the dominant acceptor in most MOCVD samples, but is hardly present in the LPE and VPE samples, whereas Si and Zn are unintentionally both present in LPE, VPE and most MOCVD samples. The Zn-doped p-type samples show only a weak exciton line but a broad donor–Zn acceptor pair transition band with a LO-phonon replica.

12.4. Shallow p$^+$ layers in GaInAs grown by MOCVD by mercury implantation

For some devices, especially for structures like junction FET, thin and abrupt p^+–n junctions are required. Diffusion is a widely practiced technique in realizing p-type layers. Zn- and Cd-diffused junctions have been obtained, but thin p-type layers are difficult to realize, since these dopants have high diffusion constants. It has been shown [Vescan et al. 1982 and Vescan et al. 1984] that Be and Cd ion implantation can also give p-type layers, but due to diffusion during annealing, it is not possible to have thin p-type layers. For example, after an anneal at 675 °C for 25 s, a 100 keV Be implantation gives a junction depth of approximately 1 μm [Tell et al. 1984]. The classical acceptor impurities in III–V materials are Be, Zn and Cd. All these impurities show a relatively high diffusion. We have tried to look for other acceptor impurities that would diffuse slowly; L'Hanidon et al. [1985] have investigated how mercury behaves in terms of the type of dopant, electrical activity and diffusion.

Growth of 1 μm depth of $Ga_{0.47}In_{0.53}As$ layers has been obtained on $\langle 1\,0\,0 \rangle$ S- and Fe-doped substrates. All epitaxial layers were unintentionally doped. The mercury beam came from a Penning source. An energy of 700 keV (Hg^{2+}) has been used with a maximum beam intensity of 100 nA·cm^{-2}. The implantation was performed at room temperature and the samples were tilted 7° from the ⟨1 0 0⟩ axis. The projected range of 700 keV Hg^{2+} ions was about 1200 Å. After implantation the samples were annealed capless in a flowing high-purity 0.4% PH_3/H_2 mixture. All the anneals were performed at 700 °C for 10 min. Annealing did not modify the chemical surface, and the mirror-like aspect was retained.

The surface carrier concentration N_s of mercury-implanted layers was measured by the Hall effect. N_s is the result of Hall experiments and is a good approximation of the surface carrier concentration. All implanted layers were found to be p-type after annealing. The results are summarized in Fig. 12.24, which is a plot of N_s as function of the implantation dose φ ranging from 10^{12} to 10^{14} ions·cm^{-2}. For low doses, $N_s = \varphi$, indicating the full electrical activity of the implanted species. For higher doses, only a part of the mercury dopant is electrically activated. For an Hg^+ dose of 10^{14} ions·cm^{-2}, we have obtained 3×10^{14} holes·cm^{-2}.

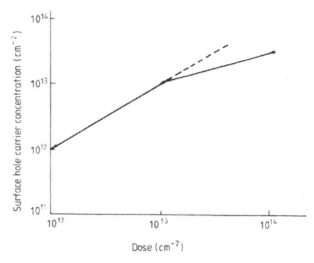

Fig. 12.24. Surface carrier concentration plotted against mercury (Hg^+) dose.

Fig. 12.25 shows typical results of the in-depth hole carrier distribution obtained by Hall measurements on a mesa etched Hall pattern in conjunction with chemical stripping of thin layers. The electrical profile has a similar shape for Hg^+ doses of 10^{13} and 10^{14} ions·cm^{-2}. Near the surface a highly p^+-doped layer was obtained, and the hole carrier concentration should be about 10^{19} holes·cm^{-3}. Hole concentration decreases as a function of depth. The junction is located at 2000 ± 200 Å for an Hg^+ dose of 10^{13} ions·cm^{-2} and at 2300 ±200 Å for an Hg^+ dose of 10^{14} ions·cm^{-2}. These electrical profiles show that impurities do not diffuse towards the bulk, but that there is a high surface pile-up after annealing at 700 °C. Electrical profiles are the only indication of the non-diffusion of mercury in GaInAs. Unfortunately, it was not easy to measure how the mercury and the matrix elements interfered. On InP layers, SIMS analysis showed that mercury diffuses towards the surface and does not diffuse towards the bulk after a 700 °C anneal. The diffusion of mercury impurities can be expected to be equivalent in InP and InGaAs.

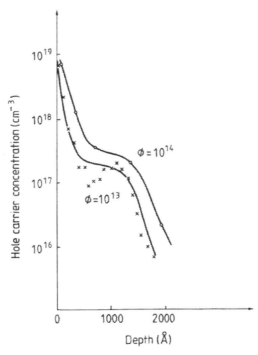

Fig. 12.25. Hole carrier distribution for Hg$^+$ doses of $\varphi = 10^{13}$ and 10^{14} ions·cm^{-2}.

A few p^+–n diodes were made from Hg-implanted epilayers grown on n^+ InP substrates. The structure was Au–Zn/Au/implanted p^+-GaInAs/8000 Å $n(10^{16})$-GaInAs/n^+-InP substrate/Au–Ge. Mesa diodes of area 3×10^{-4} cm^2 were fabricated. Fig. 12.26 shows the current–voltage characteristic of a p^+–n diode. *I–V* characteristics show a soft reverse breakdown at about 8 V. This soft breakdown can be due to tunneling, as was previously shown by Ito et al. [1981] in the case of GaInAs homojunction diodes with carrier concentrations of more than 10^{15} cm^{-3}. The leakage current is 5×10^{-3} A cm^{-2} at $V_R = -1$ V. Capacitance–voltage (C–V) data for the diodes yielded a linear plot of C^{-2} against V_R, indicating abrupt junctions.

So, mercury behaves as an acceptor dopant impurity in GaInAs and does not diffuse towards the bulk, whereas other acceptor impurities (Be, Zn, Cd) do. Mercury implantation should be a suitable method to obtain shallow p^+–n junctions on Ga$_{0.47}$In$_{0.53}$As.

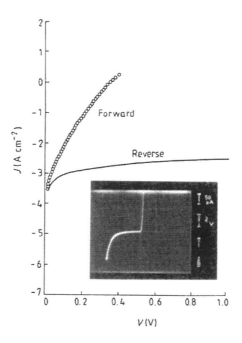

Fig. 12.26. Semilogarithmic plot of I–V characteristics of Hg-implanted n-GaInAs diode (dose = 10^{13} cm^{-2}). Inset is linear plot of I–V characteristic.

12.5. GaInAs–InP heterojunctions: Multiquantum wells and superlattices grown by MOCVD

In 1970 Esaki and Tsu [1970] envisaged solid-state superlattice semiconductors in which two materials with different electronic properties are interleaved in thin layers by periodically depositing two semiconducting materials, in contrast to natural superlattices (like SiC, which possesses various polytypical forms [Verma 1953]), whose formation is difficult to control and is not well understood. The man-made superlattice can be achieved with predetermined prescriptions in terms of thickness, carrier concentration, strength of the periodic potential and number of periods. The superlattice potential subdivides the Brillouin zone of the bulk semiconductor into mini-zones, resulting in allowed mini-bands of electron transmission separated by forbidden bands. These mini-bands are relatively narrow on both energy and wavevector scales. The superlattice structure determines the mini-bands, which in turn dictate the transport and optical properties. Superlattices consisting of thin semiconductor layers were predicted to have interesting physical properties such as the following:

(1) Quantum size effect (QSE). This effect occurs when the quantum-well thickness is of the order of the electron de Broglie wavelength. QSE becomes

a dominant feature, changing the bulk, three-dimensional, behavior into a quasi-two-dimensional one [Dingle et al. 1974].

Quantum tunneling. Tunneling phenomena across barriers open the way to many fascinating effects, the most eagerly expected being the Bloch oscillator. Esaki and Tsu [1970] predicted Bloch oscillations in a compositional superlattice before superlattices existed. The superlattice should generate microwave radiation with wavelengths of less than 1 mm [Esaki and Tsu 1970].

Experimentally the quantization of energy levels in superlattices was shown directly by the optical measurements of Dingle et al. [1974] and Chang et al. [1977]. The two-dimensionality of carrier transport in quantum wells (QW) and superlattices has been demonstrated in *n*-doped GaAs–GaAlAs superlattices by magnetoresistance quantum oscillations. Increase of two-dimensional carrier mobility by modulated doping, which separates donor atoms and charge carriers, was proposed and demonstrated by Störmer et al. [1979].

Optical modulation devices. GaAs–GaAlAs superlattices show large non-linear optical effects associated with saturation of the room-temperature exciton resonances and also large electroabsorptive effects around the band-gap energy [Razeghi et al. 1984]. A new type of optoelectronic device, SEED (self-electro-optic effect device), was demonstrated using this effect [Wood et al. 1984]. It relies on modulation and detection within the same device and one possible mode of operation gives optical bistability at very low powers.

Another recent advance is the ability to grow strained layers, superlattices in which the crystal lattices of the two materials are not very closely matched. In that case there is a built-in strain in each layer.

12.5.1. Growth technique

Semiconductor multilayers and superlattices of III–V, II–VI and IV compounds can be grown by LPE, VPE, MBE and MOCVD.

Multiple-layer lasing structures with GaInAsP ($\lambda = 1.1$ μm) layers as thin as 150 Å, separated by InP layers, have been reported using LPE [Rezek et al. 1980].

The first semiconductor strained superlattices (GaAs–GaAs$_x$P$_{1-x}$) were grown using VPE [Blakesley and Aliotta 1970 and Alferov 1967]. MBE is the most highly developed technique for multiple- and thin-layer crystal growth. Composition modulations down to alternate single atomic-layer thicknesses can be produced with MBE.

There have been a number of reports [Gossard 1983, Cho 1979, Chang et al. 1973, Tsang and Cho 1977, and Massies et al. 1987] of excellent MBE growth of heterojunctions and superlattices of III–V and II–VI compounds. MBE has been used for growing semiconductor superlattices such as GaAs–GaAlAs, InAs–GaSb, InP–InGaAs, GaSbAs–GaSbAlSb, Ge–GaAs, Si–SiGe, CdTe–HgTe and PbTe–PbGeTe.

The disadvantages of MBE are that it is an expensive and complex process, and that there are difficulties with the growth of phosphorus-containing materials.

The evolution of MOCVD as a technique for the growth of high-quality layered semiconductor structures (e.g. GaAs–GaAsP, GaAs–GaInAs, GaInAs–InP, GaInAsP–InP and GaAs–GaAlAs) has opened the way to the use of superlattice effects in new devices with many potential commercial applications. One of the major advantages of MOCVD is the ability to control solid composition easily in such systems as $Ga_xIn_{1-x}As_yP_{1-y}$–InP. The MOCVD growth of GaInAsP–InP superlattices, quantum wells (as thin as 8 Å) and heterojunctions exhibiting two-dimensional electron-gas behavior was first reported by Razeghi et al. [1983].

High-quality InP/$Ga_{0.47}In_{0.53}$As heterojunction, multiquantum-well and superlattice layers have been obtained using TEIn (or TMIn) and TEGa for In and Ga sources and pure AsH_3 and PH_3 for group V arsenic and phosphorus elements. Pure H_2 or a mixture of H_2 and N_2 is used as carrier gas. The growth temperature was between 500 and 650 °C and the growth pressure was 76 Torr. The quality of epitaxial layers depends on the purity of the starting material. The highest purity of TEIn has been supplied by SMI. Highest purities of TEGa and TMIn have been obtained from α-Ventron (see Chapter 11). Good-quality AsH_3 has been achieved using a passivated Matheson product. The growth conditions and characterization of these layers grown by MOCVD have been reported in detail elsewhere [Razeghi and Duchemin 1983].

12.5.2. *Structural characterization of GaInAs–InP quantum wells grown by MOCVD*

Many high-quality InP/$Ga_{0.47}In_{0.53}$As heterostructures such as modulation-doped heterojunctions (MDH), multiquantum wells (MQW) and superlattices (SL) have been grown. We describe now the different characterizations of these structures.

X-ray diffraction is of much use, particularly in the case of superlattices, to ensure the reproducibility of the wells and the barriers. Fig. 12.27 exhibits the x-ray (4 0 0) diffraction pattern of a 10-period $Ga_{0.47}In_{0.53}$As/InP superlattice, with InP barriers of about $L_B = 250$ Å and $Ga_{0.47}In_{0.53}$As wells of $L_z = 200$ Å [Razeghi et al. 1987]. A synchrotron beam of wavelength $\lambda = 1.2834$ Å is used as the x-ray source. Satellites corresponding to the artificial crystalline periodical $L_B + L_z$ introduced during growth have been resolved up to $n = \pm 5$. This is rarely observed and underlines the excellent structural quality of the sample under study. The $n = 0$ satellite, attributed to the (4 0 0) Bragg diffraction of the mean parameter of the superlattice, appears at the position of the InP substrate peak. This indicates that the lattice parameter of the ternary alloy is well adapted to that of the substrate.

From the angle $\Delta\theta = 360$ arcsec between two adjacent satellites, we deduce, using the classical expression

Eq. (12.12) $L_B + L_z = (\lambda / 2)\Delta\theta \cos\theta$

that the superperiodicity is $L_B + L_z = 410$ Å; this corresponds quite well to the expected value from the growth parameters.

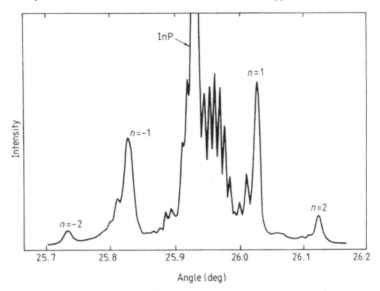

Fig. 12.27. X-ray diffraction pattern of a 10-period $Ga_{0.47}In_{0.53}As/InP$ superlattice grown by MOCVD. $L_B(InP) = 250 \pm 20$ Å, $L_z(GaInAs) = 200 \pm 20$ Å.

The hyperfine structure observed in Fig. 12.27 near the substrate peak may be attributed to the interaction between incident and reflected waves through the $L = 5000$ Å epilayer (Pendellosung oscillations). If we use the classical equation

Eq. (12.13) $\delta\theta = (\lambda / 2L)\cos\theta$

with $L = 5000$ Å, we find $\delta\theta = 30$ arcsec, in good agreement with experimental results shown in Fig. 12.27. Such behavior has rarely been observed in III–V superlattices.

SIMS analyses are performed for quantitative determination of impurities accumulated at the substrate–epilayer interfaces. The analyses are carried out on a modified Cameca IMS3F. The surfaces of the epitaxial layers are scanned with a focused mass-filtered oxygen-ion beam ($I_p = 1.5$ A at 10 keV). The scanned area is 250×250 μm and the analyzed region is 150 μm in diameter.

Fig. 12.28(a) exhibits the SIMS profiles of the four majority species (In, Ga, As and P) in a 25-period $Ga_{0.47}In_{0.53}As/InP$ superlattice with well and barrier thicknesses of 150 Å. The In concentration appears almost constant, because it is the only element present in the barriers as well as in the wells. The signals of the other species are clearly oscillating with the change in chemical composition of the layers, thus evidencing the abruptness of the interface and the perfect control of compositions and of thicknesses.

Auger spectra have been performed on chemical bevels. The samples are chemically etched by using a methanol–bromine solution (15% Br), in order to obtain a bevel with a mean amplification coefficient of 2100 (measured with a Talysurf). This means that a change of 1 μm along the surface corresponds to a change of 4.75 Å depth (*z* direction). By scanning the incident electron beam along the bevel, the phosphorus profile of a GaInAs–InP superlattice has been obtained. It is shown in Fig. 12.28(*b*). There again the modulation of composition is clearly evidenced, thus confirming the abruptness of the interfaces.

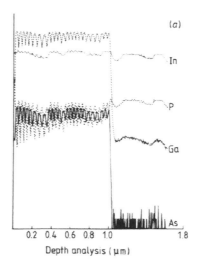

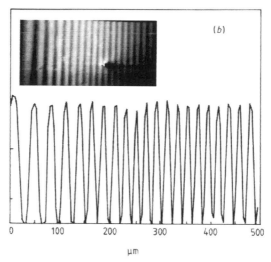

Fig. 12.28. (a) Ga (— — —), In (....), As (—) and **P** *(• • •) SIMS profiles of a 25-period Ga$_{0.47}$In$_{0.53}$As/InP superlattice (L$_z$ = 150 Å, L$_B$ = 150 Å) grown by MOCVD. (b) Phosphorus signal relative to the Auger spectrum of a Ga$_{0.47}$In$_{0.53}$As/InP superlattice grown by MOCVD.*

The etch-pit density (EPD) is also performed on a chemical bevel. The defects are revealed in an H$_3$PO$_4$–HBr solution. The etch speed of the acid solution varies with the orientation of the atomic planes. Around a dislocation, the atomic plane appears slightly disoriented so that it is etched at a higher speed. For a fixed developing time, the etched thickness is greater around the defects than in the rest of the crystal; etch patterns appear that allow us to determine the density of dislocations.

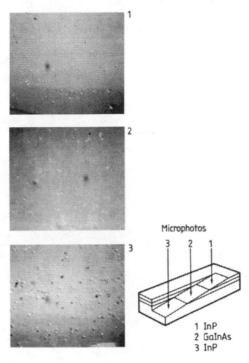

Fig. 12.29. Etch-pitch density photograph of a $Ga_{0.47}In_{0.53}As/InP$ heterojunction grown by MOCVD.

We show in Fig. 12.29 the etch-pit density photograph of a $Ga_{0.47}In_{0.53}As/InP$ heterojunction. The epitaxial layers, with a density of dislocations of about $10^4 cm^{-2}$, equal to that of the substrate, show a crystalline quality at least corresponding to that of the InP substrate. No defects are evidenced at the interface.

Edge-on transmission electron microscope (TEM) characterization has been performed on a 40-period GaInAs–InP superlattice grown by MOCVD [Razeghi et al. 1988]. Fig. 12.30 shows that the thickness of the layers is not regular and there are some dark point defects in the layers. These irregularities are due to the fact that the MOCVD reactor was operated manually, and there was a pyrolysis oven for PH_3. After elimination of the PH_3 pyrolysis oven and automation of the MOCVD reactor, the thickness homogeneity of the superlattices was improved as shown on Fig. 12.30. The dark-field images at atomic resolution of the same superlattices are shown in Fig. 12.30. It is clear from this photograph that the interfaces between these layers are perfectly smooth and flat. Looking carefully at the InP/GaInAs interface (Fig. 12.30) we can see atomic steps every 80 Å, which can be attributed to the 2° misorientation of the InP substrate. The results show that perfect interfaces on an atomic scale can be grown by MOCVD.

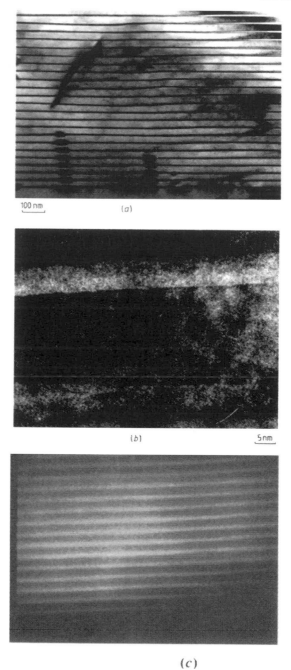

100 nm

(a)

(b)   5 nm

(c)

Fig. 12.30. (a) Transmission electron micrograph of an LP-MOCVD
$Ga_{0.47}In_{0.53}As/InP$ superlattice. (b) Dark-field image at atomic resolution of the
same superlattice. (c) TEM of a $Ga_{0.47}In_{0.53}As/InP$ superlattice grown by an
automatic MOCVD reactor.

12.5.3. Optical properties of GaInAs–InP quantum wells

(a) Photoluminescence

Quantum-well (QW) luminescence has been extensively studied in the GaAs–GaAlAs system and has provided significant information about the structural properties of heterostructure devices and about the 2D nature of elementary excitations. QW made of ternary or quaternary materials such as GaInAs or GaInAsP imbedded in InP [Razeghi et al. 1983] or InAlAs [Welch et al. 1983] barrier material have been much less studied. Standard growth conditions were used, leading to a growth rate of about 2.5 Å·s^{-1} and well defined interfaces by rapid switching of gas composition. Razeghi et al. [1984] studied multiple single-well samples, i.e. samples with a stack of quantum wells with different thicknesses grown under the same conditions. Three main subjects are addressed in what follows: (i) the origin and properties of luminescence lines; (ii) the observation of the excited states of the quantum wells by photoluminescence excitation spectroscopy (PLE) and the determination of QW parameters from these data; and (iii) the carrier-capture properties of thin quantum wells.

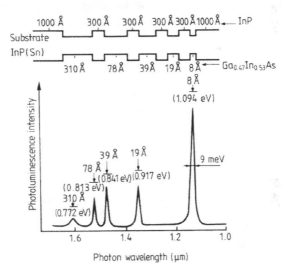

Fig. 12.31. Photoluminescence spectrum of a multi-well LP-MOCVD GaInAs–InP sample at 5 K excited by an He–Ne laser in the InP overlayer.

A typical spectrum for a five-well structure is shown in Fig. 12.31. The well thicknesses indicated ("nominal") are those deduced from the growth rate as extrapolated for thin layers from the measured growth rate of thick layers. The remarkable features are the sharpness of lines, even for QW with an 8 Å nominal thickness $(\Delta E \simeq 9 \text{ meV})$ at 5 K. The progress in sample quality can be traced by comparing these spectra with those obtained before by Razeghi et al. [1983]. Extremely large shifts of luminescence due to the quantum confinement effect of QW are observed, up to 0.320 eV for the nominal 8 Å QW.

The growth of layers with different thicknesses has allowed us to assess the origin of the variation in energy while scanning across the sample (Fig. 12.31). The recombination energy in an alloy QW can be approximated by

Eq. (12.14) $E = h\nu = E_0(x, y) + E_{conf}^e + E_{conf}^h$

where $E_0(x, y)$ is the alloy band gap (dependent on local composition), and E_{conf}^e and E_{conf}^h are the confining energies for the electron and hole ground states, respectively. The small shift between the quantized electron–hole levels and luminescence line position due to exciton binding has been neglected here. Variations of $h\nu$ according to Eq. (12.14) can have two origins: (i) a variation in $E_0(x, y)$ due to alloy *macroscopic* variations of composition across the sample; and (ii) a change of the confining energies in a quantum well due to a *macroscopic* variation of QW thickness (originating in a non-uniform growth rate across the wafer). In case (i), one expects a similar shifting of the QW energies across the wafer, independent of the thickness of the quantum well under observation. In case (ii), if υ_1 and υ_2 are the two growth rates at two points 1 and 2, and remembering that $E \sim L^{-2}$ in the infinitely deep-well limit, one finds

Eq. (12.15) $\Delta E \simeq L^{-3}\Delta L = L^{-2}(\upsilon_2 - \upsilon_1)/\upsilon_2$

In this case, the energy shift between two points on the wafer depends on the well thickness. The *constant shift* observed for all the QW in those samples which display energy shifts across the wafer allows us to conclude that this is due to the macroscopic variations of the alloy composition across the sample.

It is not possible with the present level of understanding to decide the nature of the luminescent level at low temperatures. Comparing with GaAs–GaAlAs quantum wells, where the dominance of free-exciton-mediated recombination has been established through spin orientation and lineshape analysis [Weisbuch et al. 1981], one could tentatively ascribe the same origin to the luminescence line in the GaInAs–InP system: the 2D exciton binding energy should be about 4 times larger than in 3D (as was recently well established in the GaAs–GaAlAs system by Maan et al. [1984]), i.e. should be 10 meV.

The observed linewidth of the luminescence lines (and excitation spectra) is surprisingly small. A model of PLE spectra in GaAs–GaAlAs has described this linewidth as being caused by the spatially varying well thickness of the *microscopic* island-like topology of the interface due to a layer growth mode [Weisbuch 1981 and Singh and Madhukar 1983]: assuming a half-monolayer height for the islands and a correlation length greater than or approximately equal to the exciton Bohr radius a_B, one expects variations of the exciton absorption or recombination energy given by $\Delta E \simeq L^{-3}a/2$. An $a/2$ fluctuation in interface position yields $\Delta E \simeq 42$ meV for a 20 Å thick sample, much above the observed linewidth. It is possible to conclude that the previous model, although very successful in the case of GaAs–GaAlAs, does not apply to the GaInAs–InP case. One can therefore describe the interface as rather fuzzy. However, the observation of a narrow, well defined, intense luminescence peak, even for thicknesses down to 8 Å (nominal), leads us to emphasize the very good uniformity of this fuzzy well, i.e. very small deviations from the average thickness are allowed over macroscopic (here larger than a_B) distances.

(b) Excitation spectra

Excitation spectra have been recorded for a number of GaInAs–InP quantum wells. Fig. 12.32 shows a typical PLE spectrum for a three-well structure of GaInAs–InP grown by MOCVD. A remarkable feature compared to the GaAs–GaAlAs case is that one cannot observe pairs of absorption peaks due to each confined electron level associated with either the heavy- or light-hole level. One can rather observe two series of peaks, one at lower energies and rather sharp, the other quite broad and more rightly described as a bump. Razeghi et al. [1984] tentatively ascribed these two series as due to heavy hole–electron and light hole–electron transitions. The different shapes of these transitions might be due to the different non-parabolicities of the two hole bands due to the strong valence-band mixing recently evidenced in the GaAs–GaAlAs case [Sooryakumar et al. 1984].

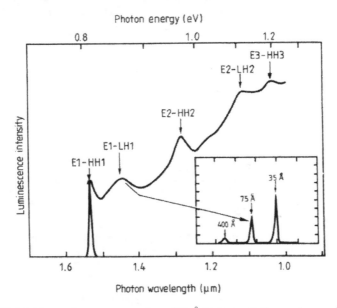

Fig. 12.32. Excitation spectrum for a 75 Å GaInAs–InP quantum well at 5 K.

From these measurements one can try to evaluate the various parameters entering Eq. (12.14). It should be remarked that the situation is much more complicated than in the GaAs–GaAlAs case, as E_0 is not known so precisely and the various band parameters (mainly the hole masses) are also not too well determined. We therefore only produce here a preliminary fit of the experimental results taking all the unknowns as fitting parameters. Using a standard quantum-well calculation of energy levels, we introduce, in a simple iterative form, the non-parabolicity of the energy bands. No provision is taken for the mixing of the valence bands, which are considered fully uncoupled (Fig. 12.33). Neglecting exciton effects, which are small on the scale of confining energies, the best-fit parameters are

E_0 = 0.742 eV
L_1 = 39 Å
L_2 = 78 Å
m_e = 0.041 m_0
m_{hh} = 0.5 m_0 (has no influence on the fit down to m_{hh}, which is certainly not
         the case)
m_{lh} = 0.042·m_0
Q = $\Delta EC/\Delta Eg$ = 0.5.

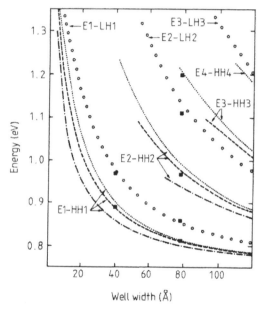

Fig. 12.33. *Fit of the transitions observed in excitation spectra for two wells with 75 and 35 Å nominal thickness. Fitting parameters: $E_0 = 0.742$ eV, $m_e = 0.041 m_0$, $m_{hh} = 0.5 m_0$, $m_{lh} = 0.042 m_0$, $L = 39$ Å, $L = 78$ Å. The various E–HH transition curves correspond to $Q = \Delta E_C/\Delta E_g, = 0.3$ (—·—), 0.5 (———) and 0.7 (······). Only one curve is traced for E–LH transitions (○), as this parameter has a negligible influence on transition energies.*

As dramatically pointed out by Miller et al. [1984] in the GaAs–GaAlAs case, it can be observed that such optical measurements of interband transitions do not lead to an accurate determination of the relative band offsets: the change in the confining energies induced by a change in Q can easily be corrected by a change in other fitting parameters, here mainly E_0 and L. A more detailed knowledge of sample structure and band parameters is therefore needed before a more careful calculation can lead to a better precision in Q.

12.5.4. Room-temperature excitons in GaInAs–InP superlattices grown by MOCVD

Excitonic absorption at room temperature in GaAs–GaAlAs multiquantum wells has attracted a great deal of interest because of the novel physical properties and the potential usefulness in optoelectronic devices. Wood et al. [1984] have shown that the application of an electric field perpendicular to the quantum-well layers can cause a strong shift in the absorption edge, and that this shift can be used to make an optical modulator. Room-temperature excitonic absorption in GaInAs–AlInAs was first reported by Wood et al. [1984]. Razeghi et al. [1986] have made the first observation of room-temperature excitonic absorption in $Ga_{0.47}In_{0.53}As$–InP superlattices grown by MOCVD. Fig. 12.34 shows the low-temperature photoluminescence spectrum of a GaInAs–InP multilayer grown by MOCVD. The details of the structure are shown in Fig. 12.34. It can be seen that the GaInAs quantum well of 100 Å thickness has a linewidth of 5 meV, while the GaInAs of 600 Å thickness has a linewidth of 3 meV, which can be attributed to free-exciton recombination, and the peak with energy 0.769 eV may be due to recombination of excitons bound to shallow impurities [Goetz et al. 1983]. The excitonic nature of the dominant high-energy peak of the 100 Å quantum well is proved by photoluminescence excitation spectroscopy (PLE). Fig. 12.35 shows the PLE spectrum of a GaInAs–InP quantum well of 100 Å thickness at 6 K. The PLE spectrum exhibits a series of peaks, which can be attributed to the heavy hole–electron and light hole–electron transitions as shown in Fig. 12.35.

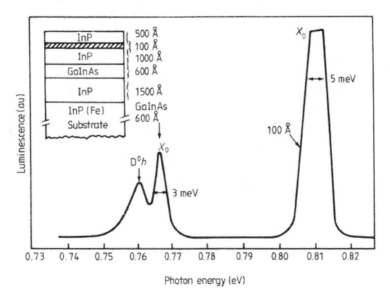

Fig. 12.34. Photoluminescence spectrum of GaInAs–InP multilayer at 6 K. Inset: schematic representation of a GaInAs–InP multilayer grown by MOCVD.
$P_{exc} = 0.8 \ W \cdot cm^{-2}$.

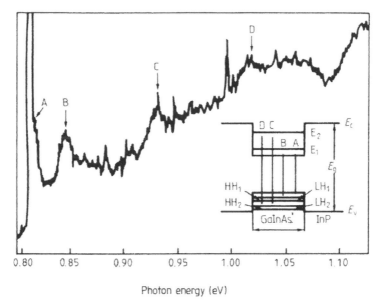

Photon energy (eV)

Fig. 12.35. PLE spectrum of a GaInAs–InP quantum well of 100 Å thickness at 6 K. Transitions: A, E_1—HH_1; B, E_1–LH_1; C, E_2–HH_2; and D, E_2–LH_2. Inset: energy levels in a GaInAs–InP quantum well.

By comparing photoluminescence (PL) and PLE spectra taken at 6 K in Fig. 12.35, we find that the energy of the E_1–HH_1 subband transition, as determined by PLE, coincides with the peak energy of the (e, h) emission in the PL spectrum. Such a coincidence is only rarely observed as it can only occur in highly pure materials with very little disorder from alloying or interfaces. Usually luminescence is due to both intrinsic and extrinsic recombination processes, but PLE probes only intrinsic states. This coincidence of luminescence and PLE spectra shows that the PL of the GaInAs quantum well of 100 Å thickness under study is intrinsic and can be attributed to free-exciton recombination. Fig. 12.36 shows the optical absorption spectrum of a GaInAs–InP superlattice at room temperature. The superlattice consists of 40 periods of approximately 100 Å thick GaInAs wells and 200 Å thick InP barriers. The room-temperature absorption spectrum exhibits the step-like behavior characteristic of the two-dimensional density of states and excitonic features at the onset of the steps. Furthermore, the 5 K photoluminescence spectrum of that superlattice shows a halfwidth of 6 meV, evidencing a good layer-to-layer reproducibility (see Fig. 12.36).

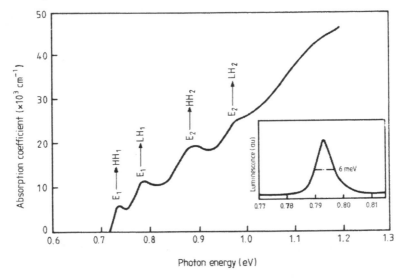

Fig. 12.36. Room-temperature absorption spectrum of a GaInAs–InP superlattice consisting of 40 periods of 100 Å thick GaInAs wells and 200 Å thick InP barriers. Inset: photoluminescence spectrum of the same superlattice; $T = 5$ K, $P_{exc} = 0.8$ $W \cdot cm^{-2}$.

12.5.5. Negative differential resistance at room temperature from resonant tunneling in GaInAs/InP double-barrier heterostructures

The principle of resonant tunneling is shown schematically in Fig. 12.37. Quasi-bound states in a quantum well confined by two narrow barriers can be brought into coincidence with the Fermi energy of the doped region outside one barrier by the application of a voltage bias. There is a resonant increase in the tunneling probability and hence current under these conditions. Further bias destroys the alignment and a negative differential resistance follows. In Fig. 12.38, we show current–voltage simulations for resonant tunneling through 3 and 6 nm wide GaInAs wells bounded by 3 nm InP barriers. The model [Davies 1988] used includes transport by tunneling and thermionic emission. Depletion effects on the low-voltage side of the barrier have also been included. The position of the negative differential resistance reflects the energies of the bound states in the wells. A conduction-band offset of 530 meV has to be assumed [Guldner et al. 1982].

The layers in order of growth were as shown in Fig. 12.39. Boothroyd and Stobbs [1986] have performed "edge-on" transmission electron microscopy (TEM) characterization on a resonant-tunneling structure of GaInAs–InP, viz. a 28 Å $Ga_{0.47}In_{0.53}As$ layer sandwiched between two InP layers of 36 Å thickness grown under the same conditions. Fig. 12.40 shows "edge-on" transmission electron microscopy characterization of this sample. The figure shows that the thickness of the triple layer is generally regular. The dark-field images at atomic resolution of another resonant-tunneling structure, i.e. a 33 Å GaInAs layer sandwiched between two InP layers of 35 Å thickness grown by MOCVD, show that very thin epitaxial layers of GaInAs–InP of accurately controlled thickness and composition can be prepared by MOCVD. The material was processed into mesa diodes of 100 μm diameter with ohmic contacts to back and front formed using Au–Ge: Au metallization and transient alloying.

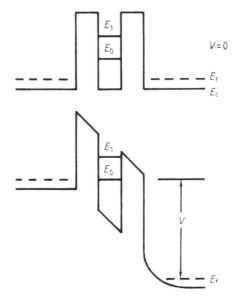

Fig. 12.37. Conduction-band profile for a double-barrier resonant-tunneling structure.

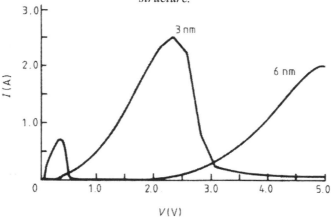

Fig. 12.38. Resonant-tunneling current–voltage simulations for 3 and 6 nm wide wells (after Razeghi [1987]).

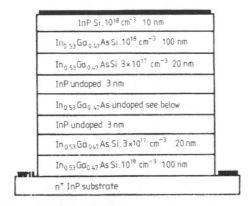

| InP Si. 10^{18} cm^{-3} 10 nm |
| In$_{0.53}$Ga$_{0.47}$As Si. 10^{18} cm^{-3} 100 nm |
| In$_{0.53}$Ga$_{0.47}$As Si 3×10^{17} cm^{-3} 20 nm |
| InP undoped 3 nm |
| In$_{0.53}$Ga$_{0.47}$As undoped see below |
| InP undoped 3 nm |
| In$_{0.53}$Ga$_{0.47}$As Si. 3×10^{17} cm^{-3} 20 nm |
| In$_{0.53}$Ga$_{0.47}$As Si. 10^{18} cm^{-3} 100 nm |
| n$^+$ InP substrate |

Fig. 12.39. Double-barrier structure grown by MOCVD.

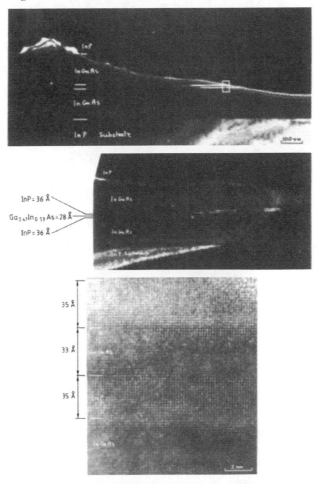

InP = 36 Å
Ga$_{0.47}$In$_{0.53}$As = 28 Å
InP = 36 Å

35 Å
33 Å
35 Å

Fig. 12.40. Edge-on transmission electron micrograph for a double-barrier resonant-tunneling GaInAs–InP structure grown by MOCVD.

For those samples which in TEM showed a marked variability of barrier and well thickness (see for example Fig. 12.40) no negative-differential-resistance effects were observed. For a wafer which TEM indicated to have more uniform layering, and a well thickness within 10% of that specified, negative-resistance effects were observed.

The results achieved with pulsed $I-V$ measurements are shown in Fig. 12.41. The presence of a thermionic emission contribution to the current at room temperature reduces the peak-to-valley ratio from the high values expected on the basis of resonant tunneling alone. It is also possible that interface roughness and alloy scattering in the GaInAs contribute to the reduction of the peak-to-valley ratio, since inhomogeneities will destroy the coherence needed for resonant tunneling. The voltage at which the negative differential resistance occurs agrees with the model calculations shown in Fig. 12.38.

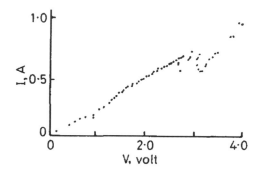

Fig. 12.41. *Pulsed current/voltage characteristic for a sample with 3 nm wells showing negative differential resistance at 3 V bias.*

12.6. Magnetotransport in GaInAs–InP heterojunctions grown by MOCVD

12.6.1. Shubnikov–de Haas and quantum Hall effects

The study of the Shubnikov–de Haas and quantum Hall effect (QHE) is one of the most useful and direct ways of characterizing semiconductor heterostructures. At low temperatures a two-dimensional gas of carriers bound in a heterostructure acts like a metal with a small Fermi energy, typically of order 10–100 meV. The magnetic field causes a quantization of the free-carrier states into a ladder of Landau levels (see Ando et al. [1982]). Changing the magnetic field sweeps the level through the Fermi energy, causing oscillations in the magnetoresistance, known as the Shubnikov–de Haas effect. Since the cyclotron motion induced by the field is in the plane perpendicular to its direction, this means that a two-dimensional system is sensitive only to the component of the field parallel to the surface normal, so that rotation of the sample relative to the field can be used to provide a direct proof of the two-dimensional nature of any system under investigation. In the presence of a magnetic field, the energy of the Landau levels will increase linearly with field:

Eq. (12.16) $E(n_L, n) = (n_L + \frac{1}{2}) \hbar \omega_c + E_n$

where the Landau quantum number n_L takes integer values 0, 1, 2, 3,... and E_n is the bottom of the subband n.

The degeneracy of a state with given quantum numbers n_L and n will also be proportional to the field. It is given by $2eB/h$.

By increasing the magnetic field, the Fermi energy will therefore remain approximately constant, and the Landau levels will be swept upwards through the Fermi energy. The density of states at the Fermi energy will thus vary, and the electrical conductivity, which depends upon the density of states within approximately kT of the Fermi energy, will show oscillations periodic in inverse magnetic field.

The density of states for a two-dimensional electron gas (2DEG) in a magnetic field is shown in Fig. 12.42. We can calculate the oscillation periodicity from the degeneracy of the Landau levels; this derivation requires only that the separation of the levels is large relative to their broadening, or that the broadening is symmetric.

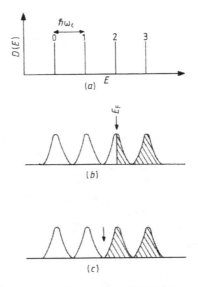

Fig. 12.42. The density of states for a 2DEG in a magnetic field.

The conductivity of the 2DEG in a high magnetic field is approximately proportional to the product of the number of electrons within about kT of the Fermi energy and the scattering rate. Both these quantities are proportional to the density of states at the Fermi energy, and so the conductivity is essentially proportional to the square of the density of states at the Fermi energy. The conductivity will therefore be a minimum when the Fermi energy lies between two Landau levels, so that an integral number of levels are completely filled. This occurs when the electron density n satisfies

Eq. (12.17) $n = N(2eB/h)$ $N = 1, 2, 3, ...$

where N is the number of filled levels, as the degeneracy of a Landau level is $2eB/h$. Conductivity minima will thus be seen when

Eq. (12.18) $B = \dfrac{1}{N} \dfrac{nh}{2e} = \dfrac{B_F}{N}$

where B_F is the fundamental field, and so the minima form a series periodic in $1/B$. For a general field direction at an angle θ to the normal to the 2DEG, the oscillation periodicity depends only upon $B \cos \theta$, and the two-dimensional electron concentration is given by

Eq. (12.19)
$$n = (2e/h) B_F \cos \theta$$
$$= (4.836 \times 10^{10}) B_F \cos \theta \ \text{cm}^{-2}$$

where B_F is in tesla.

In Fig. 12.42(b), the Fermi energy lies within a half-filled Landau level; this is analogous to a half-filled band, and so the 2DEG behaves like a metal. In Fig. 12.42(c), the Fermi energy lies between a completely filled level and a completely empty one, and the 2DEG behaves like an insulator. The difference in the conductivity in these two situations can be a factor in excess of 10^9 [Stoermer et al. 1982]. The conductivity will thus oscillate strongly as a function of magnetic field.

Under a magnetic field, plateaux in the Hall resistance of a 2DEG (ρ_{xy}) appear, corresponding with a very high accuracy to the quantized values $\rho_{xy} = h/ie^2$, where $i = 1, 2, ...$, when the minimum of the magnetoresistance becomes very small. This quantum Hall effect (QHE) has been observed in modulation-doped $In_xGa_{1-x}As$–InP HJ with two populated electric subbands at zero magnetic field and low electron density $n_s \leq 5 \times 10^{11}$ cm^{-2} [Razeghi et al. 1986].

The $In_{0.53}Ga_{0.47}As$–InP HJ used were grown by low-pressure metalorganic chemical vapor deposition on (1 0 0) semi-insulating Fe-doped substrates. The InP layer, roughly 1000 Å thick, was n-type with $N_D-N_A \sim 3 \times 10^{15}$ cm^{-3}. The $In_xGa_{1-x}As$ layer was also n-type with $N_D-N_A \sim 2 \times 10^{15}$ cm^{-3}, its thickness being equal to 1 μm. Electrons are transferred from InP into $In_xGa_{1-x}As$ and an accumulation layer is formed at the interface [Razeghi et al. 1986]. The mobility of the 2DEG and the electron density at 4 K were obtained from low-field Hall measurements and are listed in Table 12.6 for each sample. Standard Hall bridges were used to measure ρ_{xx} and ρ_{xy} between 1.3 and 4.2 K.

Sample	Hall mobility ($cm^2 \cdot V^{-1} \cdot s^{-1}$)			n_s (cm^{-2}) at 4 K
	300 K	77 K	4 K	
S1	10,000	50,000	60,000	4.2×10^{11}
S2	11,800	55,000	70,000	5.4×10^{11}
S3	9,500	41,000	55,000	5.4×10^{11}

Table 12.6. Characteristics of the heterojunctions used in this work.

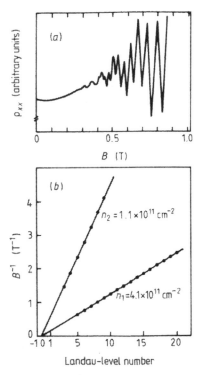

Fig. 12.43. (a) Shubnikov–de Haas oscillations at low magnetic field ($B \le 1$ T) in sample S2. $T = 1.6$ K, $\theta = 0$. (b) Reciprocal magnetic field at maxima of the high- and low-frequency oscillations versus the Landau index number.

Fig. 12.43(a) shows the B dependence of ρ_{xx} for sample S2 at low magnetic field ($B < 1$ T) applied perpendicular to the interface and at $T = 1.6$ K. Pronounced Shubnikov–de Haas oscillations are observed for $B > 0.2$ T and different oscillation periodicities appear clearly, which, we believe, originate from populated subbands E_0 and E_1 in the 2DEG. Fig. 12.43(b) gives the reciprocal field corresponding to the oscillation maxima as a function of the Landau-level index. Two linear dependences are observed and are associated with the two periodicities in B^{-1}. From the slopes of the two linear variations, the populations $n_0 = 4.1 \times 10^{11}$ cm^{-2} and $n_1 = 1.1 \times 10^{11}$ cm^{-2} are calculated for E_0 and E_1, respectively. The total electron density $n_0 + n_1 = 5.2 \times 10^{11}$ cm^{-2} deduced from the Shubnikov–de Haas data is in good agreement with the value $n_s = 5.4 \times 10^{11}$ cm^{-2} obtained from low-magnetic-field Hall measurements at 4 K in sample S2. Using $m^* = 0.047\, m_0$ for the 2DEG effective mass, one finds Fermi energies $E_{F,0} = 21$ meV and $E_{F,1} = 5.5$ meV, measured from the bottom of the two subbands E_0 and E_1, respectively. The subband separation is then approximately 15.5 meV and the higher subband E_1 starts to be occupied at $n_s \sim 3 \times 10^{11}$ cm^{-2}. Similar results are obtained in samples S1 ($n_0 = 3.5 \times 10^{11}$ cm^{-2}, $n_1 = 0.6 \times 10^{11}$ cm^{-2}) and S3 ($n_0 = 4.6 \times 10^{11}$ cm^{-2}, $n_1 = 0.9 \times 10^{11}$ cm^{-2}).

The existence of two populated subbands gives rise to inter-subband scattering as calculated [Mori and Ando 1980] and observed [Stoermer et al. 1982] in GaAs–Al$_x$Ga$_{1-x}$As HJ. The influence of inter-subband scattering can be demonstrated from magnetoresistance experiments with B parallel to the interface ($\theta = 90°$). The effect of a parallel magnetic field is to increase the subband separation and, finally, to depopulate E_1 [Ando 1975 and Beinvogl 1986]. Fig. 12.44 shows the magnetoresistance ρ_{xx} as a function of B for $\theta = 90°$ in sample S2, which exhibits a fall of about 60% in the resistance occurring at $B \sim 5$ T. This fall corresponds to the magnetic depopulation of the higher subband E_1 and to the reduction of the inter-subband scattering. Similar results were obtained [Portal et al. 1983] in Al$_x$In$_{1-x}$As–In$_x$Ga$_{1-x}$As HJ and also in the resistance of GaAs–Al$_x$Ga$_{1-x}$As HJ as a function of the electron concentration [Stoermer et al. 1982].

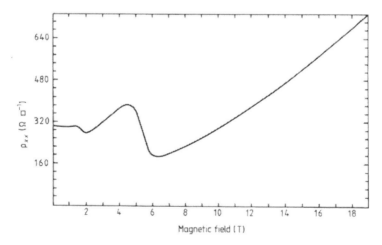

Fig. 12.44. Magnetoresistance of sample S2 with B parallel to the interface.
T = 1.6 K, θ = 90°.

Fig. 12.45 shows ρ_{xx} and ρ_{xy} in sample S1 at 15 mK as a function of B applied perpendicular to the interface. Striking features are observed in ρ_{xy} compared to the usual QHE curves reported [Briggs et al. 1983 and Guldner et al. 1982] in similar samples with only one occupied subband. While the $i = 2$, 8 and 10 plateaux are well developed at $B \sim 10.3$ and 2.25 T, the $i = 4$ and $i = 6$ plateaux are nearly missing. Moreover, the width of the $i = 5$ plateau occurring at $B \sim 4.5$ T is anomalous. Indeed, in experiments previously reported on In$_x$Ga$_{1-x}$As–InP HJ with similar values of n_s and electron mobility but with only one occupied subband, the $\rho_{xy} = h/5e^2$ plateau was not observed even at very low temperature, because the spin splitting of the $N = 2$ Landau level was not resolved. Remarkable features are also observed in the ρ_{xx} dependence as a function of the filling factor $v = n_s h/eB$. Fig. 12.45 shows that ρ_{xx} minima occur for $v = 2, 5, 8, 10$ and 12 but only smaller structures are observed at $v = 4$ and 6. Furthermore, the best vanishing ρ_{xx} minimum occurs at $v = 2$. The anomalous behavior of ρ_{xy} can be explained by the existence of two occupied electric subbands at $B = 0$. Fig. 12.46 shows a schematic diagram of the Landau levels $N_\pm$ and $N'_\pm$ arising from the two occupied subbands E_0 and E_1, respectively. The calculations are done for $\theta = 0$ using 15 meV as the subband separation, $m^* \simeq 0.047m_0$ [Guldner et al. 1982] and an enhanced Landau factor [Guldner Y et al. 1982] $g \sim 10$, independent of the Landau-level index and B. The resulting variation of the Fermi energy E_F, assuming infinitely sharp Landau levels, is also shown for a total electron concentration $n_s = 4.2 \times 10^{11}$ cm^{-2}. The middle of each ρ_{xy} plateau and the ρ_{xx} minima in Fig. 12.45 must correspond to vertical jumps of E_F from one level to the next one occurring for integral values of v. Fig. 12.46 shows that the two Landau levels $0'_+$ and 1_- overlap in the vicinity of $v = 4$, yielding no significant jump of E_F. This is consistent with the absence of the $i = 4$ plateau in ρ_{xy} and the observation of a small dip instead of a real minimum in ρ_{xx} at $v \sim 4$ (Fig. 12.45). Similarly, the crossover of the 1_+ and $0'_+$ levels in the vicinity of $v = 3$ strongly perturbs the ρ_{xy} plateau whose value is significantly larger than $h/3e^2$ (Fig. 12.46). The well developed $i = 5$ plateau appears when the $0_\pm$, $1_\pm$ and $0'_+$ Landau levels are completely filled and corresponds to the jump of E_F from $0'_+$ to $0'_-$. Note that in previous experiments [Briggs et al. 1983 and Guldner et al. 1982] in In$_x$Ga$_{1-x}$As–InP HJ with only one occupied subband, the $i = 5$ plateau would have corresponded to the jump of E_F between the two N spin components which are not experimentally [Briggs et al. 1982 and Guldner et al. 1982] resolved in samples with an electron mobility less than $70,000$ cm$^2 \cdot$V$^{-1} \cdot$s^{-1}.

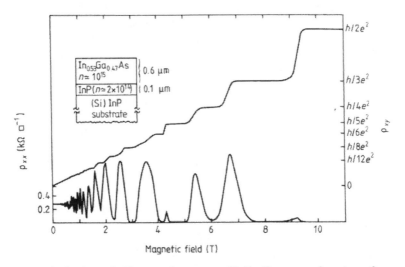

Fig. 12.45. Shubnikov–de Haas and quantum Hall effects as a function of magnetic field normal to the 2D plane measured at $T = 0.045$ K.

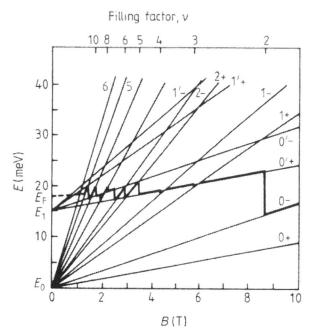

Fig. 12.46. Schematic diagram of the Landau levels N_+ and N'_+ arising from the two ground subbands E_0 and E_1. (The subscripts $+$ and $-$ refer to spin up and down, respectively.) For the sake of simplicity, the spin splitting of the levels is not presented for $N \geq 3$. The variation of the Fermi energy E_F assuming infinitely sharp Landau levels and a constant electron concentration $n_s = 4.2 \times 10^{11}$ cm^{-2} is also shown (broken line).

Fig. 12.46 shows that, in the case of S1, the $i = 6$ plateau would correspond to the spin splitting of $N = 2$ which, again, is weakly resolved. This is consistent with the weakness of the $i = 6$ plateau in ρ_{xy} and the ρ_{xx} minimum at $v = 6$ (Fig. 12.45). Finally, the $i = 8$ and $i = 10$ plateaux correspond to the usual situation when pairs of spin-split Landau levels are completely filled ($0_\pm$, $0'_\pm$, $1_\pm$, $2_\pm$ for $i = 8$ and $0_\pm$, $0'_\pm$, $1_\pm$, $2_\pm$, $3_\pm$ for $i = 10$). Note that, above 9 T, the subband E_1 is totally depopulated by the magnetic field (Fig. 12.46). The $i = 2$ plateau and the $v = 2$ minimum of ρ_{xx} in Fig. 12.45 correspond to the usual situation where only one subband is occupied. A complete redistribution of the energy levels is obtained in tilted magnetic fields ($\theta \neq 0$) since the Landau-level separation is determined only by the B component perpendicular to the interface while the spin splitting comes from the total magnetic field. Fig. 12.47 shows ρ_{xy} in sample S1 for $\theta = 45°$. The $i = 4$, 5, 6, 10 and 12 plateaux are observed, but the $i = 8$ plateau is now missing. Again the result could be qualitatively interpreted with a schematic diagram of the Landau level arising from the subbands E_0 and E_1 at $\theta = 45°$. For a more precise analysis of the results, it would be necessary to take into account the broadening of the Landau levels, the exact dependence of the effective g factor on B and N and, for $\theta \neq 0$, the shift of the E_0 and E_1 positions due to the introduction of a B component parallel to the interface.

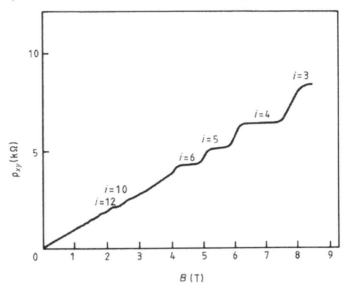

Fig. 12.47. Hall resistance ρ_{xy} as a function of B for $\theta = 45°$ at $T = 1.6$ K.

12.6.2. Observation of a two-dimensional hole gas in a $Ga_{0.47}In_{0.53}As/InP$ heterojunction grown by MOCVD

The interest in two-dimensional (2D) hole systems has been spurred on by the considerable mobility enhancement attainable in modulation-doped GaAs/AlGaAs heterostructures. The first magnetotransport measurements demonstrating the two-dimensionality of the hole gas of a GaAs/AlGaAs interface were reported by Störmer and Tsang [1980]. The integral and fractional quantum Hall effects were also observed [Stoermer et al. 1983, Mendez et al. 1984, and Cheng et al. 1984], demonstrating that these quantum phenomena are not limited to 2D electron systems and are independent of the host band structure.

One of the most outstanding features common to all 2D hole systems is the lifting of spin degeneracy due to the lack of inversion symmetry. This feature was predicted theoretically for holes in Si inversion layers [Ohkawa and Uemera 1975 and Ekenberg and Altarelli 1985], and was also demonstrated by calculations for GaAs/AlGaAs interfaces [Bangert and Landwehr 1985 and Ando 1985].

Magnetotransport studies have also been performed on a hole gas in a GaInAs–InP heterojunction [Razeghi et al. 1986].

$Ga_{0.47}In_{0.53}As/i$-InP/p^+-InP heterojunctions were grown in an i-$Ga_{0.47}In_{0.53}As/i$-InP/p^+-InP configuration on an Fe-doped semi-insulating (1 0 0) 2° off InP substrate as shown in Fig. 12.48. The wide-gap InP layer was deliberately doped with Zn to a level $N_A - N_D \simeq 10^{17}$ cm^{-3} while the narrow-gap $Ga_{0.47}In_{0.53}As$ layer was not intentionally doped. The thickness of the InP (Zn-doped) layer was 1000 Å and the GaInAs layer was 0.5 μm thick. There was a 100 Å thick undoped InP layer as a spacer between doped InP and GaInAs layers (Fig. 12.48). Growth conditions were usual.

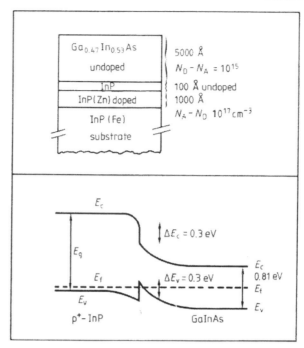

Fig. 12.48. Schematic configuration of the heterojunction i-Ga$_{0.47}$In$_{0.53}$As/i-InP/p$^+$
-InP.

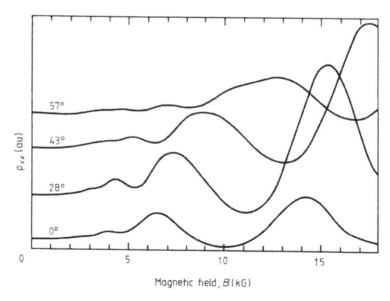

Fig. 12.49. Longitudinal magnetoresistance ρ_{xx} of LP-MOCVD GaInAs–InP versus
magnetic field for various angles at 4.2 K.

Fig. 12.49 shows Shubnikov–de Haas (SdH) oscillations for various angles θ between the field direction and the normal to the interface. The ρ_{xx} extrema appear at constant values of the perpendicular component of the magnetic field $B_\perp = B \cos \theta$ (Fig. 12.50), demonstrating the two-dimensionality of the holes at the interface.

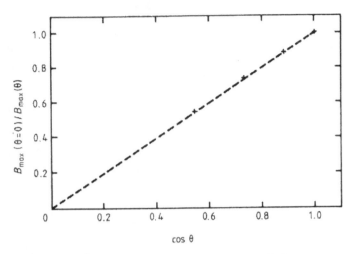

Fig. 12.50. Angular dependence of the magneto-oscillations for the maximum $B_{max}(0) = 6.5$ T. $V_{max}(\theta)$ is the total field applied as the angle increases. Broken line corresponds to the theoretical prediction for 2D behavior. LP-MOCVD GaInAs–InP, T = 4.2 K.

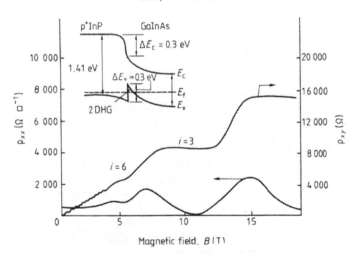

Fig. 12.51. Shubnikov–de Haas $\rho_{xx}(\Omega \cdot \square^{-1})$ and Hall effect $\rho_{xy}(\Omega')$ measurements as functions of magnetic field recorded at 4.2 K.

Fig. 12.51 shows the diagonal resistivity ρ_{xx} and the Hall resistivity ρ_{xy} as functions of magnetic field at 4.2 K. One can distinguish the steps typical for QHE analyses which also confirm the two-dimensionality of the system under study. It is worth pointing out the relatively large width of the step $\rho_{xy} = h/3e^2$ corresponding to three entirely occupied single Landau levels. In the case of an electron system, plateaux with odd indices occur at lower temperatures and develop into narrower structures than those having even indices [Von Klitzing et al. 1980 and Tsui and Gossard 1981]. This effect is attributed to a spin splitting of each Landau level. In contrast, the Hall plateaux observed in the hole system do not show this odd–even discrepancy [Stoermer et al. 1983]. This unique feature results from the lifting of the Kramers degeneracy of the 2D holes at the heterojunction interface and thus the splitting of each hole subband into two "spin subbands" even without the presence of a magnetic field.

The standard plot of the index of maxima and minima in ρ_{xx} versus inverse field positions is shown in Fig. 12.52. The slope $S = 8.96$ T gives (assuming the lifting of the Kramers degeneracy) a carrier concentration $p = (e/h)S = 2.16 \times 10^{11}\,\text{cm}^{-2}$. For $B > 10$ T ($1/B < 0.1$ T^{-1}) the slope starts to change, but unfortunately the accessible magnetic field ranging up to 18 T is not sufficient to observe the minima corresponding to two partially filled and one entirely filled Landau levels and to determine the high-field SdH periodicity.

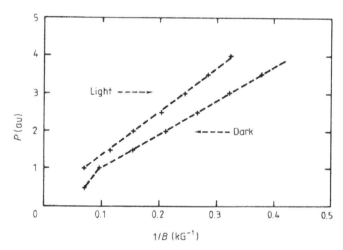

Fig. 12.52. Standard plots of indices of extrema in ρ_{xx} of LP-MOCVD GaInAs–InP versus 1/B before and after illumination. The change in slope corresponds to a change in carrier concentration displaying persistent photoconductivity. Integral indices correspond to minima in ρ_{xx}, half-integral to maxima. Broken lines are just guides for the eye.

One can try to estimate the total density of the carriers from the QHE plateau $\rho_{xy} = h/3e^2$ and take the corresponding SdH minimum as a value of the magnetic field B_3 necessary for complete filling of the three single Landau levels. Taking $p_{tot} = (e/h)NB_N$ and $NB_N = 3B_3 = 31.35$ T, we obtain $p_{tot} = 7.58 \times 10^{11}\,\text{cm}^{-2}$. The carrier concentration deduced from low-field SdH oscillations is significantly smaller than that determined from either classical or quantum Hall measurements. This observation has been reported previously for a 2D hole gas in GaAs/AlGaAs

[Stoermer et al. 1983 and Mendez et al. 1984] and has been attributed to the existence of two hole "spin subbands," degenerate at the Brillouin-zone center, with different effective masses generating independent sets of Landau levels. At low fields, only the lighter mass is resolved. At sufficiently high fields all Landau levels are resolved and, independent of their energies, carry a degeneracy eB/h. If we follow the interpretation that the hole density measured at low magnetic fields represents only a fraction of the total density populating the spin-down subband [Stoermer et al. 1983 and Takikawa et al. 1983] then the hole concentration of the spin-up subband with heavier effective mass is $p_{hh}^{+} = 5.42 \times 10^{11}$ cm^{-2} and the population ratio $p_{hh}^{+}/p_{hh}^{-} = 2.5$. In a rough approximation this ratio can be treated as the ratio of the effective masses $m_{hh}^{*}+ / m_{hh}^{*} -$. Our value $p_{hh}^{+}/p_{hh}^{-} = 2.5$ for $p_{tot} = 7.58 \times 10^{11}$ cm^{-2} is similar to those obtained for a 2D hole gas in GaAs–AlGaAs, which range between 1.8 and 3.3 for hole densities between 2.0×10^{11} and 1.15×10^{12} cm^{-2} [Stoermer and Tsang 1980, Mendez et al. 1984, and Chang et al. 1984]. This ratio has been found to increase with increasing carrier concentration, in agreement with recent calculations.

In contrast to the case of a 2D hole gas in GaAs/AlGaAs heterojunctions [Mendez et al. 1984 and Stoermer et al. 1984], a significant variation of the 2D hole concentration was found after illuminating the sample at low temperature (persistent photoconductivity effect) [Kastalsky and Hwang 1984]. Fig. 12.53 shows the SdH oscillations before and after illuminating the sample. The index extrema in SdH patterns versus $1/B$ are presented in Fig. 12.52. An increase of the 2D hole density after illumination is evident. The densities of the holes in both spin subbands before and after illuminating the sample are represented in Table 12.7.

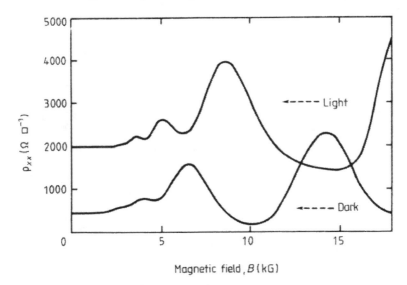

Fig. 12.53. Shubnikov–de Haas $\rho_{xx}(\Omega \cdot \square^{-1})$ measurements of LP-MOCVD GaInAs–InP at 4.2 K in the dark and after illumination with a red light-emitting diode.

	Before illuminating	After illuminating
p_H (cm^{-2})	6.31×10^{11}	8.67×10^{11}
p_{tot} (cm^{-2})	7.58×10^{11}	10.84×10^{11}
p_{hh}^{+} (cm^{-2})	5.42×10^{11}	8.00×10^{11}
p_{hh}^{-} (cm^{-2})	2.16×10^{11}	2.84×10^{11}
μ_H (cm$^2 \cdot$V$^{-1} \cdot$s^{-1})	10,450	7,320
μ_{hh}^{+} (cm$^2 \cdot$V$^{-1} \cdot$s^{-1})	6,240	4,100
μ_{hh}^{-} (cm$^2 \cdot$V$^{-1} \cdot$s^{-1})	14,850	10,770
p_{hh}^{+}/p_{hh}^{-}	2.5	2.8
$\mu_{hh}^{-}/\mu_{hh}^{+}$	2.4	2.6

Table 12.7. Densities of the holes in both spin subbands before and after illuminating the sample.

Considering the total hole concentration $p_{tot} = p_{hh}^{+} + p_{hh}^{-}$ and the value p_H obtained from low-field Hall measurements, one can see that p_H always underestimates the total hole concentration p_{tot}. Solving the equations which describe Hall density p_H and Hall mobility μ_H of a system consisting of two types of carriers with different mobilities, we can determine the relative mobilities μ_{hh}^{+} and μ_{hh}^{-}. Taking into account that for 2D hole spin subbands $p_{hh}^{+} > p_{hh}^{-}$ and $\mu_{hh}^{+} < \mu_{hh}^{-}$ we obtain

Eq. (12.20) $$\mu_{hh^+} = \frac{\mu_H p_H}{p_{hh^+} + p_{hh^-}} \left\{ 1 - \left[\frac{p_{hh^-}}{p_{hh^+}} \left(\frac{p_{hh^+} + p_{hh^-}}{p_H} - 1 \right) \right]^{1/2} \right\}$$

and

Eq. (12.21) $$\mu_{hh^-} = \frac{\mu_H p_H}{p_{hh^+} + p_{hh^-}} \left\{ 1 - \left[\frac{p_{hh^-}}{p_{hh^-}} \left(\frac{p_{hh^+} + p_{hh^-}}{p_H} - 1 \right) \right]^{1/2} \right\}$$

The values of μ_{hh}^{+} and μ_{hh}^{-} calculated for two different total hole densities (before and after illuminating the sample) are presented in Table 12.7.

It is worth noting that the mobility of the holes decreases with increasing density. This is in contrast to the behavior of the impurity-scattering mobility of 2D electrons in a single subband.

Finally, the feasibility of confining a 2D hole gas in a GaInAs/InP interface and the expectation that the valence-band discontinuity $\Delta E_C \simeq \Delta E_V \simeq 0.3$ eV offer the possibility of generating alternating layers of 2D electron gas and 2D hole gas. Such 2D electron-gas and hole-gas structures could yield new physical phenomena as well as interesting device possibilities.

12.6.3. Precise quantized Hall resistance measurements in $In_xGa_{1-x}As/InP$ heterostructures

The discovery of the quantum Hall effect offered metrologists a very remarkable means for establishing a resistance standard depending only on fundamental constants [Von Klitzing and Ebert 1985]: the Hall resistivity or resistance, ρ_{xy}, of a two-dimensional electron gas (2DEG) can take, in a high magnetic field and at low temperature, quantized values

Eq. (12.22) $R_H(i) = h/ie^2$

where h is the Planck constant, e the elementary charge and $i = 1, 2,...$ the number of filled Landau levels of the 2DEG.

Several precise measurements [Bliek et al. 1985, Cage et al. 1985, Wada et al. 1985, Endo et al. 1985, and Hartland et al. 1985] in metal-oxide semiconductor field-effect transistors (MOSFET) and/or $GaAs/Al_xGa_{1-x}As$ heterostructures have already demonstrated the excellent reproducibility, for a given laboratory and for a given experimental equipment, of the quantized Hall resistance $R_H(i)$. In Bliek et al. [1985], an agreement of $3 \times 10^{-8} \pm 3 \times 10^{-8}$ is reported for measurements of $R_H(i)$ made in the two kinds of devices.

However, the reproducibility from one laboratory to another, checked through the intercomparison of 1 Ω standards organized by the Bureau International des Poids et Mésures (BIPM, Sèvres), is not at present better than a few parts in 10^7. The observed differences are probably due mainly to random and systematic errors in the measurement of $R_H(i)$ in terms of the 1 Ω standards and, to a lesser extent, to the uncertainty of the 1 Ω intercomparison.

In this section, we present measurements of the quantized Hall resistance, made at LCIE over a 10-month time period, on an $In_xGa_{1-x}As/InP$ heterostructure grown by LP-MOCVD with a double aim:

(1) to give further experimental confirmation of the universality of the quantum Hall effect by testing InP-based heterostructures;

(2) to measure $R_H(i=2) \simeq 12,906.4$ Ω and $R_H(i=4) \simeq 6453.2$ Ω, in terms of the four 1 Ω standard resistors which comprise Ω_{LCIE}, with a low uncertainty (2.2×10^{-8}) thanks to a specially designed resistance-ratio measurement bridge using a cryogenic current comparator (CCC) [Delahayer and Reynmann 1985].

The sample consists of a 1000 Å thick InP buffer layer with carrier concentration of $N_D - N_A \simeq 3 \times 10^{15}$ cm^{-3} and a 1 μm $In_{0.53}Ga_{0.47}As$ layer with $N_D - N_A \simeq 2 \times 10^{15}$ cm^{-3}. The Hall mobility of this sample was 11,000 cm$^2 \cdot$V^{-1} s^{-1} at 300 K, 50,000 cm$^2 \cdot$V$^{-1} \cdot$s^{-1} at 77 K and 60,000 cm$^2 \cdot$V$^{-1} \cdot$s^{-1} at 4 K.

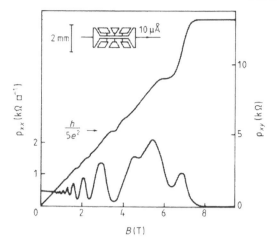

Fig. 12.54. Resistivity data ρ_{xx} and ρ_{xy} at 1.3 K for an $In_{0.53}Ga_{0.47}As/InP$
heterostructure (MOCVD) growth technique ($n = 4.3 \times 10^{11}$ cm^{-2},
$\mu = 60,000$ cm^{2}·V^{-1}·s^{-1}).

Fig. 12.54 shows resistivity data ρ_{xx} and ρ_{xy} as a function of B, at 1.3 K, for this sample ($\mu = 60,000$ cm^{2}·V^{-1}·s^{-1}, $n \approx 4.3 \times 10^{11}$ cm^{-2}). In this sample the second electric subband (E_1) is populated for $B \leq 7.5$ T. This explains the unusual well marked plateau at about $h/5e^2$; four Landau levels of the first subband E_0 and one of the second (E_1, $0\uparrow$) are entirely filled.

We note here that, for $B \geq 7.5$ T, subband E_1 is entirely depleted and the $h/2e^2$ plateau is observed. Measurements of R_H on this plateau ($B \approx 9$ T) are used for the precision measurement.

The experimental setup (a specially designed resistance-ratio measurement bridge using a cryogenic current comparator) is described by Delahaye et al. [1985] and allows experiments with a 2.2×10^{-8} uncertainty. The $i = 2$ plateau, which is associated with the smallest minimum of $\rho_{xx}(\rho_{xx}^{min} \sim 10^{-3}$ Ω·$\square^{-1}$ at 1.3 K), was investigated between 2 and 1.3 K. The plateau is flat to better than $\pm 5 \times 10^{-8}$ over a 0.4 T range. The Hall resistance as a function of ρ_{xx}^{min} shows the linear variation

Eq. (12.23) $\rho_{xy}\left(\rho_{xx}^{min}\right) = \rho_{xy}(0) + 0.07\rho_{xx}^{min}$

A similar linear law was previously observed in GaAs–Al$_x$Ga$_{1-x}$As HJ and may be due, at least partly, to the misalignment of the Hall probes. The most interesting result is that the extrapolated value $\rho_{xy}(0)$ in non-dissipative conditions, i.e. at $\rho_{xx}^{min} = 0$, is the same, within experimental uncertainty, as the value obtained on GaAs–Al$_x$Ga$_{1-x}$As HJ. The value of $\rho_{xy}(0)$ is expected to be equal to the theoretical value $h/2e^2$. At 1.3 K, the correction $\Delta\rho_{xy}/\rho_{xy}(0)$ due to the finite value of ρ_{xx}^{min} is less than 10^{-8}.

12.6.4. Persistent photoconductivity and the quantized Hall effect in $In_{0.53}Ga_{0.47}As/InP$ heterostructures grown by MOCVD

The low-field transport of the 2DEG shows a persistent photoconductivity (PPC) for $T < 200$ K, similar to that observed in the $GaAs–Al_xGa_{1-x}As$ heterostructure [Drummond et al. 1982 and Stoermer et al. 1981]. The persistent photoconductivity effect, in this case, is a consequence of increases in free electrons, due to photogeneration of electron–hole pairs in the heterostructure and trapping of the holes in the $In_{0.53}Ga_{0.47}As$ layer.

The heterostructure investigated by Tsui et al. [1981] consists of a 1500 Å thick *n*-type InP layer and a 1 μm thick undoped $In_{0.53}Ga_{0.47}As$ layer, sequentially grown on a semi-insulating InP substrate.

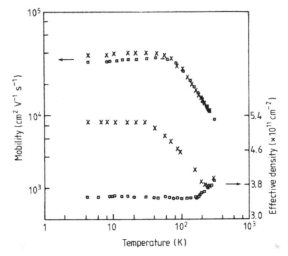

Fig. 12.55. Temperature dependence of the electron mobility, from the product of the conductivity and the Hall coefficient, and the electron density, from the low-field Hall measurements, using the one-electron model. Squares (□) are data taken in the dark. Crosses (X) are taken after exposure to light from a $GaAs_{0.6}P_{0.4}$ LED at 4.2 K.

Fig. 12.55 shows the temperature dependence of the electron density, obtained from the Hall coefficient using the one-carrier model, and the Hall mobility, obtained from the product of the conductivity and the Hall coefficient. The squares are from the dark; the crosses are from the data taken after the sample was exposed at 4.2 K to the light from a red light-emitting diode (LED). In the dark, the density is about 3.5×10^{11} cm^{-2} at 4.2 K and remains constant to better than ±2% for $T < 140$ K. Above 140 K, an exponential increase is observed, consistent with thermal activation of donor impurities in the InP, with an activation energy of approximately 7 meV. The mobility was $\mu = 3.5 \times 10^4$ cm$^2 \cdot$V$^{-1} \cdot$s^{-1} at 4.2 K and shows a slight increase with increasing T up to $T \approx 60$ K, as expected from scattering of the 2DEG by screened impurity ions close to the interface. It begins to decrease above 60 K and follows an exponential decrease, consistent with scattering dominated by LO phonons [Lin et al. 1984].

At 4.2 K, illumination with light from the red LED causes an increase of about 50% in the sample conductivity and this photoconductivity persists after the LED is turned off. As shown in Fig. 12.55, the electron density, after illumination, is increased to about 5.0×10^{11} cm^{-2} and it remains constant up to $T \simeq 40$ K. It starts to decrease at higher T and returns to its dark value for $T > 250$ K. The density decrease is due to the temperature dependence of the electron–hole recombination process. The mobility, on the other hand, shows a slight increase from its dark value for $T < 60$ K and no observable change at higher T. Thus, it is apparent that the observed persistent photoconductivity is due to an increase of free electrons in the heterostructure and the slight increase in mobility, below 60 K, is a result of the decrease in scattering of the 2DEG by screened impurity ions with increasing 2DEG density.

Similar persistent photoconductivity was previously observed in the transport of 2DEG in GaAs–Al$_x$Ga$_{1-x}$As heterostructures and the effect was attributed to photogeneration of DX centers in the Al$_x$Ga$_{1-x}$As. More recently, the importance of electron–hole pair generation in the GaAs and their subsequent separation by the junction electric field was recognized [Nathan et al. 1983]. Kastalsky and Hwang [1983] concluded that the latter mechanism is in fact the dominant mechanism for the persistent photoconductivity observed in their samples.

However, InP is not known to have DX centers and the effect is interpreted as photogeneration of electron–hole pairs in the heterostructure and subsequent trapping of holes in the In$_{0.53}$Ga$_{0.47}$As. In Fig. 12.56, we show the data on the quantized Hall effect [Tsui and Gossard 1981] at 4.2 K, under the influence of the infrared radiation from an In$_{0.53}$Ga$_{0.47}$As LED which can create electron–hole pairs in the In$_{0.53}$Ga$_{0.47}$As but not in the InP. In the dark, the $i = 4$ quantum Hall plateau (i.e. $\rho_{xy} = h/4e^2$) and its concomitant minimum in the diagonal resistivity ρ_{xx} are observed at $B \sim 3.5$ T. The 2DEG density obtained from the quantum oscillations is $n_{2DEG} = 3.5 \times 10^{11}$ cm^{-2}, in good agreement with that from the low-field Hall data. After illumination, two changes are apparent in the data shown in the lower two panels. First, the persistent photoconductivity is seen as decreases in the resistivity at $B = 0$. Secondly, an increase in the 2DEG density is seen as decreases in the period of the quantum oscillations and as corresponding shifts in the B field position of the quantum Hall plateaux. With excitation currents, $i_{LED} = 13$ and 29 μA, applied to the LED the $i = 4$ plateau is shifted to $B \sim 4$ and 4.8 T, indicating increases in the 2DEG density to $n_{2DEG} = 3.9 \times 10^{11}$ cm^{-2} and 4.8×10^{11} cm^{-2}, respectively. This increase in n_{2DEG} results from the photogeneration of electron–hole pairs in the In$_{0.53}$Ga$_{0.47}$As and their subsequent separation by the heterojunction electric field (see inset of Fig. 12.57). While the electrons are added to the 2DEG, the holes are trapped, prohibiting them to recombine. It should be emphasized that there is no observable change in either the flatness of the quantum plateau or the depth of the ρ_{xx} minima. This result is direct evidence that the illumination does not lead to any parallel conducting channels and there are no photo-induced free carriers in the InP. It differs from the 2DEG system in the GaAs/Al$_x$Ga$_{1-x}$As heterostructure, where illumination by infrared radiation of lower energy than both energy gaps is known to give rise to a parallel conduction channel in the Al$_x$Ga$_{1-x}$As.

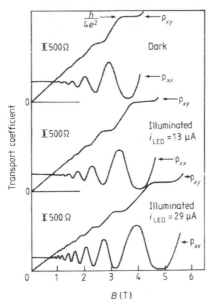

Fig. 12.56. Quantum Hall effect measured in the dark (top), after exposure to a $Ga_{0.47}In_{0.53}As$ LED with current i_{LED} = 13 μA (middle) and with i_{LED} = 29 μA (bottom).

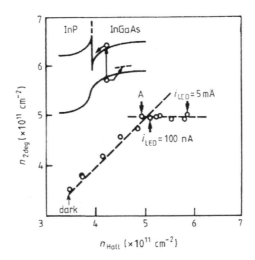

Fig. 12.57. Electron density, from quantum oscillations, versus the density, from low-field Hall measurements, with different excitation currents applied to the LED. The inset illustrates the mechanism for the persistent photoconductivity. $In_{0.53}Ga_{0.47}As$–InP, $GaAs_{0.6}P$ LED, T = 4.2 K.

The increase of n_{2DEG} with increasing excitation current applied to the LED saturates when n_{2DEG} reaches about 5×10^{11} cm^{-2}. When the sample is under illumination with $i_{LED} \geq 1$ mA, the quantum oscillations begin to show an additional period, indicative of population of the higher subband, and the quantum Hall plateau begins to have extra structures. After the LED is turned off, these extra structures disappear and the 2DEG density, calculated from the quantum oscillations, remains $n_{2DEG} \sim 5 \times 10^{11}$ cm^{-2}, independent of i_{LED}. This phenomenon is due to the fact that at saturation all the hole traps are filled and the additional electron–hole pairs generated at higher radiation intensities can contribute to charge transport only under the steady-state condition maintained by the illumination. The total number of traps in unit area obtained from the photo-induced n_{2DEG} is 1.5×10^{11} cm^{-2}. If we assume a uniform distribution in the In$_{0.53}$Ga$_{0.47}$As, we obtain a density of hole traps $n_t \sim 1.5 \times 10^{15}$ cm^{-3}.

Finally, similar results are obtained with radiation from a red LED that can generate electron–hole pairs in both the InP and the In$_{0.53}$Ga$_{0.47}$As. When the excitation current $i_{LED} < 100$ nA, only shifts in the B field position of the quantum effects are observed. At higher excitation current, ρ_{xx} shows an increasing background with increasing B and extra structures appear in the quantized Hall plateau. Again, these extra structures disappear after the LED is turned off. Since the penetration depth of the light is comparable to the thickness of the In$_{0.53}$Ga$_{0.47}$As, penetration into the InP at high intensities can create electron–hole pairs in InP. However, due to the lack of available hole traps, they cannot contribute to the persistent photoconductivity. In Fig. 12.57, we show the results on n_{2DEG}, obtained from quantum oscillations, as a function of the density, obtained from the low-field Hall data. Each point is obtained using a different excitation current to the red LED. All the points to the right of A, obtained under illumination, return to point A after the LED is turned off. Again a saturation is observed at $n_{2DEG} \sim 5 \times 10^{11}$ cm^{-2}.

12.6.5. *The effect of hydrostatic pressure on a Ga$_{0.47}$In$_{0.53}$As/InP heterojunction with three electric subbands*

The GaInAs–InP heterojunction has been studied under hydrostatic pressures up to 15 kbar [Gauthier et al. 1986]. In the range of pressure available, Shubnikov–de Haas (SdH) as well as low-field Hall and quantum Hall effects (QHE) (*Fig. 12.58* and *Fig. 12.59*) at 4.2 K have been measured. The total electron concentration is deduced from high-field SdH measurements and from the position of the $n = 2$ plateau of the quantum Hall effect. The values are very close to those obtained by low-field Hall effect measurements, pointing out the fact that no parallel conduction is introduced (Fig. 12.60). For more accurate details the reader can refer to Table 12.9, which summarizes the principal experimental results. The magnitude of the observed decrease of the total electron population agrees well with the pressure change of the conduction-band discontinuity ΔE_c if one assumes a constant conduction-band offset of 65%, i.e. $\Delta E_c = 0.65[E_{g2}(p) - E_{g1}(p)]$, and the triangular-well approximation.

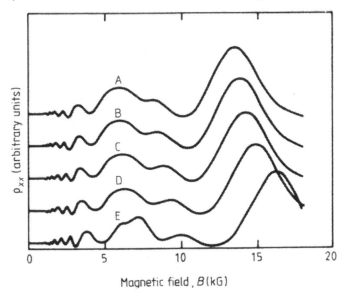

Fig. 12.58. Shubnikov–de Haas recordings for five different pressures (kbar): A, 15; B, 12.1; C, 8.8; D, 4; E, 0.

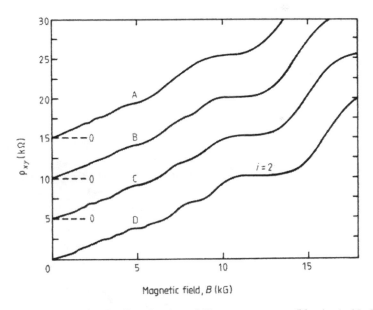

Fig. 12.59. Quantum Hall effect for four different pressures (kbar): A, 15; B, 8.85; C, 4.2; D, 0. T = 4.2 K.

518 The MOCVD Challenge

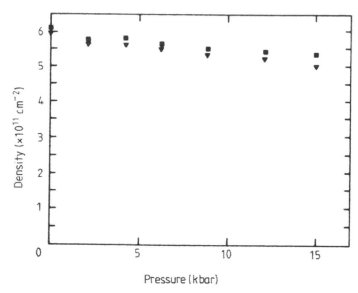

Fig. 12.60. Comparison between total carrier densities obtained by Shubnikov–de Haas (■) and low-field Hall effect (▼) measurements.

N_0 and N_1, the populations of the ground and first excited subbands, are deduced from the two sets of low-field SdH oscillations. Fig. 12.61 shows the dependence of N_S, N_0, N_1 and N_2 upon pressure.

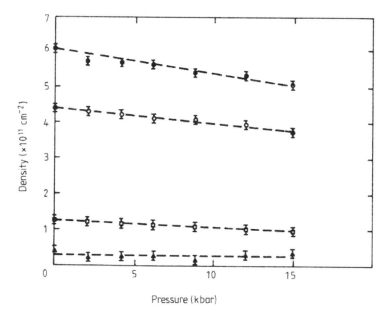

Fig. 12.61. Evolution of the population of the different subbands as well as total electron concentration versus pressure.

A decrease of 0.95×10^{11} cm^{-2} in total population leads to a decrease of 0.64×10^{11} cm^{-2} for the ground subband and 0.31×10^{11} cm^{-2} for the first excited subband, the third subband keeping roughly the same population.

In order to confirm the population of the third subband even at high pressure we have applied a parallel magnetic field ($\theta = 90°$) (Fig. 12.62). Still at 15 kbar, we see a fall in resistance at about 1.5 T due to the suppression of inter-subband scattering, indicating that the E_2 subband is not entirely empty at zero magnetic field. Taking the experimental results (Fig. 12.62) we have extrapolated to $B = 0$ with a quadratic fit the part of the curve where only one subband is populated. This can give us an order of magnitude for the importance of inter-subband scattering in this sample (for detailed results refer to Table 12.8). As can be seen, the contribution of inter-subband scattering decreases with increasing pressure.

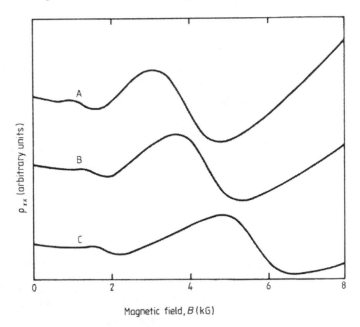

Fig. 12.62. Resistivity as a function of parallel magnetic field for different pressures (kbar): A, 15; B, 6; C, 0.

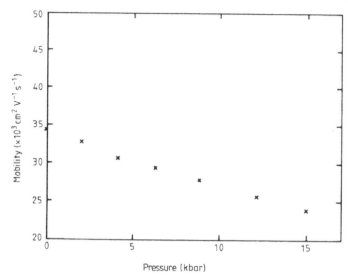

Fig. 12.63. Hall mobility versus pressure at 4.2 K up to 15 kbar.

The low-field Hall mobility of the 2DEG (Fig. 12.63) shows a decrease of about 30% for pressure up to 15 kbar. A fall in mobility of the same order of magnitude (28%) was observed earlier for pressures up to 8.8 kbar [Sotomayor et al. 1985], but in a sample displaying strong parallel conduction. The value of resistivity extrapolated to $B = 0$ for the case when only one subband is populated (taken from parallel magnetic field measurements) gives a decrease of 50%.

P (kbar)	R ($\Omega \cdot \Box^{-1}$) $B = 0$	R' ($\Omega \cdot \Box^{-1}$) (extrapolated) $B = 0$	$(R–R')/R$
0	298.4	127	0.574
2.2	357	153	0.571
4.2	393	172	0.562
6.2	432.8	193	0.554
10.1	482.4	245	0.492
12.1	535	275	0.486
15.4	587.8	310	0.471

Table 12.8. Sheet resistivity as a function of pressure.

P (kbar)	NH (10^{11} cm^{-2})	NS (10^{11} cm^{-2})	N0 (10^{11} cm^{-2})	N1 (10^{11} cm^{-2})	N2 (10^{11} cm^{-2})	EF–E0 (meV)	EF–E1 (meV)	E1–E0 (meV)	E2–E1 (meV)	$B_2^{\uparrow}$ a (T)	$B_1^{\uparrow}$ b (T)
0	6.06	6.04	4.39	1.27	0.38	21.6	6.3	15.3	4.9	1.7	5.55
2.2	5.75	5.73	4.29	1.22	0.22	20.7	5.9	14.8	4.5	1.625	4.85
4.2	5.82	5.67	4.22	1.18	0.27	20.0	5.6	14.4	4.3	1.575	4.55
6.2	5.65	5.62	4.11	1.15	0.28	19.1	5.4	13.7	4.1	1.5	4.425
8.85	5.57	5.39	4.10	1.11	0.18	18.6	5.0	13.6	3.8		
10.1										1.475	4.275
12.1	5.49	5.31	3.96	1.04	0.31	17.4	4.6	12.8	3.4	1.425	4.1
15.4	5.43	5.09	3.75	0.96	0.38	16.0	4.1	11.9	2.9	1.4	4.0

Table 12.9. Effect of pressure on a GaInAs/InP heterojunction.

a $B_2^{\uparrow}$ is the parallel magnetic field at which E_2 is depopulated.

b $B_1^{\uparrow}$ is the parallel magnetic field at which E_1 is depopulated.

12.6.6. Cyclotron resonance

Cyclotron resonance is the resonant absorption of light which occurs when the incident photon energy is equal to the separation between adjacent Landau levels. The resonance condition

Eq. (12.24) $\hbar\omega = \hbar\omega_c = heB / m*$

where ω is the frequency of the incident radiation, allows the effective mass of the electrons to be deduced.

Cyclotron resonance transitions are limited by the selection rule $\Delta n_L = \pm 1$, and so only one resonance should be seen. However, in two-dimensional systems short-range scattering can mix the Landau levels and cause additional resonances to appear [Ando et al. 1982].

Cyclotron resonance experiments described in these heterojunctions are performed by observing the transmission by the sample of fixed-frequency far-infrared (FIR) radiation as a function of magnetic field [Tsui and Gossard 1981]. The FIR sources were the H_2O (118 µm, 78.5 µm), D_2O (171.6 µm), HCN (336.6 µm, 310.9 µm) and DCN (194.8 µm, 190 µm) molecular lasers and carcinotrons ($\lambda = 1.2$ mm, 760 µm). The detectors are semiconductors which exhibit photoconductivity in the far-infrared at 4.2 K due to excitations within the impurity bands.

The far-infrared transmission is studied at fixed photon wavelength by varying the magnetic field applied tilted or normal to the surface of the sample. The finite width of the resonance peak can be explained by the fact that the transitions induced by the incident light occur between broadened Landau levels. Ando and Uemura [1974] showed that the presence of short-range scatterers in the system causes, even at $T = 0$ K, a broadening in energy Γ of the states, which can be related to the mobility μ of the electron gas, and to the magnetic field, by the relation

Eq. (12.25) $\Gamma = \dfrac{eh}{m*}\left(\dfrac{2}{\pi}\dfrac{B}{\mu}\right)^{1/2}$

In that case, we expect the cyclotron resonance halfwidth $\Delta B_{1/2}$ to depend on the resonant magnetic field inducing the Landau-level quantization according to a law of the form

Eq. (12.26) $\Delta B_{1/2} = \alpha \left(B/\mu\right)^{1/2}$

where α is a constant exhibiting the proportionality between $\Delta B_{1/2}$ and Γ. It has been found, in two-dimensional systems with mobilities similar to those of our samples, such as MOS silicium [Abstreiter et al. 1976] and a GaAs/GaAlAs heterojunction [Voisin et al. 1981], that the relation in Eq. (12.26) was remarkably verified, especially at low magnetic fields, when the assumption of short-range scatterers is fairly good. These different authors agreed in finding $\alpha = 0.63 \pm 0.02$.

Furthermore, with a detailed study of the cyclotron line for two-dimensional systems, Ando has predicted quantum oscillations in the main resonance peak [Ando 1975], which were experimentally reported [Abstreiter ct al. 1976 and Voisin et al. 1981] in the two previous systems. These beats are a consequence of the modulation of the resonance amplitude, as the Landau levels cross the Fermi level. They are clearly resolved at low magnetic fields, when many of the Ando oscillations occur along the cyclotron absorption line.

It is clear that they appear only within the region of the cyclotron resonance, because they arise from a variation in the resonance amplitude. Like Shubnikov–de Haas oscillations of the longitudinal component ρ_{xx} of the resistivity tensor, they are periodic in $1/B$ with a periodicity $\Delta(1/B)$, and still allow us to deduce the electronic density, namely

Eq. (12.27) $n = 2eh^{-1}\left[\Delta(1/B)\right]^{-1}$

In a normal configuration (magnetic field orthogonal to the surface of the sample), a cyclotron resonance experiment performed on a system exhibiting multi-subband transport cannot show hybrid transitions between Landau levels of different subbands.

The presence of a parallel magnetic field $B_\parallel$ introduces non-null terms in the overlapping matrix element, so that such hybrid transitions appear possible [Ando et al. 1982]. All spectra have been realized at $T = 1.6$ K, by varying the magnetic field in the range 0–8 T.

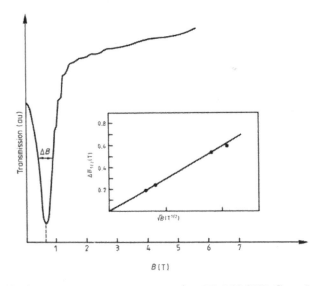

Fig. 12.64. Cyclotron resonance spectrum of an LP-MOCVD $Ga_{0.47}In_{0.53}As/InP$ heterojunction at $T = 1.6$ K, for $\lambda = 756$ μm. Inset: CR linewidth $\Delta B_{1/2}$ versus square root of the resonance magnetic field.

Fig. 12.64 shows a typical cyclotron line at $\lambda = 756$ μm with magnetic field applied normal to the sample for a $Ga_{0.47}In_{0.53}As/InP$ heterojunction exhibiting two occupied electric subbands (see following section), with respective populations $n_0 \simeq 4.2 \times 10^{11}$ cm^{-2} and $n_1 \sim 1.2 \times 10^{11}$ cm^{-2} deduced from Shubnikov–de Haas oscillations of the longitudinal resistivity ρ_{xx}. The effective mass, deduced from the plot of the resonance magnetic field, as a function of the pulsation of the incident wave, is constant. One finds $m^* = (0.047 \pm 0.001)m_0$, where m_0 is the electron mass, in good agreement with work on the same type of heterojunction [Nicholas et al. 1985 and Guldner et al. 1982].

Quantum oscillations are clearly evidenced in the region of the cyclotron resonance (CR). They are periodic in $1/B$, with a period of about $\Delta(1/B) = 0.129$ T^{-1}. The electronic density, deduced from Eq. (12.27), is $n \simeq 4 \times 10^{11}$ cm^{-2}. It corresponds quite well to the population of the ground subband.

We have shown as an inset the plot of the halfwidth at half-minimum $\Delta B_{1/2}$ of the resonance cyclotron line, as a function of the square root of the resonance field, namely $B_r^{1/2}$. The evolution appears almost linear, in good agreement with Eq. (12.26). Hall measurements gave a (relatively high) mobility, at 4 K, of about $\mu = 40,000$ cm$^2 \cdot$V$^{-1} \cdot$s^{-1}. Putting this value into Eq. (12.26) we find a proportionality constant $\alpha \simeq 0.60$, by fitting the linear relation. It is worth noting that the same factor is found in all the two-dimensional systems experienced: Si MOS, GaAs/GaAlAs and $Ga_{0.47}In_{0.53}As/InP$. It further confirms the validity of the theoretical prediction of the broadening of Landau levels with magnetic field.

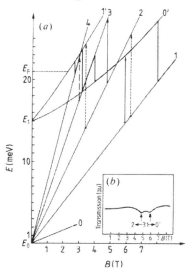

Fig. 12.65. (a) Evolution of the energy of the Landau levels, for a $Ga_{0.47}In_{0.53}As/InP$ heterojunction, with two occupied electric subbands, assuming an angle $\theta = 45°$ between the magnetic field and the normal to the surface of the sample. $\Lambda = 118$ μm (— — —), 164 μm (_____) and 255 μm (— · —). (b) Cyclotron resonance spectrum in this configuration for $\lambda = 164$ μm.

Fig. 12.65(*a*) exhibits the CR line in a tilted configuration at a wavelength of 164 μm. There is an angle of $\theta = 45°$ between the applied magnetic field and the surface of the sample.

A first resonance is resolved at $B_r \simeq 4.7$ T. It is deduced from normal cyclotron resonance, with a component of magnetic field normal to the surface of the sample: $B_\perp = B_r \cos \theta$. This further evidences the two-dimensionality of the system under study. At about $B = 5.9$ T there appears a second absorption peak attributed to a hybrid transition between the $n = 1$ Landau level of the ground electric subband and the $n = 0'$ Landau level of the first excited one. In order to explain this peak, the evolution of the Landau levels of the two subbands, as a function of magnetic field, for $\theta = 45°$, has been plotted in Fig. 12.65(*b*). The energy level E_0 of the ground subband has been taken as a reference. The energy difference $(E_1 - E_0) \simeq 15$ meV between the two subbands at zero magnetic field has been derived from their respective populations n_0 and n_1.

The effect of a parallel component of the magnetic field $B_\| = B_r \sin \theta$ is to increase this energy gap. It has been shown by Maurel et al. [1987] that this diamagnetic shift follows the equation

Eq. (12.28) $$\Delta(E_1 - E_0) = \frac{e^2 B_\|^2}{2m} \left[(\Delta z_1)^2 - (\Delta z_0)^2 \right] = \eta B_\|^2$$

where $(\Delta z_i)^2 = \langle z^2 \rangle_i - \langle z \rangle_i^2$ is the spread of the wavefunction, perpendicular to the interface, for the ith subband. The coefficient η has been determined experimentally, by observing the depopulation of the level of energy E_1 under the effect of a parallel magnetic field [Ando et al. 1982 and Razeghi et al. 1986]. Depopulation of the first excited subband manifests itself by a fall in resistance due to the elimination of intersubband scattering.

This has been retained in Fig. 12.65(*a*) to determine the evolution of the $n = 0'$ and $n = 1'$ Landau levels relative to the ground subband. The arrows represent the FIR transition experimentally observed for $\lambda = 118$, 164, and 255 μm. The evolution of the Fermi level is also shown as a darker line. The $B = 5.9$ T peak, exposed in Fig. 12.65 for $\lambda = 164$ μm, can be well explained by an $n = 0 \rightarrow n = 1'$ transition. Such a hybrid absorption has been observed, as well, for $\lambda = 118$ μm at $B_r = 3.33$ T, and can be attributed to an $n = 2 \rightarrow n = 1'$ transition, as shown in Fig. 12.65(*a*).

Cyclotron resonance experiments have been performed [Nicholas et al. 1979] using a number of different infrared laser wavelengths from 337 to 78.6 μm, over the temperature range 15 to 150 K. Direct transmission was measured as a function of magnetic field in the Faraday geometry, and some typical experimental recordings are shown in Fig. 12.66.

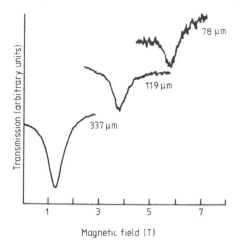

Fig. 12.66. Experimental recordings of the transmission of 337, 119 and 78 µm radiation by Ga$_{0.47}$In$_{0.53}$As as a function of magnetic field, for sample 85.

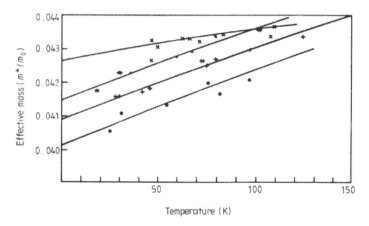

Fig. 12.67. Plots of the temperature variation of the effective mass for different laser frequencies for sample 85: (X) 79 µm; (*) 119 µm; (•) 337 µm.

The magnetic field positions of the resonances were then used to calculate the effective mass of the electrons. Fig. 12.67 shows plots of the temperature dependence of the effective mass for a number of different laser frequencies. The mass is found to increase with both increasing temperature and laser frequency. The data were then extrapolated to give the effective mass at 0 K for each different laser frequency. This procedure ensures that a value is obtained for m^* when only the lowest Landau level is populated. The variation of the 0 K effective mass as a function of laser energy is shown in Fig. 12.68. The effective mass obtained varies linearly with magnetic field. The data may again be extrapolated to zero energy to give the band-edge mass. The slope of the mass variation as a function of energy and temperature may then be used to provide a measurement of the band non-parabolicity. This is discussed in detail below, and is compared with three-band $k \cdot p$ theory [Palik et al. 1961]. The best value of the band-edge effective masses for n-$Ga_{0.47}In_{0.53}As$ are found to be in agreement with the earlier values given by Portal et al. [1978] and Nicholas et al. [1980].

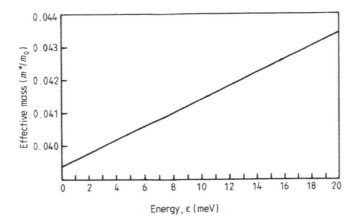

Fig. 12.68. The dependence of the effective mass for sample 85 upon laser energies, as determined from the values extrapolated to zero temperature, given in Fig. 12.67.

The same sample of $Ga_{0.47}In_{0.53}As$ was used in the magnetophonon as in the cyclotron resonance experiment. Typical experimental recordings are shown in Fig. 12.69. It is clear that oscillatory structure observed in sample 85 is not quite as simple as that observed in the components InAs and GaAs [Stradling et al. 1968]. In these materials a single series of oscillations is observed at magnetic fields satisfying the magnetophonon resonance condition

Eq. (12.29) $N h \omega_0 = h \omega_{LO}$

where ω_{LO} is the LO-phonon frequency.

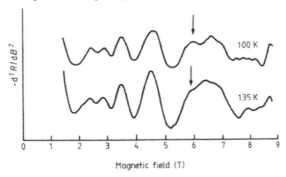

Fig. 12.69. Experimental recordings of the magnetophonon oscillations in the second derivative of the magnetoresistance in sample 85. The splitting of the higher-field peak into two LO-phonon peaks is shown by the arrow.

The oscillatory structure in sample 85 may be interpreted as due to the presence of two close series of oscillations each determined by Eq. (12.29) with fundamental fields differing by only 10%. Two series of magnetophonon oscillations in GaInAs originate from two phonon modes in the alloy system. These are "GaAs"-like and "InAs"-like modes. This is a clear contradiction of the suggestion by Pearsall et al. [1983] that only one LO-phonon mode exists in this alloy. It was not possible to resolve the 'InAs' mode in the MBE samples, which have only one series of magnetophonon oscillations originating from the GaAs-like mode.

The magnetophonon oscillations observed in $Ga_{0.47}In_{0.53}As$ obey Eq. (12.29) where $\hbar\omega_c$ is given by a suitably defined effective mass. Thus one may use this relation to calculate either the effective mass or the phonon frequencies by the assumption of the other. Results for the magnetophonon oscillations in the sample are given in Table 12.10 using the phonon frequencies measured by Brodsky and Lucovsky [1968].

Sample	T (K)	InAs phonon			GaAs phonon		
		$NB_N^*(T)$	$\hbar\omega_{LO}$ (cm^{-1})	m_{mpr}^*(BE)	$NB_N^*(T)$	$\hbar\omega_{LO}$ (cm^{-1})	m_{mpr}^*(BE)
85	97	11.25	235	0.041	12.61	271	0.040

Table 12.10. The magnetophonon series fundamental fields and effective masses. Where NB_N^* is the fundamental field measured for the series, and m_{mpr}^*(BE) is the experimental magnetophonon mass after the non-parabolic, the polaron and the second-derivative technique corrections.

The variation of the cyclotron effective mass as a function of laser energy is a consequence of the non-parabolicity of the conduction band. When the electrons are further into the conduction band, the effective mass is found to increase, as shown in Fig. 12.69, as a result of the non-linear variation of the Landau levels with applied magnetic field.

The general variation in the effective mass can be written as [Palik et al. 1961]:

Eq. (12.30)
$$\frac{1}{m^*(E)} = \frac{1}{m_0^*}\left[1 + \frac{2K_2}{E_g}\left(\tfrac{1}{2}kT + \hbar\omega_c + \frac{\hbar\omega_c}{\exp(\hbar\omega_c / kT) - 1}\right)\right]$$

where the constant K_2 is given by

Eq. (12.31) $K_2 = -[(1 + \tfrac{1}{2} x^2)/(1 + \tfrac{1}{2} x)](1 - y)^2$

with

Eq. (12.32) $x = [1 + (\Delta / E_g)]^{-1}$ and $y = m^*/m_0$

where E_g is the band gap and Δ is the spin–orbit splitting. In addition to the band non-parabolicity, the resonant polaron contribution to the effective mass increases as the laser frequency increases towards the LO-phonon frequency (ω_{LO}), as has been calculated by Larsen [1964]. This can be accounted for in Eq. (12.30) by replacing K_2 by K_2' where

Eq. (12.33) $\qquad |K_2'| = |K_2| + 3\alpha E_g / 40\hbar\omega_{LO}$

where α is the Fröhlich coupling constant. However, one can simplify Eq. (12.30) for the low- and high-temperature limits respectively. For high temperatures, one can approximate Eq. (12.30) to

Eq. (12.34) $\qquad \dfrac{1}{m*(E)} = \dfrac{1}{m_0^*}\left\{ 1 + \dfrac{2K_2}{E_g}\left[\hbar\omega_c + \tfrac{1}{2}kT + \hbar\omega_c \exp\left(\dfrac{-\hbar\omega_c}{kT} \right) \right] \right\}$

and for the low-temperature regime Eq. (12.30) has the form

Eq. (12.35) $\qquad \dfrac{1}{m*(E)} = \dfrac{1}{m_0^*}\left(1 + \dfrac{2K_2}{E_g}\left(\tfrac{1}{2}\hbar\omega_c + \tfrac{3}{2} \right) \right)$

By considering the frequency dependence of the low-temperature effective mass, an experimental value of K_2' can be derived from the slope of the plot shown in Fig. 12.68. The experimental value for K_2' is $K_2' = -2.05$ for sample 85. The theoretical value for K_2' defined in Palik et al. [1961] is $K_2' = -0.9494$, including the contribution of the resonant polaron [Larsen 1964].

The cyclotron mass is expected to be temperature-dependent and is found to be increasing with temperature. The effective band-edge mass will be decreasing with increasing temperature owing to the decreasing energy gap as, according to $k \cdot p$ theory,

Eq. (12.36) $\qquad m_0 / m_2^* \simeq 1 + 2P^2 / E_g$

While the band-edge mass is thus decreasing with increasing temperature, the cyclotron effective mass is found to be increasing, as the carriers are excited further into the band. This temperature dependence may be used to deduce a lower limit for K_2', using Eq. (12.34) in the high-temperature limit.

The experimental value K_2' obtained from the slope of the straight-line fit for the effective-mass variations with temperature, given in Fig. 12.67 and Fig. 12.68, is $K_2' = -1.47$, at fixed laser frequency. The gradient was found to become less steep for short-wavelength measurements, as these did not reach the high-temperature limit in the temperature range studied, as suggested by Eq. (12.30). The values deduced from the temperature dependence will be a lower limit on K_2' owing to the compensating effect of the decrease in band gap, as described earlier.

In conclusion, the variations of the effective mass with temperature and magnetic field yield a very important result. This is that the effect of non-parabolicity in the alloy system of $Ga_xIn_{1-x}As$ is very large, larger than would be expected from conventional three-band $k \cdot p$ theory. The $k \cdot p$ theory was originally proposed by Kane [1957].

Herman and Weisbuch [1977] found that a linear interpolation of the momentum matrix elements for the two constituent compounds could give a good description of the electronic g-factor in ternary alloys. However Nicholas et al. [1979 and 1980] and Portal et al. [1978] have shown that the variation of the effective mass as a function of alloy composition provides clear evidence for a reduction in matrix element, probably due to band mixing by disorder.

12.6.7. Shallow-donor spectroscopy and polaron coupling in $Ga_{0.47}In_{0.53}As$–InP grown by MOCVD

The energy levels associated with shallow hydrogenic impurities have been one of the most intensively studied areas in semiconductor physics. Extrinsic photoconductivity is most commonly used in these studies, frequently with the application of high magnetic fields [Stillman et al. 1976 and Yafet et al. 1956]. This method has been developed to provide a spectroscopic tool to identify residual contamination due to impurities in high-purity materials [Stillman et al. 1976 and Cooke et al. 1978].

Nicholas et al. [1984] have studied the shallow-donor energy levels in high magnetic fields, in the alloy $Ga_xIn_{1-x}As$ ($x = 0.47$), grown lattice-matched to InP. Transitions between 1s and 2p levels of the shallow hydrogenic donors have been observed in photoconductivity using Fourier-transform spectroscopy at fixed magnetic fields and also using fixed-wavelength infrared lasers in varying magnetic fields.

Epitaxial samples of high-purity n-$Ga_{0.47}In_{0.53}As$ were studied, which had been grown lattice-matched to InP. The sample characteristics are 6 μm thick GaInAs on InP substrate, with $\mu(300\ \text{K}) = 12,000\ \text{cm}^2 \cdot \text{V}^{-1} \cdot \text{s}^{-1}$ and $\mu(77\ \text{K}) = 48,000\ \text{cm}^2 \cdot \text{V}^{-1} \cdot \text{s}^{-1}$.

Some typical experimental recordings are shown in Fig. 12.70 for two wavelengths, and for comparison a recording of the sample transmission is also shown. The two transitions observed correspond to free-electron cyclotron resonance, at higher field, and 1s to $2p^+$ shallow-donor transitions at lower field. Both transitions are observed, due to the partial carrier freeze-out which occurs at 4.2 K in these samples. As the magnetic field increased, the sample resistivities rose rapidly due to a very pronounced magnetic freeze-out, and the $1s$–$2p^+$ transition became dominant at high fields. As the frequencies approached those of the optic phonons, the strength of the photoconductive signal decreased markedly, and the transitions began to broaden significantly. The optic phonon properties of $Ga_{0.47}In_{0.53}As$ are rather complex, due to its two-optic-phonon-mode behavior [Brodsky and Lucovsky 1968], which leads to the definition of two separate polaron coupling constants, α_{GaAs} and α_{InAs}. Such a two-mode polaron system has been considered by Nicholas et al. [1979] and Swierkowski et al. [1978], who showed that, when the two modes are relatively close to each other in frequency, the coupling constants are modified and become

$$\text{Eq. (12.37)} \quad \alpha_1 = \left(\frac{e^2}{h\varepsilon(\infty)}\right)\left(\frac{m^*}{2h\omega_{L1}}\right)^{1/2}\left(1-\frac{\omega_{T1}^2}{\omega_{L1}^2}\right)\left(\frac{\omega_{L2}^2-\omega_{L1}^2}{\omega_{L2}^2-\omega_{L1}^2}\right)$$

$$\text{Eq. (12.38)} \quad \alpha_2 = \left(\frac{e^2}{h\varepsilon(\infty)}\right)\left(\frac{m^*}{2h\omega_{L2}}\right)^{1/2}\left(1-\frac{\omega_{T2}^2}{\omega_{L2}^2}\right)\left(\frac{\omega_{L2}^2-\omega_{T1}^2}{\omega_{L2}^2-\omega_{L1}^2}\right)$$

where $\omega_{L1,2}$ and $\omega_{T1,2}$ represent the longitudinal and transverse optic-phonon frequencies, and $\varepsilon(\infty)$ is the high-frequency dielectric constant. For $Ga_{0.47}In_{0.53}As$ this gives $\alpha_{GaAs} = 0.05$ and $\alpha_{InAs} = 0.01$, indicating that the GaAs optic-phonon mode should dominate the coupling. The results shown in Fig. 12.71 are consistent with this conclusion, since no large discontinuity in the effectives mass is observed on passing the "InAs"-mode LO phonon at 29 meV, although there are not sufficient measurements to make a more quantitative statement.

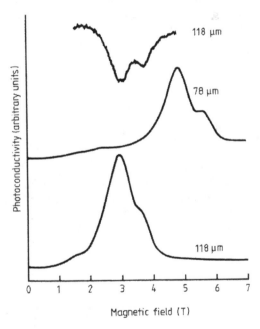

Fig. 12.70. Experimental recordings of the photoconductive response of the MOCVD sample 85, at two different wavelengths at 4.2 K. The upper trace shows a recording of direct transmission.

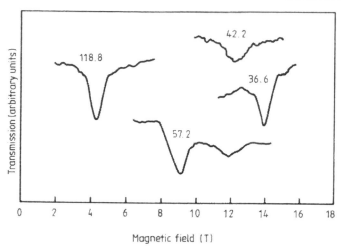

Fig. 12.71. Typical experimental recordings of the cyclotron resonance at 4.2 K in the two systems studied. The wavelengths (μm) are shown.

The shallow-donor ionization energy can now be calculated from the hydrogenic-impurity approximation. The effective mass is taken from the cyclotron resonance measurements, and the low-frequency dielectric constant is taken from a linear extrapolation of the values of 12.6 for GaAs and 14.4 for InAs [Landolt and Boernstein 1982]. Variational calculations of the shallow-donor energy levels in a magnetic field [Lee et al. 1972, Praddaude 1972, and Cabib et al. 1972], scaled with the effective Rydberg, are shown in Fig. 12.72 together with the experimental points. It may be seen that there is good agreement at low fields, but that as the transition energy approaches the optic phonons, bound-polaron coupling is observed [Cohn et al. 1972] to cause strong shifts in the levels. The data in this case are clearer than for the unbound polaron, and it is possible to distinguish separate coupling to the two different LO-phonon modes at 233 cm^{-1} (InAs) and 273 cm^{-1} (GaAs). Although it was not possible to detect the upper branch of the GaAs phonon, the much greater shifts in the resonance make it clear that this phonon dominates the coupling, as would be expected from the relative values of α_{GaAs} and α_{InAs} given above. It is interesting to note that very strongly broadened resonances were observed even at, or very close to, the LO-phonon frequencies for the InAs mode. These resonances fall at approximately the mean of the upper and lower polaron branches, and are visible as the almost vertical section of the experimental curve around 233 cm^{-1}. There is also an indication of a very weak peak in the photoconductivity, at energies corresponding to the transition 1s to 3d^{2+}, for sample 85 with 78 and 118 μm wavelengths. This extra peak does not appear at longer wavelengths owing to the low magnetic field position for this transition. The observation of the extra peak 1s to 3d^{2+} is unexpected due to the selection rule forbidding even-parity transitions. A mechanism responsible for the symmetry breaking could be disorder, either due to the relatively high doping level or to alloy fluctuations.

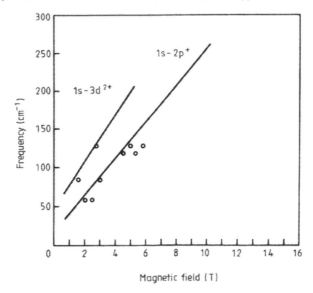

Fig. 12.72. The transition frequencies as a function of magnetic field, for the three samples studied. The full lines show the calculated values of the 1s–2p$^+$ and the possible 1s–3d^{2+} transitions. Circles (○) show experimental results grown by MOCVD.

12.7. Applications of GaInAs–InP system grown by MOCVD

12.7.1. PIN photodetector

The use of the ternary material InGaAs lattice-matched to InP for long-wavelength detectors is now well established [Stillman et al. 1982, Forrest 1982]. Among the known structures, the simple and reliable PIN has become one of the more attractive, enabling the fabrication of high-sensitivity PINFET optical receivers [Brain et al. 1984] as well as that of broad-area general-purpose photodiodes [Burrus et al. 1981]. Usually, the detectors reported in this field were either grown by LPE and hydride or chloride VPE; here we describe InGaAs PIN photodiodes prepared by MOCVD.

The structure consists of three nominally undoped layers deposited on an *n*-type InP substrate as follows: (i) an InP buffer layer 1.6 μm thick; (ii) the InGaAs absorption layer 3 μm thick; and (iii) an InP "window" layer, whose standard thickness of 1.6 μm has been decreased in some experiments to 0.5 μm in order to enhance the response in the near-infrared spectrum beyond 0.9 μm. The background impurity concentration deduced from capacitance–voltage measurements is routinely found in the 10^{15} cm^{-3} range (n-type), and is therefore suitable to achieve the low capacitance required for PINFET implementation. The *p–n* junction is next formed by Zn diffusion using the semiclosed box technique [Poulain and Cremoux 1980] and is located in the InGaAs layer typically 0.5–1 µm away from the upper hetero-interface. The ohmic contacts are made by sputtered Au–Zn alloyed at 425 °C for 30 s on the p side, and by sputtered Au on the n side. Mesas are finally denned by a chemical etch in a $Br_2/HBr/H_2O$ solution, the final device structure being schematically shown in Fig. 12.73. At this point, it should be noted that the junction is not placed at its optimum depth in order to achieve very high speeds of response. In fact, in the top illumination scheme used, the photoelectrons created near the InP window in the *p*-type part of the InGaAs layer have to diffuse down to the junction since this small region of the ternary material remains undepleted under reverse bias. Nevertheless, the increase in the response time due to this effect can be expected to be small enough for current light-wave applications [Poulain and Cremoux 1980].

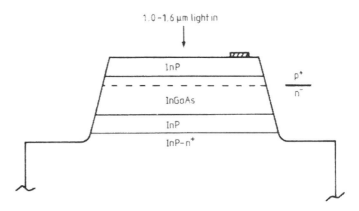

Fig. 12.73. Cross-sectional view of InGaAsP/InP mesa-type PIN photodiodes.

Three kinds of detectors have been made in the previously described wafers, differing mainly by the optically active area: (i) signal PIN with an 80 µm optical diameter $(S \sim 7 \times 10^{-5} \text{ cm}^2)$ to be used in front of 50 µm core fibers for the implementation of PINFET-type receivers; (ii) medium-sized feedback photodiodes $(S \sim 1.2 \times 10^{-3} \text{ cm}^2, 300 \text{ µm}$ optical diameter) for the monitoring of long-wavelength laser diodes; and finally (iii) large-diameter (1 mm) detectors for general-purpose both long- and short-wavelength applications.

The typical bias point found for our devices is 10 V, at which the capacitance is less than 0.3 pF, the quantum efficiency about 70% (1.3 µm wavelength), the impulse fall and rise time less than 400 ps. Fig. 12.74 gives a histogram of the dark current for one of the best wafers. The mean value is 2.3 nA and the standard deviation is 2 nA. Only the units with I_D less than 10 nA have been taken into account, which represents 237 diodes for a total number of units equal to nearly 400, including wafer-edge devices. Nearly 300 units had a dark current less than 25 nA, which represents almost all the diodes if one excludes wafer-edge ones. The performances here are in very close agreement with those measured on similar photodiodes also made in our laboratory but from LPE material. After cleaving and packaging, the performances are again checked. The quantum efficiency measured as a function of the wavelength using lock-in techniques oscillates between 60 and 70% in the 1.0–1.6 µm range, which corresponds to a nearly 100% internal efficiency for devices without an antireflection coating. The bandwidth has also been accurately determined with units mounted in a microwave package. The 3 dB cutoff frequency at the bias point is 3 GHz.

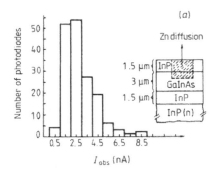 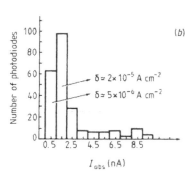

Fig. 12.74. Histogram of room-temperature dark current at −10 V bias for small-area (7 × 10⁻⁵ cm²) photodiodes: (a) LPE, I_{obs} = 2.8 nA; (b) MOCVD, I = 2.3 nA.

Concerning feedback-type and general-purpose detectors, the bias point is defined as 5 V, a voltage commonly found in electronic equipment. For the feedback type ($S \sim 1.2 \times 10^{-3}$ cm²), the mean dark current is 24 nA with a standard deviation of 16 nA. In this statistic, only the diodes for which I_D is less than 100 nA are taken into account, which represents 215 units out of 250 units including wafer-edge ones. Concerning broad-area general-purpose detectors, a poor yield of good devices was obtained since only 19 diodes out of a total number of 130 had a dark current less than 1 µA at 5 V. For these few units, the mean value is 240 nA and the standard deviation 190 nA.

12.7.2. Field-effect transistor

$In_{0.53}Ga_{0.47}As$ lattice-matched to InP has a favorable band structure configuration for the development of millimeter-wave field-effect transistors. A high electron velocity [Windhorn et al. 1982] can be achieved even at high doping levels, which will give high transconductance and high cutoff frequency. Furthermore, such a transistor could be integrated on the same chip as optoelectronic devices used for long-wavelength applications fabricated on InP substrates. This would result in an improvement of the overall performance of the circuit. Since InGaAs has too small a band gap to form a sufficiently high and leakage-free Schottky barrier, different solutions have been proposed to make the gate. A Si_3N_4 insulator [O'Connor et al. 1982] or a high-band-gap material such as InAlAs or InP [Cheng et al. 1984] can be grown on top of the channel layer; one can also use a *p–n* junction obtained either by diffusion [Onho et al. 1980 and Bandy et al. 1981], by ion implantation [Chai et al. 1985 and Vescan et al. 1982] or by etching a *p*-type epitaxial layer [Chai and Yeats 1983]. In this last case, the main difficulty is to reach a submicrometer length for the gate.

The fabrication of a junction field-effect transistor (JFET) in which the junction is formed at the interface of the *n*-type GaInAs channel and a *p*-type InP top layer has been described [Falco et al. 1987].

On an InP semi-insulating substrate, an InP buffer (0.5 μm, $N_D < 1 \times 10^{14}$ cm^{-3}), an *n*-type GaInAs channel (0.2 μm, $N_D = 1.3$–1.4×10^{17} cm^{-3}), a *p*-type InP layer (0.5 μm, $N_A = 1 \times 10^{18}$ cm^{-3}) and a *p*-type InGaAs layer (0.5 μm, $N_A > 5 \times 10^{18}$ cm^{-3}) were successively deposited. Silicon was used as the donor and zinc as the acceptor for doping. The gate metallization was evaporated first and used as a mask for the etching of the gate. This was performed using selective acid solutions in two steps: etching of the *p*-GaInAs top layer first and then the *p*-InP layer. It was found that obtaining a submicrometer gate length was much easier when the GaInAs top layer was present. Moreover, it was difficult to obtain a good ohmic contact on *p*-type InP whereas it was easy on GaInAs. A low value of gate resistance is required to achieve a low-noise microwave performance. Source and drain contacts (Au–Ge, Ti, Pt, Au) were then deposited self-aligned with respect to the gate metal. The fabrication then terminates by the etching of the isolation mesa. A cross-sectional view of the structure is shown in Fig. 12.75. A gate length of half a micrometer in a 1.5 μm source–drain spacing can be obtained by this process. The transistor width is 150 μm.

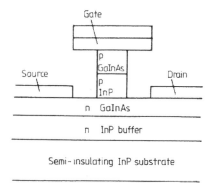

Fig. 12.75. Cross-sectional view of the JFET structure.

The DC characteristics of the junction field-effect transistor are shown in Fig. 12.76. One can observe (i) a good saturation behavior with high saturation currents, (ii) a good pinch-off, and (iii) a very high transconductance ($g_m = 235$ mS·mm^{-1}). The barrier height of the junction measured on its *I–V* characteristic is 0.35 V, which is lower than the expected value. The calculated value of the access resistance deduced from the slope of the characteristics at low drain voltage is 2.5 Ω for a 150 μm wide device (i.e. 0.4 Ω·mm). In consequence, the intrinsic transconductance is 260 mS·mm^{-1}, which is the highest value reported for a GaInAs field-effect transistor. Since the intrinsic transconductance can be approximated by the equation

Eq. (12.39) $g_m = Z \varepsilon \upsilon / \omega$

where ω is the depletion-layer thickness, one can deduce an average electron velocity of 3.6×10^7 cm·s^{-1}. Such a high value has not yet been observed in any field-effect transistor and can be attributed to an overshoot phenomenon induced by the shortness of the gate.

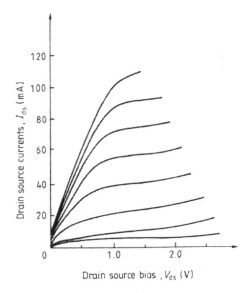

Fig. 12.76. DC characteristics of the JFET. The gate is 0.5 μm long and 150 μm wide. The transconductance is 235 mS·mm^{-1}. (Gate bias step is −0.5 V starting at 0 V)

12.7.3. GaInAs–InP optical waveguides

The development of the MOCVD method introduces possibilities for the investigation of passive and active structures at 10.6 μm wavelength.

Integrated optics geometry is particularly interesting [Falco et al. 1983] as it permits low-drive-voltage electro-optic devices, which is an important consideration at this wavelength. The large-substrate-area processing capability of MOCVD could lead to very efficient devices, especially if the propagation losses are low.

Here we describe experimental results concerning low-loss planar optical waveguides in the GaInAs–InP system obtained with the low-pressure MOCVD technique [Delacourt et al. 1987].

The waveguides are fabricated on heavily doped InP substrates (2×10^{18} cm^{-3}) and consist of a 5 μm thick undoped InP (n$^-$: 2×10^{14} cm^{-3}) buffer layer and a 5 μm thick undoped GaInAs (n$^-$: 6×10^{14} cm^{-3}) guide layer (Fig. 12.77). The dimensions of the substrates are typically 10×30 mm^2, and the losses are measured using the two-prism technique, which is very convenient at 10.6 μm.

The measurement on the GaInAs/InP waveguide is shown in Fig. 12.78, where losses as low as 0.7 dB·cm^{-1} are estimated for the TE$_0$ mode, compared with 0.3 dB·cm^{-1} theoretically expected. The small number of experimental points in

Fig. 12.78 is related to the very low losses of the waveguide, which lead to a difficulty in obtaining sufficient points for the measurements.

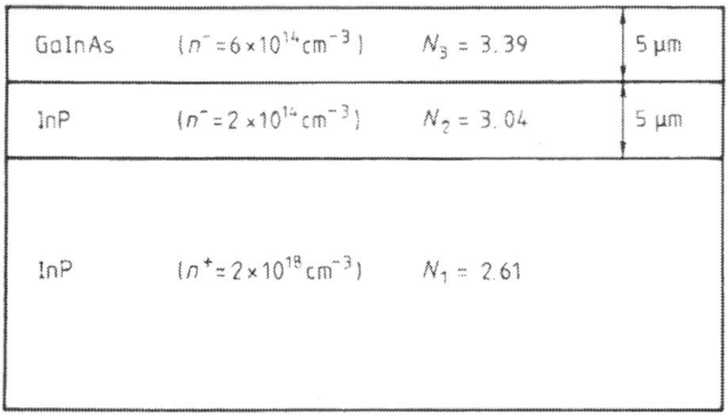

GaInAs
($n^- = 6 \times 10^{14}$ cm^{-3})
$N_3 = 3.39$
5 μm
InP
($n^- = 2 \times 10^{14}$ cm^{-3})
$N_2 = 3.04$
5 μm
InP
($n^+ = 2 \times 10^{18}$ cm^{-3})
$N_1 = 2.61$

Fig. 12.77. Schematic diagram of low-loss structure.

Experiment
3.0
2.5
log I
0
5
10
15
z (mm)

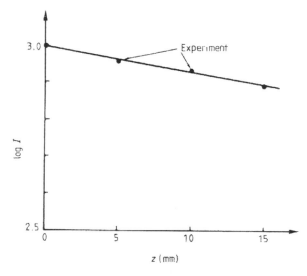

Fig. 12.78. Power coupled out against prism spacing for low-loss structure.

It must be noted that, even if this waveguide was not purely single-mode (four modes can propagate in theory), the higher-order modes present higher losses (TE_1: 4 dB·cm^{-1}; TE_2: 78 dB·cm^{-1}; TE_3: 370 dB·cm^{-1}), and as a consequence only the fundamental one can effectively be used for wave-guiding.

A very-low-loss quasi-single-mode waveguide was demonstrated using GaInAs/InP undoped epilayers on a heavily doped InP substrate. Losses as low as 0.7 dB·cm^{-1} have been measured, which demonstrates the usefulness of the MOCVD technique for this type of waveguide. These results are very encouraging for future implementation of more sophisticated devices around the 10.6 μm wavelength.

References

Abstreiter, G., Kotthaus, J.P., Koch, J.F., and Dorda, G. *Phys. Rev.* B **14** 6, 1976.
Alavi, K., Aggarwal, R.L., and Groves, S.H. *Phys. Rev.* B **21** 1311, 1980.
Alferov, Z.I. Sov. *Phys.–Semicond.* **1** 358, 1967.
Ando, T., Fowler, A.B., and Stern, F. *Rev. Mod. Phys.* **54** 2, 1982.
Ando, T. *J. Phys. Soc. Japan* **54** 1528, 1985.
Ando, T. *J. Phys. Soc. Japan* **39** 411, 1975.
Ando, T. and Uemura *J. Phys. Soc. Japan* **36** 959, 1974.
Ando, T. *J. Phys. Soc. Japan* **38** (4) 000, 1975.
Bangert, E. and Landwehr, G. *Proc. Int. Conf. on Superlattices, Microstructures and Microdevices, Urbana-Champaign, Illinois* (New York: Academic) vol 1, p 363, 1985.
Bandy, S., Nishimoto, C., Hyder, S., and Hooper, C. *Appl. Phys. Lett.* **38** 817, 1981.
Bartels, W.J. and Nijman, W. *J. Cryst. Growth* **44** 518, 1978.
Beinvogl, W., Kamgar, A., and Koch, J.F. *Phys. Rev.* B **14** 4274, 1976.
Beneking, H., Grote, N., and Selders, J. *J. Cryst. Growth* **54** 59, 1981.
Bimberg, D., Sondergeld, M., Schairer, W., and Yep, T.O. *J. Lumin.* **3** 175, 1970.
Bimberg, D., Sondergeld, M., and Grobe, E. *Phys. Rev.* B **4** 3451, 1971.
Bisaro, R., Merenda, P., and Pearsall, J.P. *Appl. Phys. Lett.* **34** 100–2, 1979.
Blakeslee, A.E. and Alliota, C.F. *IBM J. Res. Dev.* **14** 686, 1970.
Bliek, L., Braun, E., Melchert, F., Warnecke, P., Schlapp, W., Weimann, G., Ploog, K., Ebert, G., and Dorda, G. *IEEE Trans, Instrum. Meas.* **IM-34** 304–5, 1985.
Boothroyd, C.B. and Stobbs, W.M. Private communication, 1986.
Brain, M.C., Smyth, P.P., Smith, D.R., White, B.R., and Chidgey, P.J. *Electron. Lett.* **20** 894–6, 1984.
Brodsky, M.H. and Lucovsky, G. *Phys. Rev. Lett.* **21** 990, 1968.
Briggs, A., Guldner, Y., Vieren, J.P., Voos, M., Hirtz, J.P., and Razeghi, M. *Phys. Rev.* B **27** 6549, 1983.
Burrus, C.A., Dental, A.G., and Lee, T.P. *Opt. Commun.* **38** 124–6, 1981.
Cabib, D., Fabri, E., and Fioro *Nuovo Cim.* **108** 185, 1972.
Cage, M.E., Dziuba, R.F., and Field, B.F. *IEEE Trans. Instrum. Meas.* **IM-34** 301–3, 1985.
Chai, Y.G., Yuen, C., and Zdasiuk, G.A. *IEEE Trans. Electron. Devices* **ED-32** 972, 1985.
Chai, Y.G. and Yeats, R. *IEEE Electron. Device Lett.* **EDL-4** 252, 1983.
Chang, L.L., Sakaki, H., Chang, C.A., and Esaki, L. *Phys. Rev. Lett.* **38** 1489, 1977.
Chang, L.L., Esaki, L., Howard, W.E., Ludeke, R., and Schul, G. *J. Vac. Sci. Technol.* **10** 655, 1973.
Chang, L.L., Mendez, E.E., Wang, W.I., Esaki, L., and Tedrow, P.M. *Proc. 17th Int. Conf on the Physics of Semiconductors* ed J.D. Chadi (Berlin: Springer) p 299, 1984.
Chemla, D.S., Millers, D.A., Smith, P.W., Gossard, A.C., and Wiegmann, W. *Appl. Phys. Lett.* **44** 16, 1984.
Chen, Y.S. and Kim, O.K. *J. Appl. Phys.* **52** 7892, 1981.
Cheng, C.L., Liao, A.S., Chang, T.Y., Leheny, R.F., Coldren, L.A., and Lalevic, B. *IEEE Electron. Device Lett.* **EDL-5** 169, 1984.
Cho, A.Y. *J. Vac. Sci. Technol.* **16** 275, 1979.
Cohn, D.R., Larsen, D.M., and Lax, B. *Phys. Rev.* B **6** 1367, 1972.

Cooke, R.A., Hoult, R.A., Kirkman, R.F., and Stradling, R.A. *J. Phys. D: Appl. Phys.* **11** 945, 1978.

Davies, R.A., Kelly, M.J., and Kerr, T., *GEC J. Research* **4** 157-162, (1986).

Dingle, R., Wiegmann, W., and Henry, C.H. *Phys. Rev. Lett.* **33** 827, 1974.

Delacourt, D., Papuchon, M., Poisson, M.A., and Razeghi, M. *Electron. Lett.* **23** 451, 1987.

Delahaye, F. and Reymann, D. *IEEE Trans. Instrum. Meas.* **IM-34** 316–19, 1985.

Dingle, R., Wiegmann, W., and Henry, C.H. *Phys. Rev. Lett.* **33** 827, 1974.

Drummond, T.J., Kopp, W., Fischer, R., Morkoç, H., Thorne, R.E., and Cho, A.Y. *J. Appl. Phys.* **53** 1238, 1982.

Elliott, R.J. *Phys. Rev.* **108** 1384, 1957.

Ekenberg, U. and Altarelli, M. *Phys. Rev. B* **30** 3569; 1985 *Phys. Rev. B* **32** 3712, 1984.

Endo, T., Murayama, Y., Koyanagi, M., Kinoshita, J., Inagaki, K., Yamanouchi, C. and Yoshihiro, K. *IEEE Trans. Instrum. Meas.* **IM-34** 323–7, 1985.

Esaki, L. and Tsu, R. *IBM J. Res. Dev.* **14** 61, 1970.

Falco, C., Botineau, J., Azema, A., Demicheli, M., and Ostrowky, D.B. *Appl. Phys. A* **30** 23–6, 1983.

Fauri, J.P. and Million, A. *Thin Solid Films* **90** 107, 1982.

Forrest, S.R. and Kim, O.K. *J. Appl. Phys.* **53** 5738, 1982.

Forrest, S.R. *Laser Focus* **18** 81–90, 1982.

Gauthier, D., Dmowski, L., Amor, B.S., Blondel, R., Portal, J.C., Razeghi, M., Maurel, P.H., Omnes, F., and Laviron, M. *Semicond. Sci. Technol.* **1** 105, 1986.

Goetz, K.H., Bimberg, D., and Razeghi, M. Unpublished observations, 1984.

Goetz, K.H., Bimberg, D., Brauchler, K.A., Jurgensen, H., Selders, J., Razeghi, M., and Kuphal, E. *Appl. Phys. Lett.* **46** 277, 1985.

Goetz, K.H., Bimberg, D., Jurgensen, H., Selders, J., Solomonov, A.V., Glenskii, G.F., and Razeghi, M. *J. Appl. Phys.* **54** 4543, 1983.

Goetz, K.H., Bimberg, D., Solomonov, A.V., Glinski, G.F., and Razeghi, M. *J. Appl.,* 1983.

Gossard, A.C. *Thin Solid Films* **104** 279, 1983

Guldner, Y., Vieren, J.P., Voisin, P., Voos, M., Razeghi, M., and Poisson, M.A., *Appl. Phys. Lett.* **40** 877–9, 1982.

Guldner, Y., Hirtz, J.P., Vieren, J.P., Voisin, P., Voos, M., and Razeghi, M. *J. Physique Lett.* **43** L613, 1982.

Hartland, A., Davies, G.J., and Wood, D.R. *IEEE Trans. Instrum. Meas.* **IM-34** 309–14, 1985.

Herman, X. and Weisbuch, Y. *Phys. Rev. B* **15** 816, 1977.

Hilsum, C. and Rees, H.D. *Electron. Lett.* **6** 277, 1970.

L'Hanidon, H., Favenrec, P., Salvi, M., and Razeghi, M. *Electron. Lett.* **21** 122, 1985.

Ishida, K., Matsui, J., Kamejima, T., and Sakuma, I. *Phys. Status Solidi a* **31** 255, 1975.

Ito, M., Kaneda, T., Nakajima, K., Toyama, Y., and Ando, H. *Solid State Electron.* **24** 421–4, 1981.

Kane, E.O. *J. Phys. Chem. Solids* **1** 249, 1957.

Kastalsky, A. and Hwang, J.C. *Solid State Commun.* **51** 317, 1984.

Kastalsky, A. and Hwang, J.C. Unpublished, 1983.

Kawamura, Y. and Okamoto, H. *J. Appl. Phys.* **50** 4457, 1979.

Kazmierski, K., Huber, A.M., Morillot, G., and Cremoux, R. *Japan. J. Appl. Phys.* **23** 628–33, 1984.

Kordos, P., Schumbera, P., Heyen, M., and Balk, P. *Gallium Arsenide and Related Compounds, 1981* (Inst. Phys. Conf. Ser. 63) p 131, 1982.

Landolt-Börnstein *Numerical Data and Functional Relationships in Science and Technology* New Ser. III/17a, ed Madelung, O. (Berlin: Springer), 1982.

Landolt-Börnstein *Physics of Group IV Elements and III–V Compounds* vol 17a, ed. Madelung, O., Schultz, M., and Weiss, H. (Berlin: Springer), 1982

Larsen, D.M. *Phys. Rev.* **144** 679, 1964

Lee, T.P., Bunns, C.A., Dentai, A.G., and Ogawa, K. *Electron. Lett.* **16** 155, 1973.

Lorentz, M.R., Morgan, T.N., and Pettit, G.D. *Proc. Int. Conf. on the Physics of Semiconductors Moscow, 1968* ed Ryvkin, S. (Leningrad: Nauka) p 495, 1968.

Lin, B.J, Tsui, D.C., Paalanen, M.A., and Gossard, A.C. *Appl. Phys. Lett.* **45** 695, 1984.

Linh, N.T. *Adv. Solid State Phys.* 227, 1983.

Maan, J.C., Belle, G., Fasolino, A., Altarelli, M., and Ploog, K. *Phys. Rev.* B **30** 2253, 1984.

Marzin, J.Y., Benchimol, J.L., Sermarge, B., Etienne, B. and Voos, M. *Solid State Commun,* 1984.

Massies, J., Rochelle, J., Delescluse, P., Etienne, P.E., Chevrier, J., and Linh, J.J. *J. Electron. Lett.* **16** 275, 1980.

Maurel, P.H., Razeghi, M., Guldner, Y., and Vieren, J.P. *Semicond. Sci. Technol.* **2** 695, 1987.

Mendez, E.E., Wang, W.I., Chang, L.L., and Esaki, L. *Phys. Rev.* B **30** 1087, 1984.

Miller, R.C., Kleinman, D.A., and Gossard, A.C. *Phys. Rev.* B **29** 7085, 1984.

Mori, S. and Ando, T. *Surf. Sci.* **98** 101, 1980.

Nagi, H. *J. Appl. Phys.* **45** 3789, 1974.

Nahory, R.E., Pollack, M.A., Johnston, W.D., and Barns, R.L. *Appl. Phys. Lett.* **33** 659, 1978.

Nathan, M.I., Jackson, T.N., Kirchner, P.D., Mendez, E.E., Pettit, W.D., and Woodall, J.M. *J. Electron. Mater.* **12** 719, 1983.

Nicholas, R.J., Brunei, L.C., Huant, S., Kawai, K., Portal, J.C., Brummel, M.A., Razeghi, M., Cheng, K.Y., and Cho, A.Y. *Appl. Phys. Lett.* **55** 883, 1985.

Nicholas, R.J., Brummell, M.A., and Portal, J.C. *J. Cryst. Growth* **68** 356, 1984.

Nicholas, R.J., Stradling, R.A., Portal, J.C., and Askenazy, S. *J. Phys. C: Solid State Phys.* **12** 1653, 1979.

Nicholas, R.J., Sessions, S.J., and Portal, J.C. *J. Appl. Phys.* **37** 178, 1980.

O'Conner, P., Pearsall, T.P., Chang, K.K., Cho, A.Y., Hwang, J.C., and Alavi, K. *IEEE Electron. Device Lett.* **EDL-3** 64, 1982.

Ohkawa, F.J., and Uemura, Y. *Prog. Theor. Phys. Suppl.* **57** 164, 1975.

Oliver, J. and Eastman, L. *J. Electron. Mater.* **9** 693, 1980.

Onho, H., Barnard, J., Wood, C.E., and Eastman, L.F. *IEEE Electron. Device Lett.* **EDL-I** 154, 1980.

Palik, E.D., Picus, G.S., Teitler, S., and Wallis, R.F. *Phys. Rev.* **122** 475, 1961.

Pearsall, T.P., Carles, R., and Portal, J.C. *Appl. Phys. Lett.* **42** 436, 1983.

Pikhtin, A.N. *Sov. Phys. –Semicond.* **11** 245, 1977.

Pinczuk, A., Worlock, J.M., Nahory, R.E., and Pollack, M.A. *Appl. Phys. Lett.* **33** 461, 1978.

Portal, J.C., Nicholas, R.J., Brummel, M.A., Cho, A.Y., Cheng, K.Y., and Pearsall, T.P. *Solid State Commun.* **43** 907, 1982.

Portal, J.C., Perrier, P., Renucci, M.A., Askenazy, S., Nicholas, R.J., and Pearsall, T. *Physics of Semiconductors, 1978* (Inst. Phys. Conf. Ser. 43) p 829, 1978.

Poulain, P. and Cremoux, B. *Japan. J. Appl. Phys.* **19** L189–92, 1980.

Praddaude, H.C. *Phys. Rev.* A **6** 1321, 1972.

Rao, M.V. and Bhattacharya, P.K. *Electron. Lett.* **19** 196, 1983.

Raulin, J.Y., Vassilakis-Thorngren, E., Poisson, M.A., Razeghi, M., and Idomer, G. *Appl. Phys. Lett.* **50** 535, 1987.

Razeghi, M. and Duchemin, J.P. *J. Cryst. Growth* **69** 76–82, 1983.

Razeghi, M., Poisson, M.A., Guldner, Y., Vieren, J.P., Voisin, M., and Voos, M. *Appl. Phys. Lett.* **40** 877, 1982.

Razeghi, M., Maurel, P.H., and Omnes, F. *NATO Workshop on Impurity Levels in Superlattices*, 1987.

Razeghi, M., Defour, M., Omnes, F., Maurel, P.H., Bigan, E., Acher, O., Nagle, J., and Brillouet, F. *Proc. 4th MOCVD Conf.,* 1988.

Razeghi, M., Hirtz, J.P., Ziemelis, V.O., Delalande, C., Etienne, B., and Voos, M. *Appl. Phys. Lett.* **43** 585, 1983.

Razeghi, M., Nagle, J., and Weisbuch, C. *Gallium Arsenide and Related Compounds, 1984* (Inst. Phys. Conf. Ser. 74) p 379, 1984.

Razeghi, M., Maurel, P.H., Nagle, J., Omnes, F., and Pochelle. J.P. *Appl. Phys. Lett.* **49** 1110, 1986.

Razeghi, M., Duchemin, J.P., Portal, J.C., Dmowski, L., Remeni, G., Niedas, R.J., and Briggs, A. *Appl. Phys. Lett.* **48** 712, 1986.

Razeghi, M., Maurel, P.H., Tardella, A., Dimovsky, L., Gauthier, D., and Portal, J.C. *Appl. Phys. Lett.* **60** 2453, 1986.

Razeghi, M. *Semiconductors and Semimetals* vol 22, Part A ed Tsang, W.T. (New York: Academic), 1985.

Razeghi, M., Poisson, M.A., Guldner, Y., Vieren, J.P., Voisin, P., and Voos, M. *Appl. Phys. Lett.* **40** 877, 1982.

Razeghi, M., Hirtz, P., Blandeau, R., Cremoux, B., and Duchemin, J.P. *Electron. Lett.* **19** 181, 1983.

Razeghi, M., Tardella, A., Davies, R.A., Long, A.P., Kelly, M.K., Brittan, E., Boothroyd, C., and Stobbs, W.M. *Electron. Lett.* **23** 116, 1987.

Rezek, E.A., Chin, R., Holonyak, N., Kirchoefer, S.W., and Kollbas, R.M. 1980 *J. Electron. Mater.* **9** 1, 1987.

Shah, A.J., Leheny, R.F., Nahory, R.E., and Pollack, M.A. *Appl. Phys. Lett.* **37** 475, 1980.

Singh, J. and Madhukar, A. *J. Vac. Sci. Technol.* B **1** 305, 1983.

Sooryakumar, R., Chemla, D.S., Pinczuk, A., Gossard, A., Weigmann, W., and Sham, L.J. *J. Vac. Sci. Technol.* B **2** 349, 1984.

Sotomayor, C.M., Claxton, P.A., Roberts, J.S., Stradling, R.A., and Wasilewski, Z. *Proc. VIth Conf. EP2DS* Kyoto, p 558, 1985.

Störmer, H.L., Gossard, A.C., and Wiegmann, W. *Solid State Commun.* **41** 707, 1982.

Störmer, H.L. and Tsang, W.T. *Appl. Phys. Lett.* **36** 685, 1980.

Störmer, H.L., Chang, A., Tsui, D.C., Hwang, J.C., Gossard, A.C., and Wiegmann, W. *Phys. Rev. Lett.* **50**, 1983.

Störmer, H.L., Gossard, A.C., Wiegmann, W., and Baldwin, K. *Appl. Phys. Lett.* **39** 912, 1981.

Störmer, H.L., Dingle, R., Gossard, A.C., Wiegmann, W., and Stauge, M.D. *Solid State Commun.* **29** 705, 1979.

Störmer, H.L., Gossard, A.C., Wiegmann, W., Blondel, R., and Baldwin, K. *Appl. Phys. Lett.* **44** 139, 1984.

Stradling, R.A. and Wood, R.A. *J. Phys. C: Solid State Phys.* **1** 1711, 1968.

Stillman, G.E., Cook, L.W., Bulman, G.E., Tabatabaie, N., Chin, R., and Dapkus, P.D. *IEEE Trans. Electron. Devices* **ED-29** 1355–71, 1982.

Stillman, G.E., Wolfe, C.M., and Korn, D.M. *Proc. 13th Int. Conf. on Physics of Semiconductors* Rome, p 623, 1976.

Swierkowski, L., Zawadski, W., Guldner, Y., and Rigaux, C. *Solid State Commun.* **27** 1245, 1978.

Takikawa, M., Komeno, J., and Ozeki, M. *Appl. Phys. Lett.* **43** 280, 1983.

Tell, B., Leheny, R.F., Liao, A.S., Bridges, T.J., Burkhardt, E.G., Chang, T.Y., and Beebe, E.D. *Appl. Phys. Lett.* **44** 438–40, 1984.

Towe, E. *J Appl. Phys.* **53** 5136, 1982.

Trommer, R. and Albrecht, H. *Japan. J. Appl. Phys.* **22** L364, 1983.

Tsang, W.T. and Cho, A.Y. *Appl. Phys. Lett.* **30** 293, 1977

Tsui, D.C. and Gossard, A.C. *Appl. Phys. Lett.* **37** 550, 1981.

Tsui, D.C. and Gossard, A.C. *Appl. Phys. Lett.* **38** 550, 1981

Verma, A.R. *Crystal Growth and Dislocation* (London: Butterworths), 1953.

Vescan, L., Selders, J., Kräute, H., Kütt, W., and Beneking, H. *Electron. Lett.* **18** 533, 1982.

Vescan, L., Selder, J., Maier, M., Krautle, H., and Beneking, H. *J. Cryst. Growth* **67** 353–7, 1984.

Voisin, P., Guldner, Y., Vieren, J.P., Voos, M., Delescluse, P., and Linh, N.T. *Appl. Phys. Lett.* **39** (12) 982, 1981.

Von Klitzing, K., Dorda, G., and Pepper, M. *Phys. Rev. Lett.* **45** 494, 1980.

Von Klitzing, K. and Ebert, G. *Metrologia* **21** 11–18, 1985.

Wada, T., Shida, K., Nishinaka, H., and Igarashi, T. *IEEE Trans. Instrum. Meas.* **IM-34** 306–9, 1985.

Wei, H.P., Tsui, D.C., and Razeghi, M. *Appl. Phys. Lett.* **45** 666, 1984.

Weisbuch, C., Dingle, R., Gossard, A.C., and Wiegmann, W. *Solid State Commun.* **38** 709, 1981.

Weisbuch, C., Miller, R.C., Dingle, R., Gossard, A.C., and Wiegmann, W. *Solid State Commun.* **37** 219, 1981.

Welch, D.F., Wicks, G.W., and Eastmann, L.F. *Appl. Phys. Lett.* **43** 762, 1983.

Windhorn, T.H., Cook, L.W., and Stillman, G.E. *IEEE Electron. Device Lett.* **EDL-3** 18, 1982.

Wood, T.H., Burrus, C.A., Miller, D.A., Chemla, D.S., Gossard, A.C., and Wiegmann, W. *Appl. Phys. Lett.* **44** 16, 1984.

Wood, T.H., Burrus, C.A., Weiner, J.S., Chemla, D.S., Miller, D.A., Damen, T.C., Sivco, D.L., and Cho, A.Y. *Gallium Arsenide and Related Compounds, 1984* (Inst. Phys. Conf. Ser. 74) p 687, 1984.

Yafet, Y., Keys, R.W., and Adams, E.N. *J. Phys. Chem. Solids* **1** 137, 1956.

Yagi, T., Fujiwara, Y., Nishino, T., and Hamakawa, Y. *Japan. J. Appl. Phys.* **22** L467, 1983.

13. GaInAsP–InP System: MOCVD Growth, Characterization, and Applications

13.1. **Introduction**
13.2. **Growth conditions**
    13.2.1. Orientation effects
    13.2.2. Carrier-gas effects
    13.2.3. Pyrolysis-oven effects
    13.2.4. Photoluminescence spectra
13.3. **Characterization**
    13.3.1. Calorimetric absorption and photoluminescence studies of interface disorder in InGaAsP–InP quantum wells
    13.3.2. Magnetotransport
    13.3.3. Observation of quantum Hall effect in a GaInAsP–InP heterostructure grown by MOCVD
    13.3.4. Disorder of a $Ga_xIn_{1-x}As_yP_{1-y}$–InP quantum well by Zn diffusion
    13.3.5. Interface study of $Ga_xIn_{1-x}As_yP_{1-y}$–InP heterojunctions grown by MOCVD by spectroscopic ellipsometry
13.4. **Applications of GaInAsP–InP systems grown by MOCVD**
    13.4.1. Broad-area 1.2–1.6 μm $Ga_xIn_{1-x}As_yP_{1-y}$–InP DH lasers grown by MOCVD
    13.4.2. Buried-ridge-structure lasers grown by MOCVD
    13.4.3. Distributed feedback lasers fabricated on material grown completely by MOCVD
    13.4.4. CW phase-locked array $Ga_{0.25}In_{0.75}As_{0.5}P_{0.5}$–InP high-power semiconductor laser grown by MOCVD
    13.4.5. GaInAsP–InP quantum-well lasers
    13.4.6. Buried waveguides in InGaAsP–InP material grown by MOCVD

13.1. Introduction

As long-wavelength 1–1.65 μm GaInAsP electro-optical devices have become more widely used, motivated by low fiber absorption and dispersion, high transmission through water and smoke, and greatly enhanced eye safety at wavelengths greater than 1.4 μm, the low-pressure metalorganic chemical vapor deposition (MOCVD) growth technique has been used to grow $Ga_xIn_{1-x}As_yP_{1-y}$, $(0 < x < 0.47$ and $0 < y < 1)$ lattice-matched to InP for the complete compositional range between InP

($\lambda = 0.91$ μm, $E_g = 1.35$ eV) and the ternary compound $Ga_{0.47}In_{0.53}As$ ($\lambda = 1.67$ μm, $E_g = 0.75$ eV). Growth takes place in an RF induction-heated horizontal cold-wall reactor using triethylindium (TEIn) and triethylgallium (TEGa) for In and Ga sources, and AsH_3 and PH_3 for As and P sources.

13.2. Growth conditions

Growth apparatus and processes have been described in Chapter 2. $Ga_xIn_{1-x}As_yP_{1-y}$ layers lattice-matched to an InP substrate can be grown at 76 Torr and at a substrate temperature between 630 and 650 °C using TEIn, TEGa, AsH_3 and PH_3 in an $H_2 + N_2$ carrier gas. The growth rate depends linearly upon the sum of the partial pressures of TEIn and TEGa, and is independent of the arsenic and phosphorus partial pressures (see Fig. 13.1). The epitaxial-layer quality is sensitive to the pretreatment of the substrate and the alloy composition. The optimum growth conditions for $Ga_{0.23}In_{0.77}As_{0.51}P_{0.49}$ ($\lambda = 1.3$ μm) lattice-matched to InP are listed in Table 13.1.

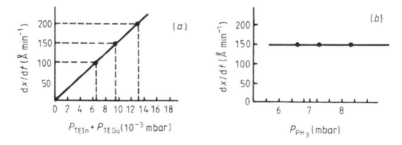

Fig. 13.1. Variation of growth rate of GaInAsP on InP ($\lambda = 1.3$ μm) with (a) TEIn + TEGa flow-rate and (b) PH_3 flow-rate at $T_G = 650°C$ and $P_{AsH_3} = 0.3$ mbar.
In (a) $P_{PH_3} = 6.6$ mbar; in (b) $P_{TEGa} = 2.5 \times 10^{-3}$ mbar, $P_{TEIN} = 7 \times 10^{-3}$ mbar.

Sources	Partial pressure (mbar)	Temperature (°C)
TEIn	8.5×10^{-3}	31
TEGa	3×10^{-3}	0
AsH_3	0.31	25
PH_3	7	25

Table 13.1. Optimum growth conditions for GaInAsP ($\lambda = 1.3$ μm) lattice-matched to the InP substrate; the growth temperature is 650°C and the growth pressure is 76 Torr; the total flow-rate ($H_2 + N_2$) is 7 l min^{-1}.

Fig. 13.2 shows the variation of lattice mismatch $\Delta a / a = (a_{InP} - a_{GaInAsP})/a_{Inp}$ for a GaInAsP epilayer with TEIn flow-rate, keeping all of the remaining growth parameters constant. By changing the TEIn flow-rate one can obtain GaInAsP layers with $\lambda = 1.3$ µm lattice-matched to the InP substrate with $\Delta a / a \leq 4 \times 10^{-4}$ (which is the limit of precision of single-crystal x-ray diffraction measurements).

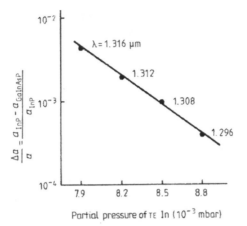

Fig. 13.2. Variation of lattice mismatch for a GaInAsP epilayer with TEIn flow-rate, keeping all of the remaining growth parameters constant, i.e. $T_G = 650$ °C, $P_{Ga} = 30$ mbar, $P_{AsH_3} = 0.31$ mbar and $P_{PH_3} = 7$ mbar.

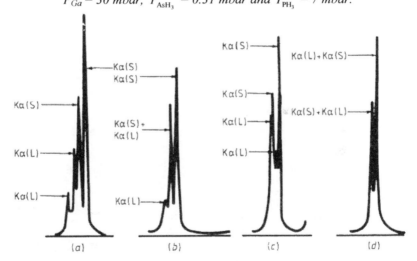

Fig. 13.3. Evolution of x-ray spectrum as a function of TEIn flow-rate. $\Delta a / a = 4.5 \times 10^{-3} (a)$, $2 \times 10^{-3} (b)$, $10^{-3} (c)$, $\leq 4 \times 10^{-4} (d)$.

Fig. 13.3 shows the evolution of the x-ray spectrum as a function of TEIn flow-rate. The full width at half-maximum (FWHM) of $K\alpha_1$ and $K\alpha_2$ reflection of the x-ray spectra of the GaInAsP epilayer is the same as the substrate, which indicates that the GaInAsP layer is homogeneous.

Fig. 13.4 shows (*a*) the ratio $R_5 = PH_3/(PH_3 + ASH_3)$ versus band-gap wavelength λ_g, between InP and $Ga_{0.47}In_{0.53}As$, (*b*) the ratio $R_3 = TEGa/(TEGa + TEIn)$ versus λ_g and (*c*) the ratio R_5/R_3 versus λ_g for $Ga_xIn_{1-x}As_yP_{1-y}$ ($0 < x < 0.47$ and $0 < y < 1$) lattice-matched to InP between InP ($\lambda = 0.9$ μm) and $Ga_{0.47}In_{0.53}As$ ($\lambda = 1.67$ μm) for a growth temperature of 650 °C, growth pressure of 76 Torr and total flow-rate of 71 min^{-1}.

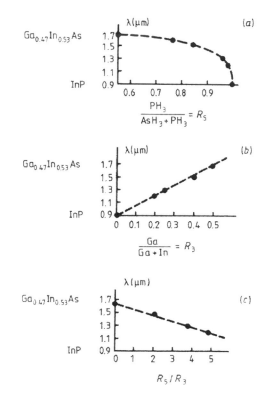

Fig. 13.4. The ratios (a) $R_5 = PH_3/(PH_3 + AsH_3)$, (b) $R_3 = TEGa/(TEGa + TEIn)$ and (c) R_5/R_3.

Fig. 13.4 shows that the incorporation behavior of group V elements is different, because both arsenic and phosphorus are volatile and a significant proportion of these elements can be lost to the gas phase. The probability of incorporation for arsenic is much higher than for phosphorus, since phosphorus has a lower sticking coefficient than arsenic. For GaInAsP ($\lambda = 1.3$ μm), one obtains the partial pressure ratios

Eq. (13.1)
$$P_{PH_3} / P_{AsH_3} = (1-y)/y = 20$$
$$P_{TEIn} / P_{TEGa} = (1-x)/x = 3$$
$$P_{(AsH_3+PH_3)} / P_{(TEIn+TEGa)} = 600$$

Using these results and the data indicated in Fig. 13.3 for a growth temperature of 650 °C and total flow-rate of 7 l·min^{-1}, it is easy to obtain $Ga_xIn_{1-x}As_yP_{1-y}$ epitaxial layers with different compositions between InP ($x = 0$, $y = 0$) and $Ga_{0.47}In_{0.53}As$ ($x = 0.47$, $y = 1$).

The MOCVD technique is attractive for its ability to grow uniform layers over a large area of substrate [Razeghi et al. 1984]. Fig. 13.5 shows the variation of x-ray spectra (lattice mismatch measured by single-crystal x-ray diffractometry, using (4 0 0) Cu K_α reflection), photoluminescence band-edge wavelength λ_g and growth rate dx/dt of $Ga_{0.28}In_{0.72}As_{0.46}P_{0.54}$—InP over an area of 10 cm^2, having a thickness of 3 μm. The layer exhibits uniform composition over a large area, the variation of lattice mismatch $\Delta a/a$ and of photoluminescence wavelength λ_g being less than 2% and the variation of thickness less than 10%.

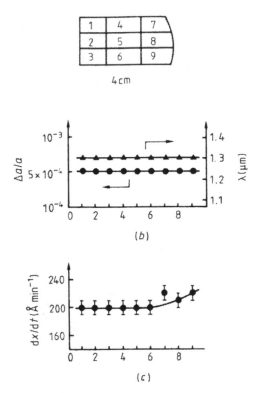

Fig. 13.5. *Variation of photoluminescence wavelength λ_g, lattice mismatch $\Delta a / a$ and growth rate dx/dt of a GaInAsP epitaxial layer over an area of 10 cm^2 of InP substrate; the thickness of the layer is 3 μm.*

13.2.1. Orientation effects

Growth of GaInAsP layers has been carried out on (1 0 0) substrates misoriented 2° off towards (1 1 0). Razeghi et al. [1984] have performed a study of MOCVD growth on (1 0 0), (1 1 1) and (1 1 5) surfaces at 650 °C in one run upon six adjacent InP substrates.

Table 13.2 lists the relative PL intensity I, PL halfwidth $\Delta h\nu$ and lattice mismatch $\Delta a/a$ of MOCVD $Ga_{0.28}In_{0.72}As_{0.46}P_{0.54}$ grown on InP substrates of various orientations at 650 °C (1 μm thickness).

The relative PL intensity (a measure of the radiative recombination efficiency within the material) is higher for (1 1 5) than for (1 0 0) orientation. Also I is higher for (1 0 0) 2° off than for (1 0 0) exact [Razeghi et al. 1984]. The PL halfwidth (a measure of crystalline quality and impurity incorporation) is higher for (1 1 5) than for (1 0 0) orientation. SIMS analysis shows that the concentration of Si impurity in (1 1 5) samples is higher than in (1 0 0) 2° off, so it is possible that higher PL intensity is due to the higher donor impurity in this layer.

Orientation	I (au)	$\Delta h\nu$ at 300 K	$\Delta a/a$
InP(Sn)(1 0 0) 2° (1 1 0)	0.1	65	$+1 \times 10^{-3}$
InP(S)(1 0 0) 2° (1 1 0)	0.05	70	$+1 \times 10^{-3}$
InP(Fe)(1 0 0) 2° (1 1 0)	0.05	60	$+1 \times 10^{-3}$
InP(Sn)(1 0 0) exact	0.01	65	$+1 \times 10^{-3}$
InP(S)(1 1 1) 2°	0.8	70	$+1 \times 10^{-3}$
InP(S)(1 1 5) 2°	1	90	$+1 \times 10^{-3}$

Table 13.2. Relative PL intensity I, PL halfwidth $\Delta h\nu$ and lattice mismatch of GaInAsP ($\lambda = 1.3 \mu m$) grown on InP substrates by MOCVD.

13.2.2. Carrier-gas effects

Razeghi et al. [1983] studied the growth of $Ga_xIn_{1-x}As_yP_{1-y}$ layers using 100% H_2 and mixtures of $H_2 + N_2$ as carrier gas, the best morphology and highest photoluminescence intensity being obtained using 50% $H_2 + 50\% N_2$.

Razeghi et al. [1983] also studied the growth of $Ga_xIn_{1-x}As_yP_{1-y}$ layers as a function of total flow-rate ($H_2 + N_2$). The best morphology and highest photoluminescence intensity were obtained for a total flow-rate between 6 and 7 $l \cdot min^{-1}$.

13.2.3. Pyrolysis-oven effects

Razeghi et al. [1983] performed a study of MOCVD growth of GaInAsP layers at a growth temperature of 650 °C as a function of temperature of the pyrolysis oven for cracking the PH_3.

Under the growth conditions mentioned above, the PL intensity decreases when the temperature of the pyrolysis oven becomes higher than 750 °C. When the parasitic reaction in the gas phase increases, the quality of the epilayer is degraded. The PL intensity and surface quality become independent of the pyrolysis-oven temperature for temperatures lower than 750 °C. This implies that, for a growth temperature of 650 °C, the pyrolysis oven for cracking of PH_3 is not necessary.

Duchemin et al. [1978] reported that it was very difficult to grow an InP layer at temperatures of 500 °C under hydrogen or nitrogen atmospheres containing a flow of PH_3 even up to 1 l·min^{-1} without damaging its surface because of preferential loss of phosphorus. The interpretation was that there was a parasitic reaction between PH_3 and $In(C_2H_5)_3$ with the formation of a complex addition compound which appeared in the form of black fumes or "aerosol." They circumvented these difficulties by introducing PH_3 into the reactor through the pyrolysis oven (whose internal temperature was 760 °C).

Razeghi et al. [1982] previously showed that the purity of starting material such as the alkyl (triethylindium and triethylgallium) or the hydride (arsine and phosphine) sources is responsible for the quality (carrier concentration and mobility) of the epitaxial layers grown by MOCVD (see also Chapter 11).

Razeghi et al. [1983] showed that, by elimination of the pyrolysis oven for PH_3, considerably better material has been obtained. Without the pyrolysis oven $Ga_xIn_{1-x}As_yP_{1-y}$–InP heterojunctions had a higher mobility, a lower carrier concentration and a higher photoluminescence (PL) intensity with sharper PL halfwidth.

In fact, the temperature of 760 °C is not enough to crack PH_3, and the optimal flow-rate of PH_3 for the growth of InP and related compounds with and without the pyrolysis oven is the same. When the temperature of the pyrolysis oven is higher than 760 °C, the epitaxial layers become compensated with the impurities coming from the hot quartz tube. Shubnikov–de Haas measurements of these samples also demonstrate the improvement of the quality of epitaxial layers after the elimination of the PH_3 pyrolysis oven.

With the pyrolysis oven present, there is a parasitic reaction between the gas compound inside the reactor near the pyrolysis oven, which induces the deposition of parasitic compounds on the reactor wall. During growth these parasitic particles fall on the epitaxial layer and create the dark spot defects in the layer, degrading its optical and electrical properties.

13.2.4. Photoluminescence spectra

The highly efficient photoluminescence of GaInAsP–InP quantum wells grown by MOCVD has been studied at various temperatures [Wei et al. 1985]. The high quantum efficiency of the material and careful optimization of the experiment for maximum sensitivity allow the detection of luminescence signals with excitation powers well below the microwatt range.

Fig. 13.6 shows a typical photoluminescence spectrum for a four-well structure of $Ga_{0.22}In_{0.78}As_{0.45}P_{0.55}$–InP. The remarkable feature is the sharpness of the lines, even for QW with a 10 Å nominal thickness (10 meV FWHM) at 5 K. Extremely large energy shifts of the luminescence due to the quantum confinement effect of QW are observed up to 1.285 eV for a nominally 10 Å thick QW. As a preliminary test of these multilayer structures, Auger scans were carried out. Samples were prepared by chemically etching a linear bevel through all layers and interfaces.

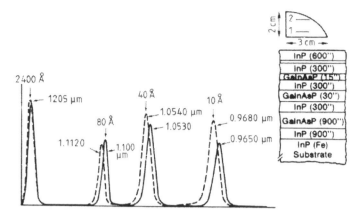

Fig. 13.6. Photoluminescence spectrum for a four-well structure of
$Ga_{0.22}In_{0.78}As_{0.45}P_{0.55}$, *at 7 K.*

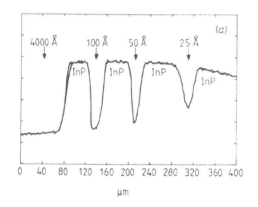

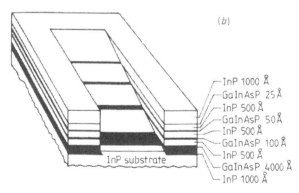

*Fig. 13.7. (a) Auger spectrum of a chemically etched bevel that cuts all four
GaInAsP layers. The abscissa indicates the distance along the bevel in micrometers,
(b) Layer assignments, MOCVD $Ga_{0.25}In_{0.75}As_{0.5}P_{0.5}$–InP.*

Fig. 13.7 indicates the nature of the bevel and also shows the Auger line scan of the bevel. The four GaInAsP layers are clearly visible, and the interfaces are abrupt.

Growing several QW with different layer thicknesses on the same substrate makes it possible to compare different QW grown under similar conditions. Especially in the case of alloy QW, it is then possible to ascertain the spatial homogeneity of composition and thickness across the wafer:

$$E = h\nu = E_g + E^e_{conf} + E^h_{conf}$$

Eq. (13.2)
$$E^{e,h}_{conf} \simeq \frac{\hbar^2}{2m*}\left(\frac{n\pi}{L_z}\right)^2 \qquad n = 1,2,3,...$$

where E_g is the alloy band gap, E^e_{conf} and E^h_{conf} are the confining energies for the electron and hole ground states, respectively, L_z is the thickness of the QW and $m*$ is the effective mass.

If the alloy composition varies across a multi-well sample, it gives rise to a constant shift of the ground-state energy (E_g) of the well, *independent* of the well thickness.

On the other hand, if R_1 and R_2 are the two growth rates at two points 1 and 2 (see Fig. 13.7) of the sample, in the infinitely-deep-well limit we find

Eq. (13.3)
$$E^{e,h}_{conf} \simeq L_z^{-2}$$
$$\Delta E \simeq L_z^{-3}\Delta L \simeq L_z^{-2}\left(R_2 - R_1\right)/R_2$$

In this case the energy shift between two points on the sample *depends* on the well thickness.

Fig. 13.6 shows that the energy shifts are due to a small variation in well thickness rather than in alloy composition.

13.3. Characterization

13.3.1. Calorimetric absorption and photoluminescence studies of interface disorder in InGaAsP–InP quantum wells

Classical semiconductor devices consist of collections of semiconductor regions separated by interfaces. The devices of the future consist of a collection of interfaces with the minimum of semiconductor between them. The characterization and control of structural, chemical and electronic properties of interfaces therefore is presently at the center of interest of semiconductor physics. A quantum well (QW) is the simplest possible model system to study the influence of structural interface disorder and variations of the chemical composition on electronic and optical properties of vertical semiconductor microstructures or heterostructure devices in general. Such disorder results from imperfect growth conditions and has a direct impact on novel devices like graded-index separate-confinement heterostructure single-quantum-well (GRINSCH-SQW) lasers or ballistic transistors. Following Bimberg et al. [1985], we show the results of an investigation of the interface disorder of $In_{0.75}Ga_{0.25}As_{0.5}P_{0.5}$/InP quantum wells grown by MOCVD. We will concentrate here on wells having a width $L_z = 10$ nm.

Calorimetric absorption spectroscopy (CAS) and photoluminescence (PL) are powerful complementary methods for characterizing the interface roughness of such materials. Luminescence spectra of quantum wells are dominated by (radiative) excitonic recombination processes. In contrast, the CAS signal is caused by single-particle interband absorption processes, since only non-radiative recombination is observed. A simple low-temperature thermometer as phonon detector is used to detect the calorimetric absorption signal.

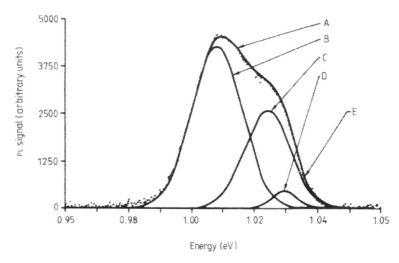

Fig. 13.8. Photoluminescence of an $In_{0.75}Ga_{0.25}As_{0.5}P_{0.5}$/InP double quantum well with $L_z = 10$ nm nominal at 1.4 K. A, experimental curve; B, Gaussian fit X(e, hh), well 1; C, Gaussian fit X(e, hh), well 2; D, Gaussian fit X (e, lh), well 1; E, superposition of all three Gaussian lines.

Fig. 13.8 shows a photoluminescence spectrum of an $In_{0.75}Ga_{0.25}As_{0.5}P_{0.5}$/InP double quantum well with $L_z = 10$ nm (nominal width, derived from growth parameters). A deconvolution of the luminescence lineshape shows that it represents a superposition of two electron-heavy hole A"(e,hh) [Bimberg and Bubenzer 1981 and Christen et al. 1984] and one electron—light hole $X(e, lh)_1$ exciton $n = 1$ subband transitions. Apparently the two wells of the sample are different, since the energetic position of the excitonic transitions are displaced with respect to each other by 18 meV. Thus the chemical composition of both wells (As/P ratio) cannot be the same, since during the MOCVD growth process the average thicknesses L_z of the two wells can be precisely controlled. The energetic positions of the different components of the luminescence can be used to calculate the compositional variation of the wells. One finds a deviation of the As/P ratio of 3% between the two wells. The energy difference between the X(e,hh) and X(e,lh) transitions—

corrected for the different exciton binding energies for heavy and light holes—
yields a real thickness of 10.1 nm, in very good agreement with 10 nm, derived
from growth parameters. The band-gap discontinuities fitting these results best are
$E_c/E_g = E_v/E_g = 0.5$. The lineshape analysis further shows that all lines have a
pronounced Gaussian shape. The halfwidth of the two dominant Gaussians is
roughly the same (19 meV). Such Gaussian profiles can be taken as indicative of
interface disorder [Bastard et al. 1984].

Unfortunately, however, thermalization of the charge carriers into the states of
lowest total energy induces a low-energy shift of the luminescence and a reduction
of halfwidth, as proved by temperature- and excitation-dependent experiments. The
minimum of total energy of the electron–hole pairs corresponds to the region with
the largest local thickness L_z. Fig. 13.9 visualizes in a qualitative way the difference
of band structure between QW with ideal and real interfaces. Thus low-temperature
luminescence does not display the true thickness distribution and can underestimate
the interface disorder of the quantum well, as will now be shown. In view of the fact
that a detailed statistical interpretation of the halfwidth of the Gaussians has still to
be developed, a quantitative determination of the interface roughness ΔL_Z is usually
based on the formula

Eq. (13.4) $\Delta L_z = \frac{1}{2}\Delta(\hbar\omega)\, L_z\, /\, \Delta E_g$

where $\Delta(\hbar\omega)$ is the full width at half-maximum of the X(e,hh) luminescence line
and ΔE_g the energy-gap difference between the QW and the three-dimensional
material. In the present case, one can derive from $\Delta(\hbar\omega) \simeq 19$ meV an interface
roughness $\Delta L_Z = 1$ nm.

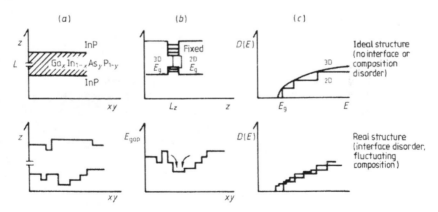

Fig. 13.9. Comparison of the ideal and real structures of quantum-well interfaces.
(a) Sample structure; (b) 2D energy gap; (c) density of states.

Fig. 13.10 shows the calorimetric absorption spectrum and additionally the
luminescence for comparison. The luminescence is shifted to lower energy by about
35 meV compared to the onset of absorption. Surprisingly the shape of the

absorption curve is well described by a $(E-E_g)^{1/2}$ dependence known from textbook theory of electron–hole interband absorption in three-dimensional semiconductors.

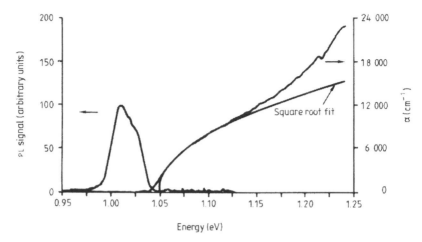

Fig. 13.10. Calorimetric absorption spectrum (right scale) and photoluminescence spectrum (left scale) of an $In_{0.75}Ga_{0.25}As_{0.5}P_{0.5}/InP$ double quantum well with $L_z = 10$ nm nominal at 1.3 K.

No influence of excitonic enhancement effects can be seen. This is due to the extremely high radiative efficiency of the excitonic recombination in quantum wells [Christen et al. 1984]. CAS in contrast reveals only non-radiative recombination processes. Structural disorder of the interface causes not only the Gaussian luminescence lineshape but also the square-root dependence of the absorption coefficient (see Fig. 13.10). For a quantum well with ideally abrupt interfaces one would expect a step-function density of states and absorption coefficient. If we take into account interface disorder or fluctuations of the chemical composition, then the two-dimensional energy gap (E_g) fluctuates in the plane parallel to the interface. Thus the ideal step-function-like density of states is smeared out and the square-root dependence is re-established. Therefore the absorption curve is a superposition of all different absorption processes according to the statistical weight of a given thickness and displays the thermal equilibrium situation. The onset of the square-root fit to the absorption gives the mean quantum-well width. The energy difference of the onset of the CAS spectrum and the onset of the dominant luminescence band $(X(e.hh))$ is 48 meV. This value has to be corrected by the binding energy of the 2D exciton (~ 0 meV) and its localization energy (~ 5 meV) [Bastard et al. 1984]. The resulting difference can be used to estimate the interface roughness. Again using $\Delta L_z = \frac{1}{2}\Delta(\hbar\omega)L_z / \Delta E_g$, but now $\Delta(\hbar\omega)$ 33 meV, one can find $\Delta L_z = 1.8$ nm. This value is by almost a factor 2 larger than that derived from the full width at half-maximum of the dominant luminescence line in Fig. 13.9.

Both the surprising square-root dependence of the CAS spectrum and the Gaussian shape of the photoluminescence (PL) prove consistently the dominant influence of interface disorder on optical and electronic properties of quantum wells. Luminescence alone, however, strongly underestimates the magnitude of the disorder. In contrast to PL, CAS displays the true interface structure of the quantum well, since no thermalization effects induce low-energy shifts.

13.3.2. Magnetotransport

Here we describe the growth of single-quantum-well (SQW), multiquantum-well (MQW) and superlattice (SL) samples from GaInAsP($\lambda = 1.3$ μm)/InP heterostructures by low-pressure metalorganic chemical vapor deposition [Razeghi et al. 1987]. Evidence of two-dimensional electron gases (2DEG) in these structures is also presented.

The optimum conditions for the reduced-pressure growth of InP and $Ga_{0.25}In_{0.75}As_{0.50}P_{0.50}$ ($\lambda = 1.3$ μm) are presented in Table 13.3. InP/GaInAsP interfaces can be obtained by switching both TEGa and AsH$_3$ flows into the reactor. GaInAsP/InP interfaces can be obtained by turning off the AsH$_3$ and TEGa flows. The growth rate is small (3 Å·s^{-1}) and it takes less than 1 s for a gas flow to reach its new steady state.

The exact composition of the QW samples is difficult to measure directly; however, x-ray diffraction and photoluminescence measurements carried out on thick (~ 1 μm) layers of GaInAsP grown under identical conditions indicate that the GaInAsP ($\lambda = 1.3$ μm) is lattice-matched to within $\Delta a/a = \pm 10^{-3}$.

Various heterojunction, SQW, MQW and SL samples were employed for this study. The SQW consisted of GaInAsP of 100 Å thickness, located between InP layers. The MQW samples consisted of 3–8 GaInAsP wells of Lz = 100–400 Å thickness with LB ~ 100–200 Å (Lz is the thickness of the GaInAsP well and Lb is the InP barrier thickness). The superlattices, consisting of 21 alternate layers of n-$Ga_{0.25}In_{0.75}As_{0.50}P_{0.50}$ of 75 Å thickness and n-InP of 50 Å thickness, were grown on a semi-insulating InP substrate.

Parameter	InP	$Ga_{0.25}In_{0.75}As_{0.50}P_{0.50}$
Growth temperature (°C)	650	650
Total flow (N$_2$+H$_2$) (l·min^{-1})	7	7
N$_2$-TEIn bubbler flow (cm^3·min^{-1})	350	350
H$_2$-TEGa bubbler flow (cm^3·min^{-1})	–	60
PH$_3$ flow (cm^3·min^{-1})	530	530
AsH$_3$ flow (cm^3·min^{-1})	–	21
Growth rate (Å·min^{-1})	100	150

Table 13.3. Optimized growth parameters.

Fig. 13.11 indicates the Auger line scan along the bevel, which had an angle of 0.02°. The Auger results indicate that three distinct GaInAsP layers with abrupt interfaces exist. Shubnikov–de Haas measurements were also performed on these samples.

The band discontinuities in lattice-matched GaInAsP–InP hetero-structures have been studied by Forrest et al. [1984] using the C–V technique: they found the conduction-band discontinuity ΔE_C to be 39% of the difference in the band gaps across the whole range of alloy compositions. This gives $\Delta E_C \simeq 170$ meV in the samples considered here, with the exception of the quantum well with higher phosphorus content, for which a discontinuity of $\Delta E_C \simeq 140$ meV is deduced.

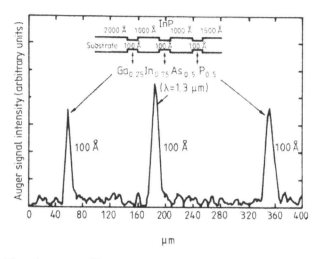

Fig. 13.11. Phosphorus profile using Auger measurements on a chemically etched bevel of MQW of GaInAsP–InP, comprising three GaInAsP ($\lambda = 1.3\ \mu m$) wells of 100 Å thickness.

Fig. 13.12(a) [Razeghi et al. 1985] shows typical Shubnikov–de Haas oscillations observed at $T = 4.2$ K, performed on MQW of $Ga_{0.25}In_{0.75}As_{0.50}P_{0.50}$/InP with the magnetic field B perpendicular to the MQW heterojunction interfaces. These oscillations disappear when the magnetic field B is applied parallel to the interfaces. This field orientation dependence of the Shubnikov–de Haas oscillation is evidence for the formation of a 2DEG in the wells of $Ga_{0.25}In_{0.75}As_{0.5}P_{0.5}$.

Fig. 13.12(*b*) shows the reciprocal magnetic field corresponding to the magneto-oscillation maxima of resistivity as a function of the Landau index for different values of θ (Fig. 13.12(*c*)), which is the angle between B and the normal to the interfaces. The oscillations are periodic in $1/B$, and they follow the expected $(\cos \theta)^{-1}$ dependence of a 2DEG. This is evidence for the two-dimensionality of the electron gas under consideration.

Fig. 13.13 gives the reciprocal magnetic field corresponding to the magneto-oscillation maxima of the resistivity as a function of the Landau index for different values of θ for a MQW with 21 alternate layers of $Ga_{0.25}In_{0.75}As_{0.50}P_{0.50}$ ($L_z = 70$ Å) and InP ($L_B = 50$ Å). The oscillations are periodic in $1/B$, and they follow the expected $(\cos \theta)^{-1}$ dependence of a 2DEG.

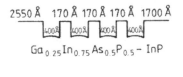

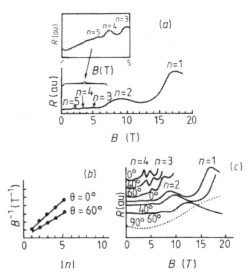

Fig. 13.12. (a) Magnetoresistance oscillations as a function of the magnetic field B. For reasons of clarity, the oscillations in the range 0–5 T are shown at higher magnification in the inset, (b) Reciprocal of magnetic field at maxima of magneto-oscillations versus Landau quantum number (n) for different values of θ for a $Ga_{0.25}In_{0.75}As_{0.50}P_{0.50}$/InP MQW grown by MOCVD. (c) Magnetoresistance oscillations as a function of the magnetic field B for different values of θ.

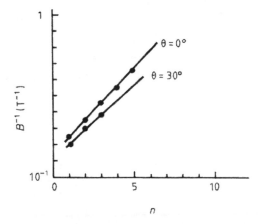

Fig. 13.13. Reciprocal of magnetic field at maxima of magneto-oscillations versus Landau quantum number for different values of θ for a $Ga_{0.25}In_{0.75}As_{0.50}P_{0.50}$/InP superlattice with 21 alternate layers (L_z = 70 Å and L_B = 50 Å).

Fig. 13.14 gives the reciprocal magnetic field corresponding to the magneto-oscillations versus Landau quantum number for different values of θ for a $Ga_{0.25}In_{0.75}As_{0.50}/InP$ single-quantum-well structure with $n_{2D} = 4.16 \times 10^{11}$ cm^{-2}.

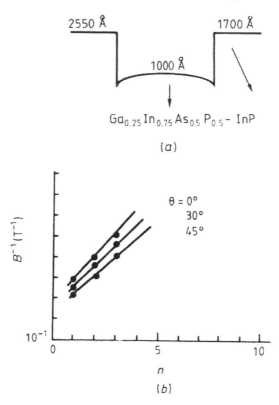

(a)

(b)

Fig. 13.14. Reciprocal field at maxima of the magneto-oscillations versus Landau quantum number for different values of θ (b), for a Ga₀.₂₅In₀.₇₅As₀.₅₀P₀.₅₀/InP heterojunction (a).

Sample	Well width Lz(Å)	Barrier width Lb(Å)	Number of wells	λ (μm) at 300 K	Y	nSdH (10^{11} cm^{-2})	m_c^* / m
A	100	—	1	1.18	0.50	5.9	0.0665
B	100	1360	3	1.32	0.69	6.7	0.0600
C	100	85	10	1.31	0.68	4.95	0.0609
D	400	200	2	1.34	0.67	5.46	0.0625
E	400	200	3	1.32	0.65	5.53	—
F	400	200	4	1.30	0.62	5.58	0.0586
G	1000	—	1	1.32	0.64	4.2	0.0616

Table 13.4. The sample characteristics.

The individual sample details and the results of Shubnikov–de Haas measurements of the two-dimensional electron concentrations are given in Table 13.4. The latter range from 4×10^{11} to 8×10^{11} cm^{-2}, and were all shown to be due to two-dimensional conduction by rotation of the samples relative to the magnetic field. Perhaps the most interesting of these measurements are shown in Fig. 13.15 for sample F, a multiquantum-well structure consisting of four 400 Å wells. Two series of Shubnikov–de Haas oscillations are seen, corresponding to electron concentrations of 5.8×10^{11} and 2.5×10^{11} cm^{-2} in either two electric subbands or two different layers. Hall effect measurements on this sample give a total sheet carrier concentration of 9×10^{11} cm^{-2}, indicating that these two populations are contributing the majority of the conduction. It is a general feature of the Shubnikov–de Haas measurements that the two-dimensional carrier concentrations deduced are consistent with conduction in a single layer, even in the multiquantum-well samples, from comparison with the doping levels. The electron concentrations are sufficiently high to require the transfer of charge from the buffer and capping layers, which may lead to the formation of what might be termed a "depleted superlattice," as shown schematically in Fig. 13.16. The band structure has tipped sideways, so that the majority of the electrons have accumulated at a single interface. This sort of behavior has also been suggested in GaInAs–InP superlattices (see Chapter 12), where the use of lightly doped buffer and capping layers leads to the apparent formation of an accumulated layer at one interface only of a multiquantum-well structure, while structures without buffer and capping layers showed uniform behavior with conduction occurring in all the quantum wells. Similar effects have also been seen in GaInAs–AlInAs superlattices [Portal et al. 1985]. The two separate populations observed in sample F might arise from the occupation of a second subband at the first heterojunction interface, from population of two adjacent layers, or from the formation of a 2DEG at the interfaces with both buffer and capping layers. The last of these is thought to be the most likely in view of the extent of the electron wavefunctions at a single heterojunction, of order 100 Å, compared with the well width of 400 Å, and the relatively small proportion of the total electron concentration found in the second subband in GaInAs–AlInAs, GaInAs–InP and GaAs–GaAlAs heterojunctions [Portal et al. 1982, Razeghi et al. 1986, and Stoermer et al. 1982]. Shubnikov–de Haas measurements on heterojunctions where two subbands are occupied also show that the two series of oscillations should interact at high fields, so that neither series is periodic in $1/B$ [Portal et al. 1982]. This is not observed in the present case.

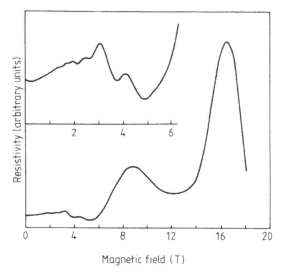

Magnetic field (T)

Fig. 13.15. The Shubnikov–de Haas oscillations in GaInAsP–InP sample F at 1.6 K. The two series of oscillations, with fundamental fields of 5.2 and 12 T, can be clearly seen in the expanded inset.

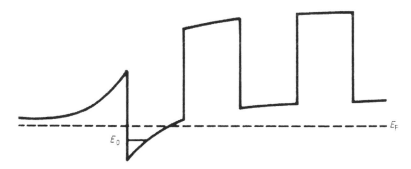

Fig. 13.16. The conduction-band edge of the depleted superlattice, with a 2DEG in one well only, proposed to explain the electron concentrations measured using the Shubnikov–de Haas effect.

Cyclotron resonance measurements have been performed [Razeghi et al. 1985] at 4.2 K using an optically pumped far-infrared laser at wavelengths between 119 and 54 μm. The electron effective masses deduced were in the range $m^*/m = 0.0585–0.0625$ except for the quantum well with higher phosphorus content, where a value of $m^*/m = 0.0665$ was measured. These values are 10–16% higher than the band-edge masses in GaInAsP of the appropriate composition [Nicholas et al. 1980] because of the non-parabolicity of the conduction band. Three-band $k \cdot p$ theory [Palik et al. 1981] gives the expression

Eq. (13.5) $m_0^* / m_{cr}^* = 1 + \left(2K_2 / E_g \right)\left(E_F + T_z \right)$

for the cyclotron mass in the limit $\hbar\omega_c < E_F$; here m_0^* is the band-edge mass and T_z is the kinetic energy associated with motion perpendicular to the layers. The coefficient K_2 (see Chapter 12) is a negative quantity, which should be related to the band structure of the material and have the value -0.8 in GaInAsP. If the 2DEG are considered to be confined in square potential wells, T_z is approximately equal to the subband energy E_0 and mass enhancements of 3–9% are calculated. However, the enhancements will be different if the multiquantum-well samples are assumed to form depleted superlattices containing one heterojunction-like 2DEG. For a heterojunction T_z can be estimated from the variational wavefunction proposed by Fang and Howard [1966], and mass enhancements in the range 6–8% are deduced. The non-parabolicity is thus greater than predicted by Eq. (13.5) regardless of whether the samples form uniform or depleted superlattices. However, recent measurements on bulk $Ga_{0.47}In_{0.53}As$ (see Chapter 12) and GaAs [Nicholas et al. 1979], and on 2DEG in the same materials [Brummel 1985 and Hopkins et al. 1986], also show higher non-parabolicities than predicted by three-band $k\cdot p$ theory and are well described by Eq. (13.5) if K_2 is taken to be greater than the calculated values by a factor of approximately 2. These results cannot be used to derive a value for the effective K_2 because of the uncertainty of the nature of the potential confining the electrons and possible errors in the measured effective masses due to the broad resonances seen, but they do suggest that, as in other III–V semiconductors, the non-parabolicity in GaInAsP is considerably greater than predicted by three-band $k\cdot p$ theory.

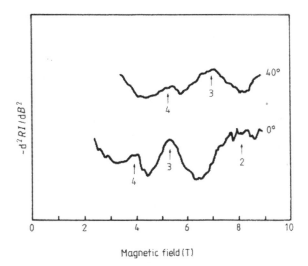

Fig. 13.17. The magnetophonon resonances in the second derivative of the magnetoresistance of sample F at 80 K. Rotation of the sample normal 40° away from the magnetic field direction shows that the oscillations are due to two-dimensional conduction.

Fig. 13.17 shows measurements of weak magnetophonon resonances in the second derivative of the magnetoresistance of sample F. Rotation of the sample relative to the magnetic field showed that the oscillations were due to two-dimensional conduction. The magnetophonon effect results from resonant scattering of the electrons by longitudinal optic (LO) phonons [Nicholas 1985], and in GaInAsP the scattering has been found to be dominated by an optic-phonon frequency of about 270 cm^{-1} associated with the $Ga_{0.47}In_{0.53}As$ ternary alloy and thought to result primarily from Ga–As vibrations [Nicholas et al. 1980]. The resonance positions are given by

Eq. (13.6) $\omega_{LO} = NeB / m^*$

in which the effective mass must be corrected for the effects of non-parabolicity [Portal et al. 1983, and Brummell et al. 1983]. The product NB_N is constant and has a value of 17.4 T in sample F. This corresponds to a phonon frequency of 269 cm^{-1}, in good agreement with the values measured in bulk material [Nicholas et al. 1980]. The oscillations observed were rather weak and showed no evidence of the long-range electron–phonon interactions and interface phonon effects seen in GaInAs–InP and GaInAs–AlInAs heterostructures [Portal et al. 1983 and Brummell et al. 1983].

So, the electron effective masses are found to be enhanced over the band-edge masses in the bulk material due to non-parabolicity, and the energy of the optic phonon dominating the electron scattering is found to be similar to that in bulk GaInAsP. The uniformly doped multiquantum-well structures display a tendency to form a depleted superlattice, in which all the carriers are confined at a single interface, and this is believed to be due to the presence of thick InP buffer and capping layers.

13.3.3. Observation of quantum Hall effect in a GaInAsP–InP heterostructure grown by MOCVD

Razeghi et al. [1986] have used high-quality trimethylindium (TMIn) instead of triethylindium (TEIn) for the growth of InP and GaInAsP by MOCVD. The growth conditions for InP and $Ga_{0.25}In_{0.75}As_{0.50}P_{0.50}$ (TMIn) determined during these investigations are presented in Table 13.5. High-quality InP epitaxial layers have been grown, with carrier concentration as low as 3×10^{13} cm^{-3} and electron Hall mobility as high as 6,300 cm^2·V^{-1}·S^{-1} at 300 K and 200,000 cm^2·V^{-1}·S^{-1} at 50 K (see Chapter 11). GaInAsP epilayers lattice-matched to InP have been grown, with a carrier concentration of 6×10^{14} cm^{-3} and electron Hall mobility as high as 6,400 cm^2·V^{-1}·S^{-1} at 300 K and 36,000 cm^2·V^{-1}·S^{-1} at 77 K.

Parameter	InP	$Ga_{0.25}In_{0.75}As_{0.50}P_{0.50}$
Growth temperature (°C)	550	550
H_2 flow (l·min^{-1})	6	6
H_2–TMIn bubbler flow (cm^3·min^{-1})	50	50
H_2–TEGa bubbler flow (cm^3·min^{-1})	_	40
PH_3 flow (cm^3·min^{-1})	270	270
AsH_3 flow (cm^3·min^{-1})	_	20
Growth rate (Å·min^{-1})	200	350
Carrier concentration (cm^{-3})	3×10^{13}	6×10^{14}
Electron mobility (77 K) (cm^2·V^{-1}·s^{-1})	150000	36000

Table 13.5. Optimized growth conditions.

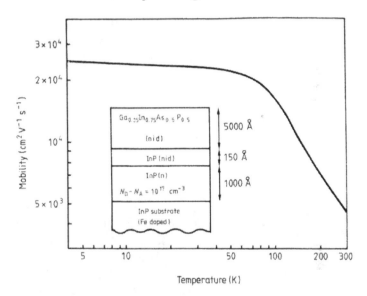

Fig. 13.18. Variation of electron Hall mobility as a function of temperature for a GaInAsP–InP epilayer grown by MOCVD (nid = non-intentionally doped).

Fig. 13.18 shows the evolution of the mobility as a function of temperature. The structure is described in the inset. It consists of a 1000 Å thick *n*-doped InP layer with an electron concentration of 10^{17} cm^{-3}, and a 150 Å thick undoped InP spacer layer, covered by 5000 Å thick undoped $Ga_{0.25}In_{0.75}As_{0.50}P_{0.50}$.

The electron Hall mobility of this modulation-doped structure was 5500 cm^2·V^{-1}·s^{-1} at 300 K, 20,000 cm^2·V^{-1}·s^{-1} at 77 K and 25,000 cm^2·V^{-1}·s^{-1} at 4 K, with a sheet carrier concentration of about 4.4×10^{11} cm^{-2} at low temperature. The variation of mobility as a function of temperature is clearly characteristic of a two-dimensional electron gas, with high mobility at low temperature due to the fact that carriers located in the quaternary compound near the interface are separated from the ionized donor impurities.

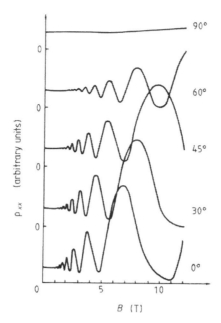

Fig. 13.19. Shubnikov–de Haas oscillations as a function of magnetic field for various tilt angles measured at 1.6 K for a GaInAsP–InP epilayer grown by MOCVD.

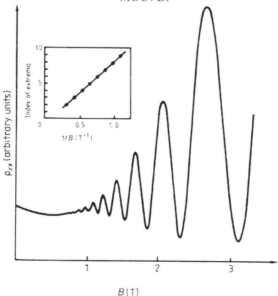

Fig. 13.20. Experimental recordings of magnetoresistance at low field (inset: standard plot of the inverse of the field position of the maxima in $\Gamma_{\chi\chi}$ versus their index), for a GaInAsP–InP epilayer grown by MOCVD.

Fig. 13.19 shows the evolution of the Shubnikov–de Haas oscillations for various angles θ between the magnetic field direction and the normal to the interface. The p_{xx} extrema appear at constant values of the perpendicular component of the magnetic field $B_1 = B \cos \theta$, and the oscillations vanish completely at $\theta = 90°$, demonstrating the two-dimensionality of the system. In Fig. 13.20, the low-field oscillations of the longitudinal resistivity ρ_{xx} at $\theta = 0$ are detailed. A plot of the index of the extrema observed, as a function of the inverse magnetic field, is shown in the inset of Fig. 13.20. It is clearly periodic, with a period $\Delta(1/B) = 0.11\ T^{-1}$ The electron density is derived from the expression:

Eq. (13.7) $$n = \frac{2e}{h\Delta(1/B)}$$

One can deduce in this case $n = 4.4 \times 10^{11} \cdot cm^{-2}$, in perfect agreement with the value of population n_H obtained from the classical Hall effect measurement.

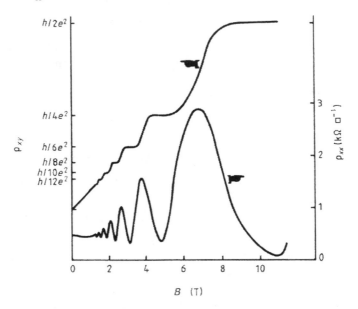

Fig. 13.21. Shubnikov–de Haas and quantum Hall effect as a function of magnetic field measured at 1.6 K for a GaInAsP–InP epilayer grown by MOCVD. The field is normal to the heterojunction plane.

Fig. 13.21 shows the diagonal resistivity ρ_{xx} and the Hall resistivity ρ_{xy} as functions of magnetic field at 4 K, for $\theta = 0$. Hall plateaux h/ie^2 with even values of i are resolved up to $h/20e^2$. The field at which only one Landau level is occupied is determined to be $B = 18.9\ T$.

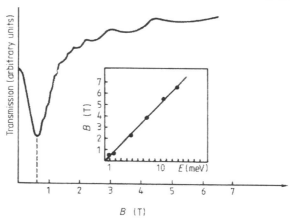

Fig. 13.22. Cyclotron resonance spectrum of a $Ga_{0.25}In_{0.75}As_{0.6}P_{0.4}$–InP heterojunction at $T = 1.6$ K, for $\lambda = 1205$ μm. Inset: plot of the resonance magnetic field versus excitation pulsation.

Fig. 13.22 shows a typical cyclotron resonance spectrum obtained at a wavelength of 1205 μm, for $\theta = 0°$ [Maurel et al. 1987]. Ando oscillations are clearly resolved. Their period in $1/B$ is about $\Delta(1/B) \equiv 0.1085$ T^{-1}, so that the electronic density derived at the heterojunction is $n \equiv 4.4 \times 10^{11}$ cm^{-2}. It remains in quite good agreement with low-field Hall measurements. The plot of the resonance field as a function of the energy of the excitation source is shown in the inset of Fig. 13.22. The dependence is clearly linear, as expected, and allowed the deduction of an electronic effective mass of $m^* = (0.059 \pm 0.001)m_0$ in the quaternary compound.

13.3.4. Disorder of a $Ga_xIn_{1-x}As_yP_{1-y}$–InP quantum well by Zn diffusion

Optical integrated circuits (IC) require that optical and electrical components can be fabricated on a single chip: for instance, monolithic optical IC with a combination of lasers and waveguides; lasers and field-effect transistors (FET); or PIN detectors and FET; etc. In this case, semi-insulating Fe-doped InP substrates should be used, and p–n junctions can be achieved by Zn diffusion.

However, the diffusion of Zn into an $Al_xGa_{1-x}As$–GaAs superlattice and a $Ga_xIn_{1-x}As$–GaAs superlattice has been shown to enhance intermixing of these SL layers greatly [Laidig et al. 1981].

Razeghi et al. [1987] have shown that Zn-enhanced intermixing is not just peculiar to the $Al_xGa_{1-x}As$–GaAs lattice-matched system or the $Ga_xIn_{1-x}As$–GaAs lattice-mismatched system. It is also effective in $In_xGa_{1-x}As_yP_{1-y}$/InP (lattice-matched) single-quantum-well and multiquantum-well structures.

For that study, various SQW and MQW were employed. The SQW consisted of a $Ga_{0.25}In_{0.75}As_{0.5}P_{0.55}$ QW of 100 Å thickness located between two 1000 Å thick InP layers. The MQW consisted of 10 $Ga_{0.25}In_{0.75}As_{0.5}P_{0.5}$ wells of $L_z = 100$ Å thickness with $L_B = 100$ Å (L_B is the InP barrier), also between two 1000 Å thick InP layers. Zn diffusion was accomplished via a semi-closed quartz box using Zn_3P_2 as the Zn source [Kamierski et al. 1984].

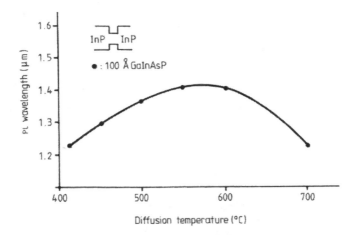

Fig. 13.23. The variation of photoluminescence wavelength of a single quantum well of GaInAsP–InP with 100 Å thickness as a function of Zn diffusion temperature for a diffusion time of 120 min.

Fig. 13.23 shows the variation of photoluminescence wavelength $\lambda(PL)$ of the SQW of GaInAsP as a function of diffusion temperature T_d, for a diffusion time of 120 min. The $\lambda(PL)$ increases when T_d increases, up to 650 °C, and then $\lambda(PL)$ decreases for $T_d > 650$ °C.

Fig. 13.24 represents the variation of $\lambda(PL)$ of this SQW as a function of diffusion time t_d for $Td = 550$ °C. The $\lambda(PL)$ increases when t_d increases.

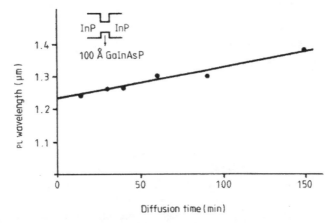

Fig. 13.24. The variation of photoluminescence wavelength of a quantum well of GaInAsP, 100 Å thick, as a function of Zn diffusion time for a diffusion temperature of 550 °C.

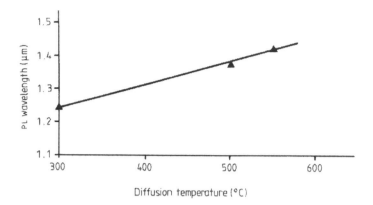

Fig. 13.25. The variation of photoluminescence wavelength of a 10-period GaInAsP–InP superlattice with $L_z = 100$ Å and $L_B = 100$ Å (L_z is thickness of GaInAsP quantum well, and L_B is thickness of InP barrier) as a function of Zn diffusion temperature for a diffusion time of 120 min.

Fig. 13.25 shows the variation of $\lambda(\mathrm{PL})$ of a 10-period GaInAsP–InP superlattice with $L_z = 100$ Å and $L_B = 100$ Å.

The morphology of SQW and MQW after Zn diffusion did not change. The intensity of the PL spectra decreases and the FWHM of the PL peak increases with diffusion time.

Thermal annealing without Zn diffusion of an SQW at 650 °C did not result in any wavelength change.

In situ Zn-doped GaInAsP–InP SQW and MQW layers did not show any variation of $\lambda(\mathrm{PL})$ after 4 h annealing at $T = 650$ °C (under PH_3 and H_2). This shows that the variation of PL wavelength is due to Zn diffusion.

Discussion

Van Vechten [1982], Burnham et al. [1986], and Laidig et al. [1983] have demonstrated that the diffusion of Zn into GaAs–GaAlAs superlattice structures promotes the intermixing of Al and Ga dramatically. They have suggested that this effect might be caused by an association between the interstitially diffusing Zn ions (Zn_i^+) and vacancies on the cation sublattice V_a (cation atoms are Ga or Al). A similar interpretation can be used for GaInAsP–InP QW layers diffused with Zn.

The increase of the photoluminescence wavelength must be attributed to the interdiffusion of In and Ga at the interfaces of the InP–GaInAsP–InP layers; interdiffusion of As and P would provide a *lower* $\lambda(\mathrm{PL})$, which is contrary to the experimental results, at least up to a diffusion temperature of $T_d = 650$ °C.

It is suggested that a closely associated Zn–vacancy pair formed by substitutional Zn moving into a neighboring interstitial site forms an intermediate link between purely interstitial and substitutional Zn according to the reactions [Laidig et al. 1984]

Eq. (13.8) $Zn_i^+ + V_a^0 \rightleftarrows (Zn_i, V)^+ \rightleftarrows Zn_s^- + 2h^+$

V_a^0 indicates a neutral cation vacancy and h a hole. Neighboring Ga or In atoms could move into the vacancy of the (Zn_i, V) pair, but the Zn_i would remain attached. In this way, interdiffusion of Ga and In would be promoted as well as provide an additional mechanism for Zn diffusion; however, the interdiffusion of As and P can be neglected.

The interdiffusion of In from the InP barriers into the $Ga_xIn_{1-x}As_yP_{1-y}$ quantum well and the opposite, Ga from the quantum well into the InP barriers, can be approximated as

Eq. (13.9) $$X(Z,t) = \frac{1}{2}X_0 \left[erf\left(\frac{L_z/2 - Z}{2(Dt)^{1/2}} \right) + erf\left(\frac{L_z/2 - Z}{2(Dt)^{1/2}} \right) \right]$$

with $X(Z, t)$ the molar fraction of Ga at time t and position Z, OZ the diffusion axis (the reference $Z = 0$ is the center of the quantum well), L_z the quantum-well thickness, t the diffusion time, X_0 the initial molar fraction of Ga in the QW and D the In–Ga interdiffusion coefficient.

The variation of the Ga molar fraction $X_d(t)$ in the QW after Zn diffusion time t is then

Eq. (13.10) $$X_d(t) = X_0 - (1/L_z) \int_{-L_z/2}^{+L_z/2} X(Z,t)\,dZ$$

Integration gives

Eq. (13.11) $$X_d(t) = X_0 \left\{ 1 - erf\left(\frac{L_z}{2(Dt)^{1/2}} \right) + \frac{2(Dt)^{1/2}}{L_z \pi^{1/2}} \left[1 - exp\left(\frac{L_z^2}{4Dt} \right) \right] \right\}$$

$X_d(t)$ can be experimentally deduced from photoluminescence measurements. The recombination energy in an alloy QW can be approximated by

Eq. (13.12) $$E = h\nu = hc/\lambda = E_g(t) + E_{conf}^e + E_{conf}^h$$
$$\lambda(\mu m) = 1.24/E(eV)$$

E_{conf}^e and E_{conf}^h are the confining energies for the electron and hole ground states respectively, λ is the photoluminescence wavelength, h is Planck's constant and c is the speed of light in a vacuum.

As we neglect the interdiffusion of As and P, y (in $Ga_xIn_{1-x}As_yP_{1-y}$) is considered constant, and the band-gap energy of a QW layer after Zn diffusion may be approximated by the formula [Olsen 1978]

$$E_g(t) = 1.35 + 0.668\big(X_w(t)\big) - 1.17Y + 0.758\big(X_w(t)\big)^2 + 0.18Y^2$$

Eq. (13.13) $$+ 0.069Y\big(X_w(t)\big)^2 - 0.069Y\big(X_w(t)\big)$$

$$- 0.322\big(X_w(t)\big)Y^2 + 0.03\big(X_w(t)\big)^3$$

with $X_w(t)$ the concentration of Ga at time t in the QW,

Eq. (13.14) $$E_g(t) = 1.24 / \lambda(t)$$

Eq. (13.15) $$\Delta\lambda = \frac{1.24}{E\big(X_w(t)\big)} - \frac{1.24}{E\big(X_0\big)}$$

Fig. 13.26 shows the experimental and calculated data of X_d as a function of diffusion time. Experimental data for a diffusion temperature of 500°C can be reasonably well fitted assuming $D = 2.5 \times 10^{-16}$ cm²·s⁻¹ in Eq. (13.11).

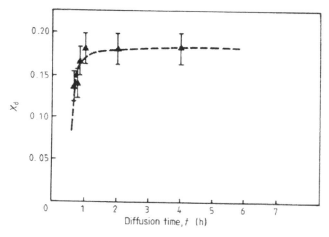

Fig. 13.26. The experimental ($\blacktriangle$) and calculated (— — —) values of X_d for a GaInAsP single QW of 100 Å thickness as a function of diffusion time, for a diffusion temperature of 500 °C.

This intermixing presents a problem for applications. For example, the transverse-junction stripe (TJS) laser is a laser structure in which the current is injected laterally into the active layer at the *p–n* junction. The *p*-type region is obtained by Zn diffusion into part of an *n*-type grown double heterostructure. The current flows laterally across the junction from the *p*-type region to the *n*-type region. In the case of a GaAs–GaAlAs system, GaAlAs has a wider band gap than GaAs, and carriers are injected predominantly across the GaAs *p–n* junction into the active layer. Disorder of a GaAs-GaAlAs quantum well by Zn diffusion increases the GaAs (active layer) energy gap so that injection into the non-diffused active layer is enhanced.

In the GaInAsP–InP system, Razeghi et al. [1987] have shown that intermixing due to Zn diffusion reduces the gap, so that injection into the non-diffused part is greatly reduced. This means that the GaInAsP system is not well suited for TJS lasers.

Moreover, the compositional change after Zn diffuses in GaInAsP wells causes a change in the lattice parameter. The deterioration of material quality when lattice mismatch is over 5×10^{-4} is well known in the case of the GaInAsP/InP double-heterostructure laser, where it increases the threshold current density with rapid degradation.

According to the Razeghi et al. [1987] model the composition of the well after Zn diffusion is near to $InAs_yP_{1-y}$, and the lattice mismatch exceeds 1%. Such an important mismatch creates dislocations and defects at the interface, which drastically increase the probability of non-radiative recombination.

The GaAlAs–GaAs system does not have this problem, since the lattice parameter hardly depends on composition. This is another difference that makes GaInAsP/InP TJS and QW lasers work less well than GaAlAs/GaAs ones.

13.3.5. Interface study of $Ga_xIn_{1-x}As_yP_{1-y}$–InP heterojunctions grown by MOCVD by spectroscopic ellipsometry

Knowledge of MOCVD material structure, on an atomic scale, in the interface region between various layers is important for understanding the electronic properties of heterojunctions and superlattices. High sensitivity makes spectroscopic ellipsometry (SE) a useful tool for addressing this problem.

Here we present a study [Drevillion et al. 1982] of $In_{0.53}Ga_{0.47}As/InP$ and $Ga_{0.25}In_{0.75}As_{0.5}P_{0.5}/InP$ heterostructures by SE. The steepness of the heterojunctions will be compared for two types of structures: those produced by the growth of InGaAs (or InGaAsP) on InP and those obtained by the reverse growth sequence.

(a) Experimental details

Two types of heterojunctions were considered for this study: $In_{0.53}Ga_{0.47}As$–InP and $Ga_{0.25}In_{0.75}As_{0.5}P_{0.5}$–InP. Growth conditions are as summarized earlier in Table 13.3. The heterostructures are obtained by depositing a 70 Å thick epilayer on top of the corresponding substrate. This thickness was chosen in order to enable SE measurements to be sensitive to the interface region. Before exposure to air, in order to prevent As and/or P effusion, the heterostructures were cooled in the deposition reactor under a gaseous atmosphere which depended on the nature of the top layer of the sample. The following gases were used: PH_3 (InP as top layer), AsH_3 (InGaAs as top layer) and an $AsH_3 + PH_3$ mixture (InGaAsP as top layer).

The spectroscopic ellipsometer is a polarization-modulation type described in detail elsewhere [Drevillion et al. 1982]. Let us recall that SE measures, as a function of the energy, the complex reflectance ratio

Eq. (13.16) $\rho = R_p R_s^{-1} = (\tan \varphi) \exp(i\Delta)$

between the reflection coefficients R_p and R_s for linearly polarized light with its polarization parallel and perpendicular, respectively, to the plane of incidence. In the case of a bulk material with a sharp surface, ρ is directly related to the dielectric function $\varepsilon(E)$ of the material through the relationship

$$\text{Eq. (13.17)} \quad \varepsilon(E) = \varepsilon_1 - i\varepsilon_2 = \left(\sin^2 \theta\right)\left[\left(\tan^2 \theta\right)\left(\frac{1-\rho}{1+\rho}\right)^2 + 1\right]$$

where θ is the angle of incidence of the light on the sample. Chemical treatments of the samples, prior to SE measurements, were performed by flowing solutions over the vertical surface of an optically pre-aligned substrate, then maintaining the sample in a dry N_2 flow while taking SE data in order to prevent a native oxide growth.

(b) Reference dielectric functions

As already mentioned, the SE study of heterojunctions requires knowledge of the dielectric functions of the bulk InP, $In_{0.53}Ga_{0.47}As$ and $In_{0.75}Ga_{0.25}As_{0.5}P_{0.5}$ materials. But, the corresponding $\varepsilon(E)$ functions cannot be deduced directly from SE measurements using Eq. (13.17) because of the presence of a native oxide surface layer. Different chemical etching and cleaning procedures were compared by x-ray photoelectron spectroscopy (XPS) and SE measurements. The SE measurements were performed, in a dry N_2 flow, immediately after XPS. XPS measurements revealed that this procedure prevents native oxide growth. The best results were obtained by using a 1:10 solution of HF:deionized H_2O. However, it has been shown, in the case of (1 0 0) GaAs surfaces, that acid solutions can produce deep micro-roughness [Hollinger et al. 1985]. These effects are illustrated below by comparing SE and XPS measurements.

The XPS measurements are shown in Fig. 13.27 to Fig. 13.31. Core-level and Auger lines P(2p), In($3d_{5/2}$), As(Auger LMM) and Ga(Auger LMM) have been recorded. In order to observe oxide removal from InP, GaInAs and GaInAsP surfaces, it is important to know the binding energies of the core levels and the kinetic energies of the Auger lines corresponding to the elements In, As, Ga and P in their oxidized forms in the III–V compounds. GaAs and InP, because they have been extensively studied, have been chosen as reference samples. The spectra shown in Fig. 13.27–Fig. 13.30 (curves A) and Fig. 13.31(a) correspond to air-oxidized InP and GaAs samples that have been chemically etched prior to air exposure. Concerning P(2p), Ga and As(Auger), oxide components are clearly visible in all XPS spectra. This is less clear for In(3d) because of the small chemical shifts existing between In in InP (binding energy BE = 444.4 eV) and In in In_2O_3 (BE = 444.7 eV) or In(OH)$_3$ (BE = 4445.3 eV). As a matter of fact, the In core level has been fitted with Gaussians in order to show the oxide component (see Fig. 13.28 (curve A)). XPS data were then taken after performing the HF:H_2O (1:10) etching treatment on three different air-oxidized substrates: InP, GaInAs and GaInAsP (Fig. 13.27–Fig. 13.31). We note that the component characteristic of overlayer oxide has disappeared from all XPS spectra, inferring oxide removal due to the etching treatment. All these results are in agreement with previous analyses, which have also shown the efficiency of HF etching solution in performing oxide removal [Krawczik and Hollinger 1984]. Immersion times from 1 s to 1 min in the etching solution did not produce any modification in the XPS spectra.

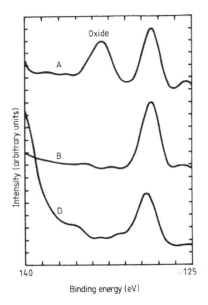

Fig. 13.27. P(2p) core-level peak recorded for (A) InP air-oxidized; and for HF:H₂O (1:10) etched samples of (B) InP and (D) InGaAsP.

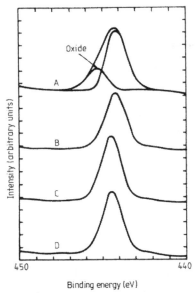

Fig. 13.28. In(3d) core-level peak recorded for (A) InP air-oxidized; and for HF:H₂O (1:10) etched samples of (B) InP, (C) InGaAs and (D) InGaAsP.

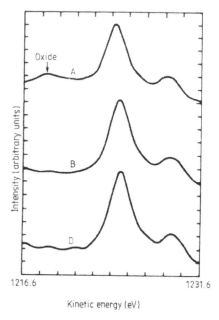

Fig. 13.29. As(LMM) Auger lines recorded for (A) GaAs air-oxidized; and for HF:H₂O (1:10) etched samples of (B) InP and (D) InGaAsP.

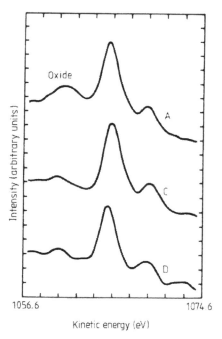

Fig. 13.30. Ga(LMM) Auger lines recorded for (A) GaAs air-oxidized; and for HF:H₂O (1:10) etched samples of (C) InGaAs and (D) InGaAsP.

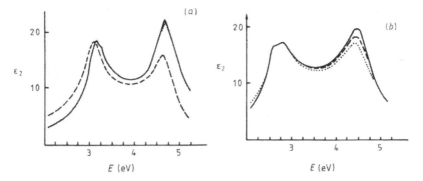

Fig. 13.31. Influence of (1:10) HF:H₂0 solution etching on the pseudo-dielectric function of InP (a) and In₀.₅₃Ga₀.₄₇As (b): broken curves, as-deposited samples; full curves, 1 s treatment; dotted curves, 3 s treatment.

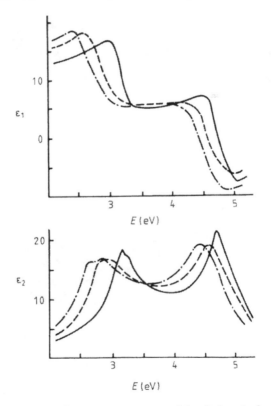

Fig. 13.32. Real part ε_1 and imaginary part ε_2 of the dielectric function spectra of InP, In₀.₅₃Ga₀.₄₇As and In₀.₇₅Ga₀.₂₅As₀.₅P₀.₅.

The SE measurements corresponding to the extreme values of y (InP and In₀.₅₃Ga₀.₄₇As) are presented in Fig. 13.32. In order to interpret the data, let us recall that the presence of an oxide layer and/or a surface micro-roughness both induce a decrease of the ε_2 spectra (together with a shift to lower energies), these effects being more sensitive in the 4.5 eV region (E_2 peak) [Aspnes 1980 and Aspnes and Studna 1985]. The first chemical treatment (full curves) performed by flowing the solution for 1 s over the surface results in an increase of the E_2 peak which corresponds to oxide removal, as evidenced by the XPS measurements. However, an extension of this chemical treatment for more than 1 s produces a decrease of the E_2 peak, revealed by SE measurements (Fig. 13.32, broken curves), which can be correlated to the presence of surface micro-roughness.

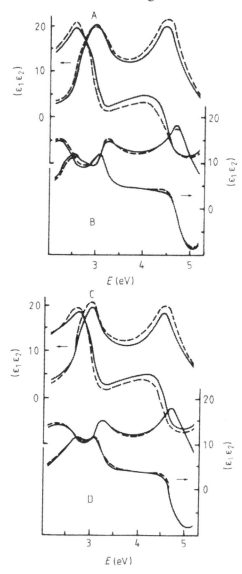

Fig. 13.33. Experiment (full curves) and best fit to the data using the physical interface model (broken curves) of the $In_{1-x}Ga_xAs_yP_{1-y}/InP$ heterostructures.

The real and imaginary parts $\varepsilon_1(E)$ and $\varepsilon_2(E)$ of the complex dielectric functions of the three reference materials are shown in Fig. 13.33 [Drevillion et al. 1982]. The two major features of the spectra are the E_1 and E_2 structures at 3.0 and 4.5 eV respectively. More precisely, all of the ε_2 spectra displayed in Fig. 13.33 have two features in the E_1 structure corresponding to the E_1 and $E_1 + \Delta_1$ transitions between the spin–orbit split upper valence band and the lowest conduction band along the [1 1 1] direction in the Brillouin zone. The three ε_1 (or ε_2) spectra are clearly distinguishable; the E_1 and E_2 structures move to lower energies with increasing y from $y = 0$ (InP) to $y = 1$ (In$_{0.53}$Ga$_{0.47}$As). The data displayed in Fig. 13.32 can be compared to previous measurements obtained on InGaAsP samples grown by liquid-phase epitaxy using a rotating analyzer ellipsometer (RAE). The overall agreement is rather good. However the dominant E_2 peaks in $\varepsilon_2(E)$ are 2–5% higher in the measurements by Kelso et al. [1982], but are 10% lower in the E_1 region (~ 3.0eV). The weak discrepancy in the E_2 region can possibly be attributed to differences in sample growth and surface preparation. As a matter of fact it was checked that a 5 Å thick In$_2$O$_3$ overlayer, or a very small surface micro-roughness (5 Å thick overlayer with 50% density deficiency), can account for the observed discrepancy in InP near 4.5 eV. But more generally, differences in $\varepsilon_2(E)$ data can reflect differences in the measurement procedures.

(c) Experimental results and discussion

Four samples of heterostructures have been examined:

> (a) a 70 Å In$_{0.53}$Ga$_{0.47}$As epilayer on an InP substrate;
> (b) a 70 Å epilayer on an In$_{0.53}$Ga$_{0.47}$As thick layer (2000 Å) on an InP substrate;
> (c) a 70 Å In$_{0.75}$Ga$_{0.25}$As$_{0.5}$P$_{0.5}$ epilayer on an InP substrate; and
> (d) a 70 Å epilayer on an In$_{0.75}$Ga$_{0.25}$As$_{0.5}$P$_{0.5}$ thick layer (2000 Å) on an InP substrate.

In each case, the cleaning procedure described above was applied before SE measurements. The $\varepsilon(E)$ functions of the heterostructures, given in Fig. 13.33, clearly differ from the data of Fig. 13.32, which correspond to the bulk materials. Because of the thinness of the epilayer (70 Å), the optical responses of the substrates clearly influence the SE measurements displayed in Fig. 13.33, this effect being more sensitive in the E_1 region (~ 3.0 eV). In order to detect the presence of an interface region between the substrate and the epilayer, the experimental $\varepsilon_2(E)$ measurements have been compared in Fig. 13.34 to a calculation using the reference dielectric functions shown in Fig. 13.32 and assuming a sharp interface between the substrate and a 60 Å (broken curves) or 80 Å (dotted curves) overlayer. The experimental ε_2 curves are only between both calculations in Fig. 13.34(d) (InP/InGaAsP). In contrast in Fig. 13.34(a), (b) and (c) the experimental curves clearly depart from the calculation in the E_2 region, showing the influence of a transition region in those cases.

This interface region may be tentatively described by assuming that the substrate and the epilayer are mixed inside a very thin layer without changing their chemical nature [Erman et al. 1983]. Such a physical mixture simulates a rough interface at the atomic scale between the substrate and the epilayer.

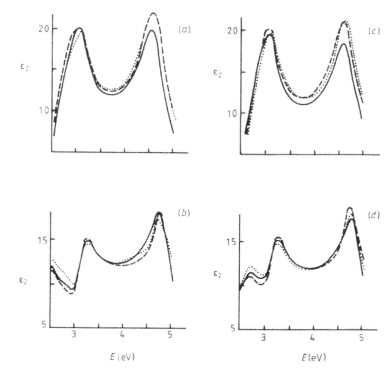

Fig. 13.34. Experiment (full curves) and modeling (broken and dotted curves) assuming a sharp interface of the $In_{1-x}Ga_xAs_yP_{1-y}InP$ heterostructures.

The dielectric function of the interface layer is computed using the standard effective-medium approximation. A least-squares fit of the data displayed in Fig. 13.33 is then performed, the free parameters of the fit being f_s, the volume fraction of the substrate inside the interface layer, and d_o and d_{int}, the thicknesses of the overlayer and the interface layer, respectively.

The results of the fits are compared to the experimental $\varepsilon(E)$ curves in Fig. 13.33 (broken curves). The best matches between calculated and experimental curves are obtained for samples (b) and (d) (InP epilayer). When using the reverse growth sequence (InP substrate), however, the physical interface model fails to reproduce the $\varepsilon_2(E)$ data in the E_2 region (see Fig. 13.33(a) and (c)).

Let us discuss first the cases corresponding to InP epilayer (samples (b) and (d)). The values of the fitted parameters obtained with these samples are displayed in Fig. 13.35. It can be seen that the overall thickness value (d_{int}, $+d_o$) obtained using the physical interface model is in very good agreement with the expected overlayer thickness (70 Å). The precision on the fitted parameter values can be estimated by varying the spectral range of the dielectric function used in the fit. The following typical values were obtained: ±0.01 on f_s and ±10 Å on d_{int} and d_o. In each case a small interface layer is obtained whose physical composition is close to InP ($f_s \simeq 0.80$). However, the sharpest interface is obtained by depositing InP on InGaAsP (sample (*d*)); in this case the existence of an interface layer is not apparent ($d_{int} = 10 \pm 10$ Å). The dielectric functions of both interface layers can be estimated by using the fitted values of f_s. The corresponding imaginary parts of the dielectric functions are compared to those of InP in Fig. 13.36. As expected, the optical responses of the interface layers of samples (*b*) and (*d*) are found to be very close to those of InP. This leads to two possible interpretations. First, a rough interface is produced in sample (*b*), but the difference between samples (*b*) and (*d*) revealed by the physical interface model (see Fig. 13.35) remains unclear. Secondly, there is an interdiffusion between the substrate and the overlayer (InP), leading to the formation of a thin layer of quaternary alloy with chemical nature different from the substrate. In this frame, the results displayed in Fig. 13.36 can be understood as an estimation of the optical response of the interface layer. An inspection of Fig. 13.36 shows that the composition of this quaternary alloy $In_{1-x}Ga_xAs_yP_{1-y}$ is probably close to InP ($y = 0$). Moreover, with this latter chemical interface model, a sharper interface is expected in sample (*d*) in which the underlayer is already a quaternary alloy ($In_{0.75}Ga_{0.25}As_{0.5}P_{0.5}$) than in sample (*b*) (substrate InGaAs). Indeed, the difference between samples (*b*) and (*d*), revealed by the physical interface model, can be correlated with the growth conditions. In both cases, before the growth of the InP layer, the introduction of $Ga(C_2H_5)_3$ and AsH_3 is interrupted, but phosphine remains present in the reactor only in the case of the deposition of InP on InGaAsP (sample (*d*)). Then, the difference between both interfaces can reflect the diffusion time of the phosphorus towards the neighborhood of the growth surface in the case of InP/InGaAs (sample (*b*)). This interpretation has recently been confirmed by Auger analysis along a chemically beveled InP/InGaAs/InP quantum well [Alnot et al. 1985].

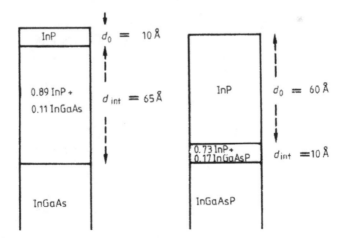

Fig. 13.35. Structures of heterojunctions (InP overlayer) as determined by using the physical interface model.

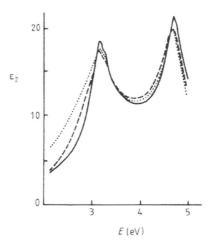

Fig. 13.36. Determination of the imaginary part of the dielectric function of the interface layers (InP overlayer) using the physical interface mode, compared to InP.———*, InP;*– – –*, InP–InGaAs; ....., InP–InGaAsP.*

Let us now consider the opposite cases in which InP is the substrate (samples (*a*) and (*c*)). In both cases very bad agreement with the physical interface model is obtained. This may be attributed to the presence of a large chemical interface. But results from Auger spectra of a chemically etched bevel through $Ga_{0.47}In_{0.53}As$–InP quantum wells do not seem to support this hypothesis. However, it has to be noted that the quantum-well structures are produced using a different growth procedure. In this latter case, the ternary layer is located between InP epitaxial layers. So, we favor another interpretation. Obviously, the optical responses of the thin 70 Å epilayers (InGaAs and InGaAsP) are incompatible with the corresponding bulk dielectric functions obtained by using 2000 Å thick InGaAs and InGaAsP grown on InP, as shown in Fig. 13.32. These differences may possibly reflect differences in sample preparation. In particular, let us recall that, during the cooling procedure in the deposition reactor before SE measurements, the introduction of group III alkyls is interrupted in both cases, group V hydrides still being resident in the reaction tube. Then, the composition of the gaseous boundary layer is expected to vary in the early stage of the cooling procedure, before complete evacuation of the group III alkyls. However, the epitaxial growth is controlled by the mass transport of group III species (see section 2.2). This can result in a modification of the composition of the InGaAsP epilayer near the surface. Furthermore, as the gaseous composition inside the deposition reactor is fixed, the chemical nature of the $In_{1-x}Ga_xAs_yP_{1-y}$ alloys is known to vary as a function of substrate temperature. The influence of these effects on the optical response of the sample is expected to be relatively stronger for a thin overlayer (70 Å) than for the bulk InGaAs or InGaAsP. Because of the lower substrate temperature dependence of the InP layer, the discrepancy between the dielectric functions of a thin layer and bulk material is not observed in samples (*b*) and (*d*) (InP epilayer).

Obviously, *in situ* ellipsometric measurements have to be performed in real time to check this hypothesis. First, spectroscopic ellipsometry appears to be a valuable tool to measure *in situ* the variation in chemical nature of $In_{1-x}Ga_xAs_yP_{1-y}$ alloys during the cooling procedure in the deposition chamber. Secondly, real-time measurements at a fixed wavelength, by kinetic ellipsometry, are known to be very sensitive to the early stage of growth on an atomic scale. This technique appears to be well adapted to perform *in situ* high-precision studies of the interface region in the MOCVD growth reactor.

13.4. Applications of GaInAsP–InP systems grown by MOCVD

There is strong pressure for the development of optoelectronic components necessary for trunk telecommunications, local area networks, cable TV distribution and avionic communications. The optoelectronic devices that are being incorporated into communications systems are quite sophisticated but further advanced components are required for future systems which make strong demands on cost, complexity and performance.

These advanced optoelectronic components exploit particular material parameters and the optimization of these parameters has led to the need for several new group III–V materials, new materials structures and alloy combinations, and novel preparation techniques.

System optimization here is heavily influenced by the need to keep the optoelectronic component and interconnection costs to a minimum and hence the drive is to employ the principles of integrated circuits as soon as possible.

Telecommunications systems have a hierarchy on the number of telephone lines multiplexed. Data rates of 8, 34, 140 and 565 Mbit·s^{-1} have emerged as the standards so far. The 8, 34 and 140 Mbit·s^{-1} systems have been implemented with LED and PINFET receivers and multimode fibers, but the emphasis has been to optimize the LED parameters for high switching speed, launch power and spectral purity, and the receiver for sensitivity so that repeater spacings up to 15 km can be achieved.

Longer repeater spacing and high levels of multiplexing requirements have led to 140 and 565 Mbit·s^{-1} laser-sourced monomode fiber systems. These systems require light control of laser and detector parameters and the early introduction of sophisticated fabrication techniques.

13.4.1. Broad-area 1.2–1.6 µm $Ga_xIn_{1-x}As_yP_{1-y}$–InP DH lasers grown by MOCVD

After optimization of growth conditions for InP and GaInAsP layers, MOCVD has been successfully used for the growth of GaInAsP–InP double heterojunctions (DH) for laser applications emitting at 1.3 and 1.5 µm. The various interfaces are produced by controlling the flow of the relevant component, as in the case of GaInAs mentioned previously.

The growth rate for these quaternary materials is small ($\sim 3 \, \text{Å·s}^{-1}$), and the gas flow is stabilized to its new steady-state value in less than 1 s after switching.

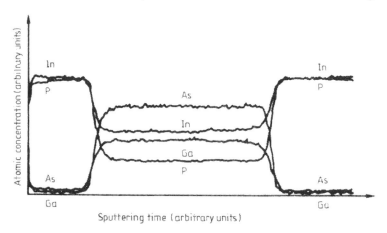

Fig. 13.37. Auger profile for GaInAsP/InP DH for laser applications (the sputter timescale is not linear in the central region).

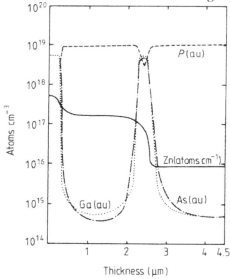

Fig. 13.38. SIMS profile of GaInAsP/InP DH for laser applications, at $\lambda = 1.3 \, \mu m$.

Fig. 13.37 and Fig. 13.38 show the Auger and SIMS profiles of a GaInAsP–InP DH emitting at 1.3 μm. The structure consists of four layers:

(1)   S-doped InP ($n \simeq 10^{18} \, \text{cm}^{-3}$), confinement layer;
(2)   GaInAsP ($\lambda \simeq 1.3 \, \mu m$), no intentional doping, active layer;
(3)   Zn-doped InP ($p \simeq 10^{17} \, \text{cm}^{-3}$), confinement layer; and
(4)   Zn doped GaInAsP ($\lambda = 1.3 \, \mu m$), cap layer.

The Zn profile shows that it is possible to control the *p*-type doping level, to avoid the diffusion of Zn into the *n*-type confinement InP layer, and to keep the *p–n* junction in the quaternary active layer.

Fig. 13.39 shows the transition region after chemical etching (see section 4.2) of a GaInAsP–InP DH laser. The etch-pit density (EPD) is the same in the substrate as in the confinement InP layers, and the interfaces are defect-free.

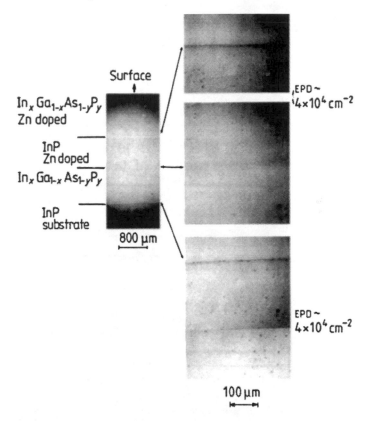

Fig. 13.39. The depth of a chemical bevel, realized on a GaInAsP/InP DH grown by LP-MOCVD, for laser applications, in a bromine solution. It is not possible to reveal the interfaces between the InP n-type confinement layer and the substrate. $In_xGa_{1-x}As_{1-y}P_y/InP/In_xGa_{1-x}As_{1-y}P_y/InP$ substrate n^+; x = 0.75, y = 0.44; bevel angle a = 0.065°; etchant H, 2 min.

(a) Laser fabrication and characteristics

Broad-area and stripe-geometry lasers have been fabricated from MOCVD grown materials. A typical example of a broad-area laser is shown schematically in Fig. 13.40. The DH structure of these laser diodes is described in Table 13.6. These layers are grown successively on (1 0 0) Sn-doped ($n \simeq 2 \times 10^{18}$ cm^{-3}) InP substrates, 2° off (1 0 0) towards (1 1 0). The growth temperature of the epilayers is 650 °C, and the working pressure is 100 mbar.

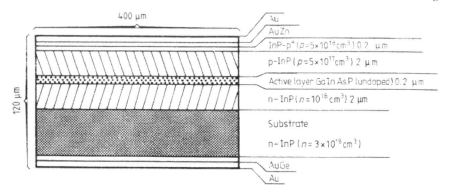

Fig. 13.40. Typical broad-area GaInAsP/InP laser diode.

Layer	Composition	Type	Thickness(μm)
Substrate	InP(Sn)	n^+, 2×10^{18} cm^{-3}	100
Buffer	InP	n^+, 10^{18} cm^{-3}	1
Active	Ga$_x$In$_{1-x}$As$_y$P$_{1-y}$	Undoped	~0.2
Confinement	InP	p, 2×10^{18} cm^{-3}	2
Contact	Ga$_x$In$_{1-x}$As$_y$P$_{1-y}$.	p^+, 5×10^{18} cm^{-3}	0.5

Table 13.6. DH structure of laser diodes.

The lattice mismatch of the Ga$_x$In$_{1-x}$As$_y$P$_{1-y}$, epitaxial layer is less than or equal to 5×10^{-4} at room temperature. The grown wafers are then lapped to a thickness of 100 μm and Au–12%Ge and Au–8%Zn contact metallizations are deposited on the n and p sides, respectively. The contacts are then annealed at 400 °C for 5 min in an argon atmosphere. The devices are cleaved and sawn, producing chips of width 150 μm with cavity lengths in the range 300–1000 μm. The laser chips are tested unmounted under pulsed conditions at a pulse repetition rate of 10^4 Hz with a pulse length of 100 ns. For chips cleaved from the same bar, the standard deviation in the lasing threshold current density is only ±5%. For a large slice area (10 cm^2), the standard deviation in lasing threshold is typically less than 20%. The emitting wavelengths can be controlled with an accuracy of 0.02 μm in the range 1.2–1.6 μm.

The dependence of threshold current density J_{th} on thickness d of the active layer of a laser diode grown by MOCVD is shown in Fig. 13.41. The threshold current density of broad-area lasers with 300 μm cavity length and 200 μm cavity width decreases linearly with decreasing thickness of the active layer down to $d = 2000$ Å and is about 500 A cm^{-2} for $d = 0.2$ μm. The threshold current density increases when the thickness is less than 1500 Å. These results are in satisfactory agreement with those obtained with LPE-grown materials [Nahory and Pollack 1978].

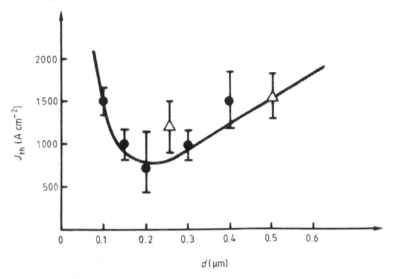

Fig. 13.41. Experimental threshold current density as a function of the thickness of the active layer for GaInAsP/InP MOCVD laser diodes emitting at wavelengths of 1.3 (•) and 1.5 (Δ) μm.

(b) Lasers emitting at 1.5 μm

Here, we describe the continuous-wave (CW) 1.5 μm room-temperature operation of $In_{1-x}Ga_xAs_yP_{1-y}$/InP double-heterostructure (DH) diode lasers grown by the low-pressure metalorganic chemical vapor deposition technique. The DH structure is given in Table 13.7.

Layer	Composition	Type	Thickness(μm)
Substrate	InP(Sn)	n^+, 2×10^{18} cm^{-3}	100
Buffer	InP	n^+, 10^{18} cm^{-3}	1
Active	$Ga_{0.37}In_{0.63}As_{0.8}P_{0.2}$	Undoped	0.5
Confinement	InP	p, 5×10^{17} cm^{-3}	2
Contact	InP	p^+, 5×10^{18} cm^{-3}	0.5

Table 13.7. DH structure of laser diode.

The composition of the $Ga_{1-x}In_xAs_yP_{1-y}$ epitaxial layer was determined by x-ray diffraction and photoluminescence. Rather thick active layers were grown (0.4–0.6 μm) in order to allow us to measure the mismatch of the active layer, which, for the data mentioned here, is $\Delta a/a < 3 \times 10^{-4}$. The compositional uniformity of the active layer was characterized by measurement of the photoluminescence wavelength at various points on an epitaxial layer of 8 cm^2 area. The wavelength varied by less than 2 nm over 95% of the area, the photoluminescence efficiency changed by less than 2%, and the thickness uniformity was better than 0.48 ±0.01 μm over the same area.

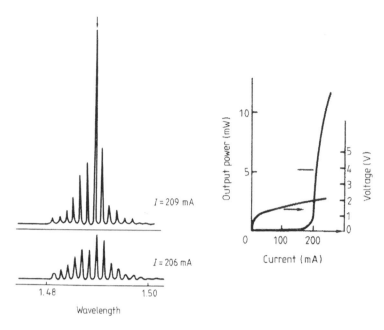

Fig. 13.42. Lasing spectrum and light–current and voltage–current characteristics of a single-transverse-mode GaInAsP/InP DH CW laser diode emitting at 1.5 μm at room temperature.

The contacting technology and fabrication of broad-area lasers are as described in the previous section. Broad-area laser chips are made by standard cleaving and sawing procedures producing Fabry–Perot cavities 400 μm in length. Fig. 13.42 shows a typical spectrum under CW operation at room temperature of a MOCVD GaInAsP/InP laser emitting at 1.5 μm.

All the chips from the same cleaved bar showed the same results. The minimum threshold current is 1.0 kA·cm^{-2}. Fig. 13.43 shows the threshold current for a 1.5 μm laser for several temperatures ranging from 10 to 70°C. As with GaInAsP/InP lasers grown by LPE, the lasing threshold varies exponentially with temperature, and the T_0 value (see Eq. (13.19)) for a 1.5 μm laser tested in pulsed operations is about 62 K in the room-temperature range.

This section gives the life test of a laser diode emitting at $\lambda = 1.5$ μm grown by MOCVD, which operated at room temperature for more than 10^4 h without significant degradation. The DH structure of this diode is described in Table 13.7.

These lasers have a stripe width of 9 μm and a cavity length of 300 μm. The localization of the current flow was obtained by shallow-proton implantation. After cleaving and sawing, individual diodes were mounted *p*-side down on gold-plated copper heatsinks.

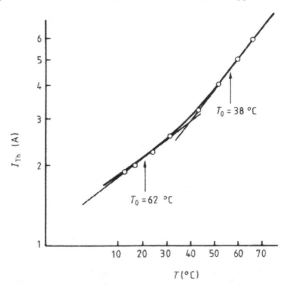

Fig. 13.43. Typical CW light–current characteristics at different temperatures for a GaInAsP/InP buried-ridge-structure (BRS) laser emitting at 1.5 µm.

Fig. 13.42 shows the continuous-wave (CW) light–current characteristic of one of these diodes with its emission spectrum before life testing. A threshold current of 200 mA and a P–I curve with fundamental transverse-mode operation up to 10 mW were obtained.

The characteristics of a laser diode were life tested under the following conditions. The diode was placed under dry nitrogen and aged in the lasing mode. The injected current was periodically adjusted to keep a constant output power of 1 mW/facet. The diode was aged at 20 °C owing to its high threshold current.

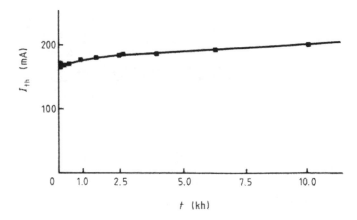

Fig. 13.44. Variation of threshold current density I_{th} for InP–GaInAsP DH lasers emitting at 1.5 µm by LP-MOCVD with time (T = 20 °C, P = 1 mW).

Fig. 13.44 shows threshold current against time for a laser soldered with indium to a gold-plated copper mount and operating at 1 mW output power at room temperature. The degradation rate is approximately $\Delta I_{th} / I_{th}\Delta t \simeq 10^{-5}\,h^{-1}$ with no detectable change in external quantum efficiency over 12,000 h.

The external quantum efficiencies of the diode after 12,000 h CW operation are illustrated in Table 13.8 in comparison with the values before life testing. Table 13.8 shows that η_{ext} after 12,000 h CW operation is comparable to the value before life testing.

	η_{ext} (W·A^{-1}) per facet	
	Before life testing	After 12,000 h
Pulsed operation	0.40	0.39
CW operation	0.23	0.20

Table 13.8. External quantum efficiencies of diode before and after life testing.

The far-field patterns of this laser diode are shown in Fig. 13.45. These patterns were measured under CW operation ($I_{th} = 254$ mA) after 12,000 h. The laser operated stably in the fundamental lateral mode without significant distortion of the field patterns. Far-field full angles at half power, both parallel and perpendicular to the junction plane, are 12° and 38°, respectively.

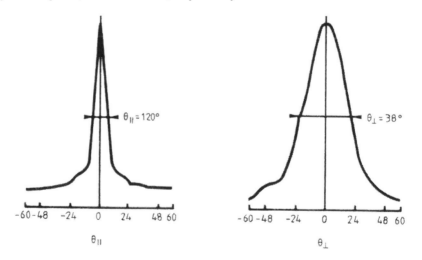

Fig. 13.45. Far-field patterns parallel and perpendicular to the junction plane for an MOCVD laser with 9 μm stripe width; λ = 1.5 μm, T = 20 °C, I_{th} = 254 mA, P = 3.5 mW.

These preliminary results on non-optimized structures are most encouraging and demonstrate that the MOCVD-grown GaInAsP/InP DH lasers have comparable degradation rates to those of LPE lasers, suggesting that the MOCVD technique is promising for large-scale production of laser diodes.

(c) $Ga_{0.25}In_{0.75}As_{0.5}P_{0.5}$–InP DH lasers emitting at 1.3 μm

A good layer uniformity was demonstrated by the experimental result that the threshold current density continued to fall with lasing cavity length for cavity lengths up to 950 μm. The variation of lasing threshold current density J_{th} with cavity length L is shown in Fig. 13.46. The value of the absorption coefficient α was $16\ cm^{-1}$ and was obtained by fitting, to the data in Fig. 13.46, the standard expression

Eq. (13.18) $\qquad J_{th} = \dfrac{1}{\beta}\left[\alpha + \dfrac{1}{2L}\ln\left(\dfrac{1}{R_1 R_2}\right)\right]$

where R_1 and R_2 are the facet reflectivities ($R_1 = R_2 = 0.32$) and β is a constant equal to $0.058\ cm \cdot A^{-1}$. In a series of 10 laser slices the average emission wavelength was 1.284 μm with a standard deviation of only ±1% for a broad-area laser of cavity length 300 μm (width 150 μm). The average threshold current density is $800\ A \cdot cm^{-2}$. The latter decreases to $500\ A \cdot cm^{-2}$ for a cavity length of 950 μm, the lowest threshold being $430\ A \cdot cm^{-2}$ for a cavity length of 950 μm.

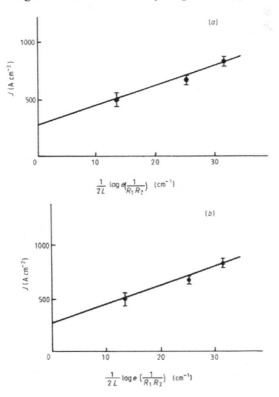

Fig. 13.46. Variation of threshold current density J_{th} with reciprocal of cavity length: (a) 1.3 μm lasers; (b) 1.5 μm lasers. In equation (13.18), $\alpha = 16\ cm^{-1}$, $\beta = 5.8 \times 10^{-2}\ cm \cdot A^{-1}$ and $R_1 = R_2 = 0.32$.

A 1.3 µm laser diode can be operated CW at room temperature at a typical current of 80–120 mA. The characteristic was kink-free up to 10 mW/facet with fundamental transverse-mode operation up to 10 mW.

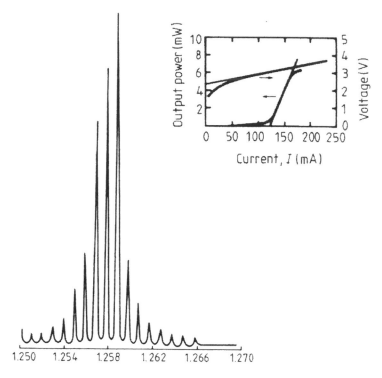

Fig. 13.47. Lasing spectrum, light–current and voltage–current characteristics of a single-transverse-mode GaInAsP/InP DH CW laser diode emitting at 1.26 µm at room temperature (T = 20 °C); output power 3.01 mW, current 143.5 mA, refractive index 3.29, laser RAZ 558 6A.

Fig. 13.47 shows a typical lasing spectrum, light–current and voltage–current characteristics of an MOCVD GaInAsP–InP DH CW laser diode emitting at 1.26 µm. The far-field patterns of this laser diode are represented in Fig. 13.48. The full widths at half-maximum (FWHM) are 15° and 22°, parallel and perpendicular to the junction plane, respectively, up to 4 mW output power, measured by scanning a Ge avalanche photodiode perpendicular to the junction. These patterns were measured under CW operation for a stripe width of 12 µm. For the same laser structure grown by LPE, $\theta_\perp$ varies between 25° and 40°. In the field of optical-fiber communication, the fabrication of laser diodes with low beam divergence is essential for coupling the emission into an optical fiber.

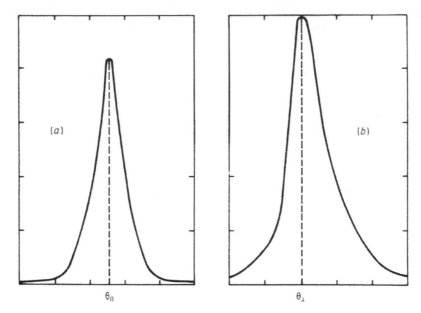

Fig. 13.48. Far-field patterns (a) parallel and (b) perpendicular to the junction plane of an LP-MOCVD GaInAsP/InP laser diode emitting at 1.26 μm; T = 20 °C, P = 3 mW, I = 143 mA, laser RAZ 558 6A.

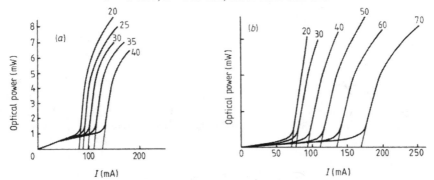

Fig. 13.49. Light output versus pulsed current at several heatsink temperatures of 1.3 μm GaInAsP/InP laser grown by LP-MOCVD: (a) under CW operation (sample RAZ 558); (b) under pulse operation ($T_0 = 67$ °C).

Fig. 13.49 shows the light power versus pulsed threshold current of this diode for various temperatures; the 20 °C average threshold current is 70 mA. The threshold current density of DH lasers depends empirically on temperature as

Eq. (13.19) $J_{th}(T) = J_{th}(T_0)\exp(T/T_0)$

where T_0 is a characteristic temperature. The average T_0 value is 67 °C. A few
diodes were aged with a constant power of 5 mW/facet at room temperature. The
results are illustrated in Fig. 13.50, which shows that GaInAsP/InP lasers emitting
in the wavelength region of 1.3 μm made by MOCVD can be operated at room
temperature for more than 6000 h without significant degradation.

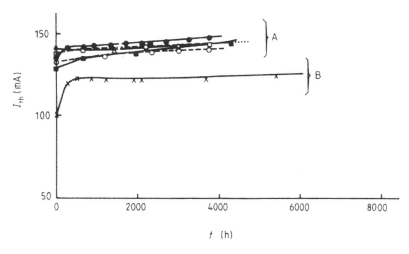

*Fig. 13.50. Current versus time aging characteristics (at room temperature) of
GaInAsP–InP laser diodes made from LP-MOCVD material: (A) $P_1 = 5$ mW, $\lambda =
1.26$ μm; (B) $P_1 = 2$ mW, $\lambda = 1.29$ μm; T = RT.*

13.4.2. Buried-ridge-structure lasers grown by MOCVD

In order to decrease the injection current a buried-ridge structure (BRS) can be
chosen. Fig. 13.51 and Fig. 13.52 show SEM micrograph and schematic diagrams
of the cross section of a BRS laser grown by two-step MOCVD. This laser was
fabricated as follows. First the following layers were grown successively on an
InP(Sn) substrate with orientation (1 0 0) exact or (1 0 0) 2° off:

(1) 1 μm InP, sulfur-doped, for confinement layer;
(2) 0.2 μm GaInAsP ($\lambda = 1.3$ or 1.55 μm), undoped, active layer; and
(3) 0.2 μm InP, Zn-doped, confinement layer.

Next, a ridge of about 1 μm width and 0.5 μm depth was etched in the InP and
GaInAsP layers through a photolithographic resist mask, using a selective etch
composed of H_2SO_4, H_2O_2 and H_2O (1:8:40).

After removing the resist mask, the ridge was then covered by a 1 μm thick Zn-
doped InP confinement layer and 0.5 μm Zn-doped GaInAsP contact layer with N_A-
$N_D = 1 \times 10^{19}$ cm^{-3} (using again the MOCVD growth technique). Fig. 13.53 shows
an electrochemical profile (polaron) of this structure before and after regrowth. In
order to localize the injection in only the buried-ridge active region, a deep-proton
implantation was realized through a 5 μm wide photoresist mask after metallization

of the contacts. The n contact is realized by the deposition of Au. The p contact is made by the deposition of Pt–Au and annealing at 450 °C for 30 s. After cleaving and scribing into chips of 300 μm length and 350 μm width, the lasers were mounted, epilayer-side down, onto nickel-plated copper heatsinks using indium bonding.

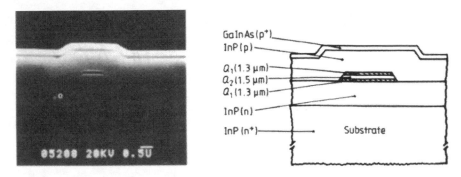

Fig. 13.51. Scanning electron micrograph and schematic diagram of the cross section of a buried-ridge structure (BRS) of a 1.3 μm laser grown by two-step LP-MOCVD.

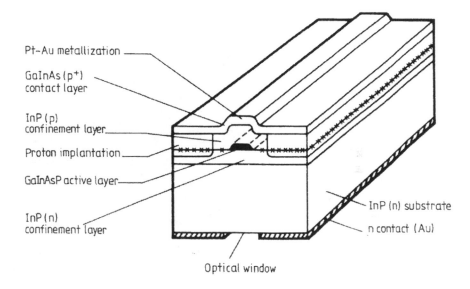

Fig. 13.52. Schematic diagram of the cross section of a BRS 1.3 μm laser grown by two step LP-MOCVD, with proton implantation.

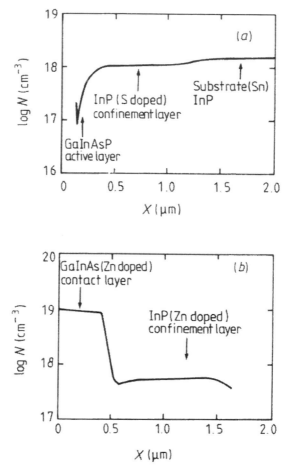

Fig. 13.53. Polaron profile of a GaInAsP–InP BRS 1.5 μm laser: (a) first step growth; (b) second step growth.

(a) BRS lasers emitting at 1.3 μm

Typical light–current, voltage–current and lasing characteristics under CW operation at 20 °C are shown in Fig. 13.54 and Fig. 13.55; CW linear output powers up to 10 mW have been measured. An external quantum efficiency $\eta_q \simeq 60\%$ has been obtained.

Typical far-field patterns parallel and perpendicular to the junction plane are shown in Fig. 13.56. These patterns were measured under CW operation for different optical output powers. A single transverse mode without significant distortion was observed at least up to 10 mW. The far-field full widths at half-maximum (FWHM) power, parallel and perpendicular to the junction plane, were about 26° and 36°, respectively.

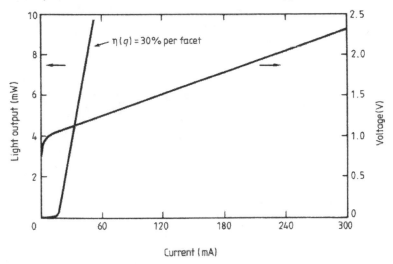

Fig. 13.54. *Light-current (P–I) and voltage–current (V–I) characteristics of LP-MOCVD BRS laser under CW operation at 20 °C.*

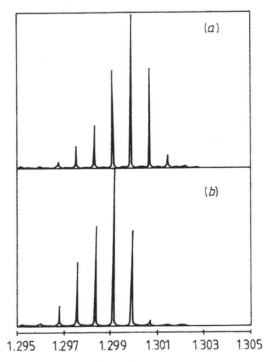

Fig. 13.55. *Lasing spectra of a BRS laser of a single traverse mode under CW operation at 20°C: (a) P = 8 mW, I = 40 mA, λ = 1.2998 μm; (b) P = 4 mW, I = 30 mA, λ = 1.299 μm.*

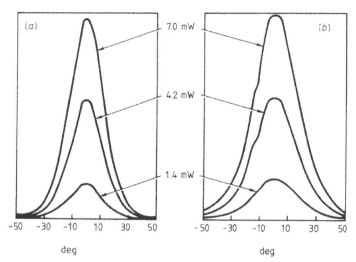

*Fig. 13.56. Far-field patterns (a) parallel and (b) perpendicular to the junction of a
BRS laser for different output powers under CW operation.*

Fig. 13.57 shows the variation of threshold current as a function of stripe width.
The threshold current increases linearly with buried-ridge width.

Excellent reproducibility, homogeneity and uniformity have been obtained over
10 cm^2. Table 13.9 shows the threshold current, quantum efficiency ηq and
wavelength of nine wafers of 8 cm^2. The standard deviation of threshold current
density over 8 cm^2 becomes less than 5%. Table 13.9 also presents the threshold
current density of diodes over 8 cm^2 using two different technologies.

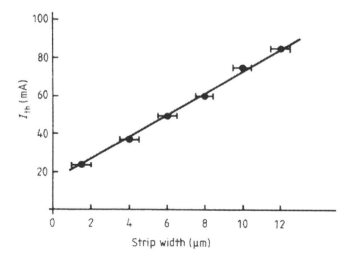

*Fig. 13.57. Variation of threshold current (I_{th}) as a function of stripe width; LP-
MOCVD BRS laser, λ = 1.3 μm, η_q = 60%.*

Wafer number	I_{th} (mA)	Number of diodes tested	T_{meas} (°C)	λ (μm)	λ_q (%)
1262–83	24.2 ± 1.9	25	27	1.32	~60
1264–91	17.9 ± 2.0	20	24	1.29	~60
1266–88	26.3 ± 5.6	25	29	1.32	~60
1267–89	37.4 ± 2.9	20	24	1.23	~60
1268–88	36.0 ± 3.1	20	27	1.3	~60
1204–06	45.7 ± 5.0	60	23	1.32	~60
1205–06	50.0 ± 2.5	60	26	1.3	~60
1218–27	36.6 ± 3.3	19	25	1.3	~60
1223–30	26.7 ± 5.4	20	28	1.3	~60

Table 13.9. Threshold current I_{th}, number of diodes tested, measurement temperature T_{meas}, lasing wavelength λ and external quantum efficiency η_q of nine wafers.

Fig. 13.58 shows the variation of threshold current (I_{th}) and external quantum efficiency (η) of a $Ga_{0.28}In_{0.72}As_{0.46}P_{0.54}$–InP BRS laser, under CW operation at room temperature, from four bars randomly selected from one wafer of 10 cm^2. Threshold currents as low as 6 mA, with an external quantum efficiency as high as 30% for a cavity length of 300 μm, have been measured.

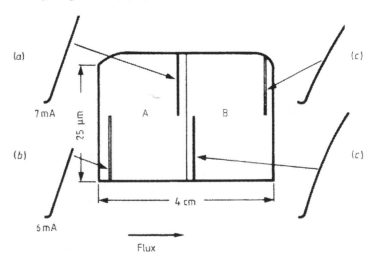

Fig. 13.58. The variation of threshold current (I_{th}) and external quantum efficiency (η) of a GaInAsP–InP BRS laser over 10 cm^2 of InP substrate; $\lambda = 1.30$ μm, 48 tested lasers.

(b) GaInAsP–InP BRS lasers emitting at 1.5 μm

Fig. 13.59 shows the variation of threshold current (I_{th}), external quantum efficiency (η) and emitting wavelengths of a GaInAsP–InP buried-ridge-structure laser emitting at 1.5 μm, fabricated on material grown by two-step MOCVD over 4 cm² of InP substrate. The measurements are done at room temperature. The fabrication process is similar to the 1.3 μm BRS laser mentioned before. Bars are taken from different parts of a wafer in order to have a typical sample representative of the wafer. The average threshold current is 14.53 mA with a standard deviation of 1.82 mA. For 79 tested lasers the average external differential efficiency is 0.17 W·A^{-1} per facet with a standard deviation of 0.01 W·A^{-1}.

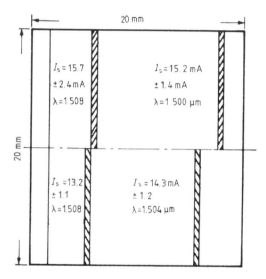

Fig. 13.59. Variation of threshold current (I_{th}), external quantum efficiency (η) and emitting wavelengths of a GaInAsP–InP BRS laser over 4 cm² of InP substrate; laser RAZ 112.120/n H⁺; 79 tested lasers; good lasers 74–94%; λ = 1.5 μm, I_{th} = 14.53 ±1.02 mA, η = 0.17 ± 0.01 W·A^{-1}.

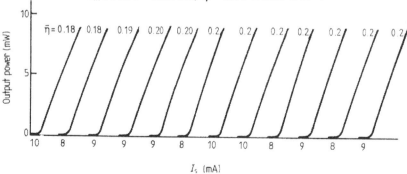

Fig. 13.60. Light–current characteristics of 12 MOCVD BRS laser diodes emitting at 1.55 μm, at T = 28 °C.

Fig. 13.60 shows light–current characteristics of 12 diodes randomly selected from another wafer emitting at 1.55 μm. External quantum efficiency as high as 20% per facet and threshold current as low as 8 mA at room temperature have been measured under pulse operation. The typical far-field full widths at half-maximum parallel and perpendicular to the junction for the lasers are 28° and 35° respectively, with excellent uniformity. The T_0, measured under pulsed operation between 20 and 70 °C, is 65 K. The maximum operating temperature is around 110 °C.

(c) Validation of MOCVD GaInAsP–InP lasers by photoluminescence imaging method

Photoluminescence imaging is done with simple equipment represented in Fig. 13.61, which consists of an argon laser and infrared PbS camera–monitor system. The laser spot is focused on the sample, the luminescence image of which is then magnified and detected by the camera.

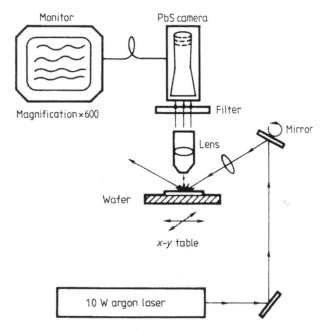

Fig. 13.61. Photoluminescence imaging system.

For practical reasons, the observed surface is 2×10^{-3} cm^2. So, x–y mechanical movements have been added in order to observe the wafers over a typical surface area of about 8 cm^2.

The image observed on the monitor screen is magnified 600 times so that the luminescence microstructure of layers is observed rather than accidental growth defects. Under these observation conditions, the image spatial resolution is limited by generation and recombination mechanisms and not by the detection method.

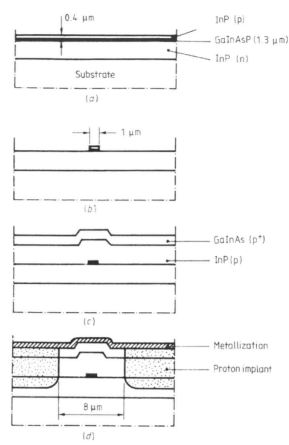

Fig. 13.62. Technology and image test steps of BRS laser fabrication. (a) First epitaxy; (b) photolithography; (c) second epitaxy; (d) completed structure.

Three validations of epitaxy and lithography steps are provided by PL imaging, two of which are in wafer processing. Fig. 13.62 represents the localization of the PL tests between the technology steps showing simultaneously, in a schematic manner, the observed structure. It is seen that the observed layers are double heterostructures consisting of an active quaternary layer (small gap) on InP buffer covered by a thin InP layer (large gap). In this structure a strong luminescence from the active layer can be observed because of the carrier confinement in the small-gap material and of the relatively smaller absorption of laser light in the surface layer. Moreover, the luminescence from different materials can be separated and identified using appropriate selective optical filters.

In addition to the tests in wafer processing, a standard electroluminescence imaging of the stripe active area in mounted lasers is provided using the same camera–monitor system.

Dark line defects (DLD) and dark spot defects (DSD) have been typically detected by PL imaging of the active layer. The DLD, attributed to deviations from lattice matching, can be eliminated by readjustments in gas composition during MOCVD growth. A correlation between DSD and chemically revealed defects has shown that some of the dark spots can correspond to dislocations, but the origin of the DSD is still not well understood.

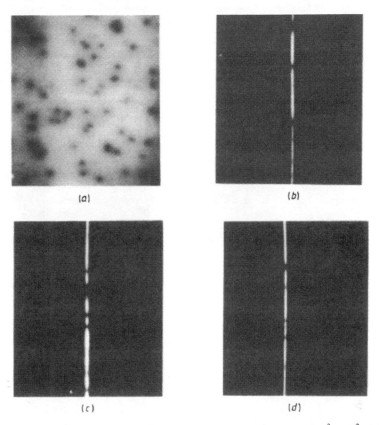

Fig. 13.63. Luminescence test images, wafer 1383/1393, DSD < 10^3 cm^{-2}, (a) After first epitaxy (wafer process); (b) after lithography (wafer process); (c) after second epitaxy (separated bar); (d) electroluminescence test (mounted laser).

Two examples of layers with DSD density below 10^3 cm^{-2} and above 3×10^4 cm^{-2} will now be considered in the light of luminescence image tests. Fig. 13.63 and Fig. 13.64 present the luminescence images of both the examples. It can be seen that a good PL image test after the first epitaxy, where the DSD were not detected ($< 10^3$ cm^{-2}), leads to good tests of stripe PL image after lithography and the second epitaxy. On the other hand, the DSD present in the first epitaxy layer image (3×10^4 cm^{-2}) could also be found in the buried stripe after lithography and the second epitaxy. Systematic tests on tenths of wafers have confirmed that the DSD are principally localized in the layers of the first epitaxy. Also, it was seen that the second epitaxy did not perturb the good PL quality of the buried stripe.

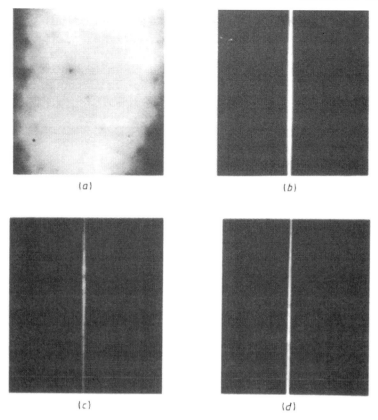

*Fig. 13.64. Luminescence test images, wafer 1262/1263, DSD $\simeq 3 \times 10^4$ cm^{-2}, (a)–
(d) as in Fig. 13.63.*

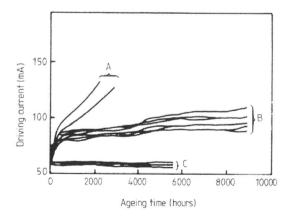

*Fig. 13.65. Aging characteristics of BRS 1.3 μm lasers with and without DLD and
DSD under accelerated test conditions (5 mW, 7 °C): A, with DSD and DLD; B,
with DSD only; C, without DSD and DLD.*

The PL and electroluminescence (EL) images have been systematically correlated with an accelerated aging test. The aging of lasers was carried out at 70 °C maintaining 5 mW of optical power. The characteristics of threshold current under aging are presented in Fig. 13.65. Good aging behavior was observed only in the absence of DSD in buried stripes. On the other hand, an important dispersion of aging characteristics has been noted if DSD over 10^4 cm^{-2} were present in the active layer.

Validations of the wafers are done after the first epitaxy, after strip lithography and after the second epitaxy by a photoluminescence imaging method. The electroluminescence of the active area of lasers mounted *p*-side down can be observed by a window in the n contact.

In order to localize the injection current in the active region, two processes have been used:

(1)   realization of a Schottky mesa of 6 μm width and 0.7 μm depth through the GaInAs cap layer (Fig. 13.66), and

(2)   realization of a deep-proton implantation of 6 μm width through the InP homojunction (Fig. 13.67).

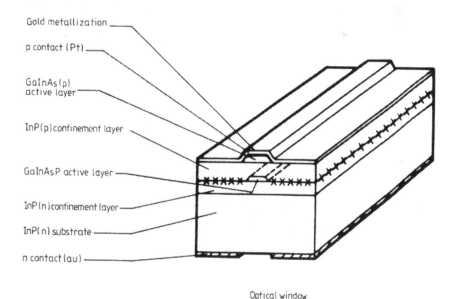

Gold metallization
p contact (Pt)
GaInAs(p) active layer
InP(p) confinement layer
GaInAsP active layer
InP(n) confinement layer
InP(n) substrate
n contact (au)

Optical window

Fig. 13.66. Diagram of a BRS laser with Schottky mesa.

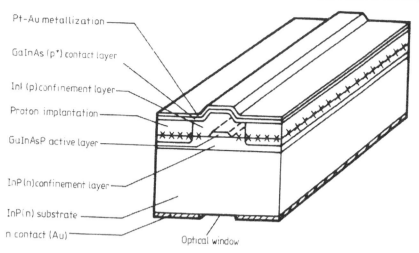

Pt-Au metallization

GaInAs (p⁺) contact layer

InP (p) confinement layer

Proton implantation

GaInAsP active layer

InP (n) confinement layer

InP (n) substrate

n contact (Au)

Optical window

Fig. 13.67. Diagram of a BRS laser with proton implantation.

Fig. 13.68 shows the histogram of threshold current under pulsed operation at room temperature for 864 chips, from the two processes, randomly selected from one wafer. The average current is 19.0 mA with a standard deviation of 2.2 mA.

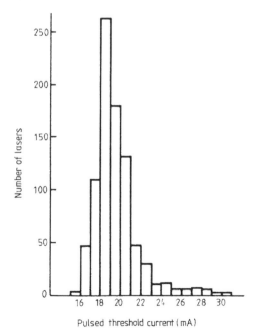

Fig. 13.68. Histogram of threshold current under pulsed operation at room temperature for 864 chips randomly selected from one wafer: $\bar{I}_{th} = 19.0 \pm 2.2$ *mA,* $\lambda = 1.3$ μm.

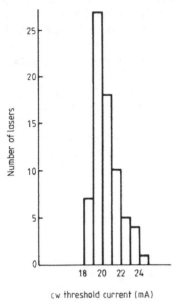

Fig. 13.69. Histogram of CW threshold current at 20 °C for 72 mounted lasers from one wafer: $\bar{I}_{th}$ = 19.9 ±1.4 mA.

Fig. 13.69 shows the histogram of CW threshold current at 20 °C for a total of 72 lasers mounted down or up, from the two processes. The average threshold current is 19.9 mA with a standard deviation of only 1.4 mA. (These parameters are respectively 45.8 and 3.5 mA at 70 °C for 58 lasers mounted down.) For the 72 lasers the average external differential efficiency is 0.22 W·A^{-1} per facet with a standard deviation of 0.01 W·A^{-1}.

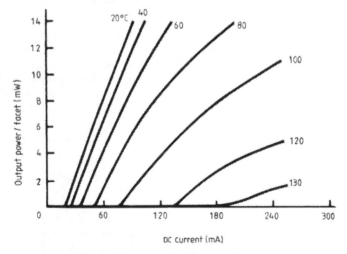

Fig. 13.70. Light output versus pulsed current at several heatsink temperatures of 1.3 μm GaInAsP–InP BRS laser grown by two-step LP-MOCVD.

Fig. 13.70 shows typical CW light–current characteristics at different temperatures. The maximum operating temperature is around 130 °C. A stable fundamental transverse mode was observed at least up to 7 mW. Typical far-field patterns parallel and perpendicular to the junction plane measured at 1.4, 4.2 and 7 mW are shown in Fig. 13.71. The far-field full widths at half-maximum parallel and perpendicular to the junction are, respectively, 27° and 35° with a good uniformity. The T_0, measured under pulsed operation between 20 and 70 °C, is 70 K. The low serial resistance of 2.5 Ω is favorable for CW operation at high temperature and also at high modulation frequency. For 50 lasers randomly selected on the wafer, the average wavelength is 1.3012 μm with a standard deviation of 68 Å.

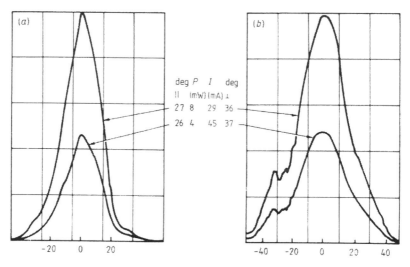

Fig. 13.71. Far-field patterns (a) parallel and (b) perpendicular to the junction of a BRS λ = 1.3 μm laser for different output powers under CW operation at 20 °C; FWHM are indicated, along with P and I, for each curve.

Table 13.10 presents statistical results on the threshold currents of 2117 chips taken randomly from the 20 wafers processed into lasers. The measurements are done at room temperature under pulsed operation. Bars are taken from different parts of the wafer (except the sides) in order to have a typical sample representative of the wafer.

The yield, which is defined as the number of good chips that work with a threshold current less than three times the standard deviation above the average divided by the total number of tested chips, gives an indication of the quality of the epitaxy, of the process and of the cleaving. The average yield obtained on these 20 wafers has the high value of 95% with a standard deviation of 4%.

The amplitude response of a laser diode is characterized with a sweep oscillator, which covers the frequency range 10 MHz–8.4 GHz, with a network analyzer, which provides magnitude and phase measurement capability from 100 MHz to 40 GHz, and with a photodiode having a bandwidth of 8 GHz.

Wafer number	Number of tested chips	Average threshold current (mA)	Standard deviation (mA)
1380	87	26.0	6.5
1383	864	19.0	2.2
1386	75	17.0	3.0
1424	62	18.4	2.3
1426	181	36.2	3.3
1427	136	45.1	8.2
1428	109	38.8	2.8
203	31	16.3	1.3
204	28	17.3	2.4
215	46	20.2	7.4
234	16	7.6	1.4
302	98	12.4	2.9
303	48	9.5	1.3
304	61	13.4	3.3
305	50	10.5	1.9
308	53	9.8	3.1
445	61	21.1	4.6
447	43	24.2	5.5
448	31	22.9	3.0
449	37	27.2	2.9
Average		20.6	3.5
Standard deviation		10.1	2.0

Table 13.10. Statistical results concerning the currents of 2117 chips.

Fig. 13.72 shows a typical modulation characteristic of a proton-implanted laser biased at 27 mA ($2I_{th}$) with a modulation current of 18 mA peak-to-peak at 20 °C.

Eleven proton lasers, 300 μm long, from five wafers have been characterized. The average bandwidths f at −3 dB and the standard deviations are the following:

at 1 mW/facet:	$f = 3.0$ GHz	$\sigma = 0.6$ GHz
at 2 mW/facet:	$f = 3.9$ GHz	$\sigma = 0.5$ GHz
at 4 mW/facet:	$f = 4.5$ GHz	$\sigma = 0.4$ GHz.

Modulation characteristics are nearly flat from 2000 MHz to 1 GHz (no roll-off limits the bandwidth). The overshoot is less than 10 dB and has a typical value of 5 dB. Results show a good reproducibility of modulation characteristics from wafer to wafer. These proton laser diodes can be used in optical transmission systems in the S band (2–4 GHz).

On the contrary, in the Schottky mesa laser diodes some roll-off limits the bandwidth at around 1 GHz. Since the two kinds of laser diodes come from the same wafers, have the same stripe structure and the same serial resistance, the difference in the modulation characteristics is attributed to a lower parasitic capacitance in the proton laser diodes (the deep-proton implantation through the InP *p–n* junction reduces the surface parasitic capacitance).

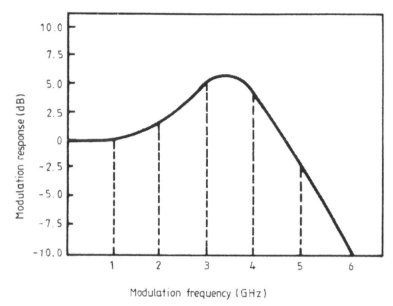

Fig. 13.72. Typical modulation characteristic of a 300 µm long BH laser with proton implantation at 20 °C (2.9 mW/facet).

Accelerated ageing tests are performed at 70 °C and at an output power of 5 mW/facet without any preliminary screening test. Twenty-two lasers, mounted down, from the two processes and from two wafers have been aging for a total of 209,630 h and have not yet shown any significant degradation. Fig. 13.73(*a*) and (*b*) show the driving current as a function of the aging time for, respectively, six and nine lasers with Schottky mesa from the two wafers. Fig. 13.74 shows aging characteristics of seven lasers with proton implantation from the two wafers. Degradation saturation appears between 500 and 1000 h. The time for stabilization is shorter for the proton lasers than for the mesa lasers. The long-term degradation rates are calculated using the values of the driving currents at 100 h and at the maximum aging time.

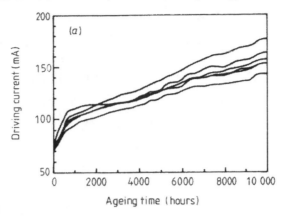

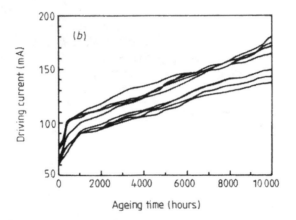

Fig. 13.73. Aging characteristics of (a) six BRS lasers with Schottky mesa from one wafer and (b) nine BRS lasers with Schottky mesa from another wafer (70 °C, 5 mW).

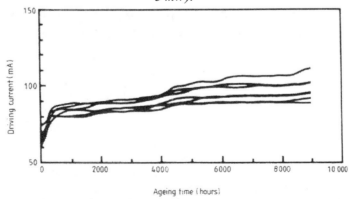

Fig. 13.74. Aging characteristics of seven BRS lasers with proton implantation from two wafers (70 °C, 5 mW).

The better reliability of the proton-bombarded lasers could be attributed to their smaller InP homojunction width (about 5 μm) compared to the 350 μm homojunction width of the Schottky lasers. But this hypothesis has not been confirmed since the different possible causes of degradation (homojunction, active layer, thermal impedance) have not yet been correlated with the observed aging.

The two wafers show identical and homogeneous degradation. With a 50% increase in driving current as a failure criterion, the above value of degradation rate for proton lasers corresponds to a lifetime of 23,800 h at 70 °C.

13.4.3. Distributed feedback lasers fabricated on material grown completely by MOCVD

Semiconductor lasers which oscillate in single longitudinal mode (SLM) under high-speed direct modulation are expected to improve the performance of optical transmission systems. In order to obtain the SLM oscillation, a semiconductor laser should include a wavelength-selective optical feedback mechanism in the device structure. In distributed feedback (DFB) lasers, a corrugation grating near the active layer is responsible for the wavelength-selective optical feedback mechanism. In the laser cavity, right-travelling waves and left-travelling waves couple through the grating. The strength of coupling, which governs the stability of SLM operation, is determined by the corrugation depth, the refractive indices and the thicknesses of the layer structure.

GaInAsP–InP DFB lasers have been studied using LPE [Suematsu et al. 1983], a hybrid of MBE and LPE [Itaya et al. 1982] and a hybrid of LPE and MOCVD [Westbrook et al. 1984]. Many important problems exist in DFB laser fabrication. Prominent among them is the phenomenon of surface deformation, which occurs during LPE regrowth. The MOCVD growth technique can be used to overcome this problem and has produced structures after overgrowth without any deformation in corrugation depth.

(a) Device fabrication

The DFB laser structure is manufactured in the following manner. First, the following layers were successively grown by MOCVD on an Sn-doped (1 0 0) 2° off or (1 0 0) exact InP substrate:

(1) 1 μm InP confinement layer, sulfur-doped, with N_D-$N_A = 10^{18}$ cm^{-3};
(2) 0.2 μm thick undoped GaInAsP (composition $\lambda = 1.55$ μm) active layer;
(3) 0.2 μm thick Zn-doped (N_A-$N_D = 2 \times 10^{17}$ cm^{-3}) GaInAsP (composition $\lambda = 1.3$ μm) guiding layer.

Next, second-order corrugation gratings with a period of 4700 Å ($\Lambda = \gamma_L/2n_{eq}$) and a depth of 1500 Å were formed in the top of the guiding layer by holographic photolithography followed by chemical etching, orienting the gratings along the (0 1 1) direction. The gratings were then covered with 1 μm of Zn-doped InP and 0.5 μm of Zn-doped GaInAsP cap layer, regrown by MOCVD.

To avoid surface deformation, great care must be exercised during the MOCVD regrowth. The corrugated InGaAsP ($\lambda = 1.3$ µm) surface is heated from room temperature to 650 °C for 3 min in an $N_2 + H_2$ carrier gas, and then the TEIn and PH$_3$ are switched into the reactor. Fig. 13.75 shows scanning electron microscope cross sections of the corrugated structure before and after MOCVD overgrowth. It appears from these photomicrographs that no significant surface deformation has occurred and the final depth of the corrugation is similar to the pregrowth height.

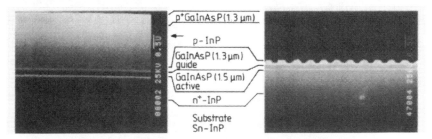

Fig. 13.75. Scanning electron microscope cross sections of the corrugated structure before and after LP-MOCVD overgrowth.

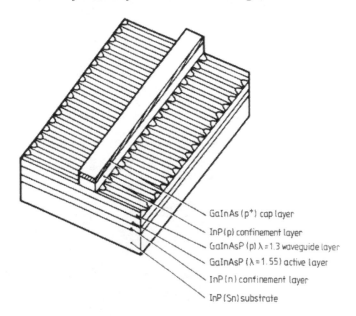

Fig. 13.76. Ridge-waveguide structures of GAInAsP–InP DFB laser.

The ridge-waveguide structures developed by Kaminow et al. [1983] have been fabricated (Fig. 13.76) from these wafers. The devices are separated by cleaving two facets and scribing the other two facets. The width of the ridge is 5 µm and the length of the cavity about 300 µm. Individual diodes are bonded with In to an Ni-plated copper heatsink.

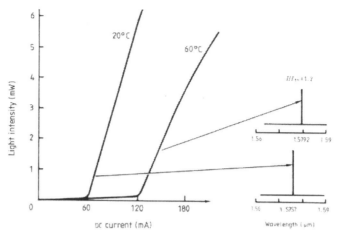

Fig. 13.77. *Light–current characteristics of* DFB *laser at 20 and 60 °C under CW operation.*

Fig. 13.77 shows typical light–current characteristics and the lasing spectrum of a DFB laser at 20 and 60 °C under CW operation. The typical threshold current I_{th} of CW operation was 60 mA. A CW output power of 6 mW/facet was measured. These results can be improved with antireflective (AR) coating on the front facet. Quantum efficiencies of 15% per facet have then been obtained. The temperature dependences of the threshold current I_{th} and the lasing wavelength under CW operation are shown in Fig. 13.78. Stable SLM operation is obtained in the temperature range from 9 to 90°C, with a temperature coefficient of the lasing wavelength $d\lambda L/dT$ of 0.9 Å·°C^{-1}.

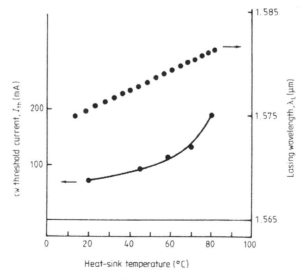

Fig. 13.78. *The variation of threshold current and lasing wavelength of a DFB laser as a function of heatsink temperature under CW operation.*

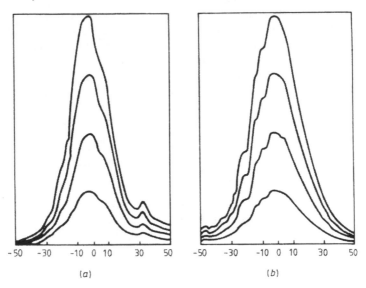

(a) (b)

Fig. 13.79. Far-field patterns (a) parallel and (b) perpendicular to the junction of a DFB 1.5 μm laser for different output powers under CW operation, at 20 °C.

Fig. 13.79 shows the far-field patterns parallel and perpendicular to the junction plane of these DFB lasers. These patterns were measured under CW operation for different optical output powers. A stable operation in the fundamental transverse mode without significant distortion was observed at least up to 8 mW. The far-field full widths at half-maximum power parallel and perpendicular to the junction planes were about 28° and 35°, respectively.

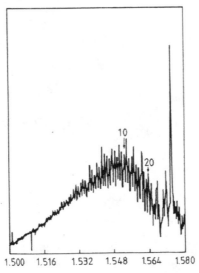

Fig. 13.80. Emission spectrum of a DFB laser just below the threshold; $\Delta\lambda = 12$ Å, $L = 300$ μm, $n_{eff} = \lambda/(2L,\Delta\lambda) = 3.4$.

The detailed spectrum of a DFB laser just below the threshold is shown in Fig. 13.80. The maximum of gain peak wavelength is 1.55 μm, which is the same as the photoluminescence wavelength of the active layer. The distance between two Fabry–Pérot modes is 12 Å. The DFB mode wavelength is 1.57 μm. The cavity length is 300 μm, so the value of the effective refractive index can be obtained by

Eq. (13.20) $n_{eff} \simeq \lambda^2 / 2L \simeq 3.4$

Kogelnik and Shank [1972] predicted that periodic structures consisting of corrugation gratings have a stopband of frequencies in which propagation is forbidden. In Fig. 13.81, the stopband of 25 Å, where no resonance-mode emission existed, is clearly observed between DFB modes. In the output spectrum of the DFB laser, most of the lasing power is concentrated in a mode next to the stopband. As current increases, the intensity of the lasing line abruptly increases and the highest satellite mode becomes more than 30 dB (see Fig. 13.82).

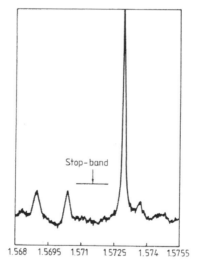

Fig. 13.81. The detailed emission spectrum of a DFB laser; $I_{th} = 78$ mA, $I/I_{th} = 0.89$.

The coupling coefficient K, per unit length, was experimentally determined by using the relation [Itaya et al. 1982]

Eq. (13.21) $K = n_{eff} \Delta \lambda_s / \lambda_B^2$

where $\Delta \lambda_s$ is the width of the "stopband" in the subthreshold laser emission spectrum, n_{eff} is the effective refractive index and λ_B is the stopband center wavelength. The coupling coefficient K of DFB lasers using this equation was 108 cm^{-1} (with two cleaved facets). Single-longitudinal-mode operation under high-speed direct modulation up to 2 ns has been obtained (Fig. 13.83). Stable SLM was also maintained under sinusoidal modulation at 1 GHz (see Fig. 13.83). This should be compared to the lasing spectrum of a conventional Fabry–Perot single-mode semiconductor laser, which becomes multimode or is broadened under rapid direct modulation; the lasing mode of these lasers sometimes hops from one to another, even when operating in single mode.

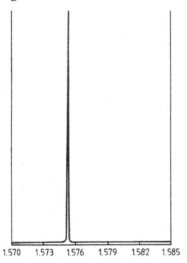

1.570 1.573 1.576 1.579 1.582 1.585

Fig. 13.82. Lasing spectrum of a DFB laser under CW operation; T = 20 °C,
dB = 10·log(P₀/P) = 30.

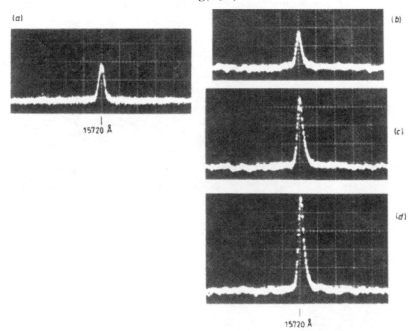

Fig. 13.83. Lasing spectra under rapid direct modulation at 1 GHz; 1.57 μm ridge MOCVD DFB laser, (a) Sinusoidal modulation at 1 GHz; pulsed modulation time-resolved spectra at (b) 0.5 ns; (c) 1 ns; (d) 2 ns.

The threshold current density in DFB lasers is found to depend on temperature, as in classical DH lasers (Eq. (13.19)): $J_{\text{th}}(T) = J_{\text{th}}(T_0) \exp(T/T_0)$ where T_0 describes the characteristic temperature. The average T_0 value is 61 °C without break point up to 75 °C (see Fig. 13.84).

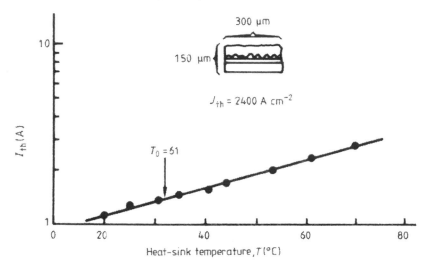

Fig. 13.84. Variation of threshold current of DFB laser as a function of temperature; $J_{th} = 2400\ A \cdot cm^{-2}$.

(b) Theoretical analysis of MOCVD DFB lasers by Correc et al. [1981]

The waveguide geometry and indices used in the calculations are shown in Fig. 13.85. The grating of period $\Lambda = 4650$ Å has a trapezoidal shape with $L_1 \simeq 0.27$ and $L_2 \simeq 0.23\Lambda$. Its depth is $g \simeq 1500$ Å. Following the usual method [Streifer et al. 1982], it is possible to compute the coupling coefficient for such gratings [Westbrook et al. 1984]. The thicknesses of the waveguide layer (T_{wg}), active layer (T_{a}) and buffer layer (T_{b}) are 0.18, 0.2 and 0.1 µm, respectively. T_{wg} has been slightly decreased from 0.2 µm (its value before the corrugation was made) to 0.18 µm to take the presence of the grating into account. Fig. 13.86 shows the dependence of the coupling coefficient K on the corrugation depth g. The Bragg order is denoted by m. Since the grating profile is rather smooth and almost sinusoidal, the higher-order coefficients are much smaller than that of the first order. This stresses the importance of abrupt tooth-shape gratings for large higher-order coupling coefficients and the necessity to avoid surface deformation during regrowth.

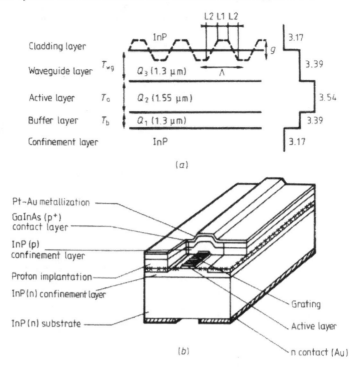

Fig. 13.85. Schematic waveguide structure of the GaInAsP/InP DFB laser showing: (a) the layers and the grating; and (b) the indices at 1.55 μm.

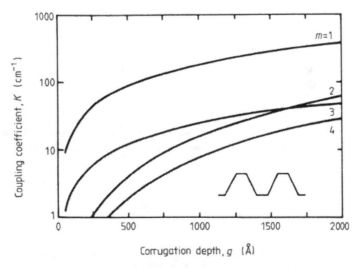

Fig. 13.86. Coupling coefficient as a function of corrugation depth for trapezoidal grating. The Bragg order is denoted by m.

Using K, the field threshold gain α_{th} is computed using the coupled-wave theory [Kogelnick and Shack 1972 and Streifer et al. 1980]. Moreover, since in actual devices there are facets where reflections occur, one has to take them into account. The position of the uncoated facet with respect to the grating will have an important impact on the properties of the laser [Matsuoka 1982, Matsuoka et al. 1984, and Itaya et al. 1982], especially on the discrimination between modes [Razeghi et al. 1985]. The field reflectivities are ρ_r and ρ_l for the right and left facets and the positions of the latter with respect to the grating are given by angles called "facet phases" θ_r, and θ_l, which are incorporated into the reflectivities in complex form. With $\rho_r = 0$ for the AR-coated facet, it is possible to compute the threshold gain as a function of the mirror phase θ_l for various KL, L being the laser length.

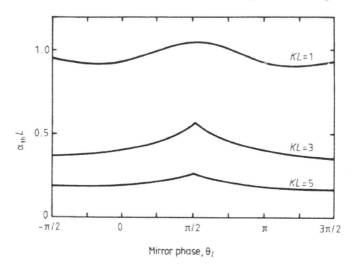

Fig. 13.87. Effect of the mirror phase θ_l on the threshold gain α_{th}; $\rho_r = 0$,
$\rho_l = 0.565 \cdot exp(i\theta_l)$.

Fig. 13.87 shows this dependence. From these curves, it appears that α_{th} is weakly dependent on θ_l and that a good estimation can be obtained with $\theta_l = 0$. Therefore, in the following calculations, $\theta_l = 0$ will be assumed. The uncertainty concerning θ_l will somewhat widen the curves obtained with $\theta_l = 0$. The threshold current I_{th} is calculated by

Eq. (13.22) $$I_{th} = e \frac{B_{eff}}{A_0^2} dWL \left| \frac{2\alpha_{th}}{\Gamma} + (\alpha_{in} + \alpha_{ac}) + \alpha_{ex} \frac{1-\Gamma}{\Gamma} \right|^2$$

The notations and numerical values are those of Utaka et al. [1982]; W is the active-region width.

Fig. 13.88 shows I_{th} as a function of the active-layer thickness T_a with the waveguide layer T_{wg} as a parameter. The broken curves illustrate the case where both facets are AR-coated and the full curves the case where only one facet is AR-coated. The threshold current corresponding to $T_{wg} = 0.3$ and $\rho_1 = 0$ is higher than 50 mA and the corresponding curve cannot be displayed. As expected, the additional feedback provided by the facet decreases I_{th} by roughly 30%. Optimum values of T_a range from 0.1 to 0.15 μm. If T_a is too small, the optical confinement factor Γ is too small and I_{th} increases; but if T_a is too large, then K is too small, hence α_{th}, is too large and I_{th} increases again. Also of interest is the fact that, for usual thicknesses ($T_a \geq 600$ Å), the thinner the waveguide layer the better.

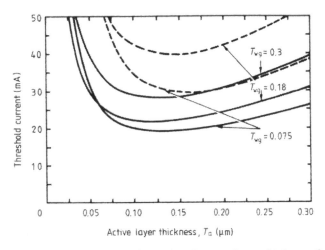

Fig. 13.88. Threshold current as a function of active-layer thickness for various waveguide-layer thicknesses; g = 1500 Å, L, = 240 μm, W = 2 μm; ρ_r = 0, ρ_l = 0 (— — — —) and ρ_r = 0, ρ_l = 0.565 (————).

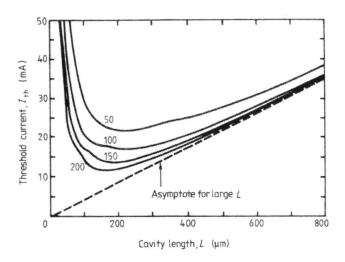

Fig. 13.89. Threshold current as a function of cavity length for various coupling coefficients; ρ_r = 0, ρ_l = 0.565.

Fig. 13.89 shows the threshold current as a function of the cavity length L for various coupling coefficients. It appears that the optimum laser length is around 200 μm for lasers where reflection has been suppressed only on one side ($\rho_r = 0$). For shorter lasers, the coupling coefficient is too small and the threshold gain α_{th} increases rapidly for decreasing L; and for longer lasers, α_{th} decreases, but then the cavity volume is too large and I_{th} increases again. The broken line indicates the asymptote for large L.

(c) Wavelength dispersion

Some particular applications, such as heterodyne detection, demand lasers with well controlled lasing wavelength λ_L, which requires excellent homogeneity over the entire wafer. Since lasing occurs in the vicinity of the Bragg wavelength λ_B, the first source of dispersion is due to λ_B. For second-order gratings, λ_B is simply the product of the effective refractive index of the guide n_e and the grating period Λ. If we consider that Λ is constant over the entire wafer, then λ_B variations $\Delta\lambda_B$ are solely due to n_e variations, which can arise from inhomogeneities in the layer thicknesses and/or stripe width. Here, we want to address the problem of the influence of waveguide parameter variations upon n_e and hence λ_B. For that, Correc et al. [1985] consider a rectangular waveguide composed of three layers ($\Omega_1, \Omega_2, \Omega_3$) embedded into InP. The unperturbed geometry corresponds to $T_{wg} = 0.18$ μm, $T_a = 0.2$ μm, $T_b = 0.1$ μm and $W = 2$ μm. Then, they vary one of the four parameters, keeping the other three constant, and evaluate the change in λ_B. The range of variation is ±5% around the nominal value. The results are displayed in Fig. 13.90. It appears that the most crucial parameter is the active-layer thickness, since a variation of only ± 5% shifts λ_B by more than 20 Å. The Bragg wavelength λ_B is less sensitive to the other parameters: ±5% variation results in a smaller shift of the order of 5 Å. Therefore, wavelength reproducibility demands excellent control of layer thickness, especially that of the active region.

The other source of dispersion is the facet phase θ_1 of the uncoated facet. So far, it has been impossible to control it by cleavage, and sophisticated methods are required to achieve phase control. It is known [Suematsu et al. 1983] that, for high enough coupling coefficient, lasing is prohibited at λ_B and that, depending on θ_1 the lowest-threshold DFB mode is located on one side or the other of λ_B. This means that the allowed λ_L will be distributed into two sets located on each side of λ_B and separated by the stopband. Within each set, λ_L variations induced by θ_1 variations are denoted by $\Delta\lambda_\theta$. Using the coupled-mode theory, Correc et al. [1985] have estimated $\Delta\lambda_\theta$ to be 2–3 Å.

In actual lasers, both sources of dispersion are present and one expects to see two sets of λ_L, separated by a stopband. Each set will have a width $\Delta\lambda_L$ that will be the sum of λ_B variation $\Delta\lambda_B$ and uncertainty due to θ_1, i.e.

Eq. (13.23) $\Delta\lambda_L = \Delta\lambda_B + \Delta\lambda_\theta$

Since we know $\Delta\lambda_\theta$, we can deduce $\Delta\lambda_B$ from $\Delta\lambda_L$ and consequently have some knowledge of the homogeneity of the wafer.

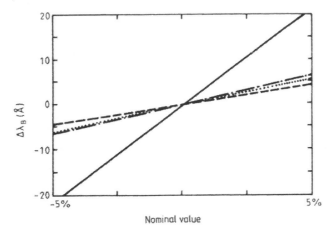

Fig. 13.90. Influence of waveguide parameters on λ_B. Nominal values are indicated in parentheses: (— · — · —) waveguide layer ($T_{wg} = 0.18 \ \mu m$); (————) active layer ($T_a = 0.2 \ \mu m$); (— — — —) buffer layer ($T_b = 0.1 \ \mu m$); (·········) waveguide width ($W = 2 \ \mu m$).

(d)   Experimental results

BRS-DFB lasers have been made by a two-step MOCVD growth technique. Typical threshold current was 30 mA for 240 μm long lasers with two cleaved facets and 40 mA for lasers with one AR-coated facet. Thresholds as low as 10 mA have been measured. Fig. 13.91 shows light–current characteristics under CW operation of a BRS-DFB laser. A single longitudinal mode is observed for temperatures up to 70 °C and output powers up to 6 mW. Beyond, power saturates. An 18–20 Å wide stopband is clearly visible in the near-threshold spectrum. From this, the coupling coefficient can be estimated at 40 cm^{-1} [Westbrook et al. 1984]. This value is in good agreement with that deduced from Fig. 13.88 for $g = 1500$ Å. The theoretical threshold current of this structure can be estimated as 25 mA from Fig. 13.88, whereas the experimental value of 40 mA is somewhat higher. This discrepancy can be explained by the fact that the DFB mode is not exactly located at the maximum of the gain curve and that it does not experience the maximum available gain.

Fig. 13.92 shows the wavelength of 10 lasers obtained from the same wafer. Obviously, there are two sets of lasing wavelengths, separated by a region where lasing is apparently forbidden. In each set, the wavelength dispersion is of the order of 8 Å. This wavelength distribution can be explained by the theoretical analysis given above. In this case, $\Delta\lambda_L = 8$ Å and $\Delta\lambda_\theta = 3$ Å. Using these, we infer that $\Delta\lambda_B \simeq 5$ Å, and from Fig. 13.90, we deduce that the control over the waveguide parameters is very good. It is better than 5% over T_{wg}, T_b and W, and better than 2% over T_a. This demonstrates the quality and homogeneity of material grown by MOCVD.

Some of these lasers have been successfully tested on an installed optical link [Le Boutet et al. 1985]. One of them has been used in a 40 km, 560 Mbit·s^{-1} link with a bit error rate of the order of 10^{-11} and another one on a 1.7 Gbit·s^{-1} link with a similar error rate [Le Boutet and Sorel 1985].

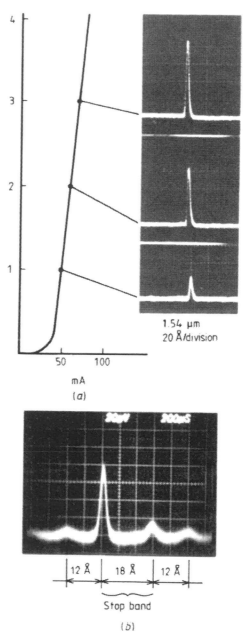

Fig. 13.91. DFB laser with SiO$_x$ AR coating: (a) light-current characteristics; (b) stopband. Coupling coefficient is 40–50 cm^{-1}, side-mode suppression is 10^3, and operating temperature is 20 °C.

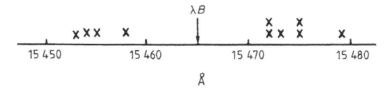

Fig. 13.92. Lasing wavelength of 10 lasers from the same wafer. Wavelength dispersion is about 3 Å due to facet phase and about 5 Å due to Bragg dispersion.

13.4.4. CW phase-locked array $Ga_{0.25}In_{0.75}As_{0.5}P_{0.5}$–InP high-power semiconductor laser grown by MOCVD

Optically coupled semiconductor laser arrays are increasingly gaining importance for applications where high CW or pulsed power in a coherent radiation pattern is needed. It has been suggested [Butler et al. 1984] that individual emitters in a phase-locked diode laser array can be expected to oscillate collectively in modes characteristic of a composite waveguide formed as a result of the optical coupling between individual waveguides of the array. These collective oscillations, called "supermodes," are determined primarily by the strength and uniformity of the optical coupling between modes of individual emitters in the array. The model characteristics of phase-locked arrays have been studied by several authors. The first direct experimental evidence confirming the existence of supermodes in an array of coupled diode lasers has been reported by Paoli et al. [1984]. An array of N lasers can be expected to possess N supermodes, each of which has an optical field whose amplitude and phase show characteristic profiles from emitter to emitter. As with any collection of identical interacting oscillators, coupling shifts the oscillating frequency of the emitters to N different values, one for each supermode.

To date, all phase-locked arrays reported in the literature have been fabricated with GaAs–GaAlAs semiconductor material. In general, most of the gain-guided or index-guided laser arrays have operated with adjacent emitters 180° out of phase, resulting in a far field with two lobes symmetrically located about the facet normal [Ackley 1983].

A fundamental problem in array design is the development of a structure that will operate with a single diffraction-limited beam.

Razeghi et al. [1985] have described the operation of a phase-locked GaInAsP–InP high-power semiconductor laser array emitting at 1.3 μm fabricated on materials grown completely by a two-step LP-MOCVD growth technique. This phase-locked array laser was manufactured by two-step MOCVD as follows. First, the following layers were grown successively on an InP(Sn) substrate with orientation (1 0 0) exact or (1 0 0) 2° off:

(1)   1 μm InP sulfur-doped ($n = 1 \times 10^{18}$ cm^{-3}) for confinement layer;
(2)   0.2 μm undoped GaInAsP lattice-matched to InP (1.3 μm wavelength composition) for active layer;
(3)   100 Å undoped InP as blocking layer;
(4)   100 Å $Ga_{0.25}In_{0.75}As_{0.5}P_{0.5}$ undoped waveguide layer.

Next, an array consisting of seven stripes of 2 μm width and stripe–stripe spacing of 1 μm was etched in the 100 Å GaInAsP layer (100 Å InP layer is used as a blocking layer) through a photolithographic resist mask, using a selective etch composed of H_2SO_4:H_2O_2:H_2O (1:8:200). After removing the resist mask, the array was then covered by a 1 μm thick Zn-doped InP confinement layer ($N_A - N_D \simeq 5 \times 10^{17}$ cm^{-3}) and a 0.5 μm Zn-doped $Ga_{0.47}In_{0.53}As$ contact layer with $N_A - N_D \simeq 10^{19}$ cm^{-3} (again using MOCVD).

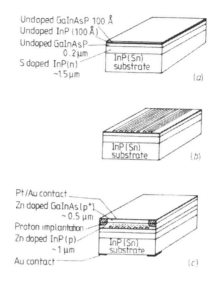

Fig. 13.93. Schematic diagram of ridge-island laser array (RILA) structure: (a) first epitaxy; (b) selective chemical etching of seven stripes; (c) after regrowth, metallization and proton implantation.

Fig. 13.93 shows a schematic diagram of the ridge-island laser array (RILA) structure before and after regrowth, with details of technology stages. In order to localize the injection only in the RILA active region, a shallow layer of protons (0.5 μm from active region) was implanted through a 20 μm wide photoresist mask after metallization of the contacts. The n contact was realized by the deposition of Au. The p contact was made by the deposition of Pt–Au and annealing at 450 °C for 30 s. After cleaving and scribing into chips 300 μm long and 350 μm wide, the lasers were mounted, epilayer-side down, onto nickel-plated copper heatsinks using indium.

Fig. 13.94 shows a Scanning electron microscope cross section of the RILA before and after regrowth. Typical pulsed and CW light–current characteristics of this RILA are displayed in Fig. 13.95. The pulsed and CW threshold was 200 mA; external quantum efficiencies were 0.19 W·A^{-1} (pulsed operation) and 0.13 W·A^{-1} (CW operation) at 20 °C. Output powers of more than 300 mW at a pulse-modulated current of 2 A (limited by the pulse generator) and 120 mW at 1.3 A (CW) were measured at 20 ° C. These results were obtained without optical surface treatment, which indicates that the radiated power can be increased by a factor of 1.5–2.

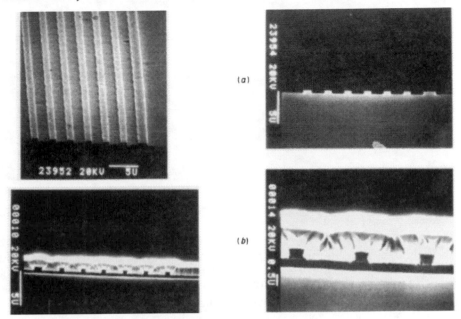

Fig. 13.94. Scanning electron micrographs of cross sections of a phase-locked high-power GaInAsP–InP laser array with seven stripes of 2 μm width (a) before and (b) after LP-MOCVD overgrowth.

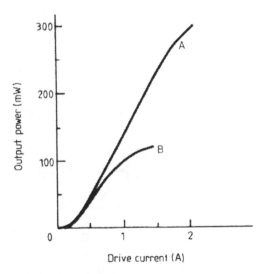

Fig. 13.95. Typical light-current characteristics of phase-locked high-power GaInAsP–InP 1.3 μm laser arrays at 20°C, under (a) pulsed and (b) CW operations. Under pulsed operation, I_{th} = 220 mA and η = 0.19 W·A^{-1}; under CW operation, I_{th} = 225 mA and η = 0.13 W·A^{-1}.

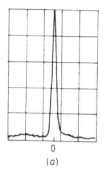
(a)

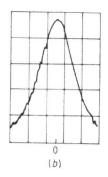
(b)

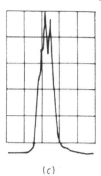
(c)

Fig. 13.96. Far-field patterns of a CW phase-locked GaInAsP–InP high-power laser array (a), (c) parallel and (b) perpendicular to the junction plane, at 20 °C.

Fig. 13.96 shows the far-field patterns parallel and perpendicular to the junction plane of these RILA. These patterns were measured under CW operation for 10 and 90 mW output powers. The far-field full widths at half-maximum power parallel and perpendicular to the junction plane were 3° and 45°, respectively, at 10 mW, which shows the phase-locked behavior of these RILA. At higher output power the far-field lobe broadens and at 90 mW it becomes about 16° parallel to the junction plane (see Fig. 13.96(c)).

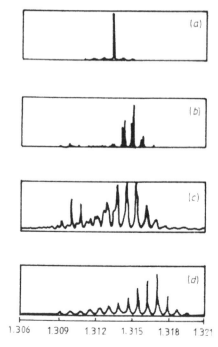

Fig. 13.97. CW longitudinal-mode spectra of a phase-locked GaInAs-P-InP laser array, at T = 20 °C. (a) P = 2 mW; (b) 10 mW; (c) 20 mW; (d) 30 mW.

Fig. 13.97 shows the appearance of two sets of longitudinal modes at 10 mW, indicating the onset of lasing on higher-order supermodes, which is consistent with the broadening of the far-field pattern.

The devices should be useful in applications requiring high CW output power in single-longitudinal-mode communications systems. Because these devices are phase-locked, it is possible to achieve even higher output power levels by adding more diodes to the array and using facet coatings.

13.4.5. GaInAsP–InP quantum-well lasers

The superiority of the quantum-well laser (QWL) over the double-heterostructure (DH) laser as regards threshold current properties has been established for the GaAs–GaAlAs system [Tsang 1982 and Hersee et al. 1983]. Comparing the DH and QWL structures it is necessary to consider three key parameters: the optical confinement factor, the cavity absorption losses and the gain properties.

The confinement factor Γ of the optical wave is the part of the light intensity that is in the active layer. It is poorer for the quantum well due to the reduced thickness of the active layer. The situation can be improved if a separate optical cavity is placed around the active layer (separate-confinement heterostructure (SCH) or graded-index separate-confinement heterostructure (GRINSCH) lasers). Typical values of Γ calculated using a five-layer model [Casey et al. 1984] are given in Table 13.11. It is interesting to note that although Γ is lower for GaInAsP/InP lasers than for GaAs/GaAlAs lasers, the ratio of the Γ values for the DH and QWL structures is similar for each material system. Values of Γ for the SCH and GRINSCH QWL structures are similar when the width of the optical confinement cavity is carefully chosen [Streifer et al. 1983 and Chinn 1984].

	$L(\text{Å})$				
Structure	50	100	200	1000	1500
GaAs/GaAlAs DH, $x_C = 40\%$	–	0.004	0.018	0.312	0.50
GaAs/GaAlAs SCH, $x_B = 20\%$, $x_C = 40\%$	0.013	0.027	0.057	0.347	0.517
GaAs/GaAlAs SCH, $x_B = 20\%$, $x_C = 60\%$	0.019	0.038	0.078	0.420	0.600
GaInAs/InP DH	–	0.002	0.008	0.163	0.305
GaAs/GaAlAsP(1.3 µm)/ InP = SCH	0.009	0.019	0.039	0.2222	0.343

Table 13.11. Values of the optical confinement factor.

Absorption losses α (free-carrier losses, scattering at interface defects, etc) affect the propagation of the optical wave in the optical cavity. Nagle et al. [1985] have measured the quantity α by extrapolating the curve of threshold current density as a function of the inverse of chip length, as described by Hersee et al. [1985]. Many DH and QWL structures were measured and it was found that the value of α was typically below 20 cm^{-1} and independent of the type of structure. L represents the width of the active layer. For GaAs/GaAlAs lasers, x_B and x_C indicate respectively the aluminum fraction in the optical cavity and in the confinement layers. The width of the optical cavity for SCH lasers is chosen for optimum confinement (2000 Å for GaAs and 3500 Å for GaInAs).

The value of the material gain at a given injection is increased for a narrow active layer (two-dimensional case) and can compensate for the deleterious effects of a small Γ in QWL structures. This can be understood if we consider the number of states that must be filled in order to achieve the population inversion that is required for lasing action. This corresponds roughly to the number of states between the band edge and kT above the edge, and at room temperature is 4.5 times higher in a DH laser (active layer = 1000 Å) than in a QWL (with GaAs effective mass $m^* = 0.067$). This ratio r is given by

Eq. (13.24) $$r = \frac{2\sqrt{2}\left(m^*\right)^{1/2}\left(kT\right)^{1/2} L}{3\pi h}$$

where L is the active-layer width of the DH laser. The above arguments should also apply to GaInAsP/InP quantum-well lasers; experimentally, however, these devices exhibit threshold currents similar to DH lasers [Dutta 1985]. The origin of this different behavior is discussed below.

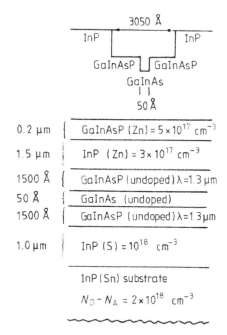

Fig. 13.98. Schematic diagram of GaInAs–GaInAsP–InP SCH QW laser

Razeghi et al. [1985] and Nagle et al. [1986] extensively studied a $Ga_{0.47}In_{0.53}As/Ga_{0.27}In_{0.73}As_{0.59}P_{0.41}/InP$ SCH laser with a 50 Å active layer and a 3000 Å optical cavity. Fig. 13.98 shows the schematic diagram of this SCH QW structure. At ambient temperature, the diodes lased at 1.30 μm for an average current of 450 mA under pulsed operation. The spontaneous emission (shown in Fig. 13.99) exhibits three main characteristics as the injection current is increased:

(1)   The maximum of the spontaneous emission shifts towards higher energy.

(2)   The integrated spontaneous emission intensity tends to saturate.

(3)   The spectral shape broadens and the slope of the high-energy tail decreases until carriers escape from the well and recombine in the optical cavity. Eventually, at high injection, population inversion and lasing take place in the cavity and emission occurs at 1.30 μm.

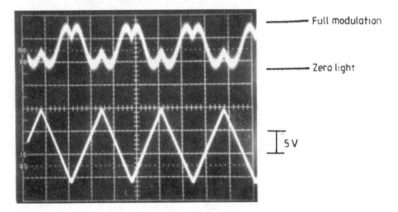

Fig. 13.99. The spontaneous emission of a GaInAs SCH QW laser. Relative intensities are significant. The spikes on the curves around 0.9 eV are due to atmospheric water absorption. $T = 295$ K, $I_{th} = 450$ mA.

It was shown by Nagle et al. [1986] that this behavior is most probably caused by carrier heating in the quantum well. The increase of the quasi-Fermi levels with injection cannot explain the observed behavior, since it is found that the maximum value of gain decreases with increasing injection current (Fig. 13.100). Equally this cannot be due to an inefficient capture of carriers by the quantum well since there was no luminescence from the cavity at low injection levels. Also studies of the photoluminescence emitted by GaInAs/InP undoped quantum wells have shown that the capture is efficient even for narrow wells [Razeghi et al. 1985].

As the temperature is lowered the maximum gain of the quantum well increases and approaches that required for lasing. At a temperature between 100 and 150 K (chip-dependent), lasing occurs in the quantum well. At 5 K, the threshold current was only 0.9 mA. Fig. 13.101 shows the lasing spectrum and light-current characteristics of this SCH QWL (lasing wavelength 1.40 μm), showing that it is not the intrinsic properties of the GaInAs quantum well or the interfaces that prevent lasing at room temperature. For the temperature range 5–120 K the T_0 value of these lasers was approximately 100 K.

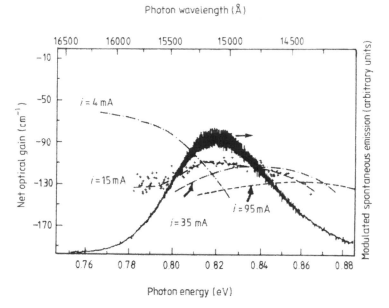

Fig. 13.100. Gain spectra of a GaInAs SCH QW laser at room temperature (TE polarization). For clarity we trace the corresponding modulated spontaneous emission and calculated values (crosses) only for I = 15 mA.

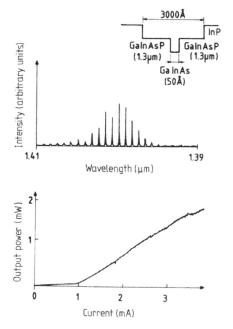

Fig. 13.101. Lasing spectrum and light–current characteristics of this SCH QWL.
$T = 5$ K, $I_{th} = 0.8$ mA.

This extremely strong temperature-dependent behavior indicates that non-radiative Auger recombination is probably responsible for carrier heating. This conclusion is further supported by the observation that, as the temperature is increased, the photoluminescence efficiency of GaInAs quantum wells decreases more rapidly than the T^{-2} variation corresponding to a band-to-band recombination with a 2D density of states [Razeghi et al. 1985].

Whereas the superiority of SQW and MQW is not well established in the GaAs–GaAlAs system, the opposite is true in the GaInAsP–InP system. A major issue in QWL is therefore to understand this outstanding difference: up to now, all the reported MQW lasers in GaInAsP–InP systems have features that are not better than those of DH lasers [Dutta et al. 1985 and Razeghi 1984].

The case of the GaInAsP/InP standard DH lasers has been a source of numerous studies [Henry et al. 1983]. Although state-of-the-art GaInAsP DH lasers have threshold current densities quite similar to those of GaAs DH lasers, their T_0 is much worse. The T_0 question in GaInAsP/InP lasers has long been a matter of controversy, but there seems now to be a consensus about the following points [Henry et al. 1983]:

(1) For optimum lasers ($d \simeq 0.15$ μm), the threshold current can be as low as 500 A·cm^{-2} [Razeghi et al. 1987]. Typical values of T_0 are 60–80 K. The density of carriers at threshold is about 2×10^{18} cm^{-3}.

(2) The threshold current has a radiative component and several non-radiative components. The radiative component is roughly 60% at 300 K [Henry et al. 1983]. The main non-radiative component is due to Auger recombination, with smaller contributions due to inter-valence-band light absorption and thermal-carrier leakage from the active region. The Auger recombination process is of the CHSH type, i.e. an electron and a heavy hole recombine by exciting a heavy hole to the spin-split-off valence band.

(3) The Auger coefficient value is $C_A = 4 \times 10^{-29}$ cm^6·s^{-1} and 2×10^{-29} cm^6·s^{-1} for GaInAs and 1.3 μm GaInAsP material, respectively.

A central issue is the value of the Auger coefficient in quantum wells. There is an abundant literature on the theoretical evaluation of the various possible processes in 3D [Etienne et al. 1982], but there seems to be a consensus now about the dominance of the CHSH process. At some point, the Auger effect in QW has been predicted to be much weaker than in 3D, but consensus is now that it should be comparable, as is also the experimental result [Sermage 1986].

Sugimura [1983] and Asada et al. [1984] have produced detailed calculations where they can optimize I_{th} and T_0 separately for the various types of QW lasers in the GaInAsP–InP system. However, the values found for the threshold current are significantly lower than those obtained up to now in MQW InGaAsP/InP lasers.

13.4.6. Buried waveguides in InGaAsP–InP material grown by MOCVD

In the case of 2D guiding in III–V materials using a classical ridge structure, the main problem to overcome is in general the necessary lateral confinement to obtain a single-mode regime and the corresponding difficulty of controlling very precisely the different etching processes. As a matter of fact, widths of the waveguides have currently to be decreased below 3 µm at $\lambda = 1.55$ µm, which makes the structure sensitive to imperfections in its optogeometrical parameters. As a result, high scattering losses may occur despite the good quality of the material. In the vertical direction, light confinement can be tailored by stacking layers of suitable composition, i.e. refractive index. Anyway, in any waveguide structure the only non-compressible cause for losses is due to interface quality.

A natural way to overcome that problem is to buy the guiding material in a cladding material whose refractive index can be adjusted to provide the desired lateral confinement. It then becomes possible to design a pair of waveguides conveniently in order to make a directional coupler switch with specified coupling strength or coupling length. A new technique relaxes the difficult control of the etching processes and permits layers whose thicknesses are well defined by the epitaxial growth. A schematic drawing of a cross section of the structure is given in Fig. 13.102. The steps for its fabrication are explained below. On a heavily doped $(2 \times 10^{18}$ cm$^{-3})$ n-type InP substrate are grown a low-doped x µm thick cladding InP layer ($<10^{15}$ cm^{-3}) and a low-doped ($<10^{15}$ cm^{-3}; $\lambda_g = 1.3$ µm) y µm thick guiding InGaAsP layer. They bear an e µm thick low-doped ($<10^{15}$ cm^{-3}) InP layer and a τ µm thick low-doped quaternary material. Then photolithographic and selective etching techniques can be used to reveal W µm wide strips in the upper quaternary material, the e µm thick InP upper layer acting as a barrier to the etching process. Then an epitaxial regrowth occurs during which an InP layer is deposited. The waveguide structure is more flexible to design than more conventional ones because of the extra parameters e and τ. Its obvious disadvantage is the need for an epitaxial regrowth.

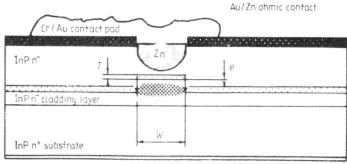

Fig. 13.102. Schematic drawing of a cross section of the modulator.

Among the many techniques available for waveguide modeling, the effective-index method (EIM) can be used in order to adjust the guiding characteristics of the structure. The goal is mainly to achieve low lateral guiding, allowing a reasonable range of guide widths W for single-mode operation. In our case, InGaAsP ($\lambda_g = 1.3$ μm) and InP have been chosen respectively for the guiding and cladding layers and a 100 Å thick etching barrier layer was assumed. For $x = 0.4$ μm, $y = 0.55$ μm and $e = \tau = 100$ Å, Table 13.12 shows the calculated results for the propagation constant β in channel waveguides. It is clear that there exists a large range of W values ($0 \le W \le 8$ μm) ensuring a single 2D mode regime with low $\partial\beta/\partial W$ values [Albrecht et al. 1983].

			W (μm)		
	3	5	7	8	10
TE	3.29183	3.29216	3.29240	3.29948	3.29261
				3.29142	3.29162
TM	3.28516	3.28552	3.28577	3.28586	3.28599
				3.28471	3.28496

Table 13.12. Variation of propagation constant β with respect to strip width.

A directional coupler switch can also be designed using the effective-index method. Once W has been chosen, the only free parameter to adjust is the inter-guide spacing d in order to reach a given value of the coupling length. With $W = 7$ μm, $e = \tau = 100$ Å and $d = 5$ μm the difference between the propagation constants of the eigenmodes of the structure is about 14×10^{-5} for TE waves. With $d = 10$ μm, it becomes 3×10^{-5}. This gives coupling lengths of 5.3 and 25.8 mm, respectively, constituting a realistic range of interaction lengths suitable for low drive voltages if the coupler is operated through the electro-optic effect [Fujiwara et al. 1984].

Two samples have been fabricated from the same substrate with the structure as described above ($e = \tau = 100$ Å; $x = 0.4$ μm; $y = 0.55$ μm). The epitaxial regrowth that occurred just after selective chemical etching consisted of 1.8 μm of n^--InP (less than 10^{15} cm^{-3}). The same mask as that used to reveal the guiding regions was used for opening windows in a top SiO_2 layer in order to diffuse Zn ions in InP for making the p–n junction. Two masks have been used during the photolithographic process: one composed of straight waveguides and one composed of directional couplers.

The straight-waveguide sample consisted of a series of waveguides with different W values ranging from 3 to 11 μm. Sample length was 22 mm and light guiding was first observed at $\lambda = 1.55$ μm through it by using an IR camera. The modal regime was studied by looking at the output light pattern when coupling was laterally displaced. A second-order mode was shown to exist for widths greater than 8 μm as predicted by theory. For the monomode waveguides, loss measurements were carried out by the cut-back method. Loss values were found to be independent of W and were all around 4 dB·cm^{-1} (best 2 dB·cm^{-1}).

Phase modulation was then observed on 7 µm wide waveguides by applying a voltage on the contacts (Fig. 13.103). By polarization interferometry, one finds a modulation efficiency of about $5° \cdot V^{-1} \cdot mm^{-1}$.

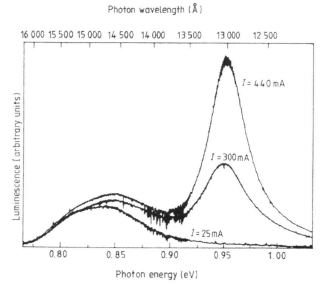

Fig. 13.103. Observation of phase modulation. Interaction length is 10 mm; phase modulation at λ = 1.5 µm.

A capacitance measurement with respect to the applied voltage V shows that, for V ranging from 0 to -10 V, the capacitance was less than 8 pF for the 4.5 mm long electrode. It then makes it possible to have $1/\tau (\tau = RC)$ of the order of 2 GHz if the load resistor is chosen equal to the conventional value of 50 Ω.

The directional coupler sample consisted of a series of couplers (using 7 µm wide guides with inter-waveguide spacing ranging from 5 to 15 µm in 0.5 µm steps). Its length was 20 mm between the two cleaved end faces and all the guides were shown monomode. Light was launched in one guide and the intensities emerging from the two guides made it possible to look at the light exchange between them. For that 20 mm long interaction, all the couplers were found to be in the parallel state for $d > 8$ µm, which is consistent with the predicted value (26 mm with $d = 10$ µm) [Mikami and Nakagome 1985].

References

Ackley, D.E. *Appl. Phys. Lett.* **42** 152, 1983.
Alnot, P., Huber, A., and Olivier, J. *Surf. Interface Anal.* **9** 283, 1985.
Albrecht, P., Bach, H.G., Bornholdt, C., Doldissen, W., Franke, D., Grote, N., Krauser, J., Niggebrugge, U., Nolting, H.P., Schalk, M., and Tiedke, I. *2nd ECIO* Firenze, p 72, 1983.

Asada, M., Kaweyama, A., and Suematsu, Y. *IEEE J. Quantum Electron.* **QE-20** 745, 1984.

Aspnes, D.E. *J. Vac. Sci. Technol.* **17** 1057, 1980.

Aspnes, D.E. and Studna, A.A. *Appl. Phys. Lett.* **46** 1071, 1985.

Bastard, G., Delalande, C., Meynadier, M.H., Frijlink, P.M., and Voos, M. *Phys. Rev.* **B 29**, 1984.

Bimberg, D., Christen, J., Steckenborn, A., Weimann, G., and Schapp, W. *J. Lum.* **30** 562, 1985.

Bimberg, D. and Bubenzer, A. *Appl. Phys. Lett.* **38** 803, 1981.

Brummell, M.A. *DPhil Thesis* University of Oxford, 1985.

Brummell, M.A., Nicholas, R.J., Portal, J.C., Cheng, K.Y., and Cho, A.Y. *J. Phys. C: Solid State Phys.* **16** L579, 1983.

Burnham, R.D., Holonyak, N., Hirch, K.C., Kalishi, R.W., Nan, D.W., Thornton, R.L., and Paoli, T.L. *Appl. Phys. Lett.* **48** 800, 1986.

Butler, J.K., Ackley, D.E., and Botez, D. *Appl. Phys. Lett.* **44** 293, 1984.

Casey, H.C., Panish, M.B., Schlosser, W.O., and Paoli, T.L. *J. Appl. Phys.* **45** 1, 1984.

Chinn, S.R. *Appl. Opt.* **23** 3508, 1984.

Christen, J., Bimberg, D., Steckenborn, A., and Weimann, G. *Appl. Phys. Lett.* **44** 84, 1984.

Correc, P., Landreau, J., Bouley, J.C., Razeghi, M., Blondeau, R., Kazmierski, K., Krakowski, M., de Cremoux, B. and Duchemin, J.P. *Optical Fiber Sources and Detectors* SPIE, p 2, 1985.

Drevillon, B., Perrin, J., Marbot, R., Violet, A., and Dalby, J.L. *Rev. Sci. Instrum.* **53** 969, 1982.

Duchemin, J.P., Bonnet, M., Beuchet, G., and Koelsch, F. *Gallium Arsenide and Related Compounds, 1978* (Inst. Phys. Conf. Ser. 45) p 10, 1979.

Dutta, N.K., Napholtz, S.G., Yen, R., Brown, R.L., Shen, T.M., Olsson, N.A. and Craft, D.C. *Appl. Phys. Lett.* **46** 19, 1985.

Erman, M., Theeten, J.B., Vodjdani, N., and Demay, Y. *J. Vac. Sci. Technol.* B **1** 328, 1983.

Etienne, B., Shah, J., Leheny, R.F., and Nahory, R.E. *Appl. Phys. Lett.* **41** 1018, 1982.

Fang, F.F. and Howard, W.E. *Phys. Rev. Lett.* **16** 797, 1966.

Forrest, S.R., Schmidt, P.H., Wilson, R.B., and Kaplan, M.L. *Appl. Phys. Lett.* **45** 1199, 1984.

Fujiwara, M., Ajisawa, A., Sugimoto, Y., and Ohta, Y. *Electron. Lett.* **20** 790, 1984.

Henry, C.H., Logan, R.A., Temkin, H., Nevit, F.R., and Luongo, J.P. *IEEE J. Quantum Electron.* **QE-19** 941, 1983.

Hersee, S.D., Baldy, M., Cremoux, B., and Duchemin, J.P. *Gallium Arsenide and Related Compounds, 1982* (Inst. Phys. Conf. Ser. 65) p 281, 1983.

Hersee, S.D., Baldy, M., Assenat, P., Cremoux, B., and Duchemin, J.P. *Electron. Lett.* **18** 870, 1982.

Hollinger, G., Gergignal, E., Joseph, J., and Robach, Y. *J. Vac. Sci. Technol.* A **3** 2082, 1985.

Hopkins, M.A., Nicholas, R.J., Brummell, M.A., Harris, J.J., and Foxon, C.T. *Superlattices Microstruct.* **2** 319, 1986.

Itaya, Y., Matsuoka, J., Nakano, Y., Suzuki, Y., Kuroiwa, K., and Ikegami, T. *Electron. Lett.* **18** 1006–8, 1982.

Itaya, Y., Wakita, K., Motosugi, G., and Ikegami, T. *IEEE J. Quantum Electron.* **QE-21** 527–33, 1982.

Kaminow, I.P., Stulz, L.W., Ko, J.S., Milles, B., Feldman, R.D., and Pollack, M.A. *Electron. Lett.* **19** 877–9, 1983.

Kazmierski, K., Huber, A.M., Morillot, G., and Cremoux, B. *Japan. J. Appl. Phys.* **23** 628, 1984.

Kelso, S.M., Aspnes, D.E., Pollack, M.A., and Nahory, R.E. *Phys. Rev.* B **26** 6669, 1982.

Kogelnik, H. and Shank, C.V. *J. Appl. Phys.* **43** 2327–35, 1972.

Krawczik, S.K. and Hollinger, G. *Appl. Phys. Lett.* **45** 871, 1984.

Laidig, W.D., Lee, J.W., Chiang, P.K., Simpson, L.W., and Bedair, S.M. *J. Appl. Phys.* **54** 6382, 1983.

Laidig, W.D., Holonyak, N., Camras, M.D., Hess, K., Coleman, J.J., Dapkus, P.D., and Barden, J. *Appl. Phys. Lett.* **38** 776, 1981.

Le-Boutet, A., Auffret, R., Claveau, G., Guibert, M., Moalic, J., Pophillat, L., Sorel, Y., and Tromeur, A. *Electron. Lett.* **20** 834–6, 1985.

Le-Boutet, A. and Sorel, Y. *IOOC-ECOC'85* Venezia, Italy, 1985.

Matsuoka, T., Nagai, H., Noguchi, Y., Suzuki, Y., and Kawaguchi, Y. *Japan. J. Appl. Phys.* **23** L138–40, 1982.

Matsuoka, T., Nagai, H., Suzuki, Y., Noguchi, Y., and Wakita, K. *Japan. J. Appl. Phys.* **23** L782–4, 1984.

Maurel, P., Razeghi, M., Guldner, Y., and Vieren, J.P. *Semicond. Sci. Technol.* **2** 695, 1987.

Mikami, O. and Nakagome, H. *Opt. Quantum Electron.* **17** 449–55, 1985.

Nagle, J., Hersee, S., Krakowski, M., Weil, T., and Weisbuch, C. *Appl. Phys. Lett.* **46** 1325, 1985.

Nagle, J., Hersee, S., Razeghi, M., Krakowski, M., Cremoux, B., and Weisbuch, C. *Surf. Sci.* **174** 148, 1986.

Nahory, R.E. and Pollack, M.A. *Electron. Lett.* **14** 727, 1978.

Nicholas, R.J., Sessions, S.J., and Portal, J.C. *Appl. Phys. Lett.* **37** 178, 1980.

Nicholas, R.J., Stradling, R.A., Portal, J.C., and Askenazy, S. *J. Phys. C: Solid State Phys.* **12** 1653, 1979.

Nicholas, R.J. *Prog. Quantum Electron.* **10** 1, 1985.

Nicholas, R.J., Sessions, S.J., and Portal, J.C. *Appl. Phys. Lett.* **37** 178, 1980.

Olsen, G.H. *IEEE Trans. Electron. Devices* **ED-2** 217, 1978.

Palik, E.D., Picus, G.S. Teitler, S., and Wallis, R.F, *Phys. Rev.* **122** 475, 1961.

Paoli, T.L., Streifer, W., and Burnham, R.D. *9th Int. Semiconductor Laser Conf.* 7–10 August, Rio de Janeiro, Brazil, 1984.

Portal, J.C., Cisowski, J., Nicholas, R.J., Brummell, M.A., Razeghi. M., and Poisson, M.A. *J. Phys. C: Solid State Phys.* **16** L573, 1983.

Portal, J.C., Nicholas, R.J., Brummell, M.A., Cho, A.Y., Cheng, K.Y., and Pearsall, T.P. *Solid State Commun.* **43** 907, 1982.

Portal, J.C., Nicholas, R.J., Brummell, M.A., Cho, A.Y., and Sivco, D.L. unpublished, 1985.

Razeghi, M., Blondeau, R., Cremoux, B., and Duchemin, J.P. *Appl. Phys. Lett.* **45** 784, 1984.

Razeghi, M., Poisson, M.A., Larivain, J.P., and Duchemin, J.P. *J. Electron. Mater.* **12** 371, 1983.

Razeghi, M., Poisson, M.A., Guldner, Y., Vieren, J.P., Voisin, M., and Voos, M. *Appl. Phys. Lett.* **40** 877, 1982.

Razeghi, M., Maurel, P., Omnes, F., Defour, M., Acher, O., Tsui, D., Wei, H.P., Vieren, J.P., and Guldner, Y. *Appl. Phys. Lett.* **51** 1821, 1987.

Razeghi, M., Duchemin, J.P., and Portal, J.C. *Appl. Phys. Lett.* **46** 46, 1985.

Razeghi, M., Duchemin, J.P., Portal, J.P., Dmowski, L., Remenyi, G., Nicholas, R.J., and Briggs, A. *Appl. Phys. Lett.* **48** 712, 1986.

Razeghi, M., Nicholas, R.J., Brunei, L.C., Huant, S., Karrai, K., Portal, J.C., Brummell, M.A., Cheng, K.Y., and Cho, A.Y. *Phys. Rev. Lett.* **55** 883, 1985.

Razeghi, M., Maurel, P., Omnes, F., Amor, B.S., Dmowski, L., and Portal, J.C. *Appl. Phys. Lett.* **48** 1267, 1986.

Razeghi, M., Blondeau, R., Krakowski, M., Bouley, J.C., Papuchon, M., Cremoux, B., and Duchemin, J.P. *IEEE J. Quantum Electron.* **QE-21** 507–11, 1985.

Razeghi, M., Nagle, J., and Weisbuch, C. *Gallium Arsenide and Related Compounds, 1984* (Inst. Phys. Conf. Ser. 74) p 379, 1985.

Razeghi, M., Blondeau, R., and Duchemin, J.P. *Gallium Arsenide and Related Compounds, 1984* (Inst. Phys. Conf. Ser. **74**) p 679, 1985.

Razeghi, M., Blondeau, R., Krakowski, M., Cremoux, B., Duchemin, J.P., Lozes, F., Martinot, M., and Bensonsson, M.A. *Appl. Phys. Lett.* **50** 230, 1987.

Razeghi, M., Acher, O., and Launay, F. *Semicond. Sci. Technol.* **2** 793, 1987.

Razeghi, M., Blondeau, R., Kazmierski, K., Krakowski, M., and Duchemin, J.P. *Appl. Phys. Lett.* **46** 131, 1985.

Sermage, B., Chemla, D.S., Sivco, D., and Cho, A.Y. *IEEE J. Quantum Electron.* **QE-22** 774, 1986.

Störmer, H.L., Gossard, A.C., and Wiegmann, W. *Solid State Commun.* **41** 707, 1982.

Streifer, W., Scifres, D.R., and Burnham, R.D. *IEEE J. Quantum Electron.* **QE-11** 867–73, 1982.

Streifer, W., Burnham, R.D., and Scifres, D.R. *IEEE J. Quantum Electron.* **QE-11** 154–61, 1980.

Streifer, W., Burnham, R.D., and Scifres, R.D. *Opt. Lett.* **8** 283, 1983.

Suematsu, Y., Arai, S., and Kishino, K. *J. Lightwave Technol.* **LT-1** 162–4, 1983.

Sugimura, A. *IEEE J. Quantum Electron.* **QE-19** 932, 1983.

Tsang, W.T. *Appl. Phys. Lett.* **40** 217, 1982.

Utaka, K., Akiba, S., Sakai, K., and Matsushima, Y. *IEEE J. Quantum Electron.* **QE-20** 236–45, 1982.

van Vechten, J.A. *J. Appl. Phys.* **53** 7081, 1982.

Wei, H.P., Chang, A.M., Tsui, D.C., and Razeghi, M. *Phys. Rev.* **B 37** 32, 1985.

Westbrook, L.D., Nelson, A.W., Fiddyment, P.J., and Evans, J.S. *Electron. Lett.* **20** 225–6, 1984.

Westbrook, L., Henning, I.D., Nelson, A., and Fiddyment, P. *IEEE J. Quantum Electron.* **QE-21** 512–18, 1984.

14.   Strained Heterostructures: MOCVD Growth, Characterization, and Applications

14.1.   Introduction
14.2.   Growth procedure and characterization
    14.2.1.   Growth procedure
    14.2.2.   Secondary-ion mass spectrometry
    14.2.3.   Auger analysis
    14.2.4.   Etch-pit density
14.3.   Growth of GaInAs–InP multiquantum wells on $Gd_3Ga_5O_{12}$ garnet (GGG) substrates
14.4.   Applications
    14.4.1.   Photonic devices based on strained layers
    14.4.2.   Electronic devices—GaInP/GaInAs/InP MESFET
    14.4.3.   Optoelectronic integrated circuits (OEIC)
14.5.   Monolayer epitaxy of $(GaAs)_n(InAs)_n$–InP by MOCVD
    14.5.1.   Multiquantum wells of $(GaAs)_2(InAs)_2$/InP

14.1. Introduction

We have seen that MOCVD is well adapted for the growth of a variety of III–V semiconductor binary, ternary and quaternary heterojunctions, multiquantum wells and superlattices on lattice-matched substrates for optoelectronic or microwave device applications.

In heterostructures, it is certainly desirable to select a pair of materials closely lattice-matched in order to minimize defect formation or stress. However, heterostructures lattice-mismatched to a limited extent can be grown with essentially no misfit dislocations, if the layers are sufficiently thin, because the mismatch is accommodated by a uniform lattice strain. Without the requirement of lattice matching, the number of available pairs for device applications and integrated circuits is greatly increased.

Strained heterostructures of high-quality InP and related compounds have been grown on alternative substrates by the low-pressure metalorganic chemical vapor deposition growth technique. Photoluminescence, SIMS and Auger measurements have shown the high-quality optical and electrical properties of these layers.

14.2. Growth procedure and characterization

14.2.1. Growth procedure

The growth apparatus has been described in Chapter 11. The optimum conditions for the low-pressure growth of these layers are presented in Table 14.1. Growth was carried out at 76 Torr. Pretreatment of the substrates was found to be crucial. The pretreatment procedure for InP and InAs was given in Chapter 11.

	InP	GaAs	InAs	GaP	GaInAs	GaInP	GaInAsP
Growth temperature (°C)	550	550	550	550	550	550	630
Total flow-rate (N$_2$ + H$_2$) (l·min^{-1})	6	6	6	6	6	6	6
N$_2$-TEIn bubbler flow (cm^3·min^{-1})	200	–	200	–	200	200	350
H$_2$-TEGa bubbler flow (cm^3·min^{-1})	–	120	–	120	120	120	120
PH$_3$ flow (cm^3·min^{-1})	300	–	–	300	–	300	530
AsH$_3$ flow (cm^3·min^{-1})	–	90	90	–	90	–	21
Growth rate (Å·min^{-1})	100	100	100	100	200	200	150

Table 14.1 Optimized growth parameters.

Smooth single-crystal films exhibiting mirror-like surfaces have been obtained, even in the presence of a large layer-substrate lattice-parameter mismatch. These layers tend, however, to be heavily dislocated, and their electrical and optical characteristics, especially those related to minority-carrier properties such as diffusion length and lifetime, are generally inferior to those of the typical lattice-matched system. These effects tend to be especially severe for thin layers, but it has been found that they can be partly eliminated by the use of thick buffer layers or special grading or superlattice techniques. Table 14.2 shows the range of heterostructures grown so far.

Featureless mirror-like surfaces have been grown over a wide temperature range of 500 to 650 °C. The x-ray diffraction rocking curve about the (4 0 0) Kα reflection from an InP epilayer on GaAs and GaAs on InP substrate is shown in Fig. 14.1. The x-ray diffraction rocking curve of a heterostructure of InAs–GaAs on InP substrate is shown in Fig. 14.2. Razeghi et al. [1986] have performed a study of LP-MOCVD growth of InP simultaneously on InP, GaAs and InAs substrates with orientations of (1 0 0) placed adjacent to one another within the reactor for a growth temperature of 550 °C.

Substrate	First epilayer	Second epilayer
InP	GaAs	
InP	InAs	
GaAs	InP	
GaAs	InAs	
InAs	InP	
InAs	GaAs	
InP	GaAs	InAs
InP	$Ga_{0.47}In_{0.53}As$	$Ga_{0.49}In_{0.51}P$
GaAs	$Ga_xIn_{1-x}As$	
GaAs	InP	$GaIn_{1-x}AS_yP_{1-y}$

Table 14.2. Strained heterostructures grown by LP-MOCVD.

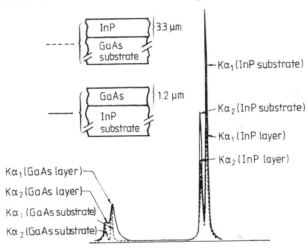

Fig. 14.1. X-ray diffraction rocking curve of (4 0 0) Cu Kα reflection from an InP epilayer on a GaAs substrate and a GaAs epilayer on an InP substrate.

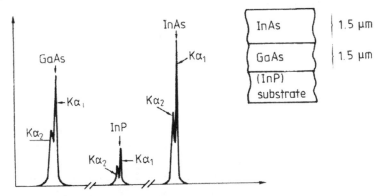

Fig. 14.2. X-ray diffraction rocking curve of (4 0 0) Cu Kα reflection from an InAs–GaAs–InP heterostructure grown by LP-MOCVD.

Fig. 14.3 shows the photoluminescence spectra of these layers at 5 K. A series of luminescence transitions, which can be attributed to recombination mechanisms such as free excitons, excitons bound to shallow impurities (such as Zn) and donor–acceptor recombination, were observed on these spectra. The exciton recombination energy of InP on InP, InP on GaAs and InP on InAs substrates are 1.419, 1.423 and 1.421 eV respectively. Considering that the lattice parameters of InP, GaAs and InAs are 5.869, 5.633 and 6.057 Å respectively, one expects the InP layer on GaAs substrate to be compressed, and the InP layer on InAs substrate to be expanded. Usually these layers are pseudomorphic (i.e. the elastic straining of the deposited lattice produces a zero misfit with the substrate). The accommodation of the mismatch by elastic strains induces changes in the magnitude of the gap. Thus in the case of an InP epilayer on an InAs substrate, one expects a lower energy for exciton recombination than for an InP epilayer on an InP substrate. This is not observed and there is no interpretation for these results yet.

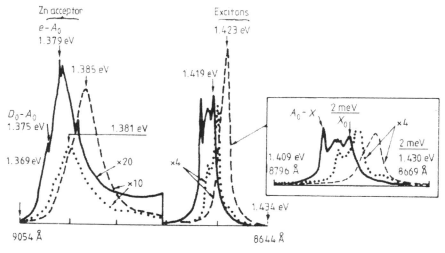

Fig. 14.3. Photoluminescence spectrum of an InP epilayer on InP, GaAs and InAs substrates at 5 K using a He–Ne laser: InP on GaAs (– – –), InP on InAs (...), InP on InP (———).

14.2.2. Secondary-ion mass spectrometry

SIMS analyses have been performed for quantitative determination of impurities accumulated at the substrate–epilayer interfaces.

Fig. 14.4 shows depth profiles of Mg, Si, Cr, Fe, Mn, As and P in InP layers grown on InP, InAs and GaAs substrates by LP-MOCVD. Each sample was analyzed in two different areas about 10 mm apart. Generally, analysis of two clean areas gives reproducible and representative results for the material. We have already shown (see Chapter 11) that the major source of impurities at the interfaces is the adsorption of atoms on the substrate surface during chemical etching prior to epitaxy. The pretreatments of the InP and InAs substrates are similar (described in Chapter 11), so their SIMS profiles are identical. But the chemical etch of the GaAs substrate prior to epitaxy is $H_2SO_4 + H_2O + H_2O_2$, and the concentration of impurities at the interface of InP–GaAs is lower than InP–InP or InP–InAs.

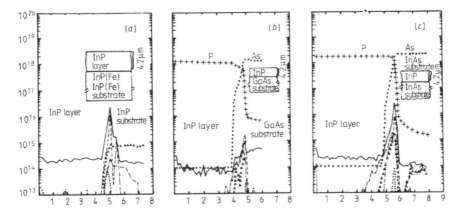

Fig. 14.4. Depth profiles of Mg, Si, Cr, Fe, Mn, P and As in the MOCVD growth of: (a) InP on InP substrate; (b) InP on GaAs substrate; and (c) InP on InAs substrate. P(+), As(...), Mg(– – –), Si(———), Cr(–·–), Fe(▲).

These results show that the quality of InP layers far from the interfaces is independent of substrate origin. Fig. 14.5 shows an automatic electrochemical profile through an InP layer grown on a GaAs substrate, with and without light (under illumination, for *n*-type semiconductors, to generate holes required for the reaction). The result shows that near the interface there are some perturbations, but far from the epilayer–substrate interfaces the quality of the epilayer and the carrier concentration become similar to those for an InP epilayer grown on InP substrate under the same conditions.

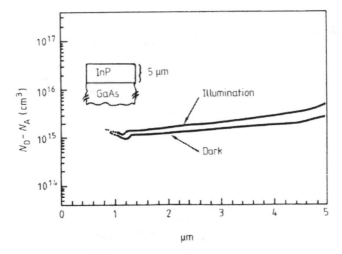

Fig. 14.5. Electrochemical polaron profile of an InP epilayer on GaAs substrate.

14.2.3. *Auger analysis*

The constituent concentration gradients at an InP/GaAs interface have been
determined by Auger analysis on a chemical bevel. The sample was chemically
etched by using a methanol–bromine solution (15% Br), in order to obtain a bevel
having a mean amplification coefficient (M) of 2100 (measured with a Talysurf).
This means that a change of 1 μm along the surface corresponds to a change of
4.5 Å in depth (z direction). Fig. 14.6 shows a schematic representation of the bevel.
By scanning the incident electron beam four times along the bevel, the successive
Auger profiles of the four elements P, As, In and Ga have been obtained. The four
profiles are shown in Fig. 14.7.

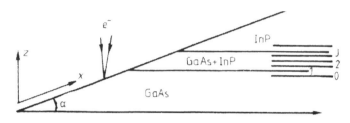

Fig. 14.6. *Schematic representation of the bevel of the InP/GaAs structure showing
how a surface analysis along the bevel is converted into depth analysis.*

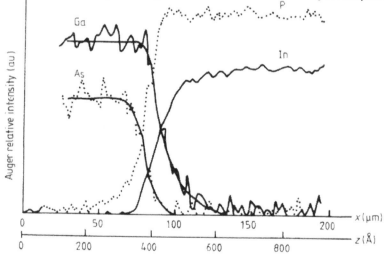

Fig. 14.7. *Corresponding Auger profiles relative to the four components: P, In, As
and Ga.*

Under certain simplifications, the relation between Auger intensity I^A and
concentration C^A of an element A can be easily obtained by assuming that the
interface region consists of n slices each of thickness a, where the slices are
numbered starting from the deepest one (at the end of the homogeneous GaAs
concentration, here) [Gazaux et al. 1986]—see Fig. 14.6.

The intensity due to the element A in the Jth slice corresponds to the intensity flowing from outside the slice, I_J^A, minus the intensity coming from all the other deeper slices, I_{J-1}^A, which is attenuated by the factor K by travelling through the Jth slice of thickness a. The interface composition (transition region) of InP–GaAs is $Ga_xIn_{1-x}As_yP_{1-y}$ where $0 \leq x, y \leq 1$.

After Auger analysis, the transition region constituting the interface can be subdivided into three parts (starting from the GaAs substrate):

(1) A region (thickness 150 Å) where In is absent. Its chemical composition is $GaAs_yP_{1-y}$, with $0.87 \leq y \leq 1$, which corresponds to the heating of the GaAs substrate under PH3 before growth. So it is possible to have the adsorption of As and adsorption of P. This can be remedied by heating the GaAs substrate under AsH3 pressure, before introducing PH3 into the reactor.
(2) The mid-region (thickness 120 Å) where all four components are present. Its chemical composition is $Ga_xIn_{1-x}As_yP_{1-y}$ with $0 \leq y \leq 0.87$ and $0.24 \leq x \leq 1$.
(3) The region (thickness 115–130 Å) where As is absent. Its chemical composition is $Ga_xIn_{1-x}P$ with $x \leq 0.24$.

Such analyses can be developed on any hetero-epitaxial structure if a good composition of the chemical etching is found [Gazaux et al. 1986].

14.2.4. Etch-pit density

Fig. 14.8 shows photomicrographs of the layers after forming a chemical bevel with very low angle and selective etching. The EPD of epitaxial layers and substrates are indicated in Table 14.3.

The results show that the EPD in InP/GaAs and InP/InAs interfaces is independent of the EPD of the InAs or GaAs substrates [Huber et al. 1984].

Epilayer/substrate	EPD(epilayer) (cm^{-2})	EPD(substrate) (cm^{-2})
InP/InP	8×10^3	8×10^3
InP/GaAs	6×10^5	2×10^3
InP/InAs	5×10^5	1×10^2

Table 14.3. Etch pit density in epitaxial layers of InP on InP, InP on GaAs, and InP on InAs substrate.

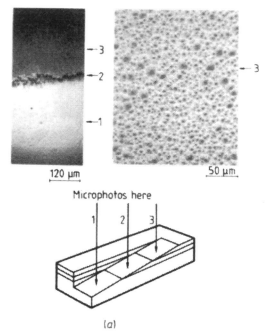

120 µm

50 µm

Microphotos here

(a)

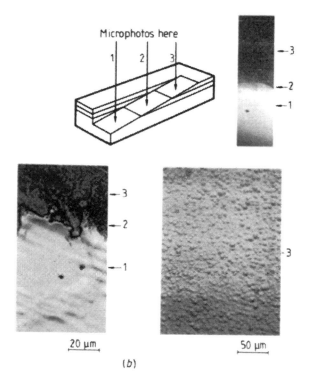

Microphotos here

20 µm

50 µm

(b)

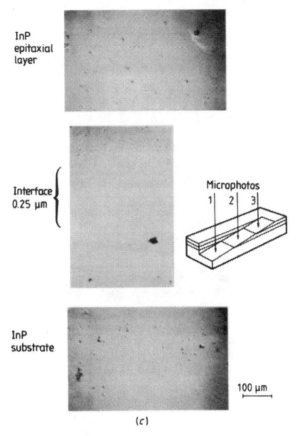

InP
epitaxial
layer

Interface
0.25 μm

Microphotos
1 2 3

InP
substrate

100 μm

(c)

*Fig. 14.8. Etch-pit density photographs of (a) InP/InAs, (b) InP/GaAs and (c)
InP/InP.*

14.3. Growth of GaInAs–InP multiquantum wells on Gd$_3$Ga$_5$O$_{12}$ garnet (GGG) substrates

The composition of the garnets is given by the formula R$_3$A$_2$B$_3$O$_{12}$, in which the trivalent R ions are rare earths such as lutetium, yttrium, gadolinium, or other large ions and the A and B ions are smaller ions, such as Ga^{3+} or Al^{3+}. The R ions are surrounded by a decahedron (not the regular one, but a distorted cube of eight oxygen ions). The B ions are surrounded by a tetrahedron of four oxygen ions and the A ions are surrounded by by an octahedron of six oxygen ions. The basic structure of GGG is cubic. The lattice parameter is 12.383 Å. The GGG is an insulator and can be used as a substrate.

Y$_3$Fe$_5$O$_{12}$ (YIG) is a magnetic garnet material having the same lattice parameter as GGG, and it can be epitaxially grown on GGG. Fig. 14.9 shows the elementary lattice cell of GGG. YIG is a potentially useful material for application to integrated optical devices, non-reciprocal elements such as isolators and circulators in planar integrated optics.

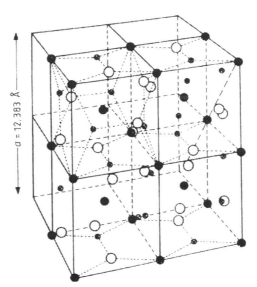

Fig. 14.9. Crystal structure of GGG.

The growth of these III–V semiconductor compounds on GGG substrate can be used for:

(1)   Optical telecommunications, such as integrated magneto-optical isolators and the integration of a laser diode with an isolator (YIG).
(2)   Integrated magneto-optical circulators with possibilities for integrated emitters, detectors, isolators, circulators, non-reciprocal phase-shifters, modulators, etc.
(3)   Magnetic recording with integrated magnetic read heads.
(4)   Magnetometry.

The growth processes have been presented in Table 14.1.

In a typical experiment to demonstrate the application of garnet substrate a GGG of (1 0 0) orientation and an InP substrate were placed on the susceptor, which was then placed inside the reactor [Razeghi et al. 1986]. The substrates were heated to 550 °C. The growth pressure was 76 Torr. The growth conditions for InP and GaInAs are indicated in Table 14.1.

The epitaxial structure was as follows: one 1.5 μm InP buffer layer, five quantum wells of GaInAs (200 Å thick) with InP barriers (200 Å thick), and one 1.5 μm InP top layer. The growth process for the GaInAs–InP multiquantum wells was the same as indicated in Chapter 12.

X-ray diffraction was used to examine the quality of the deposited layers. Fig. 14.10 presents the x-ray diffraction pattern of GaInAs–InP multilayers grown on GGG substrate indicated in Table 14.4. These data show that the orientation of the epitaxial layer is (1 1 1) while the orientation of the substrate is (1 0 0). The lattice mismatch between the substrate and epilayer is $\Delta a/a = 0.526$. However, there are many important cases where the epitaxial layer has either a different orientation from the substrate, for example CdTe(1 1 1) on GaAs(1 0 0) ($a_{GaAs} = 5.653$ Å and $a_{CdTe} = 6.481$ Å), or a totally different crystal structure (for example, silicon on sapphire) [Cullen and Wang 1978].

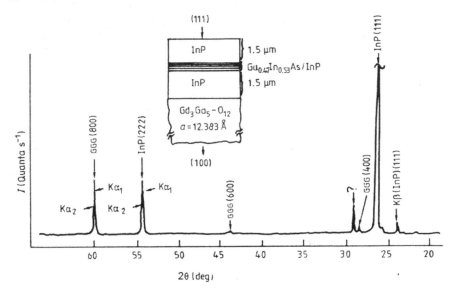

Fig. 14.10. X-ray diffraction pattern of the 3 μm thick InP/GaInAs/InP epitaxial layer grown by LP-MOCVD on a GGG(1 0 0) substrate. The inset shows a schematic view of the sample configuration.

Layers	hkl	$2\theta_{exp}$(deg)	$2\theta_{calc}$(deg)
InP	111	26.60	26.30
InP	222	54.30	54.10
InP	311	51.85	51.63
GGG	800	60.08	59.70
GGG	600	43.85	43.83
GGG	400	28.60	28.82

Table 14.4. Comparison between the experimental and reference value of 2θ (x-ray diffraction data) due to GaInAs–InP multilayers grown on GGG substrate.

In such cases the criterion of lattice match, namely, comparing the lattice parameters, is no longer applicable, and a new criterion should be used. Zur and McGill [1984] defined the concept of lattice match for any pair of crystal lattices in any given crystal direction, allowing for a periodic reconstruction of the interface. Instead of comparing the bulk lattice parameters, they compared the interface translational symmetry with that of the bulk materials on both sides of the interface.

A cut in any crystal direction through a three-dimensional lattice results in a surface which has two-dimensional (2D) translational symmetry. Any additional reconstruction of the atoms near the interface during the interface formation may, in general, reduce the symmetry of the reconstructed surface; thus the symmetry group of the reconstructed surface will be a subgroup of the symmetry group of the unreconstructed surface. Therefore, the problem of lattice match is reduced to scanning the 2D cuts in a given pair of lattices, and then comparing the two resulting 2D lattices, looking for a common superlattice.

Fig. 14.11 shows the photoluminescence (PL) spectrum of this sample at 77 K. The difference of the PL wavelengths may be due to misfit dislocations at the interface between the epilayer and the GGG substrate.

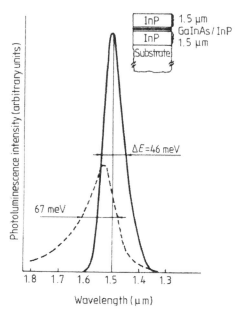

Fig. 14.11. *Photoluminescence spectra at 77 K of InP/GaInAs/InP epitaxial layer on a GGG (1 0 0) substrate (broken curve) and InP(1 0 0) substrate (full curve). The inset shows a schematic view of the sample configuration.*

Razeghi et al. [1986] have grown thick layers (7 μm) of InAs, InP, GaAs and GaInAs on GGG substrate. In these cases, all of the epilayers were polycrystalline. After growing the MQW structure they succeeded in growing single crystals of III–V compounds on top of the MQW on the GGG substrates. These results indicate that a wide range of potentially useful new phenomena based on multilayer epitaxy on alternative substrates is becoming available and with imaginative implementation may be expected to lead to further novel device structures.

14.4. Applications

14.4.1. Photonic devices based on strained layers

Long-distance optical links use lasers emitting at 1.3 μm fabricated by material grown on InP substrates. Unfortunately the technology of integrated circuits (IC) for signal treatment is a lot more difficult in this material than on GaAs substrates. A solution would be to combine the advantages of these two materials, which is against nature, owing to the large lattice mismatch.

Nevertheless, it is possible to do it with the MOCVD growth technique. $Ga_{0.25}In_{0.75}As_{0.5}P_{0.5}$–InP buried-ridge-structure (BRS) lasers emitting at 1.3 μm have been fabricated on GaAs substrates using the LP-MOCVD growth technique [Razeghi et al. 1985]. The BRS laser structure was manufactured as follows.

First the following layers were successively grown by LP-MOCVD on a Si-doped (1 0 0) 2° off GaAs substrate:

(1)   2 μm InP confinement layer, sulfur-doped with N_D–$N_A \approx 10^{18}$ cm^{-3};

(2)   0.2 μm thick undoped GaInAsP (composition 1.3 μm) active layer;

(3)   0.2 μm thick Zn-doped (N_A-$N_D \approx 2 \times 10^{17}$ cm^{-3}) InP layer in order to avoid the formation of defects near the active layer during etching.

The morphology of these layers was excellent, and the photoluminescence intensity and PL halfwidth were *the same* as for material grown on an InP substrate. The details of growth conditions were those given in Table 14.1. Next, a ridge of about 2 μm width was etched in the InP(p) and GaInAsP active layers through a photolithographic resist mask.

After removing the resist mask, the ridges were covered with 1 μm of Zn-doped InP confinement layer and 0.5 μm Zn-doped GaInAs (with N_A-$N_D \approx 10^{19}$ cm^{-3}) cap layer, also grown by LP-MOCVD. In order to localize the injection current only in the buried-ridge active region, a deep-proton implantation was performed through a 5 μm wide photoresist mask after metallization of the contacts. Further localization of the current in the buried ridge is achieved by the built-in potential difference between the *p–n* InP homojunction on each side of the active region and the *n–p* InP–GaInAsP heterojunction of the active region.

Fig. 14.12 shows schematically the resulting GaInAsP–InP BRS laser. The devices were cleaved and sawn, producing chips of width 350 μm with cavity lengths of 300 μm.

Fig. 14.13 shows the light–current characteristics of seven LP-MOCVD laser diodes obtained from the same wafer. Pulse threshold currents of 190 mA at room temperature have been measured with an output power of up to 10 mW.

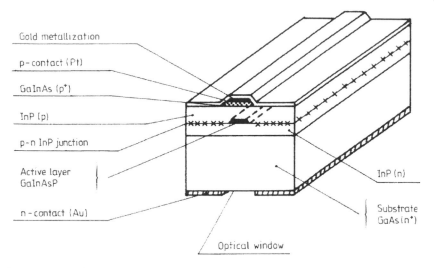

Gold metallization

p-contact (Pt)

GaInAs (p⁺)

InP (p)

p-n InP junction

Active layer
GaInAsP

n-contact (Au)

InP (n)

Substrate
GaAs (n⁺)

Optical window

Fig. 14.12. Schematic diagram of the cross section of a GaInAsP/InP BRS laser emitting at 1.3 μm grown by LP-MOCVD on a GaAs substrate.

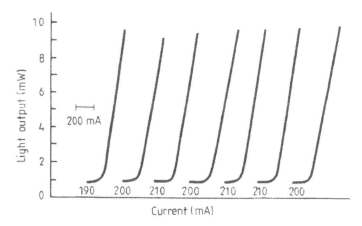

Fig. 14.13. Light–current characteristics of GaInAsP–InP laser diodes emitting at 1.3 μm grown by LP-MOCVD on a GaAs substrate.

14.4.2. Electronic devices—GaInP/GaInAs/InP MESFET

$Ga_{0.47}In_{0.55}As$ lattice-matched to InP is a potentially important material for field-effect transistors (FET) with high peak electron velocity and high electron mobility for application in optoelectronic integration.

Metal–Schottky barrier heights on GaInAs are too low to be used as MESFET gates. FET in lattice-matched $Al_{0.48}In_{0.52}As/InP$ which exploit the increased Schottky barrier height provided by AlInAs [Barnard et al. 1980 and Scott et al. 1986] and lattice-mismatched GaAs gate/GaInAs structures have also been prepared [Chen et al. 1985].

The preparation of a $Ga_{0.49}In_{0.51}P/Ga_{0.47}In_{0.53}As/InP$ MESFET fabricated from material grown by LP-MOCVD has also been reported by Razeghi et al. [1987]. The energy gap of $Ga_{0.49}In_{0.51}P$ is 1.9 eV [Hilsum 1964] and the lattice parameter is 5.65 Å. The growth conditions are given in Table 14.1.

Fig. 14.14 represents the FET device structure. The structure consists of n-type $Ga_{0.47}In_{0.53}As$ 1500 Å thick, doped to 3×10^{17} cm^{-3} with sulfur, and undoped $Ga_{0.49}In_{0.51}P$ 800 Å thick with a carrier concentration $N_D-N_A = 10^{16}$ cm^{-3} grown at 550 °C onto (1 0 0) oriented Fe-doped semi-insulating InP substrates.

Large-geometry FET with 2 μm gate lengths, 150 μm gate widths and 5 μm source–drain spacing have been fabricated. Fig. 14.14 shows source–drain current–voltage characteristics of a GaInP–GaInAs–InP FET; the gate bias step is 0.5 V. The transconductance of this device is $g_m \approx 50$ mS·mm^{-1}.

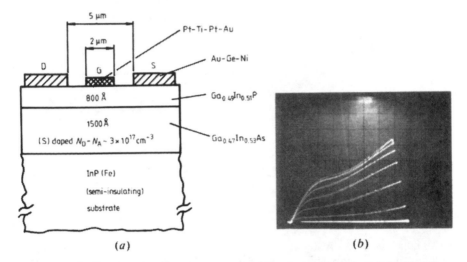

Fig. 14.14. (a) Schematic representation of the GaInP/GaInAs/InP MESEFT structure (b).

14.4.3. Optoelectronic integrated circuits (OEIC)

(a) Planar monolithic integrated photoreceiver for 1.3–1.55 μm wavelength applications using GaInAs–GaAs heteroepitaxy

Planar monolithic integrated photoreceivers are desirable devices for fiber-optic communication systems. In particular, the planar structure simplifies the problem of interconnecting the components and, for high-speed applications, such a technique reduces the parasitic capacitances. The feasibility of such devices has recently been demonstrated for 0.8 μm wavelength applications [Wada et al. 1983, Kolbas et al. 1983, and Verriele et al. 1985], and to a lesser extent for 1.3–1.55 μm wavelength applications [Hata et al. 1984, Razeghi et al. 1984, and Chen et al. 1984]. The purpose of this section is to present a planar monolithic integrated photoreceiver suitable for long-wavelength optical communications systems ($\lambda \approx 1.3 - 1.55$ μm) using $Ga_{0.47}In_{0.53}As$–GaAs strained heteroepitaxy [Razeghi et al. 1987 and Razeghi et al. 1986]. It consists of a planar $Ga_{0.47}In_{0.53}$ As photoconductive detector associated with a GaAs field-effect transistor (FET). The strained heteroepitaxy is constituted of an undoped $Ga_{0.47}In_{0.53}As$ layer grown on a classical GaAs FET epitaxy (Fig. 14.15).

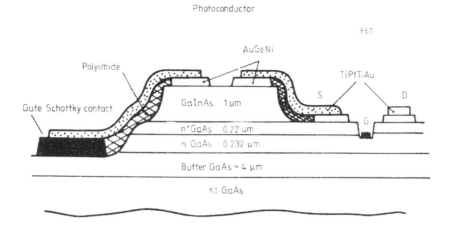

Fig. 14.15. Schematic cross section of integrated photoreceiver. The gate metallization has been deposited on the n-GaAs layer to fabricate the FET and on a part of the buffer GaAs (called gate Schottky contact on the figure) to realize an insulated interconnection between the gate and the photoconductor.

A scanning electron micrograph of the integrated circuit and its electrical circuit is shown in Fig. 14.16. The photoconductive detector has been formed by ion etching the $Ga_{0.47}In_{0.53}As$ up to the n^--GaAs layer, except for the area corresponding to the photoconductor. The lattice is constituted of two ohmic contacts (Au–Ge–Ni 2000 Å) separated from each other by 20 µm, leading to a 20×80 µm photosensitive area. The FET has been fabricated on the GaAs layers. The FET has a 2×900 µm gate (Ti 500 Å; Pt 300 Å; Ti 300 Å; Au 2000 Å), and, to produce a compact design integrated circuit, the FET surrounds the photoconductor. The interconnection between one of the photoconductor ohmic contacts and the FET source (Au–Ge–Ni 2000 Å) is made using a polyimide bridge. The same technique allows an interconnection to be made between the other photoconductor ohmic contact and the FET gate. It can be noticed that the n^+-GaAs layer reduces the dark resistance of the $Ga_{0.47}In_{0.53}As$ photoconductor; this effect will be discussed in the next paragraphs. A photoconductor was also processed on the same chip but without the transistor, in order to perform test experiments on the photodetector only. All experimental results presented on the photoconductor have been performed on the latter device.

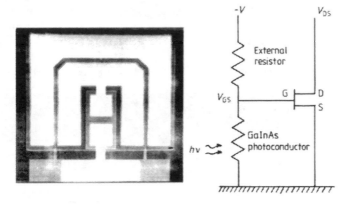

Fig. 14.16. Photograph and electrical circuit of monolithic integrated
photoreceiver.

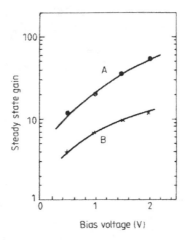

Fig. 14.17. Steady-state gain G of the GaInAs–GaAs photoconductor versus its bias
voltage V using (A) a CW 1.3 μm wavelength laser diode and (B) a CW 1.06 μm
wavelength YAG laser (P_L = 0.2 μW).

Fig. 14.17 shows typical steady-state gain (defined as the number of charges collected in the external circuit per incident photon) obtained for the photoconductor versus bias voltage. We observe that the gain value increases when the bias voltage increases; this result is due to the reduction of the transit time T_t when the electron drift velocity increases corresponding to the expression for the gain $G = T_v/T_t$ [Gammel et al. 1981 and Victor et al. 1985], where T_v is the electron–hole pair lifetime. The low value of the gain is explained by a short lifetime T_v, as can be observed on the picosecond response (Fig. 14.18). Typical dynamic gains G(ω) are given in Fig. 14.19. The experimental results can be described by the expression $G(\omega)=T_v / T_t(1+\omega^2 T_v^2)^{1/2}$ corresponding to an assumption of a generation–recombination process governed by a Poisson law. The gain–bandwidth product is close to 1 GHz, a value related to the 20 μm electrode spacing.

Obviously, this performance can easily be improved by using a shorter electrode spacing. The FET has been characterized separately. Its transconductance is close to 90 mS·mm^{-1} corresponding to the gate size (2×900 µm). A current amplification of 2 has been achieved on the picosecond response (Fig. 14.18) and the same typical amplification factors are obtained for the low-frequency response (Fig. 14.19).

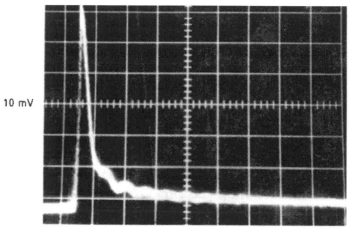

Time base (1 ns / division)

Fig. 14.18. Picosecond response using a mode-locked dye laser; λ = 0.580 µm,
P_L = 0.25 mW. The light has been focused on the photosensitive area of the
photoconductor using a microscope objective, to be sure that the light impinging on
the device creates electron–hole pairs only in the $Ga_{0.47}In_{0.53}As$ of the
photoconductor and not in the GaAs of the FET. The voltage measurements have
been performed using a 10 dB attenuator.

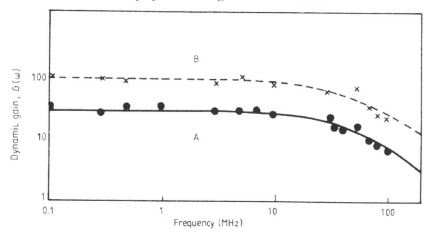

Fig. 14.19. Dynamic gains versus frequency using a sinusoidally modulated 1.3 µm
laser diode: (A) photoconductor, bias voltage 2 V; (B) integrated circuit,
V_{GS} = −2 V, V_{DS} = 2 V.

The noise properties of the photoconductor and of the integrated circuit have also been investigated in the 10 MHz–1.5 GHz frequency range. The experiments (Fig. 14.20) show a high $1/f$ photoconductor noise for frequencies lower than 100 MHz. For frequencies higher than 100 MHz this $1/f$ noise is greatly reduced, as is commonly observed in III–V photoconductors [Vilcot et al. 1985 and Chen et al. 1984]. The noise of the integrated circuit is higher; its value can be explained by the FET amplification plus the noise of the FET itself (Fig. 14.20). It can be noted that the noise under illumination is very close to the noise in darkness. This result can be explained by the low value of the gain of the photoconductor, according to the expression for the noise due to illumination of such a device [Vilcot et al. 1985].

Several improvements can be proposed to obtain better performances:

(1)   An interdigitated photoconductive detector would lead to a larger photosensitive area with a shorter electrode spacing, and obviously a higher gain–bandwidth product.

(2)   An increase in the dark resistance of the photoconductor by growing the $Ga_{0.47}In_{0.53}As$ photosensitive area directly on the GaAs buffer would constitute the main improvement of this device.

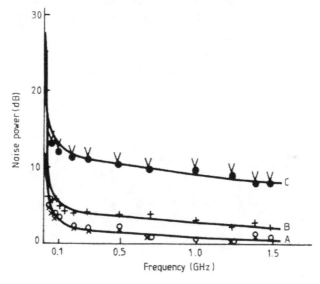

Fig. 14.20. Noise power versus frequency for CW 1.06 μm YAG laser illumination at PL = 0.2 mW. (A) Photoconductor; bias voltage 0.5 V: (X) without illumination and (○) with illumination. (B) Field-effect transistor; $V_{GS} = -0.5$ V; $V_{DS} = 2$ V. (C) Integrated circuit; $V_{GS} = -0.5$ V; $V_{DS} = 2$ V; (●) without illumination and (□) with illumination.

(b) Monolithic integration of a Schottky photodiode and a FET using a $Ga_{0.49}In_{0.51}P/Ga_{0.47}In_{0.53}As$ strained material

For the realization of optoelectronic integrated circuits (OEIC) suitable for long-wavelength optical communications systems (1.3–1.55 μm), $Ga_{0.47}In_{0.53}As$ lattice-matched to InP is a potentially important material.

Razeghi et al. [1987] have grown different epilayers by MOCVD on an Fe doped InP substrate. The $Ga_{0.47}In_{0.53}As$ epilayer ($N_D \approx 2 \times 10^{17}$ cm^{-3}; 2000 Å thick) is suitable for long-wavelength (1.3–1.55 μm) photodetection and FET channel fabrication. The $Ga_{0.49}In_{0.51}P$ epilayer is undoped (residual *n*-type) and is 1000 Å thick. A photodiode, a FET and an OEIC have been realized on the same chip. The photodiode has a 40×40 μm photosensitive area. The Schottky contact has been realized by the deposition of a 200 Å thick Ti–Pt electrode. A responsivity close to 0.1 A·W^{-1} has been measured for a 1.3 μm wavelength optical signal and the picosecond response is given in Fig. 14.21.

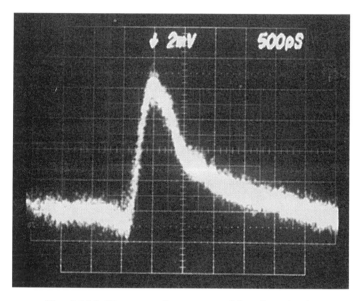

Fig. 14.21. Picosecond response of the photodiode.

The FET gate has a 3 μm length and 300 μm width. The saturation current is about 10 mA and the pinch-off voltage is close to -3 V. The average measured transconductance is 30 mS·mm^{-1} [Razeghi et al. 1988].

The OEIC (schematic cross sections in Fig. 14.22) associates the same photodiode and FET (600 μm gate width, in this case) as that previously presented in section 14.4.2. A SEM micrograph and equivalent electrical circuit are shown in Fig. 14.23. Static, dynamic and noise properties of the IC have been evaluated. As an example, a responsivity of 50 A·W^{-1} have been achieved at 1.3 μm for a 1.5 K·Ω bias resistor.

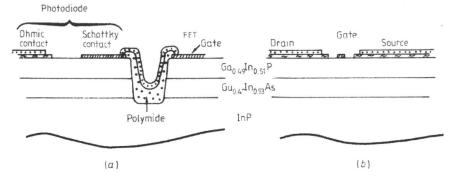

Fig. 14.22. Schematic cross sections of the device (OEIC).

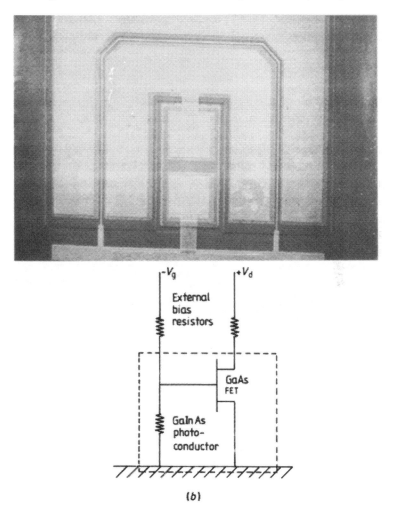

Fig. 14.23. SEM photograph and electrical equivalent circuit of the photoreceiver.

14.5. Monolayer epitaxy of $(GaAs)_n(InAs)_n$–InP by MOCVD

The atomic-layer epitaxy (ALE) proposed by Suntola [1984] and described in section 2.3.2 differs completely from the usual growth techniques. His idea is accomplished by alternative adsorption of reactant gas molecules onto the growing surface. So, in ALE, the growth thickness is only determined by the number of adsorptions. Using this method, quantum-well structures, superlattices and modulation-doped heterojunctions can be fabricated with precise thickness control and compositional homogeneity over a large surface of substrate. In modulation-doped GaInAs–InP structures, the enhancement of mobility is limited by the disorder scattering in $Ga_xIn_{1-x}As$ alloys. With monolayer epitaxy of $(GaAs)_n(InAs)_n$, one can provide ordered atomic arrangements in $Ga_{0.5}In_{0.5}As$ alloy, and improvement of the drift velocity is expected. In this section, we describe the successful growth [Razeghi et al. 1988] of $(InAs)_n(GaAs)_n$–InP multiquantum-well structures by LP-MOCVD, which exhibited an excellent uniformity of thickness and composition over a large area of InP substrate.

Growth conditions of the different monolayers are detailed in Table 14.5. Table 14.6 details the lattice parameter (a), the monatomic layer thickness (d) and the time necessary to grow this atomic layer (t).

Several growth sequences were first investigated, in order to optimize the quality of the $(GaAs)_n(InAs)_n$ bulk layer.

	InP	InAs	GaAs
Growth temperature (°C)	550	550	500
Growth pressure (mbar)	100	100	100
N_2 flow ($1 \cdot min^{-1}$)	2	2	2
H_2 flow($l \cdot min^{-1}$)	2	2	2
In flow (N_2) ($cm^3 \cdot min^{-1}$)	100	100	–
Ga flow (H_2) ($cm^3 \cdot min^{-1}$)	–	–	60
Growth rate ($Å \cdot min^{-1}$)	50	50	50
PH_3 flow ($cm^3 \cdot min^{-1}$)	200	–	–
AsH_3 flow ($cm^3 \cdot min^{-1}$)	–	45	45

Table 14.5. Growth conditions.

	a (Å)	d (Å)	t (s)
InP	5.86	2.93	2.4
GaAs	5.66	2.83	2.3
InAs	6.06	3.03	2.5

Table 14.6. Growth parameters.

Type-1 epitaxy

By taking $n = 1$ and 2000 monolayers of InAs and GaAs, a $Ga_{0.5}In_{0.5}As$ bulk layer about 1 μm thick was made, in a continuous growth, without purging between the monolayers.

The epitaxy was initiated by growing an InP buffer layer on an InP ((1 0 0) oriented) substrate. The first InAs layer was realized by switching PH_3 to the waste line and AsH_3 to the reactor. For growing the first GaAs layer, In was then switched to the waste line and Ga to the reactor. The global layer $((GaAs)_1(InAs)_1)$ was then grown, without any purging time between the monolayers.

Fig. 14.24(a) shows the photoluminescence spectrum at $T = 4$ K of the structure. The recombination peak is located at about 735 meV, at a lower value than the $T = 4$ K gap of the $Ga_{0.47}In_{0.53}As$ material lattice-matched to InP. A full width at half-maximum (FWHM) of 30 meV is determined, which clearly remains higher than the value reported in bulk $Ga_{0.47}In_{0.53}As$ (see Chapter 12).

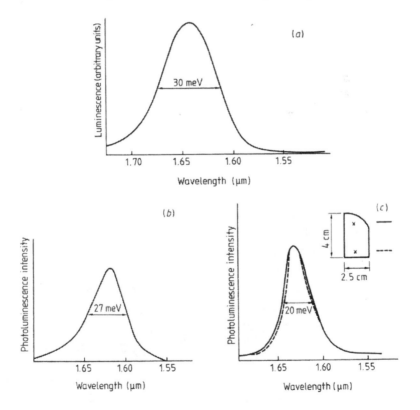

Fig. 14.24. Photoluminescence spectra at 4 K of: (a) $(GaAs)_1$ $(InAs)_1$ bulk layer grown without purge; (b) $(GaAs)_1$ $(InAs)_1$ bulk layer grown with purge; (c) $(GaAs)_2(InAs)_2$ bulk layer grown with purge.

Type-2 epitaxy

This was realized by keeping $n = 1$ with 2000 periods of $(GaAs)_1$ $(InAs)_1$, but purging times were introduced between the layers. After growing the InP buffer layer, PH_3 was switched to the waste line and AsH_3 to the reactor to realize the first InAs monolayer. In was then switched to the waste line, and Ga introduced into the reactor after waiting 1 s during which only AsH_3 was present in the reactor. This has been done to stabilize with As the growing solid surface, in order to prevent mixing between In and Ga.

Fig. 14.24(b) shows the $T = 4$ K photoluminescence spectrum of the structure. The recombination peak is located at 766 meV, still at a lower energy than the $T = 4$ K band gap of $Ga_{0.47}In_{0.53}As$. A slight decrease is found in the FWHM parameter, which is about equal to 27 meV.

Type-3 epitaxy

The purging time is maintained at 1 s, but the time of each GaAs and InAs sequence is multiplied by 2, so that $n = 2$ and the growing material is $(GaAs)_2(InAs)_2$. 1000 periods of $(GaAs)_2$ have been realized, so that the total thickness of this material is kept constant, and equal to nearly 0.6 μm.

Fig. 14.24(c) shows the $T = 4$ K photoluminescence spectrum of the structure. The peak is located at 758 meV, with a FWHM of 20 meV, which is further lowered, compared to the two previous results.

The quality of the epilayer has been improved by introducing a purging time of 1 s between the InAs and GaAs layers, and by growing two monolayers ($n = 2$) of these compounds at each sequence.

Mobility measurements, performed on this layer, using the classical Van der Pauw technique gave $\mu(300$ K$) = 7000$ cm$^2 \cdot$V$^{-1} \cdot$s^{-1} and $\mu_{(77\ K)} = 30{,}000$ cm$^2 \cdot$V$^{-1} \cdot$s^{-1}. Such results are not far away from those of bulk $Ga_{0.47}In_{0.53}As$.

It is worth adding that all these structures showed a remarkable composition uniformity, with no dispersion of the luminescence recombination peak on wafers of about 10 cm^2 area.

14.5.1. Multiquantum wells of (GaAs)₂(InAs)₂/InP

This fact was further demonstrated by studying the uniformity of a multiquantum-well structure. The wells were composed of $(GaAs)_2(InAs)_2$, which were grown as described in type-3 epitaxy. The global structure was as follows:

(1)	1000 Å InP buffer layer,
(2)	20 times $(GaAs)_2(InAs)_2$ sequence,
(3)	500 Å InP barrier,
(4)	10 times $(GaAs)_2(InAs)_2$ sequence,
(5)	500 Å InP barrier,
(6)	five times $(GaAs)_2(InAs)_2$ sequence,
(7)	500 Å InP barrier,
(8)	two times $(GaAs)_2(InAs)_2$ sequence,
(9)	500 Å InP top layer.

Fig. 14.25 shows the low-temperature ($T = 4$ K) photoluminescence spectra of the structure, using He–Ne and YAG lasers of wavelengths 0.63 μm and 1.16 μm as excitation sources. The penetration depth of the YAG laser is more important, so that it appears more sensitive than He-Ne to the wells that are situated far away from the surface of the sample.

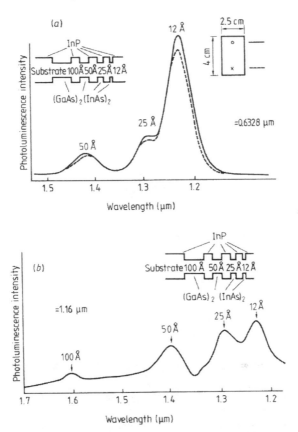

Fig. 14.25. Photoluminescence spectra at 4 K of a $(GaAs)_2(InAs)_2/InP$ multiquantum well.

The thicknesses of the quantum wells, deduced from the growth rate, are about 100, 50, 25 and 12 Å. It is noticeable that a quantum well of $(GaAs)_2(InAs)_2$ as thin as 12 Å shows a clear quantum size effect (Fig. 14.25). Another interesting feature is the remarkable uniformity in thickness and composition obtained by monatomic layer epitaxy (see Fig. 14.25). The photoluminescence spectrum has been recorded at two extremities of the sample, with no significant shift in energy in the recombination peaks.

Fig. 14.26 shows, for comparison, spectra recorded at the same points of the wafer, in a multiquantum-well structure with wells composed of classical $Ga_{0.47}In_{0.53}As$ material. Shifts of about 20 meV are seen between the peaks attributed to the same quantum wells for all the quantum wells. This indicates that the composition of the GaInAs varies across the wafer. This variation is not found in the case of the monolayer epitaxy (Fig. 14.25).

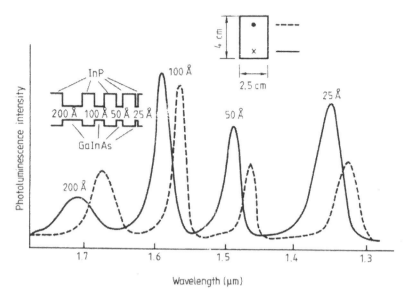

Fig. 14.26. Photoluminescence spectrum at 4 K of a GaInAs–InP multiquantum well recorded in two opposite points of the layer.

References

Asuka, N., Kolodziejski, L.A., Gunshor, R.L., Datta, S., Bicknell, R.N., and Schetzina, J.F. *Appl. Phys. Lett.* **46** 860, 1985.

Barnard, J., Ohno, H., Wood, C.E., and Eastman, L.F. *IEEE Electron. Device Lett.* **EDL-9** 174, 1980.

Cazaux, J., Etienne, P., and Razeghi, M. *J. Appl. Phys.* **59** 3598, 1986.

Chen, C.Y., Cho, A.Y., and Garlinski, P.A. *IEEE Electron. Device Lett.* **EDL-6** 20, 1985.

Chen, C.Y., Pang, Y.M., Alavi, K., Cho, A.Y., and Garbinski, P.A. *Appl. Phys. Lett.* **44** 99, 1984.

Cullen, G.W. *Hetero-epitaxial Semiconductors for Electronic Devices* ed. Cullen, G.W. and Wang, C.C. (Berlin: Springer) p 50, 1978

Gammel, J.C., Ohno, H., and Ballantyne, J.M. *IEEE J. Quantum Electron.* **QE-17** 269, 1981.

Hata, S., Ikeda, M., Amano, T., Motosugi, G., and Kurmuda, K. *Electron. Lett.* **20** 947, 1984.

Hilsum, C. *Proc. 7th. Int. Conf. Semiconductor Physics* (Paris: Dunod) p 1127, 1964.

Huber, A.M., Razeghi, M., and Morillot, G. *Gallium Arsenide and Related Compounds, 1984* (Inst. Phys. Conf. Ser. 74) p 223, 1985.

Kolbas, R.M., Abrokwah, J., Carney, J.K., Bradshaw, D.H., Elmre, H.R., and Briard, J.R. *Appl. Phys. Lett.* **43** 821, 1983.

Razeghi, M., Maurel, P., Omnes, F., and Thorngren, E. *Optical Properties of Narrow-gap Low-dimensional Structures* (New York: Plenum) pp 39–53,

Razeghi, M., Meunier, P.L., and Maurel, P. *J. Appl. Phys.* **59** 2261, 1986.

Razeghi, M. *Lightwave Technology for Communication* ed. Tsang, W.T. (New York: Academic) p 299, 1985.

Razeghi, M., Blondeau, R., and Duchemin, J.P. *Gallium Arsenide and Related Compounds, 1984* (Inst. Phys. Conf. Ser. 74) p 679, 1985.

Razeghi, M., Ramdani, J., Legry, P., Vilcot, J.P., and Decoster, D. *Gallium Arsenide and Related Compounds, 1987* (Inst. Phys. Conf. Ser. 91) p 781, 1988.

Razeghi, M., Hosseini-Therani, A., Vilcot, J.P., and Decoster, D. *Gallium Arsenide and Related Compounds, 1987* (Inst. Phys. Conf. Ser. 91) p 625, Abstracts, p 647, 1988.

Razeghi, M., Maurel, P., Omnes, F., and Nagle, J. *Appl. Phys. Lett.* **51** 2216, 1988.

Razeghi, M., Ramdani, J., Verride, H., Decoster, D., and Constant, M. *Appl. Phys. Lett.* **49** 215, 1986.

Scott, M.D., Moore, A.H., Griffith, I., Griffith, R.J, Sussman, R., and Oxley, C. *Gallium Arsenide and Related Compounds, 1985* (Inst. Phys. Conf. Ser. 79) p 475, 1986.

Suntola, T. *16th Int. Conf. on Solid State Devices and Materials* Tokyo, Extended, 1984.

Verriele, H., Maricot, S., Constant, M., Ramdani, J., and Decoster, D. *Electron. Lett.* **21** 878, 1985.

Vilcot, J.P., Constant, M., Decoster, D., and Fauquembergue, R. *Physica* B **129** 488, 1985.

Wada, D., Mivra, S., Ito, M., Fujh, T., Sakurai, T., and Hiyamizu, S. *Appl. Phys. Lett.* **42** 380, 1983.

Zur, A. and McGill, J. *J. Appl. Phys.* **55** 378, 1984.

15. MOCVD Growth of III–V Heterojunctions and Superlattices on Silicon Substrates

15.1. **Introduction**
    15.1.1. The difference in lattice parameter
    15.1.2. The difference in lattice symmetry
    15.1.3. The difference in thermal expansion coefficient
    15.1.4. Preparation and chemical etching of Si substrates
    15.1.5. The cross doping of the III–V epilayer with Si
15.2. **MOCVD growth of GaAs on silicon**
15.3. **InP grown on silicon**
15.4. **GaInAsP–InP grown on silicon**
15.5. **Applications**
    15.5.1. Room-temperature CW operation of a GaInAsP/InP ($\lambda = 1.15$ µm) light-emitting diode on silicon substrate
    15.5.2. GaInAsP/InP double heterostructure laser emitting at 1.3 µm on silicon substrate
    15.5.3. CW operation of a $Ga_{0.25}In_{0.75}As_{0.5}P_{0.5}$/InP BRS laser on silicon substrate
    15.5.4. InGaAs–InP MQW on silicon substrate
    15.5.5. GaInAs–InP PIN photodetector

15.1. Introduction

At present, there is a great deal of interest in the growth of GaAs on silicon substrate [Christou et al. 1985, Nishi et al. 1985, Sheldon et al. 1984, Soga et al. 1985, Akiyama et al. 1984, and Wang 1985]. Devices such as GaAs FET transistors [Aksum et al. 1986] or GaAs lasers [Van der Ziel 1987 and Dupuis 1988] have been realized using a silicon substrate. Long-distance optical links use semiconductor lasers emitting at 1.3 and 1.55 µm fabricated using materials such as the GaInAsP–InP lattice-matched system on an InP substrate. Unfortunately, the technology of integrated circuits for signal treatment is more difficult on this material than on GaAs substrates. Since silicon technology is more advanced than that for GaAs- and InP-based materials, the solution is to combine the advantages of InP and GaAs devices with the maturity of Si processing technology, using existing Si integrated circuit equipment.

However, high device quality for InP- and GaAs-based materials has not been obtained by direct growth on Si substrates, since the lattice parameter, crystal symmetry and thermal expansion coefficient of InP, GaAs and Si are all different. Furthermore, nucleation of InP and GaAs on Si substrate is difficult.

At the present stage of III–V/Si technology development, attention is being focused on perfecting the growth of InP, GaAs and related compounds on Si and demonstrating that devices with acceptable operating characteristics can be fabricated in these layers.

The most important problems encountered when III–V semiconductors are grown on Si substrates are as follows:

(1)   the difference in lattice parameter between the III–V semiconductor and the Si substrate;

(2)   the difference in lattice symmetry;

(3)   the difference in thermal expansion coefficient;

(4)   the preparation and chemical etching of the Si substrate to remove the oxide and carbon; and

(5)   the cross doping of the III–V epilayer with silicon.

15.1.1. The difference in lattice parameter

An 8% lattice mismatch between InP and Si and 4% between GaAs and Si produces a misfit dislocation at the interface. The formation of misfit dislocations and the introduction of tensile stress in heteroepitaxial InP-and GaAs-based materials on Si are the principal problems besetting the application of these materials for minority carrier devices, such as lasers. The dislocations at the III–V/Si interface do not degrade the III–V epilayer crystalline quality; instead, the propagation of dislocations away from the interface region (the so-called threading dislocation) appears to be primarily responsible for the non-radiative recombination centers. The threading dislocations also reduce minority carrier lifetimes, producing inferior properties for devices fabricated on III–V epilayers on Si substrates. Therefore, very low misfit dislocation densities in the III–V epitaxial overlayers are essential for laser applications.

In order to reduce the misfit dislocations in GaInAsP–InP layers grown on Si substrate, using MOCVD, Razeghi et al. [1988] used a two-step growth method. First a thin layer of 200 Å of GaAs is grown at 450 °C, using a growth rate of 50 Å·min^{-1}. Then the substrate temperature is raised to 550 °C. A five-period lattice-matched superlattice of GaInAsP–InP or GaAs–GaInP with $L_z = 100$ Å and $L_B = 100$ Å (L_B is the thickness of InP or GaInP barrier and L_z is the thickness of the GaInAsP or GaAs quantum well) is then grown. These superlattices have been shown to be effective in bending dislocation lines out toward the edges of the crystal, preventing the cross doping of the epilayer with Si, and thus preventing dislocation movement in device operation.

15.1.2. The difference in lattice symmetry

One of the problems in growing a III–V (a polar semiconductor) on Si (a non-polar semiconductor) is the possible occurrence of antiphase domains (p, g, $-a$). In certain regions of the epilayer the cation and anion sublattices have been interchanged. The interface between domains with opposite sublattice allocation forms a two-dimensional structural defect called an antiphase boundary [Kroemer 1986]. Antiphase boundaries in a III–V semiconductor containing III–III and V–V bonds represent an electrically charged defect. The III–III bonds act as acceptors, the V–V bonds as donors (see Fig. 15.1(b)). In general, antiphase boundaries which act as highly compensated doping sheets with very little net doping can also act as non-radiative recombination centers.

In order to eliminate the antiphase domain, in III–V layers grown on Si by MOCVD one can use a 4° off (1 0 0) towards (0 1 1) oriented Si substrate, and then perform the high-temperature annealing (1000 °C) under AsH$_3$–H$_2$ of the tilted substrate prior to growth, in order to form the double atomic layer steps, which help to eliminate antiphase domains (p, g, $-c$) [Kroemer 1986].

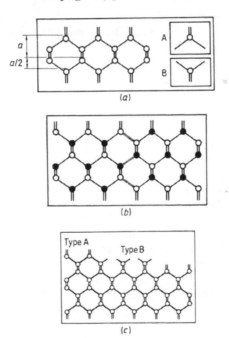

Fig. 15.1. (a) Two sublattices in diamond structure; (b) an antiphase boundary in a III–V semiconductor (○, III; ●, V); (c) two kinds of step on a Si(1 0 0) surface.

In fact, (1 0 0) Si substrates cannot be prepared to have misorientations totally along one direction, e.g., just along the direction, leading to the growth of a surface consisting of steps along the $\langle 0\ 1\ 1 \rangle$ directions and terraces having a (1 0 0) orientation. Two types of double steps can form on the misoriented (1 0 0) Si surface, the so-called "type A" surface step where the Si dangling bonds point parallel to the step edge, and the "type B" surface step in which the Si bonds are perpendicular to the step edge [Kroemer 1986 and Aspnes et al. 1987] (see Fig. 15.1(*a*) and (*c*)). As a result, the possibility exists that two distinct domains can result in the heteroepitaxial GaAs film. It has been shown experimentally [Kroemer 1986 and Sakamoto et al. 1986] that for a properly prepared (1 0 0) Si surface, the type A steps form preferentially, leading to the formation of a single-domain GaAs/Si heteroepitaxial layer (see Fig. 15.1(*c*)). Calculations by Aspnes and Ihm [Aspnes et al. 1987] have shown that the production of single-domain regions on such a surface is favored if the substrate misorientation is within a certain range of directions off the nominal (1 0 0) surface, as determined experimentally for MOCVD GaAs/Si by Akiyama et al. [1986].

Studies of the early stages of MBE GaAs/Si growth on clean (1 0 0) Si substrates slightly misoriented towards the ⟨0 1 1⟩ have shown that the deposition of GaAs begins with the nucleation of GaAs islands on the Si surface [Hull et al. 1987]. Surface preparation techniques strongly influence this process, probably because of the details of the step density and the Si atom surface mobility. Hull and co-workers have shown that, for Si substrates annealed at low temperatures ($T_{ann} \approx 730$ °C), the surface reconstruction consists primarily of steps of one or two (2 0 0) monolayers [Hull et al. 1987]. In contrast, high anneal temperatures ($T_{ann} \approx 900$ °C) can lead to the formation of large clusters of Si surface steps to form local Si facets about 6–10 (2 0 0) monolayers high. Similar detailed studies of the early stages of the MOCVD growth of GaAs/Si have not yet been reported. However, the $H_2 + AsH_3$ atmosphere used in the Si annealing pre-growth sequence for MOCVD GaAs/Si films is quite different from the high-vacuum environment used in the substrate cleaning step for MBE growth of GaAs/Si and this could influence the surface reconstruction [Dupuis et al. 1987].

15.1.3. The difference in thermal expansion coefficient

The difference in thermal expansion coefficients of GaAs, InP and related compounds with Si substrates puts the epilayer under extreme tensile stress. The difference in the lattice parameters parallel and perpendicular to the interface arises from the difference in thermal expansion between the III–V compound and Si, not from misfit accommodation by elastic strain.

Analyses of the tetragonal distortion in III–V on Si would thus be expected to yield information concerning the defect structure in this material.

Some physical parameters of Si, GaAs and InP relevant for the hetero-epitaxy of these materials are listed in Table 15.1.

15.1.4. Preparation and chemical etching of Si substrates

Preparation of atomically clean surfaces is essential in order to achieve good epitaxial growth. One has to eliminate the oxides and carbon from the Si surface before growth. Several cleaning procedures have been employed in the MOCVD growth of III–V/Si layers. One of the most commonly used techniques involves the sequential chemical oxidation and removal of a thin non-stoichiometric oxide film [Razeghi et al. 1988]. This relatively volatile oxide layer is subsequently removed by heating to a temperature of 1000 °C for 10 min. This procedure, or some variation of it, is generally employed in MOCVD heteroepitaxy on Si.

	Si	GaAs	InP
Thermal coefficient of expansion (K^{-1})	2.6×10^{-6}	6.8×10^{-6}	4.56×10^{-6}
Thermal conductivity ($W \cdot cm^{-1} \cdot K^{-1}$)	1.5	0.46	0.68
Lattice parameter (Å)	5.43	5.65	5.86
Crystal symmetry	Diamond	Zinc blende	Zinc blende
Density ($g \cdot cm^{-3}$)	2.328	5.32	4.787

Table 15.1 InP, GaAs and Si properties.

15.1.5. The cross doping of the III–V epilayer with Si

To eliminate the cross doping of the III–V epilayer with Si using a Si substrate, several types of superlattices have been evaluated. Razeghi et al. [1988] showed experimentally that GaInAsP–InP lattice-matched superlattices can prevent the cross doping of the III–V semiconductor epilayer with Si.

15.2. MOCVD growth of GaAs on silicon

The growth procedure is as follows. A Si substrate is chemically cleaned by HF dipping, and then loaded into the reactor. The substrate is then heated to approximately 1000 °C for 10 min to remove the oxide and the double-atomic-layer steps. This high-temperature cleaning procedure, 4° off substrate, is essential to obtain a single-domain structure. The substrate is then cooled to a relatively low temperature, 400–450 °C, and a 200 Å GaAs layer is deposited on the substrate. The initial growth is normally by island formation followed by coalescence. The growth temperature dramatically affects the shape and size of the islands that nucleate. At lower temperatures, the islands are smaller and closer together, resulting in a smoother surface. (However, it is necessary to have decomposition of AsH$_3$ and TEGa for growth of GaAs to occur. At temperatures lower than 400 °C decomposition of alkyls and hydrides is not complete, so with the MOCVD growth technique a temperature of 450 °C for the nucleation is adequate.)

After nucleation, the substrate temperature is raised to 550 °C for the growth of the rest of the structure. To decrease the threading dislocations, different kinds of interlayers between GaAs and Si, such as a Ge, GaP strained layer superlattice, a GaAs–GaInP superlattice, or a GaInAs–InP or GaInAsP–InP lattice-matched superlattice can be used [El-Masry et al. 1987]. Razeghi et al. [1988] showed that the strain fields associated with the lattice-matched superlattice interlayers are effective in leading dislocation lines out toward the edges of the crystal. These superlattices may also prevent dislocation movement in device operation.

GaAs is grown on silicon using GaAs–Ga$_{0.49}$In$_{0.51}$P superlattices as buffer layers in order to decrease the mismatch dislocations [Razeghi et al. 1988]. They consist of ten periods, 20 Å (GaAs) and 40 Å (GaInP). The superlattices are introduced to prevent the propagation of the dislocations present at the GaAs/Si interface. The GaAs top layer is grown at 550 °C with a growth rate of 100 Å·min^{-1}.

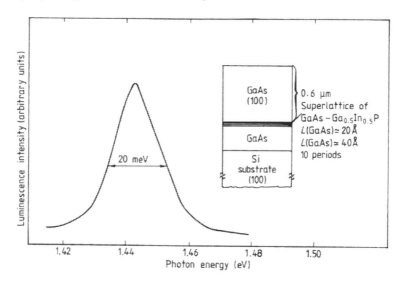

Fig. 15.2. Photoluminescence spectrum at T = 77 K of bulk GaAs grown on Si, using GaAs–Ga$_{0.49}$In$_{0.51}$P superlattices as buffer layers.

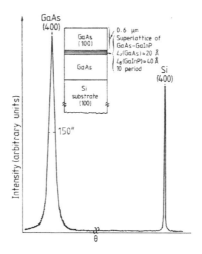

Fig. 15.3. Double x-ray diffraction pattern of GaAs grown on Si using GaAs– Ga$_{0.49}$In$_{0.51}$P superlattices as a buffer layer.

Fig. 15.2 and Fig. 15.3 show the typical optical and crystallographic properties of the 0.6 µm thick GaAs epilayer. The T = 77 K photoluminescence intensity is as high as that for GaAs grown simultaneously on a GaAs substrate (see Fig. 15.2). The full width at half maximum (FWHM) is about 20 meV.

The GaAs epilayer has the same (1 0 0) orientation as that of the substrate. Fig. 15.3 shows the double x-ray diffraction spectrum of this layer. The good crystallinity of the GaAs is shown by the relatively narrow linewidth ($\approx$ 150 arcsec).

15.3. InP grown on silicon

For the growth of InP on Si Razeghi et al. [1988] used the same preparation and heating for the Si substrate. A thin ($\approx$ 500 Å) layer of GaAs is first directly grown on the silicon substrate. A GaInAsP–InP superlattice is then deposited, to prevent the propagation of misfit dislocations. The InP layer is finally grown at a temperature of 550 °C.

The electrochemical profile of such a structure is detailed in Fig. 15.4. Note that InP has a low n-type residual doping level, about 5×10^{15} cm^{-3}. Measurements in the dark (dotted curve) and illumination (full curve) give essentially the same results, indicating that almost no deep traps are present in the layer. The $T = 2$ K photoluminescence spectrum of a 1.5 µm thick undoped InP layer grown on Si substrate is shown in Fig. 15.5, together with the corresponding electrochemical profile. The intrinsic (or shallow donor related) peak appears at about 1410 meV, which is 13 meV lower than the low-temperature gap for unstrained InP. The lattice mismatch existing between InP and Si seems to indicate that some strain is present in the layer. The peaks at 1370 and 1326 meV are believed to be due to an electron–neutral acceptor recombination (e–A^{0}) with an associated one-phonon replica. Extrinsic recombinations attributed to shallower impurities or defect-related excitons appear between 1380 and 1405 meV. Their precise origin is not clearly understood.

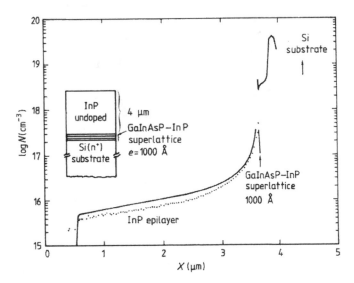

Fig. 15.4. Typical electrochemical profile of an InP epilayer grown on Si substrate.

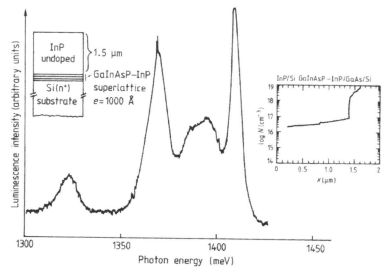

Fig. 15.5. Low-temperature PL spectrum of undoped InP grown on Si. The excitation power is around 0.8 W·cm⁻² (642.8 nm). The electrochemical profile of this layer is shown in the inset.

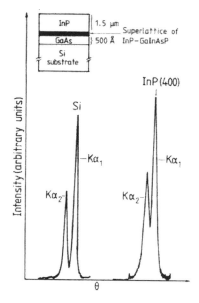

Fig. 15.6. X-ray rocking curve of InP grown on Si.

The excellent crystallographic properties of the same InP epilayer are shown in the simple crystal diffraction pattern of the structure (Fig. 15.6). The profile of the InP-related signal appears to be nearly equivalent to that of the Si substrate.

Razeghi et al. [1988] studied the problem of Si autodoping in the following way. InP was grown on a silicon substrate using different buffer layers:

(a) a simple GaAs layer;
(b) a GaAs–GaInP superlattice; and
(c) a GaInAsP–InP superlattice.

They measured the electrochemical profile of a nominally undoped InP layer, following procedures (a), (b) and (c).

In case (a) the observed doping interface is very large and has a thickness of about 2 μm, indicating that Si has efficiently diffused through the InP epilayer during growth. For case (b) the doping interface between the Si substrate and the InP layer has decreased to 0.5 μm, thus indicating that the GaAs–GaInP superlattice acts as an efficient blocking layer for Si diffusion.

Case (c) shows a very abrupt doping interface between Si and InP, with a thickness of less than 500 Å. Thus it seems that the InP–GaInAsP superlattice prevents Si diffusion and cross doping (see Fig. 15.7).

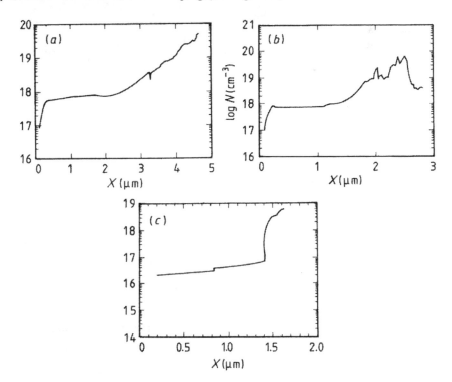

Fig. 15.7. *Electrochemical profile of InP on Si substrate using different buffer layers: (a) GaAs, (b) GaAs–Ga$_{0.49}$In$_{0.51}$P superlattice, (c) GaInAsP–InP superlattice.*

SIMS measurements were performed on sulfur-doped (2×10^{18} cm^{-3}) InP layers grown simultaneously on InP, GaAs and Si substrates, using GaAs–GaInP superlattices as buffer layers (Fig. 15.8). Fig. 15.8 shows no evidence of silicon substrate doping by P and As, because the growth temperature is low enough to avoid the deterioration of the Si substrate. This is an excellent result in view of monolithic integration of III–V compounds on Si substrate.

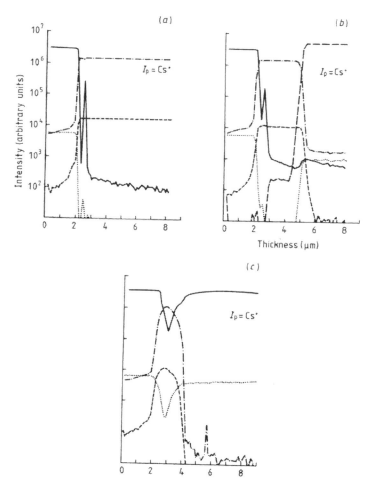

Fig. 15.8. SIMS profile of an InP layer grown simultaneously on (a) GaAs, (b) Si and (c) InP substrates: · · · · ·, In;——, P;— —, As; ---, Ga; — — —, Si.

The 20 K photoluminescence spectrum of the InP layers doped with S on Si substrate shows [Razeghi 1988] a full width at half maximum of about 28 meV, a value essentially the same as that for InP layers grown simultaneously on GaAs and InP substrates (Fig. 15.9). However, the InP/Si PL peak is shifted to a lower energy by 15 meV relative to PL the InP/InP PL, indicating that some strain is present in the InP/Si layer.

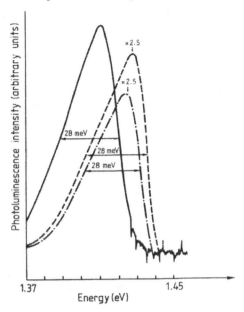

Fig. 15.9. PL spectrum at T = 20 K of the InP/GaAs–Ga$_{0.49}$In$_{0.51}$P structure grown on InP (— —), GaAs (---) and Si (——) substrates. The excitation power of the 632.8 nm He–Ne laser is around 0.8 W·cm^{-2}.

15.4. GaInAsP–InP grown on silicon

Because of the application of long-wavelength components, especially 1.3 and 1.5 μm lasers, quaternary compounds Ga$_x$In$_{1-x}$As$_y$P$_{1-y}$ ($0 \leq x \leq 0.47$ and $0 \leq y \leq 1$) lattice-matched to InP have become very important materials for the optoelectronic industry, and it would be of great interest to combine these materials with Si technology [Razeghi 1988].

The lattice parameter of the quaternary compounds with a composition Ga$_x$In$_{1-x}$As$_y$P$_{1-y}$ is given by Sharma and Purohit [1974]:

Eq. (15.1)
$$a(x,y) = a(\text{GaAs})xy + a(\text{InAs})(1-x)y + a(\text{GaP})x(1-y) + a(\text{InP})(1-x)(1-y)$$

The coefficient of linear thermal expansion, $\alpha(xy)$, of the quaternary compound may also be calculated using Vegard's law:

Eq. (15.2)
$$\alpha(x,y) = [\alpha(\text{GaAs})\alpha(\text{GaAs})xy + \alpha(\text{InAs})\alpha(\text{InAs})(1-x)y$$
$$+ \alpha(\text{GaP})\alpha(\text{GaP})x(1-y)$$
$$+ \alpha(\text{InP})\alpha(\text{InP})(1-x)(1-y)] / \alpha(x,y)$$

The thermal expansion coefficient of GaP is 5.3×10^{-6}, InAs is 5.16×10^{-6}, InP is 4.56×10^{-6}, GaAs is 6.8×10^{-6} and Si is $2.6 \times 10^{-6} \text{K}^{-1}$. The difference between the mean linear coefficients of thermal expansion of InP and Si is less than the difference between those of GaAs and Si. In addition, the deposition temperatures for MOCVD growth of GaInAsP–InP are typically 550 °C, whereas MOCVD GaAlAs–GaAs layers are grown at 750 °C. As a result of the lower thermal expansions and the lower growth temperature, the thermally-induced stress in GaInAsP–InP/Si films due to cooling from the growth temperature is less than for GaAs–GaAlAs/Si layers of comparable thickness.

The preliminary results for GaInAsP–InP/Si lasers [Razeghi 1988] show that the solution to the problem of degradation of GaAlAs–GaAs/Si lasers is to use instead GaInAsP–InP/Si lasers. It is known that InP-based long-wavelength lasers are more "immune" to the deleterious effects of stress than are GaAs-based lasers [Hayashi 1980 and Thompson 1980]. As a result, the lifetime of heteroepitaxial GaInAsP–InP lasers grown on Si substrates may be longer than that of GaAlAs–GaAs/Si lasers, although the lattice mismatch of InP/Si is 8% compared to 4% for GaAs/Si.

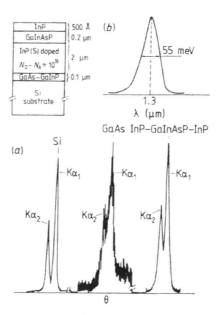

Fig. 15.10. (a) X-ray diffraction pattern of the double hetero-structure InP–GaInAsP–InP on Si; (b) 300 K PL spectrum of this sample.

A double InP/Ga$_{0.25}$In$_{0.75}$As$_{0.5}$P$_{0.5}$ heterostructure has also been grown on a silicon substrate. Fig. 15.10 shows the global structure of the layer. The growth was initiated with a 1000 Å GaAs–GaInP buffer layer, followed by 2 µm InP. The Ga$_x$In$_{1-x}$As$_y$P$_{1-y}$ active layer is about 2000 Å thick. The single diffraction pattern in Fig. 15.10(a) shows the good crystallographic quality of the InP/GaInAsP/InP structure. The Kα_1 and Kα_2 peaks also appear to be very well resolved. The GaAs buffer layer remains very thin, so its related signal is difficult to detect.

Nevertheless, $K\alpha_1$ and $K\alpha_2$ are separated, even when the epilayer lies near the silicon substrate. The quaternary compound has a wavelength of 1.3 μm. The 300 K photoluminescence spectrum in Fig. 15.10(*b*) shows the good optical properties of the quaternary. The recombination peak is situated at the expected energy, with a full width at half maximum of 55 meV, which can be favorably compared with that of the same structure grown on InP substrates.

In addition, a structure with an InP–GaInAsP superlattice as its buffer layer was also studied. The layers are composed as follows:

(1)   500 Å undoped GaAs;
(2)   five periods $InP–Ga_{0.25}In_{0.75}As_{0.5}P_{0.5}$ superlattice with barriers of 50 Å and wells of 50 Å;
(3)   3 μm InP sulfur-doped with $N_D\text{-}N_A = 2 \times 10^{18}\,cm^{-3}$ as an *n*-type confinement layer;
(4)   0.2 μm GaInAsP ($\lambda = 1.3$ μm) as an active layer;
(5)   1 *μm* InP zinc-doped with $N_A\text{-}N_D = 5 \times 10^{17}\,cm^{-3}$ as a *p*-type confinement layer; and
(6)   0.5 μm $Ga_{0.47}In_{0.53}As$ zinc-doped with $N_A\text{-}N_D = 10^{19}\,cm^{-3}$ as a contact layer.

The quality of the structure was first derived from its (4 0 0) double x-ray diffraction pattern (Fig. 15.11). The full width at half maximum (FWHM) of the 400 reflection peak of the epilayer appears only eight times higher than that of the silicon substrate. Transmission electron microscopy (TEM) shows that the superlattice blocking layer prevents most of the dislocations (Fig. 15.12). Since the InP confinement layer seems to have no dislocations, one can conclude that there are less than $10^5\,cm^{-2}$, corresponding to the TEM resolution. The optical quality of the quaternary compound active layer was derived by photoluminescence spectra at $T = 300$ and 4 K, recorded on a structure with GaInAsP covered by only 500 Å InP. The typical room-temperature photoluminescence spectrum of this layer shows the expected wavelength of GaInAsP of 1.3 μm, with FWHM of about 55 meV, which can be favorably compared with that of the same structure grown on an InP substrate. The $T = 4$ K photoluminescence of the structure is shown in Fig. 15.13. Two different peaks corresponding to extrinsic (975 meV) and intrinsic (1010 meV) recombination in the quaternary compound were noted. Their origins were then confirmed by plotting their respective intensities as a function of the excitation power of the laser. The acceptor (e–A^0) related peak saturated, while the other peak linearly increased with the excitation power, as expected from extrinsic and intrinsic recombination.

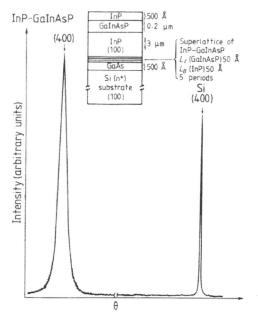

Fig. 15.11. Double x-ray diffraction pattern of GaInAsP–InP DH laser on Si substrate.

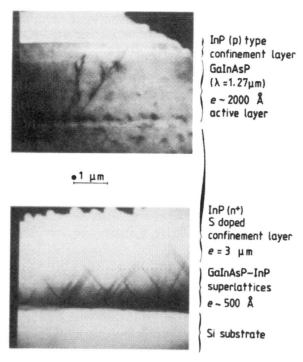

Fig. 15.12. TEM cross section of GaInAsP–InP DH laser on Si substrate.

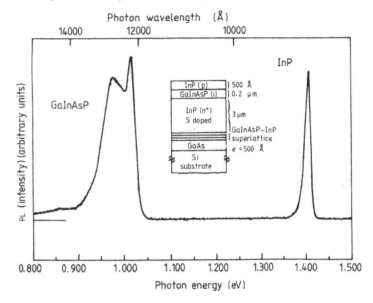

Fig. 15.13. *4 K PL spectrum of GaInAsP–InP DH laser on Si substrate.*

15.5. Applications

15.5.1. *Room-temperature CW operation of a GaInAsP/InP ($\lambda = 1.15$ μm) light-emitting diode on silicon substrate*

The epitaxial structure was grown by MOCVD on (1 0 0) silicon n^+ substrate misoriented 4° off towards (1 1 0). The growth procedure has been described in section 15.4. The structure of the devices includes:

    (a)   1000 Å GaAs–GaInP superlattice undoped as buffer layer;
    (b)   2 μm InP sulfur-doped as *n*-type confinement layer;
    (c)   2000 Å GaInAsP ($\lambda = 1.15$ μm) undoped active layer;
    (d)   500 Å InP zinc-doped as *p*-type confinement layer.

Fig. 15.14 shows the room-temperature photoluminescence spectrum of this layer, as well as its double x-ray diffraction pattern. The photoluminescence (PL) showed a composition of $\lambda = 1.15$ μm as expected. It exhibits a full width at half maximum of about 50 meV. The PL intensity is ten times lower than high-quality GaInAsP of the same composition grown on InP substrate. Furthermore, the FWHM of the (4 0 0) InP reflection peak is only six times that of the Si(4 0 0), demonstrating the good crystallographic quality of the epilayer.

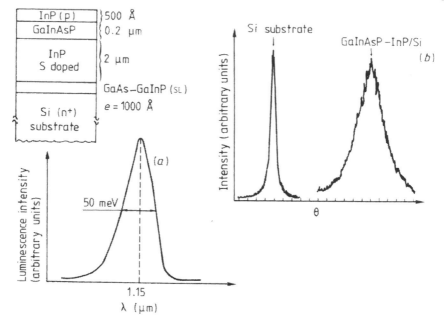

Fig. 15.14. (a) Room-temperature photoluminescence spectrum of a double heterostructure InP/GaInAsP (λ = 1.15 μm) grown on silicon substrate by MOCVD. (b) X-ray double diffraction pattern of an InP/GaInAsP (λ = 1.15 μm) heterostructure grown on silicon substrate by MOCVD.

Using this structure (Fig. 15.15(*a*)), Razeghi et al. [1988] have prepared stripes of 1 μm width by using photolithography followed by chemical etching. The stripe-to-stripe spacing is about 300 μm, as shown in Fig. 15.15(*b*).

In the second growth step, the stripes were covered with 1 μm InP Zn-doped as a *p*-type confinement layer and 0.5 μm thick $Ga_{0.47}In_{0.53}As$ Zn-doped as *p*-type contact layer (see Fig. 15.15(*c*)).

The diodes were prepared as follows: Pt and gold metallization are used for *p*-type contacts and gold has been used for *n*-type contacts on Si substrate (Fig. 15.15(*d*)). Next, proton implantation was used to localize the injection current through the stripe active layers, as shown in Fig. 15.15(*e*)). Their length is 50 μm.

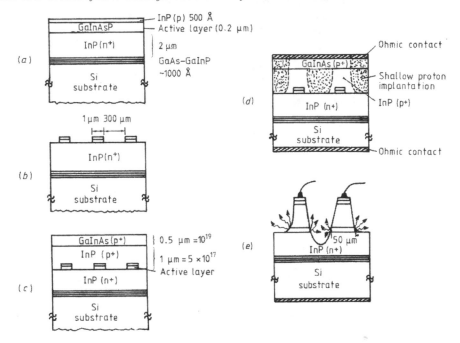

Fig. 15.15. Successive technological steps for fabricating a light-emitting diode on a Si substrate.

After bonding the contacts, an infrared camera was used to visualize the electroluminescence (Fig. 15.16). This shows light coming out from both diode sides. These diodes were placed under continuous wave operation for 24 h, with an injection current of 200 mA. No degradation was observed.

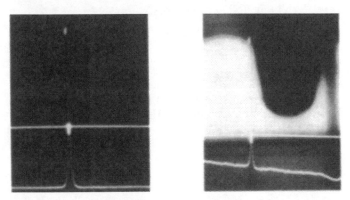

Fig. 15.16. Electroluminescence of the InP–GaInAsP ($\lambda = 1.15$ μm) heterostructure grown on a Si substrate.

Fig. 15.17 shows the output power against injection current in continuous wave operation of this stripe-buried heterostructure GaInAsP/InP on an Si substrate.

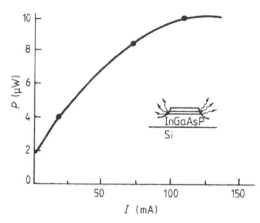

Fig. 15.17. *Output power against injected current of a stripe-buried heterostructure InGaAsP on Si under CW operation.*

15.5.2. GaInAsP/InP double heterostructure laser emitting at 1.3 μm on silicon substrate

The structure of this device is as follows:

(a)   Five periods InP/$Ga_{0.25}In_{0.75}As_{0.5}P_{0.5}$ superlattice with an InP thickness of 50 Å and a GaInAsP thickness of 50 Å, for elimination of mismatch dislocations;

(b)   3 μm InP S-doped with N_D-$N_A = 10^{18}$ cm^{-3} as an *n*-type confinement layer;

(c)   0.2 μm $Ga_{0.25}In_{0.75}As_{0.5}P_{0.5}$ undoped active layer;

(d)   1 μm InP Zn-doped $(N_A$-$N_D \simeq 5 \times 10^{17}$ cm$^{-3})$ as a *p*-type confinement layer;

(e)   0.5 μm $Ga_{0.47}In_{0.53}As$ Zn-doped $(N_A$-$N_D \simeq 10^{19}$ cm$^{-3})$ as cap layer.

Fig. 15.18 shows the electrochemical polaron profile of this structure.

Transmission electron microscopy (TEM) shows the elimination of most dislocations by superlattices of GaInAsP–InP at the interfaces (Fig. 15.12). Photoluminescence, photoluminescence excitation and double x-ray diffraction show high-quality GaInAsP material. Fig. 15.19 shows the SIMS spectrum of this structure.

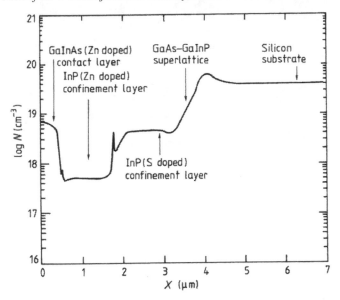

Fig. 15.18. Electrochemical polaron profile of a GaInAsP–InP DH laser on Si.

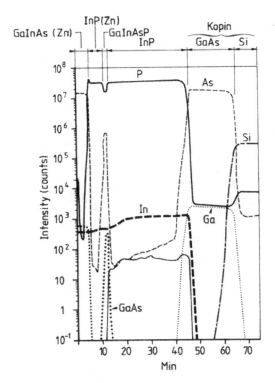

Fig. 15.19. SIMS spectrum of a GaInAsP–InP DH laser on a Si substrate.

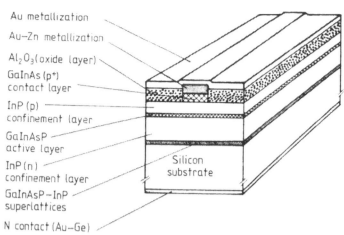

Au metallization
Au–Zn metallization
Al₂O₃(oxide layer)
GaInAs (p⁺) contact layer
InP (p) confinement layer
GaInAsP active layer
InP (n) confinement layer
GaInAsP–InP superlattices
N contact (Au–Ge)
Silicon substrate

Fig. 15.20. Oxide-defined stripe geometry of a GaInAsP–InP DH laser on a Si substrate.

Gain-guided oxide-defined stripe-geometry lasers were fabricated as follows (see Fig. 15.20). The injected current profile is defined by a self-aligned technique: Au–Zn is evaporated on the GaInAs cap layer through a photoresist mask, a selective etching is then done of the contact layer and subsequently Al₂O₃ is deposited, and lift-off is realized. After annealing Au–Zn (at $T = 420\,°C$ for 10 s), the silicon substrate is lapped down to 50 μm thickness and Au–Ge is evaporated. The wafer is then cleaved in bars along the (1 1 0) direction, which is a cleavage plane of InP but not of the Si substrate.

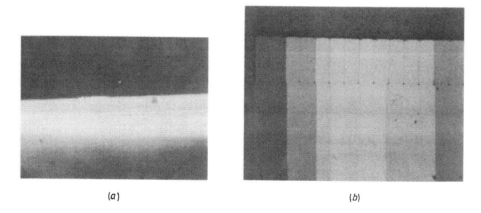

(a)   (b)

Fig. 15.21. (a) Top view of the wafer GaInAsP–InP DH laser on Si substrate cleaving in bars, (b) The cleaved facet of the stripe laser region.

Fig. 15.21(a) shows a typical top view of a wafer cleaving in bars. The cleaved facet of the stripe laser region is presented in Fig. 15.21(b). The interface between the InP and the Si substrate is delineated on this photograph by a continuous line due to different crystal parameters.

Fig. 15.22 shows the light–current characteristic in pulse operation of a laser with a cavity length of 250 μm and a stripe width of 12 μm. Measurements show a threshold current $I_{th} = 510$ mA with a differential quantum efficiency $\eta_D = 10\%$ per face and an output power up to 20 mW per facet. Threshold current density for these devices is about $J_s = 10$ kA cm^{-2}. The improvement in the quality of the epitaxial layers and laser structures will lead to a lower threshold current density and a higher differential efficiency.

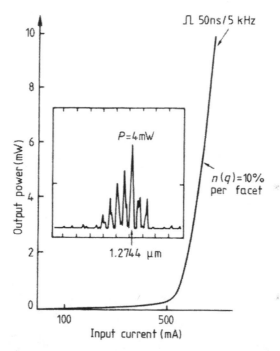

Fig. 15.22. Typical characteristics of light output–current input of a GaInAsP–InP DH laser on Si. Inset: spectral output of a gain-guided oxide-defined stripe geometry GaInAsP–InP DH laser on a Si substrate at a power output level of 4 mW.

The inset of Fig. 15.22 of the spectral output of these devices shows power output levels of 4 mW using driving pulses of 50 ns at 5 kHz. The behavior of the lasing spectra as a function of output power is similar to that obtained on InP substrates having a gain-guided structure.

Preliminary aging tests were run in pulse conditions at room temperature. After an equivalent CW aging time of 80 s (under driving pulses of 50 ns at 5 kHz), an increase of threshold current $\Delta I_{th}/I_{th} = 7\%$ is measured at an output power of 3 mW, with no degradation of the differential quantum efficiency. Such a rapid increase of threshold current may be due to the mismatch dislocation between the GaInAsP active layer and InP confinement layer. These results are very promising compared to aging tests of GaAs lasers on silicon [Van der Ziel et al. 1987].

15.5.3. CW operation of a $Ga_{0.25}In_{0.75}As_{0.5}P_{0.5}/InP$ BRS laser on silicon substrate

Fig. 15.23 shows a schematic diagram of a $Ga_{0.25}In_{0.75}As_{0.5}P_{0.5}$–InP buried ridge structure (BRS) laser on Si substrate. The structure of devices includes:

(a)   1 µm undoped GaAs as a buffer layer;

(b)   five periods $Ga_{0.47}In_{0.53}As$–InP superlattice with an InP thickness of 100 Å and a GaInAs thickness of 100 Å for decreasing the mismatch dislocations;

(c)   3 µm InP S-doped with N_D-$N_A = 10^{18}$ cm^{-3} as an *n*-type confinement layer;

(d)   0.2 µm undoped $Ga_{0.25}In_{0.75}As_{0.5}P_{0.5}$ active layer;

(e)   0.1 µm Zn-doped InP with N_A-$N_D = 2 \times 10^{17}$ cm^{-3} as a *p*-type confinement layer.

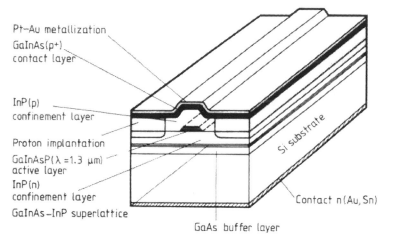

Fig. 15.23. Schematic diagram of a GaInAsP–InP BRS laser on a Si substrate.

The growth temperature was 550 °C, with a growth rate of 100 Å·min^{-1} for InP, 150 Å·min^{-1} for GaInAsP and 200 Å·min^{-1} for GaInAs.

The double x-ray diffraction pattern of this structure is shown in Fig. 15.24. The FWHM of the (4 0 0) InP–GaInAsP reflection peak is only about 250 arcsec, evidence for the good crystallinity of the InP–GaInAsP layer grown on Si substrate.

Next, the stripes of 2 µm width and stripe–stripe spacing of 350 µm were etched in the 1000 Å InP + 2000 Å GaInAsP active layer through a photolithographic resist mask, followed by chemical etching.

In the next step, after removing the resist mask, the stripes were then covered by a 1 µm thick Zn-doped InP confinement layer (with N_A-$N_D = 5 \times 10^{17}$ cm^{-3}) and a 0.5 µm Zn-doped $Ga_{0.47}In_{0.53}As$ contact layer with N_A-$N_D = 10^{19}$ cm^{-3}. In order to localize the injection in the active region, a shallow proton implantation was realized through a 5 µm width photoresist mask after the metallization of the contacts.

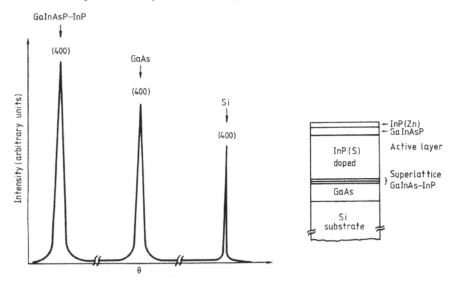

Fig. 15.24. Double x-ray diffraction of a GaInAsP–InP BRS laser on Si.

Pt and Au are used for the *p*-type contact, and Au/Sn is used for the *n*-type contact on the Si substrate. The wafer is then cleaved in bars along the (1 1 0) direction which is a cleavage plane of InP but not of the Si substrate.

After cleaving and scribing into chips 200 μm long and 350 μm wide, the diodes are bonded with In to an Au-plated copper heat sink.

Fig. 15.25 shows the light–current characterization under CW operation of a BRS laser with a cavity length of 200 μm and a stripe width of 2 μm. Threshold current $I_{th} = 80$ mA with $\eta(q) = 16\%$ per face, and an output power of 20 mW under CW operation at room temperature have been measured. The inset of Fig. 15.25 shows the spectral output of these devices at power output levels of 2 mW under CW operation.

On these lasers, a value of $T_0 = 61$ °C has been measured. The best threshold current obtained on this structure is $I_{th} = 45$ mA which corresponds to a threshold current density $J_{th} \leq 4$ kA·cm^{-2} (see Fig. 15.26).

A preliminary aging test has been processed in CW operation at room temperature, with an injection current of 94 mA corresponding to an output power of 2 mW per facet. The inset of Fig. 15.26 shows the typical driving current as a function of the aging time for this device. An increase of the injected current of less than 10% is observed after 10 hour aging time without any decrease in differential quantum efficiency.

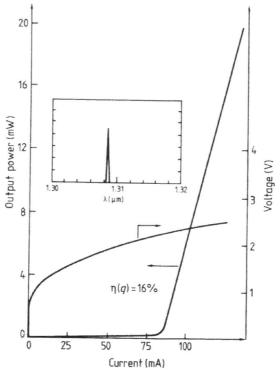

Fig. 15.25. *Light output–current input of a GaInAsP–InP BRS on a Si substrate under CW operation. Inset: spectral output of this BRS laser at an output power of 2 mW.*

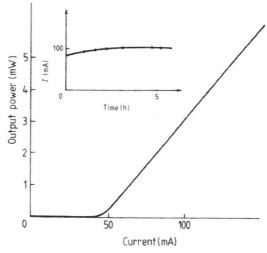

Fig. 15.26. *Light output–current input before and after ageing test. Inset: typical aging characteristics of a GaInAsP–InP BRS laser on Si substrate.*

15.5.4. *InGaAs–InP MQW on Silicon substrate*

The use of GaInAs–InP multiquantum well (MQW) layers for optical waveguides has been a subject of much interest, especially in regard to excitonic dielectric and absorption properties. Fig. 15.27 shows the scanning electron transmission micrograph of the GaInAs–InP MQW layers for 1.5 μm waveguides grown by MOCVD on a Si substrate. The structure includes:

(a)	1 μm GaAs on Si substrate as a buffer layer [Salerno et al. 1988];
(b)	1 μm undoped InP as a confinement layer;
(c)	50 GaInAs quantum well layers 30 Å thick, separated by 300 Å InP thick barriers;
(d)	1.5 μm InP undoped as a confinement layer;
(e)	1 μm InP Zn-doped;
(f)	0.5 μm GaInAs Zn-doped as a contact layer.

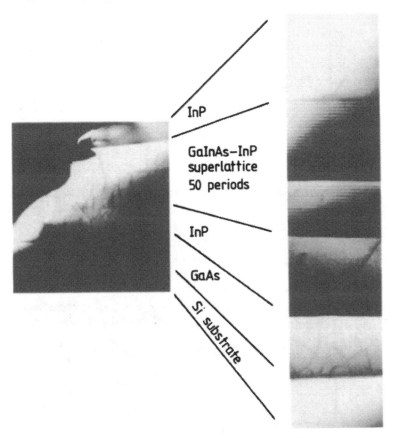

Fig. 15.27. *Scanning electron transmission micrograph of a GaInAs–InP MQW layer for a 1.5 μm waveguide grown by MOCVD on a Si substrate.*

Most of the dislocations are eliminated by superlattices. The excellent quality of the epilayers is confirmed by the single diffraction pattern of the structure: the $K\alpha_1$ and $K\alpha_2$ components of the $(4\,0\,0)$ reflection peak of Si, GaAs and InP/$Ga_{0.47}In_{0.53}$As are clearly separated. Near the InP/$Ga_{0.47}In_{0.53}$As related peak, satellites due to the superperiodicity of the 55-period superlattice are evidenced. They establish the perfect thickness control of the barriers and the wells in the superlattices (see Fig. 15.28).

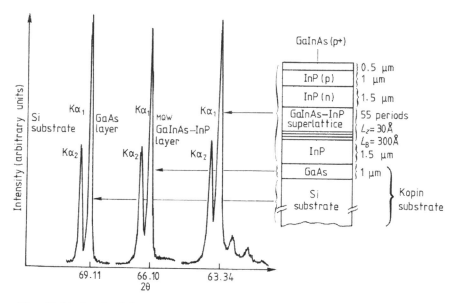

Fig. 15.28. Single diffraction pattern of a GaInAs–InP MQW layer for a 1.5 μm waveguide grown by MOCVD on a Si substrate.

15.5.5. GaInAs–InP PIN photodetector

Razeghi et al. [1988] fabricated a GaInAs–InP PIN photodetector on Si. The growth was carried out in two separate steps. First, 1 μm undoped GaAs with $N_D - N_A = 10^{14}$ cm^{-3} was grown on the $(1\,0\,0)$ 4° off η-type Si substrate [Salerno et al. 1988]. Fig. 15.29 shows the transmission electron microscope cross section of the GaAs on the Si substrate used for this structure; most of the dislocations are stopped at the interface.

Then the InP–GaInAs–InP undoped layers were deposited on the GaAs/Si layer as follows:

(a)   an InP buffer layer 1 μm thick;
(b)   a GaInAs absorption layer 4 μm thick;
(c)   an InP window layer 1 μm thick (in order to enhance the responsivity in the near-infrared spectrum beyond 0.9 μm).

The background impurity concentration deduced from a polaron electrochemical profile was found to be 2×10^{15} and 8×10^{14} cm^{-3} for GaInAs and InP, respectively.

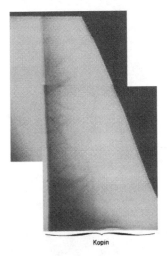

Fig. 15.29. Scanning electron microscope pattern of a GaAs layer on a Si substrate (Kopin material).

The growth temperature is 520 °C and the growth pressure 76 Torr. Fig. 15.30 shows the double x-ray diffraction pattern of this GaInAs–InP PIN photodetector on a Si substrate, evidencing the good crystallinity of the GaInAs–InP–GaAs layer grown on the Si substrate.

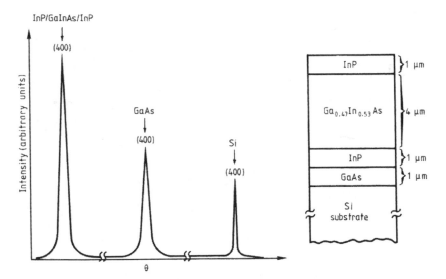

Fig. 15.30. Double x-ray diffraction pattern of InP—GaInAs—InP—GaAs on silicon substrate.

The *p–n* junction was formed next by Zn diffusion using the semiclosed box technique, it being located through the InP window in the GaInAs layer (see Fig. 15.31), using a SiO₂ mask.

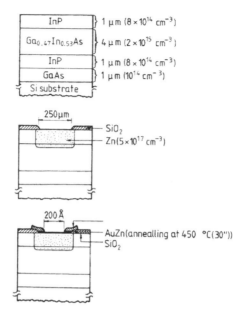

Fig. 15.31. Successive technological steps for fabricating a GaInAs–InP PIN photodiode on a Si substrate.

The ohmic contacts are made by sputtered Au–Zn alloyed at 450 °C for 30 s and then covered with Cr–Au on the *p*-side, and by sputtered Au on the *n*-side (Si substrate) (Fig. 15.31).

Fig. 15.32 shows a photograph of the GaInAs–InP PIN photodiode on Si substrate. Diodes with a diameter of 250 μm have been fabricated. Table 15.2 shows the characteristics of these devices compared with those grown on InP substrates.

	PIN GaInAs–InP on InP	PIN GaInAs–InP on Si
I_d (− 1 V)	8 nA	0.5 μA
I_d (− 10 V)	200 nA	100 μA
Capacitance	5 pF	5 pF
Responsivity	with	without
$R = \eta \dfrac{q\lambda}{hC} = \eta\lambda / 1.24$	anti-reflection coating	anti-reflection coating
at λ = 0.8 μm	0.12	0.27
at λ = 1.3 μm	0.72	0.71
at λ = 1.55 μm	0.77	0.75

Table 15.2. Characteristics of GaInAs–InP PIN photodetector on Si and InP.

|←   400 µm   →|

Fig. 15.32. Photograph of GaInAs–InP PIN photodiode on a Si substrate.

Fig. 15.33 shows the capacitance–voltage characteristic of the device. The capacitance of the device is low, evidencing the low carrier concentration of the GaInAs layer. For a good receiver performance, capacitance must be low. For PINFET optical receivers the capacitance should be ≤ 0.50 pF, a figure which could be obtained by reducing the junction area. The thickness of the active region of a PIN photodiode is the result of a trade-off between the competing effects of fast transit time requiring a narrow depletion region and the combination of quantum efficiency and low capacitance which requires a wide depletion region. The thickness of the GaInAs active layer of this device is 4 µm, so the responsivity and quantum efficiency are high. Quantum efficiency is the first consideration in a photodiode (i.e. the effective collection of photogenerated carriers by a depleted volume).

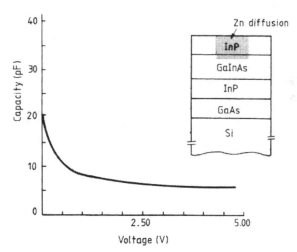

Fig. 15.33. The C(V) characteristics of a GaInAs–InP PIN photo-diode on a Si substrate.

The dark current of this device is high. Usually the reverse current of the device in the darkness arises from two sources: (i) a generation current in the depleted volume and minority carrier collection at the edge of the depletion region; and (ii) surface leakage.

Current generated in the diode can be kept low by using high-quality dislocation- and defect-free material and using care in processing to avoid introducing crystalline defects. In this structure the GaAs and InP buffer layers have very low background carrier concentration (10^{14} cm^{-3}), so the depletion region extends down to the interface which contains a lot of defects and dislocations due to the lattice mismatch between InP, GaAs and the Si substrate. To reduce the dark current one has to dope GaAs and InP buffer layers by up to 10^{16} cm^{-3}.

References

Akiyama, M., Kawarada, Y., and Kaminishi, K. *J. Cryst. Growth* **68** 21, 1984.

Akiyama, M. *J. Cryst. Growth* 77 490, 1986.

Aksum, M.I., Morloc, H., Lester, L.F., Duh, K.H.G., Smith, P.M., Chao, P.C., Longabone, M., and Erickson, L.P. *Appl. Phys. Lett.* **49** 24, 1986.

Aspnes and Ihm *Heteroepitaxy on Silicon* ed Fan, J.C., and Ponte, J.M. (Pittsburgh: Materials Research Society) p 45, 1987

Bedair, S.M. *Heterostructures on Si II*, ed. Fan, J.C., 1987.

Christou, A., Wilkins, B.R., and Tsang, W.T. *Electron. Lett.* **21** 9, 1985.

Dupuis, R.D., Van der Ziel, J.P., Logan, R.A., Brown, J.M. and Pinzone, C.J. *Appl. Phys. Lett.* **50** 407, 1987.

Dupuis, R.D. *Proc. 4th Int. Conf. on Metalorganic Chemical Vapor Phase Epitaxy, Hakone, Japan,* 1988.

El-Masry, N., Hamaguchi, N., Tarn, J.C., Kasan, N., Humphreys, T.P., Moore, D., and Hayashi, I. *Japan. J. Appl. Phys. Suppl.* **19** 23, 1980.

Hull, R., Fischer-Collrie, A., Rasner, S.J., Koch, S.M, and Harris, J.S. *Appl. Phys. Lett.* **51** 1723, 1987.

Kroemer, H. *Mat. Res. Soc. Symp. Proc.* **67** 3, 1986.

Nishi, S., Inomata, H., Akiyama, M., and Kaminishi, K. 1985 *Japan. J. Appl. Phys.* **24** 6, 1985.

Ota, Y. *J. Electrochem. Soc.* **124** 1795, 1977

Razeghi, M., Maurel, P.H., Defour, M., Omnes, F., and Acher, O. *Proc. 14th European Conf. on Optical Communication, Brighton,* 1988.

Razeghi, M., Omnes, F., Defour, M., and Maurel, P.H. *Appl. Phys. Lett.* **52** 209, 1988.

Razeghi, M. *Proc. Solid State Devices and Materials Conf., Tokyo* p 363, 1988.

Razeghi, M., Maurel, P.H., Omnes, F., and Defour, M. *J. Appl. Phys.* **63** 4511, 1988.

Razeghi, M., Defour, M., Omnes, F., Maurel, P.H., and Chazelas, J. *Appl. Phys. Lett.* **53** 725, 1988.

Razeghi, M., Defour, M., Omnes, F., Maurel, P.H., Bigan, E., Acher, O., Nagle, J., and Brillouet, F. *Proc. 4th Int. Conf. on Metalorganic Vapor Phase Epitaxy, Hakone, Japan,* 1988.

Razeghi, M., Blondeau, R., Defour, M., Omnes, F., and Maurel, P.H. *Appl. Phys. Lett.* **53** 854, 1988.

Razeghi, M., Defour, M., Omnes, F., Nagle, J., Maurel, P.H., Acher, O., and Mijuin, D. *Mtg of Materials Research Society, Reno,* 1988.
Salerno, J.P., Lee, J.W., McCullough, R.G., and Fan, J.C. *Proc. MRS Spring Meeting, April 1988, Bally's Reno, Nevada,* 1988.
Sakamoto, T. and Hashiguchi, G. *Japan. J. Appl. Phys.* **25** L57, 1986.
Sharma, L. and Purohit, R.K. *Semiconductor-heterojunctions* (Oxford: Pergamon) ch 4, 1974.
Sheldon, P., Jones, K.M., Hayes, R.E., Tsaur, B.Y., and Fan, J.C. *Appl. Phys. Lett.* **45** 3, 1984.
Soga, T., Hattori, S., Sakai, S., Takayasu, M., and Umeno, M.*J. Appl. Phys.* **57** 10, 1985.
Thompson, G.B *Physics of Semiconductor Laser Devices* (Chichester: Wiley) p 51, 1980.
Van der Ziel, J.P., Dupuis, R.D., Logan, R.A., and Pinzone, C.J. *Appl. Phys. Lett.* **51** 2, 1987.
Van der Ziel, J.P., Dupuis, R.D., Logan, R.A., and Pinzone, C.J. *Appl. Phys. Lett.* **51** 89, 1987.
Wang, W.I. *J. Vac. Sci. Technol.* B **3** 2, 1985.

16. Optoelectronic Devices Based on Quantum Structures

16.1. Introduction
16.2. GaAs- and InP-based quantum well infrared photodetectors (QWIP)
    16.2.1. GaAs-based QWIPs
    16.2.2. InP-based QWIPs
    16.2.3. InP-based nanopillar QWIPs
16.3. Self-assembled quantum dots, and quantum dot–based photodetectors
    16.3.1. Growth of InAs QDs on InP
    16.3.2. The quantum dot-in-well infrared photodetector (QDWIP)
    16.3.3. QDWIP device performance
    16.3.4. The QDIP structure
    16.3.5. Gain in QDIP and QDWIP
    16.3.6. QDIP performance
16.4. Quantum dot lasers
16.5. InP based quantum cascade lasers (QCLs)
    16.5.1. Quantum cascade laser growth
    16.5.2. Short wavelength QCLs
    16.5.3. Middle wavelength QCLs
    16.5.4. Long wavelength QCLs
    16.5.5. Photonic crystal distributed feedback QCLs

16.1. Introduction

This chapter covers some of the most recent advanced work on GaAs and InP based devices. In particular, exciting new quantum devices are discussed, including brief reviews of GaAs and InP based quantum well infrared photodetectors (QWIPs), quantum dot infrared photodetector (QDIPs), quantum dot-in-well infrared photodetectors (QDWIPs), quantum dot lasers, and quantum cascade lasers (QCLs).

GaAs based QWIPs were discussed previously in Chapter 7. However, there have been many recent improvements in GaAs based QWIPs including the development of long wavelength (~10 μm) QWIPs, and aluminum-free QWIPs. We also focus on more recent results in the InP material systems. InP based QWIPs possess several important advantages over GaAs based QWIPs.

Using the same InP material system, but by using Stranski-Krastanov growth, it is also possible to grow self-assembled quantum dots which can be integrated into

QWIP devices, either by simply replacing the quantum well (QDIPs), or by placing quantum dots inside wells (QDWIPs) and then growing the barrier layer[Lim et al. 2007]. The QDWIPs are also sometimes referred to as dot in well (DWELL) devices. The QDIP and QDWIP devices all use the physics of 3D quantum confinement, and open up exciting new quantum physics applications that can be exploited to realize sophisticated modern high performance infrared photodetectors.

Another exciting application of quantum mechanics in the context of quantum dots (QDs) is realized when one notes that electron and hole energies are both quantized inside a QD. This opens the way for using the QD as a localized light emitting center. The photon can be emitted because when an electron and a hole are injected into the confined area, radiative recombination can take place. GaAs and InP based QD-lasers have been built for a number of years and have proven to be a successful technology. QD lasers now cover the important 1.3 to 1.5 µm range, crucial to communications and are beginning to become competitive as a replacement for existing quantum well lasers.

Most QW and QD lasers operate in the infrared range, and use the interband transitions but it has also recently become possible to realize intersubband lasers based on InP. These quantum cascade lasers (QCLs) are intersubband devices that rely on precise control of the growth of complex multi-layer superlattices to create an artificial band structure within the conduction band of the material. These mini-bands can be used to realize intersubband transitions that can be harnessed to create high power infrared lasers spanning the wavelength range from 3.5 µm all the way out to the terahertz (>50 µm).

16.2. GaAs- and InP-based quantum well infrared photodetectors (QWIP)

We recall from Chapter 7 the basic mechanisms and properties which characterize a quantum well infrared photodetector (QWIP), and then we will discuss why it should matter on what type of material and with what structure the device is made. First we note that what is common to all QWIPs is the basic photoconductive mechanism which is the absorption of a photon by charges sitting in the ground states of the valence or conduction bands. The QWIP is *p*-type, if it is acceptor doped, and *n*-type if it is donor doped. In an *n*-type QWIP, the electron absorbs the photon and undergoes a transition to an excited state which is preferably very close to the continuum or in the continuum as shown in Fig. 16.1

In Fig. 16.1 the transition A to B which also exists as a thermal transition, is the one normally used for photodetection. If the excited state is too far from the continuum edge, then the excited charge cannot exit and conduct easily and this is clearly counterproductive because large bias has to be applied which also raises the dark current and the noise. The absorption wavelength, which determines in what range the detector is operating is determined mainly by the width of the quantum well, and to some extend by the height of the barrier, as we have seen before from Chapter 7.

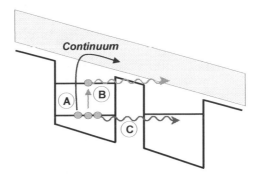

Fig. 16.1. Dark and photocurrent mechanisms in QWIPs and QDIPs. (A) Thermionic emission (B) Thermally assisted tunneling (C) Sequential tunneling. The line joining A to B can also be photon assisted, and is used for photodetection.

The challenge is to go to long wavelengths, and to beat the noise which in turn is determined by the magnitude of the dark current. The challenge is best summarized by the responsivity formula. Responsivity R is defined as the ratio of photocurrent to photonic power and can be written as

Eq. (16.1) $$R = g \frac{e(1 - \exp(-\alpha L))}{\hbar \omega} \{ \frac{v_0 e^{-E_{eff}/kT}}{v_{eg} + v_0 e^{-E_{eff}/kT}} \}$$

where e is the magnitude of the charge, α is the absorption coefficient, L the width of the active part of the device, ω is the photon frequency, g is the gain given by

Eq. (16.2) $$g = \frac{\tau_{lifetime}}{\tau_{transit}} = \frac{\text{carrier lifetime}}{\text{carrier transit time}}$$

and v_0 is an attempt frequency for carrier thermal/tunneling escape out of the excited states with effective activation energy E_{eff}. One can see that good performance in the first place; means a high absorption coefficient. The absorption efficiency normally depends on the polarization direction of the incoming light. Most focal plane array cameras operate in normal incidence mode so one would normally have to use a grating coupler to mix the incoming polarization to include the p-mode, which is the one which gives rise to the strong intersubband absorption. In the second place, we need high gain g which means long recombination times and high mobilities. In the third place it is desirable to operate the detector at low bias, so that the carrier escape barrier should be a minimum.

Now consider the material class. GaAs has the benefit of mature material growth with proven device processing technologies. In this section we will be reporting on some important new QWIP results using GaAs substrates, which use hole absorption (*p*-QWIPs) and demonstrated the importance of spin orbit coupling in determining the intersubband absorption spectrum. Later we focus on more recent results in the InP material systems. These possess several advantages.

16.2.1. GaAs-based QWIPs

QWIPs based on GaAs technology were extensively discussed in Chapter 7. More recent work on *n*-QWIPs include in particular the contributions by Jelen et al. [1997] with GaAs/Ga$_{0.51}$In$_{0.49}$P using gas phase epitaxy on GaAs. Noteworthy in this technology is the long wavelength detection achieved (~10 μm) and the aluminum-free nature of the device which thus avoids aluminum related complications such as DX centers. Remember that DX centers have a catastrophic effect on photocurrents because they lead to long lived excited states which survive for minutes and hours, and only disappear when the material is heated up again, and the defects anneal out. Basically it means that the excited carrier cannot recombine, because there is an energy barrier to go back to the ground state configuration. Working without aluminum is therefore most helpful in this respect. Also we may note the important work of [Hoff et al. 1996] this time using MOCVD growth of GaAs–Ga$_{1-x}$In$_x$P but with *p*-doping and using zinc as a dopant. The *p*-doping has the advantage of normal incidence absorption, and in principle does not need extra light coupling schemes. This would make the device fabrication easier and cheaper. Because the absorption now takes place in the valence subbands, the spin-orbit mixing of the valence bands deserves special attention. Indeed it turns out that one can only understand the measured intersubband absorption strength by including the spin-orbit mixing of the valence bands in addition to the QW induced admixtures. This has been shown in the work of Hoff et al. [1996]. The authors show that the spin-orbit mixing calculated in $k \cdot p$ theory explains their absorption data [Hoff et al. 1996]. Another type of Al-free device grown on GaAs was made in the work of Jelen et al. [1996] with *n*-QWIPs. These authors demonstrate quantum well infrared photodetectors based on a GaAs–Ga$_{0.51}$In$_{0.49}$P superlattice structure grown by gas-source molecular beam epitaxy. Wafers were grown with varying well widths. Wells of 40, 65, and 75 Å resulted in peak detection wavelengths of 10.4, 12.8, and 13.3 μm with a cutoff wavelength of 13.5, 15, and 15.5 μm, respectively. The measured and calculated absorption coefficients reach value of order 10^3 cm^{-1}. Noteworthy is the very low dark current observed which for the same parameter range, is lower than the corresponding device studied by Levine [Levine 1993].

16.2.2. InP-based QWIPs

InP based QWIPs possess several important advantages over GaAs based QWIPs. First, GaInAs–InP QWIPs use a binary barrier (InP), while GaAs–AlGaAs QWIPs use a ternary barrier (AlGaAs) which has necessarily a disordered alloy structure. Thus binary layers are ordered and have an inherently lower defect density than ternary layers which are necessarily atomically disordered. This in turn translates into lower dark current, because there are fewer defect assisted leakage paths, especially at cryogenic temperatures when these dominate. Secondly, GaInAs–InP constitutes an aluminum-free material system, which usually has less oxidation, higher device reliability, simpler device processing, no DX-centers, and reduced surface recombination. Thirdly it should be mentioned that, both GaInAs and InP, have higher electron mobilities than GaAs and AlGaAs. The effective mass of an electron is lower for GaInAs (0.041 m$_0$) than for GaAs (0.067 m$_0$). The effective mass enters the mobility and this gives rise to a higher photoconductive gain for GaInAs–InP QWIPs. Fourthly, GaInAs–InP QWIPs grown on silicon substrates, show better device performance than GaAs–AlGaAs QWIPs grown on silicon substrates. This makes the GaInAs–InP QWIP better material systems on which to build future monolithic focal plane arrays (FPAs) integrated with Si-based readout integrated circuits (ROICs). In this chapter, we report the first demonstration of LWIR FPAs based on Al-free GaInAs–InP QWIPs.

An excellent example of an InP based QWIP using MOCVD growth is provided by the work Jiang et al. [2003], who made an FPA using GaInAs–InP, with an NEDT of 29 mK at 70 K, at a wavelength of 8.5 µm. The NEDT is a measure of camera performance and is defined below

The noise equivalent difference temperature (NEDT) is a measure of FPA performance that gives us the minimum resolvable temperature difference when viewing a scene. It is defined as

$$\text{Eq. (16.3)} \qquad NEDT = \frac{1}{C_d(\Delta\lambda)} \frac{\sqrt{A\Delta f}}{D^*} = \frac{1}{\int_{\Delta\lambda} d\lambda \frac{dR}{d\lambda}} \frac{\sqrt{A\Delta f}}{D^*}$$

where a thermal variation in an object of ΔT gives a change in the blackbody emittance of $C_d(\Delta\lambda)\Delta T$ over a spectral range $\Delta\lambda$, A is the area, and $dR/d\lambda$ is the emittance.

16.2.3. InP-based nanopillar QWIPs

With the same material class, exciting innovation was made by Gin et al. [2005] who fabricated 40 nm radius nanopillar QWIPs, using e-beam lithography opening the way for truly nano-engineered photodetection structure. The nanopillar structure is shown in Fig. 16.2. Because the active absorbing area is smaller, the nanopillar photodetector has a smaller responsivity, but it has the potential of being integrated into sensors which exploit the space within the nanopillar where active polymers with selective binding sites or even energy transferring molecules or small quantum dots can be inserted. These could help gather the light arriving between the columns and transfer the exciton to the columns and produce carriers.

Also we observe that importantly, nanopillar structures no longer have the planar symmetry of a normal "flat" layered structured QWIP, and exhibit a higher grating free normal incidence light coupling. In the device studied by Gin [Gin et al. 2005], the space between the columns was filled with Polyimide but one could use a more active layer.

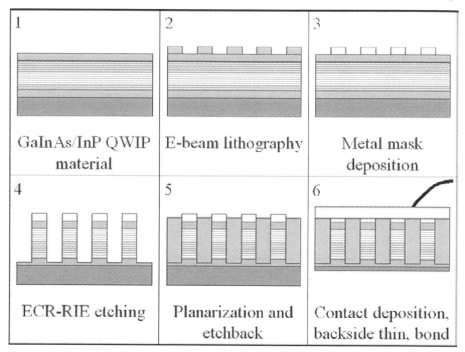

Fig. 16.2. Schematic diagram of nanopillar device fabrication

16.3. Self-assembled quantum dots, and quantum dot–based photodetectors

One of the areas where the MOCVD growth of InP-based materials has seen much recent innovation is in the creation of quantum dots. It is possible to use strained layers to grow self-assembled quantum dots by MOCVD using the Stranski–Krastanov growth technique. These self-assembled quantum dots can then be used to realize novel quantum dot (QD) based devices. QD-based infrared detectors have been the subject of intensive research because they are expected to eventually outperform current quantum well infrared photodetectors (QWIPs) [Levine 1993, Gunapala et al. 2005], due to their i) intrinsic sensitivity to normal incidence light, ii) longer lifetime of the photo-excited electrons due to the reduced recombination rate associated with a multi-phonon relaxation step, and iii) lower dark and noise currents [Ryzhii 1996]. In particular, the lower dark currents enable higher operating temperatures. Achieving higher operating temperatures will reduce the cost and complexity of detector and imaging systems by reducing the cooling requirements normally associated with QWIP-based detector systems running at cryogenic temperatures. Pure QD-based detectors, called quantum dot infrared photodetectors (QDIPs), have been extensively studied, with many showing promise for high performance [Szafraniec et al. 2003, Zhang et al. 2005,

Chakrabarti et al. 2005] however, recent work has also shifted towards quantum dot–quantum well combination devices such as the quantum dot-in-well infrared photodetectors (QDWIPs) [Krishna et al. 2005, Lu et al. 2007, Varley et al. 2007]. In this section we consider both QDIPs and QDWIPs.

16.3.1. Growth of InAs QDs on InP

When the InAs is grown on top of the InAlAs, there is a 3.5% lattice mismatch, which creates strain in the material. Beyond two monolayers the system must reconstruct itself in order to reduce the strain energy, and it does so by spontaneously transforming from flat epitaxial layers to small islands. This phenomenon is the basis of the Stranski-Krastanov growth mode.

The Stranski-Krastanov growth mode occurs when growing lattice mismatched materials where the quantum dot (QD) material is grown on a substrate or matrix layer with smaller lattice constant. The first few deposited layers of the QD material grow in a flat, layer-by-layer fashion. This flat layer is called the wetting layer. Since the QD material is lattice mismatched, the wetting layer is pseudomorphically strained and the strain builds up with increasing material deposition. Beyond a certain critical thickness the wetting layer spontaneously reorganizes, and continued deposition of material results in the growth of 3D dot features on top of the thin wetting layer. This process is illustrated in Fig. 16.3.

Fig. 16.3. Illustration of the Stranski-Krastanov growth mode.

This growth process can be understood by considering the interplay between the surface, interface, and strain energies in this situation [Ryzhii 1996, Lim 2007b]. Initially the sum of the epilayer surface energy and the interface energy is lower than the surface energy of the matrix, therefore it is favorable to have layer-by-layer growth of the QD material. After several monolayers of deposition, the buildup of strain energy makes it no longer favorable to have flat, layer-by-layer growth. With this built-up energy the material needs to relieve the strain and does so by forming 3D structures—or quantum dots.

The key factor here for optoelectronic devices is that the relaxation process in Stranski-Krastanov growth can be controlled to produce defect-free (also called coherent) QDs. This coherent relaxation was first noted in the mid-80s [Goldstein et al. 1985]. Since the dot material is experiencing a compressive strain, when the coherent relaxation takes place the lattice constant expands toward the dot edges, as illustrated in Fig. 16.4. This coherent relaxation means that the QDs grown by Stranski-Krastanov mode do not have the defects that reduce the 3D confinement in other QD fabrication methods.

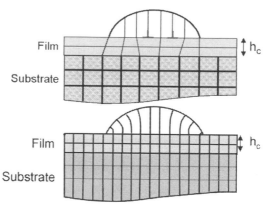

Fig. 16.4. (top) Schematic illustrating an Stranski-Krastanov dot that has relaxed via defect formation and (Bottom) schematic of a dot that has coherent relaxed without defects.

The QDs formed by this process are remarkably regularly spaced, and the array of dots is reasonably uniform, see Fig. 16.5. The precise diameter and height of these islands or QDs depends on the materials, wetting layer thickness, and growth conditions (such as temperature, deposition rate, and ripening time). Using atomic force microscopy (AFM), one can image the QD surface to determine the geometry and distribution of the dots. A typical AFM image of InAs QDs on top of an InAlAs matrix is shown in Fig. 16.5 [Lim et al. 2007a, Martyniuk et al. 2008]. The QD density and geometry are very important factors in determining the light sensing properties, as we shall see later.

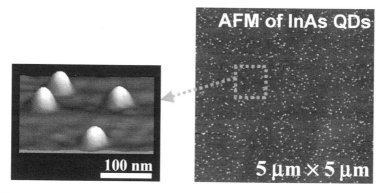

Fig. 16.5. Atomic force microscopy imaging of the InAs/InAlAs QDs used in devices.

In order to illustrate the process QD growth optimization, we show a series of selected experimental studies of QD formation as a function of various growth parameters. Fig. 16.6 and Fig. 16.7 illustrate the effects of varying the QD growth rate, while Fig. 16.8 illustrates the effect of the V/III ratio. When increasing the

TMIn flow rate from 70 to 90 sccm for 3.6 sec, the dot density increased when compared to the value at 70 sccm. This is expected as one is enhancing the effect of the mismatch. The result from the complete series of growth is shown in Fig. 16.6. In this particular set of growth rate conditions, the optimum one was 120 sccm for 2.7 sec. The dot density was 2.4×10^{10} cm^{-2}. The average lateral size was $20 \pm 4nm$ and the average height was $5 \pm 2nm$. The optimization process is a learning process. One grows at different flow rates and growth times and then selects the best material. Best material for photodetection is the one with i) highest QD density, ii) highest uniformity, iii) shapes which are closest to spheres and iv) lateral dimensions ranging from 10–15 nm.

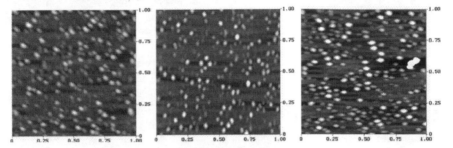

Fig. 16.6. AFM images of InAS QD on InAlAs with different growth rates (left) TMIn 120 sccm, 2.7 s; (center) TMIn 90 sccm, 3.6 s; (right) TMIn 45 sccm, 7.2 s.

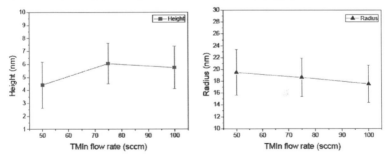

Fig. 16.7. Statistics of the height and lateral size (radius) for InAs/InP QDs (lattice mismatch ~4%) as a function of the TMIn flow rate.

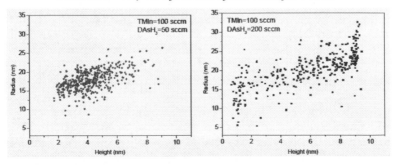

Fig. 16.8. The distribution of InAs/InP QD radius and height for different dilute AsH₃ flow.

16.3.2. The quantum dot-in-well infrared photodetector (QDWIP)

We start by discussing hybrid quantum dot (QD) in a quantum well (QW) photodetector on InP substrate. An example device structure is shown in Fig. 16.9. The energy level structure of a single unit is shown to the right. These hybrid devices are quantum dot-in-well infrared photodetectors (QDWIP), or sometimes called dot-in-well (DWELL) detectors. This name is to distinguish them from the simple QD/barrier/QD devices known as quantum dot infrared photodetectors (QDIPs). We will discuss QDIPs in section 16.3.4 and will discuss the advantages of each design.

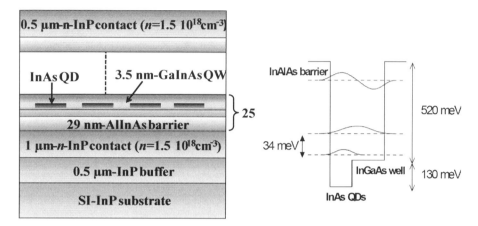

Fig. 16.9. (Left) QDWIP device structure for 4 μm detection and (Right) band structure of 1 period of the active region.

In QDWIPs (Fig. 16.9), lattice matched InAlAs and lattice matched InGaAs form the barrier and quantum well layers, respectively, and InAs is used to form the QD material [Lim et al. 2007a, Lim 2007b]. The small band-offset between InAs and InGaAs gives rise to a single QD bound state with a binding energy of 34 meV. Some examples of the atomic force microscopy (AFM) pictures of the QD are shown in Fig. 16.5 and Fig. 16.6. For the best devices, AFM gives us roughly a QD density of $\sim 10^{10}$ cm^{-2}.

The QW is n-doped approximately 10^{18} cm^{-3} with silicon dopants. This completely fills the QD levels and the remaining electrons have to occupy the QW subbands. Infrared light, impinging on the active area, excites a carrier from the second level to the upper excited state shown in Fig. 16.9 with an energy of 330 meV.

From the excited state, the carrier still has an energy barrier of 40 meV to overcome before it can conduct in the barrier layer. The escape process is assisted by the bias and by thermal activation. The formula for responsivity given by Eq. (16.1) applies here too, and one can see that the escape factor will normally give rise to a more than linear bias dependence of the responsivity at small biases. Again the important performance numbers are absorbance αL and gain g. The absorption turns out to be strongly polarization dependent and under the best circumstances with 25 repeat units and 10^{18} cm^{-3} doping level, the responsivity can reach values ~0.7 A/W as we shall now see below.

16.3.3. QDWIP device performance

The best performance so far has been achieved with the QDWIP of Fig. 16.9 [Lim et al. 2007a, Lim 2007b]. The bandstructure, photoresponse spectrum, and detectivity are shown in Fig. 16.10, Fig. 16.11, and Fig. 16.12.

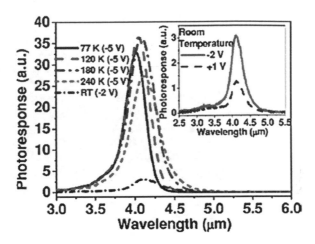

Fig. 16.10. The photoresponse versus wavelength for the QDWIP device shown in Fig. 16.9.

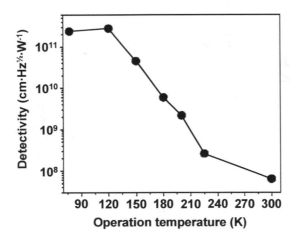

Fig. 16.11. The detectivity (D*) as a function of temperature for the QDWIP device of Fig. 16.9.

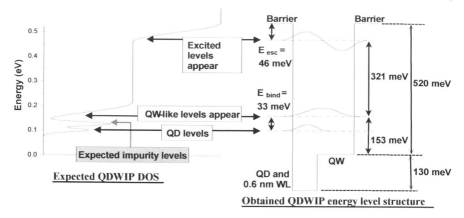

Fig. 16.12. A schematic representation of the QDWIP density of states. The QD layer only contributes 10^{10} cm^{-2} shallow bound states compared to 10^{12} cm^{-2} from the silicon dopant and 10^{14} cm^{-2} from the QW subband. The QD potentials and dopants will enhance the strong Anderson localization at the lower band edge which will now extend into the first QW "subband" up to the mobility edge which is at a density of states of about 5×10^{12} cm^{-2}.

The responsivity shown in Fig. 16.13 is remarkably stable with temperature up to about 260 K, and, were it not for the noise increase, one would have a high performance device even at room temperature. In most QDIPs the responsivity decreases rapidly as we go beyond 80 K because the "phonon bottleneck" which lowers the recombination rate [Lim et al. 2006, Lim et al. 2007a, Lim 2007b] gets removed by thermal activation. In the QDWIPs there is no phonon bottleneck; the recombination is as fast as it can be even at lower temperatures.

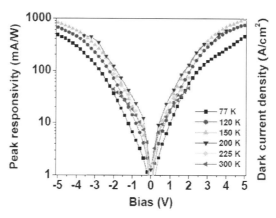

Fig. 16.13. Responsivity as a function of bias for the QDWIP device of Fig. 16.9.

The comparison of calculated and measured responsivities, and recent controlled measurements, have shown us that the absorption is not due to a single pass, as it would be in a highly absorbing geometry with p-polarized light, but is caused by multi-path, multiple scattering processes inside the system. The focal plane array (FPA) gives a reasonable image even at high temperatures as shown in Fig. 16.14 with NEDT values given in Fig. 16.15 [Tsao et al. 2007, Tsao et al. 2008].

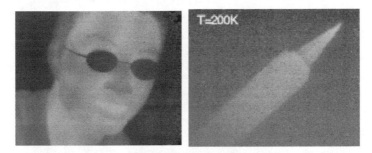

Fig. 16.14. Representative imaging at high temperature from a FPA based on the device design of Fig. 16.9.

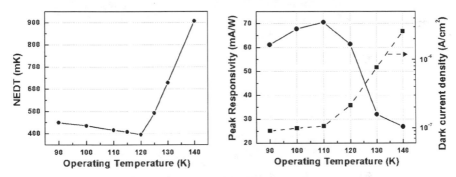

Fig. 16.15. (Left) Noise equivalent difference temperature (NEDT) as a function of operating temperature. (Right) The peak responsivity and dark current density of the FPA as a function of operating temperature.

We must remember that in this device, the QDs form essentially part of the 1 nm InAs wetting layer. The NEDT (Eq. (16.3)) is shown in Fig. 16.15; it has the interesting property that the NEDT first decreases with temperature and then goes up above about 120 K. The reason for this unusual behavior is that the responsivity is an increasing function of temperature and goes up more quickly than the noise does in this temperature range, as shown in the right Fig. 16.15. Above 120 K the opposite happens, responsivity decreases with temperature, and the NEDT increases.

16.3.4. The QDIP structure

The other way to realize a quantum dot (QD) based photodetector is to completely remove the quantum wells and replace them with layers of quantum dots thus forming a quantum dot infrared photodetector (QDIP) which consists of repeat units of QD/barrier. An example of an InP based QDIP structure is shown in Fig. 16.16, with an accompanying transmission electron microscopy (TEM) image of the structure shown in Fig. 16.17. Whereas in the first example Fig. 16.9, we are dealing with a QD embedded in a quantum-well structure, in this example (Fig. 16.16 and Fig. 16.17), we have the pure QD device or so called QDIP [Lim et al. 2005]. As mentioned before, in a QDIP, the QD layers are separated from each other only by barrier layers. So when the carriers are excited and reach the barrier band, there is no well to cross, and, as a result, recombination times are expected to be longer than in a QDWIP device [Lim et al. 2005]. This was considered to be advantageous for the realization of infrared photodetectors [Ryzhii 1996, Lim et al. 2005, Lim et al.2007a]. The problem turns out to be that as the recombination times increase, both the gain and the noise also increase; thus the overall benefit of a high gain is only marginal for detection.

The principle of the QDIP photodetection is as follows. One dopes the QD layer by adding silicon impurities to the material deposited. The electron leaves its silicon counterion and goes to occupy the lowest eigenstate in the QD region. In QDIPs, in general, one dopes so as to achieve an occupation of roughly two electrons per QD. A bias is applied and a current can flow between the electrodes. In the absence of photonic sources other than the blackbody radiation, we have the so-called "dark current" condition. The device designer tries to maximize the QD density but keep the dark current to a minimum, because it is a source of noise. As shown in Fig. 16.16, here one uses a current blocking layer (CBL) to reduce the dark current signal.

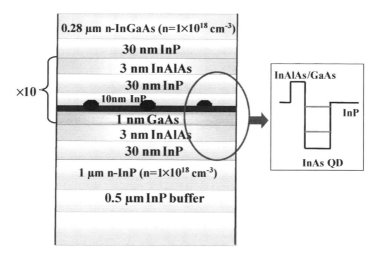

Fig. 16.16. A MWIR QDIP device structure grown with two-step barrier growth and InAlAs current blocking layers. The inset at the right shows a schematic of the conduction band alignment.(After Lim et al. [2007])

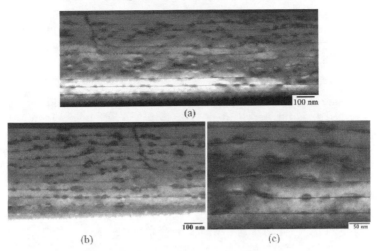

Fig. 16.17. Transmission electron microscope (TEM) images of the mid-wavelength IR QDIP device with InAs/QDs on GaAs/InAlAs/InP and 10/30 nm two step barrier growth. a) Bright field image shows overall structure; b) Dark field image shows overall structure; c) Magnified dark field image of the first few layers.

The TEM images in Fig. 16.17 show fairly good uniformity of QD formation, although there is some tilt in the QD layers, which could actually be beneficial for normal incidence absorption. When light comes in with sufficient energy to take an electron from the ground state to an excited state near the continuum levels of the barrier, the electron can escape with help from the applied bias, and we have a photocurrent. The process is shown schematically in Fig. 16.1 and is similar to the QWIP, and the resulting photocurrent responsivity is also given by the same Eq. (16.1). The parameters entering the formula are of course different in the two cases and have to be calculated correspondingly [Lim et al. 2005].

16.3.5. Gain in QDIP and QDWIP

Each time a carrier is photonically or spontaneously emitted out of the QD in a QDIP, for example see Fig. 16.1, it can only get captured to recombine by another QD [Levine et al. 1993, Lim et al. 2005, Lim et al. 2006]. The wetting layer is thin (1.5 monolayers) and only acts as a temporary trap. In a QDWIP on the other hand, see Fig. 16.9, the emitted carrier had to cross other QWs on its way to the electrode, and thus the QDWIP has a higher capture cross section and a lower gain. The gain enters the responsivity R given by Eq. (16.1) [Lim et al. 2005]. The gain in a QDIP can be $g > 1$, but it is almost always <1 in a QWIP or QDWIP [Lim et al. 2005, Movaghar et al. 2008]. Therein lies one of the most important differences.

The device performance is measured by the detectivity $D*$ which defined by

Eq. (16.4) $$D* = \frac{R\sqrt{A\Delta f}}{I_n}$$

where A is the area, and I_n is the noise current. Assuming that the noise is dominated by the generation-recombination (G-R) noise ($I_n = i_{GR}$), [Levine 1993]

Eq. (16.5) $i_{GR}^2 = 4eg\Delta f I_D$

where i_{GR} is the G-R noise, e is the electron charge, g is the gain, Δf is the noise bandwidth, and I_D is the dark current. To verify the difference in gain in the two technologies, let us compare some QDIP data with some QD in the well QDWIP data. As mentioned before, the QD in the well is expected to have a much lower gain, because the carrier cannot avoid crossing the well region where it has a strong chance of relaxing down into the bound energy levels. In Fig. 16.18 we can see the data taken for two different QDIP types.

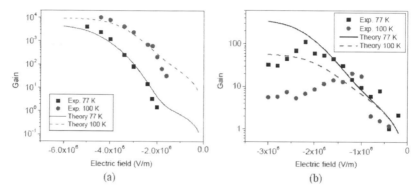

Fig. 16.18. Gain versus applied electric field, the solid lines are theoretical fits, a)
is InGaAs/InGaP and b) is InAs/InP QD structures.

Note the values of g in the QDIP are $g > 1$. This means that the carrier can go around the circuit that many times before it recombines. Now consider the data for the QD in the well or QDWIP, shown in Fig. 16.19. The difference is striking, but despite the high gain in the QDIP, which is helpful for the responsivity, the QDIP technology has a number of issues that in the present design give it a lower performance than the QDWIP. The reason is mainly the lower density of absorbing centers (around 3×10^{10} cm^{-2}) compared to 10^{12} cm^{-2}, and the fact that escape out of the QD needs a higher bias than in the QDWIP design. Also, as mentioned before, the benefit of the high gain is usually lost again by the noise term. The non-uniform nature of the QD size distribution and the fact that every QD has slightly different escape energy is another reason for the lower performance. The discrepancy between experiment and theory for gain analysis is shown in Fig. 16.18 and Fig. 16.19 and is mainly due to the fact that the theoretical model assumes that only G-R noise is operating, whereas in reality, other sources of noise are present as well (1/f, hot carrier, Johnson), [Movaghar et al. 2008] and currents can be due to tunnel ionization at high bias.

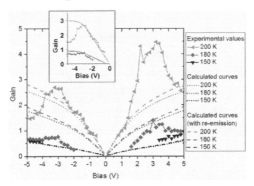

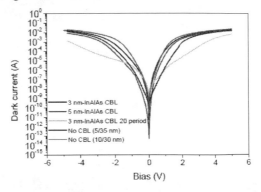

Fig. 16.19. Gain in the 35 period QDWIP as a function of bias at various operating temperatures. The inset shows the results from experiments at 180 K and 200 K (symbols) and the expected G-R gain curves extrapolated from the experimental results (solid lines).

16.3.6. QDIP performance

Let us now look at some of the other results achieved with InP-based quantum dot infrared photodetectors (QDIPs). Fig. 16.20 shows the dark current in the QDIP device of Fig. 16.16, and the corresponding detectivity D^* is shown in Fig. 16.21, and the normalized responsivity spectrum is shown in Fig. 16.22. The current blocking layer has lowered the dark current but its effect on the detectivity has been negative as can be seen in Fig. 16.21. The reason is that the obstacle created in the current path affects the photocurrent as well. So one has to be careful with current blocking layers (CBLs); they are not always beneficial. What is often beneficial is not this simple barrier but the filtering of the current through a narrow energy window [Su et al. 2004]. The example of the QDIP given above is selected because it illustrates how a current blocking layer might help. There are of course many other examples of QDIP device structures in the literature but this design provides a particularly interesting scenario.

Fig. 16.20. The dark current in the device of Fig. 16.16 at 77 K with well formed QD shown in the TEM picture Fig. 16.17, and with InAlAs current blocking layers of differing widths, and repeat period.

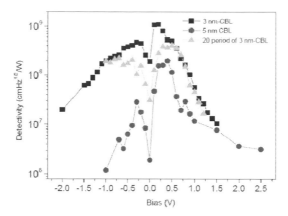

Fig. 16.21. Comparison of detectivities (D*) for the device with well-formed QDs shown in Fig. 16.16 at 77 K and with InAlAs current blocking layer.

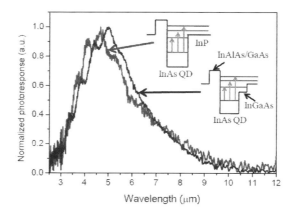

Fig. 16.22. Normalized spectral response at 77 K and a bias of 0.4 V for the device with well formed QD of Fig. 16.16. The second curve shows the response from the same structure with 3 nm InGaAs capping layers added on top of the InAs layers.

16.4. Quantum dot lasers

Stranski-Krastanov growth on GaAs and InP has also been applied with great success to make quantum dot (QD) lasers in the $1.3-1.55$ μm wavelength regime of interest for telecommunications [Poole et al. 2009]. QD lasers were first proposed by Arakawa and Sakaki in 1982 [Arakawa et al. 1982]. There are a number of reasons why one would want to apply QD technology to lasers as well. The main one is the sharp delta function like density of states which allows one to focus the gain and stimulation feedback on a very sharp wavelength region.

Quantum dot lasers have now surpassed the quantum well (QW) lasers by demonstrating lower threshold currents (critical current at which the lasing starts and which should be as small as possible to avoid heating), and higher power output [Poole et al. 2009]. Traditionally GaAs was used, but more recently the material class has been extended to InP as well. Growth on InP offers advantages similar to those already mentioned in the context of QWIPs. InP based QD lasers reach the important 1.55 μm emission range which is difficult to achieve with GaAs technology [Poole 2009]. The growth on InP also reaches a high QD density, now on the order of 4.5 10^{10} cm^{-2}.

A QD laser starts off being in the first place an electroluminescent (EL) device, which when surrounded by mirrors, recycles the light back to stimulate the emission process. The important performance criteria are high electron-hole interband radiative rates, high QD densities, and ease of carrier injection. It is also important to keep the linewidth as narrow as possible, and therefore to achieve optimal size uniformity. Remember that in QD laser fabrication, one has apart from the *p*- and *n*-injecting layers, to grow the quantum dot layers which replace the quantum well active region of a traditional GaAs or InP-based intra-band laser.

The QD laser operates on the principle that electrons and holes from *n*- and *p*-doped regions are injected into the layers and occasionally get trapped in a QD where they most often thermalize down to the lowest respective conduction and valence subband levels. When a QD is occupied by an electron and a hole, these can recombine and emit light of the corresponding energy. The light emitted can then be confined by the laser waveguide and mirrors such that it goes back to stimulate the emission process. The difference between a QW- and a QD-laser is that in a QW-laser the carriers are delocalized in the plane and one is normally dealing with only one well. There is very little chance that the carrier crosses the well and then recombines in the next one. In a QD-laser, multilayers can be used, but non-uniformity will give rise to band broadening which may or may not be beneficial. The matrix elements for radiative recombination is also different in a QD and in a QW. The 3D-localization of the wavefunction in a QD, and the absence of momentum restrictions for photon emission are most helpful to the radiative channel. Also one should note that the exciton binding energy, and importantly the concomitant electron-hole spatial correlation, is stronger in a QD as compared to a QW. This also helps the radiative channel because the electron and hole move in each other's orbit most of the time. In a QD the breaking up of an electron hole pair is more difficult than in a QW. To completely break up a pair in a QD not just temporarily, one of the carriers has to be excited outside the QD, or into a nonradiative defect channel inside the barrier. In a QW on the other hand, the electron or hole is easily thermally excited to the continuum of higher momentum states, and then is subject to momentum selection rules for recombination. Since the threshold current is given by $J = J_0 \exp[T / T_0]$, the target for high temperature operation is to minimize the transparency current J_0, while keeping T_0 high. High threshold current gives rise to high Joule heating and early breakdown of the device. Values of T realized for InAs–GaAs technology are in the range 5–75 °C [Fathpour et al. 2004].

The early QD-laser work was mainly based on established Stranski-Krastanov growth with InAs on GaAs. The large lattice mismatch of 7% gives rise to a high QD density but the wavelength is difficult to bring down to the important 1.3–1.5 μm range. So Deppe's group in Texas [Huffaker et al. 1998] tried to cover InAs with the alloy $In_xGa_{1-x}As$ and succeeded in getting 1.3 μm with a J_{th} of 45 A·cm^{-2} in continuous wave (CW) operation. Better performance than quantum wells was first achieved using the dot-in-well technology with $In_{0.15}Ga_{0.85}As$ by Liu et al. [2000] and obtained a $J_{th} = 26$ A·cm^{-2} in pulsed mode. One has to note that since 1995, the threshold current has dropped 3 orders of magnitude thanks to improvement in growth and better uniformity and higher QD densities. Critical to performance improvement is narrowing the line and increasing the QD density with optimized injection conditions. Much work has been devoted recently to fabricating lasers in the InAs–InP QD material system, and considerable improvements in device performance have been made in lasers grown on InP substrates in the last few years [Poole et al. 2009]. Whereas previously, only pulsed operation at low temperature were achieved, recent work has succeeded in making devices which work up to room temperature [Lelarge et al. 2007] (and above) in continuous wave operation. The inevitable distribution in QD size, gives rise to a broad full-width as half-maximum (FWHM) of the emission, and this results in a broad gain spectrum, which is useful if a tunable laser is required. On the other hand, this does limit the gain at a particular wavelength. To have the best gain at a given wavelength, we want a narrow distribution of emitting wavelengths. With QD densities of 10^{11} cm^{-2} and more, it has been possible to make lasers with a single QD layer only [Poole et al. 2006, Homeyer et al. 2006]. An optimized dot structure was used by Poole et al. [2009] with a QD density of 4×10^{10} cm^{-2}, as the core of a single lateral mode laser diode at 1.55 μm, demonstrating continuous wave operation at room temperature with a threshold of 25.9 mA and slope efficiency of 0.3 A·W^{-1}.

16.5. InP based quantum cascade lasers (QCLs)

Due to its unique material properties and relatively mature growth and fabrication technology, the InP-based material system is well suited for the growth of mid-infrared quantum cascade lasers (QCLs), proposed by Kazarinov and Suris [1971]. Like other semiconductor lasers, a QCL consists of a laser core, a waveguide, and a feedback mechanism. While the waveguide and the feedback mechanism is not very different compared to conventional semiconductor lasers, the laser core of a QCL is truly novel, usually consisting of 30–50 identical QCL stages. The heterostructure for one QCL stage contains tens of layers, which translates to hundreds of layers in the laser core. The uniqueness of a QCL is that the laser transition takes place between two intersubband states, whose energy difference can be engineered to produce lasers with different wavelengths using the same material system. An illustration of this process is shown in Fig. 16.23. This quantum-based transport combined with the cascaded emitter design is responsible for the name of the device. At present, QCLs have been demonstrated to emit from mid-infrared up to THz.

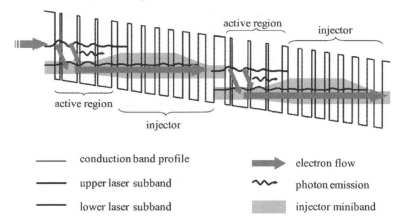

	conduction band profile		electron flow
	upper laser subband		photon emission
	lower laser subband		injector miniband

Fig. 16.23. Two emitting stages of the quantum cascade laser. This illustrates the role of the injector region and the cascaded nature of the photon emission, whereas a single electron can emit multiple photons.

16.5.1. Quantum cascade laser growth

Unlike the traditional interband lasers discussed in Chapter 8, the QCL is truly a "quantum" device. The individual layer thicknesses, as shown in Fig. 16.23, range only 1–4 nm. With tens of layers per stage, and tens of emitting stages within a laser core, this means even the simplest QCL requires hundreds of thin layers. This requires precision control of the material growth.

To form the complex conduction band profile shown in Fig. 16.23 we need to grow alternating $Ga_xIn_{1-x}As/Al_yIn_{1-y}As$ layers. Using Vegard's law, which assumes the average lattice constant varies linearly with the alloy content, the composition of a $Ga_xIn_{1-x}As/Al_yIn_{1-y}As$ heterostructure can be chosen such that the lattice constant matches that of the InP substrate. This case is called lattice matched (Fig. 16.24). It is easiest to grow crystals with the same lattice constant as the substrate. However, in a mismatched system there are many interesting effects that can occur before dislocations start to form.

Strain balanced growth

A mismatched crystal is initially pseudomorphic. In other words, it tries to take the form of the substrate by deforming itself at the interface between itself and the substrate (Fig. 16.24(b, c)). Most growth methods are two-dimensional, which results in compressive (tensile) distortion in one plane and tensile (compressive) distortion in the third dimension to compensate. This distortion is called strain, and it affects the band structure due to displacement of the atoms in the unit cell. This strain exists until a certain critical thickness, at which point the strain is relieved by the formation of dislocations.

In order to limit the carrier escape from the top of the barrier for short wavelength designs and to extend the short wavelength limit, the conduction band offset can be increased by using a strain-balanced $Ga_xIn_{1-x}As/Al_yIn_{1-y}As$

heterostructure pseudomorphically grown on an InP substrate, where x and y are chosen such that the compressive strain in the wells is balanced by equal and opposite tensile strain in the barriers [Evans et al. 2004]. Due to the absence of net strain, the strain-balanced QCL wafer can be grown free of misfit dislocations up to a certain degree. The conduction band offset of the strain-balanced heterostructure can be increased by at least 50% more than the lattice matched $Ga_{0.47}In_{0.53}As/Al_{0.48}In_{0.52}As$ heterostructure, as shown in Fig. 16.25.

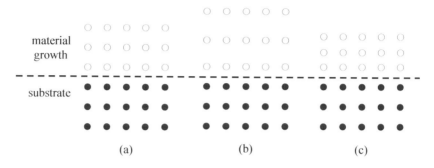

Fig. 16.24. Schematic diagram of lattice matched (a) and strained (b, c) heterostructures.

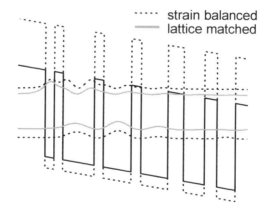

Fig. 16.25. Conduction band profiles and the corresponding upper and lower laser levels of lattice matched $Ga_{0.47}In_{0.53}As/Al_{0.48}In_{0.52}As$ and strain balanced $Ga_{0.33}In_{0.67}As/Al_{0.65}In_{0.35}As$.

To design a strain balanced QCL core heterostructure, the net strain of the system:

Eq. (16.6) $\varepsilon_{\perp} = \sum_{i=1}^{N} h_i \varepsilon_{\perp i} \Big/ \sum_{i=1}^{N} h_i$

needs to be minimized, where h_i and $\varepsilon_{\perp i}$ are the thickness and perpendicular strain of each layer, with $\varepsilon_{\perp i}$ defined by:

Eq. (16.7) $\varepsilon_{\perp i} = \dfrac{a_{\perp i}}{a_i} - 1$

where a_i is the equilibrium lattice constant, i.e., the lattice constant from Vegard's law, and $a_{\perp i}$ is the lattice constant in the direction perpendicular to the substrate and is given by:

Eq. (16.8) $a_{\perp i} = a_i \left[1 - D \left(\dfrac{a_{//}}{a_i} - 1 \right) \right]$

with $a_{//}$ denoting the in-plane lattice constant and $D = \dfrac{\varepsilon_\perp}{\varepsilon_\parallel}$ relating the ratio of the perpendicular strain to the strain in the plane of the substrate. In the (0 0 1) direction, we can define D in terms of the diagonal (c_{12}) and off diagonal (c_{11}) components of the elastic modulus for a cubic crystal:

Eq. (16.9) $D = 2 \dfrac{c_{12}}{c_{11}}$

Since the pseudomorphic system implies that the in-plane lattice constant $a_{//}$ takes the lattice constant of the substrate a_{sub} (Fig. 16.24), we can calculate the perpendicular lattice constant of each material (Eq. (16.8)), and calculate the composition of materials required to minimize the net strain.

Laser waveguide growth

A functional QCL needs a waveguide which has to provide both optical confinement and efficient heat extraction. In some cases these functions are in conflict with each other. One has to optimize the laser waveguide according to the intrinsic properties of the laser core. The absence of ternary material in the waveguide structure greatly enhances the heat removal in the transverse direction, which allows for better CW performance with only a small sacrifice of the confinement factor. Furthermore, the doping of the waveguide needs to be optimized. A lower doping level near the waveguide core favors a smaller optical loss (lower free carrier absorption) but also gives a poorer electrical conductivity, which leads to excess resistance. At the position of the cap layer, far from the laser core, where the overlap with the optical field is minimal, the doping can be made very high without introducing appreciable optical loss. This highly doped layer is necessary as it isolates the waveguide mode from coupling to the lossy surface plasmon mode at the metal-semiconductor interface. In the case of Fig. 16.26, starting from the InP substrate, the rest of the waveguide consists of a 1 μm low-doped (Si: $\sim 2 \times 10^{16}$ cm^{-3}) InP lower cladding layer, a 1.6 μm QCL core, a 3 μm low-doped (Si: $\sim 2 \times 10^{16}$ cm^{-3}) InP upper cladding layer, and a 1 μm high-doped (Si: $\sim 1 \times 10^{19}$ cm^{-3}) InP cap layer.

Quantum cascade laser growth

Even with all the design information available, the actual epitaxial growth has to be able to accurately recreate the intended structure. While most QCL growth has been done using molecular beam epitaxy (MBE), MOCVD has also shown itself to be capable of growing this type of device. Initially, MBE was used for the active layers, and MOCVD was used for the cladding and/or buried ridge processing [Evans et al. 2004]. Later, several groups began to explore MOCVD as growth technique for the entire heterostructure, active region and all [Wang et al. 2007, Green et al. 2003, Troccoli et al. 2008]. At present, the MOCVD QCL performance is very similar to MBE-grown QCLs.

The advantage of the MOCVD technique is the potential for large scale (multi-wafer) growth for manufacture, drawing on the same infrastructure built for the InP-based telecommunication industry. At present, limited work has been done in this area, but the QCL demonstration by MOCVD had a large affect on industrial interest in this technology. Regardless of the growth technique, however, the same material characterization techniques must be employed to check the quality and accuracy of the process.

Post growth characterizations need to be performed for each growth to ensure high material quality. In particular, x-ray diffraction is the main tool to characterize the layer thickness and composition for a QCL structure.

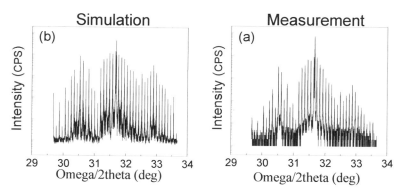

Fig. 16.26. Experimental measurement (a) and computer simulation (b) of the x-ray diffraction for a 30 stage strain-balanced QCL structure. The regularly spaced peaks are the signature of a periodic structure. The material quality is reflected by the width of the peaks. The fine position and the relative intensity of these peaks are related to the layer thicknesses and material compositions of the QCL heterostructure.

Shown in Fig. 16.27 is a comparison of the simulated and the measured x-ray diffraction curves. The close resemblance indicates precise recreation of the design parameters. In addition, optical characterization, such as photoluminescence, is used to probe the band structure. Varying layer composition or quantum well width can be detected by changes in the luminescence wavelength or lineshape. Electrical characterization is also very important. Hall effect or capacitance-voltage techniques can be used to identify doping levels and dopant distributions, which are critical to device operation.

16.5.2. Short wavelength QCLs

Due to the limited conduction band offset for the $Ga_xIn_{1-x}As/Al_yIn_{1-y}As/InP$ system, it becomes more and more difficult to extend the lower wavelength boundary for QCLs based on this material system. The limitations for short wavelength QCLs are:

1. As the upper laser level is pushed toward the conduction band of the barrier, thermally activated carrier leakage into the continuum becomes the major factor that hinders high temperature operation.
2. The intervalley scattering of electrons from the Γ valley into X and/or L valleys becomes more likely and may diminish the gain and population inversion of the emitter.
3. A short wavelength QCL generally requires a higher voltage drop than longer wavelength devices due to the larger photon energy. This increases the power density of the core for a given current density.

Therefore the thermal management for high temperature continuous wave (CW) operation is more difficult. However, with the strain-balanced technique, room temperature CW operation down to 3.8 μm has been demonstrated in GaInAs/AlInAs/InP material system [Yu et al. 2006]. Fig. 16.28 shows the CW power-current-voltage performance of a QCL at this wavelength.

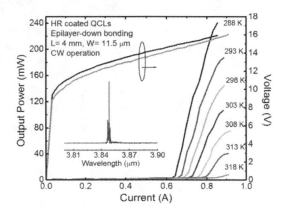

Fig. 16.27. Temperature dependent performance of a 3.8 μm QCL working in CW mode between 288 K and 318 K. The inset shows a CW spectrum taken at 298 K.

16.5.3. Middle wavelength QCLs

Mid-infrared QCLs made from the GaInAs/AlInAs/InP system are best suited for emission wavelengths between 4 μm and 9 μm. While similar limitations exist to limit the laser performance, all aspects are more manageable within this wavelength range, including electron confinement and optical losses. Although room temperature CW operation was achieved at shorter and longer wavelengths with more than 100 mW output power, the power conversion efficiency, i.e., the wall-plug efficiency (WPE), was only a few percent. Since 2007, significant effort has been devoted to optimizing the WPE of QCLs at wavelengths around 4.7 μm. Room temperature CW operation with watt-level output power [Bai et al. 2008a] and WPE exceeding 10% [Bai et al. 2008b] was demonstrated quickly. Further improving the power and WPE are extremely challenging. Shown in Fig. 16.28 is the room temperature CW output power and WPE performances of a 4.8 μm QCL. The CW maximum output power reaches 3.4 W with a CW WPE of 16.5%. This result is obtained through systematic study of the WPE behavior as a function of various parameters [Razeghi 2009]. As the QCL design, material growth, and fabrication technology have continued to improve, 4.6 μm lasers with 51% WPE at 80 K and 22% at room temperature have been achieved.

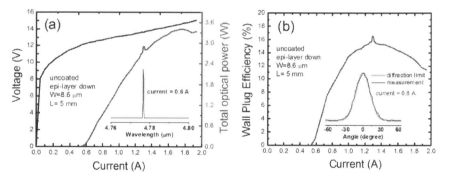

Fig. 16.28. Power-current-voltage (a) and WPE-current (b) behavior of a 4.7 μm QCL operating in CW mode at room temperature. The inset of (a) shows the lasing spectrum measured at a driving current of 0.6 A. The inset of (b) shows the lasing far field measured at a driving current of 0.8 A, together with a calculated diffraction limit.

16.5.4. Long wavelength QCLs

Unlike in the short wavelength side, where a lower boundary exists due to the finite conduction band offset, there is no boundary in the long wavelength side. In fact, the THz QCL, i.e., hundreds of micrometers in wavelength, has been demonstrated in pulsed mode up to 186 K [Kumar et al. 2009]. The following considerations need to be addressed for long wavelength QCL design. First of all, the free carrier absorption, which is the main source of waveguide loss, increases roughly as the square of the wavelength. This makes it necessary to use lower doping and thicker layers to maintain low absorption loss. Secondly, a thicker QCL core is needed to maintain the confinement factor for a longer wavelength. This increases the operating voltage and epitaxial thickness. Thirdly, the thermal resistance of the device increases as the core gets thicker. Despite all of these challenges, at 10.2 μm, room temperature pulsed and CW operation with 25 W and 0.65 W output power, respectively, have been demonstrated in the GaInAs/AlInAs/InP material system (Fig. 16.29) [Razeghi 2009].

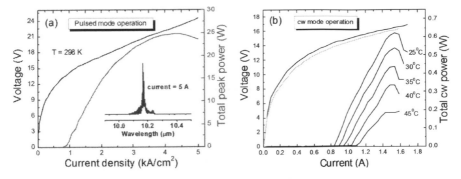

Fig. 16.29. Power-current-voltage behavior of two 10.2 μm QCLs operating in pulsed (a) and CW (b) mode at room temperature. The inset of (a) shows the lasing spectrum measured at a driving current of 5 A.

16.5.5. Photonic crystal distributed feedback QCLs

Broad area QCLs (ridge width greater than 50 μm) are natural candidates to produce extremely high peak power (greater than 100 W), however, increasing the emitter width with the conventional Fabry Perot (FP) cavity suffer from a wide emitting spectrum and a broad far field profile, due to the appearance of higher order lateral modes. The photonic crystal distribute feedback (PCDFB) mechanism was introduced to deal with the spectral and spatial problems of the conventional broad area QCLs and the first experimental demonstration of this coupling mechanism with QCLs has been realized with an output power of about 0.3 W [Bai et al. 2007]. With improved processing technology, the output power has been increased to 12 W [Bai et al. 2009]. Fig. 16.30 shows the SEM picture and the room temperature performances of three PCDFB QCLs monolithically fabricated on the same wafer. The spectra show typical index coupled DFB behavior and the far fields are diffraction limited.

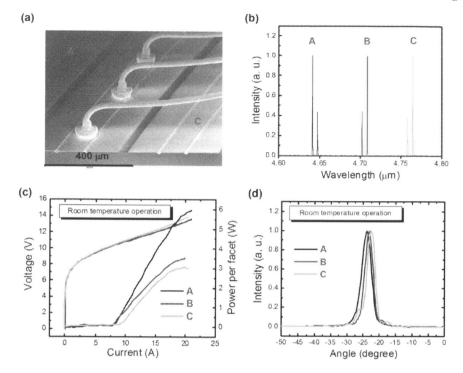

Fig. 16.30. (a) SEM image of a bonded PCDFB laser chip incorporating three different grating periods and room temperature pulsed mode performances including (b) lasing spectra, (c) power-current-voltage characteristics, and (d) far field.

References

Aivaliotis, P., Zibik, E.A., Wilson, L.R., Cockburn, J.W., Hopkinson, M. and Vinh, N.Q., *Appl. Phys. Lett.* **92** 023501-3, 2008.

Bai, Y., Darvish, S.R., Slivken, S., Sung, P., Nguyen, J., Evans, A., Zhang, W., and Razeghi, M., *App. Phys. Lett.* **91** 141123, 2007.

Bai, Y., Darvish, S.R., Slivken, S., Zhang, W., Evans, A., Nguyen, J., and Razeghi, M., *App. Phys. Lett.* **92** 101105, 2008a.

Bai, Y., Slivken, S., Darvish, S.R., and Razeghi, M., *App. Phys. Lett.* **93** 021103, 2008b.

Bai, Y., Gokden, B., Darvish, S.R., Slivken, S., and Razeghi, M., *App. Phys. Lett.* **95** 031105, 2009.

Chakrabarti, S., Stiff-Roberts, A.D., Su, X.H., Bhattacharya, P., Ariyawansa, G., and Perera, A.G., *J. Phys. D: Appl. Phys.* **38** 2135, 2005.

Chakrabarti, S., Stiff-Roberts, A.D., Bhattacharya, P., and Kennerly, S.W., *J. Vac. Sci. & Techno. B* **22** 1499-1502, 2004.

Evans, A., Yu., J.S., David, J., Doris, L., Mi, K., Slivken, S., Razeghi, M. *App. Phys. Lett.* **84** 314, 2004.

Fathpour, S., Mi, Z., Bhattacharya, P., Kovsh, A.R., Mikhrin, S.S., Krestnikov, I.L., Kozhukhov, A.V., and Ledentsov, N.N., *Appl. Phys. Lett.* **85** 5164, 2004.

Gin, A., Movaghar, B., Razeghi, M., and Brown, G.J., *Nanotechnology* **16** 1814, 2005.

Goldstein, L., Glas, F., Marzin, J.Y., Charasse, M.N. and Roux, G.L., *App. Phys. Lett.* **47** 1099-1101, 1985.

Green, R.P., Krysa, A., Roberts, J.S., Revin, D.G., Wilson, L.R., Zibik, E.A., Ng, W.H., Cockburn, J.W. *App. Phys. Lett.* **83** 1921, 2003.

Gunapala, S.D., Bandara, S.V., Liu, J.K., Hill, C.J., Rafol, S.B., Mumolo, J.M., Trinh, J.T., Tidrow, M.Z., and LeVan, P.D., *Semicond. Sci. Technol.* **20** 473, 2005.

Hoff, J., Razeghi, M., and Brown, G., *Phys. Rev. B* **54** 10774, 1994

Homeyer, E., Piron, R., Caroff, P., Paranthoen, C., Dehaese, O., Corre, A.L., Loualiche, S., *Phys. Status Solidi C* **3** 407, 2006.

Huffaker, D.L, Park, G., Zou, Z., Shchekin, O.B., and Deppe, D.G., *Appl. Phys. Lett.* **73** 2564, 1998.

Jelen, C., Slivken, S., Razeghi, M., and Brown, G., *App. Phys. Lett.* **70** 360, 1997.

Jiang, J., Mi, K., McClintock, R., Razeghi, M., Brown, G.J., and Jelen, C., *IEEE Photonics Technol. Lett.* **15** 1041, 2003.

Kazarinov, R.F. and Suris, R.A., *Sov. Phys. Semicond.-USSR* **5** 707, 1971.

Krishna, S., Forman, D., Annamalai, S., Dowd, P., Varangis, P., Tumolillo, T., Gray, A., Zilko, J., Sun, K., Liu, M., Campbell, J., and Carothers, D., *Appl. Phys. Lett.* **86** 193501, 2005.

Kumar, S., Hu, Q., and Reno, J.L., *App. Phys. Lett.* **94** 131105, 2009.

Lelarge, F., Dagens, B., Renaudier, J., Brenot, R., Accard, A., Dijk, F., Make, D., Gouezigou, O.L., Provost, J.G., Poingt, F., Landreau, J., Drisse, O., Derouin, E., Rousseau, B., Pommereau, F., Duan, G.H., *IEEE J. Quantum Electron.* **13** 111, 2007.

Levine, B.F., *J. App.Phys.* **74** R1-R81-R1-R81, 1993.

Lim, H., Tsao, S., Zhang, W., and Razeghi, M., *Appl. Phys. Lett.* **90** 131112-3, 2007a.

Lim, H., *PhD Dissertation*, Northwestern University, 2007b.

Lim, H., Movaghar, B., Tsao, S., Taguchi, M., Zhang, W., Quivy, A.A., and Razeghi, M., *Phys. Rev. B* **74** 205321-8, 2006.

Lim, H., Zhang, W., Tsao, S., Sills, T., Szafraniec, J., Mi, K., Movaghar, B., and Razeghi, M., *Phys. Rev. B* **72** 085332-085332, 2005.

Liu, G.T.; Stintz, A.; Li, H.; Newell, T.C.; Gray, A.L.; Varangis, P.M.; Malloy, K.J.; Lester, L.F. *J. of Quantum Electronics* **36** 1272, 2000.

Lu, X., Vaillancourt, J., and Meisner, M.J., *Appl. Phys. Lett.* **91** 051115, 2007.

Martyniuk, P., Krishna, S., and Rogalski, A., *J. App. Phys.* **104** 034314-6, 2008.

Movaghar, B., Tsao, S., Pour, S.A., Yamanaka, T., and Razeghi, M., *Phys. Rev. B* **78** 115320-10, 2008.

Poole, P.J, Kaminska, K, Barrios, P., Lu, Z., and Liu, J., *J. Crystal Growth* **311** 1482-1486, 2009.

Razeghi. M., *IEEE J. Selec. T. Quant. Electron.* **15** 941-951, 2009.

Ryzhii, V., *Semi. Sci. &Techno.* **11** 759-765, 1996.

Szafraniec, J., Tsao, S., Zhang, W., Lim, H., Taguchi, M., Quivy, A.A., Movaghar, B., and Razeghi, M., *Appl. Phys. Lett.* **88** 121102, 2006.

Su X.H, Chakrabarti S., Bhattacharya, P., Ariyawansa, G., and Perera A.G., *IEEE Journal of Quantum Electronics*, **41** 974, 2005.

Troccoli, M., Diehl, L., Bour, D.P., Corzine, S.W., Yu, N., Wang, C.Y., Belkin, M.A., Höfler, G., Lewicki, R., Wysocki, G., Tittel, F.K., Capasso, F. *J. Lightwave Tech.* **26** 3534, 2008.

Tsao, S., Hochul, L., Hosung, S., Wei, Z., and Razeghi, M., *IEEE Sensors J.* **8** 936-941, 2008.

Tsao, S., Lim, H., Zhang, W., and Razeghi, M., *App. Phys. Lett.* **90** 201109-3, 2007.

Varley, E., Lenz, M., Lee, S.J., Brown, J.S., Ramirez, D.A., Stintz, A., Krishna, S., Reisinger, A., and Sundaram, M., *Appl. Phys. Lett.* **91** 081120, 2007.

Wang, X.J., Fan, J.Y., Tanbun-Ek, T., Choa, F.-S. *App. Phys. Lett.* **90** 211103, 2007.

Yu, J.S., Evans, A., Slivken, S., Darvish, S.R., and Razeghi, M., *App. Phys. Lett.* **88** 251118, 2006.

Zhang, W., Lim, H., Taguchi, M., Tsao, S., Movaghar, B., and Razeghi, M., *Appl. Phys. Lett.* **86** 191103 2005.

Appendices

A.1.   Effect of substrate miscut on the measured superlattice period
A.2.   Optimization of thickness and indium composition of InGaAs wells for 980 nm lasers
A.3.   Energy levels and laser gains in a quantum well (GaInAsP): The effective mass approximation
A.4.   Luttinger–Kohn Hamiltonian
A.5.   Infrared detectors
A.6.   Physical properties and safety information of metalorganics
A.7.   Original frontmatter:
       MOCVD Challenge Volume 1: Original foreword
       MOCVD Challenge Volume 1: Original introduction
       MOCVD Challenge Volume 2: Original foreword
       MOCVD Challenge Volume 2: Original introduction

731

A.1. Effect of substrate miscut on the measured superlattice period

A miscut substrate is widely used in crystal growth since the miscut surface provides a step where minimum growth energy is required, and thus growth is easier and better interfacial quality is obtained. The relation between the thickness of the layer D, the wavelength of the x-rays λ, the separation of the superlattice satellites $\Delta\theta$, and the Bragg diffraction angle of the substrate θ_B can be described by

Eq. (A.1)
$$D = \frac{\lambda}{2\Delta\theta\cos\theta_B}$$

It appears that the effect of the miscut of the substrate on the measurement result is noticed when about 5% difference in the measured superlattice period between $(0\,0\,2)$ diffraction and $(0\,0\,4)$ diffractions is observed. The observed discrepancy results from the fact that Eq. (A.1) is only valid for symmetric geometry. For a miscut substrate, the diffraction results from asymmetric geometry. The real thickness D_{asym} is given by

Eq. (A.2)
$$D_{asym} = \lambda\frac{\sin(\theta_B + \alpha)}{\Delta\theta\sin(2\theta_B)}$$

when the separation of the neighboring peaks, $\Delta\theta$, is not too large. In the above equation, α is the cross-section angle in the beam plane between the diffraction plane and the surface plane, as shown in Fig. A.1. A relation among α, the miscut angle β, and the azimuthal angle φ can be obtained from Fig. A.1 as

Eq. (A.3) $\tan\alpha = \tan\beta\cos\varphi$

Combining Eq. (A.2) and Eq. (A.3) under the limit where the miscut angle β is small, we can obtain the relation between the real superlattice period D_{asym} and the period D calculated from Eq. (A.1)

Eq. (A.4) $D_{asym} = D(1 + \beta\cot\theta_B\cos\varphi)$

When $\beta\cot\theta \ll 1$, then Eq. (A.4) can be approximated as

Eq. (A.5) $D = D_{asym}(1 - \beta\cot\theta_B\cos\varphi)$

Effect of Substrate Misorientation

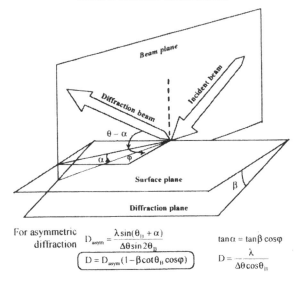

For asymmetric diffraction

$$D_{asym} = \frac{\lambda \sin(\theta_H + \alpha)}{\Delta\theta \sin 2\theta_H}$$

$$\boxed{D = D_{asym}(1 - \beta \cot\theta_H \cos\varphi)}$$

$$\tan\alpha = \tan\beta \cos\varphi$$

$$D = \frac{\lambda}{\Delta\theta \cos\theta_H}$$

Fig. A.1. A schematic drawing of the x-ray diffraction geometry.

Effect of Misorientation —— Measurement

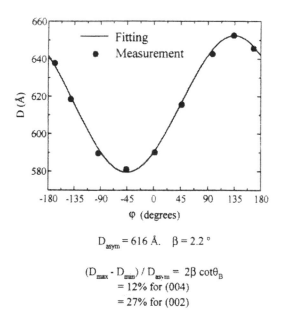

$$D_{asym} = 616 \text{ Å.} \quad \beta = 2.2°$$

$$(D_{max} - D_{min}) / D_{asym} = 2\beta \cot\theta_B$$
$$= 12\% \text{ for (004)}$$
$$= 27\% \text{ for (002)}$$

Fig. A.2. The relation between measured periods of a GaAs/Ga$_{0.51}$In$_{0.40}$P superlattice and the azimuthal angle.

Fig. A.2 shows the variation of D with azimuthal angle φ measured by HRXRD for a 10-period GaAs/Ga$_{0.51}$In$_{0.49}$P superlattice grown by LPMOCVD on a GaAs(0 0 1) substrate misoriented by $2°$ towards $\langle 1\,1\,0 \rangle$ (miscut $2°$). The nominal well thicknesses and barrier thicknesses are 90 Å and 100 Å, respectively. Fig. A.2 indicates that a satisfactory agreement between the measured value of D and theoretical predictions is obtained, β and D_{asym} from the fitting are $2.2°$, and the measured superlattice period is also close to the thickness measured from the TEM picture.

The maximum error in D from incorrect alignment estimated from Eq. (A.5) is 25% and 11% at (0 0 2) and (0 0 4) diffraction, respectively, for a sample grown on a $2°$ miscut GaAs substrate. This is an extremely significant error for most measurements.

A.2. Optimization of thickness and indium composition of InGaAs wells for 980 nm lasers

An InGaAs/GaAs strained quantum well is used as the active layer for lasers lasing at the wavelength 980 nm.

In the biaxially compressed InGaAs quantum well (QW) the normally cubic unit cell deforms elastically, matching to the smaller GaAs lattice in the plane of the QW, while in the plane perpendicular to the QW the unit cell stretches. Such a deformation affects the InGaAs energy band structure.

The strain Hamiltonian for a band at $k = 0$ is

Eq. (A.6) $\qquad H_\varepsilon = -a\varepsilon_{ii} - 3b\left(L_i^2 - \frac{1}{3}L^2 \right)\varepsilon_{ij}\delta_{ij} - \sqrt{3}dL_iL_j\varepsilon_{ij}$

where a is the hydrostatic deformation potential and b and d are the shear deformation potentials corresponding to strains of tetragonal and rhombohedral symmetries, respectively, ε is a strain tensor.

In the [1 0 0] direction

Eq. (A.7) $\qquad \varepsilon_{xx} = \varepsilon_{yy} = \varepsilon$

Eq. (A.8) $\qquad \varepsilon_{zz} = -2\frac{C_{12}}{C_{11}}\varepsilon$

Eq. (A.9) $\qquad \varepsilon_{ij} = 0 \qquad$ for $i \neq j$

where C_{ij} is the ij^{th} component of the stiffness tensor and $-\varepsilon$ is equal to the relaxed mismatch, $\varepsilon > 0$ for tensile strain and $\varepsilon < 0$ for compressive strain.

Applying the wavefunction of the appropriate band to the Hamiltonian in Eq. (A.6), the energy of each band under strain is obtained:

Eq. (A.10) $\qquad E(\text{hh}) = E_g + E_{hydr} - E_{shear}$

Eq. (A.11) $\qquad E(\text{ll}) = E_g + E_{hydr} + E_{shear}$

Eq. (A.12) $\qquad E(\Delta) = E_g + \Delta + E_{hydr}$

where E_g is the bandgap energy, Δ is the split-off energy, and

Eq. (A.13) $E_{hydr} = 2a\dfrac{C_{11} - C_{12}}{C_{11}}\varepsilon$

Eq. (A.14) $E_{hydr} = b\dfrac{C_{11} + C_{12}}{C_{11}}\varepsilon$

are the hydropressure energy shift and shear strain energy shift, respectively.

The critical thickness of a strained layer is an important parameter for the strained layer. The force balancing model [Matthews and Blakeslee 1974] gives the following equation for critical thickness:

Eq. (A.15) $h_c = \dfrac{2}{k\sqrt{2}\pi\varepsilon}\dfrac{1 - 0.25\gamma}{1 + \gamma}\left[\ln\left(\dfrac{h_c\sqrt{2}}{a}\right) + 1\right]$

where a is the lattice constant of the substrate, ε is the mismatch of the strained layer, and γ is the Poisson ratio, which is defined as

Eq. (A.16) $\gamma = \dfrac{C_{12}}{C_{12} + C_{11}}$

The coefficient k has a value of unity for a strained layer superlattice, two for a single quantum well, and four for a single strained layer.

The energy balancing model [People and Bean 1985] gives the following equation for the critical thickness of a single strained layer:

Eq. (A.17) $h_c = \dfrac{a}{32\sqrt{2}\pi\varepsilon^2}\dfrac{1 - \gamma}{1 + \gamma}\ln\left(\dfrac{h_c\sqrt{2}}{a}\right)$

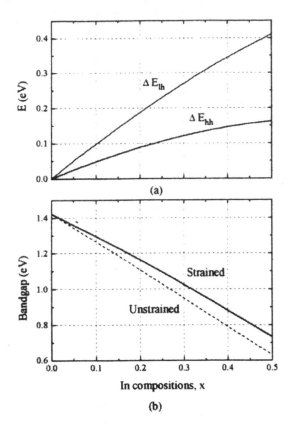

Fig. A.3. (a) Energy shifts in the light-hole valence band edge and heavy-hole valence band edge, (b) The bandgap energies of the strained (solid line) and unstrained $In_xGa_{1-x}As$ (dashed line) on a GaAs substrate.

Fig. A.3(a) shows the energy shifts in the light-hole valence band edge and heavy-hole valence band edge calculated for strained InGaAs on GaAs using Eq. (A.10)–Eq. (A.14). The necessary parameters in the calculation are listed in Table A.1.

		InAs	GaAs
Lattice constant (Å)	a_0	6.0583	5.6533
Elastic coefficient (10^{12} dyn·cm^{-2})	C_{11}	0.8329	1.188
	C_{12}	0.4526	0.538
Hydrostatic deformation potential (eV)	a	5.9	9.8
Shear deformation potential (eV)	b	1.7	1.8
Electron effective mass	m_e^*/m_0	0.023	0.0665
Heavy-hole effective mass	m_{hh}^*/m_0	0.32	0.34
Light-hole effective mass	m_{lh}^*/m_0	0.027	0.074

	$In_xGa_{1-x}As$	
Bandgap energy (eV)	E_g	$1.424 - 1.614x + 0.54x^2$
Exciton binding energy (meV)	E_{ex}	8
Band offset	$\Delta E_c/\Delta E_g$	0.65

Table A.1. Material parameters for GaAs, InAs, and $In_xGa_{1-x}As$.

Since the energy shift in the light-hole band edge with In composition is much larger than the shift in the heavy-hole valence band edge, recombination in the bulk strained InGaAs layer is dominated by transition from the conduction band to the heavy-hole valence band. There is a strain-induced increase in the bandgap energy with indium composition for InGaAs, which partly offsets the decrease in bandgap energy with composition in unstrained InGaAs. The bandgap energies of strained (solid line) and unstrained InGaAs (dashed curve) on a GaAs substrate are shown in Fig. A.3(b). Fig. A.3 does not take into account the quantum confinement effect, which introduces an extra shift energy. A single-quantum-well calculation is performed to decide the quantum confinement energy shift using the parameters listed in Table A.1.

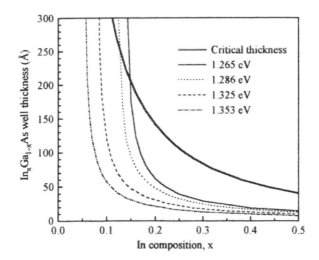

Fig. A.4. Equal-energy lines as functions of well thickness and In composition for strained In$_x$Ga$_{1-x}$As layers on a GaAs substrate.

The next figure (Fig. A.4) shows the equal-energy lines on a composition-well thickness plane together with the critical thickness calculated using Eq. (A.15). There are roughly three regions, where the dependence of energy on the In composition or well width varies. At high In composition, peak energy (e.g. observed in photoluminescence spectrum) is sensitive to the variation in well thickness and relatively insensitive to the In composition. At low In composition, the peak energy is sensitive to the variation in In composition and relatively insensitive to the well thickness. The best region for control of the lasing wavelength is between those two regions, where the peak energy is not sensitive to the In composition or well thickness. For the material lasing at 980 nm (1.265 eV), the best In composition is around 0.19.

Avoidance of misfit dislocation is another important issue. It is observed that being farther away from the critical thickness will improve laser performance. The best In composition is in the range of 0.17–0.22 for 980 nm lasers, where maximum difference between required well width and critical thickness can be maintained.

References

Matthews, J.W. and Blakeslee, A.E. *J. Crystal Growth* **27** 118, 1974.
People, R. and Bean, J.C. *Appl. Phys. Lett.* **47** 322, 1985.

A.3. Energy levels and laser gains in a quantum well (GaInAsP): The "effective mass approximation"

Energy levels for the conduction band for quantum well structures are obtained as below. Wavefunctions in a well (well width, L_z; well potential barrier, ΔE_c) are given by

Eq. (A.18) $\quad \varphi_{cn}^- = A \left\{ \begin{matrix} \cos \\ \sin \end{matrix} \right\} \left[\sqrt{2m_{c1}E_{c1}}\, y / \hbar \right] \qquad$ for $|y| \le L_z / 2$

Eq. (A.19) $\quad \varphi_{cn}^+ = B \exp\left[-\sqrt{2m_{c2}E_{c2}}\, y / \hbar \right] \qquad$ for $|y| \ge L_z / 2$

where the upper part of the curly bracket of Eq. (A.18) is for odd n, the lower being for even n. One can derive the eigenvalue equation for the energy level E_{cn} for each subband n using the boundary conditions (continuity of wavefunctions and their derivatives) as

Eq. (A.20) $\quad (m_{c1} / m_{c2})\left[(\Delta E_c - E_{cn}) / E_{cn}\right]^{1/2} = \left\{ \begin{matrix} \tan \\ -\cot \end{matrix} \right\} \left(L_z \sqrt{2m_{c1}E_{cn}} / 2\hbar \right)$

where m_{c1} and m_{c2} are the effective masses of the conduction electrons in the active region and the waveguide region. Energy bands for heavy holes and light holes are obtained in a similar way. Energy levels for intersubbands of InGaAsP($\lambda = 808$ nm)/GaAs are shown in Table A.2.

QW width (L_z)	Subbands	Energy level (meV)				
		$N=1$	$n=2$	$n=3$	$n=4$	$n=5$
	E_c	15.06	57.08			
150 Å	E_{hh}	2.82	11.24	25.20	44.30	67.68
	E_{lh}	12.14	14.83	93.62		
	E_c	4.98	19.74	48.70	75.09	
300 Å	E_{hh}	0.79	3.15	7.06	12.58	19.62
	E_{lh}	3.86	15.44	34.39	59.96	89.64

Table A.2. Energy levels for intersubbands of InGaAsP($\lambda = 808$ nm)/GaAs

Material constants such as E_g, m_0, m_{hh} and m_{lh}, tabulated in Table 8.3 are used for this calculation with the help of the following equations:

Eq. (A.21) $\dfrac{m_0}{m_{hh}(100)} = \gamma_1 - 2\gamma_2$ $\dfrac{m_0}{m_{lh}(100)} = \gamma_1 + 2\gamma_2$

Eq. (A.22) $\dfrac{m_0}{m_{hh}(111)} = \gamma_1 - 2\gamma_3$ $\dfrac{m_0}{m_{lh}(111)} = \gamma_2 + 2\gamma_3$

The quasi-Fermi energy level is the chemical potential for the electron or hole gas in the active layer at quasiequilibrium during current injection. Since the quasi-Fermi energy level determines not only injection carrier density, but also the laser gain, we calculate the quasi-Fermi energy level from the relation between the carrier density in the active layer (n_{act}) and the quasi-Fermi energy level (E_{fc}).

Eq. (A.23) $n_{act} \equiv \int f_D(E_k, E_{fc})dk = \sum_n kT\rho_n \ln\left[1 + \exp\left(\beta\left(E_{fc} - E_{cn}\right)\right)\right]$

where $f_D(E_k, E_{fc})$ is the Fermi-Dirac distribution for energy E_k, when the chemical potential is E_{fc}, and ρ_n, the density of states for 2D, is given by

Eq. (A.24) $\rho_n = qmkT / \pi\hbar^2 L_z$

Quasi-Fermi levels for holes (E_{fv}) are obtained in the same way. Once E_{fc} and E_{fv} have been determined, the gain spectrum can be calculated from the Fermi golden rule:

Eq. (A.25) $g(\hbar\omega) = \dfrac{q^2|M|^2}{\hbar\omega\varepsilon_0 m^2 c_0 \hbar n L_z} \sum_{ij} m_{r,ij} A_{ij} C_{ij}\left[f_c - (1 - f_v)\right]\theta\left(\hbar\omega - E_{ij}\right)$

where

Eq. (A.26) $A_{ij} = \dfrac{3}{4}\left(1 + \cos^2 \Theta_{ij}\right)$ (for TE)

Eq. (A.27) $A_{ij} = \dfrac{3}{2}\sin^2 \Theta_{ij}$ (for TM)

The angular factor Θ_{ij} is given by $\cos^2 \Theta_{ij} = E_{ij}/E$. Since Θ_{ij} is close to zero at the band edge, TM modes give a very small contribution compared to TE modes and therefore will be neglected below for simplicity. The calculation results for GaAs/AlGaAs lasers when an optical loss for the laser cavity of $\alpha_{loss} = 10 \cdot cm^{-1}$ is assumed are shown in Fig. A.5

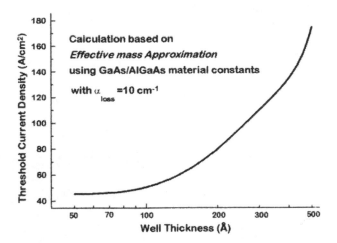

Fig. A.5. The threshold current change as a function of the quantum well thickness L_z when the total loss of the laser cavity α_{loss} is 10 cm^{-1} and material constants for GaAs/AlGaAs were used for Eq. (A.18)–Eq. (A.22).

A.4.   Luttinger–Kohn Hamiltonian

k·p Theory

The general Hamiltonian for electrons in a solid can be written as

Eq. (A.28) $$\left[\frac{p^2}{2m_0}+V_0\right]\Psi_{nk}(r)=\varepsilon_{nk}\Psi_{nk}(r)$$

By Bloch's theorem, $\Psi_{nk}(r)=e^{ik\cdot r}\,\varnothing_{nk}(r)$, where $\varnothing_{nk}(r)$ is invariant in crystal symmetry translation or rotation. Hence, (Eq. (A.28)) can be written as

Eq. (A.29) $$\left[\frac{p^2}{2m_0}+V_0+\frac{\hbar}{m_0}k\cdot p\right]\phi_{nk}(r)=\varepsilon_{nk}\phi_{nk}(r)$$

This is the basic Hamiltonian for *k·p* perturbation theory.

Luttinger–Kohn Hamiltonian

The Luttinger–Kohn Hamiltonian is a Hamiltonian describing degenerate valence bands. According to the degenerate second-order perturbation theory, energies for $|X\rangle,|Y\rangle,|Z\rangle$ states (hole eigenstates with *x, y, z* symmetry) for the *k·p* Hamiltonian (Eq. (A.29)) are given by

Eq. (A.30) $$H_{ij}=\frac{\hbar^2k^2}{2m_0}\delta_{ij}+\sum_m\frac{|\langle i|H_1|j\rangle|^2}{\varepsilon_1-\varepsilon_m}$$

where *i, j* indicate *X, Y, Z*. H_1 is the *k·p* term ($(h/m_o)\,k\cdot p$) in (Eq. (A.29)). Using the equation

Eq. (A.31) $$\frac{m_0^2}{\hbar^2}|\langle X|H_1|m\rangle|^2=|\langle X|p_x|m\rangle|^2k_x^2+|\langle X|p_y|m\rangle|^2k_y^2+|\langle X|p_z|m\rangle|^2k_z^2$$

Eq. (A.30) is rewritten as

Eq. (A.32) $H = \begin{pmatrix} E_1 + Ak_x^2 + B(k_y^2 + k_z^2) & Ck_xk_y & Ck_xk_z \\ Ck_xk_y & E_1 + Ak_y^2 + B(k_x^2 + k_z^2) & Ck_yk_z \\ Ck_xk_z & Ck_yk_z & E_1 + Ak_z^2 + B(k_x^2 + k_y^2) \end{pmatrix}$

This is not yet an accurate perturbation theory since the spin–orbit interaction is not included in this way (we have included only crystal potentials, the spin–orbit interaction is a self-interaction of the electrons). The spin–orbit interaction decouples the six states (2 (spin number) × 3 (*X*, *Y*, *Z*)) into $\left|J = \frac{3}{2}\right\rangle$ and $\left|J = \frac{1}{2}\right\rangle$.

Since the $\left|J = \frac{1}{2}\right\rangle$ state has much lower energy (by $\Delta_0 \sim 300$ meV), we consider only $\left|J = \frac{3}{2}\right\rangle$ states. Translating the $|L = 1\rangle$ states (such as $|X\rangle$) into $\left|J = \frac{3}{2}\right\rangle$ states (using Clebsch–Gordan coefficients), Eq. (A.32) can be rewritten as

Eq. (A.33) $H = \begin{pmatrix} H_{hh} & -c & -b & 0 \\ -c^* & H_{lh} & 0 & b \\ -b^* & 0 & H_{hh} & -c \\ 0 & b^* & -c^* & H_{lh} \end{pmatrix}$

where

Eq. (A.34) $H_{hh} = \dfrac{\hbar^2}{2m_0}\left\{(\gamma_1 + \gamma_2)k_\rho^2 + (\gamma_1 - 2\gamma_2)k_z^2\right\}$

Eq. (A.35) $H_{lh} = \dfrac{\hbar^2}{2m_0}\left\{(\gamma_1 + \gamma_2)k_z^2 + (\gamma_1 - 2\gamma_2)k_\rho^2\right\}$

Eq. (A.36) $c = \dfrac{\sqrt{3}\hbar^2}{2m_0}\left\{\gamma_2\left(k_x^2 - k_y^2\right) - 2i\gamma_3 k_x k_y\right\}$

Eq. (A.37) $b = \dfrac{\sqrt{3}\hbar^2}{2m_0}\gamma_3 k_z\left(k_x - ik_y\right)$

Further simplification is possible if we assume γ_2 and γ_3 are close to each other ($\gamma_2 \sim \gamma_3$) (called the axial approximation). Then by a unitary transformation [Brodio and Sham 1985], Eq. (A.33) is transformed into decoupled 2 × 2 matrices as

Eq. (A.38)

$$H_v = \frac{1}{2}\left(\frac{\hbar^2}{m_0}\right)\begin{pmatrix} (\gamma_1+\gamma_2)k_\rho^2+(\gamma_1-2\gamma_2)k_z^2 & \sqrt{3}\gamma_2 k_\rho(k_\rho+i2k_z) \\ \sqrt{3}\gamma_2 k_\rho(k_\rho+i2k_z) & (\gamma_1-\gamma_2)k_\rho^2+(\gamma_1+2\gamma_2)k_z^2 \end{pmatrix}+V(z)$$

where $V(z)$ is the potential profile of the valence band. This is the Luttinger–Kohn Hamiltonian with the axial approximation.

References

Brodio, D.A. and Sham, L.J. *Phys. Rev.* B **31** 888, 1985.

A.5.   Infrared detectors

Classification

In general, infrared detectors fall into two categories: photon detectors (also named photodetectors) and thermal detectors (see Fig. A. 6). In photon detectors the incident photons are absorbed within the material by interaction with electrons, either to lattice atoms, or to impurity atoms or with free electrons. The observed electrical signal results from the changed electronic energy distribution. The photon detectors show a selective wavelength dependence of the response per unit incident radiation power. In thermal detectors, the incident radiation is absorbed and raises the temperature of the material. The output signal is observed as a change in some temperature-dependent property of the material. In pyroelectric detectors a change in the internal electrical polarization is measured, whereas in the case of bolometers a change in the electrical resistance is measured. The thermal effects are generally wavelength independent since the radiation can be absorbed in a "black" surface coating.

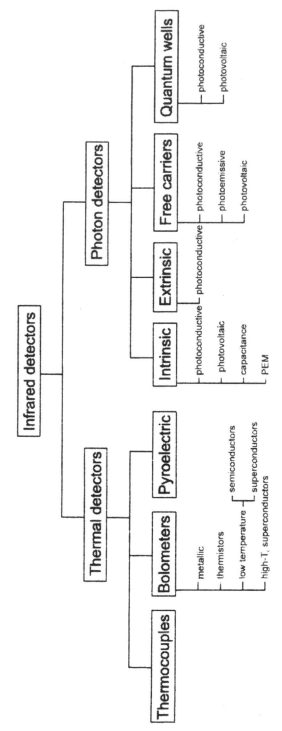

Fig. A. 6. Classification of infrared detectors

General theory of photodetectors

The photodetector is a slab of homogeneous semiconductor with the actual "electrical" area A_e which is coupled to the beam of the infrared radiation by its optical area A_o (Fig. A.7). Usually, the optical and electrical areas of the device are the same or close. The use of optical concentrators can increase the A_o/A_e ratio, however.

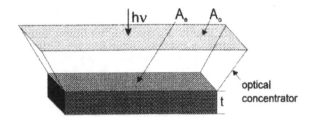

Fig. A.7. Model of a photodetector.

The current responsivity of the photodetector is determined by the quantum efficiency η and by the photoelectric gain g. The quantum efficiency value describes how well the detector is coupled to the radiation to be detected. It is usually defined as the number of electron–hole pairs generated per incident photon. The photoelectric gain is the number of carriers passing contacts per pair generated. This value shows how well the generated electron–hole pairs are used to generate the current response of a photodetector. Both values are assumed here as constant over the volume of the device.

The spectral current responsivity is

Eq. (A.39) $\qquad R_i = \dfrac{\lambda\eta}{hc}qg$

where λ is the wavelength, h is the Planck constant, c is the light velocity, q is the electron charge, and g is the photoelectric current gain. Assuming that the current gains for photocurrent and noise current are the same, the current noise due to generation and recombination processes is

Eq. (A.40) $\qquad I_n^2 = 2(G+R)A_e t\, \Delta f\, q^2 g^2$

where G and R are the generation and recombination rates, Δf is the frequency band, and t is the thickness of the slab.

The detectivity D^* is the main parameter characterizing the normalized signal to noise performance of detectors and can be defined as

Eq. (A.41) $\qquad D^* = \dfrac{R_i(A_o\Delta f)^{1/2}}{I_n}$

According to Eq. (A.39) to Eq. (A.41)

$$\text{Eq. (A.42)} \qquad D^* = \frac{\lambda}{hc}\left(\frac{A_o}{A_e}\right)^{1/2} \eta\left[2(G+R)t\right]^{-1/2}$$

For a given wavelength and operating temperature the highest performance can be obtained by maximizing $\eta/[t(G+R)]^{1/2}$ which corresponds to the condition of the highest ratio of the sheet optical generation to the square root of sheet thermal generation-recombination.

The effects of a fluctuating recombination can frequently be avoided by arranging for the recombination process to take place in a region of the device where it has little effect due to low photoelectric gain: for example, at the contacts in sweep-out photoconductors or in the neutral regions of diodes. In this case, the noise can be reduced by a factor of $2^{1/2}$ and detectivity increased by the same factor. The generation process with its associated fluctuation, however, cannot be avoided by any means.

The total generation rate is a sum of the optical and thermal generation

$$\text{Eq. (A.43)} \qquad G = G_{th} + G_{op}$$

The optical generation may be due to the signal or thermal background radiation. Usually thermal background radiation is higher compared to the signal radiation. The optical generation rate is

$$\text{Eq. (A.44)} \qquad G_{op} = \frac{A_o \eta \Phi_B}{A_e t}$$

where $\Phi\beta$ is the background photon flux density.

If the thermal generation is reduced much below the optical one, the performance of the device is determined by the background radiation (BLIP conditions, for background-limited infrared photodetector). T_{BLIP} is the temperature at which the thermally generated current is equal to the background-generated current in the detector. Assuming that recombination does not contribute to the noise

$$\text{Eq. (A.45)} \qquad D^*_{BLIP} = \frac{\lambda}{hc}\left(\frac{\eta}{2\Phi_B}\right)^{-1/2}$$

It should be noted that D^*_{BLIP} does not depend on the ratio A_o/A_e.

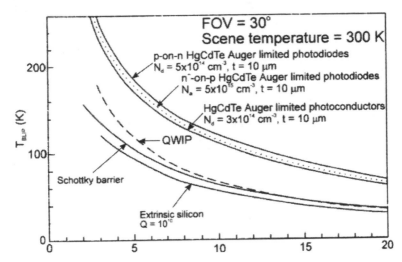

Fig. A.8. An estimation of the temperature required for background-limited operation of different types of photon detector. In the calculations FOV = 30° and T_B = 300 K are assumed (after [Rogalski 1995]).

In Fig. A.8 plots of the calculated temperature required for BLIP operation in a 30° field of view (FOV) are shown as a function of cutoff wavelength. We can see that the operating temperature for HgCdTe detectors is higher than for other types of photon detector. HgCdTe detectors with background-limited performance operate with thermoelectric coolers in the MWIR range, whereas the LWIR detectors ($8 \leq \lambda_c \leq 12$ μm) operate at $\approx$100 K. HgCdTe is characterized by high optical absorption coefficient and quantum efficiency and relatively low thermal generation rate compared to extrinsic detectors, suicide Schottky barriers, and GaAs/AlGaAs quantum well infrared photodetectors (QWIPs). However, the cooling requirements for QWIPs with cutoff wavelengths below 10 μm are less stringent in comparison with extrinsic detectors and Schottky barrier devices.

References

Rogalski, A. (ed) *Infrared Photon Detectors* (Bellingham, WA: SPIE Optical Engineering), 1995.

A.6. Physical properties and safety information of metalorganics

Table A.3 and Table A.4 summarize some of the basic thermodynamic properties of metalorganic sources commonly used in MOCVD, including their chemical formula and abbreviation, boiling point, melting point, and the expression of their vapor pressure as a function of temperature.

Additional information on their other important physical properties is also provided for a number of important metalorganic sources, including diethylzinc (Table A.5), trimethylindium (Table A.6), triethylindium (Table A.7), trimethylgallium (Table A.8), and triethylgallium (Table A.9).

In the rest of this Appendix, general information about the safety of metalorganic compounds will be given. This will be helpful in developing safety and health procedures during their handling.

Chemical reactivity

Metalorganics catch fire if exposed to air, react violently with water and any compound containing active hydrogen, and may react vigorously with compounds containing oxygen or organic halide.

Stability

Metalorganics are stable when stored under a dry, inert atmosphere and away from heat.

Fire hazard

Metalorganics are spontaneously flammable in air and the products of combustion may be toxic. Metalorganics are pyrophoric by the paper char test used to gauge pyrophoricity for transportation classification purposes [Mudry 1975].

Firefighting technique

Protect against fire by strict adherence to safe operating procedures and proper equipment design. In case of fire, immediate action should be taken to confine it. All lines and equipment which could contribute to the fire should be shut off. As in any fire, prevent human exposure to fire, smoke or products of combustion. Evacuate non-essential personnel from the fire area.

The most effective fire extinguishing agent is dry chemical powder pressurized with nitrogen. Sand, vermiculite or carbon dioxide may be used. CAUTION: re-ignition may occur. DO NOT USE WATER, FOAM, CARBON TETRACHLORIDE OR CHLOROBROMOMETHANE extinguishing agents, as these materials react violently and/or liberate toxic fumes on contact with metalorganics.

When there is a potential for exposure to smoke, fumes, or products of combustion, firefighters should wear full-face positive-pressure self-contained breathing apparatus or a positive-pressure supplied-air respirator with escape pack and impervious clothing including gloves, hoods, aluminized suits and rubber boots.

Human health

Metalorganics cause severe burns. Do not get in eyes, on skin or clothing.

Ingestion and inhalation. Because of the highly reactive nature of metalorganics with air and moisture, ingestion is unlikely.

Skin and eye contact. Metalorganics react immediately with moisture on the skin or in the eye to produce severe thermal and chemical burns.

First aid

If contact with metalorganics occurs, immediately initiate the recommended procedures below. Simultaneously contact a poison center, a physician, or the nearest hospital. Inform the person contacted of the type and extent of exposure, describe the victim's symptoms, and follow the advice given.

Ingestion. Should metalorganics be swallowed, immediately give several glasses of water but do not induce vomiting. If vomiting does occur, give fluids again. Have a physician determine if condition of patient will permit induction of vomiting or evacuation of stomach. Do not give anything by mouth to an unconscious or convulsing person.

Skin contact. Under a safety shower, immediately flush all affected areas with large amounts of running water for at least 15 minutes. Remove contaminated clothing and shoes. Do not attempt to neutralize with chemical agents. Get medical attention immediately. Wash clothing before reuse.

Eye contact. Immediately flush the eyes with large quantities of running water for a minimum of 15 minutes. Hold the eyelids apart during the flushing to ensure rinsing of the entire surface of the eyes and lids with water. Do not attempt to neutralize with chemical agents. Obtain medical attention as soon as possible. Oils or ointments should not be used at this time. Continue the flushing for an additional 15 minutes if a physician is not immediately available.

Inhalation. Exposure to combustion products of this material may cause respiratory irritation or difficulty with breathing. If inhaled, remove to fresh air. If not breathing, clear the victim's airway and start mouth-to-mouth artificial respiration which may be supplemented by the use of a bag-mask respirator or manually triggered oxygen supply capable of delivering one liter per second or more. If the victim is breathing, oxygen may be delivered from a demand-type or continuous-flow inhaler, preferably with a physician's advice. Get medical attention immediately.

Industrial hygiene

Ingestion. As a matter of good industrial hygiene practice, food should be kept in a separate area away from the storage/use location. Smoking should be avoided in storage/use locations. Before eating, hands and face should be washed.

Skin contact. Skin contact must be prevented through the use of fire-retardant protective clothing during sampling or when disconnecting lines or opening connections. Recommended protection includes a full-face shield, impervious gloves, aluminized polyamide coat, hood and rubber boots. Safety showers—with quick-opening valves which that stay open—should be readily available in all areas where the material is handled or stored. Water should be supplied through insulated and heat-traced lines to prevent freeze-ups in cold weather.

Eye contact. Eye contact with liquid or aerosol must be prevented through the use of a full-face shield selected with regard for use-condition exposure potential. Eyewash fountains, or other means of washing the eyes with a gentle flow of tap water, should be readily available in all areas where this material is handled or stored. Water should be supplied through insulated and heat-traced lines to prevent freeze-ups in cold weather.

Inhalation. Metalorganics should be used in a tightly closed system. Use in an open (e.g. outdoor) or well ventilated area to minimize exposure to the products of combustion if a leak should occur. In the event of a leak, inhalation of fumes or reaction products must be prevented through the use of an approved organic vapor respirator with dust, mist and fume filter. Where exposure potential necessitates a higher level of protection, use a positive-pressure, supplied-air respirator.

Spill handling

Make sure all personnel involved in spill handling follow proper firefighting techniques and good industrial hygiene practices. Any person entering an area with either a significant spill or an unknown concentration of fumes or combustion products should wear a positive-pressure, supplied-air respirator with escape pack. Block off the source of spill, extinguish fire with extinguishing agent. Re-ignition may occur. If the fire cannot be controlled with the extinguishing agent, keep a safe distance, protect adjacent property and allow product to burn until consumed.

Corrosivity to materials of construction

This material is not corrosive to steel, aluminum, brass, nickel or other common metals when blanketed with a dry inert gas. Some plastics and elastomers may be attacked.

Storage requirements

Containers should be stored in a cool, dry, well ventilated area. Store away from flammable materials and sources of heat and flame. Exercise due caution to prevent damage to or leakage from the container

Compound	Formula	Abbreviation	Melting point (°C)	Boiling point (°C)	$\log_{10}P$ (mmHg) (T in K)	Temperature Range (°C)
Group II sources						
Dimethylberyllium	$(CH_3)_2Be$	DMBe				
Diethylberyllium	$(C_2H_5)_2Be$	DEBe	12	194	$7.59-2200/T$	
Bis-Cyclopentadienyl Magnesium	$(C_5H_5)_2Mg$	Cp_2Mg	176		$25.14-2.18\ln T-4198/T$	
Group IIB sources						
Dimethylzinc	$(CH_3)_2Zn$	DMZn	−42	46	$7.802-1560T$	
Diethylzinc	$(C_2H_5)_2Zn$	DEZn	−28	118	$8.280-2190/T$	
Dimethylcadmium	$(CH_3)_2Cd$	DMCd	−4.5	105.5	$7.764-1850/T$	
Group III sources						
Trimethylaluminum	$(CH_3)_3Al$	TMAl	15.4	126	$7.3147-1534.1/(T-53)$	17–100
Triethylaluminum	$(C_2H_6)_3Al$	TEAl	−58	194	$10.784-3625/T$	110–140
Trimethylgallium	$(CH_3)_3Ga$	TMGa	−15.8	55.7	$8.07-1703/T$	
Triethylgallium	$(C_2H_5)_3Ga$	TEGa	−82.3	143	$8.224-2222/T$	50–80
Ethyldimethylindium	$(CH_3)_2(C_2H_5)In$	EDMIn	5.5	133.8		10–38
Trimethylindium	$(CH_3)_3In$	TMIn	88.4	184	$10.520-3014/T$	
Triethylindium	$(C_2H_5)_3In$	TEIn	−32		1.2	44
					3	53
					12	83

Table A.3. Physical properties of some organometallics used in MOCVD. [Ludowise 1985, and http://electronicmaterials.rohmhaas.com]

Compound	Formula	Abbreviation	Melting point (°C)	Boiling point (°C)	$\text{Log}_{10}P$ (mmHg) (T in K)	Temperature Range (°C)
Group IV sources						
Tetramethylgermanium	$(CH_3)_4Ge$	TMGe	−88	43.6	139	0
Tetramethyltin	$(CH_3)_4Sn$	TMSn	−53	78	$7.495\text{-}1620/T$	
Tetraethyltin	$(C_2H_5)_4Sn$	TESn	−112	181		
Group V sources						
Diethylarsine Hydride	$(C_2H_5)_2AsH$	DEAs			$7.339\text{-}1680/T$	
Tertiarybutylarsine	$(C_4H_9)AsH_2$	TBAs			$7.5\text{-}1562.3/T$	
Tertiarybutylphosphine	$(C_4H_9)PH_2$	TBP			$7.586\text{-}1539/T$	
Trimethylphosphorus	$(CH_3)_3P$	TMP	−85	37.8	$7.7329\text{-}1512/T$	
Triethylphosphorus	$(C_2H_5)_3P$	TEP	−88	127	$7.86\text{-}2000/T$	18−78.2
Trimethylarsenic	$(CH_3)_3As$	TMAs	−87.3	50−52	$7.7119\text{-}1563/T$	
Triethylarsenic	$(C_2H_5)_3As$	TEAs	−91	140	15.5	
Trimethylantimony	$(CH_3)_3Sb$	TMSb	−86.7	80.6	$7.7280\text{-}1709/T$	37
Triethylantimony	$(C_2H_5)_3Sb$	TESb	−98	116	17	75
Group VI sources						
Diethylselenide	$(C_2H_5)_2Se$	DESe		108		
Dimethyltellurium	$(CH_3)_2Te$	DMTe	−10	82	$7.97\text{-}1865/T$	
Diethyltellurium	$(C_2H_5)_2Te$	DETe		137~138	$7.99\text{-}2093/T$	

Table A.4. Physical properties of some organometallics used in MOCVD. [Ludowise 1985, and http://electronicmaterials.rohmhaas.com]

Acronym	DEZn
Formula	$(C_2H_5)_2Zn$
Formula weight	123.49
Metallic purity	99.9999 wt% (min) zinc
Appearance	Clear, colorless liquid
Density	1.198 $g \cdot ml^{-1}$ at 30 °C
Melting point	−30 °C
Vapor pressure	3.6 mmHg at 0 °C 16 mmHg at 25 °C 760 mmHg at 117.6 °C
Behavior towards organic solvents	Completely miscible, without reaction, with aromatic and saturated aliphatic and alicyclic hydrocarbons. Forms relatively unstable complexes with simple ethers, thioethers, phosphines and arsines, but more stable complexes with tertiary amines and cyclic ethers.
Stability in air	Ignites on exposure (pyrophoric).
Stability in water	Reacts violently, evolving gaseous hydrocarbons, carbon dioxide and water.
Storage stability	Stable indefinitely at ambient temperatures when stored in an inert atmosphere.

Table A.5. Chemical properties of diethylzinc. [Razeghi 1989]

Acronym	TMIn
Formula	$(CH_3)_3In$
Formula weight	159.85
Metallic purity	99.999 wt% (min) indium
Appearance	White, crystalline solid
Density	1.586 $g \cdot ml^{-1}$ at 19 °C
Melting point	89 °C
Boiling point	135.8 °C at 760 mmHg 67 °C at 12 mmHg
Vapor pressure	15 mmHg at 41.7 °C
Stability in air	Pyrophoric, ignites spontaneously in air.
Solubility	Completely miscible with most common solvents.
Storage stability	Stable indefinitely when stored in an inert atmosphere.

Table A.6. Chemical properties of trimethylindium. [Razeghi 1989]

Acronym	TEIn
Formula	$(C_2H_5)_3In$
Formula weight	202.01
Metallic purity	99.9999 wt% (min) indium
Appearance	Clear, colorless liquid
Density	1.260 g·ml^{-1} at 20 °C
Melting point	−32 °C
Vapor pressure	1.18 mmHg at 40 °C 4.05 mmHg at 60 °C 12.0 mmHg at 80 °C
Behavior towards organic solvents	Completely miscible, without reaction, with aromatic and saturated aliphatic and alicyclic hydrocarbons. Forms complexes with ethers, thioethers, tertiary amines, phosphines, arsines and other Lewis bases.
Stability in air	Ignites on exposure (pyrophoric).
Stability in water	Partially hydrolyzed; loses one ethyl group with cold water.
Storage stability	Stable indefinitely at ambient temperatures when stored in an inert atmosphere.

Table A.7. Chemical properties of triethylindium. [Razeghi 1989]

Acronym	TMGa
Formula	$(CH_3)_3In$
Formula weight	114.82
Metallic purity	99.9999 wt% (min) gallium
Appearance	Clear, colorless liquid
Density	1.151 $g \cdot ml^{-1}$ at 15 °C
Melting point	−15.8 °C
Vapor pressure	64.5 mmHg at 0 °C 226.5 mmHg at 25 °C 760 mmHg at 55.8 °C
Behavior towards organic solvents	Completely miscible, without reaction, with aromatic and saturated aliphatic and alicyclic hydrocarbons. Forms complexes with ethers, thioethers, tertiary amines, tertiary phosphines, tertiary arsines, and other Lewis bases.
Stability in air	Ignites on exposure (pyrophoric).
Stability in water	Reacts violently, forming methane and Me_2GaOH or $[(Me_2Ga)_2O]_x$.
Storage stability	Stable indefinitely at ambient temperatures when stored in an inert atmosphere.

Table A.8. Chemical properties of trimethylgallium. [Razeghi 1989]

Acronym	TEGa
Formula	$(C_2H_5)_3Ga$
Formula weight	156.91
Metallic purity	99.9999 wt% (min) gallium
Appearance	Clear, colorless liquid
Density	1.0586 $g \cdot ml^{-1}$ at 20 °C
Melting point	−82.3 °C
Vapor pressure	16 mmHg at 43 °C 62 mmHg at 72 °C 760 mmHg at 143 °C
Behavior towards organic solvents	Completely miscible, without reaction, with aromatic and saturated aliphatic and alicyclic hydrocarbons. Forms complexes with ethers, thioethers, tertiary amines, tertiary phosphines, tertiary arsines and other Lewis bases.
Stability in air	Ignites on exposure (pyrophoric).
Stability in water	Reacts vigorously, forming ethane and Et_2GaOH or $[(Et_2Ga)_2O]_x$.
Storage stability	Stable indefinitely at room temperatures in an inert atmosphere.

Table A.9. Chemical properties of triethylgallium. [Razeghi 1989]

References

Ludowise, M., "Metalorganic chemical vapor deposition of III-V semiconductors," *Journal of Applied Physics* 58, R31-R55, 1985.

Mudry, W.L., Burleson, D.C., Malpass, D.B., and Watson, S.C., *Journal of Fire Flammability* 6, p. 478, 1975.

Razeghi, M., *The MOCVD Challenge Volume 1: A Survey of GaInAsP-InP for Photonic and Electronic Applications*, Adam Hilger, Bristol, UK, 1989.

Sze, S.M., *Physics of Semiconductor Devices*, John Wiley & Sons, New York, 1981.

MOCVD Challenge Volume 1:
Original foreword

More than 35 years have elapsed since Welker, in the Erlangen Laboratory of Siemens, discovered the semiconducting properties of III–V compounds. This immediately created a flurry of experimental and theoretical research, showing clearly that physicists and engineers were well aware that this new class of semiconductors might, in time, have important, maybe revolutionary, applications.

Yet, except for some very specialized, low-volume devices, such as Hall effect devices, many years were to pass before some of these materials, mostly GaAs and InP, began to be used in significant amounts in the information technology industry. And, at first, they were almost entirely used in devices which could not be made using the, by then, workhorse of the semiconductor industry, silicon. Injection lasers, made possible by the 'direct gap' character of most of the III–V compounds, or the Gunn effect, are typical examples.

The reasons behind this long lead time are clear: silicon technology was mature and comparatively easy to apply. III–V technologies, on the other hand, had to be developed almost from scratch. New types of 'compositional structures' were, in many cases, necessary for the full advantage of III–V compounds over silicon, in some applications, to be realized fully. The early semiconductor lasers, based on GaAs, could only operate in a pulsed mode, with a low duty factor, and at reduced temperature. 'Separate confinement' and later quantum wells made it possible to operate semiconductor lasers in continuous wave mode above room temperature. But that implied that not only the dopant concentration, but also the composition of the materials be varied in a controllable way over very short distances, for example in the GaAs–(Ga, Al)As system. And the same type of structures proved necessary to take full advantage of the specific properties of III–V compounds (direct gap, high electron mobility, etc) in microwave or digital applications.

All this was only made possible through the development of new technologies for the epitaxial growth of these materials. Liquid phase epitaxy (LPE) first, molecular beam epitaxy (MBE) a little later, and finally metalorganic chemical vapor deposition (MOCVD) have been as important for III–V technology as diffusion has been for a long time for silicon.

This is so well recognized by the scientific and technical community that the very important IBM Europe Science and Technology Prize was awarded jointly to the three European scientists who had made decisive breakthroughs in developing these technologies: Mme Bauser for LPE, Mr Joyce for MBE, and Mme Razeghi for MOCVD. They were dubbed, on the occasion, the Queens and King of epitaxy!

Needless to say, when one of the Queens writes as thorough a treatise on this very subject, this is bound to become a major event. At a time when, at last, III–V compounds, mostly InP (covered in Volume 1) and GaAs (covered in Volume 2) based materials, are becoming of major importance to industry, this treatise will be a must for every scientist and engineer in the field.

Manijeh Razeghi has been recognized all over the world for her ability to achieve astonishingly brilliant results with MOCVD technologies. Simply reading the book may not be quite enough for every engineer to do fully as well as she does. But it is a necessary step in that direction.

Pierre Aigrain
Paris, France
October 1989

MOCVD Challenge Volume 1:
Original introduction

The secret of accomplishing meaningful basic or applied research on semiconductors is to have control over the preparation of the materials. This has proved particularly so for compound semiconductors, and for many compounds progress has been minimal for long periods because synthesis and purification were not mastered. The III–V compounds have not been exceptions, though their potential applications have drawn many scientists into their study, and a great deal of effort has been expended on developing growth techniques. Certainly we have progressed from the days when the preparation of GaAs was invariably accomplished by an explosion, but even 20 years after the discovery of GaAs and InP, the manufacture of pure crystals was best described as occasional. These problems had led many laboratories to study epitaxial deposition, because it was believed, correctly, that the lower temperature of preparation would give both higher purity and better control. A range of methods were tried, exploiting hydrides, chlorides, and, later, liquid phases. All were marginally successful, but the practice still remained an art. The introduction of metalorganic chemical vapor deposition in 1972 was not greeted universally with acclaim. Few workers were enthusiastic about its advantages over the established methods, and it was regarded as an unpleasant necessity for dealing with compounds incorporating aluminum. Acceptance came slowly, with the realization that the modern forms of device, with many layers containing a variegated selection of elements, could not be produced in numbers by the older methods. Today it is an established and standard technique, available throughout the world.

Science and engineering move forward through the concerted efforts of many people. The history of progress in GaAs and the other III–V compounds is studded with many distinguished names following the original discovery by Professor Heinrich Welker in 1952. Various scientists have made important discoveries on the long road since then, and their contributions must not be minimized. Nevertheless, one name stands out in the field of MOCVD, that of Dr Manijeh Razeghi. Her achievements in material and device preparation are too many to list. That is why this is an unusual book. It sets out to describe growth, characterization and applications of heterojunctions, quantum wells and superlattices based on InP and its derivatives. It draws heavily, perforce, on Dr. Razeghi's own work because she has contributed so much. It is a pleasure for me to acknowledge her success in this introduction.

Cyril Hilsum
Wembley, UK
August 1988

MOCVD Challenge Volume 2:
Original foreword

History has often marked its progress in terms of materials, starting with the stone age, then the iron age, then the bronze age and on to the present age. Indeed, materials continue to be a limiting factor in enabling many areas of technology even today. For example, the development of optical fibers for communications has been a key factor in constructing the information superhighway.

In the future when historians look back on our current stage of development they will probably describe it as the age of artificial materials, a period when we learned how to create new forms of materials tailored specifically to our needs, not found in nature or by refining nature. As a material scientist, I find that composites come to mind, but just as pervasive now is the development of layered semiconductors by those interested in electronic and optical devices. How marvelous it is to think that we are now able to build, atom by atom, different optoelectronic devices constructed of a whole series of materials many of which never existed before. Professor Manijeh Razeghi's efforts in this regard are almost legendary at this point; she has been one of a handful of pioneers in developing the techniques and the devices themselves.

For me, as an engineering Dean, it has been a distinct adventure—a ride into the future—to encourage Manijeh to leave Paris and to come to our University to join our Department of Electrical Engineering and Computer Science. In no time we have been able to help Manijeh to establish her new laboratory facilities and the Center for Quantum Devices. (Indeed help is the correct word, because often she had to look back and wait for some of us to catch up.)

Now many new things are coming from these laboratories, but Manijeh's life also now has a very new dimension beyond the research that she aggressively pursues. Generations of "students," high-schoolers, undergraduates, graduates, postdocs and visitors are all being influenced by her philosophy and approach to work, an influence she could not fulfill in her previous roles.

Those of us who are in academic research have a secret: in our careers we may indeed discover something grand (and Manijeh has certainly done so and more can be expected) but most of all we have inspired and launched many young people on their way to successful careers. In many senses, this results in a greater satisfaction than our own splendid discoveries. Manijeh is now firmly entrenched in academia and clearly enjoying both sides of such a career. This book is another example of her ability to do cutting-edge research and her desire to educate. It is a text of the fundamentals of semiconductor crystal growth and material characterization leading into many of the important concepts in advanced device design and fabrication.

Jerome B. Cohen
Robert R. McCormick School of Engineering and Applied Science
Northwestern University, Evanston, Illinois
April 1995

MOCVD Challenge Volume 2:
Original introduction

Professor Razeghi's *The MOCVD Challenge* Volume 1 appeared in 1989 and was extremely well received by the scientific and engineering communities working on semiconductors. This volume complements Volume 1 in that it has a focus on GaAs and other arsenides grown by MOCVD, whereas Volume 1 covered mainly the phosphide-based materials.

Naturally this volume conveys much more than MOCVD growth of GaAs. Based on her world-class research work carried out at Thomson-CSF and more recently at Northwestern University, the author explores thoroughly the issue of *in situ* MOCVD growth monitoring of both arsenides and phosphides by means of differential reflectivity. This pioneering work is explained in detail supported by a solid theoretical analysis of its principles. The potentials for studying and understanding the initial stages of growth are clearly indicated and this will be extremely useful for future developments in, for example, self-organized growth of semiconductor nanostructures by MOCVD. Moreover, the author gives a clear example of the dynamic interaction between an *in situ* growth characterization technique, the following modifications of a reactor design and the impact of these on the resultant sample design and properties. A full cycle of improvements, which constitutes a marvelous example of how progress can be made in an experimental multi-variable problem.

Reflecting Manijeh Razeghi's trajectory in recent years, the book offers welcome perspectives from both industrial and academic points of view; whereas most topics covered are treated at the level of final year/graduate students in material sciences, there is a consistent treatment of the relevant science issues which have a bearing on device applications. This is most obviously exemplified by the chapter on *ex situ* characterization and the last four chapters on device application. Device-relevant issues are addressed from the very beginning of the book in the unique breathtaking style of the author.

Another fascinating aspect offered by Professor Razeghi's book is the journey offered, which starts by putting atoms together with MOCVD, via the growth and characterization of semiconductor layers, then through discrete devices and finally reaching an overview of an optoelectronic system. And at most stops there is the possibility of an in-depth study tour.

It is now almost three years since Professor Razeghi was appointed to a prestigious chair at Northwestern University. Within two years her research team and laboratory achieved critical mass, with many projects flowing in and high-impact novel results on photonic devices physics and engineering coming out. Some of these are based on another family of promising materials, the antimonides, and I very much hope that in a few years' time we shall see Volume 3.

Clivia M. Sotomayor Torres
Glasgow

Index

1/f noise 293
2D electron-gas 510

Absorption loss 354
Absorption spectra 139, 282, 286, 469
Acceptor binding energies 472
Acceptor ionization 172
Acceptors 137
Acoustic phonon scattering 150
Activation energy 117
Adducts 35
AlGaAs/GaAs 368, 400
Alkyls 36
Aluminum-free 704
Ando formula 230
Atomic layer epitaxy (ALE) 13, 42
Auger analysis 646
Auger coefficient 633
Auger electron spectroscopy (AES) 27, 255
Auger lines 574
Auger measurements 558
Auger recombination 350
Auger spectrum 247, 436, 552

Ball polishing 157
Band offsets 212, 259
Band structure of GaAs 168
Bandgap energy 3, 292
Bath revelation 108
Bevel revelation 106
Binding energy 283
Bipolar devices 5
Bipolar transistor 387, 388
Bloch oscillator 483
Bond method 122
Bound-exciton 137
Bowing parameter 3
Bragg peaks 120, 251
Bragg's law 120
Bruggeman theory 86, 89, 91
Buried waveguides 634

Buried-ridge-structure lasers 594
Burstein shift 205

Calorimetric absorption 553
Capacitance 112
Capacitance transients 112, 180
Capacitance–voltage (C–V) 139, 210
Capture cross-section 118, 119, 715
Capture rate 120
Carrier–carrier scattering 434
CARS 46
Cathodoluminescence (CL) 336
Cavity length 338, 361
Chemical beam epitaxy (CBE) 13, 40
Chemical etching 672
Chloride atomic layer epitaxy 43
Cladding layer 348, 353
Cleavage of lateral epitaxial films for
    transfer (CLEFT) 307
CLEFT 307
Coherent QDs 707
Collector current 388
Complex dielectric functions 579
Conduction band discontinuity 18, 261, 366
Conduction band offset 260, 264, 725
Conductivity 146
Conversion efficiency 306
Coulomb interaction 280, 282, 472
Cross doping 673
Current blocking layer 714
C–V method 431
Cyclotron mass 529
Cyclotron resonance 220, 522

D^* 291
Dark current 303
DC conductivity 239
Deep levels 111, 118, 476
Deep-level transient spectroscopy
    (DLTS) 17, 111, 114, 180, 199, 436, 477

Defect scattering ... 434
Deformation potential ... 150
Depth profiles ... 425
Detectivity ... 297
DFB lasers ... 618
Dielectric functions ... 86, 581, 582
Differential efficiency ... 357
Direct energy gap ... 7
Direct-bandgap ... 7
Dislocations ... 109, 111
Distributed feedback lasers ... 612
Donor ... 137
Donor ionization energy ... 172, 435
Dopants ... 443
Double heterostructure (DH) ... 7
Double-heterostructure laser ... 323
Drain–source current ... 370
DX centers ... 16, 194, 231, 233, 236, 238, 267, 371, 374, 514, 704

EER spectrum ... 261, 265
Effective mass ... 360, 418, 466, 528
Effective medium theories (EMTs) ... 85
Effective-index method (EIM) ... 634
Electroabsorption ... 283
Electroabsorption spectra ... 288
Electrochemical (polaron) profile ... 202, 246, 333, 444, 447, 645, 675
Electrochemical C–V (ECV) ... 140
Electroluminescence ... 685
Electroluminescence imaging ... 602
Electron Hall mobilities ... 76
Electron spectroscopy for chemical analysis (ESCA), ... 27
Electron spin resonance ... 223
Electronic devices ... 5
Electro-optical modulators ... 279
Electroreflectance ... 259
Ellipsometry ... 47, 48
Emitter current ... 388
Energy band diagram ... 319
Energy band structure ... 416
Energy gaps ... 3
Erbium-doped GaAs ... 190
Etch-pit density ... 647
Excitation spectra ... 491
Exciton resonances ... 281
Excitonic absorption ... 243, 493
Excitonic luminescence ... 181
Excitons ... 280, 466
External quantum efficiency ... 321, 332, 600

Fabry-Pérot cavity ... 72, 315, 616
Fe trap ... 478
Field-effect transistor ... 535
FIR emission spectra ... 242
FIR magnetoemission study ... 242
Focal plane arrays ... 269, 299, 713
Fourier transform spectroscopy ... 134
Fractional quantum Hall ... 505
Frank exciton ... 136
Free excitons ... 137
FTS ... 134
FWHM ... 138

$Ga_{0.25}In_{0.75}As_{0.5}P_{0.5}$–InP DH lasers ... 591
$Ga_{0.47}In_{0.53}As$ ... 454
Ga0.47In0.53As–InP heterojunction ... 458
Ga0.49In0.51P ... 163
GaAs/AlGaAs QWIPs ... 298
GaAs/$Ga_{0.51}In_{0.49}P$ ... 210
GaAs/GaInP superlattice ... 260
GaAs–GaAlAs systems ... 16
GaAs–GaInP systems ... 16
Gain ... 290, 359, 703, 715
GaInAlP–GaAs systems ... 18
GaInAs–GaAs photoconductor ... 657
GaInAs–InP multiquantum wells ... 649
GaInAsP–InP systems ... 17
GaInP/GaAs ... 392
GaInP/GaAs superlattice ... 244
GaInP/GaAs/GaInP ... 76
GaInP/GaInAs/InP MESFET ... 654
GaInP/InGaAs/GaAs ... 383
GaInP–GaAs ... 236
Garnet (GGG) substrates ... 649
Gas source molecular beam epitaxy (GSMBE) ... 13
Gate capacitance ... 370
Gate–source voltage, ... 370
$Ga_xAl_{1-x}As$ ... 163
$Ga_xIn_{1-x}As_yP_{1-y}$ ... 168
$Ga_xIn_{1-x}P$ alloys ... 197
g-factor ... 227, 530
Graded index (GRIN) ... 337
Graded-index (GRIN) ... 326
Gunn diode ... 6, 444, 448
Gunn oscillator ... 399

Hall coefficient ... 152
Hall constant ... 152
Hall measurement ... 149, 480

Hall mobility 150, 178, 219, 229, 374, 433, 434, 510, 511, 564
Hall resistance 505
Hall voltage 151
HBT 17, 386, 390, 392
Heavy-hole–light-hole 253
Helmholtz layer 141
HEMT 17, 367
Heterojunction 4, 18, 366
Heterojunction bipolar transistor (HBT)
.... 16, 391
Heterojunctions 14
Heterostructure laser 331, 340
$Hg_{1-x}Cd_x$ Te 292
HgTe–CdTe-superlattice 293
HIGFET 17, 383
High-mobility 219, 441, 454
Hydrides 36
Hyperfine structure 485

III–V binary compounds 1
III–V quaternary compounds 3
III–V ternary alloys 3
impact ionization avalanche transit time
(IMPATT) diode 6
Impurities 182, 183
Impurity scattering 149
InAs/InP 79
$InAs_{1-x}Sb_x$ materials 294
infrared photodetectors 289, 751
InP epitaxial layers 448
$InP/Ga_{0.47}In_{0.53}As$ heterojunction 484
InP/GaAs/Si 79
Integrated photoreceiver 655
Internal quantum efficiency 337
Interstitials 111
Intersubband 259, 296
intersubband absorption 703
Intersubband hole absorption 271
Intersubband scattering 525
intersubband states 720
Inter-valley scattering 434
Intraband 259
Ionization energy 183
Ionized impurity scattering 235

JFETs 5

$k \cdot p$ perturbation method 165
$k \cdot p$ theory 165, 530, 562, 747

Landau levels 223, 498, 504, 509, 522, 567
Laser diodes 341, 354
Laser fabrication 585
Lasing spectrum 617
Lattice constant 3, 4
Lattice mismatch 124, 456, 546, 721
Lattice-matched heterojunctions 62
Lattice-mismatched structures 79
Leakage current 376
Light-emitting diodes (LEDs) 6
$LiNbO_3$ 280
Liquid phase epitaxy (LPE) 10, 25
Lorentz force 151
Low threshold current 16
Low-energy electron diffraction (LEED)
.... 27
Luttinger–Kohn Hamiltonian 356, 358, 747

Magnesium 189
Magnetic freeze-out 530
Magneto-optic 223
Magnetophonon resonance 527, 563
Magnetophotoluminescence (MPL) ... 442
Magnetoresistance 217, 225, 527, 559
Magnetotransport 216, 222, 498, 557
Maxwell equation 128
Maxwell–Garnett 85
MCT 267
Mercury implantation 479
MESFET 400, 411
Metalorganic atomic layer epitaxy
(MOALE) 13, 42
Metalorganic chemical vapor deposition
(MOCVD) 10
Microwave diodes 448
Migration-enhanced epitaxy (MEE) 14, 43
Misfit dislocations 18, 670
Misfit strain 241
Mobility 146, 149, 246, 247, 353, 434
MODFET 367, 371, 376, 382
Modulation-doped structures 15
Molecular beam epitaxy (MBE) 10, 27
Momentum relaxation rate 349, 353
Monolithic focal plane arrays 705
Monolithic integration 412, 660
MOSFET 5
Multiquantum wells (MQW) 28

Nanopillar QWIPs 705
Negative differential resistance .. 449, 495
Noise equivalent difference temperature
      (NEDT) 299, 705
Noise power 659
Noise-equivalent power (NEP) 290
Non-radiative recombination 343
n-type doping 201

Optical confinement factor 318, 629
Optical diagnostics 46
Optical gain 315
Optical models 95
Optical phonon scattering 150, 352
Optical waveguides 537
Optic-phonon mode 531
Optoelectronic devices 6
Optoelectronic integrated circuit (OEIC)
      .... 399, 655
Oscillator strength 280

Pendellosung oscillations ... 124, 130, 254
Persistent photoconductivity 231, 371,
      508, 513
Phase modulation 636
Phase-modulated ellipsometry (PME) .. 49
Phonon bottleneck 712
Phonon replicas 470
Photoconductor 659
Photocurrent 305
Photodetectors 289, 292, 411, 753
Photoluminescence 133, 139, 181, 249,
      268, 271, 352, 423, 439, 465, 489,
      549, 554, 665
Photoluminescence excitation spectrum
      252
Photonic crystal distribute feedback ... 727
Photovoltage spectroscopy 144, 269
Piezoelectric 434
Piezoelectric scattering 152
PIN photodetector 400, 533, 694
Pinch-off 376
Pockels effect 279
Poisson equation 351, 430
Polar semiconductors 434
Polariton 181
Polarization interferometry 636
Polaron 117
Precipitates 111
Pseudomorphic 721
p-type doping 207

p-type GaAs 188
p-type QWIPs 300

Quantum cascade laser (QCL) 720
Quantum dot growth 707
Quantum dot infrared photodetector
      (QDIP) 713
Quantum dot laser 718
Quantum dot-in-well infrared
      photodetector (QDWIP) 710
Quantum dots 706
Quantum efficiency 289, 343, 534, 753
Quantum Hall effect ... 218, 498, 511, 516
Quantum Hall plateau 514
Quantum size effects 321
Quantum transport 234
Quantum tunneling 483
Quantum well infrared photodetector
      (QWIP) 267, 269, 300, 702
Quantum well infrared photodetectors
      (QWIP) 297
Quantum well lasers 328, 629
Quantum wire 330
Quantum-confined Stark effect 283
Quantum-size effects 138

Raman effect 470
Raman scattering 46
RCA solution 158
Reactor 30
Readout integrated circuit (ROIC) 705
Recombination processes 136
Reduced mass 466
Reflectance anisotropy (RA) 47
Reflection coefficient 89
Reflection difference spectroscopy (RDS)
      .... 47, 57, 66, 69, 76, 89, 93, 97
Reflection high-energy electron
      diffraction (RHEED) 27, 71
Refractive index 325
Relaxation time 179, 353
Resonant-tunneling 496
Responsivity 703
Reynolds number 33
Ridge-waveguide structures 613
RMS noise 290
Rocking curve 643

Saturation current 660
Saturation velocity 370
Scanning electron micrograph 83, 85,
      595, 626

Scattered radiation effect. 126
Scattering mechanisms........................179
Schottky contact..................................112
Schottky photodiode............................660
Secondary-ion mass spectroscopy
  (SIMS) 27, 644
Second-harmonic generation................47
Selenium...186
Self-assembled quantum dots..............701
Self-electro-optic effect device (SEED)
 483
Semiconductor lasers...................313, 332
Shallow-donor spectroscopy 530
Sheet carrier concentration.................373
Sheet resistance 146
Shubnikov–de Haas oscillations 216, 217,
  221, 228, 240, 498, 507, 561
Si δ-doped GaAs/AlGaAs SL..............110
Silane..184
Silicon 184
Solar cell 6, 304
Spectroscopic ellipsometry.........573, 582
Spin subbands.....................................510
Spin–orbit interaction.........................165
Spin–orbit splitting.....................418, 529
Spontaneous emission 315, 351
Stacking faults....................................111
Stark effect 280, 282, 412
Stimulated emission 315
Strain-balanced...................................721
Strained GaInP/GaAs heterostructures
 380
Strained heterostructures 643
Strained layers....................................653
Strained *p*-channel.............................380
Strained quantum well........................737
Strained-layer superlattices 295
Stranski–Krastanov growth 99, 706
Sulfur..187
Superlattices (SL)..................................28

TEGFET..367

Terahertz...726
Ternary compounds................................3
Thermal broadening 354
Thermal conductivity 33
Thermal expansion............. 169, 672, 680
Threshold current 316, 329, 337, 348,
  355, 599
Threshold current density ... 338, 342, 357
Threshold voltage.......................369, 372
Thyristor..5
Transconductance....... 374, 376, 404, 537
Transit time 370, 389, 390
Transmission electron microscopy
  (TEM) 198, 487, 686
Trap...112, 116
Trap activation energy........................116
Trap concentration 118
Trimethylindium (TMIn)......................55
Tunnel diode 6

Vacancies..111
Van der Pauw's method 148
Vapor phase epitaxy (VPE)............10, 26
Vegard's law 3, 120
Viscosity 33
Visible lasers......................................339
VLSI..5

Wall-plug efficiency...........................726
Wannier–Stark localization 285, 286, 288
Waveguide 319, 723
Wetting layer......................................707

X-ray diffraction. 120, 250, 462, 485, 650
X-ray photoelectron spectroscopy (XPS)
  ....27, 574
X-ray reflectivity 125, 130, 131
X-ray refraction..................................125
X-ray rocking curve 676

Zinc..188